BIOCHEMISTRY: FUNDAMENTALS AND APPLICATIONS

BIOCHEMISTRY:
Fundamentals and Applications

S. S. Purohit, Ph.D.
Ex-Head
P. G. Department of Botany
Dungar College, Bikaner

AGROBIOS (INDIA)

Published by:
AGROBIOS (INDIA)
Agro House, Behind Nasrani Cinema
Chopasani Road, Jodhpur 342 003
Phone: 91-0291-2643993, 2642319
E. mail: agrobiosindia@gmail.com; info@agrobiosonline.com
Web Site: agrobiosonline.com

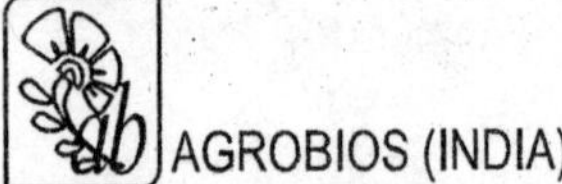

First Published: 2010

ISBN No. (10): 81-7754-395-4
ISBN No. (13): 978-81-7754-395-7

Published by: Dr. Updesh Purohit for Agrobios (India), Jodhpur
Lasertypeset at: Yashee Computers & Printers, Jodhpur
Cover Design by: Reena
Printed at: Agrobios Digitals, Jodhpur

Preface

Biochemistry is the organic chemistry of compounds and processes occurring in organisms. It is the study of the chemical processes in living organisms. It deals with the structure and function of cellular components such as proteins, carbohydrates, lipids, nucleic acids and other biomolecules. Although there are a vast number of different biomolecules many are complex and large molecules (called polymers) that are composed of similar repeating subunits (called monomers). Each class of polymeric biomolecule has a different set of subunit types. Biochemistry also deals with the studies of chemical properties of important biological molecules, like proteins, in particular the chemistry of enzyme-catalyzed reactions. Other areas of biochemistry include the genetic code (DNA, RNA), protein synthesis, and signal transduction.

Biochemistry also encompasses a study of the chemical nature of all living matter from the smallest virus and microorganism to the most complex and highly evolved human being. The relationship of the living to their environment; the processes by which an exchange of chemical substances takes place between the living organism and its environment through digestion, absorption and excretion, the processes by which the absorbed materials are utilized for synthetic reactions leading to growth and replenishment of tissues and multiplication of the cell and the species; the metabolic breakdown of the materials to supply energy for all the above; the mechanisms which regulate with precision all these processes by means of hormonal regulatory stimuli- all these come under the purview of biochemistry.

The book covers structure, function, and regulation of biologically active molecules; gene structure and expression; biochemical mechanisms; protein biosynthesis; protein folding; membrane structure-function relationships; bioenergetics; and immunochemistry. The book includes chapters like Biochemistry: Historical Development; Water: The Solvent of Life; pH and Buffers; The Atom and Chemical Bonds; Bioenergetics; Organic Components: Proteins; Organic Components: Carbohydrates; Organic Components: Lipids; Organic Component Nucleic Acids: DNA; Organic Component Nucleic Acids: RNA; Enzymes; Metabolism: General; Metabolism: Carbohydrates; Metabolism: Protein; Metabolism Lipids; Metabolism: Nucleic Acid; Pheromones; Animal Hormone; Plant Hormones/ Growth Regulators; Plant Hormone: Auxins, Gibberellins, Cytokinins, Ethylene, Abscisic Acid; The Vitamins; Photosynthetic Pigments; Hemoglobin; Plasma Proteins and Immunoglobulins; Glycoproteins; Hemostatis and Thrombosis; Metabolism of Xenobiotics; Muscle; The Cytoskeleton; Water and Human Metabolism; Nutrient Metabolism; Recombinant DNA Technology; Genomics; Gene Expression; Biotechnology for New Drug Discovery; New Therapeutic Approaches; Kidney and Liver Function Tests; Alkaloids; Appendix 1: Uric Acid - A Uremic Toxin?; Glossary; Subject Index.

In this presentation, efforts have been made to elucidate the different aspects, which are directly or indirectly are linked students of life sciences and pharmacy.

This book is self-reliant record, since all the salient features are extensively covered in an elaborate and self-explanatory manner. The book will serve as a source book for students of cellular and molecular life sciences.

If one has conviction of presentation of any thought and matter, that presentation may be having scope of improvement and refinement. I humbly accept all suggestions, comments, criticism in order to better my presentation.

Author

Contents

CHAPTER 1

Biochemistry Historical Development

Biochemistry is the study of the substances found in living organisms, and of the chemical reactions underlying life processes. This science is a branch of both chemistry and biology; the prefix bio- comes from *bios*, the Greek word for *life*. The chief goal of biochemistry is to understand the structure and behavior of biomolecules. These are the carbon-containing compounds that make up the various parts of the living cell and carry out the chemical reactions that enable it to grow, maintain and reproduce itself, and use and store energy.

A vast array of biomolecules is present in the cell. The structure of each biomolecule determines in what chemical reactions it is able to participate, and hence what role it plays in the cell's life processes. Among the most important classes of biomolecules are nucleic acids, proteins, carbohydrates, and lipids.

PROTEINS

Proteins are large molecules built up of small subunits called amino acids. Using only 20 different amino acids, a cell constructs thousands of different proteins, each of which has a highly specialized role in the cell. The proteins of greatest interest to biochemists are the enzymes, which are the *worker* molecules of the cell. These enzymes serve as promoters, or catalysts, of chemical reactions.

CARBOHYDRATE

Carbohydrate is a group of organic compounds that includes sugars, starches, celluloses, and gums and serves as a major energy source in the diet of animals. These compounds are produced by photosynthetic plants and contain only carbon, hydrogen, and oxygen, usually in the ratio 1:2:1.

Carbohydrates are mainly sugars and starches, together constituting one of the three principal types of nutrients used as energy sources (calories) by the body. Carbohydrates can also be defined chemically as neutral compounds of carbon, hydrogen and oxygen.

Carbohydrates, as a class, are the most abundant organic compounds found in nature. They are produced by green plants and by bacteria using the process known as photosynthesis, in which carbon dioxide is taken from the air by means of solar energy to yield the carbohydrates as well as all the other chemicals needed by the organisms to survive and grow..

The carbohydrate group consists principally of sugar, starch, dextrin, cellulose, and glycogen, substances that constitute an important part of the human diet and that of many animals. The simplest of them are the simple sugars, or monosaccharides, which contain either an aldehyde or a ketone group. The most important is glucose. Two monosaccharide molecules joined together by an oxygen atom, with the elimination of a molecule of water, yield a disaccharide, of which the most important are sucrose (ordinary cane sugar), lactose, and maltose. Polysaccharides have enormous molecules made up of one type or several types of monosaccharide units–about 10 in glycogen, for example; 25 in starch; and 100 to 200 in cellulose.

In living organisms, carbohydrates serve both essential structural and energy-storage functions. In plants, cellulose and hemicellulose are the main structural elements. In invertebrate animals, the polysaccharide chitin is the main component of the exoskeletons of arthropods. In vertebrate animals, the cell coatings of connective tissues contain carbohydrates. Cell membranes are rich in glycoproteins, and so forth. Plants use starch and animals use glycogen to store energy; when the energy is needed, the carbohydrates are broken down by enzymes.

LIPIDS

Lipids, diverse group of fatty substances found in all living organisms. Lipids are distinguished from other classes of organic compounds in that they do not dissolve in water but are soluble in alcohol, ether, or other organic solvents. Among the most important lipids are the phospholipids, which are major components of the cell membrane. Phospholipids limit the passage of water and water-soluble compounds through the membrane, enabling the cell to keep its contents separate from the outside environment.

FATS AND OILS

Fats and oils, which are composed of triglycerides, serve as stored energy reserves in plant and animal cells. Each triglyceride is composed of three fatty acid molecules bonded to one glycerol molecule. When an organism has excess energy available from food or from photosynthesis, it may use that energy to form stores of triglycerides. These can later be broken down to yield energy when the organism needs it. Fats and oils contain twice as much stored energy per unit of weight, as carbohydrates or proteins.

HISTORY OF BIOCHEMISTRY

Biochemistry as a science is rather young. Any attempts to trace its history must go back to the beginnings of organic chemistry from which it mainly took root.

Theophrastus Bombastus von Hohenheim (1493-1541) better known as Paracelsus – was the earliest to refer to some chemical aspects of medicine in his writings. Early chemists were mostly seized with the task of finding out the chemistry of plant and animal materials. The Swedish pharmacist Scheele (1742-1786) isolated and studied a number of substances like citric acid, lactic acid and uric acid from biological materials.

Lavoisier (1743-1794) was the earliest to notice that the "*burning*" inside the body is similar to the combustion and oxidation of organic materials; he realized that oxygen was consumed and carbon dioxide was given out in both instances. This may be said to be the beginning of the modern animal calorimetry. Leibig (1803-1873) showed that plant material depended for its nutrition on simple inorganic substances.

The theory of 'vitalism' indicates that animal materials fundamentally differed from lifeless substances in that only living matter can synthesise them – proved to be only a myth when Wohler (1800-1882) synthesised urea by simple chemical processes in the laboratory. The earliest book relating to biochemistry was Lectures in Animal Chemistry published by the Swedish chemist Berzelius in

1806.

During the later part of the nineteenth century Chevreul, Kossel and Emil Fischer contributed a great deal to the elucidation of the chemistry of fats, proteins and mucoproteins and carbohydrates, the principal constituents of protoplasm. The foundations for enzyme chemistry were laid by the studies on yeast fermentation by Pasteur (1822-1895). Buchner ably pursued these concepts further into the early part of the present century. It is to Michaelis we owe the present theory of enzyme-substrate interactions.

The concepts developed by Arrhenius vant Hoff and Oswald about osmotic pressure and electrolytes and the concept of pH developed by Sorenson enabled better understanding of the functioning and homeostasis of the body fluids. Several pioneers have contributed in bringing the subject to its present stage. They used relatively simple tools and analytical methods for their studies and succeeded in unravelling many of the fundamental processes of biochemistry. Meyerhoff and Hill discovered the interrelationship between muscle contraction and lactic acid production; Warburg and others extensively studied the mechanisms of cellular oxidation and helped in the understanding of hydrogen and electron transferring systems in biological oxidations; Lippman recognized the high energy stored in the terminal pyrophosphate linkages of adenosine triphoshpate (ATP). Szent-Gyorgi and Krebs elucidated the aerobic oxidation reactions of citric acid cycle.

While these fundamental discoveries were being made, advances in other sciences, notably physics (including optics) and physical chemistry led to the development of highly sophisticated tools of study and analysis. The electron microscope, the ultracentrifuge, chromatographic and electrophoretic methods of analysis, labelling with radioactive and mass isotopes, x-ray diffraction and a host of others have all contributed to the rapid advance of the subject to its present stage during the last few decades. The scientific and enquiring outlook have ousted superstition and blind belief and research has become and honourable career and got its due recognition from the State and the public paving the way to research by thousands of enthusiastic workers and rapid unravelling of many a hitherto difficult problem. The elucidation of the structure of insulin by Sanger, synthesis of the posterior pituitary hormones by Du Vigneaud, elucidation of the structure and enzymic synthesis of nucleic acids by Watson and Crick, Kornberg, Nirenberg and Khorana and the unravelling of the structure of the gene by Jacob and Monod are but a few of the more recent achievements of modern biochemistry.

SCOPE OF BIOCHEMISTRY

Biochemistry encompasses a study of the chemical nature of all living matter from the smallest virus and microorganism to the most complex and highly evolved human being. The relationship of the living to their environment; the processes by which an exchange of chemical substances takes place between the living organism and its environment through digestion, absorption and excretion, the processes by which the absorbed materials are utilized for synthetic reactions leading to growth and replenishment of tissues and multiplication of the cell and the species; the metabolic breakdown of the materials to supply energy for all the above; the mechanisms which regulate with precision all these processes by means of hormonal and neuro regulatory stimuli-all these come under the purview of biochemistry.

The chemical substances involved are both inorganic and organic and the processes in their entry into and removal from the organism invoke the principles familiar to physical chemistry. Thus the study of biochemistry requires a proper basic knowledge of inorganic, organic and physical chemistry since it has its roots in all the three.

So far no RNA molecules that direct the replication of other RNA molecules have been identified in nature. But ingenious techniques devised by Cech and Jack W. Szostak (Massachusetts, USA) modified naturally occurring **ribozymes** so that they can carry out some of the most important subreactions of RNA replication, such as stringing together nucleotides or oligonucleotides (short sequences of nucleotides).

TRENDS IN BIOCHEMISTRY

Any working biochemist should know what is currently going on in the field of biochemistry because the biochemical literature is still growing faster than the scientific literature as a whole. Progress has been made in the field of several areas which are useful for mankind. These areas are Immunology, Cell and molecular Biology, Plant biochemistry, Biotechnology, Environmental and Biochemical Toxicology, Microbial and Fermentation Technology, Bio-informatics, Molecular Medicine, Nutrition in Health and Disease and Environmental Biochemistry.

In several teaching institutes, new approaches of biochemistry teaching have been developed specially in the field of metabolic pathways, designed to engage students with the material, and its complexities. Based on a novel way of presenting metabolic pathways, in which the focus is placed on proteins rather than metabolites, simple tutorial-based exercises and mini-projects are described, bringing metabolism to life, yet requiring very few resources.

Recent developments in mass spectrometry and protein arrays provide opportunities to derive systematically proteomic information from small samples of cellular material. Relative quantification among samples can be achieved with either gel-based or gel-free approaches. Furthermore, the adaptation of specific techniques facilitates absolute quantification. The relative quantification in two-dimensional gel electrophoresis may be used in non-gel-based approaches such as isobaric tagging of peptides, pre-labelling of living cells with isotopomeric forms of essential amino acids and protein array platforms. In addition, novel flow-cytometry-based approaches are also in use. These technologies are used to determine accurately the levels of proteins or biomarkers in a wide range of samples.

The studies of topological properties of the genetic material including DNA under winding and overwinding, knotting, and tangling profoundly influence virtually every major nucleic acid process. Such studied helps to visualize three-dimensional relationships, offer practical approaches and knowledge of topoisomerases (the enzymes that regulate the topological state of DNA in the cell). These ubiquitous enzymes perform a number of critical cellular functions by generating transient breaks in the double helix. During this catalytic event, topoisomerases maintain genomic stability by forming covalent phosphotyrosyl bonds between active site residues and the newly generated DNA termini. Topoisomerases are essential for cell survival. However, because they cleave the genetic material, these enzymes also have the potential to fragment the genome. The enzyme is exploited by some of the most widely prescribed anticancer and antibacterial drugs currently in clinical use. In addition to curing cancer, topoisomerase action also has been linked to the induction of specific types of leukemia.

CHAPTER 2

Water: The Solvent of Life

Water (from the Anglo-Saxon and Low German word, *wœter*) is a colourless, tasteless, and odourless substance that is essential to all forms of life that we know of. Can you think of anything that has no taste, colour or smell, comes in the form of a liquid, solid and gas, and is necessary for the survival of people, plants and animals? You guessed it–water! Water is remarkable. You can't taste it, smell it, and you can almost see right through it. It is the only thing on earth that can exist naturally as a solid, a liquid and a gas. Liquid water is found in rivers, lakes and oceans. Solid water is water that is frozen, like snow and ice. Water is a gas when it is in the air of our atmosphere as vapor or steam.

Water is known in science as H_2O. That means the smallest possible amount of water, **one molecule, is made from two hydrogen atoms that attach to one oxygen atom**. The hydrogen atoms have a positive electrical charge and the oxygen atom has a negative charge. Because opposite charges attract, the charges on either end of the water molecule attract other water molecules and group together, forming water drops. The charges act like magnets. That is how water sticks together in raindrops, rivers, oceans, ponds, and your drinking glass.

Water has multifarious uses, hence it is one of the most precious gifts of nature without which no life could survive on earth. When our population was limited, its bountiful supplies seemed endlessly renewable. We could then afford to foul one water source, abandon it, and move to another. This, however, is no longer possible since the exponential growth rates of human population, industries, and farms have already reduced the availability of water to below its minimum *per capita* daily requirement of about 23 litres. Consequently, any further expansion of our activity will depend upon how well are we able to prevent the loss and contamination of the available water resources.

The physical properties of a substance are properties that have everything to do with the substance's appearance. The chemical properties are properties that are often used in chemistry, to address the state of a substance. Physical and chemical properties can tell us something about the behaviour of a substance in certain circumstances. There are several different physical and chemical properties, which are often used alternately.

CHEMICAL

1. **Light absorption:** This is the amount of light a certain amount of water can absorb over time.
2. **The pH:** The pH has its own scale, running up from 1 to 14. The pH shows whether a substance is acid (pH 1-6), neutral (pH 7) or basic (pH 8-14). The number of hydrogen atoms in the substance determines the pH. The more hydrogen atoms a substance contains, the lower the pH will be. A substance that contains many hydrogen atoms is acid. We can measure the pH by dipping a special colouring paper in the substance, the colours shows which pH the

substance has.

3. **Alkalinity:** This is the capacity of water to neutralize an acid or a base, so that the pH of the water will not change.

PHYSICAL

The properties of water arise from the hydrogen bonding and tetrahedral arrangement of electron pairs around the oxygen atom. The size of water molecule is 3 Å (1Å=10^{-10} m)

FIG. 2.1: Snowflake crystal.

- **Capillary rise:** The rise of water in a capillary tube (tube having an approximate diameter of a hair) to a level higher than that in the container is called capillary rise. This rise is due to the combined action of cohesion, adhesion and surface tension.
- **Cohesion and adhesion**: The attraction of like molecules (here water molecules) is known as cohesion and the attraction of unlike molecules (like soil and water or water and its contents) is known as adhesion. Both of these forces are responsible for retention, movement and utilization.
- **Conductivity:** This means the amount of electricity that water can conduct. It is expressed in a chemical magnitude.
- **Density:** The density of water means the weight of a certain amount of water. It is usually expressed in kilograms per cubic meter.
- **Hydrogen bonding**: The phenomenon by which hydrogen atoms act as connecting linkages between water molecules is called hydrogen bonding.
- **Polarity:** Water molecules are dipolar in nature for which they cling to each other through H^+ bonding rather than acting individually.
- **Surface tension:** It is a phenomenon occurring typically at the interface of a liquid and gas and results from the greater attraction of water molecules for each other than for the air above.
- **Thermal properties:** This refers to what happens to water when it is heated; at which temperature it becomes gaseous and that sort of thing.
- **Viscosity:** This means the syrupiness of water and it determines the mobility of water. When the temperature rises, the viscosity degrades; this means that water will be more mobile at higher temperatures. When a fluid is moved in shear, the force required is proportional to the velocity of the shear. The proportionality factor is called viscosity. For water, viscosity decreases by about 3 per cent per 10 °C rise in temperature.

CHEMICAL STRUCTURE

A water molecule consists of three atoms; an oxygen atom and two hydrogen atoms, which are bond together like little magnets. The atoms consist of matter that has a nucleus in the centre. The difference between atoms is expressed by atomic numbers. The atomic number of an atom depends on the number of protons in the nucleus of the atom. Protons are small positively charged particles. Hydrogen has one proton in the nucleus and oxygen has eight. There are also uncharged particles in the nucleus, called neutrons. Next to protons and neutrons, atoms also consist of negatively charged electrons, which can be found in the electron cloud around the nucleus. The number of electrons in an atom equals the number of protons in the nucleus. The attraction between the protons and electrons is what keeps an atom together.

Molecule Weight

The weight of a molecule is determined by the atomic masses of the atoms that it is built of. The atomic mass of an atom is determined by the addition of the number of protons and neutrons in the nucleus, because the electrons hardly weigh anything. When the atomic masses of the separate atoms are known, one simply has to add them up to find the total atomic mass of a molecule, expressed in grams per mol. A mol is an expression of the molar weight of a molecule, derived from the weight of a hydrogen molecule, which is 1 mol.

Hydrogen has a relative atomic mass of 1 g/ mol and oxygen has a relative atomic mass of 16 g/ mol. Water consists of one oxygen atom and two hydrogen atoms. This means that the mass of a water molecule is 1 g + 1 g + 16 g = 18 g / mol.

When the number of moles of water is known, one can calculate how many grams of weight this is, by using the molar weight of water.

The molar weight of separate atoms can be found in the periodic table of Mendelejef.

Water is often called the universal solvent. Water molecules in contact with a crystal orient themselves to neutralize the attractive forces between the ions in the crystal structure. The liberated ions are then hydrated by these water molecules as shown in Fig. 2.2, preventing them from recombining and re-crystallizing. This solvency and hydration effect is shown quantitatively by water's relatively high dielectric constant.

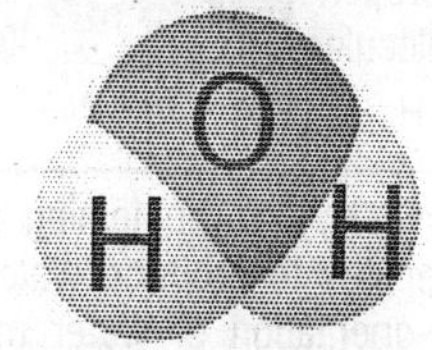

FIG. 2.2: A water molecule.

Water ionizes so very slightly, producing only 10^{-7} moles of hydrogen and 10^{-7} moles of hydroxyl ions per liter that it is an insulator - it cannot conduct electrical current. As salts or other ionizing materials dissolve in water, electrical conductivity develops. The conductivity of naturally occurring waters provides a measure of their dissolved mineral content.

Another important phenomenon occurring in water solutions related to dissolved matter (solutes), rather than to water (solvent), is osmotic pressure. If two aqueous solutions are separated by a membrane, water will pass from the more dilute into the more concentrated one. This important process controls the performance of all living cells. It explains the effectiveness of food preservation by salting; the salt creates a strong solution, disrupting the cells of organisms that might cause food spoilage, as the water inside their bodies leaves in an attempt to dilute the external salt solution. In specially designed membrane cells, the osmotic flow of water across the membrane can be reversed by applying a sufficiently high pressure to the more concentrated solution. This process of *reverse osmosis* is a practical one for desalination of water.

The oceans alone constitute 97 per cent of the hydrosphere, and the polar ice caps, ice fields and glaciers make up another 2 per cent in the form of ice. Inland surface waters (lakes and rivers) account for barely 0.02 per cent.

The chemical substances are usually classified into groups of compounds with similar properties such as salts, alcohol, water, it would be solid (ice at room temperature, with a regular crystal structure. Clearly, liquid water is not solid at room temperature and, therefore, it is generally thought that is made up of a large number of small clusters (or *icebergs*) or hydrogen-bonded water molecules separated by volumes containing free bonded molecules only. Each of these clusters, which have a rather open cage-like cluster structure, is therefore dynamic, in contrast to the static structure of crystalline solids or sugars. Water, however, is unique; there is nothing else like it. To understand the properties of water, it is best to start with its molecule, which consists of two hydrogen atoms bonded to an oxygen atom. The water molecule is planar and angled that, due to the high electronegativity of oxygen, there is a partial separation of charge leading to the formation of hydrogen bonds between molecules.

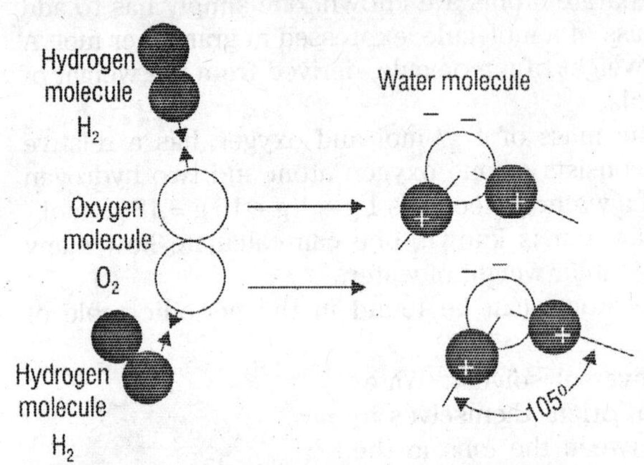

FIG. 2.3: (A) The joining of diatomic hydrogen and oxygen molecules to produce water molecules of a polar nature. (B) The orientation of water molecules tends to keep ions from recombining and thus precipitating from solution. This accounts of water's capabilities as a solvent.

It is important to note that although these hydrogen bonds are strong compared with other intermolecular forces (van der Walls bonds), they are only about 1/24th as strong as O-H covalent bond. These hydrogen bonds are responsible for joining several water molecules together and forming a three-dimensional structure. In the liquid water, there are average 2 to 3 hydrogen bonds per molecule of water, but in the ice four bonds per molecule of water exist. The presence of four bonds per molecule in ice gives the three-dimensional figure with each water molecule is surrounded by four other water molecules in a tetrahedral fashion to a form a tridymite arrangement with the oxygen atoms forming layers of *puckered* hexagonal rings.

Electrolytes

Water is known as the universal solvent. When it dissolves a mineral, new materials are produced from the atoms released by the mineral. These are fundamental particles called ions. The break-up of a chemical compound by dissolution in water forms cations, which are positively charged, and anions, which are negatively charged. Generally, an increase in water temperature causes an increase in the solubility of most salts. Important and notable exceptions are $CaCO_3$, $CaSO_4$, $MgCO_3$, and $Mg(OH)_2$, all of the which become less soluble as the temperature increase.

There are a number of ways in which this phenomenon of ionization can be demonstrated. The simplest is an experiment in which an electric light is connected to an electrical circuit having two separated poles inserted in a beaker of water. In pure water, the lamp does not light when the switch is on because pure water is an insulator, not a conductor. Salt is added to the water in small increments, and gradually the light brightens, the intensity being proportional to the amount of salt added. In pure water, the circuit is open; current begins to flow only after sodium and chloride ions from the salt are present to transport electrons through the solution. If crystals of sugar are added to the water instead of salt, nothing happens; salt is an electrolyte and sugar is not.

It was observed many years ago that impurities in water lower its freezing point. Upon careful investigation it was discovered that 1 mole/L of a non-electrolyte such as sugar or alcohol dissolved in water lowers the freezing point by 1.86°C. It was found that 1 mole/L of sodium chloride lowered the freezing point by almost twice the value found for sugar and alcohol, and 1 mole/L of sodium sulphate produced almost 3 times this depression. This is due to the production of two ions, sodium and chloride, from one molecule of sodium chloride salt, and three ions, two sodium and one sulphate, from one molecule of sodium sulphate (Na_2SO_4).

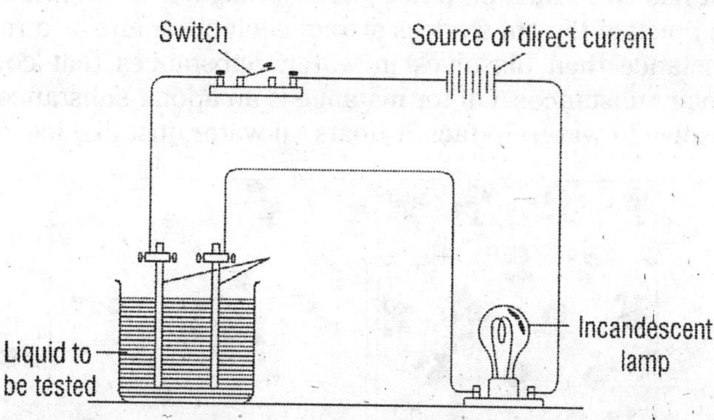

FIG. 2.4: Pure water is nonconductive, but addition of an ionizing salt allows current to pass.

PHYSICAL STATES OF WATER

Water exists in three states: solid, liquid and gaseous. At a normal temperature of about 25°C it is liquid, but below 0°C it will freeze and turn to ice. Water can be found in the gaseous state above 100°C, this is called the boiling point of water, at which water starts to evaporate. The water turns to gas and is then odourless and colourless. How fast water evaporates depends on the temperature; if the temperature is high, water will evaporate sooner.

What Happens if Water Changes Phase?

The changes from a liquid to a solid or to a gas are called phase changes (when a substance, such as water changes phase, its physical appearance changes, but not its chemical properties). This is because the chemical structure remains the same, but the molecules of which it consists will float a little further apart. In the solid state, the water molecules are fairly close together, but in the liquid state they are a bit further apart. The water becomes liquid as a result of parting molecules. When water changes from liquid to gas the molecules will part even further, that is why we cannot detect it.

Why Does Ice Float on Water?

When substances freeze, usually the molecules come closer together. Water has an abnormality there: it freezes below 0°C, but when temperatures fall to 4°C, water starts to expand and as a result the density is lower. Density of a substance means the weight in kilograms of a cubic metre of a substance. When two substances are mixed but do not dissolve in one another, the substance with the lowest density floats on the other substance. In this case, that substance is ice, due to the increased density of water.

How Come Not All Substances are Water-Soluble?

Polarity determines if a substance is water-soluble. A polar substance is a substance that has two kinds of 'poles', as in a magnet. When another substance is also polar, the poles of the substances attract each other and as a result, the substances mix. A substance then dissolves in water. Substances that contain no 'poles' are called apolar substances. Oil for instance is an apolar substance, which is why oil does not dissolve in water. In fact, it floats on water, just like ice, due to its smaller density.

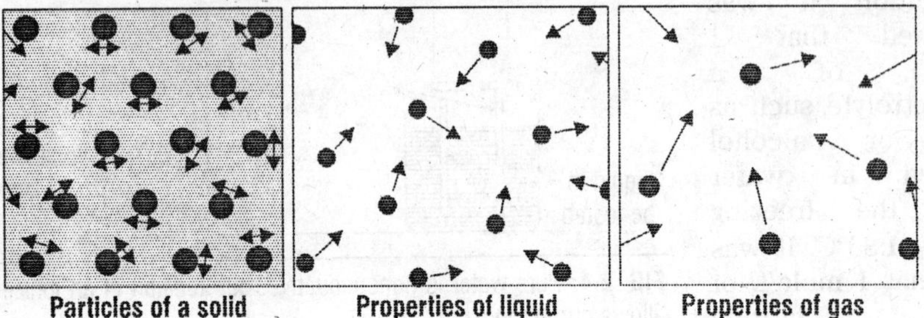

Hard Water

When water is referred to as 'hard' this simply means, that it contains more minerals than ordinary water. These are especially the minerals calcium and magnesium. The degree of hardness of the water exceeds, when more calcium and magnesium dissolve.

Magnesium and calcium are positively charged ions. Because of their presence, other positively charged substances will dissolve less easy in hard water than in water that does not contain calcium and magnesium. This is the cause of the fact that soap does not really dissolve in hard water.

PHYSICAL PROPERTIES OF WATER

Dissolving Ability

Water can dissolve more substance than other solvent so it is called universal solvent. It is a polar molecule so it is very efficient in dissolving particularly the substance in which the atoms are held by ionic bonds. Besides, many gases and organic molecules can also be dissolved in water. Water is chemically inert which does not change the properties of solutes, nor does it changed itself. Most biochemical reactions take place in the liquid phase which is provided by the water. Solvent properties of water are not only essential to life processes, but also important for domestic and industrial purposes.

Specific Heat

Specific heat of water is 1.0 calorie which is the highest among all substances except lithium and ammonia. Because of high specific heat, water heats and cools more slowly than other substances. Water has a greater capacity to store heat than other substances. Large bodies of water do not fluctuate much in temperature as compared to large bodies of land, because so much heat is required to raise the water temperature.

Latent Heat

A great deal of heat energy is required to change water from liquid to vapour phase which is known as latent heat of vaporization. The latent heat of vapourisation of water is 539.55 cal. per gram which is quite high than other substances. Similarly, extra energy is required to melt the ice into liquid. It is called latent heat of fusion. It is 79.71 cal per gram for water. The value of two types of latent heat is high for water because of presence of hydrogen bonding. These properties help in moderating the surface temperature of large bodies of water. It also helps in maintaining the body temperature of body by evaporating the perspiration during heat periods.

Density

Ice is about 9 per cent less dense than water, whereas most other substances in nature have their densities in the order: solid > liquid > gas. The maximum density of water is reported at 4°C at atmospheric pressure due to which ice or water at any other temperature will float on it. This property of water is due to hydrogen bonding. The density properties of water cause the temperature stratification, and vertical mixing of lakes which are helpful in many ways to the aquatic organisms. Due to this property, complete freezing of the bodies of water is also checked. In cold seasons when the atmospheric temperature reaches at 4°C the surface water sinks to the bottom, so only surface of lakes is freezed while the bottom temperature remains at 4°C keeping the organisms alive.

Cohesion and Adhesion

The ability of stick water itself is known as cohesion whereas the ability of water to stick to some solid substance is called adhesion. Cohesion property gives water a very high surface tension which is caused by the strong attraction of polar water molecules for each other by strong hydrogen bonds on the surface of water. Several insects are able to move on the top of water without breaking the surface film because of the surface tension of water. Water in soil rises to several feet up by capillary action. It is due to adherence to soil particles where oxygen atoms of silica attract the hydrogen atoms of water molecule.

Water Retention Forces

Soil serves as a water reservoir but a leaky one. When too much water is added, the excess runs off over the surface or into deeper layers. Water is held in soils because of the attraction between unlike charges- a positive ion attracted to a negatively charged ion. The positively charged hydrogens of water are attracted to nearby negatively charged ions, such as oxygen, even to the oxygen of another adjacent water molecule.

Water Potential

Water potential is defined as the potential energy of water per unit mass of water in the system. The total water potential of a sample is the sum of four component potentials: gravitational, matric, osmotic, and pressure. Gravitational potential depends on the position of the water in a gravitational field. Since, it refers to work, it is an energy term, not a pressure term. The soil water potential is a combination of the effects of the surface area of soil particles and small soil pores that adsorb water (metric potential), the effects of dissolved substances (solute or osmotic potential), and the atmospheric or gas pressure effects (pressure potential). In non-salty well-drained soil, the metric potential is almost equal to the water potential.

The water potential of a solid or liquid sample can be found by relating the sample water potential reading to the vapor pressure of air in equilibrium with the sample. The relationship between the sample's water potential (Y) and the vapor pressure of the air is:

$$\frac{RT}{M}\ \text{In}\ \frac{}{0}$$

Where p is the vapor pressure of the air, p_0 is the saturation vapor pressure at sample temperature, R is the gas constant (8.31 J/mol K), T is the Kelvin temperature of the sample, and M is the molecular mass of water. The vapor pressure of the air can be measured using a chilled mirror, and ρ_0 is computed from sample temperature.

WATER QUALITY

Water is essential to human life and to the health of the environment. As a valuable natural resource, it comprises marine, estuarine, freshwater (river and lakes) and groundwater environments, across coastal and inland areas. Water has two dimensions that are closely linked - quantity and quality. Water quality is commonly defined by its physical, chemical, biological and aesthetic (appearance and smell) characteristics. A healthy environment is one in which the water quality supports a rich and varied community of organisms and protects public health.

Water quality in a body of water influences the way in which communities use the water for activities such as drinking, swimming or commercial purposes. More specifically, the water may be used by the community for:

1. Supplying drinking water.
2. Recreation (swimming, boating)
3. Irrigating crops and watering stock.
4. Industrial processes.
5. Navigation and shipping.
6. Production of edible fish, shellfish and crustaceans.
7. Protection of aquatic ecosystems.
8. Wildlife habitats.
9. Scientific study and education.

In a recent EPA survey the NSW community rated water quality one of the highest priority environmental issues, and has expressed its wish that water quality be improved or, at the very least, not further degraded.

Importance Water Quality

Our water resources are of major environmental, social and economic value to NSW, and if water quality becomes degraded this resource will lose its value. Water quality is important not only to protect public health - water provides ecosystem habitats, is used for farming, fishing and mining, and contributes to recreation and tourism.

If water quality is not maintained, it is not just the environment that will suffer - the commercial and recreational value of our water resources will also diminish.

Factors Affecting Water Quality

Water quality is closely linked to the surrounding environment and land use. Other than in its vapour form, water is never pure and is affected by community uses such as agriculture, urban and industrial use, and recreation. The modification of natural stream flows by dams and weirs can also affect water quality. The weather too can have a major impact on water quality, particularly in a dry country like Australia which is periodically affected by droughts.

Groundwater is a major source of water in NSW, with reserves estimated to be 200 times more than the water in dams. Groundwater is an integral part of our water supply. At times of low river flow groundwater enters the rivers, maintaining river flow. Although data on groundwater quality is limited, it is clear that, like other bodies of water, groundwater close to urban or industrial development is vulnerable to contamination.

Generally the water quality of rivers is best in the headwaters, where rainfall is often abundant. Water quality often declines as rivers flow through regions where land use and water use are intense and pollution from intensive agriculture, large towns, industries and recreation areas increases.

There are of course exceptions to the rule and water quality may improve downstream, behind dams and weirs, at points where tributaries or better quality groundwater enter the mainstream, and in wetlands.

Rivers frequently act as conduits for pollutants by collecting and carrying wastewater from catchments and, ultimately, discharging it into the ocean. Storm water, which can also be rich in nutrients, organic matter and pollutants, finds its way into rivers and oceans mostly via the storm water drain network. Beach water quality in NSW may also be affected by bacteria from sewer overflows or other runoff into storm water drains.

Measurement of Water Quality

The presence of contaminants and the characteristics of water are used to indicate the quality of water. These water quality indicators can be categorized as:

1. **Biological:** bacteria, algae.
2. **Physical:** temperature, turbidity and clarity, colour, salinity, suspended solids, dissolved solids.
3. **Chemical:** pH, dissolved oxygen, biological oxygen demand, nutrients (including nitrogen and phosphorus), organic and inorganic compounds (including toxicants)
4. **Aesthetic:** odours, taints, colour, floating matter.
5. **Radioactive:** alpha, beta and gamma radiation emitters.

Measurements of these indicators can be used to determine, and monitor changes in, water quality, and determine whether the quality of the water is suitable for the health of the natural environment and the uses for which the water is required.

The design of water quality monitoring programs is a complex and specialized field. The range of indicators that can be measured is wide and other indicators may be adopted in the future. The cost of a monitoring program to assess them all would be prohibitive, so resources are usually directed towards assessing contaminants that are important for the local environment or for a specific use of the water.

This water quality information can then be used to develop management programs and action plans to ensure that water quality is protected.

CHAPTER 3

pH and Buffers

pH is the acronym for *potential hydrogen*. It is a measure of the degree of saturation of the hydrogen ion in a substance or solution. The molecule of water (H_2O, H = Hydrogen and O = Oxygen). If water and water is combined we get $H_2O + H_2O \rightarrow H_3O + OH$. H_3O (the hydronium ion$^+$) is the acid element and OH^- (the hydroxyl ion$^-$) is the base or alkaline element. In pure water these are balanced and upon measuring with a pH meter the reading would be 7. Seven (7) is neutral on a pH scale which goes from 0 to 14. This scale corresponds to the hydrogen ion concentration from 10^0 to 10^{-14} moles per liter. This is a huge range which sensitive instruments can measure.

When the H_3O and OH^- are out of balance a pH meter will detect this and the reading will move above or below 7. Like a teeter-totter, if one goes up the other goes down and vice versa. In the human body a pH balancing act is continuously going on to maintain homeostasis. When defining measurement values of certain pH levels of human fluids, there are no absolutes that can be written in stone because the value that *should be here* has to be balanced against other values *that should be there*. In essence, in the human body things never happen in a vacuum and you need to be ever mindful of these things as you make your measurements.

Acidic and basic are two extremes that describe chemicals, just like hot and cold are two extremes that describe temperature. Mixing acids and bases can cancel out their extreme effects; much like mixing hot and cold water can even out the water temperature. A substance that is neither acidic nor basic is neutral.

The pH scale measures how acidic or basic a substance is. It ranges from 0 to 14. A pH of 7 is neutral. A pH less than 7 is acidic, and a pH greater than 7 is basic. Each whole pH value below 7 is ten times more acidic than the next higher value. For example, a pH of 4 is ten times more acidic than a pH of 5 and 100 times (10 times 10) more acidic than a pH of 6. The same holds true for pH values above 7, each of which is ten times more alkaline–another way to say basic–than the next lower whole value. For example, a pH of 10 is ten times more alkaline than a pH of 9.

Pure water is neutral, with a pH of 7.0. When chemicals are mixed with water, the mixture can become either acidic or basic. Vinegar and lemon juice are acidic substances, while laundry detergents and ammonia are basic.

Chemicals that are very basic or very acidic are called "reactive". These chemicals can cause severe burns. Automobile battery acid is an acidic chemical that is reactive. Automobile batteries contain a stronger form of some of the same acid that is in acid rain. Household drain cleaners often contain lye, a very alkaline chemical that is reactive.

Therefore, pH is the quantitative measure of the acidity or basicity of aqueous or other liquid solutions. The term is widely used in biochemistry. pH translates the values of the concentration of the hydrogen ion–which ordinarily ranges between about 1 and 10-14 gram-equivalents per litre–into numbers between 0 and 14. In pure water, which is neutral (neither acidic nor alkaline), the concentration of the hydrogen ion is 10^{-7} gram-equivalents per litre, which corresponds to a pH of 7. A solution with a pH less than 7 is considered acidic; a solution with a pH greater than 7 is considered basic, or alkaline.

FUNDAMENTAL ACID-BASE CONCEPTS

An **acid** is a chemical species that can donate a proton (H^+), and a **base** is a species that can accept (gain) a proton. The conjugate base of an acid is the species formed after the acid loses a proton; the base can then gain another proton to return to the acid. In solution, these two species (the acid and its conjugate base) exist in equilibrium. Therefore, pH is:

$$pH = -\log [H^+]$$

Where $[H^+]$ is the molar concentration of protons in aqueous solution. When an acid is placed in water, free protons are generated according to the general following reaction in which HA and A^- are generic symbols for an acid and its de-protonated form, the conjugate base:

$$HA + H_2O \qquad H_3O^+ + A^+$$

$$\text{Acid + Base} \qquad \text{Conjugate Acid + Conjugate Base}$$

The above equation is useful because it shows that HA giving up a proton to become A^- and water acts as a base (accepting the proton released by HA). However, the nomenclature H_3O^+ is somewhat misleading, because the proton is actually solvated by many water molecules. Hence, the equilibrium is often written as follows, where H_2O is the base:

$$HA \qquad HA^+ + A^-$$

The Law of Mass Action and Equilibrium Constants

Using the Law of Mass Action, which says that for a balanced chemical equation of the type:

$$aA + bB \overset{K}{} \quad cC + dD$$

in which A, B, C, and D are chemical species and a, b, c, and d are their stoichiometric coefficients, a constant quantity, known as the equilibrium constant (K) can be found from the expression:

$$K = \frac{C^c \; D^d}{A^a \; B^b}$$

Where the brackets indicate the concentrations of species A, B, C, and D at equilibrium.

Equilibrium Constant for an Acid-Base Reaction

Using the Law of Mass Action, one can also define an equilibrium constant for the acid dissociation equilibrium reaction. This equilibrium constant, known as K_a, is defined as follows:

$$K = \frac{H^+ \quad A^-}{HA}$$

Equilibrium Constant for the Dissociation of Water

One of the simplest applications of the Law of Mass Action is the dissociation of water into H^+ and OH^-shows as follows:

$$H_2O\ (I) \rightleftharpoons H^+_{(aq)} + OH^-_{(aq)}$$

The equilibrium constant for this dissociation reaction, known as K_w, is given as follows:

$$Kw = [H^+][OH^-]$$

(H_2O is not included in the equilibrium-constant expression because it is a pure liquid.) Hence, increasing the OH^- concentration of an aqueous solution has the effect of decreasing the H^+ concentration, because the product of these two concentrations must remain constant at a given temperature. Thus, in water, the equilibrium underlies the equivalency.

Definition of pH, pOH, and pKw.

The concentrations of hydrogen ions and indirectly hydroxide ions are given by a pH number. *pH is defined as the negative logarithm of the hydrogen ion concentration.* The equation is:

$$pH = -\log [H^+]$$

Similarly, $$pOH = -\log [OH^-]$$

and $$pKw = -\log [Kw]$$

Logarithms of numbers that are multiples of ten are merely the exponents of the number including the sign.

BRONSTED-LOWRY (BL) ACIDS AND BASES

The Bronsted-Lowry definition is named for Johannes Bronsted and Thomas Lowry. They independently proposed it in 1923. A Bronsted-Lowry (BL) **acid** is defined as *any substance that can donate a hydrogen ion (proton)* and a Bronsted-Lowry **base** is *any substance that can accept a hydrogen ion (proton)*. Thus, according to the BL definition, acids and bases must come in what is called *conjugate pairs.*

The most common acid-base theory used by medicinal chemists is the Bronsted-Lowry definition, which is concerned with a molecule's ability to donate or accept protons. The strength of an acid is defined as the position of the following equilibrium expression:

$$HA \rightleftharpoons H^+ + A^-$$

For example, consider acetic acid dissolved in water:

$$CH_3COOH\ (aq) + H_2O \rightleftharpoons H_3O^+\ (aq) + CH_3OO{=}\ (aq)$$

Notice that it has been written as $H_2O(l)$ explicitly in these reactions. The reason is that acid/base dissociation occurs by a *proton transfer reaction* from an acid species to a specific water molecule. The transfer occurs through a hydrogen bond between the acid molecule and a solvating water molecule.

Here, CH_3COOH is a BL acid because it can donate a proton, and CH_3COO^- its *conjugate* base because it can accept a proton. Note that H_2O and H_3O^+ also form such a conjugate pair.

Note that the H_3O^+ rather than H^+ has been used to denote the nature of H^+ ions in water in the above reaction. This is really only a *very crude* representation of the true nature of solvated H^+ ions. Although we will use it in the context of our discussion of acids and bases, the more modern understanding of the true nature of H^+ in water will be given in a later section in the context of how acidic solutions conduct electricity.

Similarly when ammonia is dissolved in water, one has:

$$H_2O\ (l) + NH_3\ (aq) \rightleftharpoons NH_4^+\ (aq) + OH^-\ (aq)$$

Here, NH_3 is the BL base and its conjugate acid is NH_4^+. Similarly, H_2O acts as a BL acid and OH^- acts as a BL base.

Another important advantage of the BL definition is that we are not limited to water as the solvent. Consider the reaction that occurs when HCl is dissolved in ammonia:

$$HCl\ (in\ NH_3) + NH_3\ (l) \rightleftharpoons NH_4^+\ (in\ NH_3) + Cl^-\ (in\ NH_3)$$

Here, HCl acts as a BL acid with Cl^- as its conjugate base. Also, NH_3 acts as a BL base with NH_4^+ as its conjugate acid.

An interesting ambiguity comes up within the BL definition, namely, that some species can act either as a BL acid *or* a BL base. Such beasts are called *amphoteric*. An example is the hydrogen carbonate ion, HCO_3 When dissolved in water, two possible reactions can occur:

$$HCO_3\ (aq) + H_2O\ (l) \rightleftharpoons H_3O^+\ (aq) + CO_3^{2-}(aq)$$

Or

$$HCO_3^-\ (aq) + H_2O\ (l) \rightleftharpoons H_2CO_3^+\ (aq) + OH^-(aq)$$

In the first of these, HCO_3^- acts as a BL acid with CO_3^{2-} as its conjugate base, while in the second it acts as a BL base with H_2CO_3 as its conjugate acid. The treatment of amphoteric reactions is mathematically a little more hideous than you might think, so we will return to such reactions at the end of our acid-base section, when we have developed the necessary machinery.

HENDERSON–HASSELBALCH EQUATION

Lawrence Joseph Henderson (1908) described an equation for the use of carbonic acid as a buffer_solution. Karl Albert Hasselbalch later re-expressed that formula in logarithmic terms, resulting in the Henderson–Hasselbalch equation. Hasselbalch was using the formula to study metabolic acidosis, which results from carbonic acid in the blood.

Henderson–Hasselbalch describes the derivation of pH as a measure of acidity (using pK_a, the acid dissociation constant) in biological and chemical systems. The equation is also useful for estimating the pH of a buffer solution and finding the equilibrium pH in acid-base reactions (it is widely used to calculate isoelectric point of the proteins).

Two equivalent forms of the equation are:

$$pH = pK_a + \log \frac{[A^-]}{[HA]}$$

and

$$pH = pK_a + \log\left(\frac{[base]}{[acid]}\right)$$

Here, pKa is – log(Ka) where Ka is the acid dissociation constant, that is:

$$pK_a = -\log(K_a) = -\log\left(\frac{[H_3O^+][A^-]}{[HA]}\right)$$

For the non-specific Bronsted acid-base reaction:

$$HA + H_2O \rightleftharpoons A^- + H_3O^+$$

In these equations, A – denotes the ionic form of the relevant acid. Bracketed quantities such as [base] and [acid] denote the molar concentration of the quantity enclosed.

In analogy to the above equations, the following equation is valid:

$$pOH = pK_b + \log\left(\frac{[BH^+]}{[B]}\right)$$

Where, BH^+ denotes the salt of the corresponding base B.

Limitations

There are some significant approximations implicit in the Henderson–Hasselbalch equation. The most significant is the assumption that the concentration of the acid and its conjugate base at equilibrium will remain the same as the formal concentration. This neglects the dissociation of the acid and the hydrolysis of the base. The dissociation of water itself is neglected as well. These approximations will fail when dealing with relatively strong acids or bases (pKa more than a couple units away from 7), dilute or very concentrated solutions (less than 1 mM or greater than 1 M), or heavily skewed acid/base ratios (more than 100 to 1).

IONIZATION OF WATER

Water molecules exist in equilibrium with hydrogen ions and hydroxide ions.

$$H_2O \rightleftharpoons H^+ + OH^-$$

The water equilibrium constant is written as:

$$Kw = [H^+]\,[OH^-]$$

Experimentally, it has been found that the concentration of:

$$H^+ = OH^- = 10^{-7}$$

Therefore: $Kw = [10^{-7}][10^{-7}] = [10^{-14}]$

(To multiply exponential numbers - simply add the exponents.)

The values for Kw, H^+, and OH^- concentration all indicate that the equilibrium favors the reactant (water molecules). In other words, only very small amounts of H^+ and OH^- ions are present.

Effect of Acids and Bases on Water Equilibrium

If an acid (H^+) is added to the water, the equilibrium shifts to the left and the OH^- ion concentration decreases.

Water Equilibrium: $H_2O \longleftrightarrow H^+ + OH^-$

If base (OH^-) is added to water, the equilibrium shifts to left and the H^+ concentration decreases.

MEASUREMENT OF pH

The measurement was originally used by the Danish biochemist S.P.L. Sorensen to represent the hydrogen ion concentration, expressed in equivalents per litre, of an aqueous solution: $pH = -\log[H^+]$ (in expressions of this kind, enclosure of a chemical symbol within square brackets denotes that the concentration of the symbolized species is the quantity being considered).

The pH is usually measured with a pH meter, which translates into pH readings the difference in electromotive force (electrical potential or voltage) between suitable electrodes placed in the solution to be tested. Fundamentally, a pH meter consists of a voltmeter attached to a pH-responsive electrode and a reference (unvarying) electrode. The pH-responsive electrode is usually glass, and the reference is usually a mercury-mercurous chloride (calomel) electrode, although a silver-silver chloride electrode is sometimes used.

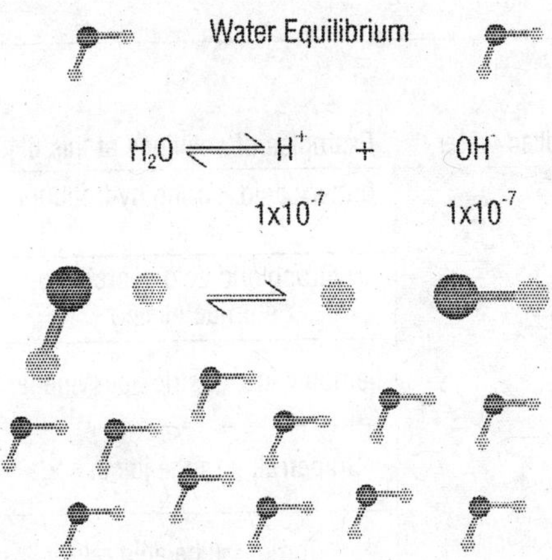

When the two electrodes are immersed in a solution, they act as a battery. The glass electrode develops an electric potential (charge) that is directly related to the hydrogen-ion activity in the solution, and the voltmeter measures the potential difference between the glass and reference electrodes.

The meter may have either a digital or analogue (scale and deflected needle) readout. Digital readouts have the advantage of exactness, while analogue readouts give better indications of rates of change. Battery-powered portable pH meters are widely used for field tests of the pH of soils; such tests may also be performed, less accurately, by mixing indicator dyes in soil suspensions and matching the resulting colours against a colour chart calibrated in pH.

pH Scale

The concentration of hydrogen ions is commonly expressed in terms of the pH scale. Low pH corresponds to high hydrogen ion concentration and *vice versa.* A substance that when added to water increases the concentration of hydrogen ions (lowers the pH) is called an acid. A substance that reduces the concentration of hydrogen ions (raises the pH) is called a base. Finally some substances enable solutions to resist pH changes when an acid or base is added. Such substances are called buffers. Buffers are very important in helping organisms maintain a relatively constant pH.

Study the pH chart given below carefully. Note that each decrease in pH by one pH unit means a tenfold increase in the concentration of hydrogen ions.

BUFFER SOLUTION

A buffer, as defined by Van Slyke (1922), is "a substance which by its presence in solution increases the amount of acid or alkali that must be added to cause unit change in pH". Buffers are important components in experiments designed to study biological reactions by maintaining a constant concentration of hydrogen ions within the physiological range. The pH of mammalian blood is maintained close to 7.38 by buffer systems such as:

$$H_2PO_4^- \Leftrightarrow HPO_4^{2-},\ CO_2 \rightleftharpoons H_2CO_3,\ H_2CO_3 \Leftrightarrow HCO_3^-$$

In living plants, the normal range of pH in tissues is about 4.0-6.2. It is not as narrowly defined as in mammalian tissues. Buffers based on organic and inorganic salts, acids and bases are widely used in biochemical and biological research. Universally applicable buffers for biochemistry have following characteristics:

1. Water solubility.
2. No interference with biological processes.
3. Known complex-forming tendency with metal ions.
4. Non-toxicity.
5. No interference with biological membranes (penetration, solubilisation, adsorption on surface etc.)
6. Very low UV absorption at wavelength >260 nm.

Buffer solutions are solutions that resist change in Hydronium ion and the hydroxide ion concentration (and consequently pH) upon addition of small amounts of acid or base, or upon dilution. Buffer solutions consist of a weak acid and its conjugate base (more common) or a weak base and its conjugate acid (less common). The resistive action is the result of the equilibrium between the weak acid (HA) and its conjugate base (A^-):

$$HA_{(aq)} + H_2O(l) \rightarrow H_3O^+_{(aq)} + A^-_{(aq)}$$

Any alkali added to the solution is consumed by the Hydronium ions. These ions are mostly regenerated as the equilibrium moves to the right and some of the acid dissociates into Hydronium ions and the conjugate base. If a strong acid is added, the conjugate base is protonated, and the pH is almost entirely restored.

Concentration of hydrogen ions compared to distilled water	Examples of solutions at this pH
10,000,000	Battery acid, strong hydrofluoric acid
1,000,000	Hydrochloric acid secreted by stomach lining
1,00,000	Lemon juice, gastric acid vineger
10,000	Grapefruit, orange juice, soda
1,000	Tomato juice acid rain
100	Soft drinking water black coffee
10	Urine saliva
1	"Pure" water
1/10	Sea water
1/1,00	Baking soda
1/1,000	Great salt lake milk of magnegsia
1/10,000	Ammonia solution
1/100,000	Soapy water
1/1000,000	Bleaches oven cleaner
1/10,000,000	Liquid drain cleaner

This is an example of Le Chatelier's principle and the common ion effect. This contrasts with solutions of strong acids or strong bases, where any additional strong acid or base can greatly change the pH. This may be easier to see by comparing two graphs when an strong acid is titrated with a strong base the curve will have a large gradient throughout showing that a small addition of base/acid will have a large effect compared to a weak acid/strong base titration curve which will have a smaller gradient near the pKa.

When writing about buffer systems they can be represented as salt of conjugate base/acid, or base/salt of conjugate acid. Buffer solutions are presented in terms of the Bronsted-Lowry notion of acids and bases, as opposed to the Lewis acid-base theory.

COMMON-ION EFFECT

The common-ion effect is a term used to describe the effect on a solution of two dissolved solutes that contain the same ion. The presence of a common ion suppresses the ionization of a weak acid or a weak base.

For example, if both sodium acetate and acetic acid are dissolved in the same solution they both dissociate and ionize to produce acetate ions.

1. **Sodium acetate** is a strong electrolyte and it dissociates completely in solution.
2. **Acetic acid** is a weak acid and it only ionizes slightly.

According to Le Chatelier's principle, the addition of acetate ions from sodium acetate will suppress the ionization of acetic acid and shift its equilibrium to the left. Thus the per cent dissociation of the acetic acid decreases and increases pH of the solution.

$$NaC_2H_3O_{2(s)} \rightarrow Na^+{}_{(aq)} + C_2H_3O_2^-{}_{(aq)}$$

$$HC_2H_3O_{2(l)} \rightleftharpoons H^+{}_{(aq)} + C_2H_3O_2^-{}_{(aq)}$$

This will decrease the hydrogen ion concentration and thus the common-ion solution will be less acidic than a solution containing only acetic acid.

CALCULATING pH

Titration of a weak acid with a strong base the flat region at pKa is the buffering region.

Titration of a strong acid with a strong base. Note the sharp rise in pH: this solution cannot buffer.

The equilibrium above has the following acid dissociation constant:

Simple manipulation with logarithms gives the Henderson-Hasselbalch equation, which describe pH in terms of pK_a:

In this equation.

1. $[A^-]$ is the concentration of the conjugate base. This may be considered as coming completely from the salt, since the acid supplies relatively few anions compared to the salt lyn.
2. [HA] is the concentration of the acid. This may be considered as coming completely from the acid, since the salt supplies relatively few complete acid molecules (A^- may extract H^+ from water to become HA) compared to the added acid.

Maximum buffering capacity is found when pH = pKa, and buffer range is considered to be at a pH = pKa ± 1.

CHEMISTRY ALLOWS BUFFERS

Illustration of buffering effect: Sodium acetate/acetic acid.

The acid dissociation constant for acetic acid-sodium acetate is given by the equation:

Since this equilibrium only involves a weak acid and base, it can be assumed that ionization of the acetic acid and hydrolysis of the acetate ions are negligible. In a buffer consisting of equal amounts of acetic acid and sodium acetate, the equilibrium equation simplifies to:

$$K_a = [H^+],$$

and the pH of the buffer as is equal to the pK_a.

For determination of the effect of addition of a strong acid such as HCl, the following mathematics would provide the new pH. HCl is a strong acid and is completely ionized in solution. This increases the concentration of H^+ in solution and neutralizes the acetate. The consumed hydrogen ions change the effective number of moles of acetic acid and acetate ions:

After accounting for volume change to determine concentrations, the new pH could be calculated from the Henderson-Hasselbalch equation. Any neutralization will result in a small change in pH, since it is on a logarithmic scale.

THE CHOICE OF THE CORRECT BUFFER

The choice of the correct buffer for a particular biochemical system or technique depends on a number of additional factors. For example: undesired interaction of the buffer with the biopolymer, redox stability, metal ion complexing properties and purity. One way to solve the difficult problem of selecting the right buffer is to evaluate as many buffers as possible. Reviews on the use of buffers in various areas are available [7-9]. However they do not provide detailed information and a comprehensive treatise on the subject should be consulted.

TABLE 3.1: Common buffer compounds used in biology

Common Name	pKa at 25°C	Buffer Range	Temp Effect (pH / °C)**	Mol. Weight	Full Compound Name
TAPS	8.43	7.7 – 9.1	–0.018	243.3	3-{[tris(hydroxymethyl)methyl]amino} propanesulphonic acid
Bicine	8.35	7.6 – 9.0	–0.018	163.2	N,N-bis(2-hydroxyethyl)glycine
Tris	8.06	7.5 – 9.0	–0.028	121.14	tris(hydroxymethyl)methylamine
Tricine	8.05	7.4 – 8.8	–0.021	179.2	N-tris(hydroxymethyl) methylglycine
HEPES	7.48	6.8 – 8.2	–0.014	238.3	4-2-hydroxyethyl-1-piperazineethanesulphonic acid
TES	7.40	6.8 – 8.2	–0.020	229.20	2-{[tris(hydroxymethyl)methyl]amino} ethanesulphonic acid
MOPS	7.20	6.5 – 7.9	–0.015	209.3	3-(N-morpholino) propanesulphonic acid
PIPES	6.76	6.1 – 7.5	–0.008	302.4	piperazine-N,N'-bis(2-ethanesulphonic acid)
Cacodylate	6.27	5.0 – 7.4		138.0	dimethylarsinic acid
MES	6.15	5.5 – 6.7	–0.011	195.2	2-(N-morpholino) ethanesulphonic acid
Acetate	4.76	3.8 – 5.8		59.04	Deprotonated Ethanoic Acid (Non-IUPAC Name);

** Values are approximate.

TABLE 3.2: Citric acid-phosphate buffers

pH	0.2 M Na_2HPO_4	0.1 M Citric Acid
3.0	20.55 ml	79.45 ml
4.0	38.55 ml	61.45 ml
5.0	51.50 ml	48.50 ml
6.0	63.15 ml	36.85 ml
7.0	82.35 ml	17.65 ml
8.0	97.25 ml	2.75 ml

MAKING BUFFER SOLUTIONS

In general, preparing a buffer solution requires either:

1. A weak acid and a salt of the acid's conjugate base.
2. Or a weak base and a salt of the base's conjugate acid.

Both of which in sufficient amounts to maintain the ability to buffer.

Example: Citric acid-phosphate buffer.

Make up 0.1 M citric acid and 0.2 M Disodium hydrogen phosphate solutions then mix as follows to make a 100 ml solution:

Good's Buffers

The use of buffers based on inorganic or organic salts is limited because of the interference of buffer cations and anions with the biological reaction under study. The development and introduction of the Zwitterionic Biological Buffers by Good (1966) did much to change this situation. This type of buffer displays the desired characteristics: Low interference with biological processes is due to the fact that anionic and cationic sites are present as non-interacting carboxylate or sulphonate and cationic ammonium groups. Good's buffers or Good buffers are twelve buffering agents selected. Good buffers are based on following criteria which helped biochemists and biologists in modern laboratories. These are as follows:

- **Ease of preparation:** Buffers should be easily prepared and should be purified from inexpensive materials.
- **Membrane impermeability:** Ideally, a buffer is not readily pass through cell membranes, which reduces the accumulation of buffer compound within cells.

TABLE 3.3: Important twelve Good's buffers

Buffer	pKa at 20°C	ΔpKa/°C
MES	6.15	-0.011
ADA	6.6	-0.011
PIPES	6.8	-0.0085
ACES	6.9	-0.020
Cholamine chloride	7.1	-0.027
BES	7.15	-0.016
TES	7.5	-0.020
HEPES	7.55	-0.014
Acetamidoglycine	7.7	-
Tricine	8.15	-0.021
Glycinamide	8.2	-0.029
Bicine	8.35	-0.018

- **Minimal salt effects:** Highly ionic buffers may cause problems or complications in some biological systems.
- **Optical absorbance:** Buffers should not absorb visible or ultraviolet light at wavelengths longer than 230 nm so as not to interfere with commonly used spectrophotometric assays.
- **pK$_a$:** Because most biological reactions take place at near-neutral pH between 6 and 8, ideal buffers would have pK$_a$ values in this regime to provide maximum buffering capacity there.
- **Solubility:** The biological systems are in aqueous systems and need good solubility in water. Low solubility in non-polar solvents (fats, oils, and organic solvents) was also considered beneficial, as this would tend to prevent the buffer compound from accumulating in non-polar compartments in biological systems like cell membranes and other cell compartments.
- **Stability:** The buffers should be chemically stable, resisting enzymatic and non-enzymatic degradation.
- **Well-behaved cation interactions:** If the buffers form complexes with cationic ligands, the complexes formed should remain soluble. Ideally, at least some of the buffering compounds will not form complexes.

How to Use pH Buffer

Buffer solutions of known pH value allow the user to adjust the system to read accurate measurements.

For best accuracy

1. Standardization should be performed with fresh buffer solutions.
2. Buffers used should frame the range of pH for the samples being tested.
3. Buffers should be at the same temperature as the samples. (For example; if all your samples are at 50°C, warm your buffers to 50°C using a beaker in a water bath.)
4. Buffer values are dependent upon temperature. The following table lists the NIST values of pH buffers at various temperatures.

TABLE 3.4: NIST Values of pH Buffers at Various Temperatures

Temperature (°C)	4.01 Buffer 0.05 M Potassium Acid Phthalate	6.86 Buffer 0.025 M KH_2PO_4 0.025 M Na_2HPO_4	9.18 Buffer 0.01 M Borax
0°	4.00 pH	6.98 pH	9.46 pH
10°	4.00 pH	6.92 pH	9.33 pH
20°	4.00 pH	6.88 pH	9.22 pH
25°	4.01 pH	6.86 pH	9.18 pH
30°	4.02 pH	6.85 pH	9.14 pH
40°	4.04 pH	6.84 pH	9.07 pH
50°	4.06 pH	6.83 pH	9.01 pH
60°	4.09 pH	6.84 pH	8.96 pH
70°	4.13 pH	6.85 pH	8.92 pH
80°	4.16 pH	6.86 pH	8.88 pH
90°	4.21 pH	6.88 pH	8.85 pH

Primary and Secondary Standard Buffer Solutions

The primary and secondary standard buffer solutions are prepared from the standard buffer substances as indicated in the table. The standard pHs [pH(S)] of the buffer solution in the temperature range 0-95°C are indicated and can be used for calibration purposes. Buffer compositions are on the molal scale. Only freshly prepared solutions should be used, as the initial buffer composition may change very rapidly. Tartrate, phthalate, and phosphates may be dried at 110°C for 1-2 hours prior to use. Potassium tetraoxalate should not be dried above 60°C, and borax should not be dried at all.

- **Molality (m):** Dimension: Mole per kilogram of solvent.
- **Molarity (M):** Dimension: Mole per litre of solution.
- **Dilution value (ΔpH ½):** Change of pH value observed by dilution of a buffer solution with an equal volume of pure water. It is positive when pH increases and negative when pH decreases with increasing dilution.
- **Buffer value β (equiv./pH):** (also Buffer capacity, Van Slyke Buffer value) $\beta = d[B]/dpH$, where d[B] is the increment (in equivalents) of a strong base required to produce a certain pH change of the buffer solution. Strong acids effect negative (-d[B]) increment and thus lower the pH.
- **Temperature coefficient (pH unit/°C):** dpH(S)/dT standard change of pH value per degree centigrade. It can be positive or negative.

Primary Standards

Composition and properties of the five primary standard buffers at 25°C (see notes on preparation, below).

TABLE 3.5: Buffer solution

	Tartrate	Phthalate	Phosphate D	Phosphate E	Borate
Buffer substance	$KHC_4H_4O_6$	$KHC_8H_4O_4$	KH_2PO_4 + Na_2HPO_4	KH_2PO_4 + Na_2HPO_4	$Na_2B_4O_7 10 H_2O$
g/l soln. at 25 °C	Saturated at 25°C	10.12	3.39 [b] 3.53 [c]	1.179 [b] 4.30 [c]	3.80
Molality (m)	0.0341	0.05	0.025 [a]	0.008695 [b] 0.03043 [c]	0.01
Molarity (M)	0.034	0.04958	0.02490 [a]	0.008665 [b] 0.03032 [c]	0.009971
Density (g/ml)	1.0036	1.0017	1.0028	1.0020	0.9996
pH at 25°C	3.557	4.008	6.865	7.413	9.180
Dilution value, D pH½	+0.049	+0.052	+0.080	+0.07 [d]	+0.01
Buffer value, b, equiv./pH	0.027	0.016	0.029	0.016	0.020
Temp. coeff., dpH(S)/dt, units/°C	-0.0014	+0.0012	-0.0028	-0.0028	-0.0082

[a] Concentration of each phosphate salt. [b] KH_2PO_4. [c] Na_2HPO_4. [d] Calculated value.

TABLE 3.6: Recommended standard values of pH(S) for primary standard buffers (+/- 0.005 at 0 - 60°C and +/-0.008 from 60-90°C)

	Buffer pH				
Temp. (°C)	Tartrate	Phthalate	Phosphate D	Phosphate E	Borate
0		4.003	6.984	7.534	9.464
5		3.999	6.951	7.500	9.395
10		3.998	6.923	7.472	9.332
15		3.999	6.900	7.448	9.276
20		4.002	6.881	7.429	9.225
25	3.557	4.008	6.865	7.413	9.180
30	3.552	4.015	6.853	7.400	9.139
35	3.549	4.024	6.844	7.389	9.102
38	3.548	4.030	6.840	7.384	9.081
40	3.547	4.035	6.838	7.380	9.068
45	3.547	4.047	6.834	7.373	9.038
50	3.549	4.060	6.833	7.367	9.011
55	3.554	40.75	6.834		8.985
60	3.560	4.091	6.836		8.962
70	3.580	4.126	6.845		8.921
80	3.609	4.164	6.859		8.885
90	3.650	4.205	6.877		8.8850
95	3.674	4.227	6.886		8.833

*N.B.S. National Bureau of Standards.

Secondary Standards

TABLE 3.7: Recommended standard values of pH(S) for the secondary buffer standards

	Buffer pH	
Temp. (°C)	**Tetraoxalate**	**Calcium hydroxide**
0	1.666	13.423
5	1.668	13.207
10	1.670	13.003
15	1.672	12.810
20	1.675	12.627
25	1.679	12.454
30	1.683	12.289
35	1.688	12.133
38	1.691	12.043
40	1.694	11.984
45	1.700	11.841
50	1.707	11.705
55	1.715	11.574
60	1.723	11.449
70	1.743	
80	1.766	
90	1.792	
95	1.806	

MECHANISM OF BUFFER ACTION

Buffer solutions are solutions that resist change in Hydronium ion and the hydroxide ion concentration (and consequently pH) upon addition of small amounts of acid or base, or upon dilution. Buffer solutions consist of a weak acid and its conjugate base (more common) or a weak base and its conjugate acid (less common). The resistive action is the result of the equilibrium between the weak acid (HA) and its conjugate base (A–):

$$HA_{(aq)} + H_2O(l) \rightarrow H_3O^+_{(aq)} + A^-_{(aq)}$$

Any alkali added to the solution is consumed by the Hydronium ions. These ions are mostly regenerated as the equilibrium moves to the right and some of the acid dissociates into Hydronium ions and the conjugate base. If a strong acid is added, the conjugate base is protonated, and the pH is almost entirely restored. This is an example of Le Chatelier's principle and the common ion effect. This contrasts with solutions of strong acids or strong bases, where any additional strong acid or base can greatly change the pH. This may be easier to see by comparing two graphs when an strong acid is titrated with a strong base the curve will have a large gradient throughout showing that a small addition of base/acid will have a large effect compared to a weak acid/strong base titration curve which will have a smaller gradient near the pKa.

- Titration of a weak acid with a strong base the flat region at pKa is the buffering region.
- Titration of a weak acid with a strong base the flat region at pKa is the buffering region.
- Titration of a strong acid with a strong base. Note the sharp rise in pH: this solution cannot buffer.
- Titration of a strong acid with a strong base. Note the sharp rise in pH: this

solution cannot buffer.

When writing about buffer systems they can be represented as salt of conjugate base/acid, or base/salt of conjugate acid. It should be noted that here buffer solutions are presented in terms of the Bronsted-Lowry notion of acids and bases, as opposed to the Lewis acid-base theory (see acid-base reaction theories.

BUFFER APPLICATIONS

Their resistance to changes in pH makes buffer solutions very useful for chemical manufacturing and essential for many biochemical processes. The ideal buffer for a particular pH has a pK_a equal to the pH desired, since a solution of this buffer would contain equal amounts of acid and base and be in the middle of the range of buffering capacity.

Buffer solutions are necessary to keep the correct pH for enzymes in many organisms to work. Many enzymes work only under very precise conditions; if the pH strays too far out of the margin, the enzymes slow or stop working and can denature, thus permanently disabling its catalytic activity. A buffer of carbonic acid (H_2CO_3) and bicarbonate (HCO_3^-) is present in blood plasma, to maintain a pH between 7.35 and 7.45.

Industrially, buffer solutions are used in fermentation processes and in setting the correct conditions for dyes used in colouring fabrics. They are also used in chemical analysis and calibration of pH meters.

With few exceptions, studies of biochemical systems require the use of a buffer in order to control the pH value. Therefore the action of the buffer is of prime importance. Factors influencing the action of buffer solutions and pH are:

1. **Activity effects:** concentration and electrical charge of the species involved.
2. **Salt effects:** added "indifferent" electrolytes.
3. **Dilution effect:** pH-variation on dilution of buffer solutions buffer capacity: added base or acid.
4. Temperature dependence.

CHAPTER 4

The Atom and Chemical Bonds

The atom is a unit of matter, the smallest unit of an element, consisting of a dense, central, positively charged nucleus surrounded by a system of electrons, equal in number to the number of nuclear protons, the entire structure having an approximate diameter of 10^{-8} centimeter and characteristically remaining undivided in chemical reactions except for limited removal, transfer, or exchange of certain electrons.

Though the word *atom* originally denoted a particle that cannot be cut into smaller particles, in modern scientific usage the atom is composed of various **subatomic particles**. The constituent particles of an atom consist of the electron, the proton and, for atoms other than hydrogen, the neutron.

There are 116 known elements, or types of atoms. Elements are chemical, substances that cannot be decomposed or broken into more elementary substances by ordinary chemical means. Elements were at one time believed to be the fundamental substances but are now known to consist of a number of different elementary particles. More than 116 chemical elements are known to exist in the universe, although several of these, the so-called transuranium elements, have not been found in nature, and can only be produced artificially.

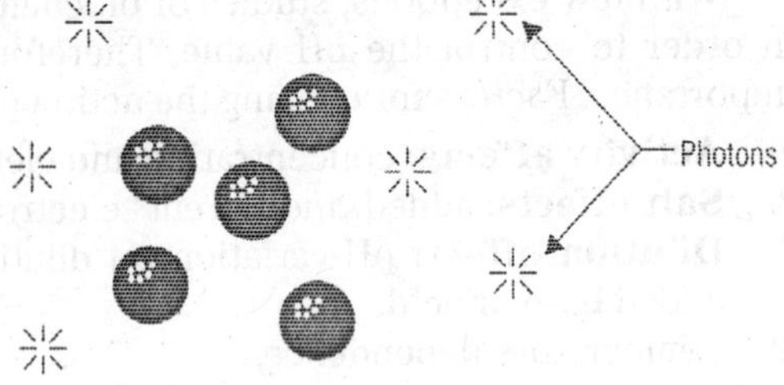

The world and everything in it is made up of atoms of one or some combination of those elements. Atom is the tiny basic building block of matter. All the material on Earth is composed of various combinations of atoms. Atoms are the smallest particles of a chemical element that still exhibit all the chemical properties unique to that element. A row of 100 million atoms would be only about a centimeter long. The arrangement and combination of these atoms determines what a thing will be. Consider the element of carbon: arrange a gaggle of carbon atoms one way, and you have a worthless lump of graphite; arrange them a bit differently, and you have a diamond. Combined with oxygen atoms, they become a gas floating through the atmosphere. When arranged in yet another manner, and combined with several additional elements - those same atoms form a human being.

HISTORICAL ASPECTS

In the 1800's an English chemist, John Dalton performed experiments with various chemicals that showed that matter, indeed, seem to consist of elementary lumpy particles (atoms). Although he did not know about their structure, he knew that the evidence pointed to something fundamental.

The English physicist J.J. Thomson (1897) discovered the electron and proposed a model for the structure of the atom. The electron is by far the least massive of these particles at 9.11×10^{-28} g, with a negative electrical charge and a size that is too small to be measured using available techniques. Protons have a positive charge and a mass 1,836 times that of the electron, at 1.6726×10^{-24} g, although this can be reduced by changes to the atomic binding energy.

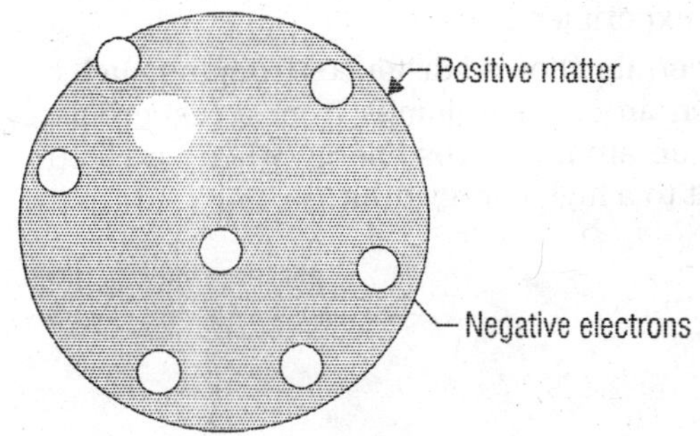

FIG. 4.1: Thomson's 'Resin in the Pudding' model of the atom.

Max Planck (1900) showed that when you vibrate atoms strong enough, such as when you heat an object until it glows, you can measure the energy only in discrete units. He called these **energy packets, quanta**.

Physicists at the time thought that light consisted of waves but, according to Albert Einstein, the quanta behaved like discrete particles. Physicists call Einstein's discrete light particle, a **photon**. Atoms not only emit photons, but they can also absorb them.

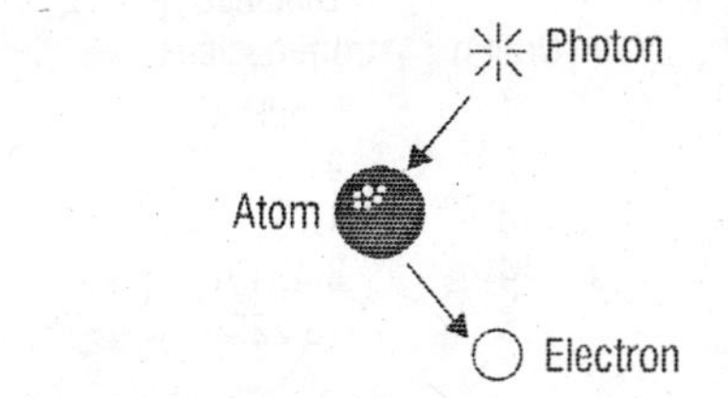

FIG. 4.2: Photoelectric effect.

In 1905, Albert Einstein (1805) explained that light absorption can release electrons from atoms, a phenomenon called the **photoelectric effect**.

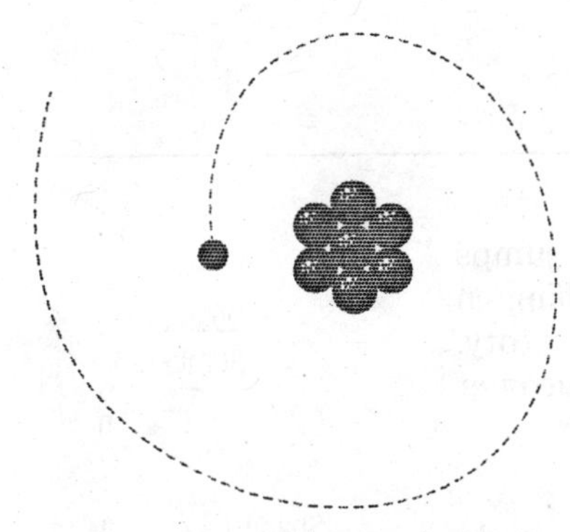

Atoms consist of a compact positively charged nucleus. The mass of the nucleus following a fusion reaction is less than the sum of the masses of the separate particles. The difference between these two values is emitted as energy, as described by Albert Einstein's mass–energy equivalence formula, $E = mc^2$, where m is the mass loss and c is the speed of light. This deficit is the binding energy of the nucleus.

All of the bound protons and neutrons in an atom make up a tiny atomic nucleus, and are collectively called nucleons. The radius of a nucleus is approximately equal to fm, where *A* is the total number of **nucleons**. The nucleus occupies less than one thousand million millionth of the atomic volume, but contains almost all of the atom's mass. If an atom had the size of the earth, the nucleus would have the size of a football stadium. This is much smaller than the radius of the atom, which is on the order of 10^5 fm. The nucleons are bound together by a short-ranged attractive potential called the residual strong force. At distances smaller than 2.5 fm, this force is much more powerful than the electrostatic force that causes positively charged protons to repel each other.

Rutherford identified the particles of the nucleus as discrete positive charges of matter. Using alpha particles as bullets, Rutherford knocked hydrogen nuclei out of atoms of six elements: boron, fluorine, sodium, aluminium, phosphorus, and nitrogen.

Danish physicist, Niels Bohr (1912) proposed a theory that electrons do not spiral into the nucleus and came up with some rules for what does happen. Bohr came up with two following rules of the experiment:

1. **Rule 1:** Electrons can orbit only at certain allowed distances from the nucleus.
2. **Rule 2:** Atoms radiate energy when an electron jumps from a higher-energy orbit to a lower-energy orbit. Also, an atom absorbs energy when an electron gets boosted from a low-energy orbit to a high-energy orbit.

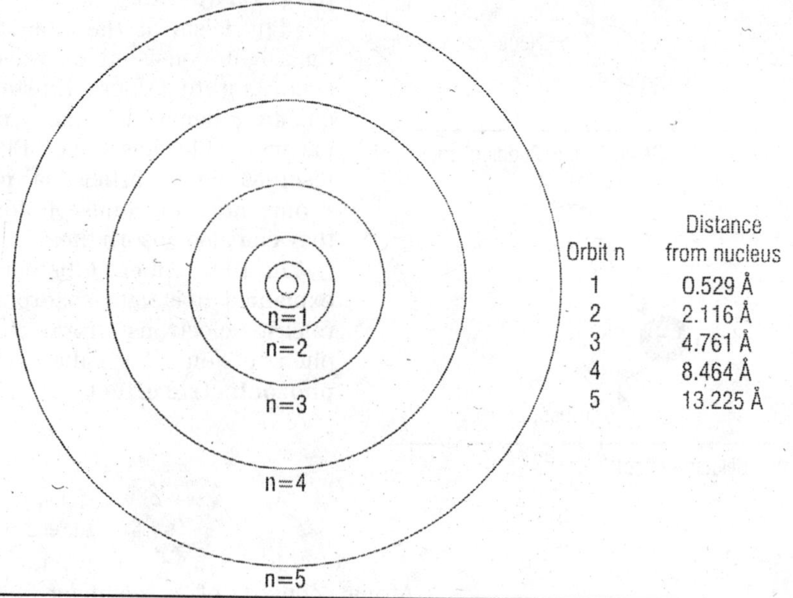

FIG. 4.4: Bohr's atom for Hydrogen.

Light (photons) emits whenever an electron jumps from one orbit to another. The jumps seem to happen instantaneously without moving through a trajectory. Bohr's atom seemed too simple to describe the heavier elements.

Bohr-Sommerfeld model

According to the Bohr-Sommerfeld model, not only do electrons travel in certain orbits but the orbits have different shapes and the orbits could tilt in the presence of a magnetic field. Orbits can appear circular or elliptical, and they can even swing back and forth through the nucleus in a straight line.

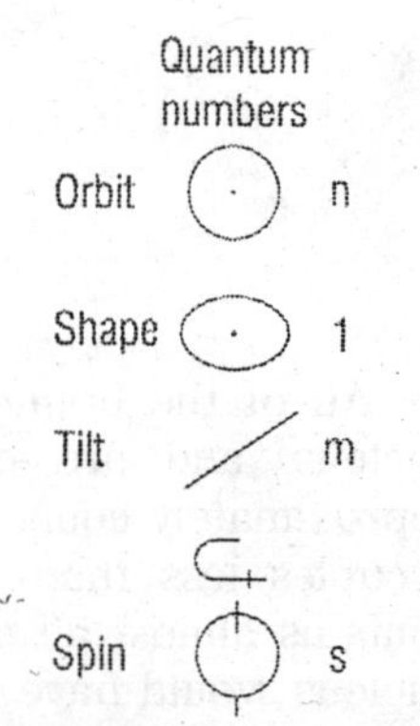

FIG. 4.6: Quantum numbers.

The orbit shapes and various angles to the magnetic field could only have certain shapes, similar to an electron in a certain orbit. As an example, the fourth orbit in a hydrogen atom can have only three possible shapes and seven possible traits. These added states allowed more possibilities for different spectral lines to appear. This brought the model of the atom into closer agreement with experimental data.

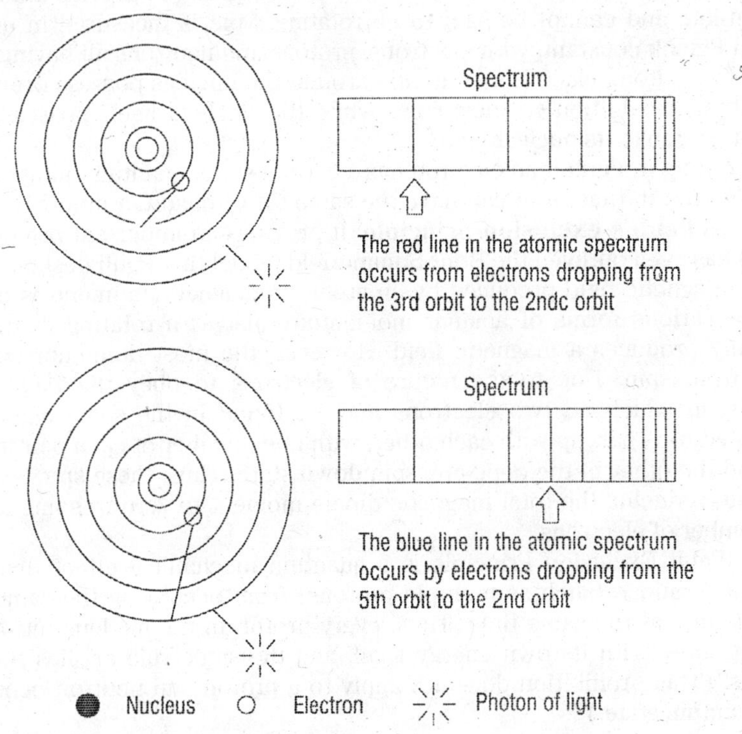

FIG. 4.5: The electron can exist in only one of the orbits. The diagram shows only five orbits, but any number of orbits can theoretically exist.

Wolfgang Pauli (1924) predicted that an electron should spin (kind of like a top) while it orbits around the nucleus. The electron can spin in either of two directions. This spin consisted of a fourth quantum number: electron spin (s).

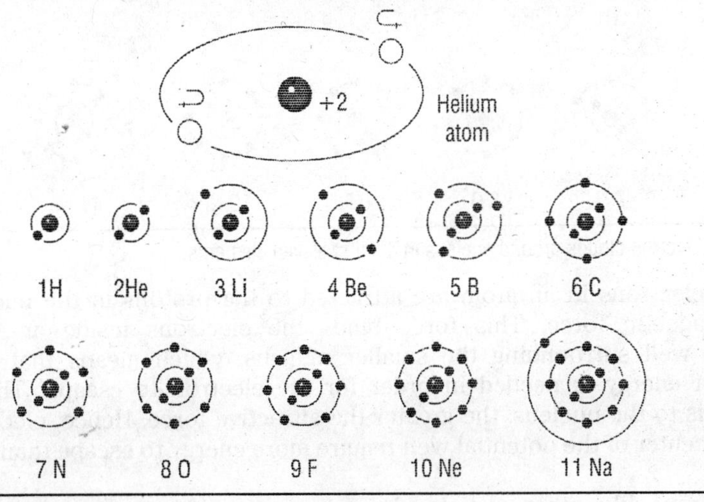

FIG. 4.7: Pauli's Exclusion principle.

Elementary particles possess an intrinsic quantum mechanical property known as spin. This is analogous to the angular momentum of an object that is spinning around its center of mass, although strictly speaking these particles are believed to be point-like and cannot be said to be rotating. Spin is measured in units of the reduced Planck constant, with electrons, protons and neutrons all having spin ½, or spin-½. In an atom, electrons in motion around the nucleus possess orbital angular momentum in addition to their spin, while the nucleus itself possesses angular momentum due to its nuclear spin.

According to Pauli if an electron has a certain set of quantum numbers, then no other electron in that atom can have the same set of quantum numbers. Physicists call this as **Pauli's exclusion principle**. It provides an important principle to this day and has even outlived the Bohr-Sommerfeld model that Pauli designed it for.

The magnetic field produced by an atom- its magnetic moment- is determined by these various forms of angular momentum, just as a rotating charged object classically produces a magnetic field. However, the most dominant contribution comes from spin. Due to the nature of electrons to obey the Pauli exclusion principle, in which no two electrons may be found in the same quantum state, bound electrons pair up with each other, with one member of each pair in a spin up state and the other in the opposite, spin down state. Thus, these spins cancel each other out, reducing the total magnetic dipole moment to zero in some atoms with even number of electrons.

The Pauli Exclusion Principle is a quantum mechanical effect that prohibits *identical* fermions (such as multiple protons) from occupying the same quantum physical state at the same time. Thus, every proton in the nucleus must occupy a different state, with its own energy level, and the same rule applies to all of the neutrons. (This prohibition does not apply to a proton and neutron occupying the same quantum state.)

The visual concept of the atom now appeared as an electron cloud which surrounds a nucleus. The cloud consists of a probability distribution map which determines the most probable location of an electron as follows:

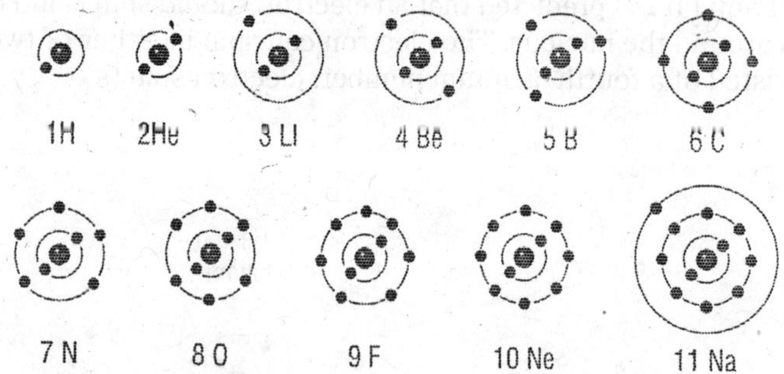

FIG. 4.8: Electron clouds around nucleus in the first eleven elements.

The electrons in an atom are attracted to the protons in the nucleus by the electromagnetic force. This force binds the electrons inside an electrostatic potential well surrounding the smaller nucleus, which means that an external source of energy is needed in order for the electron to escape. The closer an electron is to the nucleus, the greater the attractive force. Hence, electrons bound near the center of the potential well require more energy to escape than those at the exterior.

Electrons, like other particles, have properties of both a particle and a wave. The electron cloud is a region inside the potential well where each electron forms a type of three-dimensional standing wave- a wave form that does not move relative to the nucleus. This behavior is defined by an atomic orbital, a mathematical function that characteristics the probability that an electron will appear to be at a particular location when its position is measured. Only a discrete (or quantized) set of these orbitals exist around the nucleus, as other possible wave patterns will rapidly decay into a more stable form. Orbitals can have one or more ring or node structures, and they differ from each other in size, shape and orientation.

Chemists usually describe the atom as a simple solar system model similar to Bohr's model but without the different orbit shapes. The important emphasis for chemistry attempts to show the groupings of electrons in orbital shells.

Creation of molecules

Chemical behaviour of the elements become together and creates molecules. Molecules may share electrons as the hydrogen and water molecules as shown below. *The atoms which share electrons have the name ions.* The outer electron shell of an atom actually does the sharing and bonding of the atoms. This in turn allows chemists to describe the interactions of chemistry. Even though the orbit model of the atom does not provide an accurate model, it works well for describing chemistry.

The nucleus contains most of the atom's mass as well as the positive charge. James Chadwick (1932) discovered the neutron which is heavier than the proton with a mass of 1840 electrons and with no charge (neutral). The proton-neutron together called **nucleon**.

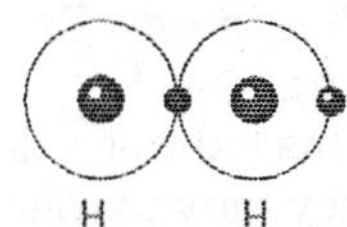

Hydrogen molecule

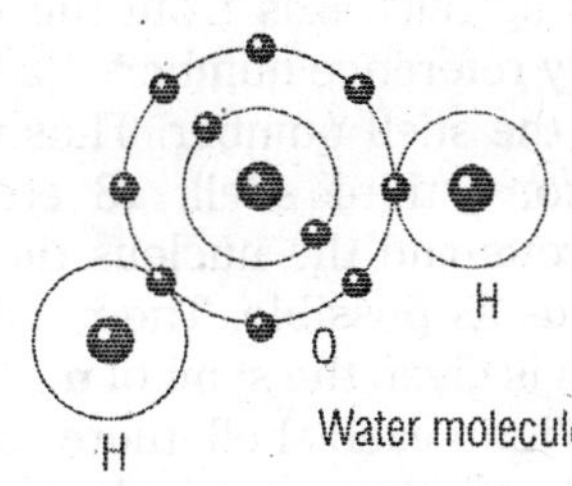

Water molecule

FIG. 4.10: A helium atom with two electrons orbiting a nucleon made of two protons and two neutrons.

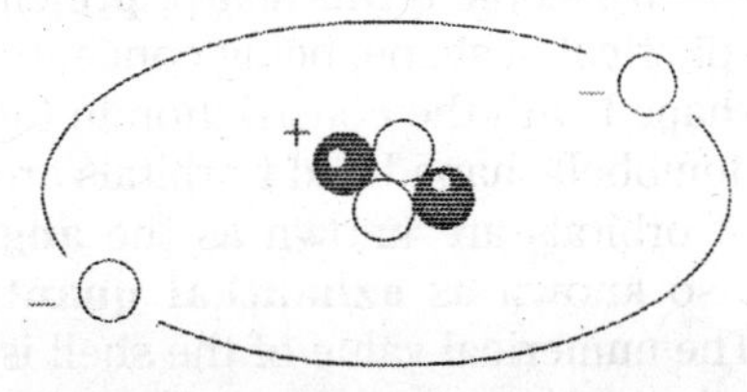

FIG. 4.9: Creation of molecule from element.

Atoms of the same element have the same number of protons, called the atomic number. Within a single element, the number of neutrons may vary, determining the isotope of that element. The number of neutrons relative to the protons determines the stability of the nucleus, with certain isotopes undergoing. The variations in mass result, more or less, from the number of neutrons in the nucleus of the atom. Atoms of an element having the same atomic number but different atomic masses get called **isotopes** of that element.

For each variety of matter there should exist a corresponding **opposite** or **antimatter**. Physicists now know that antimatter exists. There also exist anti-electrons, anti-proton, and anti-neutron. This allows the existence for anti-atoms, a true form of antimatter.

To conclude, the atoms are made of smaller particles, called electrons, protons, and neutrons. An atom consists of a cloud of electrons surrounding a small, dense nucleus of protons and neutrons. Electrons and protons have a property called electric charge, which affects the way they interact with each other and with other electrically charged particles. Electrons carry a negative electric charge, while protons have a positive electric charge. The negative charge is the opposite of the positive charge, and, like the opposite poles of a magnet, these opposite electric charges attract one another. Conversely, like charges (negative and negative, or positive and positive) repel one another. The attraction between an atom's electrons and its protons holds the atom together. Normally, an atom is electrically neutral, which means that the negative charge of its electrons is exactly equaled by the positive charge of its protons.

The nucleus contains nearly all of the mass of the atom, but it occupies only a tiny fraction of the space inside the atom. The diameter of a typical nucleus is only about 1×10^{-14} m (4×10^{-13} in), or about 1/100,000 of the diameter of the entire atom. The electron cloud makes up the rest of the atom's overall size. If an atom were magnified until it was as large as a football stadium, the nucleus would be about the size of a grape.

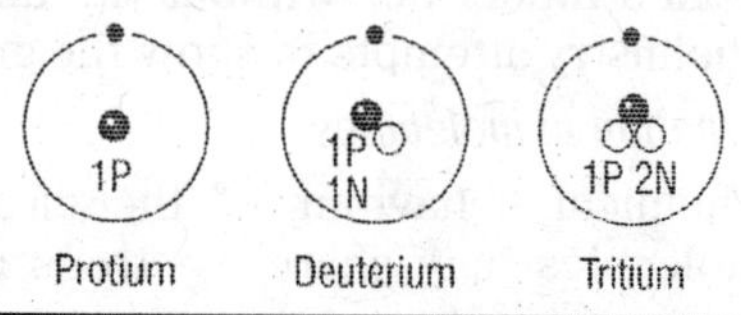

FIG. 4.11: Isotopes of Hydrogen.

Electrons

Electrons are tiny, negatively charged particles that form a cloud around the nucleus of an atom. Each electron carries a single fundamental unit of negative electric charge, or $^{-1}$.

Electrons exhibit a dual character, that of a particle and a wave, and are similar to light. They move around the atomic nuclei in concentric, successive and discrete volumes, the shells. The electrons of each shell possess a definite energy and are positioned at specific distances from the nucleus. The electrons of the inner shell possess lower energies than those of the outer shells. The shells are designated, counting outwards from the nucleus, as K, L, M, N, O, P, Q with corresponding energy reference numbers 1,2,3,7. Each shell can accommodate $2n^2$ electrons where 'n' is the shell number. Thus the first shell contains two electrons, second shell 8 electrons, third shell, 18 etc. Due to the electrostatic attraction between the electrons and the nucleus the electrons tend to occupy the shells as close to the nucleus as possible. The number of the shell is the **Principal quantum number**, which is given the symbol **n.**

Within each shell, there are regions in which the electrons are most likely to be found and these particularly shaped regions in space are known as the orbitals. Four types of orbitals are recognized based on the geometric shape of the orbital and the angular (the orbital motion) momentum, the electron exhibits when located in a given type of orbital. They are designated as **s, p, d, f** orbitals (from the spectroscopic terms **s**harp, **p**rincipal, **d**iffuse, and **f**undamental). The **s** orbitals are spherical in shape, being concentric about the nucleus. The **p** orbitals are dumbbell-shaped with the constriction in the region of the nucleus. The **d** orbitals are double dumbbell-shaped and **f** orbitals are complex with varied shapes. The type and shape of orbitals are known as the **angular momentum quantum number**, **1**. This is also known as **azimuthal quantum number** or **subsidiary quantum number**. The numerical value of the shell is the number of types of orbitals assigned for that principal quantum level. Thus principal quantum number i possesses only the **Is** orbital; the second level has **2s** and **2p** orbitals and the third level has **3s**, **3p** and **3d** orbitals. For any shell there can be only one **s** orbital, three **p** orbitals, five **d** orbitals and seven **f** orbitals. The orbitals of the same shell vary slightly in energy.

A third quantum, the **magnetic quantum number, m;** designated by subscripts

x, y and **z** denotes the direction of the orbitals when there are several **p, d,** and **f** orbitals in a specific energy level. The three **p-orbitals** are mutually perpendicular as the cartesian co-ordinates. The **d** orbitals can have five orientations in space and **f** orbitals, 7 orientations. A fourth quantum number denotes the spin of the electron. The electron can spin clockwise and counter clock-wise. The value given to clockwise spin is + ½ (↑) and to counter-clockwise spin is –½ (↓). This is **spin quantum number** represented by s. No two electrons of a given atom are allowed to possess the same four quantum numbers at a time (Pauli's exclusion principle).

At any time, and atomic orbital may contain a maximum of two electrons and these two should differ in the fourth quantum number i.e. must be of the opposite spin. The existence of an electrostatic attraction between the nucleus (positively charged) and the electrons (negatively charged) facilitates the electrons occupying the orbitals as close to the nucleus; the orbitals of the lowest energy. When there are many orbitals, orbitals of each type are first filled singly. Only after all these orbitals contain one electron each, a second electron is added to the orbitals to complete a pair and thereby fill the orbital.

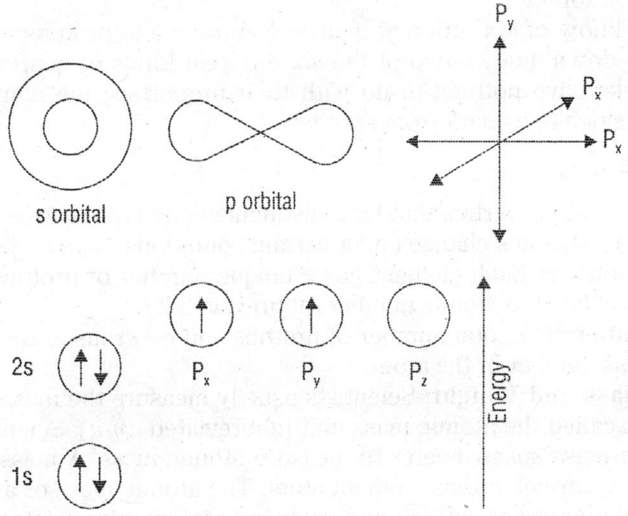

FIG. 4.12: Schematic representation of energy levels from 1s to **2p** and electronic configuration of carbon in the ground state.

Protons and Neutrons

Protons carry a positive charge of $^{+1}$, exactly the opposite electric charge as electrons. Protons have a positive charge and a mass 1,836 times that of the electron, at 1.6726 × 10^{-24} g, although this can be reduced by changes to the atomic binding energy. The number of protons in the nucleus determines the total quantity of positive charge in the atom. In an electrically neutral atom, the number of the protons and the number of electrons are equal, so that the positive and negative charges balance out to zero. The proton is very small, but it is fairly massive compared to the other particles that make up matter. A proton's mass is about 1,840 times the mass of an electron.

In 1932 British physicist Sir James Chadwick discovered the subatomic particle known as the **neutron,** filling in a key missing piece in science's understanding of the atom. Neutrons are about the same size as protons but their mass is slightly greater. Without neutrons present, the repulsion among the positively charged protons would cause the nucleus to fly apart. Neutrons have no electrical charge and have a free mass of 1,839 times the mass of electrons, or 1.6929×10^{-24} g. Neutrons and protons have comparable dimensions - on the order of 2.5×10^{-15} m - although the 'surface' of these particles is not sharply defined.

Consider the element helium, which has two protons in its nucleus. If the nucleus did not contain neutrons as well, it would be unstable because of the electrical repulsion between the protons.

In the Standard Model of physics, both protons and neutrons are composed of elementary particles called **quarks**. The quark is a type of fermion, one of the two basic constituents of matter- the other being the lepton, of which the electron is an example. There are six types of quarks, and each has a fractional electric charge of either $+^{2/3}$ or $-^{1/3}$. Protons are composed of two up quarks and one down quark, while a neutron consists of one up quark and two down quarks. This distinction accounts for the difference in mass and charge between the two particles. The quarks are held together by the strong nuclear force, which is mediated by gluons. The gluon is a member of the family of bosons, which are elementary particles that mediate physical forces.

Physicists know of six different quarks. Neutrons and protons are made up of up quarks and down quarks–two of the six different kinds of quarks. The fanciful names of quarks have nothing to do with their properties; the names are simply labels to distinguish one quark from another.

Properties of Atoms

Atoms have several properties that help distinguish one type of atom from another and determine how atoms change under certain conditions.

Atomic Number: Each element has a unique number of protons in its atoms. This number is called the atomic number (abbreviated Z).

Mass Number: The total number of protons and neutrons in the nucleus of an atom is the mass number of the atom.

Atomic Mass and Weight: Scientists usually measure the mass of an atom in terms of a unit called the atomic mass unit (abbreviated *amu*). Scientists can use a device called a mass spectrometer to measure atomic mass. A mass spectrometer removes one or more electrons from an atom. The atomic mass of an atom, which depends on the number of protons and neutrons present, also relates to the atomic weight of an element. Weight usually refers to the force of gravity on an object, but atomic weight is really just another way to express mass. An element's atomic weight is given in grams. It represents the mass of one mole (6.02 × 1023 atoms) of that element. Numerically, the atomic mass and the atomic weight of an element are the same, but the first is expressed in grams and the second is in atomic mass units. So, the atomic weight of hydrogen is 1 gram and the atomic mass of hydrogen is 1 *amu*.

Forces Acting Inside Atoms

In physics, a force is a push or pull on an object. There are four fundamental forces, three of which - the **electromagnetic force**, the **strong force**, and the **weak force**– are involved in keeping stable atoms in one piece and determining how unstable atoms will decay. The electromagnetic force keeps electrons attached to their atom. The strong force holds the protons and neutrons together in the nucleus. The weak force governs how atoms decay when they have excess protons or neutrons. The fourth fundamental force, gravity, only becomes apparent with objects much larger than subatomic particles.

Electromagnetic Force: The most familiar of the forces at work inside the atom is the electromagnetic force. This is the same force that causes people's hair to stick to a brush or comb when they have a build-up of static electricity. The electromagnetic force causes opposite electric charges to attract each other. The electromagnetic force also causes like charges to repel each other.

Strong Force: Protons and neutrons in the nuclei of atoms are held together by the strong force. This force must overcome the electromagnetic force of repulsion the protons in a nucleus exert on one another.

Particle physicists explain the behavior of the strong force by introducing another type of particle, called a pion. Protons and neutrons interact in the nucleus by exchanging pions. Pions in the nucleus exist only for the briefest instant of time, no more than 1 × 10-23 seconds, but even during their short existence they can provide the attraction that holds the nucleus together.

Weak Force: The weak force lives up to its name - it is much weaker than the electromagnetic and strong forces. Like the strong force, it only acts over a short distance, about .01 fm.

The Quantum Atom

Scientists of the early 20th century developed a new view of matter and energy to accurately describe how atoms behave. They called this theory quantum theory, or quantum mechanics. Quantum theory describes matter as acting both as a particle and as a wave. In the visible objects encountered in everyday life, the wavelike nature of matter is too small to be apparent. Wavelike nature becomes important, however, in microscopic particles such as electrons.

Valence

The outermost electron shell of an atom in its uncombined state is known as the valence shell, and the electrons in that shell are called valence electrons. The number of valence electrons determines the bonding behavior with other atoms. Atoms tend to chemically react with each other in a manner that will fill (or empty) their outer valence shells.

Rearranging Atoms Leads to New Properties

The second stage, which began in 2005, focuses on active nanostructures that change their size, shape, conductivity or other properties during use. New drug-delivery particles could release therapeutic molecules in the body only after they reached their targeted diseased tissues. Electronic components such as transistors and amplifiers with adaptive functions could be reduced to single, complex molecules.

Starting around 2010, workers will cultivate expertise with systems of nanostructures, directing large numbers of intricate components to specified ends. One application could involve the guided self-assembly of nanoelectronic components into three-dimensional circuits and whole devices. Medicine could employ such systems to improve the tissue compatibility of implants, or to create scaffolds for tissue regeneration, or perhaps even to build artificial organs.

After 2015-2020, the field will expand to include molecular nanosystems-heterogeneous networks in which molecules and supramolecular structures serve as distinct devices. The proteins inside cells work together this way, but whereas biological systems are water-based and markedly temperature-sensitive, these molecular nanosystems will be able to operate in a far wider range of environments and should be much faster. Computers and robots could be reduced to extraordinarily small sizes. Medical applications might be as ambitious as new types of genetic therapies and antiaging treatments. New interfaces linking people directly to electronics could change telecommunications.

Over time, therefore, nanotechnology should benefit every industrial sector and health care field. It should also help the environment through more efficient use of resources and better methods of pollution control. Nanotech does, however, pose new challenges to risk governance as well. Internationally, more needs to be done to collect the scientific information needed to resolve the ambiguities and to install the proper regulatory oversight. Helping the public to perceive nanotech soberly in a big picture that retains human values and quality of life will also be essential for this powerful new discipline to live up to its astonishing potential.

CHEMICAL BOND

A chemical bond is a mechanism which chemically combines atoms. There are three types of bonds: **metallic**, **ionic**, and **covalent**. The mechanism is an electrostatic force of attraction between areas of positive and negative charge. The locations of the areas of charge will differ from system to system. These differences are responsible for the various types of bonds generally found in compounds. Bonds form as an attempt to stabilize a chemical system by releasing energy. The greater the amount of energy released during the formation of a bond, the more stable the bond will be. All bond formation processes involve the use of valence level electrons.

Bond Formation

Atom rarely remains in free existence. Most of atoms form clusters or aggregates which are electrically neutral, the molecules. The attraction between atoms within a molecule is called a chemical bond. Inert elements are the stables of elements because they resist from entering into chemical combination. This optimum stability is due to the octet of outer electrons they possess (Helium is an exception to this, with only two electrons). Kossel as well as Lewis in 1916 independently proposed that elements try to acquire the electronic configuration of the nearest inert element to become stable. Thus Sodium, having a single electron in the outermost orbital, is eager to give away its solitary electron to achieve the closed shell distribution or the octet of outer electrons like that of Neon. The loss of one electron imparts a unit positive charge to the nucleus resulting in the formation of an ion in a similar way Chlorine, having seven electrons in the outermost shell, tries to acquire one more electron to attain an electronic configuration as that of the nearest inert element Argon. This addition of electron results in the formation of a chlorine ion of a unit negative charge. When these two elements, sodium and chlorine, are allowed to come together sodium releases its electron and chlorine accepts it resulting in the formation of Na^+ and Cl^- ions. These two ions, being oppositely charged, will be attracted towards each other by coulombic attraction and form the stable salt NaCl. This type of bonding is known as electrovalent bonding or ionic bonding because the ionic charges are formed by the gain or loss of electrons. Ionic bonding takes place between an electropositive element (which can release an electron) and an electro-negative element (which can accept an electron).

$$\underset{2,8,1}{Na} - 1e^- \longrightarrow \underset{2,8,}{Na^+} \quad \text{(Neon configuration)}$$

$$\underset{2,8,7}{Cl} + 1e^- \longrightarrow \underset{2,8,8,}{Cl^-} \quad \text{(Argon configuration)}$$

$$Na^+ + Cl^- \rightarrow NaCl$$

Atoms achieve the octets by sharing their electrons on a part time basis also. For example, two hydrogen atoms, each with only one electron, can attain a stable configuration (as that of Helium) by pooling the electrons together and sharing the pair of electrons between them. At this time the **s** orbitals of both atom overlap the two electrons pair and move into a new orbital where the pair of electrons are electrostatically attracted towards both the nuclei. This new orbital is known as the **molecular orbital** and the bond– **a covalent bond.** This type of bond formation takes place between two electronegative elements. Since each atom has to contribute one electron each to produce a shared pair, the number of orbitals containing an unpaired electron determines the number of shared pairs, the valences. Thus nitrogen with three unpaired orbitals shares electrons with three hydrogen atoms to form NH_3 and oxygen with two incomplete orbitals combines with two hydrogen atoms to form H_2O, Thus the number of half filled orbitals of an atom decides its valency, which is 1 for hydrogen, 2 for oxygen and 3 for nitrogen.

$$H + H \rightarrow H : H \text{ or } H - H$$

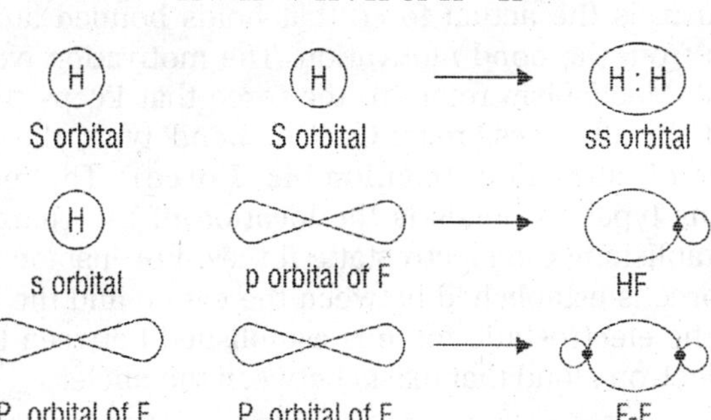

FIG. 4.13: Various types of orbitals.

The molecular orbitals are formed by the overlapping of the atomic orbitals. The more the overlap of orbitals the higher will be the bond strength. The molecular orbital (MO) shows the region where the electrons attracted by two nuclei are most likely to be. Those molecular orbitals where the electrons are concentrated between the bonded nuclei are symmetrically distributed around the inter-nuclear distance produce "Sigma bonds" (bonds). These bonds, also known as single bonds, appear circularly symmetrical when viewed along the bond axis because overlapping of the orbitals along the same axis forms them. Bonds are formed when: (1) two **s** orbitals overlap, (2) one **s** and one **p** orbital overlap, and (3) two **p** orbitals overlap head-on in a linear fashion. The resulting Mo will be intermediate in shape to the participating atomic orbitals.

In some other cases two atoms share a pair of electrons where both the electrons are donated by a single atom. This is essentially a covalent bond and is known as **coordinate covalent bond.**

Bond Energy

Bond Energy is the amount of energy released when a bond forms. It is a direct measure of the amount of stability gained when two atoms establish a chemical bond. Inversely, bond energy corresponds to the amount of energy that is required to break a bond. The amount that is required to break a bond is exactly equal to the amount released when the bond formed. The magnitude of the bond energy corresponds to the vertical drop that appears in a potential energy well at the lowest point, or most stable position, in the curve. An equation that represents a bond formation process would appear as follows:

$$A + B = AB + Energy$$

The energy term corresponds to the energy released, or bond energy.

Bond Length

Bond Length is the average distance between the centers of two bonded atoms. Because bonded atoms experience some vibration, moving towards and away from each other, the distance between bonded atoms will vary slightly over a period of time. The term Bond Length specifically refers to the average positions of the two atoms during the harmonic vibrations that they undergo. On the Potential Energy Well, the Bond Length is the position on the horizontal coordinate that corresponds to the bottom of the well, or the position of lowest energy for the system.

Bond Mechanism

The bond mechanism is the actual force that holds bonded atoms together. The mechanism differs from the bond motivation. The motivation explains why atoms would like to bond. Mechanism refers to the force that keeps them together after they have bonded. In all cases, regardless of bond type, the mechanism is an electrostatic force of attraction (**Coulombic Force**). The primary difference between the various types of bonds is the location of the charged areas that are responsible for establishing the electrostatic forces. For instance, in ionic systems the electrostatic force is established between the cation and the anion. In covalent types of systems, the electrostatic force is established between the positive nuclei and the negative electron cloud that exists between the nuclei.

Bond Motivation

Bond Motivation refers to the reasons behind why bonds are established. At a very fundamental level all bonds form in order to release energy. This loss of energy, or a negative change in Enthalpy, corresponds to fulfilling one of the two driving forces All bond formation processes are exothermic.

At another level, the motivation is interpreted as the desire of an atom to achieve some specific types of electronic configuration that is more stable that the configuration of the unbonded atom. In these case there is a special electronic configuration and is more stable because of their lower energy content. In other words, motivation still returns to the idea that all atoms want to loss energy by bonding.

Bond Order

Bond Order is the term used to distinguish between the numbers of bonds that exist between two atoms. The three primary bond orders are one, two and three. They would correspond to a single bond, double bond and triple bond. Under some conditions, atoms may find themselves with an average bond that appears to be half-way between a single and a double. In that case, the bond order would by 1.5. Other more exotic systems may have bond orders such as 1.33. These may be a result of bonded electrons switching between positions through a process known as resonance. As a result, the average appearance of the bond over a period of time will be the bond order.

TYPES OF BONDS

Ionic Bond

A loosely bound electron from one atom transfers to a deeper shell in another atom. The two ions are held together by their electrical forces. Ionic bond is also called **electrovalent bond**, type of linkage formed from the electrostatic attraction between oppositely charged ions in a chemical compound. Such a bond forms when the valence (outermost) electrons of one atom are transferred permanently to another atom. The atom that loses the electrons becomes a positively charged ion (cation), while the one that gains them becomes a negatively charged ion (anion).

Typical of ionic bonds are those in the alkali halides such as sodium chloride, NaCl.

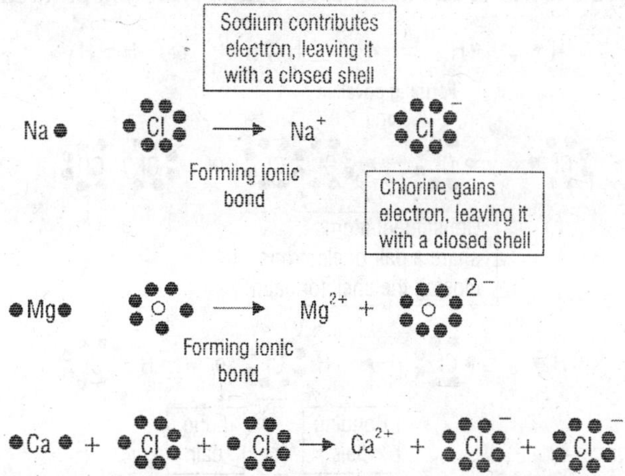

FIG. 4.14: Presentation of Ionic bonding according to Lewis diagrams.

Ionic bonding results in compounds known as ionic, or electrovalent, compounds, which are best exemplified by the compounds formed between non-metals and the alkali and alkaline-earth metals. In ionic crystalline solids of this kind, the electrostatic forces of attraction between opposite charges and repulsion between similar charges orient the ions in such a manner that every positive ion becomes surrounded by negative ions and vice versa. In short, the ions are so arranged that the positive and negative charges alternate and balance one another, the overall charge of the entire substance being zero. The magnitude of the electrostatic forces in ionic crystals is considerable. Accordingly, these substances tend to be hard and non-volatile.

An ionic bond is actually the extreme case of a polar covalent bond, the latter resulting from unequal sharing of electrons rather than complete electron transfer. Ionic bonds typically form when the difference in the electronegativities of the two atoms is great, while covalent bonds form when the electronegativities are similar. Compare covalent bond.

Covalent Bond

The idea that two electrons can be shared between two atoms and serve as the link between them was first introduced by G.N. Lewis (1916). He described the formation of such bonds due to the tendencies of certain atoms to combine with one another in order for both to have the electronic structure of a corresponding noble-gas atom. Covalent chemical bonds involve the sharing of a pair of valence electrons by two atoms, in contrast to the transfer of electrons in ionic bonds. Such bonds lead to stable molecules if they share electrons in such a way as to create a noble gas configuration for each atom.

The electron clouds of one atom are distorted by the presence of another atom. In the new cloud pattern, the outer electrons are more concentrated in the region between the two atoms. Thus, the atoms share their electrons. This allows atoms of the same element to form chemical bonds, which could not happen with ionic bonds. Chemical-bond energies typically measure several electron volts.

Covalent bond is the **interatomic linkage** that results from the sharing of an electron pair between two atoms. The binding arises from the electrostatic attraction of their nuclei for the same electrons. A covalent bond forms when the bonded atoms have a lower total energy than that of widely separated atoms.

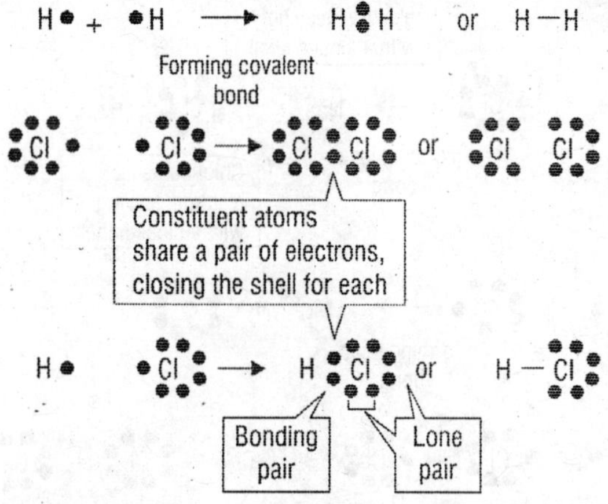

FIG. 4.15: Presentation of covalent bonding according to Lewis diagrams.

Molecules that have covalent linkages include the inorganic substances hydrogen, nitrogen, chlorine, water, and ammonia (H_2, N_2, Cl_2, H_2O, NH_3) together with all organic compounds. In structural representations of molecules, covalent bonds are indicated by solid lines connecting pairs of atoms.

A single line indicates a bond between two atoms (i.e., involving one electron pair), double lines (=) indicate a double bond between two atoms (i.e., involving two electron pairs), and triple lines (≡) represent a triple bond, as found, for example, in carbon monoxide (C≡O). Single bonds consist of one sigma (σ) bond, double bonds have one σ and one pi ( ) bond, and triple bonds have one σ and two bonds.

Covalent bonds are directional: It means that atoms so bonded prefer specific orientations relative to one another; this in turn gives molecules definite shapes, as in the angular (bent) structure of the H_2O molecule. Covalent bonds between identical atoms (as in H_2) are non-polar- i.e., electrically uniform–while those between unlike atoms are polar- i.e., one atom is slightly negatively charged and the other is slightly positively charged. This partial ionic character of covalent bonds increases with the difference in the electronegativities of the two atoms.

Hydrogen gas forms the simplest covalent bond in the diatomic hydrogen molecule. The halogens such as chlorine also exist as diatomic gases by forming covalent bonds. The nitrogen and oxygen which makes up the bulk of the atmosphere also exhibits covalent bonding in forming diatomic molecules.

Polar Covalent Bonds

Covalent bonds, in which the sharing of the electron pair is unequal, with the electrons spending more time around the more non-metallic atom, are called polar covalent bonds. In such a bond there is a charge separation with one atom being slightly more positive and the other more negative, i.e., the bond will produce a dipole moment. The ability of an atom to attract electrons in the presence of another atom is a measurable property called **electronegativity**.

Metallic Bonds

The properties of metals suggest that their atoms possess strong bonds, yet the ease of conduction of heat and electricity suggest that electrons can move freely in all directions in a metal. The general observations give rise to a picture of "positive ions in a sea of electrons" to describe metallic bonding.

Hydrogen Bonding

When a hydrogen atom is covalently linked with a strongly electronegative atom such as oxygen or nitrogen, the bond develops a considerable ionic nature with the hydrogen forming the positive end of a dipole and the electronegative atom, the negative end. This positively charged hydrogen atom gets attracted towards the negative end of another molecule and this attraction weakly binds the two molecules. A number of molecules attracted to each other by these electrostatic forces may associate together to form large clusters of molecules with hydrogen acting as a bridge between the electronegative atoms. This attractive force, which binds the hydrogen atom of one molecule with an electronegative atom of another molecule, is known as the hydrogen bond or the proton bond. In other words, hydrogen bonding is the interaction occurring between covalently bonded hydrogen atom of a proton donor group A–H and a region of high electron density of a proton acceptor group B; where A is an electronegative atom O, N, S, X (F, Br, I) or C and the acceptor group is a lone pair of an electronegative atom or a π orbital of a multiple bond (unsaturated) system. Generally the hydrogen bond is characterized as a proton shared by low electron pairs. Hydrogen bonding, being an association phenomenon, causes a decrease in the total number of free molecules and an increase in the average molecular weight (the cases of intramolecular hydrogen bonding are exceptions.) The formation of H bonds is an exothermic process while the dissociation of H bonds is endothermic.

Hydrogen bond is best explained in relation to the water molecule, which is a typical electrical dipole. The hydrogen atom of a H–O–H molecule is attracted to the oxygen atom of another molecule and this process in continued with almost all the molecules present in the vicinity. A proton donor group possesses positively charged hydrogen available for hydrogen bonding and the acceptor group has an unshared pair of electrons.

Pimentel and McClellan (1960) have classified all molecules into four groups with respect to their ability to participate in H bonding.

Molecules with one or more donor groups and no acceptor groups	Haloforms, highly haolgenated compounds, acetylenes.
Molecule with one or more acceptor groups but no donor groups.	Ketones, ethers, esters, olefins, aromatics, tertiary amines, nitriles, isonitriles.
Molecules with both donor and acceptor groups.	Alcohols, water, phenols inorganic and carboxylic acids, primary and secondary amines.
Molecules with neither donor nor acceptor groups.	Saturated hydrocarbons CCl_4, CS_2.

TABLE 4.5: Donor and acceptor groups of Hydrogen (Vinogradov and Linnel, 1971)

Donor group A–H	Examples
O–H	H_2O, ROH (alcohols, phenols), R–C"° OH (carboxylic acids), (RO)-P"°–OH (phosphoric acids), ROOH (hydroperoxides)
N–H	RNH_2 (primary amines), R_2 R_2 NH (secondary amines), $RCONHR_1$ (AMIDES, PEPTIDES)
S–H	RSH thiophenols), Cysteine
X–H (X=F,Cl, Br or I)	FH, HCl.
C–H	R≡CH (acetylenes), $CHCl_3$, HCN
P–H	R_1R_2PH (phosphines) $(RO)_2$ –P"°–H (Phosphonates)
Acceptor groups	Examples
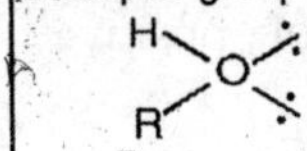 H, O, R	H_2O, alcohols, phenols, ketones, esters, carboxylic acids
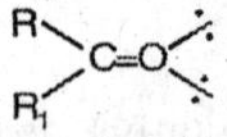 R, C=O, R_1	(R=aryl or alkyl, X=N, S, P, Se or As) Sulphoxides pyridine–N–oxide, ethers
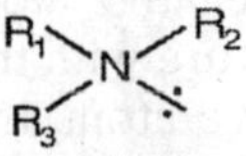 R_1, R_2, N, R_3	Primary, secondary and tertiary amines, pyridines
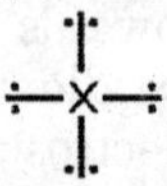 X	Oxygen and halogen anions, quaternary ammonium salts
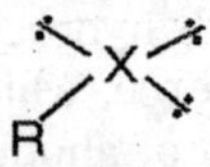 X, R	R–X (organic halides), HF
π Electron systems	Benzene, acetylenes, nitriles, cyanides

Molecules belonging to 1–3 groups can participate in hydrogen bonding while that of the fourth group is incapable of doing so. Solvents of 4th group are used in studying H bonded molecules. Intermolecular H bonding occurs between the groups 1, 2 and 3. The molecules of the third group can self-associate by H bonding with them. This electrostatic attraction is conventionally designated by dotted lines and the stronger covalent bonds by solid lines.

The energy of the H bond (2–8 Kcal/mole) is intermediate between that of van der Waal's forces (1 Kcal/mole) and that of a chemical bond (OH = 110 Kcal/mole). Some of the energy values of the various H bonds are H–F.H = 10 Kcal/mole; H–O H = 7.2 Kcal/mole; H–N.H = 2 Kcal/ mole. The bond length varies from 2.3 to 3.0 Å (Table 4.6).

TABLE 4.6: The average lengths of H bonds in Angstroms (Å)

O–H....N	=	2.78
O–H....O	=	2.72
N–H....N	=	3.07
N–H....O	=	2.93
N–H....S	=	3.40
C–H....O	=	3.23

There are two types of hydrogen bonding, intermolecular and intramolecular. The intermolecular hydrogen bonding results in association of molecules and is influenced by the shape of the molecules. This type of H bonding changes the number, mass, shape and electronic structure of the participants in a system. Due to the intermolecular association the effective molecular weight of a structure is increased leading to higher melting and boiling points.

The decrease in freedom of molecular rotation due to the increased molecular structure (associations) increases the viscosity. Due to increased polarity of the bond A–H in H-bonded system A–H.B, a larger dielectric constant (Permittivity) and greater dipole moment are experienced by the molecules.

Intramolecular H bonding takes place in molecules of the 3rd group; between the proton donor group and the proton acceptor group of the same molecule. This type of bonding occurs only when (1) the proton donor and one proton acceptor sites on the same molecule is in a favourable spatial configuration, that is, the distance between the hydrogen of the donor group and the acceptor site is in between 1.4 and 2.5 Å, (2) the angular orientation of the acceptor site does not differ much from the bond axis of the donor group A–H and (3) the molecules are in the cis or a gauche (Skew) conformation. Intramolecular H bonding gives rise to chelation i.e. formation of 5,6,7-heterocyclic rings. This reduces the solubilities of these molecules in water because these compounds cannot form H bonds with water. Other physical properties of the system remain unaltered since only the electronic structure of the molecule is modified and no molecular association takes place within the system Intramolecular H bonding is also formed with π electrons as seen in o-allyl phenol and β-phenyl ethanol.

The hydrogen bonds formed with C–H groups are very weak when compared to the H bonds involving F, O, N, Cl, S etc. It is found that H in the methyl group of many purines forms fairly strong H bonds with oxygen atoms present in the DNA molecule.

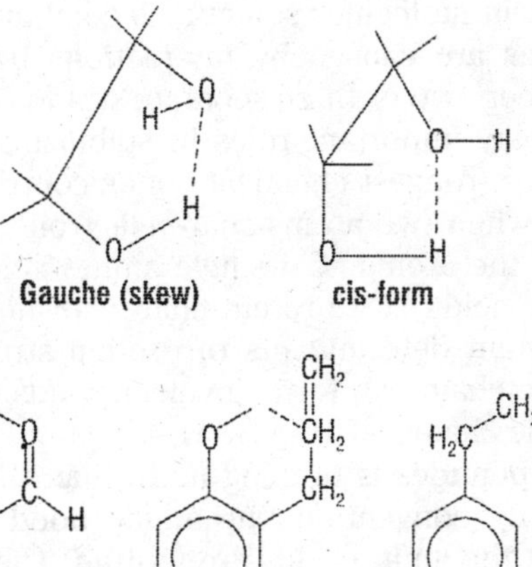

Gauche (skew) cis-form

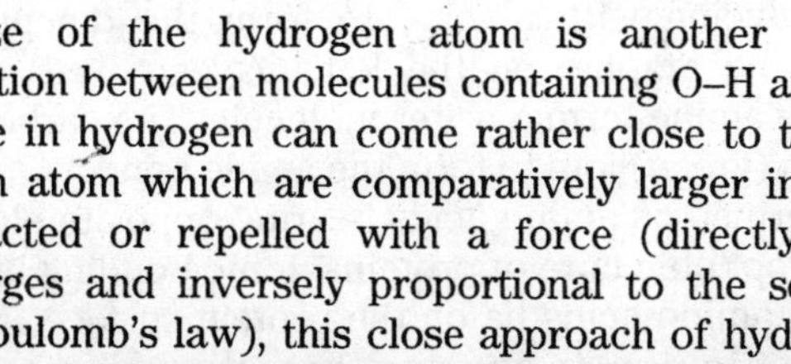

The small size of the hydrogen atom is another factor favouring the electrostatic attraction between molecules containing O–H and N–H groups in that the positive charge in hydrogen can come rather close to the negatively charged oxygen or nitrogen atom which are comparatively larger in size. As the charged particles are attracted or repelled with a force (directly proportional to the magnitude of charges and inversely proportional to the square of the distance separating them Coulomb's law), this close approach of hydrogen atom to oxygen or nitrogen atoms results in a much stronger electrical attractive force than that found between large particles of a similar magnitude of charge difference.

The presence of hydrogen bonding is detected by infrared absorption spectra, X-ray analysis, electro diffraction examination of boiling point, melting point and solubility etc.

Hydrogen Bonds in Biological Systems

The fact that life originated in water– a labyrinth of H bonds– ensures the prevalence of H bonds everywhere in the living system. The living cell contains 70–90% water. Almost all the biological molecules occur in aqueous environments, hence the role of the nearest neighbour water molecules is of great importance in the structure, conformation and arrangements of biopolymers. These biopolymers enter into intermolecular H bonding among them and with adjoining water molecules and almost all the polar molecules are interlinked through an extensive network of H bonded water molecules in a living system. The exhaustive intramolecular H bonding results in considerable folding in their skeletal structures of these macromolecules and produces microenvironments whose polarities range from non-polar and highly polar depending on the number and arrangements of water molecules in each in each microregion. The thermodynamic parameters (free energy, enthalpy and entropy) of the H bonds between the proton donor and acceptor side chain groups will vary with the polarity of their specific microenvironments. Though not directly, the extensive H bonding within a cell is responsible for the proper alignments of the non-polar molecules like lipids to form associations, the membranes.

BONDING AND PROTEIN STRUCTURE

The specific sequence of the amino acids along the protein chain is referred to as the primary structure of the protein. However, these chains are not rigid, but rather they are long and flexible like string. The strands of protein can twist to form helixes or fold into sheets. They can bend and fold back on themselves to form globs and several protein molecules sometimes combine into a larger molecule. All of these configurations are caused by interactions both within a single protein strand as well as between two or three separate strands of proteins.

Chemical bonds play important roles in stabilizing 3-D protein structure and covalent bonds are the strongest chemical bonds contributing to protein structure. Covalent bonds arise when two atoms share electrons. In addition to the covalent bonds which connect the atoms of a single amino acid and the covalent peptide bond that links amino acids in a protein chain, covalent bonds between cysteine side chains are important determinants of protein structure. Cysteine is the sole amino acid whose side chain can form covalent bonds, yielding disulphide bridges with other cysteine side chains: $-CH_2-S-S-CH_2-$.

The formation of peptides is nothing more than the application of the **amide synthesis reaction**. By convention, the amide bond in the peptides should be made in the order that the amino acids are written. The amine end (N terminal) of an amino acid is always on the left, while the acid end (C terminal) is on the right.

The reaction of glycine with alanine to form the dipeptide glyclalanine is written as shown in the graphic on the left. Oxygen (red) from the acid and hydrogens (red) on the amine form a water molecule. The carboxyl oxygen (green) and the amine nitrogen (green) join to form the amide bond.

Ionic bonds are formed as amino acids bearing opposite electrical charges are juxtaposed in the hydrophobic core of proteins. Ionic bonding in the interior is rare because most charged amino acids lie on the protein surface. Although rare, ionic bonds can be important to protein structure because they are potent electrostatic attractions that can approach the strength of covalent bonds.

Electrically charged amino acids are usually found on protein surfaces. They promote appropriate folding by interacting with the water solvent. Polar water molecules can form shells around charged surface residue side chains, helping to stabilize and solubilize the protein.

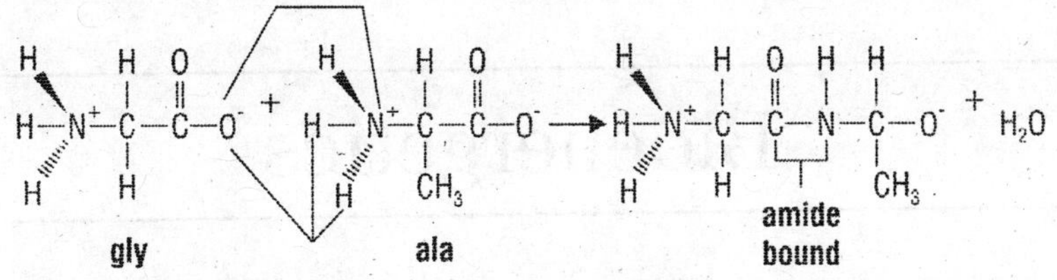

gly + ala → amide bound + H_2O

FIG. 4.16: Peptide bond formation.

When two atoms bearing partial negative charges share partially positively charged hydrogen, the atoms are engaged in a hydrogen bond (H-bond). The correct 3-D structure of a protein is often dependent on an intricate network of H-bonds. These can occur between a variety of atoms, involving:

1. Atoms on two different amino acid side chains.
2. Atoms on amino acid side chains and water molecules at the protein surface.
3. Atoms on amino acid side chains and protein backbone atoms.
4. Backbone atoms and water molecules at the protein surface.
5. Backbone atoms on two different amino acids.

Amine terminal end — Repeat nnits — Repeat units — Repeat units — Amide bond — Carboxyl acid terminal end

gly ala leu

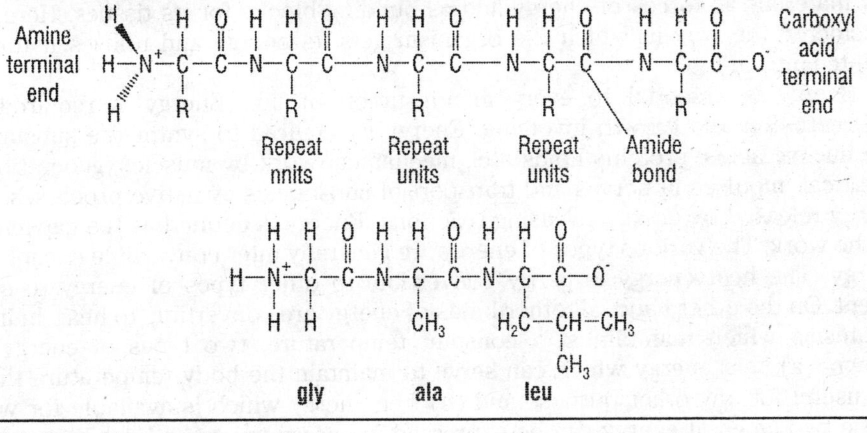

FIG. 4.17: Backbone peptide structure.

Hydrophobic bonds are a major force driving proper protein folding. They juxtapose hydrophobic side chains by minimizing lost energy caused by the intrusion of amino acids into the H_2O solvent, which disrupts lattices of water molecules. Hydrophobic bonding forms an interior, hydrophobic protein core, where most hydrophobic side chains can closely associate and are shielded from interactions with solvent H_2O's. However, not all hydrophobic amino acids are in the interior of proteins, however. When found at the surface, exposed to polar water molecules, hydrophobic side chains are usually involved in extensive hydrophobic bonding.

Van der Waals attractions, although transient and weak, can provide an important component of protein structure because of their sheer number. Most atoms of a protein are packed sufficiently close to others to be involved in transient Van der Waals attractions. Van der Waals forces can play important roles in protein-protein recognition when complementary shapes are involved.

CHAPTER 5

Bioenergetics

Bioenergetics in biochemistry is concerned with the mechanisms whereby the energy made released from the oxidation of substrates or the absorption of light can be used to drive thermodynamically unfavorable reactions such as the synthesis of ATP from ADP and Pi, or the accumulation of ions across membranes.

Living system is a highly organized arrangement of matter. The synthesis and maintenance of the system is possible only by the availability of certain reactions. These reactions used by the living systems are the degradation of complex molecules of food materials to simple molecules and photosynthesis of materials from the simple compounds. Such spontaneous reactions produce useful energy for the performance of various vital activities of the organisms. The organism uses the raw materials as forces of energy and as building blocks for its tissues. Here it is considered the way in which the organism gets its energy and utilizes it for the maintenance of life.

Energy is essential to every manifestation of life. Energy is required for differentiation and growth involving. Energy is required to synthesise substances like nucleic acids, proteins lipids etc., mechanical work by muscles, generation of electrical impulses in nerves and transport of substances by active processes. The energy released by energy releasing reactions. Energy is defined as the capacity to do the work. The various types of energy are generally inter convertible except heat energy. The heat energy is partly convertible to other types of energy to some extent. On the other hand, all other kinds of energy are convertible to heat. In living organisms which maintain at a constant temperature, two types of energy are known- (a) heat energy which can serve to maintain the body temperature this is not useful for any other purpose and (b) free energy which is available for work. These two types of energy can be expressed in energy units (Calorie, Joules). It is described here about the free energy, which is liberated in the chemical reactions.

NATURE OF ENERGY AND ITS MEASUREMENTS

Energy is a fundamental physical concept. As mentioned above it is the capacity to do work. Many kinds of energy are known. The potential and kinetic energies, surface energy, chemical energy, electrical energy and radiant energy are examples. These may be converted from one form to another. When one form is created equal amount of another form disappears. This is the law of conservation of energy.

Energy can be understood clearly, if it is measured exactly. It is measured by various ways. The Table (5.1) gives the different kinds of energies, units and their inter-conversion. The ultimate source of energy for life process is radiant energy is converted into chemical energy by biological mechanisms of transformations to vital kinds of work, mechanical, chemical biosynthesis and osmotic (transport). However, it is important to note that only a fraction of energy is released for vital processes. This is known as free energy.

Chemical Energy

Different organic compounds yield different amounts of energy when they are completely oxidized. This heat of combustion is determined by the elementary composition and structure of the substances and also its energy content. The total energy liberated is represented by the difference in energy contents of the substance oxidized and products of oxidation. The total amount of energy liberated is dependent upon the pathway of oxidation.

TABLE 5.1: Energy units and their inter conversion

Unit	Definition	Equivalent in other unit
Erg	Work performed by a force of one dyne moving its point of application one cm.	
Joule		10^7 ergs
Joule (electrical)	Energy developed when one coulomb of electrons (10.364×10^{-6} moles) passes through a potential of one volt.	0.2390 calorie
Calorie	Energy (heat) required to warm one g of H_2O from 25° to 26°C	4.1840 joules
Electron volt	Energy gained by an electron passing through a potential of one volt	1.6×10^{-12} ergs 1.6×10^{-19} joules 23.06 Kcal per mol
Quantum	Energy of an electromagnetic quantum calculated from its wave length expressed in millimicrons, m: eV= 1,239.8/m	1,239.8 m^{-1} ev
Kilo calories per Einstein	Energy of mol of photons (one Einstein) of wavelength in m: Kcal per mol = 28,589.7/m	28,589.7 m^{-1} Kcal per mol

The different compounds yield different quantities of energy on combustion. When combustion takes place at constant pressure, the heat change is called the enthalpy change. The heat liberated in the combustion of ethane is higher than its derivative like alcohol, aldehyde and acetic acid. The later can be derived from ethane by one stage of oxidation. Heats of combustion represent the maximum energy from a substance by breaking and oxidizing its elements. The combustion of glucose in the living system is carried out by a series of oxidation process and ultimately liberates energy water and carbon dioxide. The energy is liberated step by step.

Chemical reactions are always associated with energy change. Some reactions proceed with heat liberation and these are called exothermic. Other reactions absorbed heat from the surrounding and these are called endothermic. Reactions that absorbed energy proceed only when energy (heat) is supplied. Many processes associated with breakdown of foodstuffs are exothermic.

Free Energy

Free energy is the energy of the system capable of doing work. The capacity of the chemical system to do useful work often depends on the downhill flow of electrons; ΔF, the free energy available for such work can be measured in potential energy unit just as the ability to do work of falling water and just as the potential difference in the case of the aqueous system depends on the height of the water fall. So in the chemical system the free energy difference ΔF is the difference between the sum of the free energies of all the reactants ΣF and those for all the products, ΣFp.

$$\Delta F = \Sigma F_p -$$

Free energy F is a property of a system just is as volume and mass; once the size and certain environmental conditions have been fixed, F is accurately defined. Neither the previous history of the system not the route, where by a chemical system of energy ΣF_R approaches another one of energy ΣFp, plays a part in determining the magnitude of ΔF. The change $\Sigma F_R \longrightarrow \Sigma F_p$ with consequent change ΔF in the free energy, is the thermodynamic quantity of much significance. Because its sign indicates, whether a reaction theoretically able to proceed spontaneously. If the reaction is exergonic the magnitude of $-\Delta F$ indicates the maximum energy available for the work. In addition the driving force $-\Delta F$ of reversible reaction, declines as the mixture of reactants products move towards the equilibrium.

Although all chemical reactions are reversible in theory, the term irreversible is used in thermodynamics to designate those reactions, which proceedes spontaneously, the reactants being virtually converted to the products. As such reactions are usually characterized by large equilibrium constant ΔF becomes a large negative number. The chief biological significance of the determination of the free energy change lies in the fact that the negative ΔF of an exergonic reaction provides a measure of the maximum amount of work obtainable from this reaction. The positive ΔF of an endergonic reaction provides an estimate of the minimum amount of work to drive this type of reaction.

REQUIREMENTS OF BIOENERGETIC SYSTEM

The fundamental requirement of a bioenergetic system is the sources of energy. This energy is a compound capable of being degraded to products of lower potential energy. As the most energy yielding reactions are oxidative, the materials are oxidisable foodstuffs. Secondly, the enzymatic mechanisms are needed to degrade the foodstuffs. There should be a facility for the collection and storage of the liberated free energy. The phosphorylated compounds store and distribute the energy. The final mechanism of utilization of stored energy is the conversion of free energy into such type of work as muscle contraction, secretion, synthesis etc.

Scheme of Bioenergetics

The Fig. 5.1 gives the scheme of bioenergetics in the total metabolism of organism. Individual reactions are also shown in the figure. Degradation of foodstuffs of high potential energy to products of low energy is shown in coupled to a mechanism for collection and storage energy. This mechanism consists of the conversion of inorganic phosphate into high-energy phosphate compound. This high-energy phosphate compound can in turn be degraded back into inorganic phosphate after releasing energy. Another coupling mechanism ensures that this energy is made into useful energy.

Catabolism of foodstuffs may be divided into three phases. Firstly, food materials of high molecular weight (Polysaccharide, lipids) are digested to products more easily absorbed form into the alimentary canal. These reactions are hydrolytic and the small amount of energy liberated is not recoverable but is lost as heat. Products of this stage of catabolism include mainly common food sources like glycerol, fatty acids and many amino acids.

Second phase of catabolism converts this diverse collection of compounds into small number of intermediates for final oxidation in the next stage. Hexose and fatty acids are firstly activated by the formation of esters with phosphate and co-enzyme-A respectively. At this stage some free energy is liberated. In addition to the anaerobically producing lactic acid or alcohol depending upon the organism, lactic acid formation (glycolysis) makes available about 50000 calorie and alcohol formation (fermentation) about 62000 calories and the rest of the energy is available during later oxidation. Higher animals in general can oxidize glucose completely to carbon dioxide and water and liberate much free energy. Under aerobic conditions pyruvic acid rather than lactic acid is the main product and this is further oxidized to acetyl coenzyme A. Acetyl coenzyme-A is also produced from fatty acid. The glycerol of fat joins the carbohydrate path at the triose state.

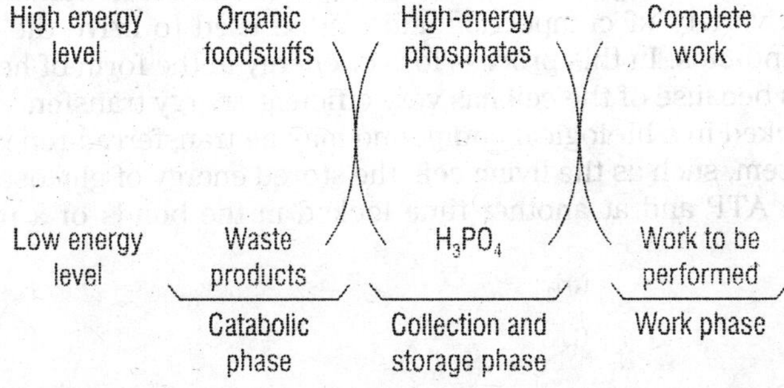

FIG. 5.1: Scheme of bioenergetics.

Certain amino acids are catabolized via special pathway, but most are initially deaminated to form the corresponding keto acid. The nitrogen of amino group follows its own pathway to eventually appear in the urine has urea and ammonia. Sulphur contained in the amino acid (e.g. methionine) is oxidized to inorganic sulphate. The carbon skeletons of a few amino acids are converted to ketone bodies or acetyl co-enzyme-A. But many amino acids either yield pyruvic acid, α-keto glutaric acid or oxaloacetic acid directly on deamination or form products readily convertible to these α-ketoacid. During this second phase a small amount of recoverable free energy is liberated.

During third phase all of the products of the second phase are channelled into a single mechanism known as tricarboxylic acid or Krebs cycle. This is a sequence of reaction, which releases the carbon dioxide, while it feeds pairs of their hydrogen atom into oxidative chain to form water.

Free energy liberated or consumed in any reaction is the difference between the free energies of formation and reactants in the reaction. It has been possible to calculate the free energy change associated with much biochemically important reaction. Depending upon the type and number of coupling system, it is possible to estimate the amount of free energy recoverable by the cell from a given oxidation.

ATP - THE ENERGY CURRENCY

Both energy yielding and energy consuming reactions occur with the living cell. The potential or stored energy of one compound (e.g. glucose) is released and utilized, in a most efficient manner, to drive the synthesis of other compounds (e.g. proteins). This energy, now stored in the newly synthesised compound, can, in turn, be made available for other synthetic reactions. What is described here is the coupling of energy yielding and energy consuming reactions. However, energy-yielding reactions of the cell occur in many instances in the absence of energy consuming reactions. The energy released in such a situation would be in the form of heat and loss to the organism. However, nature has provided the cell with a means of temporary energy storage in the form of adenosine triphosphate (ATP). Thus, the energy released in the oxidation of compounds such as carbohydrates, lipids, proteins, etc., is immediately utilized in the synthesis of ATP from adenosine diphosphate (ADP) and inorganic phosphate (iP). The chemical energy transferred to ATP can be used to drive synthetic reactions, ADP and iP being released in the process. The bond joining the last phosphate group to ATP is referred to as a *high-energy bond* (~). This last phosphate group of ATP is readily transferred from one compound to another. In this way energy is transferred from one compound to another compound through ATP. This is of obvious since ATP can be formed in the oxidation of a variety of compounds and can be used to drive the synthesis of a variety of compounds. In this process loss of energy in the form of heat occurs very less and this is because of the cell has very efficient energy transfer.

Energy locked in a biological compound may be transferred repeatedly. Thus in a dynamic system, such as the living cell, the stored energy of glucose may be found at one time in ATP and at another time locked in the bonds of a protein or lipid molecule.

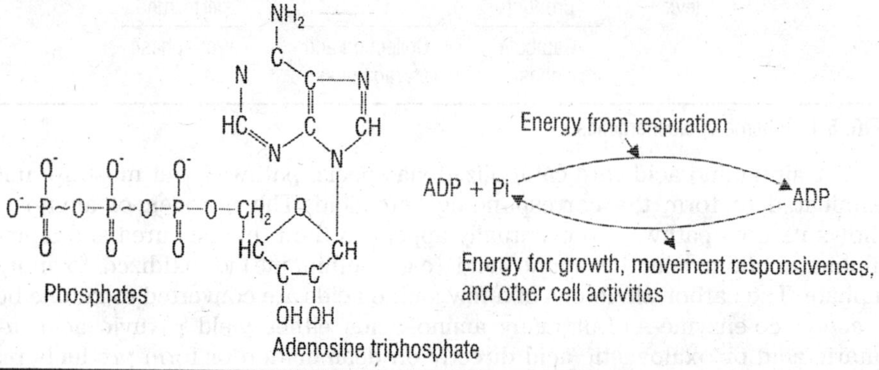

Phosphates

Adenosine triphosphate

Energy from respiration

ADP + Pi

ADP

Energy for growth, movement responsiveness, and other cell activities

FIG. 5.2: An illustration of the role of ATP as an intermediate energy transfer compound.

ATP works by losing the endmost phosphate group when instructed to do so by an enzyme. This reaction releases a lot of energy, which the organism can then use to build proteins, contact muscles, *etc.* The reaction product is adenosine diphosphate (ADP), and the phosphate group either ends up as orthophosphate (HPO_4) or attached to another molecule (e.g. an alcohol). Even more energy can be extracted by removing a second phosphate group to produce adenosine monophosphate (AMP).

$$ATP + H_2O \rightarrow ADP + HPO_4$$

When the organism is resting and energy is not immediately needed, the reverse reaction takes place and the phosphate group is reattached to the molecule using energy obtained from food or sunlight. Thus the ATP molecule acts as a chemical 'battery', storing energy when it is not needed, but able to release it instantly when the organism requires it.

Release of Energy through Biological Oxidation

The most important feature of respiration is the release of usable energy. In this process of release of energy two processes i.e. oxidation and reduction go simultaneously. In its simplest sense, oxidation refers to the removal of electrons from a compound, a process usually accompanied in the cell by the removal of hydrogen. Conversely, the reduction of a compound refers to the addition of electrons to that compound, usually accompanied in the cell by the addition of hydrogen.

Biological oxidation proceeds in three stages if we introduce a carbohydrate, such as ubiquitous glucose, into the process.

A Metabolic Degradation

- **Glycolysis:** Degradation of glucose to pyruvate.
- **Oxidative Decarboxylation of Pyruvate:** A relatively short step during which a 2-C body arises from the 3-C body pyruvate with the elimination of CO_2.
- **The Citric Acid Cycle.** Degradation of the 2-C body furnished by stage B to CO_2. It is also known as the Krebs cycle after Krebs who, together with other scientists, elucidated this pathway; a third name is tri carboxylic acid cycle (TCA cycle), which it owes to the participation of acids with three carboxyl functions.
- **Terminal Oxidation in the Respiratory Chain.** The hydrogen, which is extracted from the substrates in stages A to C, is finally combined with oxygen to form water. For this to occur a hydrogen transport takes place along a chain of redox systems, i.e., via an electron transport chain. Energy set free by this transport is utilized for the formation of ATP.

The Phosphorus Cycle

The fact that ATP is Nature's 'universal energy store' explains why phosphates are a vital ingredient in the diets of all living things. Modern fertilizers often contain phosphorus compounds that have been extracted from animal bones. These compounds are used by plants to make ATP. We then eat the plants, metabolise their phosphorus, and produce our own ATP. When we die, our phosphorus goes back into the ecosystem to begin the cycle again.

OXIDATION-REDUCTION POTENTIAL

The importance of oxidations-reductions lies in the fact that they comprise all known energy yielding reactions, which living organisms have at the disposal when compared to other reactions (hydrolysis, condensation etc.). Oxidations furnish the most energy per mole of compound. This is how the organisms of all levels have retained oxidations for reductions as source of energy during the long evolutionary period. The hydrogen atoms locked up in the food material, are oxidized from various linkages to carbon, oxygen and other elements in the organic compounds. In the oxidation-reduction process the hydrogen and oxygen are brought together in gradual step by step fashion; the energy is liberated in each step. The study of biological oxidation-reduction is largely a study of individual step by which hydrogen is brought together with oxygen to form water ultimately. Thus, biological oxidation is essentially a reverse process of photosynthesis as follows:

$$6CO_2 + 6H_2O \xrightarrow[\text{Biochemical}]{\text{Photosynthesis}} C_6H_{12}O_6 + 6CO_2$$

Contrary to the earlier held view of the direct combustion of oxygen with the carbon, the intake oxygen is used primarily to form water with the hydrogen atom of the substrate. The oxygen contained in the CO_2 is derived from the substrate itself or from the body water by an indirect route.

Oxidation-reduction potential and electron transfer: Bio-chemically oxygen has the highest oxidation potential (electron affinities) of all systems in the living cell while the lowest values are assigned to the hydrogen atoms attached to the various substrates. Other substances have potential intermediate between these two extremes. The electrons are transferred from systems of lower potential to high potential. A coupled oxidation-reduction reaction can occur between two systems are follows:

System 2 has greater potential than system 1, hence can accept electrons from it. The reductant system 1 provides these electrons by changing to its oxidant form, while the oxidant system 2 accepts the electrons and in so doing changes to its reductant form. It should be noted that the effective oxidation potential of a given system may be higher or lower than its standard potential. This is due to variation in the oxidant-reductant ratio. Thus, if conditions within a cell maintain a high ratio NAD/NADH, then this system may well act as an electron acceptor (Oxidant). The amount of energy liberated in transferring electrons from one system to another is determined by the difference in oxidation potential between two systems. Electrons derived from metabolites such as malate or succinate move from one carrier to the next along the electron transport assembly to oxygen, which is the final electron acceptor. The flow of electrons is always towards the oxygen, because electrons in the water molecule can be regarded at a lower energy level. As the electrons are transferred in steps from malate to oxygen, they are descending an energy gradient like a downhill task.

The tendency for substances to undergo oxidation-reduction reactions differs considerably and depends on the relative with which they accept or give up electrons. This tendency can be measured and can be referred the ability of hydrogen gas is oxidized to hydrogen ions.

$$H_2 \longleftrightarrow 2H^+ + 2e$$

TABLE 5.2: Oxidation-reduction potentials of important cellular reactions

Oxidation-reduction reaction	E'o (25 C; pH 7.0) (V)
Oxidized form ⟶ reduced form	
Succinate ⟶ α-ketoglutaric acid	-0.67
CO_2 + α-ketoglutarate ⟶ isocitrate	-0.48
Ferredoxin (Ox) ⟶ (red)	-0.43
NAD ⟶ $NADH_2$	-0.32
Riboflavin ⟶ reduced riboflavin	-0.21
Acetaldehyde ⟶ ethanol	-0.20
Pyruvate ⟶ lactate	-0.18
Oxaloacetate ⟶ malate	-0.17
Fumarate ⟶ Succinate	+0.03
Cytochrome b Fe^{3+} ⟶ Fe^{3+}	+0.04
Ubiquinone (COQO ⟶ $COQH_2$)	+0.10
Cytochrome C Fe^{3+} ⟶ Fe^{2+}	+0.22
Cytochrome a Fe^{3+} ⟶ Fe^{2+}	+0.29
Cu^{2+} ⟶ Cu^+	+0.35
Oxygen, ½O_2 ⟶ H_2O	+0.82

The compounds in the upper part of the Table (5.2) lose electrons more readily than the hydrogen gas, while the compounds in the lower part have lesser capacity to become oxidized. For each substance the tendency of its reduced form, to give up electrons and so become oxidized is called as the standard redox potential E'o and can be measured in Volts. The standard conditions are needed, because actual redox potential varies with concentration and pH, E'o refers to a molar solution containing equal quantities of the oxidized and reduced forms of the compound at pH7. When E'o is negative the substance tends to become oxidized more readily than hydrogen gas. A high negative E'o indicates a high electron donating or reducing capacity. On the contrary the substances with positive values of E'o tend to become reduced because they readily accept electrons and act as oxidizing agents.

The values of E'o for a number of donors and carriers of electrons are presented in the Table 5.2. The reduced forms of compounds at the upper side of the table loose electrons to the oxidized forms of those lower forms i.e., to those with more positive value of E'o. As electrons move down to low every levels of carriers with more and more redox potentials, a small amount of energy is dissipated as heat; but much energy is transferred to other systems.

Riboflavin as flavoproteins has wide range of oxidation-reduction potentials

The energy change is associated with the movement of electrons down the redox gradient towards oxygen. This can be calculated in the following manner. When the electrons move from NADH to oxygen the distance in the redox scale is from -0.32 to +0.82 V, so the potential difference in E'o is 1.14 V. The energy change can be said in calories. The equation:

$$\Delta G^\circ \text{ (in kcal)} = \frac{n \times 96.500}{4.18 \times 1000} \text{ X } \Delta E^\circ$$

n is 2, two electrons are transferred therefore,

$$-\Delta G^\circ = \frac{2 \times 96.500 \times 1.14}{4.18 \times 1000} = 52.6 \text{ kcal.}$$

Since the reaction is exergonic, 52.6 kcal of energy is made available when one mole of NADH transfers its electrons.

Another important example is succinate oxidation and the oxygen is the final electron acceptor.

$$\Delta E^\circ = 0.82\text{-}0.03 \text{ or } 0.79 \text{ V.}$$

The number of electrons involved is again 2, and the amount of energy released:

$$\Delta G^\circ = \frac{2 \times 96{,}5008 \times 0.79}{4.11 \times 1000} = 36.5 \text{ Kcal.}$$

The reaction is exergonic and small amount of energy is released compared to NADH is the electron donor. It is therefore expected that a fewer molecules of ATP are produced when succinate is oxidized.

The malate can reduce NAD, since its standard redox potential is lower than that of NAD on the redox scale. The malate, lactate and ethanol are able to reduce NAD in the presence of suitable dehydrogenases. The actual redox potential operative *in vivo* is slightly greater than the standard values. Thus when there is more reduced form (e.g. malate) than oxidized form (oxaloacetate), the redox potential is more negative than the standard of E'o. In the case of succinate/fumarate system, the redox potential is too low on the scale to reduce the limited amount of NAD^+. This carrier has a power potential than the system NAD^-/NADH.

CHAPTER 6

Organic Components: Proteins

All proteins contain C, H, O, N and generally S. Many proteins contain P. Elements such as I, Fe, Cu, and Zn are also occasionally present. The average elementary composition of proteins is: C = 50, H -= 7, O = 23, N = 16, S = 3, P = 3 per cent.

The term protein is derived from the Greek word *proteios* meaning Primary or holding first place. Proteins are the first recognizable and distinctive expressions of genetic information. All basic functions in living organisms depend on specific proteins. Many proteins are enzymes. The specific enzymes in a particular cell determine its metabolic capabilities. There are variations in enzyme content that account for the metabolic differences between cell types. Proteins form much of the structural materials of the cell. Thus, they are directly responsible for the structural variability between different cells and different organisms. They are the contractile machinery of the cell. Some proteins play an important role in the organization of the extra cellular frame work of tissues. They are present in the blood and other intercellular fluids.

Proteins are also responsible for a number of specialized physiological functions. As hormones, they serve to regulate and coordinate various physiological activities. Antibodies are important in protecting animals from disease. Proteins are responsible for muscular contraction and for oxygen transport from the lungs to the tissues. They are involved in electron transport during respiration and in photosynthesis.

Proteins have a variety of other functions as well. These examples illustrate the enormous functional diversity of the proteins. This diversity of proteins is a result of their chemical structure. Proteins are polymers of 20 common amino acids. Each protein has a unique amino acid sequence which endows it with a specific shape and function.

AMINO ACIDS

Amino acids are the building blocks (monomers) of proteins. 20 different amino acids are used to synthesise proteins. The **shape** and other properties of each protein is dictated by the precise sequence of amino acids in it.

Each amino acid consists of an **alpha carbon atom** to which is attached to:

- a hydrogen atom
- an amino group (hence *amino* acid)
- a carboxyl group (-COOH). This gives up a proton and is thus an acid (hence amino *acid*)
- one of 20 different *R* groups. It is the structure of the R group that determines which of the 20 it is and its special properties. The amino acid shown here is **Alanine**.

About twenty amino acids have been occurring in proteins. Most proteins contain a large proportion of these amino acids.

Amino acids are the building blocks (monomers) of proteins. 20 different amino acids are used to synthesise proteins. The **shape** and other properties of each protein are dictated by the precise sequence of amino acids in it.

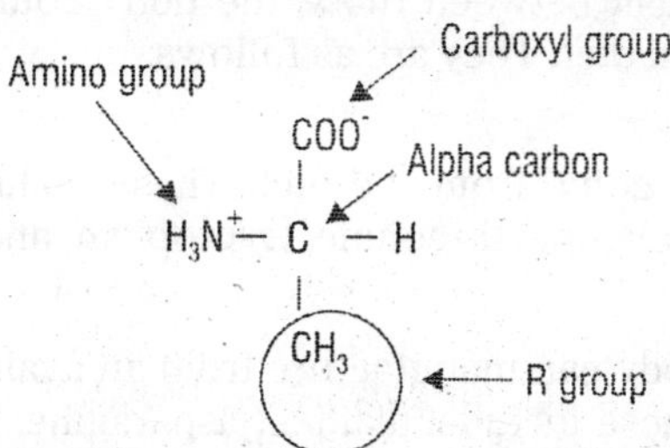

FIG. 6.1: Amino acids.

TABLE 6.1: Different kind of amino acids

Alanine	**Ala**	**hydrophobic**
Arginine	Arg	free amino group makes it basic and hydrophilic
Asparagine	Asn	carbohydrate can be covalently linked ("N-linked) to its -NH
Aspartic acid	Asp	free carboxyl group makes it acidic and hydrophilic
Cysteine	Cys	oxidation of their sulphydryl (-SH) groups link 2 Cys (S-S)
Glutamic acid	Glu	free carboxyl group makes it acidic and hydrophilic
Glutamine	Gln	Moderately hydrophilic
Glycine	Gly	so small it is amphiphilic (can exist in any surroundings)
Histidine	His	basic and hydrophilic
Isoleucine	Ile	hydrophobic
Leucine	Leu	hydrophobic
Lysine	Lys	strongly basic and hydrophilic
Methionine	Met	hydrophobic
Phenylalanine	Phe	very hydrophobic
Proline	Pro	causes kinks in the chain
Serine	Ser	carbohydrate can be covalently linked ("O-linked") to its -OH
Threonine	Thr	carbohydrate can be covalently linked ("O-linked") to its -OH
Tryptophan	Trp	scarce in most plant proteins
Tyrosine	Tyr	a phosphate or sulphate group can be covalently attached to its -OH
Valine	Val	hydrophobic

Humans must include adequate amounts of 9 amino acids in their diet. These **essential** amino acids cannot be synthesised from other precursors. However, cysteine can partially meet the need for methionine (they both contain sulphur), and tyrosine can partially substitute for phenylalanine.

The Essential amino acids are: Histidine, Isoleucine, Leucine, Lysine, Methionine (and/or cysteine), Phenylalanine (and/or tyrosine), Threonine, Tryptophan and Valine. Two of the essential amino acids: **lysine** and **tryptophan** are poorly represented in most plant proteins. Thus strict vegetarians should ensure that their diet contains sufficient amounts of these two amino acids.

Classification

Amino acids have been classified in various ways. One system classifies them according to the number of amino and carboxyl groups present in the molecule as monoamino mono carboxylic acids and monoamino dicarboxylic acids, etc. Another system classifies them as aliphatic, aromatic, and heterocyclic amino acids.

Depending on the source, eight to ten amino acids are considered essential for life. The reason for the classifications of amino acids into "essential" and "non-essential" was to distinguish between those the body could manufacture and the ones that had to come from diet. They are as follows:

Essential Amino Acids

They are those that must come from the diet. These include: isoleucine, leucine, lysine, methionine, phenylalanine, threonine, tryptophan, and valine.

Non-essential Amino Acids

They are those that the body can manufacture from an available source of nitrogen and a carbon skeleton. These include: alanine, asparagine, aspartic acid, carnitine, citrulline, cysteine, cystine, GABA, glutamic acid, glutamine, glutathione, glycine, hydroxyproline, ornithine, proline, serine, taurine, and tyrosine.

Amino acid	Characteristics
Taurine	Essential amino acid (conditionally) for adults and essential for normal infant development
Lysine	Essential amino acid because it cannot be synthesised in the body and its breakdown is irreversible
Arginine	Essential amino acid for children and possibly for adults
Histidine	Essential amino acid for children, but usually not for adults
Methionine	Essential amino acid that cannot be synthesised in the body
Phenylalanine	Essential amino acid that is converted to tyrosine in the body
Threonine	Essential amino acid, serving as a carrier for phosphate in phosphoproteins
Tryptophan	Essential amino acid, the only one with an indole nucleus responsible for the peculiar odor of feces
Leucine	Essential branched chain amino acid classified as semi-essential by some
Valine	Essential or semi-essential amino acid (a branched chain)
Isoleucine	Essential or semi-essential amino acid because it sometimes cannot be made in the body
Asparagine	Non-essential amino acid
Alanine	Non-essential amino acid whose main function is the metabolism of tryptophan and pyridoxine
Tyrosine	Non-essential amino acid (aromatic) produced from phenylalanine
Proline	Non-essential amino acid (aromatic) that requires Vitamin C for its synthesis
Cystine	Non-essential amino acid created when two cysteine molecules bond together
Serine	Non-essential amino acid derived from glycine, contributing to the formation of cystine from homocysteine
Gaba	Non-essential amino acid formed from glutamic acid with the help of Vitamin B6
Ornithine	Non-essential amino acid found free in the body tissues, but not used as a protein building block
Citrulline	Non-essential amino acid involved in the urea cycle
Glutamic Acid	Non-essential amino acid that can be synthesised from a number of amino acids
Aspartic acid	Non-essential amino acid which aids in the formation of RNA and DNA
Glycine	Non-essential glucogenic amino acid that readily converts to serine
Cysteine	Non-essential, sulphur-containing amino acid
Carnitine	Not a true amino acid but sometimes referred to as Vitamin BT
Glutathione	Not considered a true amino acid but a tripeptide of glutamic acid, cysteine, and glycine
Glutamine	Semi-essential amino acid

Semi-essential Amino Acids

They are ones that can sometimes be made internally if conditions are right. Arginine and histidine can be converted from other amino acids if needed.

Methionine can be converted to cystine, but cystine cannot be converted to methionine. Phenylalanine can be converted to tyrosine, but not the other way around. Therefore, when cystine and tyrosine are present in the diet, the requirements for methionine and phenylalanine are reduced. Thus, cystine and tyrosine are sometimes classified as "semi-essential." The liver is able to produce 80% of the amino acids it needs for protein construction, while the remaining 20% must be consumed.

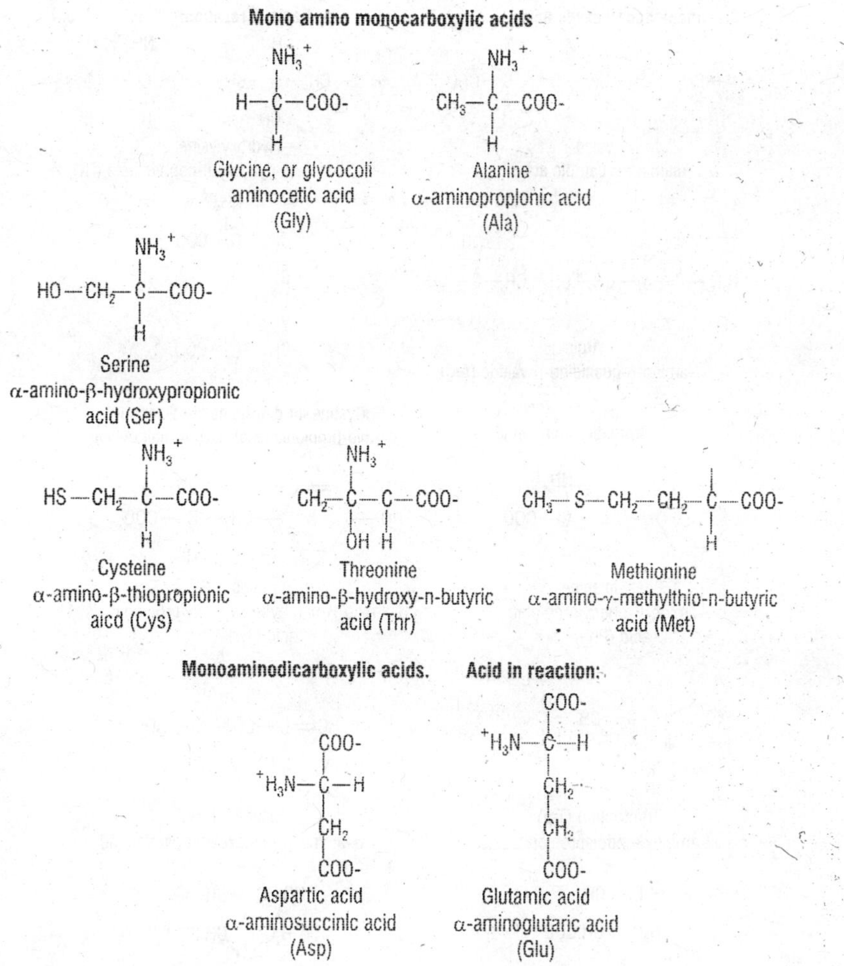

FIG. 6.2: Amino acids.

Glutamine: It is technically a semi-essential amino acid because there are times when it cannot be manufactured by the body. When the body is subjected to such metabolic stress situations as trauma, including surgeries, cancer, sepsis and burns, Glutamine becomes an essential amino acid. Officially, however, it is classified as a non-essential amino acid. Glutamine has a unique structure with each molecule containing two atoms of nitrogen instead of one - an aspect that enables it to clear ammonia from the tissues, especially the brain. Glutamine is the most abundant amino acid in the blood and cerebrospinal fluid, and is the only amino acid that easily passes the blood-brain barrier.

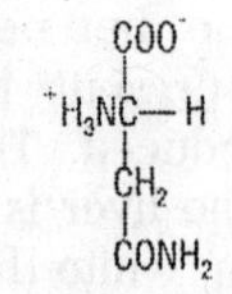

COO^- / $^+H_3NC-H$ / CH_2 / $CONH_2$

Asparagine (Asn)
α-amino-γ-succinamide

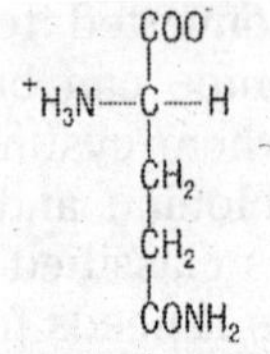

COO^- / $^+H_3N-C-H$ / CH_2 / CH_2 / $CONH_2$

Glutomine (Gln)
α-amino-δ-glutaramide

Diaminomonacarboxylic acids. Basic in reaction:

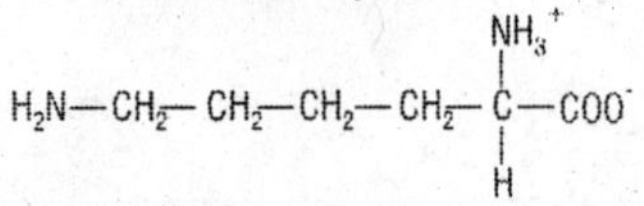

$H_2N-CH_2-CH_2-CH_2-CH_2-C(NH_3^+)(H)-COO^-$

lysine
α-ε-diamino-n-caproic acid (Lsy)

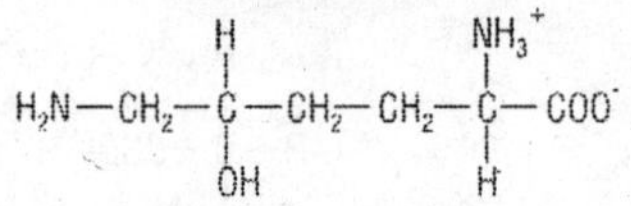

$H_2N-CH_2-C(H)(OH)-CH_2-CH_2-C(NH_3^+)(H)-COO^-$

Hydroxylysine
α-ε-diaminio-δ-hydroxy-n-copreic acid (Hyl)

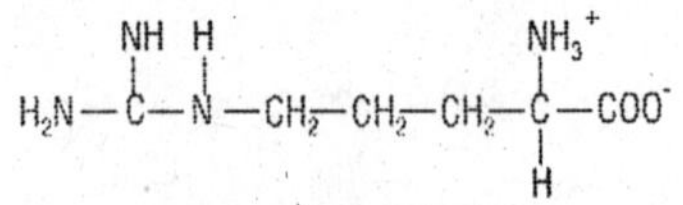

$H_2N-C(NH)-N(H)-CH_2-CH_2-CH_2-C(NH_3^+)(H)-COO^-$

Arginine
α-amino-δ-guanidino-n-valeric (Arg)

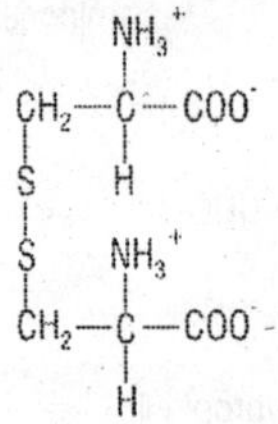

$CH_2-C(NH_3^+)(H)-COO^-$ / S / S / $CH_2-C(NH_3^+)(H)-COO^-$

Cystine, or dicysteine di-(a-amino-β -thio-propionic acid) (Cys = half cystyl)

Aromatic Amino Acid

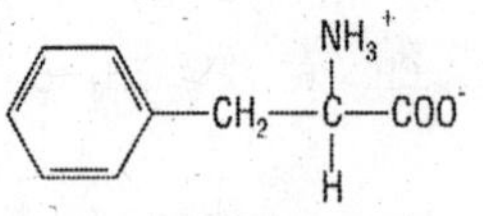

$C_6H_5-CH_2-C(NH_3^+)(H)-COO^-$

Phenylalanine
α-amino-β-phenylpropionic acid (Phe)

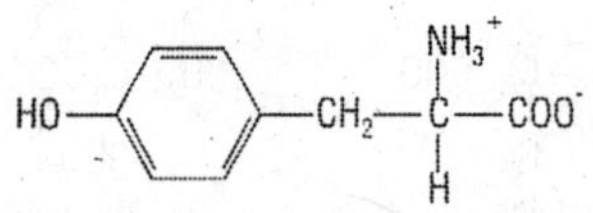

$HO-C_6H_4-CH_2-C(NH_3^+)(H)-COO^-$

Tyrosine
α-amino-β-parahydroxyphenylpropionic acid (Tyr)

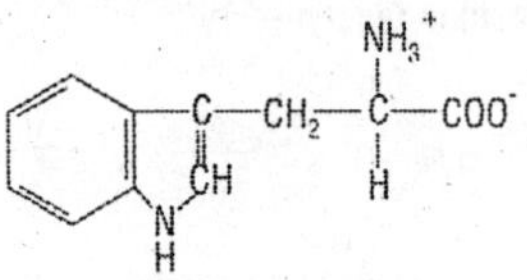

Tryptophan (Trp)
α-amino-β-indolepropionic acid

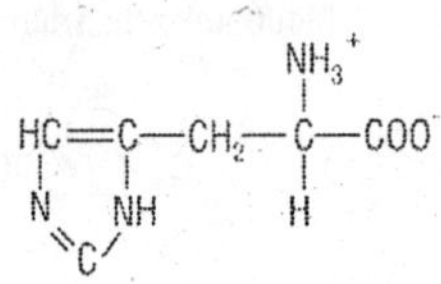

$HC=C-CH_2-C(NH_3^+)(H)-COO^-$ (N, NH, C ring)

Histidine (His)
α-amino-β-imidozolepropionic acid

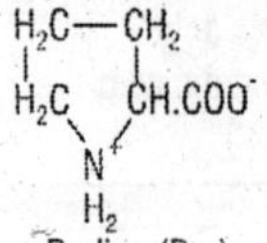

H_2C-CH_2 / H_2C $CH.COO^-$ / N^+ / H_2

Proline (Pro)
α-pyrrolidinecarboxylic acid

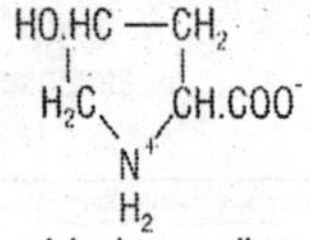

$HO.HC-CH_2$ / H_2C $CH.COO^-$ / N^+ / H_2

4-hydroxyproline
γ-hydroxy-α-pyrrolidingcorboxylic acid

FIG. 6.3: Amino acids.

Amino acids are also classified as neutral, acidic, and basic amino acids.

Aliphatic Amino Acids

1. Monoamino monocarboxylic acids. Neutral in reaction: glycine (gly), alanine (Ala), serine (ser), cysteine (cys), threonine (Thr), methionine (Met), Valine (val), Leucine (Leu), isoleucine (Ile).
2. Monoamino dicarboxylic acids. Acid in reaction: aspartic acid (Asp), glutamic

acid (Glu). Two derivatives of aspartic and glutamic acids: asparagine (Asn), glutamine (Glu).

3. Diaminomonocarboxylic acids. Basic in reaction: lysine (Lys), hydroxylysine (Hyl), arginine (Arg)

Diamino dicarboxylic acids. Cystine (cys).Aromatic Amino Acids

Monoamino monocarboxylic acids are neutral in reaction. Examples: Phenylalanine (Phe), tyrosine (Tyr).

Heterocyclic Amino Acids

Monoamino monocarboxylic acids are neutral in reaction, histidine is slightly basic. Examples: tryptophan (Trp), histidine (His), Proline (Pro), hydroxyproline (Hyp).

Amino acids as their name suggests, possess both an acidic region (a carboxyl group, -COOH) and a basic region (an amino group, $-NH_2$), as well as a third region referred to as a side chain. This third region is usually denoted in chemical notation by the letter R, indicating that its chemical composition varies from one amino acid to the next. Thus the general structure of the naturally occurring amino acids as follows:

The carbon immediately adjacent to the carboxyl group is called the α-carbon. With one exception, amino acids occurring in proteins are all α-amino acids, *i.e.*, the carboxyl and amino groups are both attached to the α-carbon.

Since the structures of different amino acids vary only with respect to their R-groups (side chain). The specific chemical properties that distinguish one amino acid from another must be dependent on the chemical nature of the particular side chain of each amino acid. The amino acids are categorized according to the general characteristics of their R-groups.

1. **Neutral amino acids**: The neutral amino acids have uncharged R-groups at neutral pH. Examples- valine, leu, ileu, met, phe, ser, thr, tyr, try, cys, gly, ala, pro.
2. **Acidic amino acids**: The side chain of these amino acids terminates in a carboxyl group. Hence these possess an extra negative charge at physiological pH. For example, Aspartic acid (Asp), Glutamic acid (Glu).
3. **Basic amino acids**: They possess side chain nitrogen containing groups that act as bases. Examples- lysine, histidine and arginine.

Structure: Amino acids, as their name suggests, possess both an acidic region (a carboxyl group, -COOH) and a basic region (an amino group, $-NH_2$), as well as a third region referred to as a side chain. This third region is usually denoted in chemical notation by the letter R, indicating that its chemical composition varies from one amino acid to the next. Thus the general structure of the naturally occurring amino acids as follows:

The carbon immediately adjacent to the carboxyl group is called the α-carbon. With one exception, amino acids occurring in proteins are all α-amino acids, *i.e.*, the carboxyl and amino groups are both attached to the α-carbon.

Since the structures of different amino acids vary only with respect to their R-groups (side chains). The specific chemical properties that distinguish one amino acid from another must be dependent on the chemical nature of the particular side chain of each amino acid. The amino acids are categorized according to the general characteristics of their R-groups.

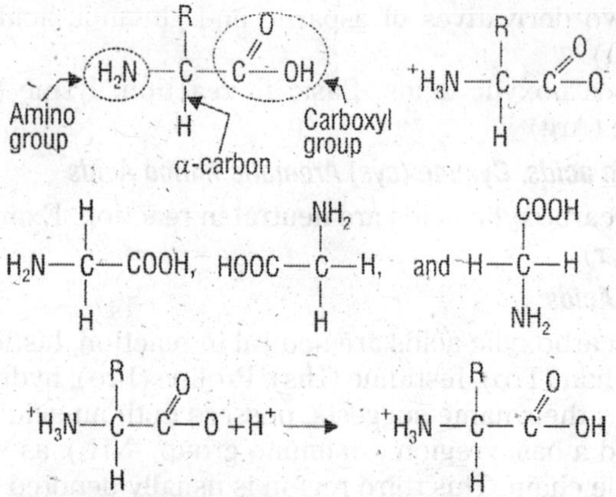

Amino group — $H_2N-C(R)(H)-COOH$ — Carboxyl group; α-carbon

$$^{+}H_3N-C(R)(H)-COO^{-}$$

$$H_2N-C(H)(H)-COOH,\quad HOOC-C(NH_2)(H)-H,\quad \text{and}\ H-C(COOH)(NH_2)-H$$

$$^{+}H_3N-C(R)(H)-COO^{-} + H^{+} \longrightarrow {}^{+}H_3N-C(R)(H)-COOH$$

FIG. 6.4: Structure of amino acid.

The amino acids found in proteins are all α-amino acids, *i.e.*, the amino and carboxyl groups are both attached to the same carbon atom (the α-carbon atom). The α-carbon atom is a potential chiral center, and except when the R group (or side chain) is H, amino acids display optical activity. All amino acids found in proteins are of the L configuration.

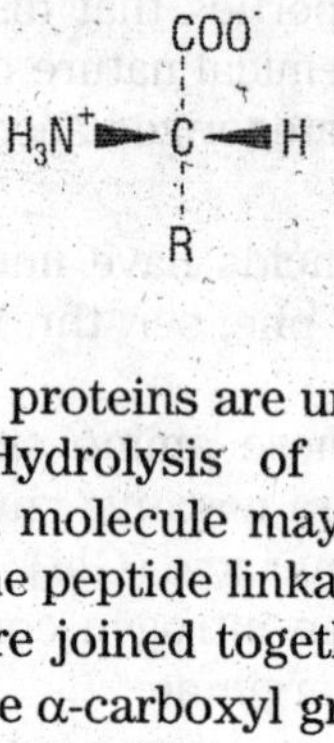

$$H_3N^{+} \blacktriangleright C(COO^{-})(R) \blacktriangleleft H$$

General structure of alpha-amino acids in the L configuration; R in one of over 20 different chemical groups.

Peptide Bonds

The amino acids in proteins are united through an acid-amide type of bond called a peptide linkage. Hydrolysis of proteins involves hydrolysis of many peptide linkages. A protein molecule may contain hundreds or thousands of amino acids. The formation of the peptide linkage between two amino acids is as follows:

Amino acids are joined together in chains by peptide bonds (amide linkages) formed between the α-carboxyl group of one amino acid and the α-amino group of the adjacent amino acid. This polymerization reaction is sometimes referred to as a dehydration condensation because one molecule of water is removed.

Several amino acids linked together by peptide bonds are called peptides. Long chains, consisting of more than eight or ten amino acids, are referred to as polypeptides. Amino acids in a peptide or polypeptide chain are called amino acid residues. The peptides and polypeptides have a free amino group at one end and a free carboxyl group at the other end. The amino acid carrying the free amino group is called the N-terminal residue. The amino acid carrying the free carboxyl group is called the C-terminal residue. The terminal amino and carboxyl groups are ionized in the intracellular pH range. All other amino or carboxyl groups, except those present in side chains, are involved in peptide bonds.

The bond lengths and bond angles involved in part of a polypeptide chain, the peptide bond is shorter than the normal C-N single bond and therefore has double bond character. The C-N linkage of the peptide bond has a length of 0.132 nm. The normal C-N single and double bond lengths are 0.147 nm and 0.128 nm, respectively. The partial double bond character of the peptide bond holds the groups of six atoms in a rigid, planar configuration. The only rotation allowed is the twisting of this unit about the bonds connecting it to the two carbons. This restriction is particularly important in determining the possible configuration of a protein.

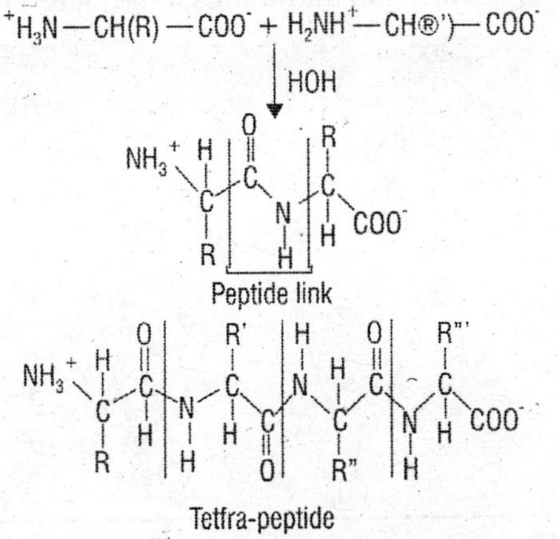

$$^{+}H_3N-CH(R)-COO^{-} + H_2NH^{+}-CH(R')-COO^{-} \xrightarrow{HOH}$$

Peptide link

Tetfra-peptide

FIG. 6.5: Amino acid showing peptide like and tetra-peptide.

Also as a consequence of the peptide bond structure, the electron distribution is such that both the carboxyl ($-\overset{|}{C}=O$) oxygen and the amide ($-\overset{|}{N}-H$) hydrogen are particularly good for hydrogen bonding, since the oxygen carries a slight negative charge and the hydrogen carries a slight positive charge.

PHYSICAL PROPERTIES OF AMINO ACIDS

Solubility

In general, the amino acids are readily soluble in water, insoluble or slightly soluble in alcohol, and insoluble in ether. The amino acids are generally soluble in dilute acids and bases in which they form the amino acid salts.

Melting Point

The amino acids possess high melting points (200°C- 300°C). Many of the amino acids undergo decomposition at or near the melting point.

Taste

Amino acids are usually either sweet or tasteless or bitter. For example, glycine, alanine, valine are sweet. isoleucine and arginine are biter, and leucine is taste less. The sodium glutamate is a flavouring agent for certain sauces and foods.

Amphoteric (or Zwitterions)

Simultaneous presence of acidic and basic groups, amino acids can have both positive and negative charges. They are therefore, amphoteric molecules or zwitterions. In solution the following relations of the zwitterion structure to H^+ ions exist:

$$^{+}H_3N\text{-}CH(R)\text{-}COOH = H^{+} + {}^{+}H_3N\text{-}CH(R)\text{-}COO^{-} \longleftrightarrow H^{+} + H_2N\text{-}CH(R)\text{-}COO^{-}$$

amino acid cation in acid — zwitterions amino acid — anion solution alkaline solution

Optical Activity

All of the amino acids except glycine show optical activity. The α-amino carbon atoms of all the amino acids except glycine are asymmetrical. Naturally occurring amino acids almost always have the L configuration. This L form is important in the three dimensional conformation of protein molecules. D-amino acids do exist naturally in bacterial cell walls and antibiotics, but not as constituents of proteins.

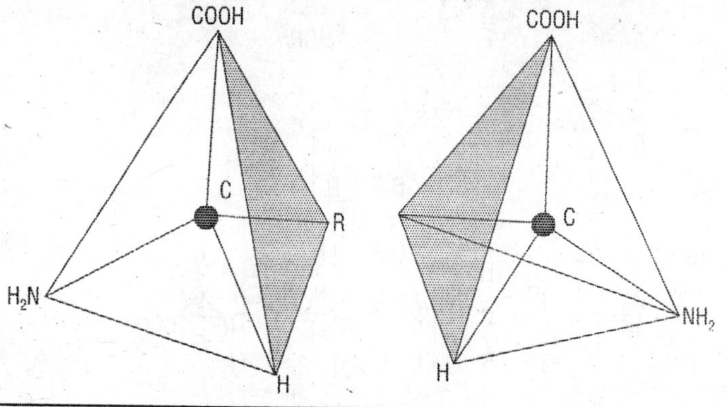

FIG. 6.6: Arrangement of the groups attached to the asymmetrical carbon atom in the optical isomers.

Isoelectric Point

An amino acid can carry a net positive or negative charge, depending upon the pH. It can behave both as a weak acid and a weak base. For example in alanine, both the amino group and the carboxyl group are ionized between pH 4 and pH 9, and it exists as dipolar ions. At one pH, the isoelectric point, an amino acid bears no net charge.

Determining the amino acid sequence is very difficult. Sanger and his co-workers (1955) established the sequence in insulin. It contains fifty-one amino acids. Later, the primary structure of ribonuclease, an enzyme containing 124 amino acids, was elucidated by stein and his co-workers. The amino acid sequence was proved by the total chemical synthesis of insulin in 1967 and ribonuclease in 1969. The amino acid sequences of many other proteins such as myoglobin, hemoglobin, lysozyme, chymotrypsin and others are known.

The first step in the process of sequences determination is protein purification owing to the large size and instability of proteins. The techniques usually employed to purify organic compounds can not be used. Usually low temperatures and neutral pH are maintained during isolation to avoid denaturation of the protein. One or more of the following techniques are used to purify proteins.

1. **Isoelectric precipitation**: Most proteins have minimal solubility at the iso electric point. This method takes advantage of difference in iso-electric points between proteins.
2. **Salting out**: This method takes advantage of differences in solubilities in concentrated salt solutions (usually ammonium sulphate).
3. **Precipitation by organic solvents**: This method takes advantage of difference in solubility of proteins when organic solvents are added to their aqueous solutions. Ethanol, methanol, and acetone are used, frequently below (0°C).
4. **Electrophoresis**: Proteins in solution can be separated as a result of their different mobilities in an applied electric field. Large quantities can be separated when the procedure is carried out using blocks of starch gel as inert supporting media, or feeding the solution continuously into a filter paper curtain.
5. **Molecular sieves**: Carbohydrates and acryl amide polymers in the form of

sponge-like networks are used. The cavities in the network are the same size or smaller than the protein. Smaller molecules retain in the interstices, while larger molecules pass through.

6. **Chromatography**: This method utilizes ion-exchange resins or adsorptive materials in columns to which proteins are adsorbed with different avidity. Proteins may be eluted sequentially by changing the pH or ionic strength of the solvent. In many cases these methods have very great power of resolution.
7. **Crystallization**: This is frequently employed as a final purification step after high enrichment of protein solutions.

The second step in the process of sequence determination requires the preparation of fragments containing several amino acids. Such fragments are obtained by the use of highly purified proteolytic enzymes which hydrolyze polypeptide chains only at specific sites. For example, tyrosine hydrolyses peptide bonds at arginine or lysine carboxyl groups. There are also a few specific chemical methods for cleavage, such as the use of cyanogen bromide to cleave at methionine residues. The amino acid sequence in each fragment is determined by use of successive analysis for end groups. The entire sequence is deducted by comparing regions of overlap.

A number of techniques can be used to degrade polypeptide fragments, one amino acid at a time. The enzyme leucine aminopeptidase carboxy peptidases, phenyl isothiocyanate (Edman reagent) and potassium cyanate reaction are used for this purpose.

RAMACHANDRAN PLOT

A **Ramachandran plot** (also known as a **Ramachandran Map** or a **Ramachandran diagram**), developed by Gopalasamudram Narayana Ramachandran who visualized dihedral angles of amino acid residues in protein structure. It shows the possible conformations and angles for a polypeptide.

FIG. 6.7: Gopalasamudram Narayana Iyer Ramachandran (1922–2001), an outstanding figure in the field of protein structure. His discovery of the triple helical structure of collagen in 1955 and his analysis of the allowed conformations of proteins through the use of the Ramachandran plot rank among the most outstanding contributions in structural biology, along with Pauling's description of the α-helix and Watson and Crick's discovery of the double helical structure of DNA. He worked on structural biophysics at the Department of Crystallography and Biophysics, University of Madras and later at Indian Institute of Science, Bangalore. He correctly proposed the triple helical structure of the fibrous protein.

The Ramachandran plot displays the *psi* and *phi* backbone conformational angles for each residue in a protein. The distance between two succession alpha carbon atoms in the backbone chain of a protein is approximately constant, as are the angles between the two bonds of such atoms. The proteins have only conformational freedom to rotate around the bonds in the backbone and in the side chain.

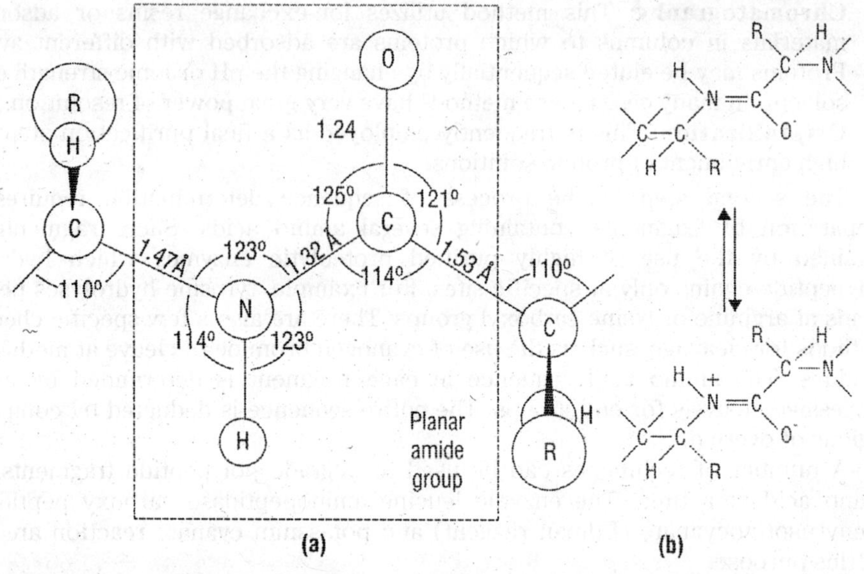

FIG. 6.8: (a) Dimensions of the peptide bond derived from X-ray analysis of simple peptides (b) Resonance structures in the peptide bond.

The conformational angles show preferences for values that are expected based on simple energy considerations and deviations from these angles may be used as indicators of potential error in crystallographic projects. Phi and psi angles are also used in the classification of some secondary structure elements such as beta turns.

In a Ramachandran plot, the core or allowed regions are the areas in the plot show the preferred regions for psi/phi angle pairs for residues in a protein. Presumably, if the determination of protein structure is reliable, most pairs will be in the favoured regions of the plot and only a few will be in "*disallowed*" regions.

There are multiple definitions of the so-called core or allowed areas in Ramachandran plots. The results of analysis can heavily depend on the definition used. As the number of protein structures that have been solved has increased, our knowledge concerning the empirical distribution of phi/psi angles in proteins has improved. In 2003, Lovell and co-workers published updated definitions of the *Preferred* and *Allowed but Disfavoured* regions of the Ramachandran plot. Nearly 100,000 residues from 500 structures with resolution better than or equal to 1.8 Angstroms were used to define the empirical distribution.

One would expect that larger side chains would result in more restrictions and consequently a smaller allowable region in the Ramachandran plot. In practice this does not appear the case; only the methylene group at the ® position has an influence.

Glycine has a hydrogen atom, with a smaller van der Waals radius, instead of a methyl group at the ® position. Hence, it is least restricted and this is apparent in the Ramachandran plot for glycine for which the allowable area is considerably larger.

In contrast, the Ramachandran plot for proline shows only a very limited number of possible combinations of and. One can also plot the dihedral angles in polysaccharides in this fashion.

It is still difficult predict the three-dimensional structure of a protein from is amino-acid sequence, protein structure could be reduced to a few elements of motifs, which are arranged in different ways in different proteins. Each of the motifs could be understood with reference to the conformational possibilities of the amino acids. In the structure (given below) the peptide group is indicated by the dashed lines.

Many chemical characteristics determine the motifs. For example, the six atoms of the peptide group should all be in the same plane. This characteristic has been explained as planarity of the peptide bond. Therefore, some degree of rotation could be allowed about the bonds, which join the peptide groups to the α-carbon atom. Such rotations are called torsion angles ϕ and ψ . The values of ϕ and ψ are influenced to different degrees due to steric hindrance from the side chains. The permissible range of ϕ and ψ could be plotted against each other to give what are popularly described as Ramachandran plots.

Because the peptide group itself is rigid and planar, there is no rotation around the bond between the carbonyl carbon atom and the nitrogen atom (C-N bond).

Glycylalanine

In the structure the peptide group is indicated by the dashed lines. Because the peptide group itself is rigid and planar, there is no rotation around the bond between the carbonyl carbon atom and the nitrogen atom (C'-N bond). However, free rotation is possible around the bond between the -carbon and the carbonyl carbon atom (the C-C' bond) and about the bond between the nitrogen atom and the alanyl _-carbon atom (the N-C- bond). Thus, for every peptide group in a protein, there are two rotatable bonds, the relative angles of which define a particular backbone confirmation.

The conformation of an amino acid residue can be defined by specifying the angles of rotation (around the N-Cα bond) and (around the Cα-C bond). The zero position for is defined with the N-H group *trans* to the Cα-N bond trans to the C = O bond.

Not all combinations of ϕ and ψ angles are possible, however, as many lead to clashes between atoms in adjacent residues. For all residues except glycine, the existence of such steric restriction involving side-chain atoms reduces drastically the number of possible conformations. The possible combinations of ϕ and angles that do not lead to clashes can be plotted on a conformation map. It is also known as a Ramachandran plot, named after the chemist who did much of the pioneering work in this field, a Ramachandran plot for the allowed conformations of alanylalanine. The double-hatched areas represent conformations (Combinations of ϕ and ψ) for which no hindrance exists. The single-hatched areas represent conformations for which some hindrance exists, but which may be possible if the distortion can be compensated for by interactions elsewhere in the protein.

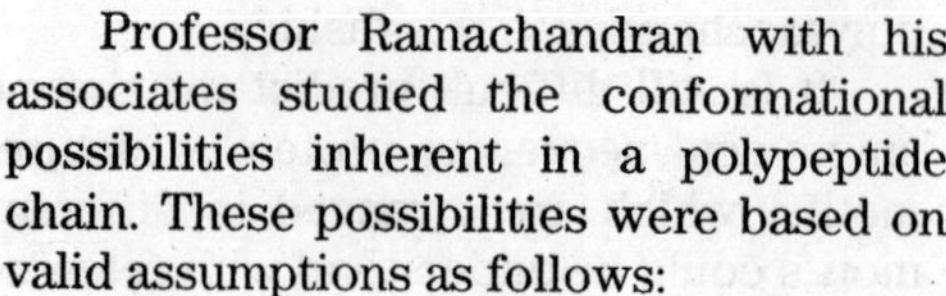

Professor Ramachandran with his associates studied the conformational possibilities inherent in a polypeptide chain. These possibilities were based on valid assumptions as follows:

1. The peptide bond is rigid and rotation is not allowed.
2. The molecule may rotate about the other single bonds in the backbone, *i.e.* the N-Cα bond and Cα-C' bonds.
3. All atoms are represented as hard spheres, there being no chance of inter-penetration,
4. No other attractive or repulsive forces present.

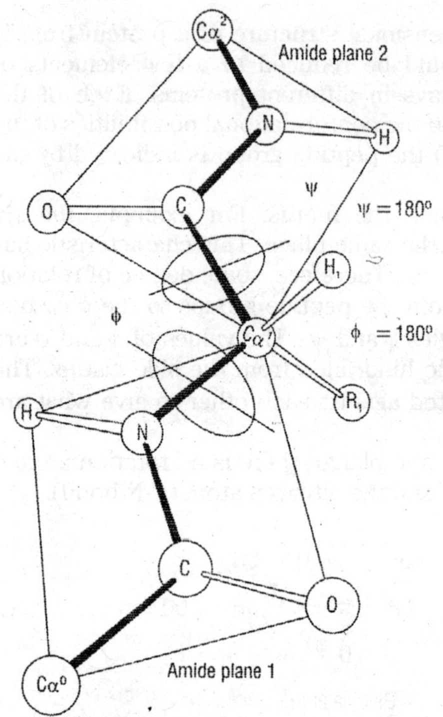

FIG. 6.9: Planar structure of the atoms involved in a peptide bond. Rotation of adjacent amide planes in a polypeptide.

According to above cited assumptions the conformational possibilities of a dipeptide can be represented in the form of a Ramachandran map. The map indicates that the hatched areas represent the combinations of φ and ψ torsion angles allowed for any dipeptide. The values of φ and ψ which if repeated in consecutive residues (and thus leading to regular structures), are also marked in the map (α helix and β sheets). Some of these secondary structures are described in terms of φ and ψ.

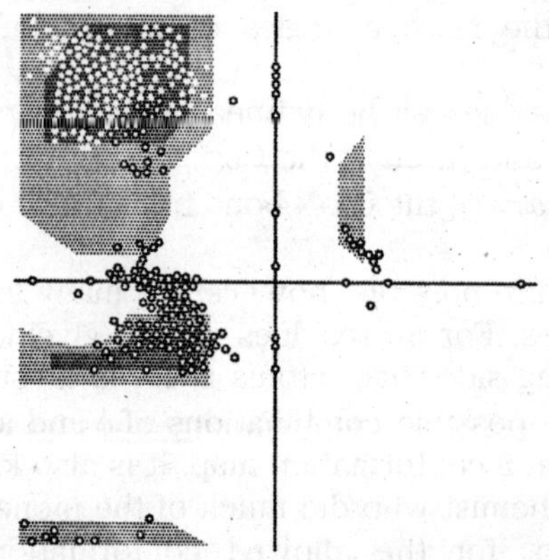

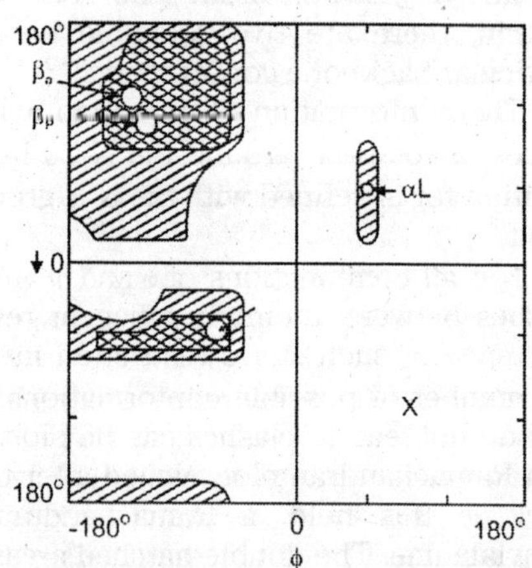

FIG. 6.10: (A) A Ramachandran plot generated from the protein PCNA, a human DNA clamp protein that is composed of both beta sheets and alpha helices (PDB ID 1 AXC). Points that lie on the axes indicate N- and C-terminal residues for each subunit. Ramachandran plot for alanylalanine, showing the fully allowed regions (double hatched) and partially allowed regions (single-hatched) of φ and ψ angles. The coordinates for the parallel and anti-parallel ß structures (ßp and ßa, respectively) and for the left-handed and right-handed α-helices (αL and αR, respectively) are indicated.

α **helix:** When (ϕ, ψ) values (-57°,-47°) are repeated, the polypeptide backbone chain of the protein is folded in a **right-handed** manner around an axis giving it the appearance of a helix with side chains extending out. Then α helix is stabilized by hydrogen bonds.

The β strand and β pleated sheet: The backbone of a polypeptide chain when extended or stretched out, it forms, what is described as β strand.

β turns (or reverse turns) and ω (omega) turns). The polypeptide chain sometimes reverses its direction, and it does so with the help of a structural motif called β turn or reverse turn or omega turn.

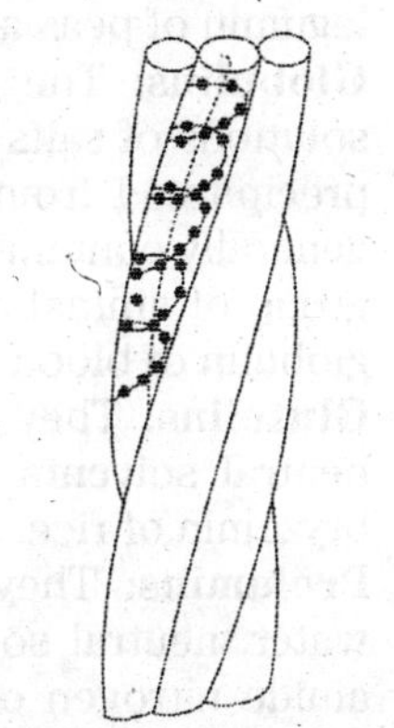

FIG. 6.11: Presentation of triple helix showing collagen.

CLASSIFICATION OF THE PROTEINS

Proteins are classified on the basis of their functions or of their physical and chemical properties. On the functional basis they are classified according to their biological role. The physical and chemical properties for classification of proteins are solubility and heat coagulability.

Functional Classification of Proteins

Catalytic proteins or enzymes

1. **Structural proteins**- collagen, elastin, keratins, lipoproteins.
2. **Contractile proteins**- myosin, actin, flagellar proteins.
3. **Transport proteins**- hemoglobin, hemocyanin, serum proteins, myoglobin.
4. **Genetic proteins**- nucleoproteins, histones.
5. **Regulatory or hormonal proteins**- ACTH, insulin, MSH, thyroxin.
6. **Immune proteins**- gamma globulins.

Classification on the Basis of the Shape of the Molecule

1. **Globular proteins:** those which have a relatively spherical or ovoid shape. These are relatively soluble.
2. **Fibrous proteins:** those which resemble long fibers in nature. These are insoluble and are usually found as components of the tougher types of tissues. Examples are the keratins of skin, hair, and feathers, the collagens of cartilage and tendons, the elastin of ligaments and silk fibroin.

Classification Based on their Composition

This classification is recommended by a joint committee of the American society of Biological chemists and the American physiological society, proteins are classified into the three main groups of simple, conjugated and derived proteins and each of these groups is sub-divided into a number of classes.

Simple Proteins

Simple proteins are those proteins that upon hydrolysis (6 M HCl) yield only amino acids or their derivatives. They are classified on their solubility in water, acids and alkalies into following types.

1. **Albumins:** The albumins are soluble in water, coagulated by heat, and usually deficient in glycine. They may be precipitated from solution by saturation with ammonium sulphate. They are products of both plants and animals. Examples- egg albumin, serum albumin of blood, myogen of muscle, lactalbumin of milk,

legumin of peas and leucosin of wheat.

2. **Globulins:** The globulins are insoluble in water. They are soluble in dilute solutions of salts of alkalies and acids. They are also heat coagulable. They are precipitated from solution by half saturation with ammonium sulphate. They generally contain glycine. The globulins are an important and widely distributed group of animal and plant proteins. Examples- ovoglobulin of egg yolk, serum globulin of blood, myosin of muscle, legumin of peas and amandin of almonds.
3. **Glutelins:** They are soluble in very dilute acids and alkalies but insoluble in neutral solvents. They are plant proteins. Examples: glutenin of wheat and oryzamin of rice.
4. **Prolamins:** They are soluble in 70 to 80 per cent alcohol, but insoluble in water, neutral solvents or absolute alcohol. The prolamines yield proline and amide nitrogen on hydrolysis, but deficient in lysine. They are plant proteins found in seeds. Examples: gliadin of wheat, hordein of barley, and zein of corn.
5. **Albuminoids:** They are also called as scleroproteins. The albuminoids are the least soluble of all the proteins. They are insoluble in water, salt solutions, dilute acids and alkalies, and alcohol. The albuminoids form a very diverse group of proteins. Their members show different physical and chemical properties. These proteins are found in animals only. They form exo-skeletal structures (*e.g.*, hair, horn, hoofs, and nails), supporting and connecting fibrous tissues, and organic material of cartilage and bone. Examples are as follows:
   a) Keratins of hair, horn, hoofs, nails,
   b) Elastin of connective tissue and ligaments.
   c) Collagen of bones, cartilage, and tendons.
   d) Fibroin and sericin of silk.
   e) Spongin of sponges.
6. **Histones:** They are soluble in water and insoluble in dilute ammonia. The histones are readily soluble in dilute acids and alkalies. They are not readily coagulated by heat. Histones are basic proteins, yield basic amino acids (arginine or lysine) on hydrolysis. They usually occur in tissues in salt combinations with nucleic acids. Examples: scomborne of mackerel sperm, gadus histone of codfish sperm.
7. **Protamines:** They are the simplest of the proteins and are large polypeptides. Protamines are strongly basic, yield basic amino acids on hydrolysis (arginine). They are soluble in water, dilute ammonia, acids and alkalies and are not coagulated by heat. They precipitate other proteins from their solutions. They also occur in tissues in salt combination with nucleic acids. Examples: salmine of salmon sperm, sturine of sturgen sperm, scombrine of mackerel sperm, and cyprinine of carp sperm.
8. Kossel has sub divided the protamines into four groups. This subdivision is based on their content of the basic amino acids:
   a) **mono protamines:** contain only arginine.
   b) **lysine diprotamines:** contain arginine and lysine.
   c) **histidine diprotamines:** contain arginine and histidine.
   d) **triprotamines:** contain arginine, lysine, and histidine.

Conjugated Proteins

Conjugated proteins are composed of simple protein (apoprotein) and some non protein substance. The non protein group is called as the prosthetic group. The prosthetic group may be inorganic or organic component.

Nucleoproteins

They are composed of simple basic proteins (protamine or histone) combined with nucleic acids (prosthetic group). They are the proteins of nucleus and are the main constituents of chromatin. They are abundant in tissues such as yeast, asparagus tips, thymus and other glandular organs, and sperm. Examples: nucleohistone and nucleoprotamine.

Mucoproteins

They are composed of simple proteins combined with mucopolysaccharides. Prosthetic group may be hyaluronic acid, chondroitin sulphate, N-acetyl hexosamine, uronic acid, sialic acid, and monosaccharides. Water soluble mucoproteins have been obtained from serum, egg white (ovomucoid). Their prosthetic groups are composed of hexosamine and hexose sugars.

Insoluble mucoproteins have been obtained from egg white (ovomucoid β), egg chalazae, vitreous humour, and submaxillary glands. Mucoproteins are important constituents of the ground substance of connective tissues. They are generally found in all types of animal mucins, the blood group substances, gonadotropic hormones. Examples: ovomucoid of egg white, seromucoid and seroglycoid of the blood serum, plasma glycoproteins, thyroglobulin, immunoglobins.

Chromoproteins

They are composed of simple proteins combined with a coloured prosthetic group. This group belongs many proteins of biological importance. Examples are as follows:

1. **Hemoglobin:** It is a respiratory protein in which the prosthetic group is the iron- containing, porphyrin complex heme.
2. **Chromoprotein** is similar to the hemoglobin in function and in chemical composition, are helicorubin (in milluscs), chlorocruorin (in marine worms), and actiniohematin (in anemones).
3. **Cytochromes:** These are cellular oxidation-reduction proteins, in which prosthetic group is heme.
4. **Flavoproteins:** These are also cellular oxidation-reduction proteins in which the prosthetic group is riboflavin.
5. **Chromoproteins** of certain animal fibers, such as black wool and hair, in which the prosthetic group is melanin.
6. **Visual purple** of the retina, a chromoprotein in which the prosthetic group is a carotenoid pigment.
7. **Catalase**, the enzyme that decomposes hydrogen peroxide into water and oxygen, a chromoprotein in which the prosthetic group is heme. The oxidative enzyme, peroxidase has a similar composition.

Phosphoproteins

Phosphoric acid is the prosthetic group of the phosphoproteins. Examples: casein of milk and vitellin of egg yolk.

Lipoproteins or Phospholipid Protein

They are composed of protein combined with a lipid, such as lecithin, fatty acid. They are found in milk, blood, cell nuclei, egg yolk, cell membranes, and chloroplasts. They also occur in bacterial antigens and viruses.

Metalloproteins

A large group of enzyme proteins contain metallic elements such as Fe, Cu, Mg, Zn, Mn, Co, which are parts of their essential structures. The heme proteins are also metalloproteins.

Derived Proteins

These are formed from simple and conjugated proteins. These proteins are of two types:

Primary Derived Proteins

These are formed by mild changes in the protein molecule. There is no cleavage of peptide bonds. These proteins are similar to denatured proteins.

1. **Proteins**: These are insoluble formed by the action of water, very dilute acids, and enzymes. Examples- myosan from myosin, edestan from edestin, and fibrin from fibrinogen.
2. **Metaproteins**: These are formed by further action of acids and alkalies on proteins. They are generally soluble in very dilute acids and alkalies but insoluble in neutral solvents. Examples- acid and alkali albuminates.
3. **Coagulated proteins**: These are formed by the action of heat or alcohol on natural proteins. These are insoluble products. They are also formed by action of ultra violet light, x-rays, very high pressure, and mechanical shaking on protein solutions at the isoelectric pH. Examples- cooked egg albumin, cooked meat and other proteins, and alcohol precipitated proteins.

Secondary Derived Proteins

These are formed by hydrolytic cleavage of the peptide bonds. They represent molecules of different sizes and amino acid composition.

1. **Proteoses**: These are hydrolytic products of proteins. These are soluble in water and are not coagulated by heat. They are precipitated from their solutions by saturation with ammonium sulphate.
2. **Peptones**: These are hydrolytic products of simpler structured proteins. They are soluble in water, and are not coagulated by heat. They are not precipitated by saturation with ammonium sulphate. They are precipitated by phosphotungstic acid.
3. **Peptides**: These are composed of a few amino acids united through peptide bonds. They are dipeptides, tripeptides, tetrapeptides etc. They are water soluble, are not coagulated by heat, are not salted out of solution, and are precipitated by phosphotungstic acid.

Protein Folding

As they are being synthesised, proteins must adopt the correct conformation for their function. Proteins may either fold spontaneously or they may need the assistance of **chaperone proteins** so that they do not get trapped in stable folding intermediates but rather fold into the correct final conformation. There are 3 major classes of chaperones:

- **The Hsp70 family:** DnaK is the best known member of this family in *E. coli*. It is a very highly conserved protein.
- **The Hsp 60 family:** The *E. coli* member of this family, GroEL forms a double-layered cylindrical structure which provides a protected environment in which proteins can fold. The structure is capped at one end by another heat-shock protein: GroES, a member of the Hsp 10 family.
- **The Hsp 90 family:** All are named because these proteins were first identified as Heat Shock Proteins - their synthesis increased in response to a sudden rise in temperature. In this role, chaperones protect important cell proteins from denaturation as a result of a sudden rise in temperature.

Post-Translational Modifications

Some proteins must be modified in one or more of a number of ways before they realize their final functional form. The following are some of the modifications that have been found to occur to proteins after they have been synthesised:

Dealing with the N-terminal Residue

In bacteria, the N-terminal residue of the newly-synthesised protein is modified in bacteria to remove the formyl group. The N-terminal methionine may also be removed. In eukaryotes, the methionine is also subject to removal.

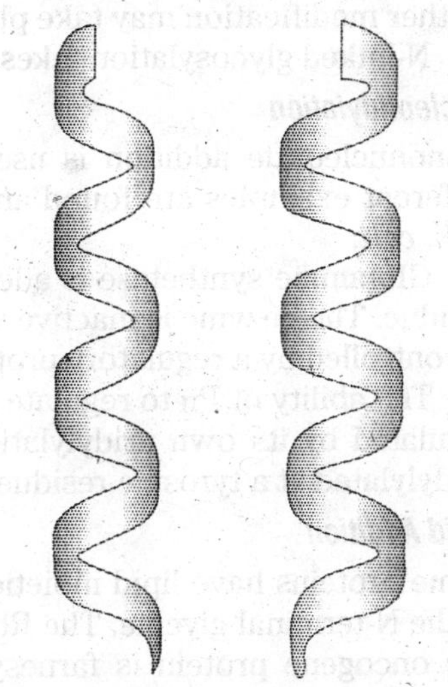

FIG. 6.12: Presentation of left and right handed helix.

Amino Acid Modifications

Many of the amino-acid side-chains can be modified. Here are some examples:

Acetylation

The amino-terminal residues of some proteins are acetylated. For example, the N-terminal serine of histone H4 is invariably acetylated as are a number of lysine residues.

Phosphorylation

Phosphorylation of proteins (at Ser, Thr, Tyr and His residues) is an important regulatory mechanism. For example, the activity of glycogen phosphorylase is regulated by phosphorylation of Serine 14.

Phosphorylation of tyrosine residues (in particular) is an important aspect of signal transduction pathways. Bacterial cells sense and respond to environmental signals through histidine phosphorylation.

Methylation

The activity of histones can be modified by methylation. Lysine 20 of histone H4 can be mono- or di- methylated.

Carboxylation

The blood coagulation factor, prothrombin, contains a large number of carboxylated glutamic acid residues in the N-terminal 32 amino acids. These modified residues are essential for activity. The modification requires vitamin K.

Hydroxylation

The conversion of proline to hydroxyproline in collagen is the classical example of a post-translational modification.

This conversion is catalysed by **prolyl-4-hydroxylase** which is a tetramer of two α and two β subunits; the β subunits are multifunctional and also carry disulphide isomerase activity.

Glycosylation

Many extracellular (but not intracellular) proteins are glycosylated. Mono- or Oligo-saccharides can be attached to asparagine (N-linked) or to serine/threonine (O-linked) residues. O-linked sugars are first assembled by serial addition to a polypeptide chain. O-linked glycosylation takes place in the Golgi apparatus.

N-linked sugar chains are first assembled as a dolichol phosphate derivative.

The sugars are then transferred to the appropriate site on a protein where some further modification may take place.

N-linked glycosylation takes place in the endoplasmic reticulum;

Nucleotidylation

Mononucleotide addition is used to regulate the activity of some enzymes. Two different examples are found among the system that regulates Nitrogen utilization in *E. coli*:

Glutamine synthetase is adenylylated (i.e. AMP is added) at a specific tyrosine residue. The enzyme is inactive when it is adenylylated. The degree of adenylylation is controlled by a regulatory protein, $\mathbf{P_{II}}$.

The ability of P_{II} to regulate the adenylylation of glutamine synthetase is in turn regulated by its own uridylylation (i.e. the covalent addition of UMP). P_{II} is also uridylylated at a tyrosine residue.

Lipid Addition

Some proteins have lipid moieties attached to the viral *src* protein is myristoylated at the N-terminal glycine. The Rhodopsin is palmitoylated at a cysteine residue. The *ras* oncogene protein is farnesylated as are some G proteins. Some eukaryotes, notably parasitic protozoa, have glycosylphosphatidylinositol-linked proteins.

Others

The protein, thyroglobin, is iodinated during the synthesis of thyroxine.

Adding Prosthetic Groups: Proteins that require a prosthetic group for activity must have this group added. For example, the haem (heme) group must be added to globins and cytochromes; Fe-S clusters must be added to ferredoxins.

Forming Disulphide Bonds: Many extracellular proteins contain disulphide cross-links (intracellular proteins almost never do). The cross-links can only be established after the protein has folded up into the correct shape.

The formation of disulphide bonds is aided by the enzyme protein disulphide isomerase in eukaryotes and by the DsbA protein in bacteria. Protein disulphide isomerase is one of a number of activities found in the beta subunit of the enzyme prolyl-4-hydroxylase. Other activities of this subunit are:

1. Cellular thyroid hormone-binding protein.
2. P4HB (glutathione-insulin transhydrogenase)
3. Smaller subunit of the heterodimeric microsomal triglyceride transfer protein.

Proteolytic Processing

Some proteins are synthesised as inactive precursor polypeptides which become activated only after proteolytic cleavage of the precursor polypeptide chain. Two well-known examples are:

Chymotrypsin and Trypsin

Chymotrypsin and trypsin are both synthesised as **zymogens.** Cleavage of chymotrypsinogen between Arg15 and Ile 16 by trypsin yields the enzymatically active pi-chymotrypsin. Two further proteolytic cleavages catalysed by chymotrypsin removes the dipeptides Ser^{14}-Arg^{15} and Thr^{147}-Asn^{148} to yield alpha-chymotrypsin. Trypsin is activated by the removal of the N-terminal seven amino acids.

Insulin

Insulin is synthesised as a precursor polypeptide. The initial **preproinsulin** contains a signal sequence since the protein is targeted for secretion. The signal sequence is removed as described in the next section.

The resulting precursor, **proinsulin**, is converted into active insulin by specific peptidases which remove amino acids 31 - 63. The final form of the hormone has two polypeptide chains held by 2 interchain and 1 intrachain disulphide bonds.

Protein Targeting

Proteins that must be targeted within the cell must be intercepted early during synthesis so that this can happen correctly. As a protein is being synthesised, decisions must be taken about sending it to the correct location in the cell where it will be required. The information for doing this resides in the nascent protein sequence itself. Once the protein has reached its final destination, this information may be removed by proteolytic processing.

Targeting in Bacteria

In bacterial cells, the targeting decision is relatively straightforward: is the protein destined to be an **intracellular** protein or an **extracellular** one?

However, even here nuances are possible. In a gram-negative bacterium, such as *E. coli,* there must be some way of knowing whether a protein is destined to go to the cell membrane, the periplasmic space, or the outer membrane.

Secreted proteins contain a signal sequence. This is a short (6 - 30) stretch of hydrophobic amino acids, flanked on the N-terminal side by one or more positively charged amino acids such as lysine or arginine, and containing neutral amino acids with short side-chains (such as glycine or alanine) at the cleavage site. Some examples are:

As proteins with signal sequences are synthesised, they are bound by the **SecB** protein. This prevents the protein from folding. **SecB** delivers the protein to the cell membrane where is secreted through a pore formed by the **SecE** and **SecY** proteins. Secretion is driven by the **SecA** ATPase. After the protein has been secreted, the signal sequence is removed by a membrane bound **leader peptidase.**

Targeting in Eukaryotes

In eukaryotic cells, the situation is more complex. Extracellular proteins can be targeted for secretion or to the cell membrane, or to one of the many internal organelles such as the lysosome. Intracellular proteins can be targeted for the cytoplasm, to the nucleus or to special organelles such as the mitochondrion or the chloroplast.

The Signal Sequence hypothesis was first enunciated by **Gunther Blöbel** who was awarded the Nobel Prize in Medicine in 1999 for his work.

The following diagram summarizes the choices/fates available to newly synthesised proteins in a eukaryotic cell:

Protein secretion in eukaryotic cells also involves a signal sequence. As in bacteria, the signal sequence is a short (6 - 30) stretch of hydrophobic amino acids, flanked on the N-terminal side by one or more positively charged amino acids such as lysine or arginine, and containing neutral amino acids with short side-chains (such as glycine or alanine) at the cleavage site:

Signal sequences are recognized by Signal Recognition Particles (SRPs) which are ribonucleoprotein particles containing a stable **305** nt **7S** RNA and 6 polypeptides.

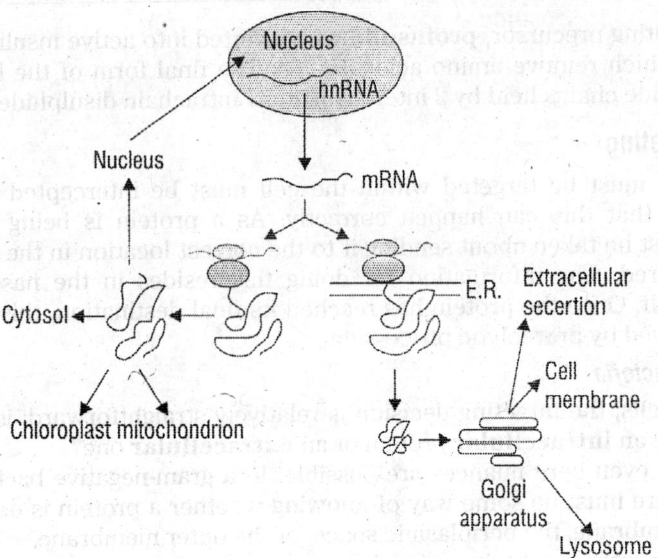

The SRP Cycle

The signal recognition particle (SRP) associates with ribosomes that are in the process of translating the mRNA for a secretory protein. The protein has a signal at the N-terminus. Subsequently, the ribosome-bound SRP interacts with the SRP-receptor, a component of the ER membrane. Finally, SRP recycles to associate with another ribosome, and translation continues with the secretory protein transferring the membrane through a channel called the translocon. The molecular details of this essential biological process have been unknown for many years, but significant progress has been made.

Components of the SRP: The RNA helices are numbered according to the nomenclature of Larsen and Zwieb and define the two functional domains of the SRP: the small domain (left; helices 2, 3, 4, and a portion of helix 5), and the large domain (right; a portion of helix 5, and helices 6, 7, and 8). (Helix 1 extends the 5'- and 3'-ends of the RNA and is present only in the SRPs of the archaea and certain bacteria.) There are six proteins [redish-green] in the mammalian SRP, named according to their approximate molecular weight in kD_a. The translation-regulatory function resides in the small domain, while protein translocation and signal recognition are mediated through the large domain. Protein SRP54 recognizes the signal peptide of the secretory protein. Overall, SRP is shaped like a dumbbell, about 240 Å long. The SRP database has more information about the SRP components.)

The 7S RNA has 2 domains - the Alu domain (so called because it is related to the Alu family of repeating sequences) and the S domain. The 7S RNA has 3 activities - a **signal sequence recognition** activity, an **SRP receptor binding** activity, and an **elongation arrest** activity.

As proteins are being synthesised, the N-terminal signal sequence is bound by the **SRP** particle. Protein synthesis is temporarily halted (**elongation arrest**) until the particle -- with the nascent polypeptide and the ribosome -- attaches to an **SRP receptor** complex in the **endoplasmic reticulum** membrane.

The SRP receptor is a heterodimer formed by a larger 68 kD alpha subunit and a smaller 30 kD subunit. The α subunit is a GTP-binding protein. Attachment of the ribosome triggers the release of GDP and binding of GTP. The GTP form of the SRP receptor binds the ribosome very tightly.

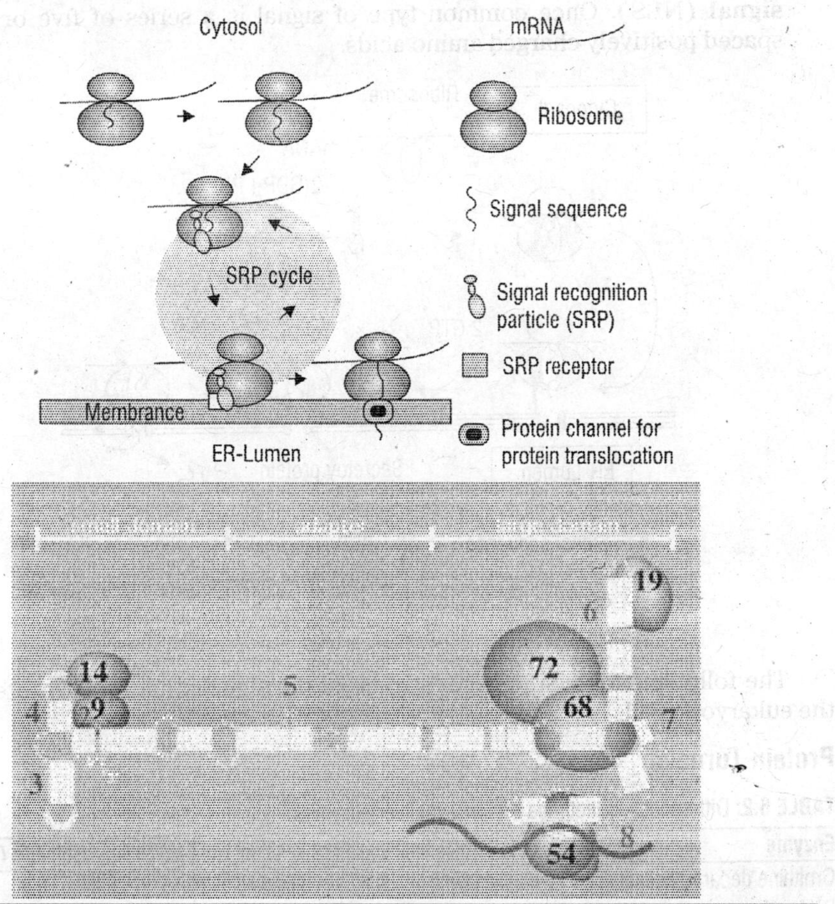

FIG. 6.13: Small domin, adopter and large domin of SRP cycle.

The ribosome is then transferred to a Ribosome receptor. Once again, the relatively slow hydrolysis of GTP allows time for this transfer to take place. Once transfer has taken place, GTP is hydrolysed and the SRP receptor is free to interact with another SRP particle-ribosome complex.

Protein synthesis then resumes but now the nascent polypeptide is secreted into the lumen of the **ER** as it is being synthesised.

Shortly after the polypeptide enters the **ER**, signal peptidase cleaves the signal peptide. The finished polypeptide is then targeted through the **ER** and **Golgi apparatus** to the cell surface or to the lysosome. During this process, the proteins are carried in coated vesicles which are enclosed either by a coat protein skeleton (ER → Golgi) or by a **clathrin** skeleton (Golgi → exterior).

Proteins that are synthesised on free ribosomes may also be targeted within the cell:

- Proteins that are targeted for organelles have their own N-terminal uptake-targeting sequence(s) that determines whether the protein must cross one or two membranes. In the former case, proteins destined for the inter-membrane space of the chloroplast or mitochondrion are first transported into the stroma (chloroplast) or matrix (mitochondrion) and then re-transported back through the inner chloroplast or mitochondrial membrane to the inter-membrane space.

- Proteins that must be targeted to the nucleus have a **nuclear localization signal (NLS).** Once common type of signal is a series of five or so closely spaced positively charged amino acids.

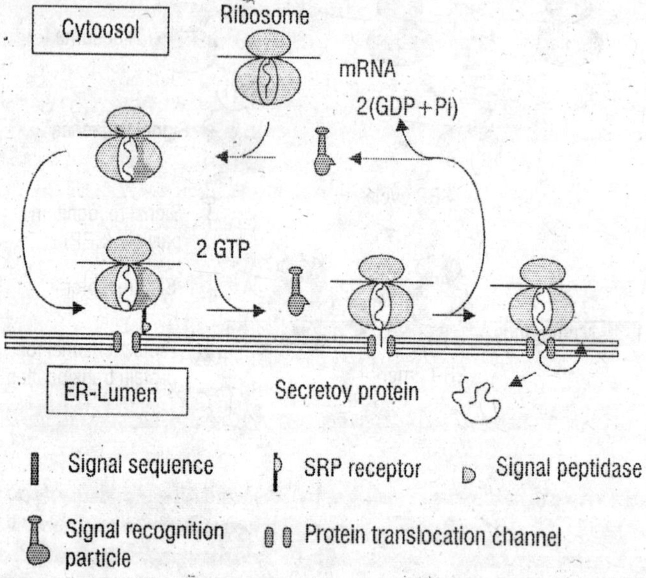

The following table summarizes some of the signals that target proteins within the eukaryotic cell:

Protein Turnover

TABLE 6.2: Differences in lifetimes of some rat liver enzymes

Enzyme	Half-life (h)
Ornithine decarboxylase	0.2
RNA polymerase I	1.3
Tyrosine aminotransferase	2.0
Serine dehydratase	4.0
PEP carboxylase	5.0
Aldolase	118
GAPDH	130
cytochrome c	150

Protein lifetimes must also be regulated. Some proteins are needed for only very short times – and could be harmful is present for too long. Others are needed all the time and it would be unnecessarily wasteful to keep re-synthesizing them.

The lifetime of proteins in eukaryotes appears to be determined by the nature of the **N-terminal amino acid**. Some amino acids (e.g. Ala, Cys, Gly, Met, Pro, Ser, Thr, Val) stabilize proteins (at least in yeast); others (e.g. Arg, His, Ile, Leu, Lys, Phe, Trp, Tyr) destabilize proteins.

In bacteria, the **C-terminal amino** acid also has an effect on the half-life of a protein.

Protein turnover in the eukaryotic cell cytoplasm involves the very highly conserved protein **ubiquitin**. This small (76 aa) protein is covalently attached to any protein selected for degradation. Such proteins are then degraded by the 26S proteosome.

PROTEIN STRUCTURE

A protein may be defined as one or more polypeptide chains folded into a characteristic three-dimensional configuration. The specific configuration of a protein is required for its biological activity. It follows inevitably from the sequence of amino acids and their individual chemical properties. The Danish protein chemist Kai Linderstrom-Lang suggested that one could consider the structure of a protein on several levels. Protein configuration is generally described in terms of four levels of organization: (1) Primary (2), Secondary (3), Tertiary (4) Quaternary.

Primary Structure

The primary structure of a protein refers to the order or sequence of amino acids in its polypeptide chain. This structure denotes the number, kinds, and sequence of amino acids, it provides no information about the proteins three-dimensional configuration.

Primary structure comprises the order of amino acids, the stereochemistry of the amino acid residues, and the properties of the covalent bonds joining the amino acids.

Three types of polypeptide chains are possible, open, cyclic and branched. Open chains contain a free α-amino group at one end of the chain and a free α-carboxyl group at the other.

An open polypeptide chain may produce a cyclic chain through peptide bond formation from the terminal α-amino and α-carboxyl groups. The antibiotics tyrocidine and gramicidin S are cyclic polypeptides. No free α-amino groups have been found in ovalbumin and certain muscle proteins because these are composed of cyclic chains.

The presence of two amino groups in lysine and of two carboxyl groups in aspartic and glutamic acids suggests the peptide bond branching from the additional amino and carboxyl groups in proteins. There is peptide bond branching between the ß and carboxyl groups of aspartic and glutamic acids in a peptide chain and a free α-amino group at the end of another peptide chain. There is branching by union of two peptide chain through the disulphide linkage-S-S- formed from cysteine groups.

In the protein molecule, amino acids are arranged like beads on a string. The sequence of amino acids is of great biological importance. The sequence of amino acids is identical in all molecules of any one specific type of protein. A change in even one amino acid residue may adversely affect the biological activity of the protein. The relationship between primary structure and biological activity is well illustrated in studies of hemoglobin. In both ß-chains a glutamic acid residue (the sixth from the N-terminal end) is replaced by a valine residue. This minor structural change greatly influences the biological function of the hemoglobin molecule. It causes an inherited disease in humans, called sickle cell anaemia.

Secondary Structure

Several weak interactions are very important in determining the secondary and tertiary structure of proteins. All these weak bonds are non-covalent, the main types are:

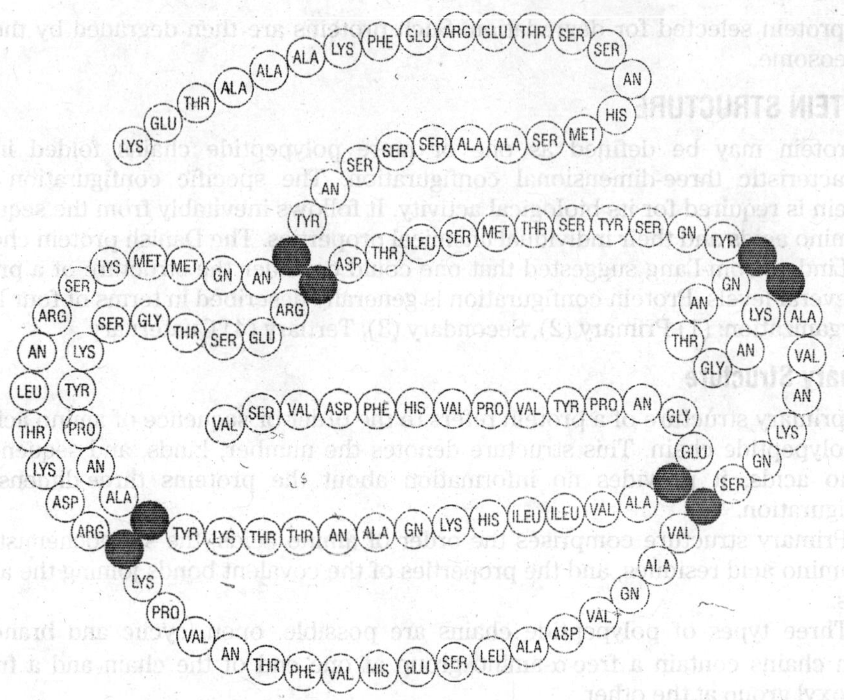

FIG. 6.14: The primary structure of bovine pancreatic ribonuclease.

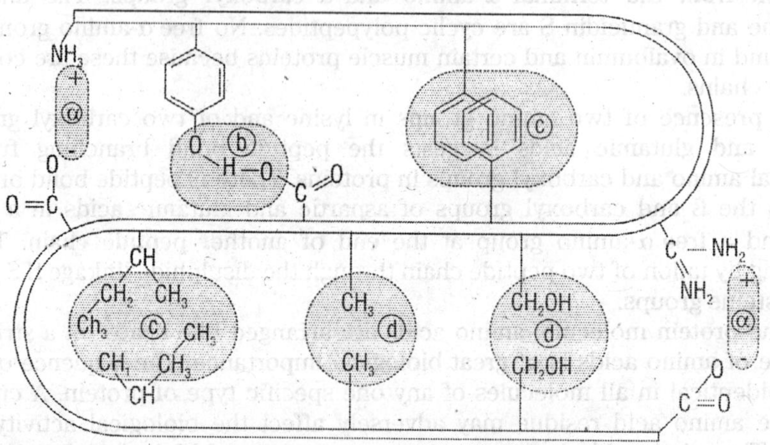

FIG. 6.15: Types of non-covalent bonds that stabilize protein structure. (a) ionic bonds; (b) hydrogen bonds; (c) hydrophobic interactions; (d) Van der Waals interactions.

1. **Ionic or electrostatic bonds**: These result from the attractive force between ionized groups of opposite change.
2. **Hydrogen bonds**: These result from a H^+ (proton) that is shared between two neighbouring electro-negative atom. The H^+ can be shared between nitrogen or oxygen atoms which are close to each other.
3. **Hydrophobic interactions**: In these water is excluded by non-polar groups. These associate with each other so that they are not in contact with water. In globular proteins the side chains of the most hydrophobic amino acids (long

non-polar side chains, such a leucine, isoleucine, valine, and pheny alanine) aggregate inside the molecule, and the polar (charged) groups protrude from the surface of the tertiary structure. The hydrophobic residues repel the water molecules that surround the protein, thereby causing the globular structure to be more compact.

4. **Van der Waals interactions**: They occur only when two atoms come very close together. The closeness of two molecules can induce charge fluctuations which may produce dipoles and mutual attraction at very short range.

The secondary structure of a protein defines the degree to which the polypeptide chain is stabilized by hydrogen bonds in the form of an α-helix or pleated sheet. Secondary structure is a direct consequence of the sequential arrangement of the amino acids in the polypeptide chain. Pauling and co-workers suggested two categories of polypeptide backbone configurations which would satisfy their findings of X-ray diffraction studies on crystalline amides: sheet structures and the α-helix.

Sheets, in which the hydrogen bonding exists between polypeptide chains (inter-chain hydrogen bonding). Coils or helices in which the hydrogen bonding exists between the peptide bonds of the same polypeptide chain (intra chain hydrogen bonding).

X-ray diffraction technique has facilitated classification of proteins into three structural types or groups- the α-helix structure (α-keratin type), the pleated sheet structure (ß-keratin type), the collagen group.

α-Helix Structure

In the α-helix, the oxygen of each α-carboxyl group is hydrogen bonded to the amide nitrogen of the third amino acid from it along the polypeptide chain. In this way, the first amino acid in a polypeptide chain is hydrogen bonded to the fourth amino acid in the chain, the second amino acid is hydrogen bonded to the fifth amino acid, and so on. Each peptide bond remains planar, and the side chain of the amino acids project outward from the axis of the helix.

In the α-helix, the hydrogen bonds form rings of atoms with the general formula where n = 3. There are 3.6 amino acid residues per turn, and 1.5 Å along the axis of the helix between one residue and the next. The peptide groups are strictly planar and the bond angle at the Cα is tetrahedral. The α-helix may be regarded as a spiral staircase in which the amino acid residues form the steps. The height of each step is 1.5 Å, and the height of each complete turn 5.4 Å, making 3.6 steps per turn. It takes 18 steps to return to a point exactly above the starting point.

The helix can turn either clockwise or counter clockwise when viewed from one end. That is it can be either right-handed or left-handed. If Cα is optically active (in the L-series of amino acids), this leads to some difference in stability of the two forms. The right handed α-helix is more stable.

The helical content of proteins can be estimated by measurements of optical rotatory dispersion. Owing to its asymmetrical character, the α-helix contributes to power of polypeptides to rotate polarized light. Individual proteins vary in α-helix content from 0 to almost 100 per cent.

The α-helix structure occur in many proteins. For example, the fibres of hair and wool (keratin proteins), myosin of muscles, fibrinogen and fibrin.

When α-keratin is treated with water, the hydrogen bonds hold the peptide chain in the α-helical configuration are broken. The fiber may be stretched out with disruption of α-helices, drying the fibres the α-helices reform with reversal of α-keratin.

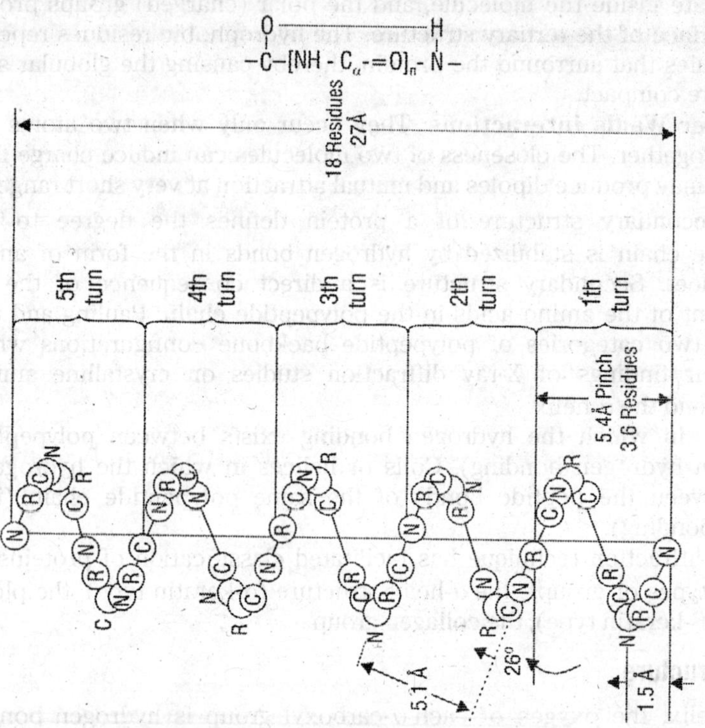

FIG. 6.16: A conventional representation of the α-helix.

Pleated Sheet Structure

This type of structures are formed by regular hydrogen bonding either between extended, separate polypeptide chains, or between extended portions of the same chain folded back on itself. In this case, the amino acid side chains project above and below hydrogen bonded sheets. Two conformations of pleated sheets are possible. One is parallel pleated sheet, in which the hydrogen bonded polypeptide chains, or portions of a chain, run in the same direction relative to their N-terminal and C-terminal ends. In the anti-parallel pleated sheet, the two hydrogen bonded chains run in opposite directions.

Pleated sheet or ß-structures result from the formation of hydrogen bonds between (across) polypeptide chain. Adjacent amino acids in a polypeptide chain occurs every 3.6 Å, and every amino acid residue is rotated 180 Å with respect to the adjacent ones. Polyglycine has a configuration of this type, the measured repeat unit of 7'2 Å, the chains are almost extended. Chains of polyglycine can form hydrogen bonds by arranging themselves in a sheet in which alternate chain lie antiparallel.

A configuration of this type is not possible with amino acids other than glycine because side chains (instead of hydrogen atoms) attached to the carbon could not be accommodated. Silk fibroin is a member of the class of ß-keratins rich in alanine or serine. Silk fibronin is an example of a protein with a related structure. The sheet in silk fibronin is pleated to accommodate the side chains on the α-carbon atoms.

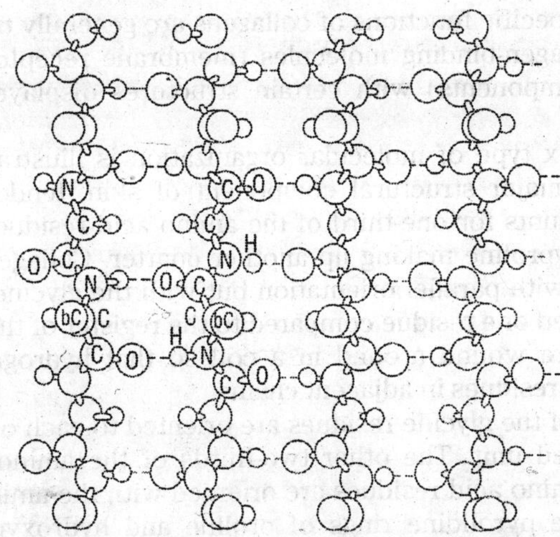

FIG. 6.17: The pleated sheet conformation of extended polypeptide chains (antiparallel).

Triple Helix

Collagens, characterized by a unique triple-helical structure, are the predominant component of extracellular matrices (ECMs) existing in all multicellular animals.

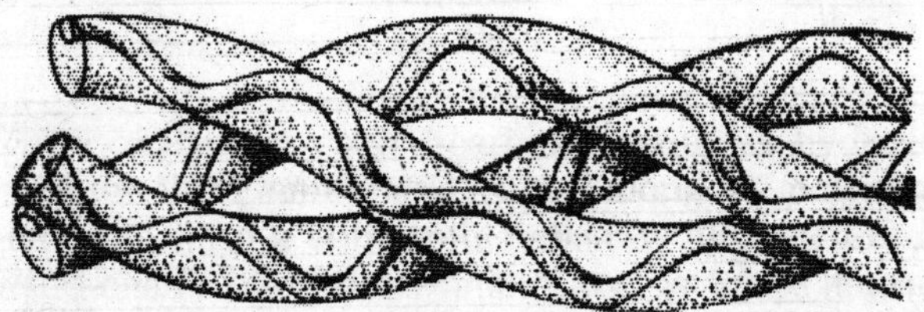

FIG. 6.18: The molecular structure of collagen, consisting of three polypeptide chains (Each polypeptide chain is in the form of the left-handed helix) forming a right-handled super helix.

The alignment contains 20 copies of the G-X-Y repeat that forms a triple helix. The first position of the repeat is glycine, the second and third positions can be any residue but are frequently proline and hydroxyproline. Collagens are post translationally modified by proline hydroxylase to form the hydroxyproline residues. Defective hydroxylation is the cause of scurvy. Some members of the collagen super family are not involved in connective tissue structure but share the same triple helical structure.

The triple helix of the collagen molecule is resistant to most proteinases, and the matrix metalloproteinases that do proteolyse collagen are affected by the architecture of collagen fibrils, which are notably more resistant to collagenolysis than lone collagen monomers. Full or limited proteolysis of the collagen fibril is known to be a key process in normal growth, development, repair, and cell differentiation, and in cancerous tumor progression and heart disease.

Collagen Group

Collagens are generally extracellular structural proteins involved in formation of connective tissue structure. Collagens not only maintain structural integrity of tissues and organs, but also regulate a number of biological events, including cell attachment, migration and differentiation, tissue regeneration and animal development. The specific functions of collagens are generally triggered by specific interactions of collagen-binding molecules (membrane receptors, soluble factors and other ECM components) with certain structures displayed on the collagen triple helices.

A more complex type of molecular organization is illustrated by the protein collagen. It is the major structural component of skin, tendon, and connective tissue. Glycine accounts for one-third of the amino acid residues of collagen, with proline and hydroxyproline making up another quarter. Collagen consists of three polypeptide chains with parallel orientation but with the glycine of every triplet of amino acids displaced one residue compared to the register of the other two chains. The three chains are wound around in a coil so that hydrogen bonding occurs between the glycine residues in adjacent chains.

The α-carbons of the glycine residues are oriented to each other and the center of the three stranded unit. The other two-thirds of the amino acid residues are oriented with the amino acid residues are oriented with the amino acid side chains facing outward. The pyrolidine rings of proline and hydroxyproline are closely packed. These produce a helical ridge on the outer surface of the molecule. Adjacent amino acid residues in a chain, occurs 2'86° apart and are rotated 108°.

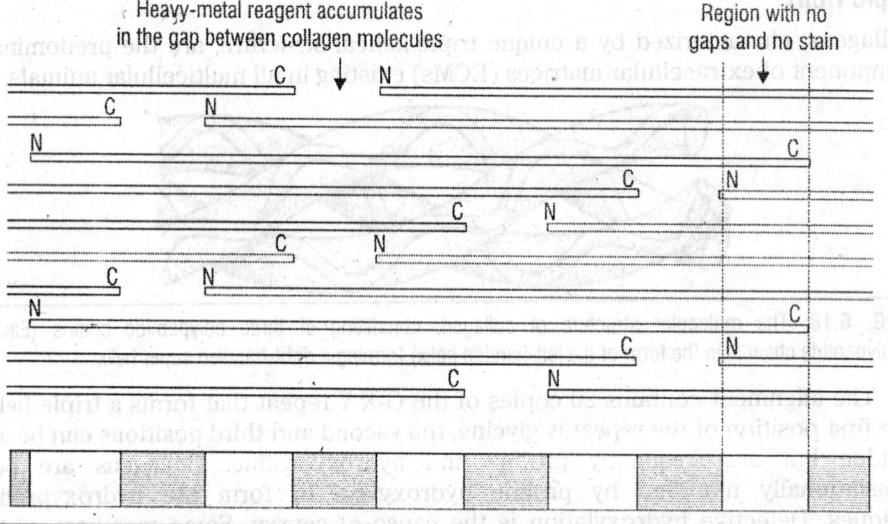

FIG. 6.19: Diagram showing the staggered arrangement of tropocollagen molecules.

The basic collagen molecule consists of a stiff, inextensible, triple-stranded helix. Eighteen genetically different individual polypeptide chains have now been described. These combine in triplets to form ten collage types. Of these, type I, the most abundant, has been most fully investigated.

Type I: 2α1 (I) and 1α2 (I), 300 nm broad fibrils with 67 nm bonds. They are most abundant and found in skin, bone, tendon and cornea. Type I collage has three chains, with a mass of 95,000 Daltons. Each chain contains about 1,000 amino acids, of which the glycine content is one-third and proline one-fifth.

For electron microscopy, collagen fibers are fixed and stained with heavy metal reagents. They show cross-striations of dark and light bonds every 67 nm. The individual tropocollagen molecules assemble in such a way that adjacent molecules are staggered, *i.e.*, displaced longitudinally be approximately one-quarter of their length (67 nm) and leaving a gap between the ends of each molecule.

It is difficult to explain how the tropocollagen molecules pack three dimensionally in the quarter-staggered array to produce a cylindrical fibril. One favoured model is a pentamicrofibril consisting of groups of five tropocollagen molecules packing together in the 67 nm staggered pattern.

The size and arrangement of the collagen fibrils vary considerably from tissue to tissue. The diameter ranges from 10 to 300 nm. The packing is random in mammalian skin and in parallel bundles in tendon.The collagen fibrils are further strengthened to withstand high tensile forces through the presence of cross-links. Covalent cross-links are formed both within a tropocollagen molecule and between different molecules.

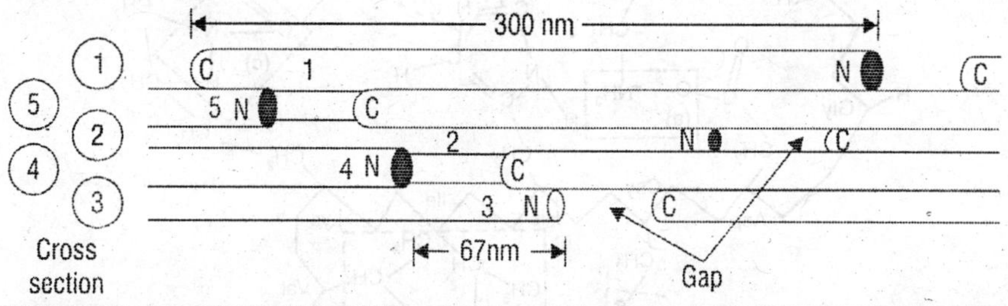

FIG. 6.20: A model for the three-dimensional packing arrangement of collagen molecules in a pentamicrofibril.

Collagen types I, II, III, V and VI form fibrillar structures. The others all aggregate to form supramolecular structures.

Tertiary Structure

Tertiary structure is stabilized by various weak molecular forces as well as by covalent linkages in the form of disulphide bonds or occasional isopeptide bonds. These are peptide linkages formed between a side chain amino group and a side chain carboxyl group. Some of the forces which determine tertiary structure are: (1) ionic bond, (2) hydrogen bond, (3) isopeptide bond (4) hydrophobic interactions, and (5) disulphide bonds.

The three-dimensional configuration of a protein is worked out by X-ray diffraction or X-ray crystallography. X-ray diffraction analysis of protein structure is not easy and consequently the tertiary structure of only a few proteins has been determined, e.g., lysozyme.

In globular proteins, the polypeptide chain is held together in a definite way to form a compact structure in three dimensions, called the **tertiary structure**. This configuration is required for biological activity. Enzymatic activity is detected in partially formed nascent polypeptide chains. They are folded into their final three-dimensional conformation as being synthesised. The helix itself possesses considerable flexibility. In the process of folding into the native conformation, many arrangements would occur until a thermodynamically stable conformation is attained. The total change in free energy during the folding process is close to zero or slightly negative.

Several soluble and globular proteins in crystal form have been analysed and the structural characteristics are summarized as follows:

1. Protein molecules are very compact, globular in form and with space for not more than a few solvent molecules in their interior.
2. Non polar (hydrophobic) side chains of amino acid residues are largely oriented toward the interior of the molecules.

3. Polar residues face the external surface of the molecule, and are exposed to the solvent. Water bounds to many polar groups at the surface.
4. The helical portions of the polypeptide chain are separated by non helical segments.

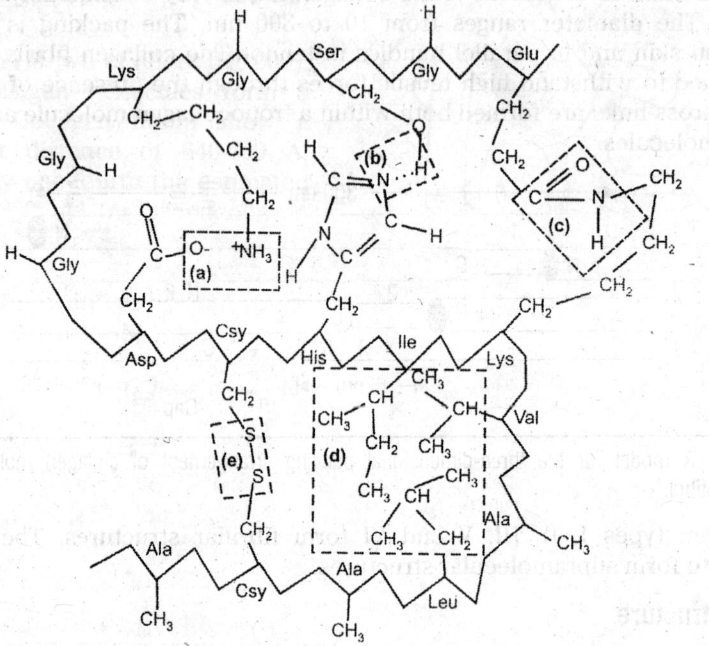

FIG. 6.21: A schematic representation of the types of forces which stabilize the three-dimensional structure of a protein.

Globular proteins have regions of α-helix or β-configurations and other parts exist as a random coil, *i.e.*, in a flexible structure that may change at random. The tertiary structure is given by the way in which these helical or random-coil segments arrange with respect to each other.

The special arrangement is predetermined by the sequence of amino acids in the primary structure and by the bonds. A series of biological properties of proteins, such as enzyme activity and antigenicity, are related to the tertiary structure.

The denaturation of a protein is caused by high temperatures or other non physiologic conditions, and disrupts the tertiary structure. This is usually accompanied by loss of biological activity.

The tertiary structure of ribonuclease is held together by four disulphide bonds which are present between pairs of cysteines. This amino acid contains -SH group. Denaturation of ribonuclease can be achieved by treatment with β-mercaptoethanol and high concentrations of urea. Mercaptoethanol is a reducing agent which can disrupt -S-S-bridges (disulphide bonds), reducing them to-SH groups.

Urea disrupts weak interactions. After denaturation, the enzyme can be refolded into its natural conformation (renaturation) by removing the urea and mercaptol ethanol by dialysis, and enzymatic activity is recovered. There are 105 possible combinations in which eight cysteines can pair to produce four disulphide bridges. The biologically active natural conformation is produced after renaturation, is a thermodynamically the most stable structure. The information for protein folding is contained in the amino acid sequence of proteins *i.e.*, primary structure.

Lysozyme is a proteolytic enzyme, derived from hen egg white. It contains 129 amino acid residues including four disulphide bridges. X-ray diffraction studies have produced an electron density distribution map. In this map, the structure of the enzyme can be clearly seen. The molecule is roughly ellipsoidal, with dimensions about 45 × 30 × 30 Å. The general arrangement is more complicated than myoglobin. The α-helical content is relatively low. Lysozyme has a hydrophobic spine, but a number of hydrophilic residues are also on the outside and several polar residues are internal. A very interesting feature of the enzyme is that it is complexed with inhibitors. The inhibitor molecules lie imbedded in a crevice in the surface of the enzyme molecule.

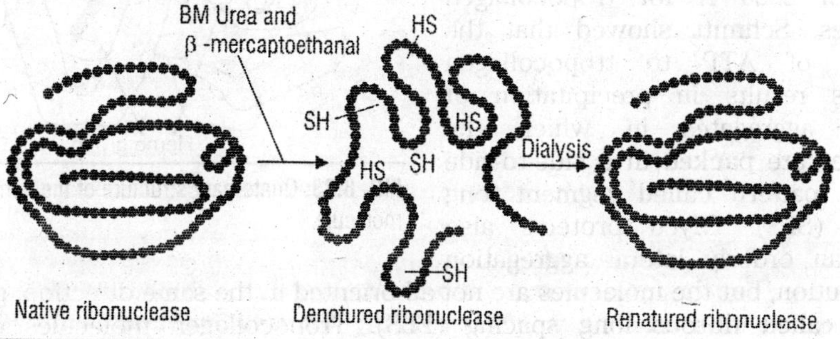

FIG. 6.22: Denaturation and renaturation of ribonuclease.

Quaternary Structure

A protein is said to have quaternary structure if it is composed of several polypeptide chains and are not covalently linked to one another. These chains may or may not be identical but in both cases they are linked by weak bonds. In such a protein, each polypeptide chain is called a subunit and has its own tertiary structure.

Hemoglobin was the first protein for which a complete quaternary structure was determined. The hemoglobin molecule consists of two α-chains and two ß-chains, each possessing a characteristic tertiary structure. These four subunits are oriented in a specific configuration relative to one another to form the quaternary structure of the hemoglobin molecule.

Separation and association of the subunits may occur spontaneously. Hemoglobin may be broken into two half molecules (two α- and two ß) by urea. When urea is removed, they reassembled forming complete functional molecules. This binding is highly specific and takes place only between the half molecules. This is called the principle of self assembly. This principle also applies to the building up of more complex cellular structures, such as the cell membrane, microtubules and so forth. Most enzymes exist as multi-subunit structures, *i.e.*, in more than one structural form (called isozymes) in the same species. They differ on the quaternary level of organization by having various associations of polypeptide chains. Frequently isozymes differ in the characteristics of their catalytic activity. For example the enzyme lactic dehydrogenase (LDH) occurs as five isozymes. They differ in amino acid composition, electrophoretic mobility, and other properties. Each lactate dehydrogenase isozyme differs in the rate of lactate hydrolysis versus substrate concentration. Lactic dehydrogenase is composed of four polypeptide chains of two possible types, called A and B. Thus, five different combinations can occur- AAAA (the most common LDH isozyme in the heart muscle), AAAB, AABB,

ABBB and BBBB (the most common LDH isozyme in adult skeletal muscle). The LDH isozymes have been separated by acrylamide gel electrophoresis.

Collagen also displays a complex organization at the quaternary level of structure collagen fibrils are composed of long, thin, rod like molecular units, tropocollagen (M. wt. of about 340,000). The tropocollagen molecules form a triple helix. Collagen fibers show a major repeat distance of 640-700 Å, approximately one-fourth the estimated length of 2820 Å for tropocollagen molecules. Schmitt showed that the addition of ATP to tropocollagen solutions results in precipitation of ordered aggregates in which the molecules are packed in a side-to-side giving a pattern called segment long spacing (SLS). Glyco proteins also induce an orderly lateral aggregation from solution, but the molecules are not all oriented in the same direction, giving a pattern called fibrous long spacing (FLS). Tropocollagen molecules can be precipitated in the form of native collagen, as a laterally oriented polarized structure.

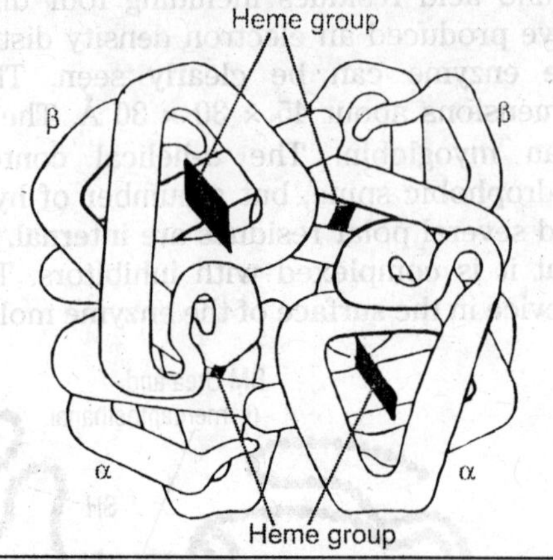

FIG. 6.23: Quaternary structure of the hemoglobin molecule.

PROPERTIES OF THE PROTEINS

Taste

Pure proteins are generally tasteless. The taste of protein hydrolysates (proteoses, peptones, peptides, amino acids) is bitter. Pure proteins are odourless. When heated they turn brown and char and give off the odour of burning feathers of hair.

Shape

The shapes vary widely from spherical (in horse heart cytochrome C) to very elongated spheroids, as in fibrinogen. The methods used to determine the shape of molecules are based on properties such as rate of diffusion, viscosity, double refraction of flow or birefringence and orientation in an alternating current field.

Molecular Weight

Protein molecules are exceedingly complex in structure and are very large. The most important method for the determination of protein molecular weights is based on the ultracentrifuge. The molecular weights of proteins also may be obtained from osmotic pressure measurements and light scattering. The amino acid content of a protein may also be used for calculating minimal molecular weights.

1. Casein varies in molecular weight from 75,000 to 375,000. The molecules of gelatin vary in mass from 10,000 to 100,000. Myoglobin of muscle is composed of 17000, 34000 and 68,000 molecular weight fractions. The globulin fraction of blood serum is complex.
2. The molecular weights of proteins cannot be determined with a high degree of accuracy in most cases, and the values reported are more or less approximate.

Amphoteric (or Zwitterions) and the Isoelectric Point

Because of the simultaneous presence of acidic (carboxyl) and basic (amino) groups proteins can have both positive and negative charges and are therefore amphoteric molecules or zwitterions.

In addition to the terminal -NH_3^+ and -COO^- charged groups, proteins contain dicarboxylic and diamino-amino acids which dissociate as follows.

1. The acidic groups lose protons and become negatively charged. For example in aspartic and glutamic acids, the free carboxyl group dissociates into -COO^- + H^+.
2. The basic groups, by gaining protons, become positively charged -NH_2 + H^+ - NH^{3+} this is found in amino acids with two basic groups, such as lysine or arginine.

All these ion-producing groups contribute to the acid-base reactions and to the electrical properties of proteins.

The actual charge of a protein molecule is the result of the sum of all single charges. Dissociation of the different acidic and basic groups takes place at different hydrogen ion concentrations of the medium. pH greatly influences the total charge of the molecule. In an acid medium, amino groups capture hydrogen ions and react as bases (-NH_2 + $H^+ \rightarrow NH_3^+$). And in an alkaline medium the reverse takes place and carboxylic groups dissociate (-$COOH \rightarrow COO^-$ + H^+). Proteins exists as complex cations in acid solution and in alkaline solution show successive over lapping stages of H^+ dissociation with the formation of zwitterions, and finally protein anions.

For every protein there is a definite pH at which the sum of positive and negative charges is zero. This pH is called the isoelectric point (P^I). The isoelectric pH is of great importance in relation to the physical and chemical properties of the protein. At this pH the net charge on the protein is least and the electrical conductivity is least. Also, at this pH the osmotic pressure, swelling capacity, viscosity, and the solubility is remains minimum. Proteins exist as cations on the acid side and as anions on the alkaline side of the isoelectric pH.

At the isoelectric point, proteins placed in an electric field do not migrate to either of the poles. At a lower pH they migrate to the negative pole (cathode) and at a high pH, to the positive pole (anode). Migration ceases or is minimal when the pH of the buffer solution is the same as the isoelectric pH of the protein. This migration is called electrophoresis. It provides a very useful technique for the separation and purification of proteins, and in the determination of isoelectric pH values.

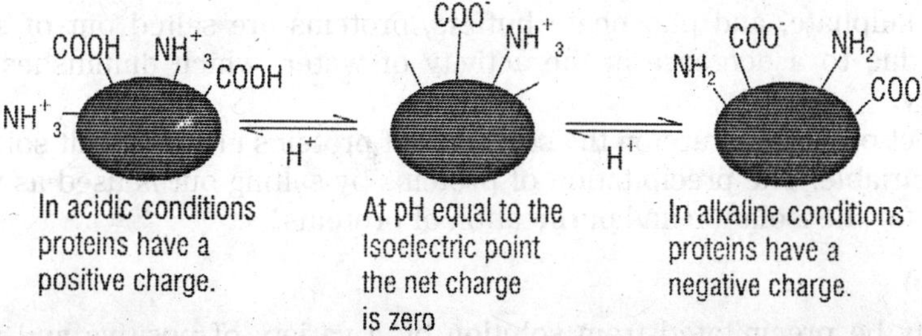

FIG. 6.24: The ionization of proteins depends on the pH. In acidic conditions proteins migrate to the cathode, in alkaline conditions to the anode.

At this pH the protein exists as the zwitterion form in which the total positive charges equal the total negative. So the net charge is zero, though the total charge may be high. The isoionic point of a protein is the pH at which the number of H^+ ions dissociated from the protein is equal to the number taken up from the solution by the protein. The isoionic and isoelectric pH values are the same when the protein does not combine with ions other than the H^+ ion.

Proteins act as buffers on both sides of the isoelectric pH. The isoelectric pH values of most of the body proteins are below the pH of blood and tissue fluids. The body proteins are generally held at a pH alkaline to their isoelectric points. They exist as negative ions or salts with the cations Na^+, K^+, Ca^{+2} and Mg^{+2}. In general the imidazole groups of histidine and the α-ammonium groups are responsible for the buffer action of proteins at physiologic pH.

Crystallization

Many of the proteins have been obtained in crystalline condition, such as hemoglobin, enzymes, and vegetable proteins, the crystallization of proteins may be enhanced by the addition of a salt (ammonium sulphate or sodium chloride) and adjustment toward the isoelectric pH. The added substances and adjustment to the isoelectric pH decrease the solubility of the protein. The proteins do not dissociate at the isoelectric pH and crystallize very readily.

Viscosity

The viscosity of protein solutions varies greatly. The viscosity is closely related to molecular shape. Long molecules have higher viscosity than those of globular proteins viscosity is minimum at the isoelectric point. Viscosity is of great value in calculating the axial length of protein molecules.

Solubility and Salting Out

The solubility of a protein depends on proportions and distribution of polar and non-polar groups and the dipole moment. The ionic polar groups interact electrostatically to form aggregates and oppose solubility. The polar water molecules interact with the polar groups of proteins to increase solubility. The addition of organic solvent (acetone or alcohol) decreases the solubility of the protein. The addition of small amounts of salt to a protein dissolved in pure water, its solubility is increased. Globulins, which are insoluble in pure water becomes soluble in the presence of small amounts of salts. This phenomenon is called salting in due to the forces of attraction between the protein ions and the salt ions. At low salt concentrations the increase in protein solubility (s) is proportional to the ionic strength of the solvent.

- At high concentrations of very soluble salts such as ammonium sulphate, sodium sulphate, and phosphate buffers, proteins are salted out of solution. This is due to a decrease in the activity of water, which diminishes protein solubility.
- The effect of temperature on the solubility of proteins in strong salt solutions is highly variable. The precipitation of proteins by salting out is used as valuable method for the isolation and purification of proteins.

Precipitation

Proteins may be precipitated from solution by a variety of positive and negative ions. Such precipitations are of importance in the isolation of proteins, in the deproteinisation of blood and fluids, extraction for analysis, and in the preparation of useful protein derivates.

The positive ions most commonly used are Zn^{+2}, Cd^{+2}, Fe^{+3}, Cu^{+2} and Pb^{+2}. The pH of the protein solution is important for the complete precipitation of proteins by heavy metal ions. These ions precipitate proteins from solutions alkaline to the isoelectric pH. At this pH protein is dissociated as protein which combines with the positive metal ion to give an insoluble precipitate of metal proteinate.

Negative ion protein precipitants are tungstic acid, phosphotungstic acid, trichloroacetic acid, picric acid, tannic acid, ferrocyanic acid, and sulphosalicylic acid. Negative ions combine with protein$^+$ (solution is acid to the isoelectric pH) to form insoluble precipitate of protein salts. Tungstic acid, picric acid, etc. are used for the preparation of protein-free filtrates of blood and other biological materials prior to analysis. Hides are converted to leather by treatment with tannic acid, which reacts with skin proteins to form insoluble protein tannates.

The injection of a foreign protein into an animal generally forms antibodies against the protein injected. The precipitation of proteins by their specific immune sera antibody proteins provides a highly sensitive and specific method for the detection of proteins.

Heat Coagulation

Coagulation of proteins from their solutions by heating is most commonly used process of deproteinisation of tissue extracts. Heat coagulation occurs best near the isoelectric pH of the protein. The presence of water is necessary for heat coagulation. Coagulated protein is generally soluble in acids, alkalies, strong urea and salicylate solutions. It is reprecipitated when the acid or alkali is neutralized and the urea or salicylate is removed by dialysis.

Proteins possess characteristic coagulation temperatures which may be altered by changing the pH. Heating an isoelectric solution of the protein in the presence of salt, a temperature is reached at which the protein is rapidly denatured. This temperature is the coagulation temperature of the protein. The coagulation temperature is lowered by the addition of small amounts of acid or alkali. It can be raised by the addition of much salt.

Denaturation

Under adverse conditions the secondary and tertiary structure characteristic of the native conformation will be lost. The protein will assume a more disordered and less compact conformation. This process is called denaturation. It is accompanied by a loss of specificity, catalytic activity and changes in physical properties such as solubility. The polypeptide backbone is not broken during denaturation. The intra molecular hydrogen bonds and other weak interactions which hold the protein in its native conformation are disrupted.

The most obvious change is that the denatured protein is insoluble at the isoelectric pH of the original protein and coagulates and precipitates in neutral solution. Proteins may be denatured by numerous agents and processes. Denaturation may be effected by solution in acid or alkali and standing for some time at ordinary temperature, or by heating. Treatment with alcohol, acetone, and other organic solvents leads to denaturation. Solution of proteins in strong urea, salicylate, and guanidine salts causes denaturation. Sodium dodecyl sulphate is an efficient agent for the denaturation of proteins. The initial action of proteolytic enzymes may cause denaturation before peptide hydrolysis occurs.

Denaturation is caused by exposure to X-rays or ultra violet light and by visible light in the presence of a photosensitizer. Violet shaking of a protein solution leads to denaturation. Proteins may also be denatured subjected to very high pressures.

The peptide chains are unfolded and uncoiled as a result of denaturation. There is a increase in viscosity of denatured proteins in urea solution. The increase is due to change from a globular to a fibrous structure.

The chemical characteristics of denatured proteins are different from those of the original protein. Denaturation does not involve the hydrolysis of peptide linkages. It leads to an increase in the number of reactive -SH and -S-S-groups, due to uncoiling of the peptide chains. It also leads to an increase in the number of reactive phenolic hydroxyl groups (of tyrosine). Denaturation generally changes the isoelectric pH of a protein to be higher. It causes no change in the combination of proteins with acids and alkalies and less capacity to combine with water. It changes the solubility of proteins. Denatured proteins do not crystallize. Denaturation of protein enzymes leads to their in activation. The digestibility of proteins by proteolytic enzymes may be markedly increased. The properties of antibodies are destroyed by denaturation. The antigenic properties are decreased but not destroyed by denaturation.

Enzymes are stabilized against denaturation by the presence of substrate or coenzyme. The substrate or coenzyme fits into a cavity at the active site of the protein. The enzyme-substrate complex is thermodynamically more stable because it has a more closed structure than the enzyme alone with an open cavity. Some proteins are stabilized against denaturation by the presence of fatty acids or sugars. For example, plasma albumin is much more stable in the presence of trace amounts of fatty acids. Some proteins are stabilized against denaturation by low concentration of non ionic detergents.

The denatured proteins have a less compact conformation than native proteins is supported by the following evidences:

1. Denatured proteins have a higher intrinsic viscosity than native proteins.
2. Native proteins contain functional groups buried in the interior which are unreactive until they become exposed during denaturation.
3. Shifts in the ultraviolet absorption of tryptophan and tyrosine residues indicate that their environment has changed from a hydrophobic one characteristic of the interior of tightly folded globular proteins to an aqueous one characteristic of extended chains.
4. Changes in optical rotatory dispersion indicate a loss of α-helices.
5. The denatured state of a protein is not always entirely disordered, but depending upon the conditions, may be intermediate between the native conformation and complete random conformation.

Some general principles of denaturation are:

1. The denaturation is an all or none phenomenon.
2. Protein denaturation is a reaction of the first order.
3. The point of equilibrium in a denaturation process is independent of the protein concentration.
4. The temperature coefficient (Q 10 value) of denaturation is generally very high.

Anson and Mirsky first demonstrated reversibility of denaturation in their work on hemoglobin. When hemoglobin is denatured in acid solution and allowed to stand before neutralization in a acid or alkaline solution to dissolve the hemoglobin. The soluble hemoglobin crystallizes in the same form and has the same absorption spectrum and affinity for oxygen as native hemoglobin. Adjustment of the pH of the denatured hemoglobin permits reorganization (recoiling) of the peptide chains as they are in the native hemoglobin.

One of the most extensively studied renatured proteins is the enzyme ribonuclease. The four disulphide bonds in native ribonuclease can be reduced, leading to denaturation of the protein. The disulphide bonds of the reduced, extended chain can be oxidized which lead to a random mixture of disulphide bonds. This protein does not possess enzymatic activity. Under conditions that favour disulphide bonds rearrange rapidly with the formation of correct disulphide pairs and reformation of the native enzymatically active enzyme. An enzyme present in microsomal membranes catalyses the disulphide interchange and speeds up the process of renaturation. A number of other enzymes have been renatured after treatment with urea.

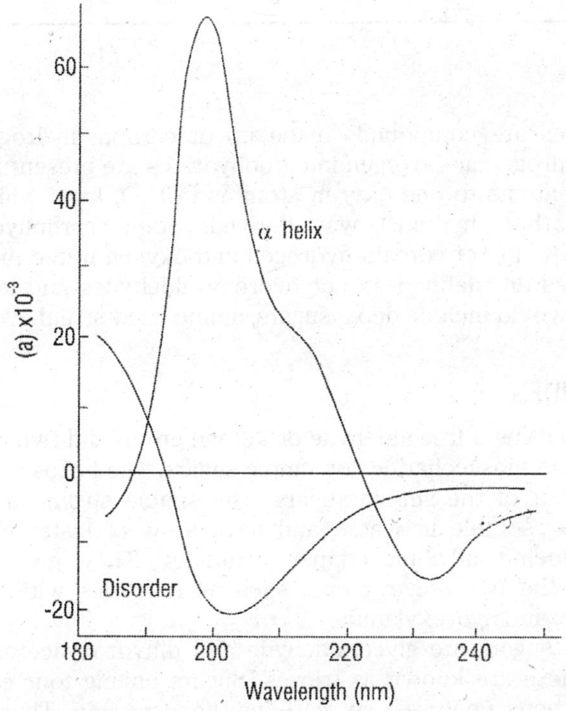

FIG. 6.25: ORD spectrum of poly-D- lysine showing the α-helical and disordered conformation regions.

So far it has been possible to re-nature entirely or partly the denaturation of only a limited number of proteins.

Optical Activity

Asymmetrical molecules such as carbohydrates, amino acids, and proteins rotate the plane of polarization of plane-polarized light. The amount of rotation depends on the concentration of the substance and the path length of light in the sample, in much the same way as for optical absorbances. The amount of rotation (even the direction of rotation) also depends on the wavelength of the light. The dependence of the specific rotation (α) (the measured rotation per unit concentration and path length) on the wavelength of light is known as the **optical rotatory dispersion** (ORD).

The conformation of a protein introduces an additional source of asymmetry that affects the ORD spectrum. Helical regions in soluble proteins give rise to a particular ORD spectrum quite distinct from that of disordered regions. By comparing the ORD spectrum of an unknown protein with suitable standards of known conformation, it is possible to calculate the approximate proportions of the different types of structure in the protein.

CHAPTER 7

Organic Components: Carbohydrates

The carbohydrates are compounds made up of carbon, hydrogen, and oxygen. Generally, the hydrogen and oxygen in carbohydrates are present in the proportion of two hydrogen atoms to one oxygen atom as in H_2O, from which fact the term carbohydrate (carbon hydrate) was derived. Some carbohydrates, such as rhamnose, $C_6H_{12}O_5$, do not contain hydrogen and oxygen in the proportion of H_2O. The carbohydrates are defined as polyhydroxy aldehydes and ketones and their derivatives. This would include deoxysugars, amino sugars, and even sugar alcohols and acids.

MONOSACCHARIDES

Compounds possessing a free aldehyde or ketone group and two or more hydroxyl groups are called monosaccharides or simple sugars. The hexoses and pentoses are the most important of the simple sugars. The simple sugars are generally well-crystallized solids, soluble in water, and have a sweet taste. They all have the property of reducing alkaline copper solutions. They give other reactions characteristic of the free sugar group, such as reactions with phenylhydrazine, hydrogen cyanide, and hydroxylamine.

The simplest sugars are glyceraldehyde and dihydroxyacetone. Having three carbon atoms, these are known as trioses. Sugars having four carbon atoms are tetroses, five carbons pentoses, six carbons, hexoses, etc. They may be further differentiated on the basis of the functional group involved into aldo-or keto sugars.

Structure of D-Glucose

D-Glucose is the most common of the monosaccharides, occurring in the free state in the blood of animals and in the polymerized state, inter alia, as starch and cellulose.

The Fischer projection formula for D-glucose is also known as the open or straight chain structure. This structure occurs only in solution. There are two crystalline forms of D-glucose, known as α and β which also have different optical activities when dissolved. X- ray diffraction studies have confirmed chemical evidence that α and β-D-glucose are structures containing a ring of five carbon atoms and one oxygen atom.

These structures are known as the Haworth projection formulas. They do not represent the true shapes of the molecules, but they do show the configuration at each of the chiral atoms. In this system, the basic structure of the two ring forms is pyranose and furanose.

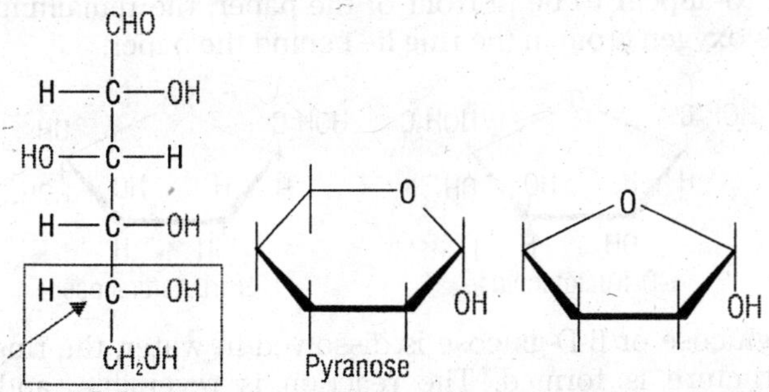

α -D Glucose

Pyranose

When the rings are viewed in this manner, the attached groups will be found to lie either above or below the plane of the ring. There are certain rules to obtain the correct Haworth form.

The first rule for translation of Fischer projections to Haworth forms is that any group to the right of the carbon chain is written down, those to the left are written up *i.e.*, above the plane of the ring. Thus, for example:

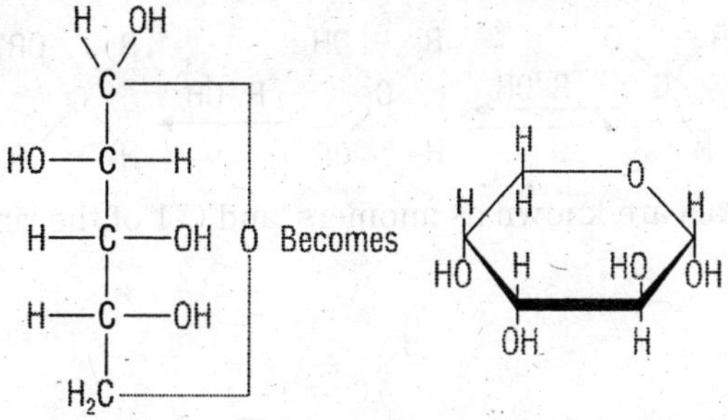

Becomes

When there are more carbon atoms in the sugar than are involved in ring formation. The rule is that if the ring is to the right, the extra carbon or carbons will be up *i.e.*, above the plane. For example, in the case of a ketohexose in the furanose form, the position of the 6-carbon will be determined as above, the 1-carbon will be up or down depending on whether the sugar is in the α- or β- form:

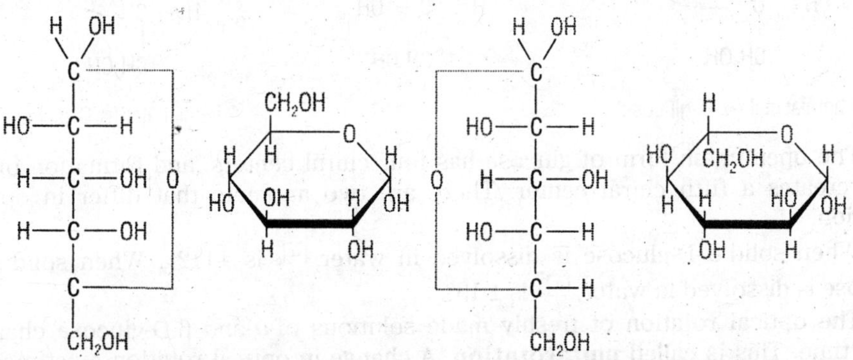

α-D-glucopyranose

β-l-glucopyranose

Haworth projection formulas do not represent the true shapes of the molecules, but they do show the configuration at each of the chiral atoms. The formulas represent the ring as a generalized plane standing at right angles to the plane of the paper. An element is given to the structure by thickening the three bonds of the ring that are meant to appear to be in front of the paper, the remaining three bonds of the ring and the oxygen atom in the ring lie behind the paper.

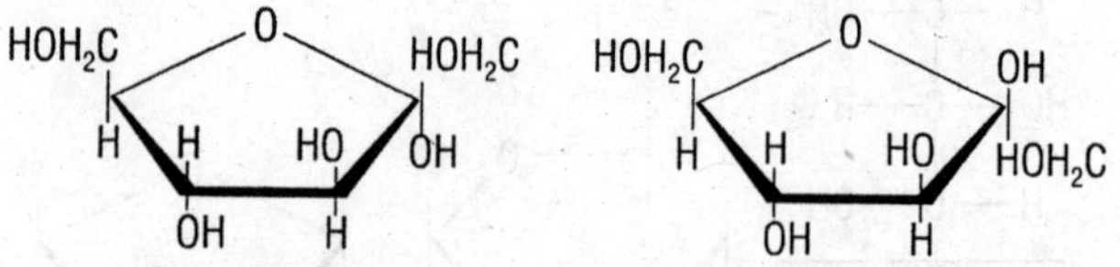

α-D-fructofuranose β-D-fructofuranose

When α-D-glucose or β-D-glucose is dissolved in water, the ring opens and the spent-chain structure is formed. The reaction is reversible, and equilibrium is established between the open form and the two ring forms. The aldehydes react reversibly with alcohols to give hemiacetals and then, in the presence of an acid catalyst, acetals:

The formation of a ring by the open-chain form of D-glucose can be the result of a reaction between the hydroxyl group on C-5 and the aldehyde group to give a hemiacetal. The aldehyde carbon becomes chiral as a result, thus giving rise to two hemiacetals, α- and β-D-glucose.

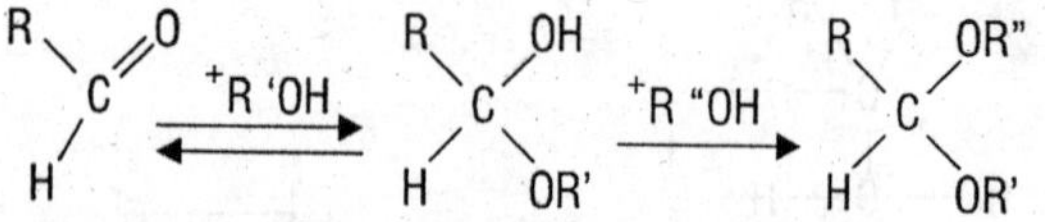

These hemiacetals are known as anomers, and C-1 of the ring form is called the anomeric carbon.

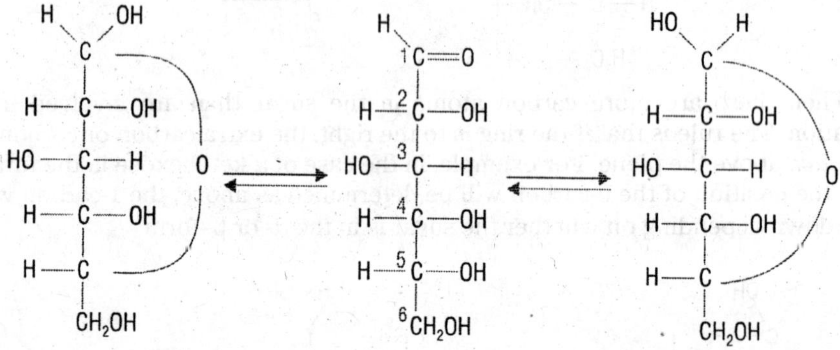

A hemiacetal (α-D-glucose) A hemiacetal (β-D-glucose)

The open chain form of glucose has four chiral centers, and formation of the ring creates a fifth chiral center. There are two anomers that differ in optical rotation.

When solid α-D-glucose is dissolved in water, $[\alpha]_D^{25}$ is +112°. When solid β-D-glucose is dissolved in water, $[\alpha]_D^{25}$ is ± 19°.

The optical rotation of freshly made solutions of α-and β-D-glucose changes with time. This is called **mutarotation**. A change in optical rotation must mean a change in structure. When α-D-glucose is dissolved in water, the α-D-anomer is the only structure present at the instant of dissolution. The opening of the ring is a slow reaction, but since the reaction is reversible, some ß-D anomer as well as the open-chain form will appear. Ultimately, an equilibrium between the open-chain form and the two anomer is established with an $[\alpha]_D^{25}$ of +52°.

β anomer α anomer

It is possible for two other anomeric forms to arise from the open-chain form by the addition of the hydroxyl group on C-4 to the carboxyl group of the aldehyde. These are five membered ring forms.

The Conformation of Glucose

The Haworth structures for the anomers of D-glucopyranose do not represent the true shape of the rings. The carbon atoms of glucose are all saturated, and the most stable form of a ring will be one that is strain free, *i.e.*, where the angles formed by the bonds at each carbon atom are 109°, the tetrahedral angle.

Consider cyclohexane, only two strains free conformations are possible: a chair and a boat, which are shown without H atoms.

C_6H_{12} Chair Boat

C-2 - C-3 (eclipsed) (b) C-3 - C-4 (staggered) (c)

There are 12 H atoms in the structure, 6 are in the general plane of the rings and are called equatorial (e), 6 are perpendicular to the general plane of the rings and are called axial (a). The chair and boat forms are inter convertible by rotation of the C-C bonds. Therefore, they are not isomers, and the term conformer is used to describe the various shapes a molecule can possess.

Although both conformers of cyclohexane are strain-free, the chair conformer is more stable than the boat. There are two reasons for this.

Two of the six C-C bonds in the boat are eclipsed, where as none are in the chair conformation. This can be illustrated by imagining the eye placed to look along the C-2 - C-3 bond, as shown in figure below. The eye, positioned as in (a), will see the configuration shown in (b). C-2 with its two H atoms and the bond leading to C-1, however, the eye cannot see C-3 or the two H atoms attached to it, because these three atoms are immediately behind C-2 and its two H atoms. Likewise, the C-5 - C-6 bond is eclipsed, but all other C-C bonds in the boat and all

C-C bonds in the chair are staggered, *e.g.*, (c) shows the eye looking along C-3 - C-4. Thus, in the boat, the four axial H atoms on C-2, C-3, C-5 and C-6 are as close as possible, and interactions between them distort the ring are maximal.

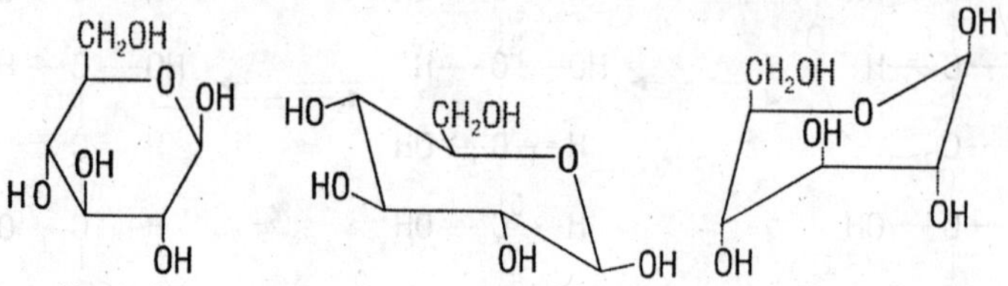

Haworth C1 1C

The axial H atoms on C-1 and C-4 of the boat come closer together (0.18 nm) than the sum of their van der Walls radii (0.24), and with these two atoms there is an unfavourable interaction tending to distort the ring.

The difference in energy between the boat and chair conformers of cyclohexane is about 25° KJ mol^{-1}, which means that 25 C only in 1,000 molecules it exists in the boat conformation.

The Haworth structure tells that a substituent that is above the general plane of the monosaccharide ring must also appear above the general plane of the chair, whether this is axial or equatorial will depend on the carbon atom being considered.

The preferred conformer is C_1, where all the bulky substituents are equatorial, on the edge of the ring. There will be considerable distortion in the 1 C conformer, where three bulky groups on the upper face of the ring produce considerable steric hindrance, as do two bulky groups on the lower face.

If the preferred forms of α and ß-D-glucopyranose (both C_1) are composed, it is seen that the ß form is the more stable since it has no bulky axial substituent, whereas α form has one. This explains ß-D-glucopyranose is the dominant anomer in an aqueous solution of glucose.

Glycosides

All the sugars in the ring forms are hemiacetals. They react with another alcohol to form acetals. **These are given the name of glycosides.** When a particular sugar is involved, a more specific name is applied, *e.g.*, glucoside, galactoside, fructoside. Further, it is possible to have α- and ß- forms and pyranose of furanose ring structures.

All monosaccharides and their derivatives that possess aldehyde or ketone groups (except alditols and aldonic acids) will have reducing properties. The reactivity of the anomeric hydroxyl group is illustrated with which monosaccharides react with alcohols and with amines. The normal hydroxyl groups in the molecule do not react, but the anomeric hydroxyl does. The process is known as glycosylation (of the alcohol or amine), and the products as O-glycosides and N-glycosides.

The R group in these glycosides is referred to as the aglycone.

As with the parent monosaccharide, the anomeric carbons have either the α- or the ß configuration because of chiral center. A glycoside is an acetal in chemical terms.

α-D-Methylglucopyranoside β-D-Methylglucofuranoside

Any monosaccharide

ROH

RNH$_2$

H_2O

O-Glycoside

N-Glycoside

Aldehyde (open-chain monosaccharide)

+R'OH

Hemiacetal (ring form)

+R"OH

H Acetal OR' (glycoside)

The glycosidic bond is found in a very wide range of biological compounds. There is an important group of O-glycosides in which the glycosidic bond links two monosaccharides, *i.e.*, the aglycone is a sugar. Such compounds are called disaccharides. The anomeric hydroxyl group of the second monosaccharide can itself glycosylate a hydroxyl group in a third monosaccharide to give a trisaccharide. Polysaccharides are polymers in which a large number of monosaccharides are linked by glycosidic bonds.

Properties of Monosaccharides

Monosaccharides are sweet in taste. These are colourless solids, soluble in water but sparingly soluble in alcohol and insoluble in ether.

Optical Isomerism

All natural monosaccharides are optically active. Most of the monosaccharides occurring in living cells are in the D-configuration.

The phenomenon of optical rotation is due to asymmetrical carbon atoms in the molecules. These asymmetric carbon atoms have their valence bonds attached to four different atoms or groups. In the tetrahedral model of the carbon atom, the carbon is in the center of the tetrahedron with its four valence bonds directed to the corners of the tetrahedron.

Two isomers have following properties:

1. All atoms are connected to the same atoms in each case and each atom has the same group of nearest neighbours. Only the 3-dimensional arrangement of the atoms in space differs.
2. The two isomers are mirror images of each other.

3. The molecules are non-superimposable.
4. Either one of these isomers as a crystal or in solution, will cause the plane of polarized light to be rotated, the other an equal amount to the left.
5. Isomers having these properties are called enantiomorphs.

D- and L- optical isomers are designated with defined configurations of glyceraldehyde, a triose which has only a single asymmetric carbon atom.

The D-isomer of glyceraldehyde is written with the -OH group on the right of the asymmetric carbon atom and the hydrogen on the left. Sugars related to dextro-glyceraldehyde are designated D compounds, and related to laevo-glyceraldehyde are L compounds.

The C_2 is the plane of the paper, the aldehyde and hydroxymethyl groups are behind the paper (as denoted by the dashed bond lines), and the hydrogen and hydroxyl groups are in front of the paper (as denoted by the solid bond lines). Enantiomorphs are mirror images of each other. When placed in front of a plane mirror, one structure will give an image that is identical with the structure of the other.

CHO, H—C—OH, CH_2OH and CHO, HO—C—H, CH_2OH CHO, H—C—OH, CH_2OH and CHO, HO—C—H, CH_2OH

In the pairs of figures above, the structure on the left is called D-glyceraldehyde, and that on the right L-glyceraldehyde.

D-Glyceraldehyde	Types of D-aldoses	Types of D-ketoses
CHO H—C—OH CH_2OH	CHO R H—C—OH CH_2OH	CH_2OH CO R H—C—OH CH_2OH

L-Glyceraldehyde	Types of L-aldoses	Types of L-ketoses
CHO HO—C—H CH_2OH	CHO R HO—C—H CH_2OH	CH_2OH CO R HO—C—H

The property of optical activity is referred to the ability of a solution of an enantiomer to rotate the plane of polarized light. One of a pair of enantiomers will rotate the plane in a clockwise direction and is given the symbol (+). The other will rotate the plane in a counter clockwise direction and is given the symbol (-). The D enantiomer of glyceraldehyde is (+) and is described as D-(+) glyceraldehyde. The other is L (-) glyceraldehyde. Mixtures of D and L enantiomers will have a net rotation depending on the proportions of the enantiomers. Equal proportions give a net rotation of zero, in which case the solution is said to be racemic.

Optical activity is measured in a polarimeter. The magnitude of the optical activity is measured as an angle of rotation, given the symbol: α. The units are degrees or radians (SI).

Whereas glyceraldehyde has one chiral center and aldotetroses, pentoses, and hexoses have two, three, and four respectively. Each chiral center gives rise to optical activity. The net optical activity of an aldose will depend on contribution from each chiral center and will be (+) or (-).

Specific Rotation

The actual measurement of the rotation of polarized light involves the use of an instrument called the polariscope or polarimeter.

Monochromatic light is used because different wavelengths of polarized light are rotated differently. The light from incandescent sodium or mercury is used. Polariscopes have special filters to purify the beam of light as it passes through. The sodium -D spectral line is commonly used. It is composed of two wavelengths D_1 (5896 Å) and D_2 ((5890 Å). The mercury line gives slightly greater rotation than sodium -D light because of its shorter wavelength.

All the principles of optical rotation were formulated by Biot. These are:

1. An optically active substance has a definite rotation for a given wavelength.
2. Rotation is proportional to the concentration of optically active substance.
3. Rotation is proportional to the thickness of solution through which the light passes.
4. Rotation varies with temperature, some substances showing an increase and others a decrease with rising temperature.
5. Rotation varies with wavelength of light, being greater the shorter the wavelength.
6. Rotation varies with the nature of the solvent in which the substance is dissolved.

Specific rotation is defined as the rotation in degrees of 1 g of substance per milliliter in a tube 1 dm (10 cm) in length.

The molecular rotation of a substance is the product of its specific rotation and molecular weight. The different rotation of a substance toward different wavelengths of light is referred to as optical rotatory dispersion.

Tautomerism

Monosaccharides are chain compounds containing the aldehyde or ketone group. In the crystalline state the monosaccharides containing five or more carbon atoms exist in tautomeric ring forms. Even in solution these sugars exist almost entirely in ring forms, which are in dynamic equilibria with very small amounts of the chain forms.

Mutarotation

Fischer treated glucose with methyl alcohol and hydrogen chloride. The product consisted of two isomeric methyl derivatives of glucose. These substances had specific rotations of +159° and -34°. They have ring structures as follows:

Methyl α-D. glucosideMethyl β-D- glucoside
(α)D = + 159° (α)D = -34°

Carbon 1 is asymmetric, this accounts for the two isomers. Dubrunfant noticed that the optical rotation of a freshly prepared solution of glucose gradually decreases and finally becomes constant. This change in the rotation of sugar solutions upon standing is a general property of reducing sugars and is called mutarotation. Tanret prepared two isomeric forms of D-glucose by crystallization under different conditions. When D-glucose is crystallized from water or dilute alcohol at room temperature, a form separates having an initial specific rotation of +112 ° which changes to +52.5 °. If crystallization takes place from water at temperature above 98° C, a different form of glucose, having an initial specific rotation of +19°, which changes to 52.5°, is obtained. The first of these isomers of glucose was called α-D-glucose and the second ß-D-glucose.

a-D-glucose ß-D-glucose
+112° ⟶ + 52.5° ⟵ + 19°°

Tanret showed that glucose exists in isomeric forms. These forms in solution change into the same equilibrium mixture regardless of which form is dissolved.

The equilibrium for fructose in aqueous solution is as follows:

- Both ring forms of fructose are ß forms but one ring contains five carbon atoms and the other four. The α- form of fructose containing a five carbon ring has been prepared.
- The sugar osazones also exist in both chain and ring forms.

HOMOPOLYSACCHARIDES

These are Complex carbohydrates formed from at least six identical monosaccharides; examples include starch, glycogen, cellulose, and insulin.

Homopolysaccharides have a well-defined chemical structure. Their molecular weight depends on the source. The molecules from a single source also may vary in size, because most polysaccharides are formed biologically by an enzyme-catalysed process lacking genetic information regarding size. Following are the examples of homopolysaccharides:

Cellulose

The sugar component is glucose with a ß, 1 → 4 linkage. It has structural function. It is found throughout plant kingdom. Cellulose is found in plants as microfibrils (2-20 nm diameter and 100 - 40 000 nm long). These form the structurally strong framework in the cell walls. Cellulose is mostly prepared from wood pulp. Cellulose is also produced in a highly hydrated form by some bacteria (for example, *Acetobacter xylinum*).

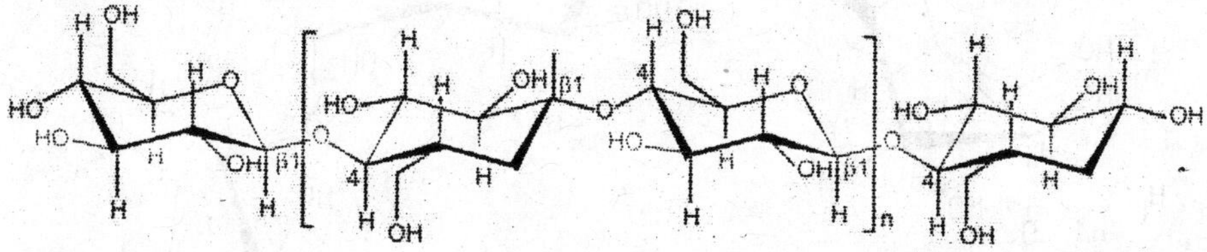

Cellulose is a linear polymer of ß-(1→4)-D-glucopyranose units in $^4C_1$ conformation. The fully equatorial conformation of ß-linked glucopyranose residues stabilizes the chain structure, minimizing its flexibility (for example, relative to the slightly more flexible α-linked glucopyranose residues in amylose). Cellulose preparations may contain trace amounts (~0.3%) of arabinoxylans.

Molecular Structure

Cellulose is an insoluble molecule consisting of between 2000 - 14000 residues with some preparations being somewhat shorter. It forms crystals (cellulose I_α) where intra-molecular (O3-H →O5' and O6→H-O2') and intra-strand (O6-H→O3') hydrogen bonds holds the network flat allowing the more hydrophobic ribbon faces to stack. Each residue is oriented 180° to the next with the chain synthesised two residues at a time. Although individual strand of cellulose are intrinsically no less hydrophilic, or no more hydrophobic, than some other soluble polysaccharides (such as amylose) this tendency to form crystals utilizing extensive intra- and intermolecular hydrogen bonding makes it completely insoluble in normal aqueous solutions (although it is soluble in more exotic solvents such as aqueous N-methylmorpholine-N-oxide (NMNO, ~0.8 mol water/mol, then up to 30% by wt cellulose at 100°C), CdO/ethylenediamine (cadoxen), LiCl/*N,N'*-dimethylacetamide or near-supercritical water). It is thought that water molecules catalyse the formation of the natural cellulose crystals by helping to align the chains through hydrogen-bonded bridging.

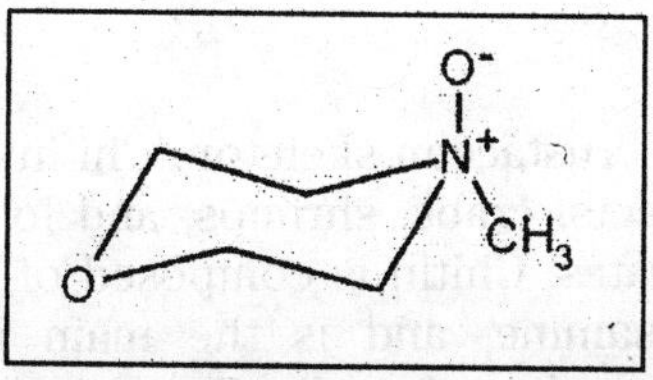

Cellulose has many uses as an anticake agent, emulsifier, stabilizer, dispersing agent, thickener, and gelling agent but these are generally subsidiary to its most important use of holding on to water. Water cannot penetrate crystalline cellulose but dry amorphous cellulose absorbs water becoming soft and flexible.

Amylose

Amylose molecules consist of single mostly-unbranched chains with 500-20,000 α-(1→4)-D-glucose units dependent on source (a very few α-1→6 branches and linked phosphate groups may be found, but these have little influence on the molecule's behavior). Its function is in food storage. It is found in starch, especially in corn, potatoes and rice.

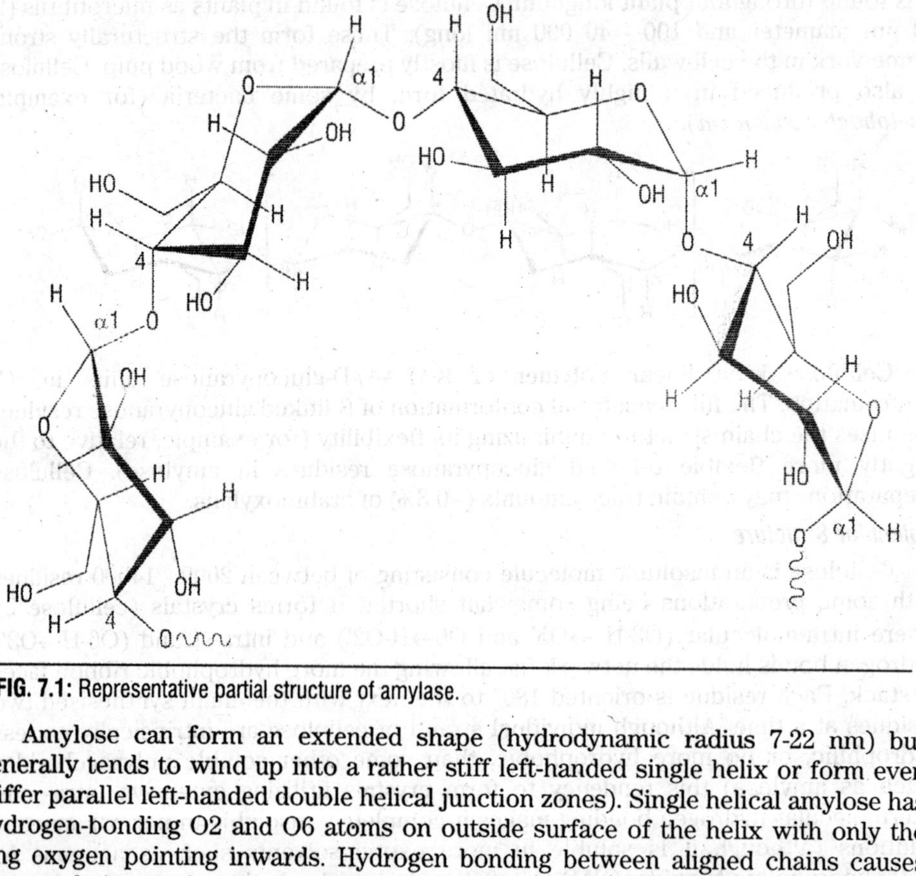

FIG. 7.1: Representative partial structure of amylase.

Amylose can form an extended shape (hydrodynamic radius 7-22 nm) but generally tends to wind up into a rather stiff left-handed single helix or form even stiffer parallel left-handed double helical junction zones). Single helical amylose has hydrogen-bonding O2 and O6 atoms on outside surface of the helix with only the ring oxygen pointing inwards. Hydrogen bonding between aligned chains causes retrogradation and releases some of the bound water (syneresis). The aligned chains may then form double stranded crystallites that are resistant to amylases. These possess extensive inter- and intra-strand hydrogen bonding, resulting in a fairly hydrophobic structure of low solubility. The amylose content of starches is thus the major cause of resistant starch formation.

Chitin

Chitin is found in insect and crustacean skeleton. Chitin is a polysaccharide found in the outer skeleton of insects, crabs, shrimps, and lobsters and in the internal structures of other invertebrates. Chitin is composed of ß(1-4) linked units of the amino sugar N-acetyl-glucosamine, and is the main source of production of chitosan, which is used in a number of applications, such as a flocculating agent, a wound healing agent, a sizing and strengthening agent for paper, and a delivery vehicle for pharmaceuticals and genes.

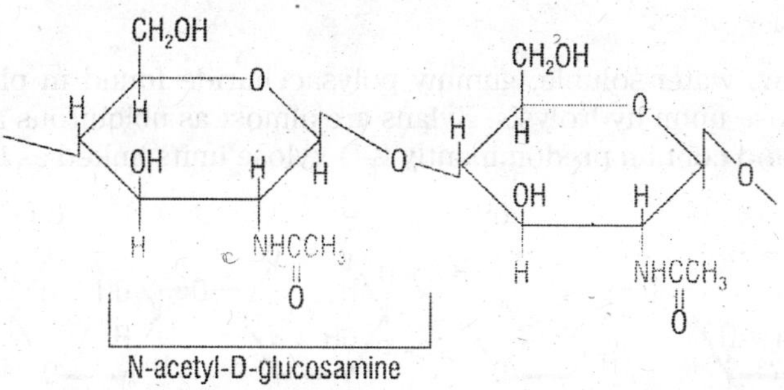

N-acetyl-D-glucosamine

Chitin.

The combination of chitin's remarkable properties makes it extremely versatile. Chitin is natural, non-toxic, non-allergenic, anti-microbial, and biodegradable. It has a strong positive charge, which allows it to bind with negatively charged surfaces or materials, including metals, skin, and macromolecules such as proteins. Chitin is also used as anti-bacterial sponges and hospital dressings, artificial blood vessels, contact lenses, tumor inhibition, dental plaque inhibition, and blood cholesterol control. Household products include sponges, diapers, feminine napkins, and tampons.

Inulin

The sugar component is fructose with a ß, 2 → 4 linkage. Its function is in food storage. It is found in atrichckes and chicory. Inulins are a group of naturally occurring polysaccharides (several simple sugars linked together) produced by many types of plants. They belong to a class of fibers known as fructans. Inulin is used by some plants as a means of storing energy and is typically found in roots or rhizomes. Most plants which synthesise and store inulin do not store other materials such as starch.

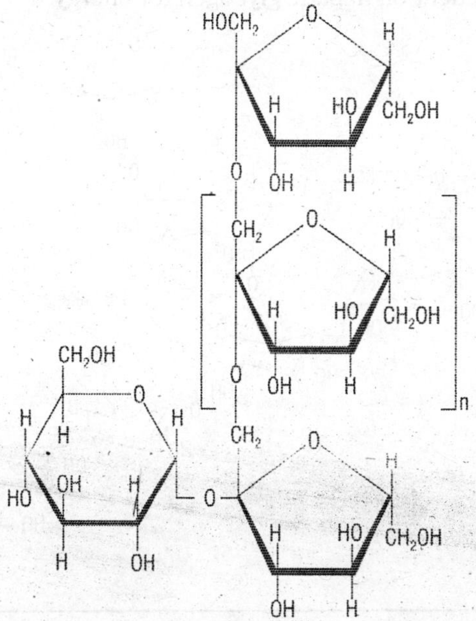

Inulins.

Xylans

Xylan is a yellow, water-soluble, gummy polysaccharide found in plant cell walls and yielding xylose upon hydrolysis. Xylans are almost as ubiquitous as cellulose in plant-cell walls and contain predominantly ß-D-xylose units linked as in cellulose.

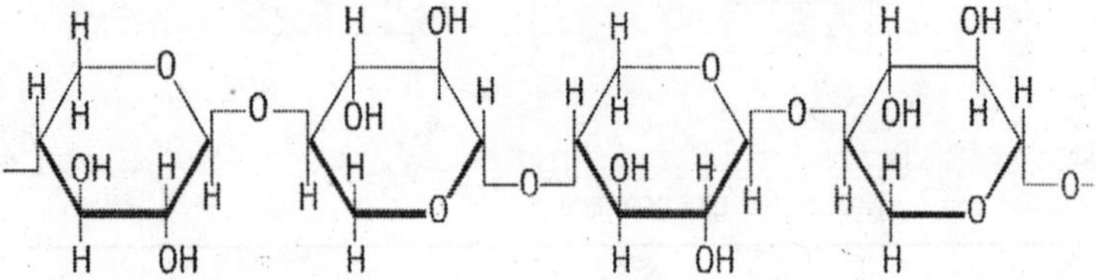

Xylan.

The sugar component is xylose with a ß, 1 → 4 linkage. Some xylans contain other sugars, such as L-arabinose, but they form branches and are not part of the main chain. It has structural function. It is found throughout land plants. Xylan is also found in the cell walls of some green algae, especially macrophytic siphonous genera, where it replaces cellulose. Similarly, it replaces the inner fibrillar cell-wall layer of cellulose in some red algae. Xylans are of little commercial importance.

Glycogen

It is white, amorphous, tasteless polysaccharide. Glycogen is found in two forms, proglycogen (with a molecular weight of around 400 kDa) and macroglycogen (with a molecular weight in the millions). It is stored primarily in the liver and muscle tissue and readily converted to glucose as needed by the body to satisfy its energy needs. It also is found in various species of microorganisms (bacteria and fungi, including yeasts). Glycogen serves as an energy reservoir, being broken down to glucose when needed. It supplies energy during heavy work. It is stored with water (1 gram of carbohydrates stored with 3 grams of water). The central nervous system (CNS) is dependent on hepatic glycogen for energy.

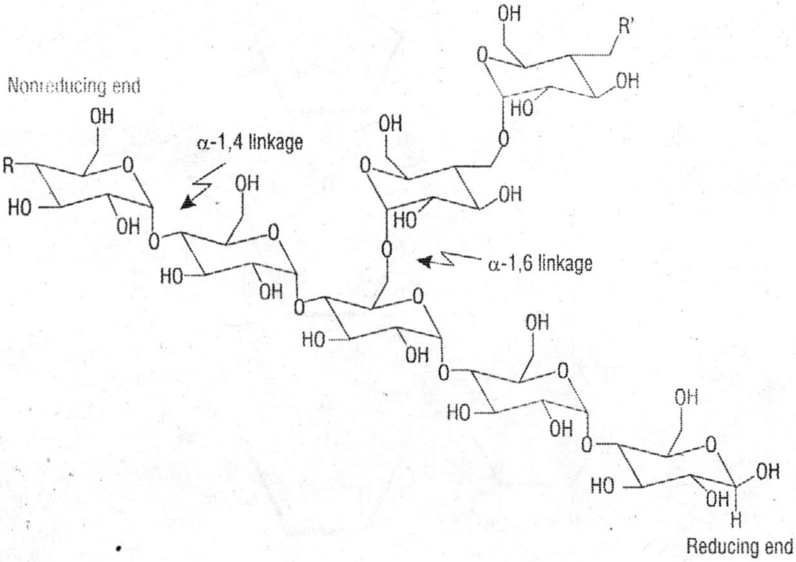

Glycogen.

The sugar component is glucose with a α, 1 → 4 and σ ← 1, α linkage. It is a polysaccharide, (C-6, H-10, O-5)$_n$. It is a highly branched polymer. It described as a dendrimer of about 60,000 glucose residues and has a molecular weight between 10^6 and 10^7 Daltons (~4.8 million). Most of Glc units are linked by α-1,4 glycosidic bonds, approximately 1 in 12 Glc residues also makes alpha-1,6 glycosidic bond with a second Glc, which results in the creation of a branch. Glycogen does not possess a reducing end: the 'reducing end. The glycogen granules contain both glycogen and the enzymes of glycogen synthesis (glycogenesis) and degradation (glycogenolysis).

Amylopectin

Amylopectin is a highly branched polymer of glucose found in plants. It is one of the two components of starch, the other being amylose. Starch is made of about 70% amylopectin by weight. It is found in grain rice and waxy potatoes.

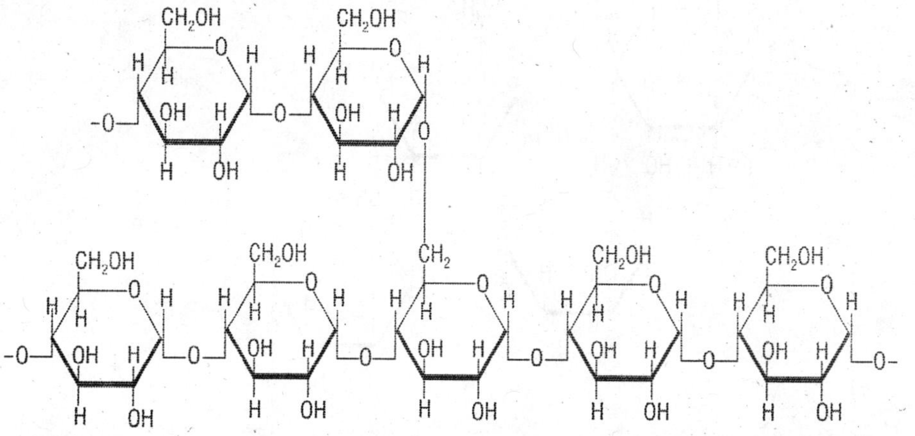

Amylopectin.

The amount is less in long-grain rice and russet potatoes. It is soluble in water. It is composed of linearly linked alpha 1,4 linked glucose units (coiled into tubular sections) with occasional alpha 1-6 glycosidic bonds which provide branching points. Each amylopectin molecule may contain 100,000-200,000 glucose units, and each branch is about 20 or 30 glucose units in length, so that these molecules are bushy and nearly spherical in shape. The many exposed ends can have more glucose units added to them by enzyme action for storage purposes or removed from them for use in respiration. Amylopectin is one fraction of starch (typically 80-90%), the other fraction being amylose (10-20%). Glycogen ("animal starch"), found in the liver and muscles, is effectively similar in structure to amylopectin, but it has shorter branches: 8-12 glucose units.

Dextran

It is a high-molecular-weight polymer of d-glucose and is a complex, branched glucan (polysaccharide made of many glucose molecules) composed of chains of varying lengths (from 10 to 150 Kilo Daltons). The straight chain consists of α-1,6 glycosidic linkages between glucose molecules, while branches begin from α-1,4 linkages (and in some cases, α-1,2 and α-1,3 linkages as well). It is produced by enzymes on the cell surface of certain lactic acid bacteria.

Dextrans formed from sucrose by bacteria in the mouth adhere to the tooth surfaces and produce dental plaque. Dextran is synthesised from sucrose by certain lactic-acid bacteria, the best known being *Leuconostoc mesenteroides* and *Streptococcus mutans*. Dextran is also formed by the lactic acid bacterium *Lactobacillus brevis* to create the crystals of *tibicos*, or water kefir fermented beverage which supposedly has some health benefits. It is used medicinally as an antithrombotic (anti-platelet), to reduce blood viscosity, and as a volume expander in anemia.

It is used as blood volume expander used in treatment of shock or impending shock when blood or blood products are not available; dextran 40 is also used as a priming fluid in cardiopulmonary bypass and for prophylaxis of venous thrombosis and pulmonary embolism in surgical procedures associated with a high risk of thromboembolic complications.

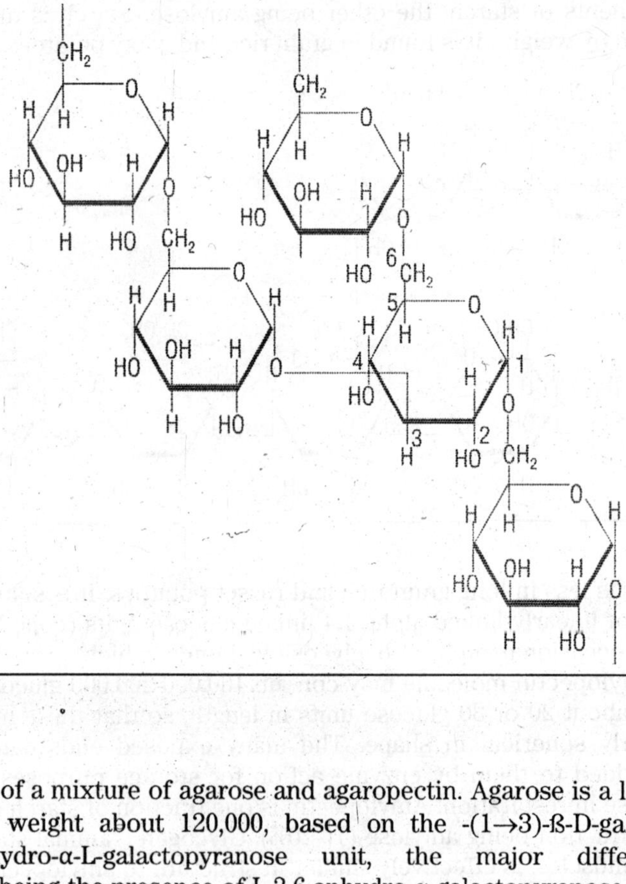

Dextran.

Agar

Agar consists of a mixture of agarose and agaropectin. Agarose is a linear polymer, of molecular weight about 120,000, based on the -(1→3)-ß-D-galactopyranose-(1→4)-3,6-anhydro-α-L-galactopyranose unit, the major differences from carrageenans being the presence of L-3,6-anhydro-α-galactopyranose rather than D-3,6-anhydro-α-galactopyranose units and the lack of sulphate groups. Agaropectin is a heterogeneous mixture of smaller molecules that occur in lesser amounts. Their structures are similar but slightly branched and sulphated, and they may have methyl and pyruvic acid ketal substituents. They gel poorly and simply removed from the excellent gelling agarose molecules by using their charge. The quality of agar is improved by alkaline treatment that converts of any L-galactose-6-sulphate to 3,6-anhydro-L-galactose.

HETEROPOLYSACCHARIDE

Complex carbohydrates formed by combining carbohydrates with non-carbohydrates or carbohydrate derivatives; examples include pectin, lignin, glycoproteins, glycolipids, and mucopolysaccharides.

Pectin

Pectin is a heterogeneous grouping of acidic structural polysaccharides, found in fruit and vegetables and mainly prepared from 'waste' citrus peel and apple pomace. It makes up between about 2% and 35% of plant cell walls and is important for plant growth, development and defense. Pectin has a complex structure. The functionality of pectin is dictated by the chemical fine structure, and the majority of the pectin used currently comes from citrus peel and apple pomace. Other sources of the ingredient have remained largely unexploited because of certain undesirable structural properties.

Pectin.

The characteristic structure of pectin is a linear chain of α-(1-4)-linked D-galacturonic acid that forms the pectin-backbone, a homogalactoronan.

The majority of the structure consists of homopolymeric partially methylated poly-α-(1→4)-D-galacturonic acid residues ('smooth', see right) but there are substantial 'hairy' non-gelling areas (see below) of alternating α -(1→2)-L-rhamnosyl-α -(1→4)-D-galacturonosyl sections containing branch-points with mostly neutral side chains (1 - 20 residues) of mainly L-arabinose and D-galactose (rhamnogalacturonan I).

Pectins may also contain rhamnogalacturonan II sidechains containing other residues such as D-xylose, L-fucose, D-glucuronic acid, D-apiose, 3-deoxy-D-*manno*-2-octulosonic acid (Kdo) and 3-deoxy-D-*lyxo*-2-heptulosonic acid (Dha) attached to poly-α-(1→4)-D-galacturonic acid regions. The neutral sugars are mainly D-galactose, L-arabinose and D-xylose, the types and proportions of neutral sugars varying with the origin of pectin.

Lignin

It is a complex oxygen-containing organic substance that, with cellulose, forms the chief constituent of wood. It is second only to cellulose as the most abundant organic material on Earth, though it has found relatively few industrial uses other than as a fuel.

Lignin is a cross-linked racemic macromolecule with molecular masses in excess. It is relatively hydrophobic and aromatic in nature. The degree of polymerisation in nature is difficult to measure, because it is fragmented during extraction and the molecule consists of various types of substructures, which appear to repeat in a haphazard manner. Various types of lignin have been described depending on the means of isolation.

There are three monolignol monomers, methoxylated to various degrees: p-coumaryl alcohol, coniferyl alcohol, and sinapyl alcohol. These lignols are incorporated into lignin in the form of the phenylpropanoids *p*-hydroxyphenyl (H), guaiacyl (G), and syringal (S) respectively.

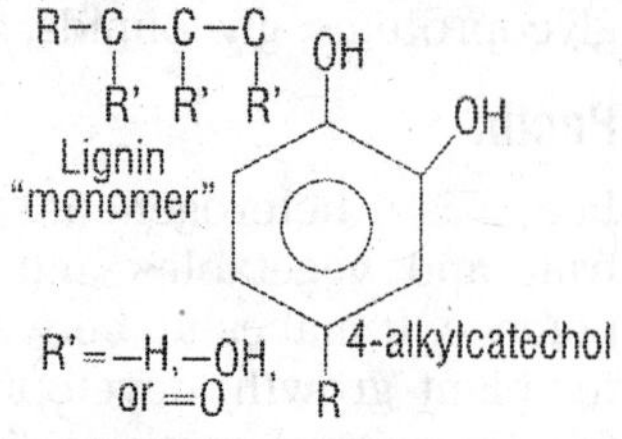

The basic monomer of lignin can be thought of as 4-alkylcatechol. The alkyl ligand is a 1 to 3 carbon chain which may be substituted with hydroxy or keto ligands at any or all positions. Both phenol groups and the alkyl substitutions may cross-link to other monomers or their side chains, yielding a fairly light, but strongly cross-linked and randomly ordered mass, *i.e.*, wood.

Lignin monomer.

Lignin is formed by removal of water from sugars to create aromatic structures. These reactions are not reversible. There are many possible monomers of lignin, and the types and proportions depend on the source in nature. Lignin is found in all vascular plants, mostly between the cells, but also within the cells, and in the cell walls. It makes vegetables firm and crunchy, and gives us what we call "fiber" in our food. It functions to regulate the transport of liquid in the living plant (partly by reinforcing cell walls and keeping them from collapsing, partly by regulating the flow of liquid), and it enables trees to grow taller and compete for sunshine and researchers see it as a disposal mechanism for metabolic wastes.

Glycoproteins

Glycoprotein is a protein containing one or more covalently linked carbohydrates of various types, i.e. from monosaccharides to branched polysaccharides, including glycosylphosphatidylinositol (GPI), glycosaminoglycans (GAG). They contain oligosaccharide chains (glycans) covalently attached to their polypeptide side-chains. The formation of glycoprotein is called glycosylation. In proteins that have segments extending extracellularly, the extracellular segments are often glycosylated. Glycoproteins are often important integral membrane proteins, where they play a role in cell-cell interactions. Glycoproteins also occur in the cytosol, but their functions and the pathways producing these modifications in this compartment are less well-understood.

Glycoconjugates also play important roles in developmental processes, as revealed by the pathology of human diseases caused by abnormal glycosylation and genetic studies in model organisms. Glycosylation is the enzymatic process that links saccharides to produce glycans, either free or attached to proteins and lipids. This enzymatic process produces one of four fundamental components of all cells (along with nucleic acids, proteins, and lipids). Glycosylation is also present in the cytoplasm and nucleus as the O-GlcNAc modification. There are two types of glycosylation:

- In N-glycosylation, the addition of sugar chains can happen at the amide nitrogen on the side chain of the asparagine.
- In O-glycosylation, the addition of sugar chains can happen on the hydroxyl oxygen on the side chain of hydroxylysine, hydroxyproline, serine, or threonine.

Glycoproteins are important compontents in various functions like, structural molecule of body (Collagens), lubricant and protective agent (Mucins), transport molecule (transferrin, ceruloplasmin), immunologic molecule(immunoglobins, histocompatibility antigens), hormone [(Chorionoic gonadotropin, thyroid-stimulating hormone (TSH)], enzyme (alkaline phosphatase), cell attachment-recognition site (various proteins involved in cell-cell), etc.

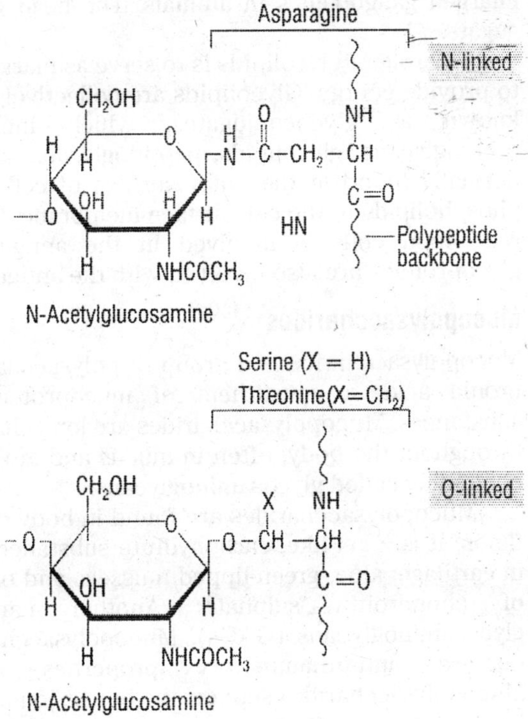

Asparagine
N-linked
CH_2OH
$NHCOCH_3$
Polypeptide backbone
N-Acetylglucosamine

Serine (X = H)
Threonine(X=CH_3)
O-linked
CH_2OH
$NHCOCH_3$
N-Acetylglucosamine

Glycolipids

The term glycolipid designates any compound containing one or more monosaccharide residues bound by a glycosidic linkage to a hydrophobic moiety such as an acylglycerol, a sphingoid, a ceramide (*N*-acylsphingoid) or a prenyl phosphate. Glycolipids are lipids that are attached to a short carbohydrate chain (mono- or oligosaccharides).

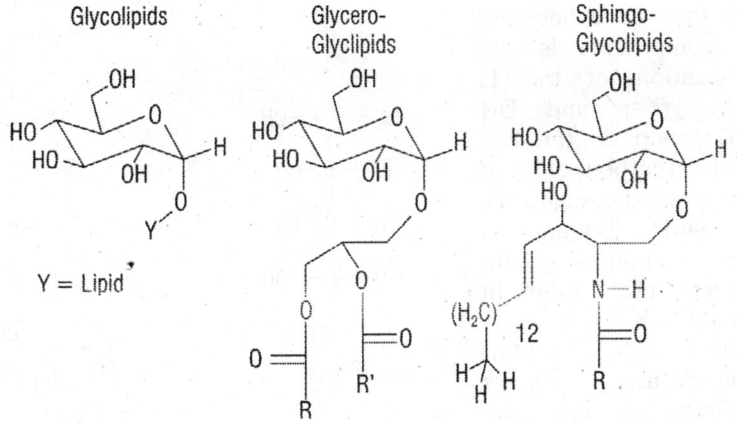

Glycolipids
Y = Lipid
Glycero-Glyclipids
Sphingo-Glycolipids

Glycolipids.

Glycolipid are of different types e.g., Glyceroglycolipids (Galactolipids and Sulpholipids (SQDG)) and Glycosphingolipids (Cerebrosides which include Galactocerebrosides and Glucocerebrosides), Gangliosides (the most complex animal glycolipids; contain negatively charged oligosacchrides with one or more sialic acid residues; more than 40 different ganglidosides have been identified; they are most abundant in cells), Globosides, Sulphatides and Glycophosphosphingolipids (complex glycophospholipids from fungi, including yeasts, and in plants, where they were originally called "phytoglycolipidds" by dCdarter, et al., may comprise as complicated add of compounds as the negatively charged gangliosides in animals. The head group of a glycolipid is composed of sugars.

The role of glycolipids is to serve as markers for cell to cell communication and to provide energy. Glycolipids are collectively part of a larger family of substances known as glycoconjugates, which include glycoproteins, glycopeptides, proteoglycans, glycolipids, peptidoglycans and lipopolysaccharides. Glycolips are normally found at the outer surface of cell membranes and are associated with phospholipids in the cell surface membrane. One of the glycolipids found in human red blood cells is involved in the antigens of the ABO blood type. Some glycoproteins are also involved with the antigens as well.

Mucopolysaccharides

Mucopolysaccharide is a group of polysaccharides containing an amino sugar and uronic acid; a constituent of mucoproteins, glycoproteins, and blood-group substances. Mucopolysaccharides are long chains of sugar molecules that are found throughout the body, often in mucus and in fluid around the joints. They are more commonly called glycosaminoglycans.

Mucopolysaccharides are found in body cells, mucous secretions, and synovial fluids. It is a gel-like, carbohydrate substance. Mucopolysaccharides can be found in cartilage, aloe, green-lipped mussels, and other great sources. They are a source of chondroitin sulphate. Another name for mucopolysaccharides is glycosaminoglycans (GAGs). Mucopolysaccharides contribute to longevity. They possess anti-inflammatory properties and other intrinsic properties. Mucopolysaccharides support the bones, joints, and cartilage.

EPIMERS

Epimer is a type of isomer in which the difference between the two compounds is the relative position of the H (hydrogen) group and OH (hydroxyl) group on the last asymmetric C (carbon) atom of the chain, as in the sugars D-glucose and D-mannose. Therefore, epimers are diastereomers that differ in configuration of only one stereogenic center. Diastereomers are a class of stereoisomers that are non-superposable, non-mirror images of one another, unlike enantiomers which are non-superposable mirror images of one another.

Epimers

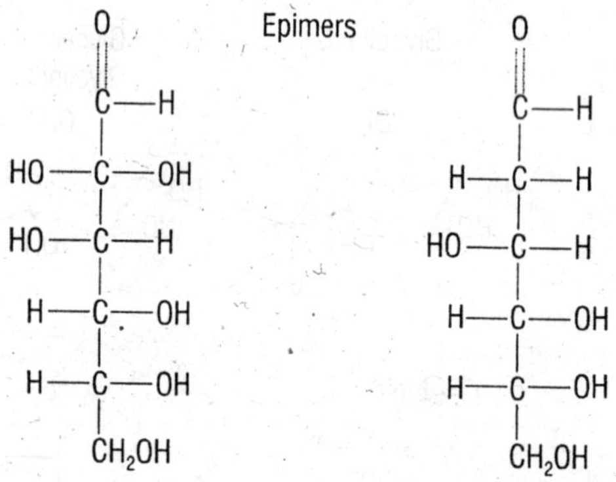

Glucose Mannose

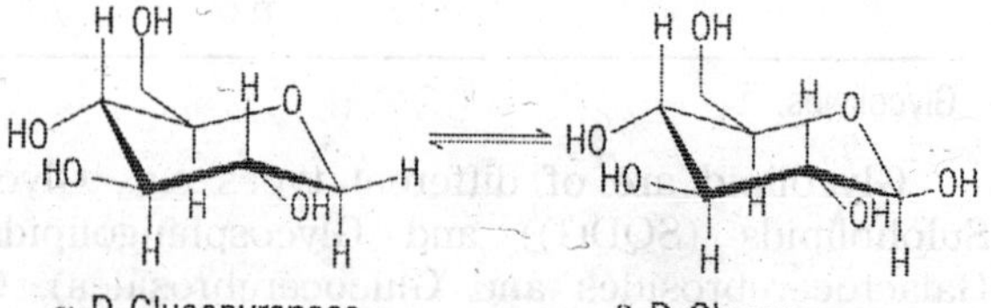

α-D-Glucopyranose β-D-Glucopyranose

For example, the sugars α-glucose and ß-glucose are epimers. In α-glucose, the -OH group on the first (anomeric) carbon is in the direction opposite the methylene group (in the axial position). In ß-glucose, the -OH group is oriented in the same direction as the methylene group (in the equatorial position). These two molecules are both epimers and anomers.

ANOMER

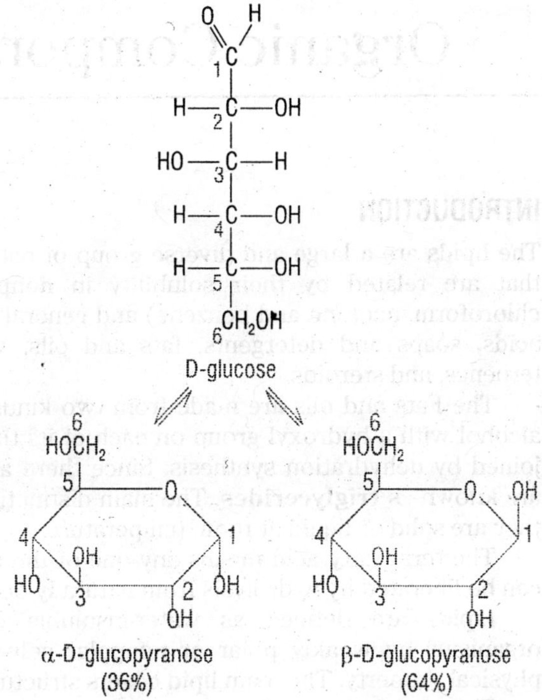

Anomer is either of a pair of cyclic stereoisomers (designated α or ß) of a sugar or glycoside, differing only in configuration at the reducing carbon atom. Anomer is a special type of epimer. It is a stereoisomer (diastereomer, more exactly) of a saccharide (in the cyclic form) that differs only in its configuration at the hemiacetal (or hemiketal) carbon, also called the anomeric carbon. If the structure is analogous to one with the hydroxyl group on the anomeric carbon in the axial position of glucose, then the sugar is an alpha anomer. If, however, that hydroxyl is in the equatorial position, then the sugar is a beta anomer. For example, α-D-glucopyranose and ß-D-glucopyranose, the two cyclic forms of glucose, are anomers.

D-glucose

α-D-glucopyranose (36%)

β-D-glucopyranose (64%)

The cyclic forms of carbohydrates can exist in two forms, α- and ß- based on the position of the substituent at the anomeric center. The two forms are sometimes described as anomers since they are isomers at the anomeric center. To assign the cyclic form of a carbohydrate as the α- or ß- form look at the relative positions of the $-CH_2OH$ group and -OH (or -OR) group at the anomeric center. In the α- form, the exocyclic O group at the anomeric center is on the opposite face to the $-CH_2OH$ group, and in the ß- form, the exocyclic O group at the anomeric center is on the same face as the $-CH_2OH$ group. If a mixture of α- and ß- anomers are present, then this is often represented by using a “wavy” line to represent the bond. In general the two forms are stable solids, but in solution they rapidly equilibrate.

CHAPTER 8

Organic Components: Lipids

INTRODUCTION

The lipids are a large and diverse group of naturally occuring organic compounds that are related by their solubility in nonpolar organic solvents (e.g. ether, chloroform, acetone and benzene) and general insolubility in water. They are fatty acids, soaps and detergents, fats and oils, waxes, phospholipids, eicosonoids, terpenes, and steroids.

The Fats and oils are made from two kinds of molecules: **Glycerol** (a type of alcohol with a hydroxyl group on each of its three carbons) and three **fatty acids** joined by dehydration synthesis. Since there are three fatty acids attached, these are known as **triglycerides**. The main distinction between fats and oils is whether they are solid or liquid at room temperature.

The term fatty acid means any one of the aliphatic monocarboxylic acids that can be liberated by hydrolysis from naturally occuring fats and oils.

Lipids are defined as water-insoluble compounds extracted from living organisms by weakly polar or nonpolar solvents. This definition is based on a physical property. The team lipid covers structurally diverse groups of compounds.

The term fats and oils are commonly used to denote crude lipid mixtures which are obtained from natural sources in specific ways (such as land and cotton seed oil). The distribution between the two terms is solely on the basis of the physical state at room temperature, fats are solids, oils are liquids.

A common feature of all lipids is that, biologically, their hydrocarbon content is derived from the polymerization of acetate followed by reduction of the chain so formed. Polymerization of acetate can give rise to the following:

- **Long, linear hydrocarbon chains:** The products are fatty acids, $CH_3\,(CH_2)_4$ COOH. Lipids containing fatty acids include the glycerol lipids the sphingolipids, and waxes.
- **Branched-chain hydrocarbon** via a fire-carbon intermediate, isopentene (isoprene), Terpenes, Steroids and caroteroids.
- **Linear or cyclic structures** that are only partially reduced. These are called a cetogenins (or polyketides), many of these compounds are aromatic and not all are lipids.

FATTY ACIDS (DERICED LIPIDS)

Fatty acids are merely carboxylic acids with long hydrocarbon chains. The hydrocarbon chain length may vary from 10-30 carbons (most usual is 12-18). The non-polar hydrocarbon alkane chain is an important counter balance to the polar acid functional group. In acids with only a few carbons, the acid functional group dominates and gives the whole molecule a polar character. However, in fatty acids, the non-polar hydrocarbon chain gives the molecule a non- polar character. Over 100 fatty acids are known to occur naturally. They vary in chain length and degree of unsaturation. Nearly all have an even number of carbon atoms. Most consist of linear chains of carbon atoms, but a few have branched chains. Fatty acids occur in very low quantities in the free states. They are found mostly in an esterified state as components of other lipids. The pKa of the carboxylic acid group is about 5, and under physiological conditions, this group will exist in an ionized state called an acylate ion, *e.g.*, the ion of palmitic acid is palmitate, $CH_3\,(CH_2)_{14}\,COO^-$.

Name	Structure	Source
Lauric acid	$CH_3(CH_2)_{10}COOH$	Lauric acid is the main acid in coconut oil and in palm kernel oil (not to be confused with palm oil). It has antimicrobial properties. It is also found in human milk (5.8% of total fat), cow's milk (2.2%), and goat's milk (4.5%).
Palmitic acid	$CH_3(CH_2)_{14}COOH$	It is a major component of the oil from palm trees (palm oil and palm kernel oil).
Stearic acid	$CH_3(CH_2)_{16}COOH$	It occurs in many animal and vegetable fats and oils. One important source is cocoa.
Oleic acid	$CH_3(CH_2)_7CH{=}CH(CH_2)_7COOH$	It makes up 55-80% of olive oil, though there may be less than 2.0% as actual free acid in the virgin olive oil, while higher concentrations make the olive oil inedible.
Linoleic acid	$CH_3(CH_2)_4(CH{=}CHCH_2)_2(CH_2)_6COOH$	It is found in corn oil, cottonseed oil, soybean oil, safflower oil, poppy seed oil, hemp oil, canola oil, walnut oil, grain-fed cow milk, olive oil, palm oil, sunflower oil, egg yolks, etc.
Linolenic acid	$CH_3CH_2(CH{=}CHCH_2)_3(CH_2)_6COOH$	α-Linolenic acid – an ω-3 fatty acid found in many vegetable oils.
Arachidonic acid	$CH_3(CH_2)_4(CH{=}CHCH_2)_4(CH_2)_2COOH$	It is the counterpart to the saturated arachidic acid found in peanut oil.

The naturally occurring fatty acids include almost all of the saturated acids having an even number of carbon atoms in the range from 4 to 30 carbons. Other prominent fatty acids are the 18- and 20- carbon unsaturated fatty acids, various hydroxy acids, and a few cyclic fatty acids.

Saturated Fatty Acids

Saturated fatty acid is a fatty acid, such as stearic acid, whose carbon chain contains no unsaturated linkages between carbon atoms and hence cannot incorporate any more hydrogen atoms. The general formula for these acids is $CH_3\,(CH_2)_n^-\,COOH$. Among those occurring most commonly are the following.

Butyric acid Cu, lauric acid C_{12}, myristic acid C_{14}, Palmitic acid C_{16}, Stearic acid C_{18}. The saturated aliphatic acids occur in two series, *i.e.*, the even-numbered and the odd-numbered. Even carbon acids from C_2 to C_{26} occur in both plant and animal fats, while acids as large as C_{34} have been found in waxes. The most common member is palmitic acid. This acid is found in practically every fat stearic acid occurs to some 10-30 per cent in animal depot fats.

Unsaturated Fatty Acids

Unsaturated fatty acids have one or more alkenyl functional groups exist along the chain. The two next carbon atoms in the chain that are bound to either side of the double bond can occur in a cis or trans configuration. A cis configuration means that adjacent carbon atoms are on the same side of the double bond. The rigidity of the double bond freezes its conformation and, in the case of the cis isomer, causes the chain to bend and restricts the conformational freedom of the fatty acid. The more double bonds the chain has in the cis configuration, the less flexibility it has. When a chain has many cis bonds, it becomes quite curved in its most accessible conformations. For example, oleic acid, with one double bond, has a "kink" in it, while linoleic acid, with two double bonds, has a more pronounced bend.

Alpha-linolenic acid, with three double bonds, favours a hooked shape. The effect of this is that in restricted environments, such as when fatty acids are part of a phospholipid in a lipid bilayer, or triglycerides in lipid droplets, cis bonds limit the ability of fatty acids to be closely packed and therefore could affect the melting temperature of the membrane or of the fat. A trans configuration, by contrast, means that the next two carbon atoms are bound to opposite sides of the double bond. As a result, they don't cause the chain to bend much, and their shape is similar to straight saturated fatty acids.

In most naturally occurring unsaturated fatty acids, each double bond has 3n carbon atoms after it, for some n, and all are cis bonds. Most fatty acids in the trans configuration (trans fats) are not found in nature and are the result of human processing (eg, hydrogenation). The differences in geometry between the various types of unsaturated fatty acids, as well as between saturated and unsaturated fatty acids, play an important role in biological processes, and in the construction of biological structures (such as cell membranes).

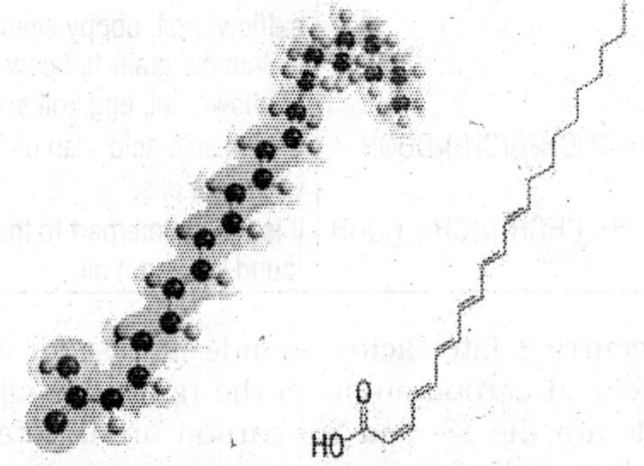

FIG. 8.1: Unsaturated Fatty Acids ($C_{22}H_{32}O_2$).

The following are the most important of the unsaturated fatty acids:

- Oleic acid (C_{18}, one double bonds)
- Linoleic acid (C_{18}, two double bonds)
- Linolenic acid (C_{18}, three double bonds)
- Arachidonic acid (C_{20}, four double bonds)

```
      O   H   H   H   H   H   H   H   H   H   H   H
      ‖   |   |   |   |   |   |   |   |   |   |   |
H—O—C—C—C—C—C—C—C—C—C—C—C—C—H
          |   |   |   |   |   |   |   |   |   |   |
          H   H   H   H   H   H   H   H   H   H   H
```

Lauric acid.

```
      O   H   H   H   H   H   H   H   H   H   H   H   H   H   H   H
      ‖   |   |   |   |   |   |   |   |   |   |   |   |   |   |   |
H—O—C—C—C—C—C—C—C—C—C—C—C—C—C—C—C—C—H
          |   |   |   |   |   |   |   |   |   |   |   |   |   |   |
          H   H   H   H   H   H   H   H   H   H   H   H   H   H   H
```

Palmitic acid.

```
    O H H H H H H H H H H H H H H H H H
    ‖ | | | | | | | | | | | | | | | | |
H-O-C-C-C-C-C-C-C-C-C-C-C-C-C-C-C-C-C-H
      | | | | | | | | | | | | | | | | |
      H H H H H H H H H H H H H H H H H
```

Stearic acid.

```
    O H H H H H H H     H H H H H H H H
    ‖ | | | | | | |     | | | | | | | |
H-O-C-C-C-C-C-C-C-C-C=C-C-C-C-C-C-C-C-C-H
      | | | | | | | | | | | | | | | | |
      H H H H H H H H H H H H H H H H H
```

Oleic acid.

```
   O H H H H H H H H   H H   H H H H H
   ‖ | | | | | | | |   | |   | | | | |
HO-C-C-C-C-C-C-C-C-C=C-C-C=C-C-C-C-C-C-H axis
     | | | | | | |   | |   | | | | | |
     H H H H H H H   H H   H H H H H H
```

Linoleic acid (trans isomer).

```
   O H H H H H H H H H H H H H H H H H
   ‖ | | | | | | | | | | | | | | | | |
HO-C-C-C-C-C-C-C-C-C=C-C-C=C-C-C-C-C-C-H axis
     | | | | | | |     |     | | | | |
     H H H H H H H     H     H H H H H
```

Linoleic acid (cis isomer).

```
  H H H H H H H H H H H H H H H H H O
  | | | | | | | | | | | | | | | | | ‖
H-C-C-C=C-C-C=C-C-C=C-C-C-C-C-C-C-C-C-O-H
  | |     |     |     | | | | | | |
  H H ↑   H     H     H H H H H H H
```

Linolenic acid.

This group of acids is characterized by having one or more double bonds or ethylenic groups. These acids are: monoethenoic, dienoic, trienoic and polyenoic.

In these acids geometric isomerism has been observed. Freedom of rotation around the C = C bond is prevented. The *cis* or *trans* geometrical isomer has been designated. 9-octadecenoic acid is an example of this type of isomerism. The cis isomer is oleic acid, while the *trans* isomer is elegiac acid.

A number of positional isomers can exist when ever a double bond is present. For the mono-ene-octadecenod acid, 16 positional isomers are possible. Fifteen of these can also exist as geometric isomers (*cis-trans*).

Branched - Chain Acids

Branched-chain fatty acids are common constituents of the lipids of bacteria and animals, although they are rarely found in the integral lipids of higher plants. Normally, the fatty acyl chain is saturated and the branch is a methyl-group. However, unsaturated branched-chain fatty acids are found in marine animals, and branches other than methyl may be present in microbial lipids. The most common branched chain fatty acids are mono-methyl-branched, but di- and poly-methyl-branched fatty acids are also known. Their main function in membranes may be to increase the fluidity of lipids as an alternative to double bonds, which are more liable to oxidation. The following discussion is not intended to be comprehensive. Branched chain acids found in nature are: Isobutyric, C_3H_7COOH; Isovaleric, C_4H_9COOH; Tuberculostearic, $C_{18}H_{37}COOH$; Phthioic, $C_{25}H_{51}$ COOH. These are more often found in waxes than in fats.

Neo fatty acids have a terminal tertiary butyl group or with two iso-methyl groups. They are found in certain microorganisms, algae, plants and marine invertebrates. For example, 13,13-dimethyltetradecanoic acid or neopalmitic acid is found as minor component of bark and resins from conifers and other plants. It is also found in the shell, chitin and chitosan of a species of crab.

HOOC

13,13-dimethyl-tetradecanoic acid

***Iso*-Methyl branched fatty acids:** They have the branch point on the penultimate carbon (one from the end).

***Anteiso*-methyl-branched fatty acids:** They have the branch point on the *ante*-penultimate carbon atom.

HOOC CH_2 CH CH_3 CH_3 HOOC CH_3 CH CH_2 CH_3

iso- *anteiso-*

Fatty acids with structures of this type and with 10 to more than 30 carbons in the acyl chain are found in nature, but those most often encountered have 14 to 18 carbons. They are common constituents of bacteria but are rarely found in other microorganisms. These fatty acids are produced biosynthetically via the conventional mechanisms for the synthesis of saturated fatty acids in bacteria. In animal tissues, the biosynthesis of these fatty acids *de novo* is normally a very minor process. Lanolin, the waxy material produced as a protective coating for the fleece of sheep, contains a high proportion of *iso* and *anteiso* fatty acids from C_{10} to C_{34} in chain-length.

Saturated Mid-Chain Methyl-Branched Fatty Acids

10-*R*-Methyloctadecanoic acid or tuberculostearic acid is a major component of the lipids of the tubercle bacillus and related bacterial species. Indeed its presence in bacterial cultures and sputum from patients is used in the diagnosis of tuberculosis. It is also found in Corynebacterium and many other species.

HOOC

Tuberculostearic acid

A number of fatty acids with a single methyl branch of this type have been isolated from specific bacteria. For example, 10-methylhexadecanoic and 11-methyloctadecanoic acids are relatively common. 12-Methylhexadecanoic acid and 14-methyloctadecanoic acid are major components of the halotolerant bacterium *Rubrobacter radiotolerans*.

Biosynthesis

The branched chain fatty acids are biosynthesised by methylation of oleic acid esterified as a component of a phospholipid, with

$CH_3(CH_2)_7CH = CH(CH_2)_7COOR$

↓ S-adenosylmethionine

$CH_3(CH_2)_7\overset{CH_2}{\overset{\|}{C}} - CH_2(CH_2)_7COOR$

↓ NADPH

$CH_3(CH_2)_7\overset{CH_3}{\overset{|}{C}H} - CH_2(CH_2)_7COOR$

S-adenosylmethionine as the methyl donor. The resulting 10-methyleneoctadecanoyl residue is reduced to the 10-methyl compound (in *Corynebacterium* bacterium) with NADPH as the cofactor. A related mechanism is in used for biosynthesis of cyclopropane fatty acids in bacteria.

A wide range of mono-methyl branched fatty acids occur in a few disparate mammalian tissues while non-isoprenoid dimethyl-branched fatty acids are frequently reported from halophilic *Bacillus* sp. The multi-branched fatty acids with the methyl branches in positions 2-, 4-, 6- and 8- are found in certain Mycobacteria. Dimethyl fatty acids are occasionally reported from sponges. Dimethyl, dibasic acids such as 14, 15-dimethyltriacontanedioic acid are sometimes found in plant waxes and in bacteria.

Isoprenoid Fatty Acids

A number of isoprenoid fatty acids occur naturally in animal tissues that are derived from the metabolism of phytol (3,7,11,15-tetramethylhexadec-*trans*-2-en-1-ol), the aliphatic alcohol moiety of chlorophyll. These range from 2,6-dimethylheptanoic to 5,9,13,17-tetramethyloctadecanoic acids.

Normally, these fatty acids occur at low levels only in tissues, with the concentrations being highest in herbivores. For example, phytanic acid is found at levels of up to 1% normally in milk fat and adipose tissue from cows. Isoprenoid branched-chain fatty acids, such as phytanic and pristanic acids, appear to be signalling molecules that function by regulating the expression of those genes that affect the catabolism of lipids in animal tissues.

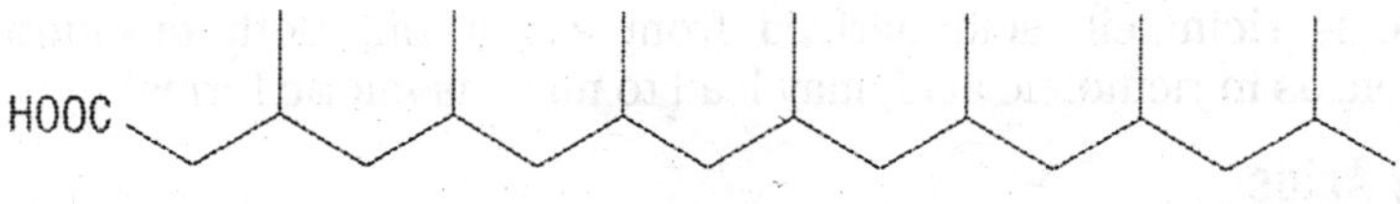

HOOC

3,7,11,15- tetramethylhexadecanoic (phytanic) acid

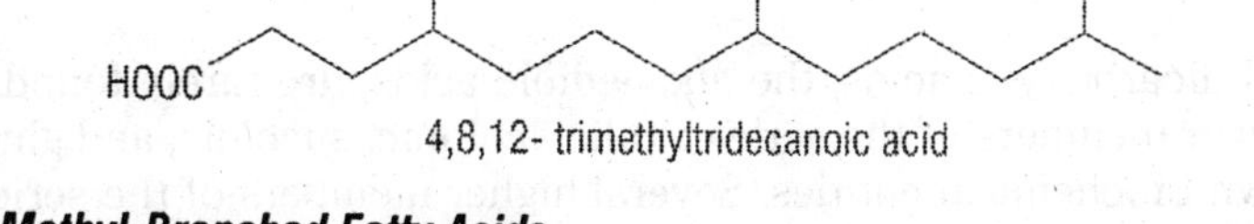

HOOC

2,6,10,14- tetramethyylpentadecnoic (phytanic) acid

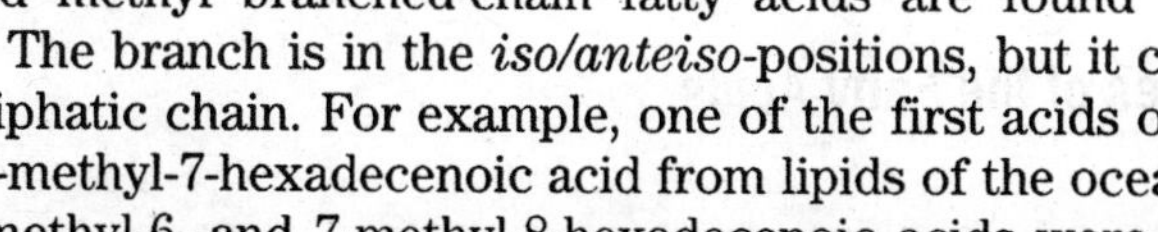

HOOC

4,8,12- trimethyltridecanoic acid

Unsaturated Methyl-Branched Fatty Acids

Monounsaturated methyl branched-chain fatty acids are found in bacteria and marine animals. The branch is in the *iso/anteiso*-positions, but it can also be more central in the aliphatic chain. For example, one of the first acids of this type to be described was 7-methyl-7-hexadecenoic acid from lipids of the ocean sunfish (*Mola mola*), while 7-methyl-6- and 7-methyl-8-hexadecenoic acids were later found in a sponges. Branched-chain fatty acids are uncommon in plants, but small amounts of 16-methyl-*cis*-9-octadecenoic and 16-methyl-*cis*-9, *cis*-12-octadecadienoic acids have been found in wood and seeds of certain Gymnosperm species.

Mycolic Acids: The mycolic acids, major components of the Mycobacteria and related species, are ß-hydroxy-α-alkyl branched structures of high molecular weight (60 to 88 carbons or more). Depending on species, these can contain a variety of functional groups, including double bonds of both the *cis-* and *trans*-configuration (but when the latter they also possess an adjacent methyl branch) and cyclopropane rings, which can also be of the *cis-* and *trans*-configuration. In addition, they can contain, methoxy-, epoxy- and keto groups, which are also adjacent to a methyl branch normally. Some representative structures are illustrated below.

alpnh'-myclic acid

$$CH_3(CH_2)_{17}CH{=}CH(CH_2)_{17}\underset{\displaystyle OH}{CH}{-}\overset{\displaystyle CH_2(CH_2)_{20}CH_3}{CHCOOH}$$

Mycobacteric acids: Mycobacteria also contain multi-methyl-branched fatty acids (C_{14} to C_{32}), often with the methyl branches in positions 2, 4, 6 and 8, and sometimes with hydroxyl groups in position 3, together with long-chain (C_{36} to C_{47}) fatty acids, termed mycobacteric acids, which are structurally related to mycolic acids and contain cyclopropyl rings, double bonds (*trans* and *cis*) and/or oxygenated functions.

$$CH_3{-}(CH_2)_{15}{-}\underset{\displaystyle CH_3}{CH}{-}CH{=}CH{-}(CH_2)_{12}{-}CH{=}CH{-}\underset{\displaystyle CH_3}{CH_2}{-}CH_2{-}COOH$$

Hydroxy and Keto Derivatives

An example is ricinoleic acid derived from castor oil. Both unsaturation and hydroxylation, as in ricinoleic acid, may lead to many isomeric forms.

Cyclic Fatty Acids

Fatty acids containing cyclic structures are rare. These have been identified in the lipids of lactobacilli and in certain seed oils. In the lactobacilli the ring structure is a cyclopropene system. In the seed oils of the tropical family Flacourtiaceae dextrorotatory fatty acids with a cyclopentenyl ring at the end of the fatty acid chain have been identified.

The best examples are chaulmoogric and hydrocarpic acids, which occur in chaulmoogra oil (used in the treatment of leprosy).

Other Acids

The saturated dicarboxylic acids, the alkanedioic acids, are rarely found in natural lipids. The lower members of the series-oxalic, malonic, succinic, and glutaric acids are well known biochemical entities. Several higher members of the series, such as azelaic acid (nonanedioic), $(CH_2)_7$ $(COOH)_2$, are important degradation products of the natural acids.

Physical Properties of the Fatty Acids

Melting Point

The melting points of the acids with even numbers of carbon atoms increase regularly with increasing chain length from- 7.9° C for C_4 to 90.0° C for C_{28}. The odd and even acid series each has its own line of melting points. The introduction of a double bond causes lowering of the melting points. The cis isomer has lower melting point than the *trans*.

Polymorphism

The higher fatty acids exist in two or three crystalline forms. It depends on the temperature and the solvent of crystallization. These forms are interconvertable.

Solubility

Acetic and butyric acid are very soluble in water. Acids higher than C_6 are only slightly soluble in water. Solubility decreases with increasing chain length. The fatty acids are soluble in a variety of organic solvents such as C_6H_6, CH_3OH, C_2H_5OH, $CHCl_3$.

Boiling Point

The boiling points of the saturated fatty acids increase regularly with increasing chain length.

Absorption Spectra

The carboxyl group in the free fatty acids absorbs light in the short ultraviolet, non conjugated double bonds do not show appreciable absorption in the ultra violet, conjugate bonds show more absorption at longer wave lengths in the ultraviolet.

Salts

The fatty acids dissociate in solution and then salts can be prepared. The alkali metal salts (Na and K) of the higher fatty acids are the emulsifying and cleansing agents of commercial soaps. The salts of calcium and magnesium are insoluble in water. They precipitate from soluble soap solutions in hard water containing Ca^{++} and Mg^{++}. The calcium soaps of the higher acids are used in greases. The aluminium salts form stable gets and are used in industry.

Detergents and Wetting Agents

Several derivatives of the fatty acids have important uses as detergents and for wetting agents. The cleansing action of soap is related to the presence of both water-soluble and oil-soluble groups in the same molecule. Catalytic reduction of the carboxyl group produces alkyl alcohols which can be sulphated or sulphonated. These derivatives are efficient detergents and wetting agents. The aerosols contain the sulphonate group and are used mainly as wetting agents.

Esters

Esters of the fatty acids are generally made by direct esterification with the chosen alcohol in the presence of concentrated H_2SO_4 or anhydrous HCl. The preparation of esters of polyhydric alcohols is more involved, but monoglycerides, diglycerides, and triglycerides may be synthesised.

Distribution in Nature

Aquatic fats, characterized by the wide range of acids present, by the low content of saturated acids, and by the high unsaturation of the C_{20} and C_{22} acids.

- **Higher land animals:** Depot fats contain mainly palmitic acid (25-30 per cent) and oleic acid (40-50 per cent), frequently with high stearic acid content.
- **Milk fats:** These are characterized by the presence of low saturated acids (C_4-C_{10}), palmitic and oleic acids are always the major components.
- **Fruit coat fats:** They contain palmitic, oleic, and sometimes linoleic acid in varying proportions independently of their source, there by differing from seed fats.
- **Seed fats:** They contain palmitic and C_{18} monoene, diene and triene acids.

CLASSIFICATION OF LIPIDS

There is no single internationally accepted system of classification for the lipids. It is more common to classify the lipids with regard to the structure of the molecules. Since most of the lipids are esters, it is most convenient to do this on the basis of the alcohol portion of the molecule.

Several classifications have been proposed for classification of lipids: Lipids have been classified as follows:

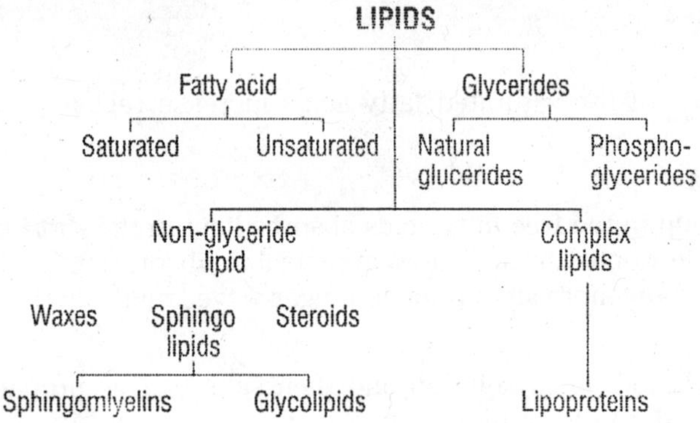

They are also classified as:

Simple Lipids	**Triglycerides, neutral fats:** Found in adipose tissue, butterfat, lard, suet, fish oils, olive oil, corn oil, etc. Esters of three molecules of fatty acids plus one molecule of glycerol; the fatty acid may all be different. **Waxes:** beeswax, head oil of sperm whale, cerumen, carnauba oil, and lanolin. Composed of esters of fatty acids with alcohol other than glycerol; of industrial and medicinal importance.
Compound Lipids	**Phospholipids (phosphatides):** Found chiefly in animal tissues. Substituted fats, consisting of phosphatidic acid; composed of glycerol, fatty acids, and phosphoric acid bound in ester linkage to a nitrogenous base. **Lecithin:** Found in brain, egg yolk, and organ meats. Phosphatidyl choline or serine; phosphatide linked to choline; a lipotropic agent; important in fat metabolism and transport; used as emulsigying agent in the food industry. **Cephalin:** Occurs predominantly in nervous tissue. Phosphatidyl ethanolamine; phosphatide linage to serine or ethanolamine; plays a role in blood clotting. **Plasmalogen:** Found in brain, heart, and muscle. Phosphatidal ethanolamine or choline; phosphatide containing an aliphatic aldehyde. **Lipositol:** Found in brain, heart, kidneys, and plant tissues together with phytic acid. Phosphatidyl inositol; phosphatide linked to inositol; rapid synthesis and degradation in brain; evidence for role in cell transport processes. **Sphingomyelin:** Found in nervous tissue, brain, and red blood cells. Sphingosine-containing phosphatide; yields fatty acids, choline, sphingosine, phosphoric acid, and no glycerol; source of phosphoric acid in body tissue. **Glycolipids:** ***Cerebroside:*** myline sheaths of nerves, brain, and other tissues. Yields on hydrolysis of fatty acids, sphingosine, galactose (or glucose), but not fatty acids; includes kerasin and phrenosin. ***Ganglioside:*** brain, nerve tissue, and other selected tissues, notably spleen; contains a ceramide linked to hexose (glucose or galactose), neuraminic acid, sphingosine, and fatty acids.

	Sulpholipid: white matter of brain, liver, and testicle; also plant chloroplast. Sulphur-containing glycolipid; sulphate present in ester linkage to galactose. ***Proteolipids:*** brain and nerve tissue. Complexes of protein and lipids having solubility properties of lipids.
Terpenoids and Steroids	**Terpenes:** Found in essential oils, resin acids, rubber, plant pigments such as caotenese and lycopenes, Vitamin A, and camphor. Large group of compounds made up of repeating isoprene units; Vitamin A of nutritional interest; fat soluble Vitamin E and K, which are also related chemically to terpenes. **Sterols:** ***Cholesterol:*** found in egg yolk, dairy products, and animal tissues. A consituent of bile acids and a precursor of Vitamin D. ***Ergosterol:*** found in plant tissues, yeast, and fungi. Converted to Vitamin D2 on irradiation. ***7-dehydrocholesterol:*** found in animal tissues and underneath skin. Converted to D3 on irradiation. **Androgens and estrogens:** (Sex hormones) found in ovaries and testes. Adrenal corticosteroids: adrenal cortex, blood.
Derived lipids	**Fatty acids:** occur in plant and animal foods; also exhibit in complex forms with other substances. Obtained from hydrolysis of fats; usually contains an even number of carbon atoms and are straight chain derivatives. Classification of fatty acids is based on the length of the carbon chain (short, medium, or long); the number of double bonds (unsaturated, mono-, or polyunsaturated); or essentiality in the diet (essential or non-essential). A current designation is based on the position of the endmost double bond, counting from the methyl (CH_3) carbon, called the omega end. The most important omega fatty acids are: Omega 6 - linolein and arachidonic acids and Omega 3 - linolenic, eicosapentaenoic, and docosahexaenoic acids.

Simple Lipids

Simple lipids (syn. Homolipids) contain only C, H, and O. They can be segregated into structural types, which are fatty acids (FA), waxes, triglycerides (TG), and sterols. A fatty acid is a long-chain monocarboxylic acid, and a wax is the ester of a long-chain alcohol and a fatty acid. A triglyceride is the ester of a glycerol that contains three FA molecules. sterols are a special class of alcohols, containing a fused four-ring structure, or steroid nucleus. Sterols may combine with a fatty acid to form sterol esters. Simple lipids are present in animal and vegetal cells, or chemically made. Several lipids from bacteria, algae, fungi and plants are omitted as they are not found frequently in natural extracts. They are further classified as:

Triglycerides

Triglycerides are formed from single molecule of glycerol, combined with three fatty acids on each of the OH groups, and make up most of fats digested by humans. Ester bonds form between each fatty acid and the glycerol molecule. It is naturally occurring ester of three fatty acids and glycerol that is the chief constituent of fats and oils. They are adipose tissue, butterfat, lard, suet, fish oils, olive oil, corn oil, etc. Esters of three molecules of fatty acids plus one molecule of glycerol; the fatty acid may all be different.

Natural Fats

These are group of oily substances of natural occurrence, as butter, lard, tallow, etc., as distinguished from certain fatlike substance of artificial production, as paraffin. Most natural fats are essentially mixtures of triglycerides of fatty acids. Natural fats are triglycerides, that is, esters of glycerine (1,2,3-propanetriol) and three fatty acids.

Waxes

Wax components

Compounds	General structure
n-alkanes	$H_3C(CH_2)_n CH_3$
Ketones	R_1CPR_2
Secondary alcohols	$R_1CH(OH)R_2$
-Diketones	$R_1COCH_2COR_2$
Monoesters	R_1CPPR_2
Primary alcohols	RCH_2OH
Aldehydes	RCOH
Alkanoic acids	RCOOH
Docarboxylic acids	$HOOC(CH_2)_nCOOH$
ω-Hydtoxy acids	$HOCH_2(CH_2)_nCOOH$

Waxes is a general term used to refer to the mixture of long-chain apolar lipids forming a protective coating (cutin in the cuticle) on plant leaves and fruits but also in animals (wax of honeybee or beeswax, cuticular lipids of insects, spermaceti of the sperm whale, skin lipids, uropygial glands of birds, depot fat of planktonic crustacea), algae, fungi and bacteria. Some waxes are of mineral origin. Montan wax originates from mob or lignite, that fossilised compound representing a late step of the transformation of vegetal into hydrocarbons. Following are the wax components:

Triacontanylpalmitate is the main component of bee wax. Palmitic acid (C16:0) is esterified by a C30 chain, triacontanol (or melissyl alcohol).

$$\underbrace{CH_3(CH_2)_{14}-\overset{\overset{\displaystyle O}{\|}}{C}}_{\text{Palmitic acid}}-O-\underbrace{CH_2-(CH_2)_{28}-CH_3}_{\text{Triacontanol}}$$

In plants the outer covering consists of a hydroxy fatty acid polymer called cutin. These acids may be saturated, unsaturated, mono- or di-hydroxylated. The underground parts and healed wound surfaces of plants are covered with an analogous substance, suberin. These substances are frequently mixed with other lipids and form a complex mixture called epicuticular wax.

In bacteria, *Mycobactorium* spp., characteristic waxes (**phthiocerol waxes**) were described in the wall cell lipids. These waxes are diesters of the phthiocerols, including phthiodiolone and phthiotriol, with mycocerosic acids. Thus, they are named **dimycocerosate esters**.

Carnauba wax is obtained from the fronds of a species of palm tree in Brazil, is a particularly hard wax used for high gloss finishes, such as in automobile wax, boat wax, floor wax, and shoe polish.

Compound Lipids

Compound lipids also called heterolipids. These lipids contain N and P atoms in addition to the usual C, H, and O. These lipids can be hydrolyzed under alkali conditions to generate smaller constituents.

Phospholipids (Phosphatides)

These are lipids containing a phosphate group. They are mainly found in animal tissues. These are substituted fats, consisting of phosphatidic acid; composed of glycerol, fatty acids, and phosphoric acid bound in ester linkage to a nitrogenous base. A phospholipid molecule has a strongly nonpolar and hydrophobic (water insoluble) tail region represented by fatty acid chains and a strongly polar or hydrophilic (water soluble) head region represented by the phosphate group. Due to this property, a phospho-lipid placed in water forms a lipid bilayer in a characteristic manner. Such an arrangement forms a basic component of plasma membranes.

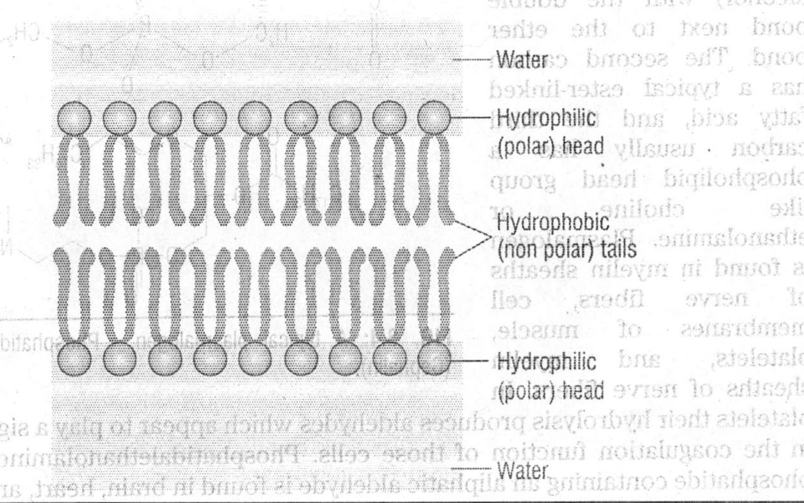

FIG. 8.2: A Lipid Bilayer in an aqueous medium.

Lecithin

Lecithin is yellow-brownish fatty substances. In this phosphatidyl choline or serine; phosphatide linked to choline; a lipotropic agent; important in fat metabolism and transport composed of phosphoric acid, choline, fatty acids, glycerol, glycolipids, triglycerides, and phospholipids (e.g., phosphatidylcholine, phosphatidylethanolamine, and phosphatidylinositol). Lecithin is sometimes used as a synonym for pure phosphatidylcholine, a phospholipid that is the major component of its phosphatide fraction. It may be isolated either from animal and plant tissues, egg yolk or from soy beans, from which it is extracted chemically (using hexane) or mechanically. It is used as emulsifying agent in the food industry.

FIG. 8.3: A phosphatidylcholine, a type of phospholipid in lecithin.

Cephalin

It occurs predominantly in nervous tissue. Cephalin is phosphatidylethanolamine like lecithin. It consists of a combination of glycerol esterified with two fatty acids and phosphoric acid. Whereas the phosphate group is combined with choline in Lecithin, it is combined with the ethanolamine in Cephalin. The two fatty acids may be the same, or different, and are usually in the 1,2 positions (though can be in the 1,3 positions). Phosphatidylethanolamine; phosphatide linage to serine or ethanolamine; plays a role in blood clotting.

Plasmalogen

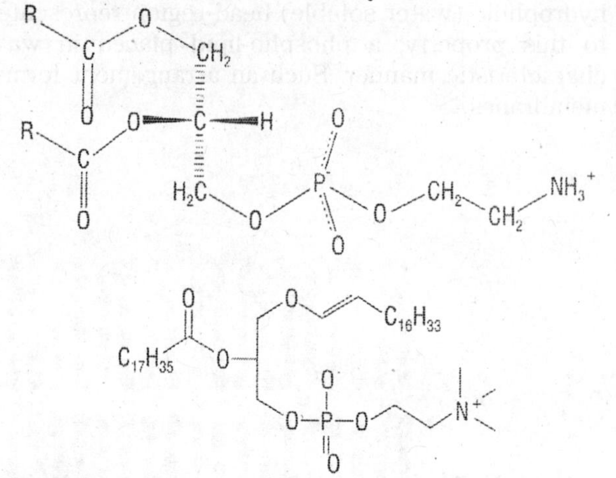

Plasmalogen is an ether lipid where the first position of glycerol binds a vinyl residue (from a vinyl alcohol) with the double bond next to the ether bond. The second carbon has a typical ester-linked fatty acid, and the third carbon usually has a phospholipid head group like choline or ethanolamine. Plasmalogen is found in myelin sheaths of nerve fibers, cell membranes of muscle, platelets, and myelin sheaths of nerve fibers. In platelets their hydrolysis produces aldehydes which appear to play a significant part in the coagulation function of those cells. Phosphatidalethanolamine or choline; phosphatide containing an aliphatic aldehyde is found in brain, heart, and muscle.

FIG. 8.4: A typical plasmalogen. Phosphatidylethanolamine (Cephalin).

Lipositol

Found in brain, heart, kidneys, and plant tissues together with phytic acid. Phosphatidyl inositol; phosphatide linked to inositol; rapid synthesis and degradation in brain; evidence for role in cell transport processes. A member of the vitamin B complex necessary for growth of yeast and of mice, absence from the diet causes hair loss and dermatitis in mice.

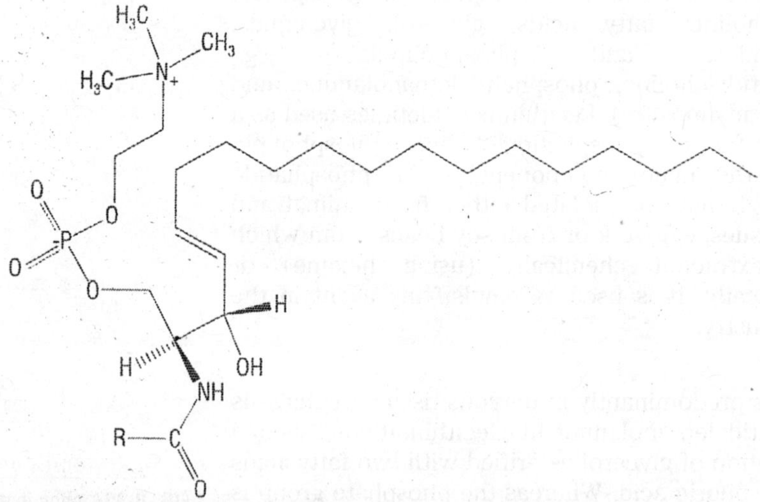

Sphingomyelin.

Sphingomyelin

In humans, sphingomyelin is believed to be the only cell membrane (the plasma membrane of cells is highly enriched in sphingomyelin and is considered largely to be found in the exoplasmic leaflet of the cell membrane) phospholipid not derived from glycerol. Sphingosine-containing phosphatide yields fatty acids, choline, sphingosine, phosphoric acid, and not glycerol; source of phosphoric acid in body tissue. Found in nervous tissue, brain, and red blood cells.

Glycolipids

The term *glycolipid* designates any compound containing one or more monosaccharide residues bound by a glycosidic linkage to a hydrophobic moiety such as an acylglycerol, a sphingoid, a ceramide (*N*-acylsphingoid) or a prenyl phosphate. Glycolipids are glycosyl derivatives of lipids such as acylglycerols, ceramides and prenols. They are collectively part of a larger family of substances known as glycoconjugates.

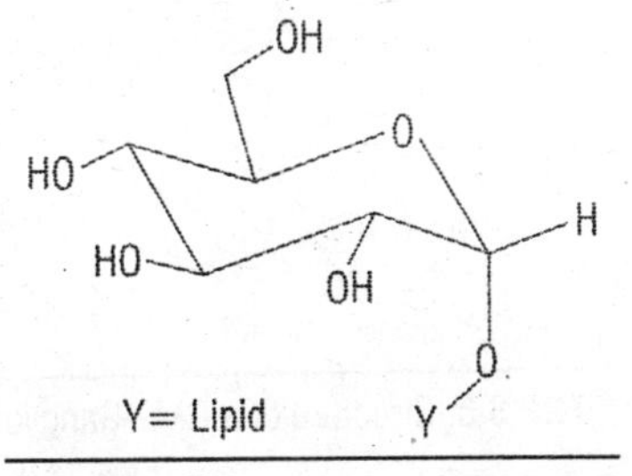

OH HO HO O H OH O Y= Lipid Y

Glycolipids.

Cerebroside: They are found in myline sheaths of nerves, brain, and other tissues. They are yield on hydrolysis of fatty acids, sphingosine, galactose (or glucose), but not fatty acids; includes kerasin and phrenosin. Cerebrosides are glycosphingolipids which are important components in animal muscle and nerve cell membranes. Myelin is the most well known cerebroside. They consist of a ceramide with a single sugar residue at the 1-hydroxyl moiety. The sugar residue can be either glucose or galactose; the two major types are therefore called glucocerebrosides and galactocerebrosides. Galactocerebrosides are typically found in neural tissue, while glucocerebrosides are found in other tissues. A defect in the degradation of glucocerebrosides is Gaucher's disease. The corresponding defect for galactocerebrosides is Krabbe disease.

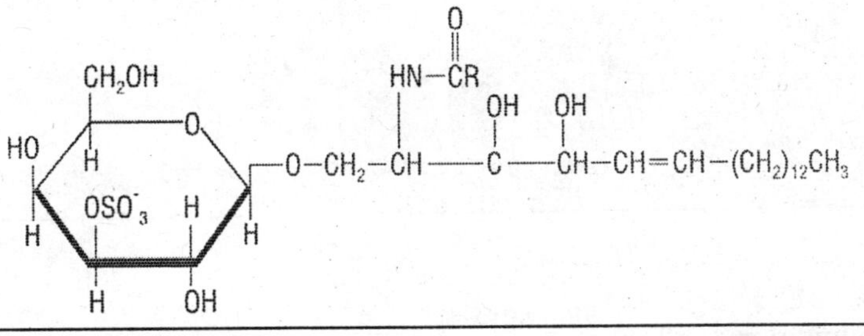

CH_2OH HO O H OSO_3^- H H H H OH — O—CH_2—CH(HN—C(=O)R)—C(OH)—CH(OH)—CH=CH—$(CH_2)_{12}CH_3$

Cerebroside.

Ganglioside: Gangliosides are the group of glycosphingolipids that show the greatest structural variation and also the more complex structure. Glycosphingolipids consist of a ceramide backbone with three or more sugars esterified, one of these being a **sialic acid** such as ***N*-acetylneuraminic acid.** These latter compounds are referred to as *acidic glycosphingolipids*, and they have a net negative charge at neutral pH.

Except some echinoderms, gangliosides are characteristic of vertebrate nervous tissues. This group includes molecules composed of ceramide linked by a glycosidic bond to an oligosaccharide chain containing hexose **and N-**acetylneuraminic acid (NANA, acidic sugar known also as sialic acid) units. It is found in brain, nerve tissue, and other selected tissues, notably spleen; contains a ceramide linked to hexose (glucose or galactose), neuraminic acid, sphingosine, and fatty acids. Gangliosides are present in nerve endings and appear to be important in nerve impulse transmission.

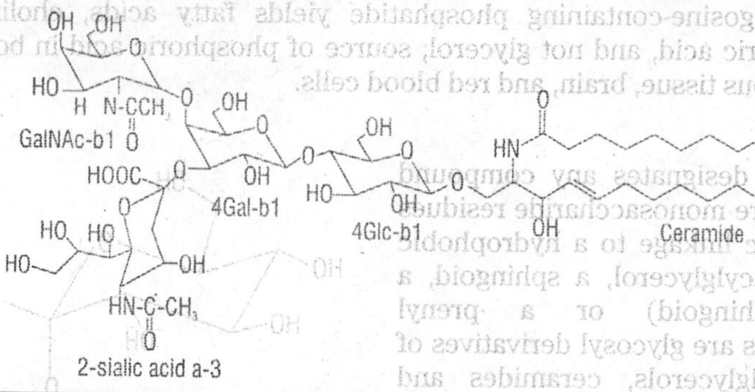

FIG. 8.5: Structure of a GM2 Ganglioside.

Sulpholipids

In sulpholipids, glycolipid and sulphate are present in ester linkage to galactose. It has molecular formula $C_{43}H_{78}O_{12}S$. Sulpholipid-1 (SL-1) is an abundant sulphated glycolipid and potential virulence factor found in *Mycobacterium tuberculosis*. SL-1 consists of a trehalose-2-sulpaate (T2S) disaccharide elaborated with four lipids. It is white matter of brain, liver, and testicle; also plant chloroplast. It is also found in *Lyngbya lagerheimii* and *Phormidium tenue*. It has anti-HIV activity.

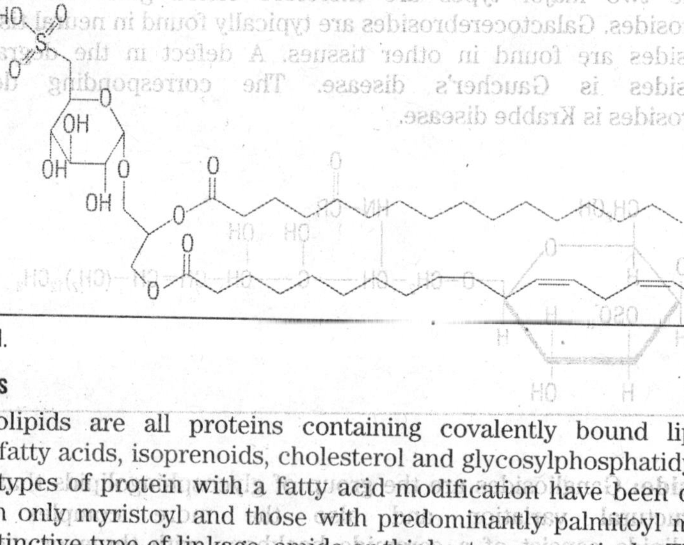

Sulpholipid.

Proteolipids

Proteolipids are all proteins containing covalently bound lipid moieties, including fatty acids, isoprenoids, cholesterol and glycosylphosphatidylinositol. Yet only two types of protein with a fatty acid modification have been described, i.e. those with only myristoyl and those with predominantly palmitoyl moieties, each with a distinctive type of linkage, amide or thiol ester, respectively. The prenylated lipids contain an isoprenoid group linked via a sulphur atom (thiol ether bond) to the protein. The term 'lipoprotein' is also used to describe such compounds on occasion, but to avoid confusion this might be better reserved for the non-covalently linked lipid-protein complexes of the type found in plasma. Lipoproteins are complex aggregates of lipids and proteins that render the lipids compatible with the aqueous environment of body fluids and enable their transport throughout the body of all vertebrates and insects. They are found in brain and nerve tissue. Such complexes of protein and lipids have solubility properties of lipids.

Glycolipids

Glycolipids are lipids that are attached to a short carbohydrate chain (mono- or oligosaccharides). The role of glycolipids is to serve as markers for cell to cell communication and to provide energy. Glycolipids are collectively part of a larger family of substances known as glycoconjugates. Their role is to provide energy and also serve as markers for cellular recognition. They occur where a carbohydrate chain is associated with phospholipids on the exoplasmic surface of the cell membrane. The carbohydrates are found on the outer surface of all eukaryotic cell membranes.

Glycolipids are present ubiquitously in cell surface membranes. The most common glycolipids in mammals are a class of sphingolipids, glycosphingolipids (GSLs). Major GSLs including gangliosides are derived from glucosylceramide (GlcCer) that is synthesised by a specific glycosyltransferase, ceramide glucosyltransferase (GlcT-1, UGCG). Galactosylceramide (GalCer) formed by ceramide galactosyltransferase (CGT) is also present as a dominant glycolipid in myelin sheath. The class of glyceroglycolipids is found in particular tissues: galactosylalkylacylglyceride (seminolipid) in the testis, and phosphatidyglucose in human cord blood cells. GSLs in cell membranes are preferentially distributed into lipid domains, so-called **rafts.**

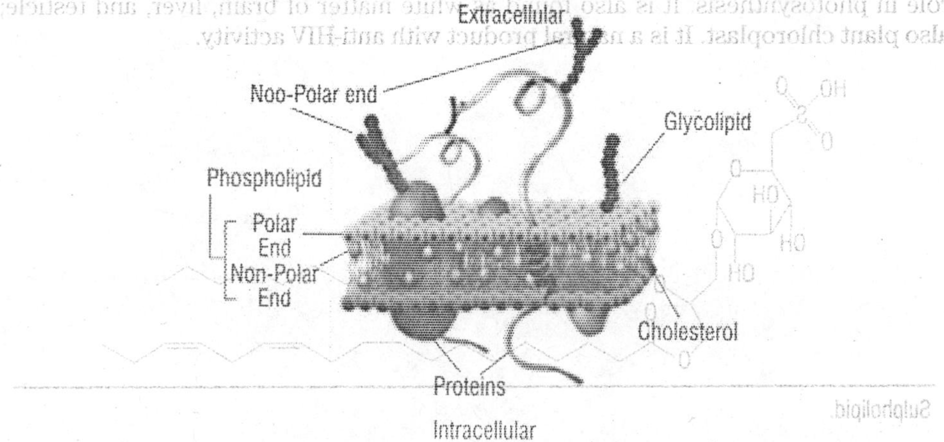

FIG 8.6: Presentation of cell membrane showing position of glycolipid.

Cerebroside

It is a lipid that contains galactose. They consist of a ceramide with a single sugar residue at the 1-hydroxyl moiety. The sugar residue can be either glucose or galactose; the two major types are therefore called glucocerebrosides and galactocerebrosides. They are found in found normally in brain and myline sheaths of nerves, brain, and other tissues.

Ganglioside

Gangliosides are the group of glycosphingolipids that show the greatest structural variation and also the more complex structure. Except some echinoderms, gangliosides are characteristic of vertebrate nervous tissues. They are found in brain, nerve tissue, and other selected tissues, notably spleen; contains a ceramide linked to hexose (glucose or galactose), neuraminic acid, sphingosine, and fatty acids. This group includes molecules composed of ceramide linked by a glycosidic bond to an oligosaccharide chain containing hexose and **N-acetylneuraminic** acid

(NANA, acidic sugar known also as **sialic acid**) units.

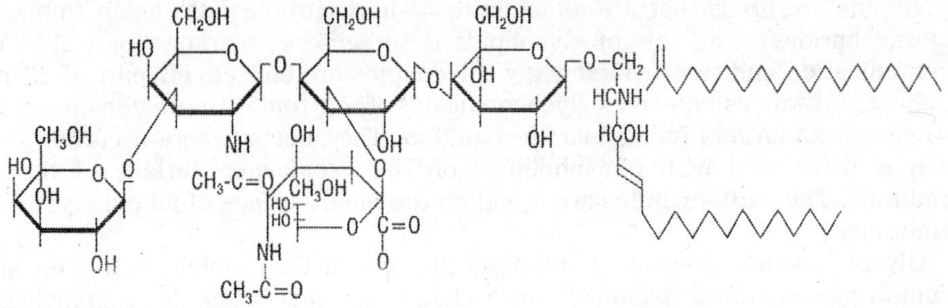

Ganglioside.

Sulpholipid

They are sulphur-containing glycolipid; sulphate present in ester linkage to galactose. It has molecular formula $C_{43}H_{78}O_{12}S$. All photosynthetic organisms, with the exception of several species of photosynthetic bacteria, are thought to contain the sulpholipid 6-sulpho-alpha-D-quinovosyldiacylglycerol. The association of this lipid with photosynthetic membranes has led to the assumption that it plays some role in photosynthesis. It is also found as white matter of brain, liver, and testicle; also plant chloroplast. It is a natural product with anti-HIV activity.

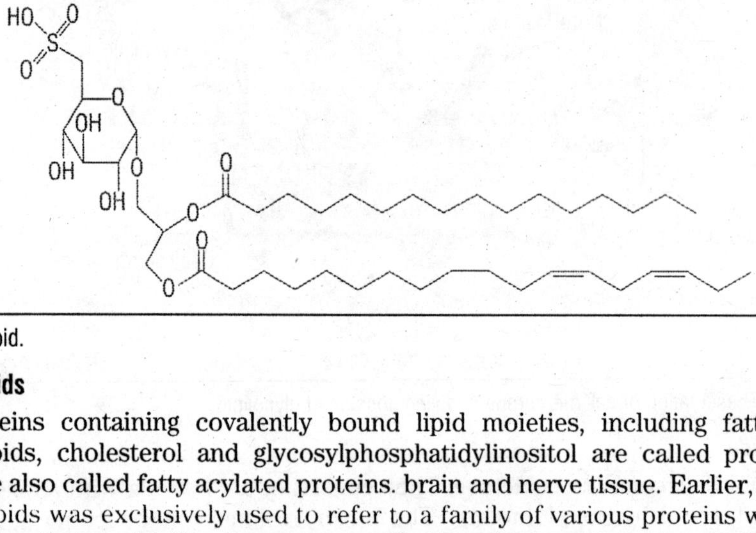

Sulpholipid.

Proteolipids

All proteins containing covalently bound lipid moieties, including fatty acids, isoprenoids, cholesterol and glycosylphosphatidylinositol are called proteolipid. They are also called fatty acylated proteins. brain and nerve tissue. Earlier, the term proteolipids was exclusively used to refer to a family of various proteins which are related by their solubility in mixtures of chloroform and methanol. Thus, the archetypal proteolipid found initially in myelin is now known as proteolipid protein (PLP). Protein-lipid combinations are abundant in brain tissue, but also present in a wide variety of animal and plant tissues. In contrast to lipoproteins, they are insoluble in water, but soluble in a chloroform-methanol mixture. The protein moiety has a high content of hydrophobic amino acids.

They are soluble in chloroform-methanol mixtures and insoluble in water. Folch considered that the proteolipids consist of a protein core covered by lipids combined by non covalent bonds. Proteolipids A, B, C, etc. were isolated and identified on the basis of the amount of protein present. Proteolipids contain protein with small amounts of aspartic and glutamic acids, arginine, lysine, and histidine, and large amounts of neutral amino acids. The lipids were found to be primarily phosphatides with phosphatidyl serine. In addition, di and tri phosphoinositides have been found associated with the protein fraction. The associated lipids consist of a mixture of phosphoglycerides, cerebrosides, and sulphatides, while lipoproteins contain phospholipids, cholesterol, and triglycerides.

TERPENOIDS AND STEROIDS

Terpenoids

The name terpene was given originally to the steam-distillable oils obtained from turpentine (an extract of pine). The terpenoids sometimes called isoprenoids and are a large and diverse class of naturally occurring organic chemicals similar to terpenes. The name terpene was given originally to the steam-distillable oils obtained from turpentine (an extract of pine). Terpenes with more than 10 carbons exist, the number of carbons being usually a multiple of five. The structures are extraordinarily diverse.

They may be assembled, and modified in various manners. The multicyclic structures differ from one another with respect to their functional groups and basic carbon skeletons. Many similar water-insoluble compounds are distributed very widely. These lipids are found in living organisms and represent the largest group of natural products.

Terpenes with 10 C atoms are known as monoterpenes. Examples are limonene and geraniol. Sesquiterpenes and diterpenes have 15 and 20 C atoms, respectively. Examples are Farnesol and Vitamin A. Triterpenes (30 C atoms) and tetraterpenes (40 C atoms) require special attention because they give rise to steroids and cerotenoids (ß-carotene). Squatene is an example of triterpene. Polyisoprenoid compounds exist, *e.g.*, rubber, but in a biochemical context, the obiquinones and dolichols are particularly important.

They are found in essential oils, resin acids, rubber, plant pigments such as caotenese and lycopenes, Vitamin A, and camphor. Large group of compounds made up of repeating isoprene units; Vitamin A of nutritional interest; fat soluble Vitamin E and K, which are also related chemically to terpenes. Plant terpenoids have aromatic properties and used in traditional herbal remedies mainly as antibacterial, antineoplastic and pharmaceutical functions. For example, contribute scent of eucalyptus terpenoids, the flavors of cinnamon, cloves, and ginger, and the color of yellow flowers. The important terpenoids are citral, menthol, camphor and cannabis (*Salvia divinorum*).

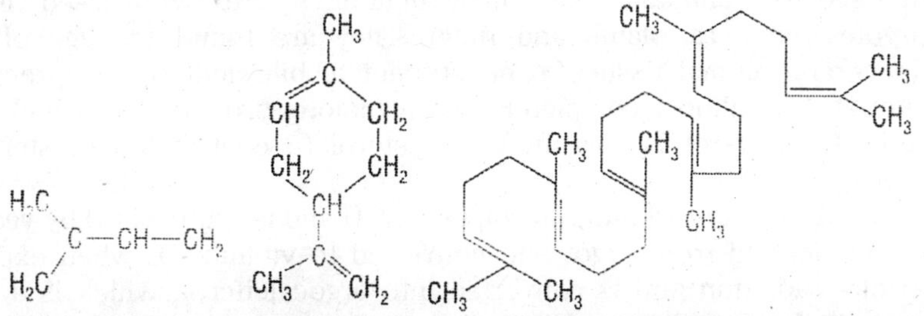

Isoprene Limonene Squatene

Steroids

The steroids and sterols in animals are produce biologically from terpenoid precursors. Sometimes terpenoids are added to proteins, e.g., to enhance their attachment to the cell membrane; this is known as isoprenylation. Sterols are also known as steroid alcohols.

Structurally, steroids are derivatives of the reduced aromatic hydrocarbon perhydro cyclo pentano phenanthrene. Although related structurally to phenanthrene, these compounds are true terpenes that are synthesised in living systems from isoprene via squalene. Sterols are steroids containing one or more hydroxyl groups. Some examples are cholesterol, a component of the cytoplasmic membrane of animal cells, testosterone, a hormone, and cholic acid, a constituent of bile.

The fused-ring system is a planar structure. Bonds shown and indicate the substituents that are, respectively, in front (ß) and behind (δ) the general plane of the sterol. The configuration of substituents off the two carbon atoms shared between two fused rings determines whether the rings are fused in the cis or trans conformation. Trans-fused rings are flat structures, but cis-fused rings are bent structures.

Classification

Sterols of plants are called phytosterols and sterols of animals are called zoosterols. Important zoosterols are cholesterol and some steroid hormones; notable phytosterols include campesterol, sitosterol, and stigmasterol. Ergosterol is a sterol present in the cell membrane of fungi, where it serves a role similar to cholesterol in animal cells.

Chemically they are classified as subgroup of steroids with a hydroxyl group at the 3-position of the A-ring.[3] They are amphipathic lipids synthesised from acetyl-coenzyme A via the HMG-CoA reductase pathway. The overall molecule is quite flat. The hydroxyl group on the A ring is polar. The rest of the aliphatic chain is non-polar. They may be Cholesterol, Phytosterol, Ergosterol and Hopanoids.

Steroids are classified as anabolic (or anabolic-androgenic) and corticosteroids. Corticosteroids, such as cortisone, are drugs that doctors typically prescribe to help control inflammation in the body. They're often used to help control conditions like asthma and lupus. They're not the same as the anabolic steroids that receive so much media attention for their use by some athletes and bodybuilders.

Steroids are also classified as:

1. **Cholesterol:** Cholesterol is a lipidic, waxy alcohol found in the cell membranes and transported in the blood plasma of all animals. It is an essential component of mammalian cell membranes where it is required to establish proper membrane permeability and fluidity. Cholesterol is the principal sterol synthesised by animals, but small quantities are synthesised in other eukaryotes, such as plants and fungi. They are found in egg yolk, dairy products, and animal tissues. A constituent of bile acids and a precursor of Vitamin D. It is almost completely absent among prokaryotes, which include bacteria. Cholesterol is classified as a sterol (a contraction of steroid and alcohol).
2. **Ergosterol:** It is a crystalline sterol, $C_{28}H_{43}OH$ and is synthesised by yeast from sugars or derived from ergot and converted to vitamin D_2 when exposed to ultraviolet radiation and is converted into ergocalciferol, which is a form of Vitamin D. Ergosterol is a biological precursor (a provitamin) to Vitamin D_2. Ergosterol is a component of fungal cell membranes, serving the same function that cholesterol serves in animal cells. The presence of ergosterol in fungal cell membranes coupled with its absence in animal cell membranes makes it a useful target for antifungal drugs. Ergosterol is also present in the cell membranes of some protists, such as trypanosomes. This is the basis for the

use of some antifungals against West African sleeping sickness. It is responsible for maintaining cell integrity and function.

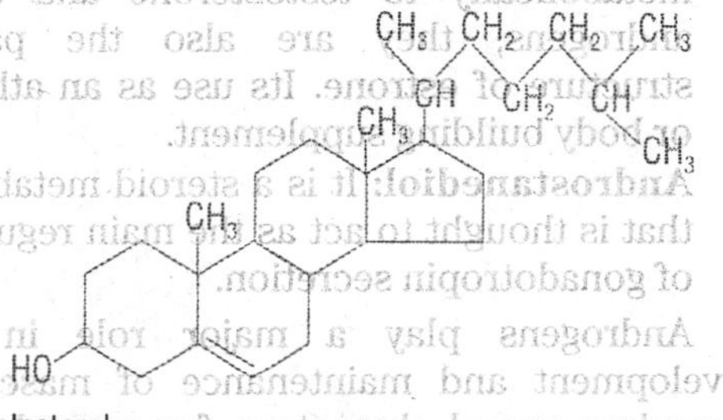

3. **7-dehydrocholesterol:** It is a zoosterol that functions in the serum as a cholesterol precursor, and is converted to vitamin D_3 in the skin, therefore functioning as provitamin-D_3. The presence of this compound in human skin enables humans to manufacture vitamin D_3 from ultra-violet rays in the sun light, via an intermediate isomer pre-vitamin D_3. It is also found in the milk of several mammalian species. They are found in animal tissues and underneath skin. They are converted to D_3 on irradiation.

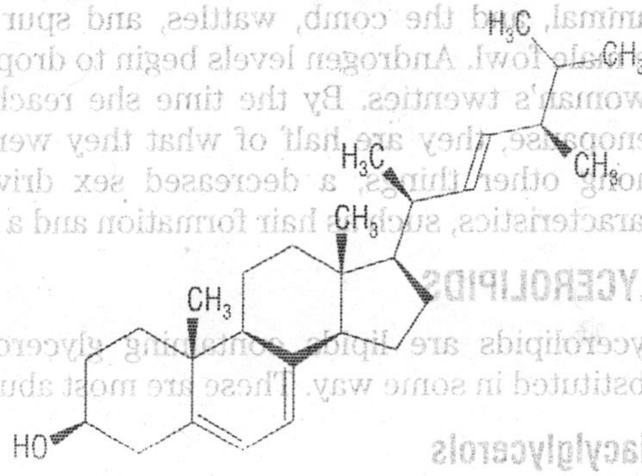

Cholesterol.

Ergosterol.

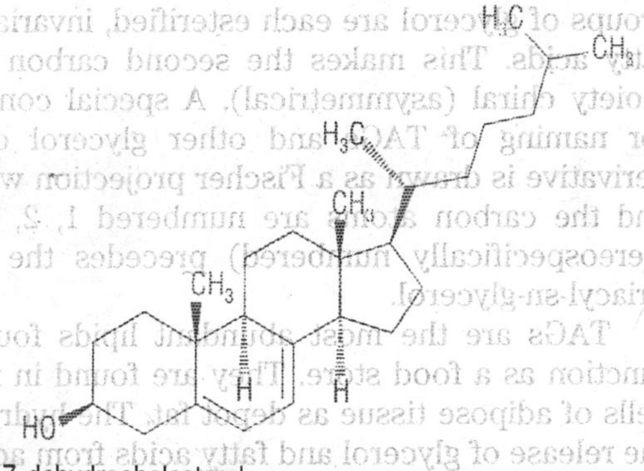

7-dehydrocholesterol.

Androgens and Estrogens

- **Androgen:** Androgen is the generic term for any natural or synthetic compound. It is a naturally occurring sex hormone and is produced by the ovaries, adrenal glands and other tissues. One of the most important members in this class of steroidal hormones is testosterone. Testosterone has 19-carbons and in general such a steroidal skeleton is called androgenic skeleton. Another member is androstenedione affectionately called andro. Andro and testosterone are interconvertable in our body.

  A subset of androgens, adrenal androgens, includes any of the 19-carbon steroids synthesised by the adrenal cortex, an adrenal gland, that function as weak steroids or steroid precursors, including dehydroepiandrosterone (DHEA), dehydroepiandrosterone sulphate (DHEA-S), and androstenedione. Besides testosterone, other androgens include:

- **Dehydroepiandrosterone (DHEA):** It is a steroid hormone produced from cholesterol in the adrenal cortex, which is the primary precursor of natural estrogens. DHEA is also called dehydroisoandrosterone or dehydroandrosterone.
- **Androstenedione (Andro):** It is an androgenic steroid, which is produced by the testes, adrenal cortex, and ovaries. While androstenediones are converted

metabolically to testosterone and other androgens, they are also the parent structure of estrone. Its use as an athletic or body building supplement.

- **Androstanediol:** It is a steroid metabolite that is thought to act as the main regulator of gonadotropin secretion.

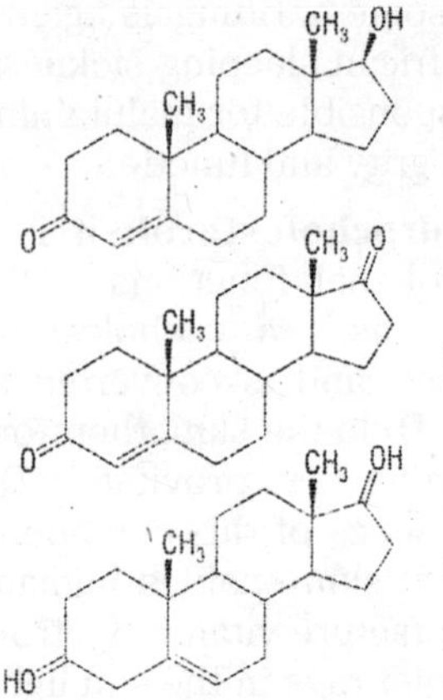

Testosterone

Androstenedione

Androst-5-ene-3β, 17β-diol

Androgens play a major role in the development and maintenance of masculine secondary sexual characters, for example, the seminal vesicle and prostate gland of the male mammal, and the comb, wattles, and spur of the male fowl. Androgen levels begin to drop in a woman's twenties. By the time she reaches menopause, they are half of what they were in her early 20s. This can result in, among other things, a decreased sex drive. Androgen also can promote male characteristics, such as hair formation and a deepening voice.

GLYCEROLIPIDS

Glycerolipids are lipids containing glycerol in which the hydroxyl groups are substituted in some way. These are most abundant lipids in animals.

Triacylglycerols

Triacylglycerols (TAGs) are neutral glycerolipids and are also known as triglycerides. In the TAGs the three hydroxyl groups of glycerol are each esterified, invariably by different fatty acids. This makes the second carbon of the glycerol moiety chiral (asymmetrical). A special convention is used for naming of TAGs and other glycerol derivatives. The derivative is drawn as a Fischer projection with the secondary hydroxyl to the left, and the carbon atoms are numbered 1, 2, and 3 from the top. The prefix (for stereospecifically numbered) precedes the name glycerol. For example, 1,2,3-Triacyl-sn-glycerol.

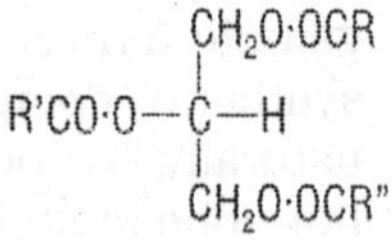

$$\begin{array}{c} CH_2O{\cdot}OCR \\ | \\ R'CO{\cdot}O-C-H \\ | \\ CH_2O{\cdot}OCR'' \end{array}$$

1,2,3.-Triacyl-sn-glycerol

TAGs are the most abundant lipids found in animals. This is because they function as a food store. They are found in most cells, particularly present in the cells of adipose tissue as depot fat. The hydrolysis of the ester bonds of TAGs and the release of glycerol and fatty acids from adipose tissue is called fat mobilization. Depot fat is a water-free mixture of TAGs differing from each other in the nature of the three fatty acyl groups. Depot fat has a relatively high content of unsaturated fatty acids. The fat exists in a liquid state. Solid fat would present only a small surface area for enzymes in the cytoplasmic water that mobilize it. Also, solid fat would render the adipose tissue rigid and unyielding during mechanical stress.

The melting point of depot fat is only a few degrees below body temperature. The fatty acid composition is such that keeps the fat liquid and the ability to store as much energy as possible. Unsaturated fatty acids yield less energy than saturated ones when oxidized.

All glycerides are soluble in ether, chloroform and benzene. They are only slightly soluble in cold methyl and ethyl alcohol and acetone.

The specific gravity of fats is generally lower than that of water. The melting point of a fat depends upon its component glycerides. Glycerides of the higher saturated fatty acids have the highest melting points, and those of unsaturated acids have the lowest. The presence of a large proportion of unsaturated acids or of short-chain acids in the glycerides of a fat lowers the melting point and causes the fat to be liquid (an oil) at room temperature.

Pure glycerides have no colour, odour, or taste. In a mixed glyceride, the molecules are asymmetric and show optical rotation. Natural triglycerides do not possess optical rotation power.

Saponification

The glyceroides of the fats may be readily decomposed into glycerol and salts of the constituent fatty acids (soaps) by boiling with strong bases (sodium or potassium hydroxide) hydrogenation. The unsaturated glycerides of fats may be hydrogenated by treatment with hydrogen in the presence of nicked catalyst.

The hydrogenation of oils such as cotton seed to prepare commercial cooking fats and margarines is an important food industry. Certain iodized oils, such as iodized poppy seed oil, are used in medicine for injection into body cavities to be X-rayed. The iodine stops the ray and causes the cavities to be outlined in the photographs.

Rancidity

The unpleasant odour and taste developed by most natural fats upon aging is referred to as rancidity. Rancidity may be due to hydrolysis of component glycerides of a fat into free fatty acids and glycerol or monoglycerides and diglycerides. This is often hastened by the presence of lipolytic enzymes in the fats.

Phosphoglycerides

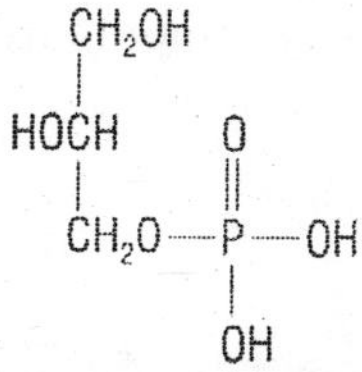

Phosphoglycerides are polar glycerolipids and are often called as phospholipids. Any lipid containing phosphorus is included in the general class of phospholipids. Glycerol phospholipids are sometimes referred to as phosphatides and are the most common. The different types of phospholipids have certain biological functions in common phospholipids are all good emulsifying agents. Large concentrations are usually found in cell membranes and in subcellular structures where lipids and water soluble materials must be capable of intimate interaction. The phospholipids may promote the transportation, absorption, and metabolic interactions of the other lipid substances.

$$\begin{array}{l} CH_2OH \\ | \\ HOCH \quad\quad \overset{O}{\|} \\ | \\ CH_2O-P-OH \\ \quad\quad\quad\;\; | \\ \quad\quad\quad\; OH \end{array}$$

sn-Clycerol-3-phosphoric acid

Glycerol phospholipids have the same essential structures, varying only in the nature of one of the groups linked to the phosphate position of the molecule. Most of the compounds are of phosphodiester nature.

All phosphoglycerides are derived from Sn-glycerol-3-phosphoric acid in which the phospheric acid moiety is esterified with certain alcohols and the hydroxyl groups on C-1 and C-2 are esterified with fatty acids.

Sn-glycerol-3-phosphoric acid: In most phosphoglycerides, the fatty acid substituted on C-1 is saturated and that on C-2 is unsaturated.

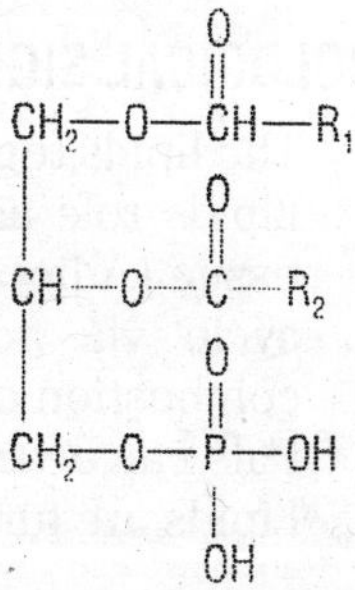

- **Phosphatidic Acids:** These are not found in tissue extracts in any significant concentration. However, they have important roles as intermediates in the biosynthesis of the other glycerol lipids.
- **Phosphatidyl cholines (Lecithines):** The choline derivatives of phosphatidic acid are commonly known as the lecithins. Lecithins contain the acidic OH of phosphoric acid and the very basic OH^- of choline, so they exist as internal

$$\begin{array}{l} CH_2-O-\overset{O}{\overset{\|}{C}}H-R_1 \\ | \\ CH-O-\overset{O}{\overset{\|}{C}}-R_2 \\ | \\ CH_2-O-\overset{O}{\overset{\|}{P}}-OH \\ \quad\quad\quad\quad | \\ \quad\quad\quad\quad OH \end{array}$$

salts or zwitterious. Variations in the fatty acid components give rise to different lecithins. Various combination of saturated and unsaturated acids occurring in leathins.

The lecithins are waxy, white substances but soon become brown when exposed to air and light. They are soluble in ordinary fat solvents. They are hygroscopic and mix well with water to form cloudy colloidal solutions. They do not have definite melting points but decompose when heated.

Lecthin molecules are highly polar structure. These associate with proteins and other substances. They are active and valuable agents for the emulsification of fats and oils.

Lecithin is the most common form of phospholipid at least in animals. The lecithins are required for normal transport and utilization of other lipids, especially in the liver. Any interruption of the normal transportation of lecithins to and from the liver, usually results in the accumulation of lecithin in the liver, giving rise to the condition called fatty liver.

1. **Phosphatidyl ethanolamines (cephalins):** The cephalins are structurally similar with the lecithins, except that the base ethanolamine, instead of choline, is present. The cephalins also form zwitterious. Cephalins are differentiated by the presence of different fatty acid groups in the molecules.
2. **Other glycerophosphatides phosphatidyl serines and phosphatidyl inositols:** There are many other phosphatides found in nature and may have important functions. These contain a nitrogenous substituent on the phosphate group. The chief of these are serine and thrionine. The alcohol group of each of the N compounds is united in the ester link.

```
            CH2—O—CO—R
            |
R—CO—O—CH       O
            |       ||
            CH2—O—P—O—CH2—CH2—NH3+
                    |
                    O-
```

In addition to these nitrogenous substances, carbohydrate-like substituents can occur. Inositol can exist in some nine stereochemical forms, two of which are optical enantiomorphs. The chief form present in the phosphatides is the optically inactive myoinositol. This form has the chain-shaped ring, and five of the hydroxyls are equatorial, one is axial. In animals, the serine and inositol phospholids are found in highest concentration in nerve tissues.

3. **Plasmalogens:** The plasmalogens make up an appreciable amount of the total phosphatides in brain and muscle.

These phospholipids yield on hydrolysis one mole each of long-chain aliphatic aldehyde, fatty acid, glycerol phosphate, and a nitrogenous base. The aldehydogenic group replaces one of the fatty acids in a typical phospholipid structure. One of the common plasmalogens is ethanolamine plasmalogen.

BIOLOGICAL SIGNIFICANCE OF LIPIDS

1. The lipids together with the carbohydrates and proteins are important as food.
2. Lipids role as the chief storage form of energy. These can be stored in fatty tissue (Adipose tissue). Lipids are energy rich molecules. They enter the Krebs cycle via acetyl CoA and involved in the synthesis of ATP. Complete combustion of 1S of fat yields 9.0 K Cal of energy.
3. Fats can be transformed into carbohydrates and vice versa.
4. Lipids are fundamental structural building blocks of protoplasm. These are also

starting materials in the synthesis of more complex protoplasmic components, for example phospholipid constitute the important components of cell membranes.

5. Subcutaneous fat serves as insulation of heat.
6. Lipids are also electrical insulation *e.g.* in myclin sheath of nerve fibres.
7. Fatty tissues protect vital organs from injuries.
8. Many hormones are lipids *e.g.* steroids.
9. Some vitamins are lipids fat soluble vitamins such as vitamins. A, D, E and K.
10. Waxes are as protective coatings of the skins and furs of animals and the leaves and fruits of plants. They also protect animals and plants against bacteria and insects.
11. Lipids are commercially important in the form of soaps, detergents, greases, and the various oils of the paint industry.

CHAPTER 9

Organic Component Nucleic Acids: DNA

Meischer (1868) isolated a substance from the nucleus of pus cells. He separated this substance from pus cells by digesting them for a week with dilute hydrochloric acid and then shaking the mixture with ether in a separatory funnel. A heavy solid layer separated at the bottom of the funnel and consisted of pure nuclear material. Meischer then obtained this material by digesting pus cells with artificial gastric juice. He considered the product the characteristic constituent of the cell nucleus and called it *nuclein.* The nuclein possessed much stronger acidic properties than proteins. It was soluble in dilute alkalies, but insoluble in dilute acids, water, and organic solvents. It contained a relatively large amount of phosphorus.

Meischer also studied the nuclear material of ripe salmon sperm. He found this material to consist of a salt of the acidic nuclein and a basic substance, called protamin. Nuclein was shown to be a mixture of a basic protein and a phosphorus-containing organic acid of high molecular weight.

Altmann (1889) introduced the term **nucleic acid**. He developed general methods for the isolation of nucleic acids from many sources. He established the existence of two main types of nucleic acids, namely ribonucleic acid (RNA) and deoxyribonucleic acid (DNA).

The discovery of the building block units of the nucleic acids was followed by some generalizations about structure, such as the trinucleotide hypothesis of Stendel and of Jones and the tetranucleotide hypothesis of Levene. The modern era of structural determination was initiated by analytical research of Chargaff and his co-workers by the development of methods of synthesis of nucleosides, nucleotides, and polynucleotides, first by Todd and later by H.G. Khorana, and finally by X- ray and physicochemical studies of the tertiary and quaternary structure of nucleic acids and their successful interpretation by Wilkins, Watson, Crick, and others.

Nucleic acids are macromolecules of the at most biological importance. All living organism contain nucleic acids in the form of deoxyribonucleic acid (DNA) and ribonucleic acid (RNA). Some viruses may contain only RNA, while others have only DNA.

In higher cells DNA is localized mainly in the nucleus as part of the chromosomes. A small amount of DNA is present in the cytoplasm and found within mitochondria and chloroplasts. RNA is found both in the nucleus and in the cytoplasm.

According the Watson and Crick, a DNA molecule has double helix, *i.e.*, it is composed of two long and parallel polynucleotide chains or strands helically coiled around the same axis like spiral staircase or ladder.

IMPORTANT FEATURES OF DNA

1. Four different deoxynucleotides, or nucleotides, the structural units of DNA, are assembled into long polymers of DNA strands, or nucleic acids. Prior to assembly, they are in the form of nucleotide triphosphates similar to ATP.
2. **DNA Nucleotides:** Each nucleotide contains three parts: a phosphate group, the sugar deoxyribose, and one of four nitrogen bases. The four bases of DNA, their designations and their triphosphate form are adenine (dATP), guanine (dGTP), thymine (dTTP), and cytosine (dCTP).
3. In 1950, Chargaff developed the principle of base-pairing. He determined the relative amounts of A, T, C, and G in a variety of cells, proving that A = T and C = G and that there is exactly as much purine (adenine and guanine) in the nucleus as there is pyrimidine (thymine and cytosine).
4. Through the use of X-ray crystallography, Wilkins and Franklin determined that DNA was double stranded and could form a helix.
5. In 1953, having used critical information from the work of others (Rosalind Franklin and Linus Pauling) and by constructing models of their own, Watson and Crick determined the double helictical structure of DNA, including its phosphate-sugar backbone, specific (A-T, G-C) base-pairing of purines and pyrimidines, and the meaning of the intramolecular distances.
6. In the double helix, the two polymers run in opposite directions (5'-to-3' and 3'-to-5'). The deoxyribose sugar is the hub of this numbering system -- the 3' carbon contains a hydroxyl group and the 5' carbon has a phosphate. In forming the phosphate sugar backbone of DNA the free 3' -OH group of deoxyribose in the first nucleotide reacts with the first 5' phosphate in the second nucleotide. Many millions of nucleotides may be present in a single DNA molecule.

THE PHYSICAL STRUCTURE OF DNA: THE DOUBLE HELIX

In 1953, James D. Watson and Francis H.C. Crick, published a paper in which they proposed a model for the physical and chemical structure of the DNA molecule. According to their model, most DNA consists of two polnucleotide chains wound around each other in a right-handed (clockwise) helix. In generating their model, Watson and Crick used three main pieces of evidence.

1. The DNA molecule was known to be composed of bases, sugars, and phosphate groups linked together as a polynucleotide (deoxyribonucleotide) chain.
2. Erwin Chargaff hydrolyzed the DNA of a number of organisms by chemical treatment. He had quantified the purines and pyrimidines released. His studies showed that in all the DNAs the amount of the purines was equal to the amount of the pyrimidines. The amount of adenine (A) was equal to that of thymine (T), and the amount of guanine (G) was equal to that of cytosine (C). These equivalencies are known as Chargaff's rules. In comparisons of DNA's from different organisms, the A/T and G/C ratios are always the same, although the (A + T) (G + C) ratio (presented as % GC) varies.

Rosalind Franklin, working with Maurice H.F. Wilkins, studied isolated fibers of DNA by using the X- ray diffraction technique.

The simplest way of looking at the structure of DNA is as two strings of molecular units, almost like beads on a string. This wind around each other in what has been called the double helix. The structure of an individual string of beads or DNA strand is made of repeating units, the **beads**. However, each of these is one of only four types, and each of these types is related and is, in fact, very similar. Each of the simplified units shown is referred to as a **monomer**, and many of these will join together, or polymerize, to make the **polymer**. This is called DNA.

Double-stranded, non denatured DNA in a preparation that has been extracted from a particular known biological source-viral, bacterial, or eukaryotic tissue as distinct from any additional material added as a probe or marker is called **native DNA**. The technique used to determine the precise sequence of basses in a length of DNA is called **DNA sequencing**.

The **sugar** (five-carbon sugar or pentose sugar) is also involved in making DNA structure. Sugar is of two forms, either as a **ring** structure or as a **long**, five-carbon chain. In DNA ring form of sugar is found. One of the important points to see here is that the 2' carbon (pronounced **two prime carbon**) has no OH group attached to it, unlike the 1' and 3' carbons around the ring, and hence this sugar is called **2'-deoxyribose**. In RNA, the 2' carbon does attach to an OH group. In this way DNA differ from RNA.

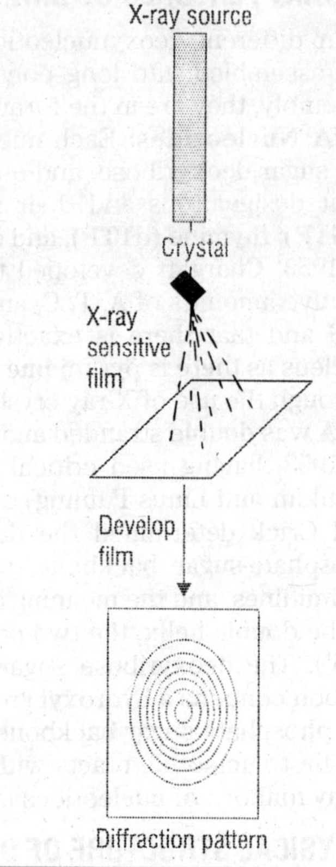

FIG. 9.1: Simplified diagram of X-ray diffraction.

Sugar

A sugar unit is attached to a second unit known as a **base** (a nitrogenous base due to its nitrogen atom content). Together, the sugar and the base are collectively called a **nucleoside**. In DNA four types of base and so four types of monomer units are found. The information, which is stored in the DNA molecule, is found according to base arrangement that is it is the order of these bases in the polymer, which ultimately determines the protein, which is being encoded by a particular stretch of DNA polymer.

Bases

The four bases can be separated into two groups, each group having two members. The simplest bases are the **pyrimidines**, which have a single ring, structure. The two forms found in DNA are *thymine* (T) and *cytosine* (C). The other bases are the **purines**, where there are two rings found in the structure. The forms found in DNA are *adenine* (A) and *guanine* (G).

The Phosphates

To complete the monomer structure, addition of a phosphate unit to the nucleoside ios required. This whole monomer unit is called a **nucleotide.** In a nucleotide three phosphate groups are joined in a long chain. The phosphate groups are labelled as α, β and γ.

Therefore DNA is a polymer of nucleotide units. The four bases, which are incorporated into the nucleotide are:

- 2'-deoxythymidine 5'-triphosphate, using thymine as the base (dTTP)
- 2'-deoxyguanosine 5'-triphosphate, using guanine as the base (dGTP)

- 2'-deoxyadenosine 5'-triphosphate, using adenine as the base (dATP)
- 2'-deoxycytidine 5'-triphosphate, using cytosine as the base (dCTP)

where *d* denotes deoxy sugar while the first letter refer to the base incorporated.

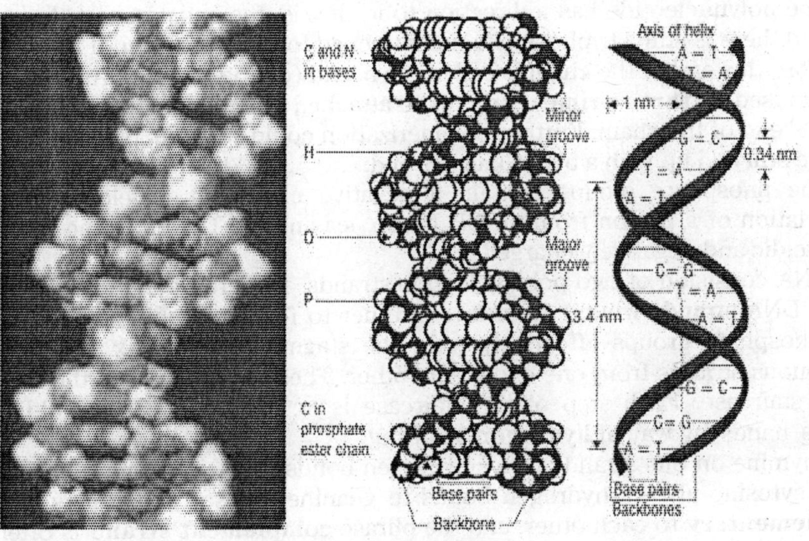

FIG. 9.2: DNA molecule -the basic tool for genetic engineering.

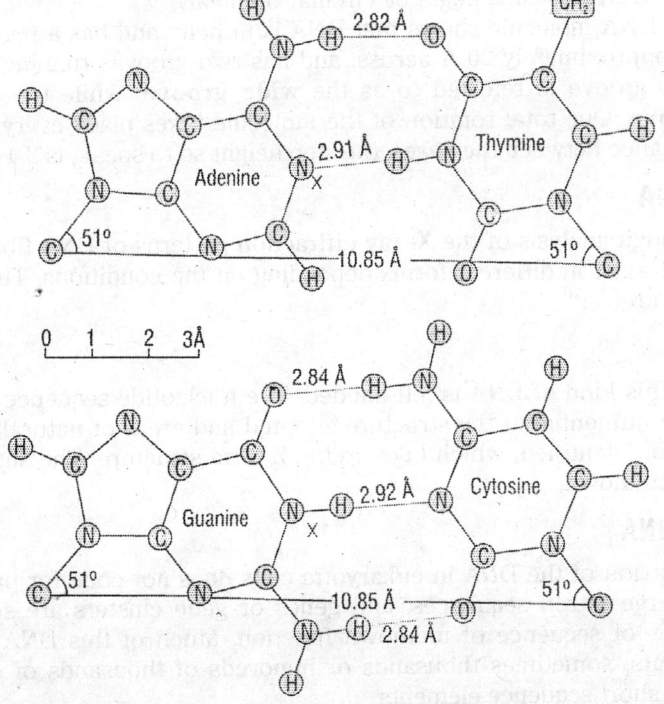

FIG. 9.3: Watson-Crick pairing of bases by hydrogen bonding. The position of the helix axis of DNA is shown by the X.

The above building blocks when polymerize they form a long chain, or **polynucleotide**. Here, the nucleotides are held together by the bonding of the **3' carbon** of the deoxyribose sugar to the first or α phosphate group of the next nucleotide. This bond is called **phosphodiester** bond, and results in β and γ phosphate groups of the bonding phosphate chain.

The polynucleotide has a direction to it, that is, the ends are not the same. At one end there is still a triphosphate unit attached to the **5' carbon** of the sugar, and therefore this end of the chain is called the 5' end (or **five prime**). At the other end is an unused 3' (**three prime**) carbon still attached to an untouched OH group. This is the 3' end of the chain. Further polymerization could still take place but, even if it did, we still end up with a 5' end and a 3' end.

The phosphate groups contain a negative charge. It develops due to the dissociation of a proton from on OH group, leaving O^-. Therefore the molecule is both acidic and negatively charged.

DNA consisted of two polynucleotide strands. They run in opposite directions. These DNA strands wound around each other to form a double helix. The sugars and phosphate groups effectively form the staging of the helix and make two continuous strands from one end to the other. These bases are like the steps on a spiral staircase. Each step of the staircase is actually two bases, holding each other's 'hands' by way of **hydrogen bonding.**

Thymine on one strand always hydrogen bonds to adenine on the other strand while cytosine always hydrogen bonds to guanine. These two strands are called **complementary** to each other, and the phrase **complement strand** is often used. There are **three** hydrogen bonds between guanine and cytosine, while there are only **two** hydrogen bonds between adenine and thymine. **DNA topology** is the configuration of DNA, which might be circular of linear.

Model of DNA molecule shows that DNA is in helix and has a regular pattern. The helix is approximately 20 Å across, and has two grooves running around and along it. One groove is referred to as the **wide groove**, while the other is the **narrow groove**. One total rotation of the molecule takes place every 34 Å along, while the distance between the bases, the step height so to speak, is 3.4 Å.

TYPES OF DNA

A more thorough analysis of the X- ray diffraction patterns of DNA fibers revealed that DNA can exist in different forms depending on the conditions. The important types of DNA are:

Z DNA

The helix of this kind of DNA is left-handed. The nucleotide sequence of the DNA appears to be influential in the structure adopted and areas of naturally occurring DNA have been identified, which take up the Z form structure. The significance of this is as yet unknown.

Non-Coding DNA

A large proportion of the DNA in eukaryotic cells does not code for protein; most genes have large intron sequences, and genes or gene clusters are separated by large stretches of sequence of unknown function. Much of this DNA consists of multiple repeats, sometimes thousands or hundreds of thousands of copies, of a few relatively short sequence elements.

Genomic DNA

DNA of the genome, which in bacteria, is a single, circular loop, and in eukaryotic cells is the entire haploid nuclear complement of DNA. Non-genomic DNA includes, in bacteria, the DNA of plasmids, and in eukaryotes the DNA of viruses or other contaminants. Note that in eukaryotes the mitochondrial genome and chloroplast genome are often considered separately from the nuclear genome.

Unique Sequence DNA

The fraction of DNA which reassociates most slowly is unique sequence, essentially comprising single – copy genes, or those repeated a few times, *Escherichia coli* DNA is essentially all unique sequence.

Dispersed Repetitive DNA

Much of the moderately repetitive DNA comprises a number of DNA sequences of a few hundred base pairs (SINES or short interspersed elements) or between one and five thousand base pairs (LINES, or long interspersed elements), each repeated more than 100000 times and scattered throughout the genome. The most prominent examples in humans are the *Alu* and the *L1* elements, which may be parasitic DNA sequences, replicating themselves by transposition.

Satellite DNA

Satellite DNA, which occurs mostly near the centromeres of chromosomes, and may be involved in attachment of the mitotic spindle, consists of huge numbers of tandem repeats of short (up to 30 bp) sequences. Hypervariability in satellite DNA is the basis of the DNA fingerprinting techniques.

When humidity is high, DNA is in the B form (the double-helical form deduced by Watson and Crick). When the humidity is relatively low, DNA is found in the A form. Both of these DNA forms involve right handed helices. In A-DNA the base pairs are tilted and pulled away from the axis of the double helix. In B-DNA the helix axis passes through the base pairs, which are oriented perpendicular to that axis.

An inherent problem in analysing DNA fibers by X- ray diffraction is that the DNA strands are generally not regularly ordered around and along the fiber axis. Thus the DNA models generated from the resulting data show averaged helical structures. These can not show local structural variations that might occur because of the presence of particular base sequences. This limitation has been overcome by methods to synthesise short DNA molecules (called oligomers) of defined sequences. The pure DNA oligomers can be crystallized and analysed by single-crystal X- ray diffraction. Some sequences produced the B helix while others produced the A helix. Unexpectedly certain sequences produced a third DNA form called Z- DNA, which has a left-handed helix and a zigzag sugar-phosphate backbone. The latter property gave this DNA form its **Z** designation.

FEATURES OF A-DNA AND B-DNA

A-DNA and B-DNA are right-handed double helices. They have 10.9 and 10.0 base pairs per 360- degree turn of the helix, respectively. These values correspond closely with the data derived from DNA fiber studies. The bases are inclined away from the perpendicular from the helix axis by 13 degrees in A-DNA and 2- degrees in B-DNA.

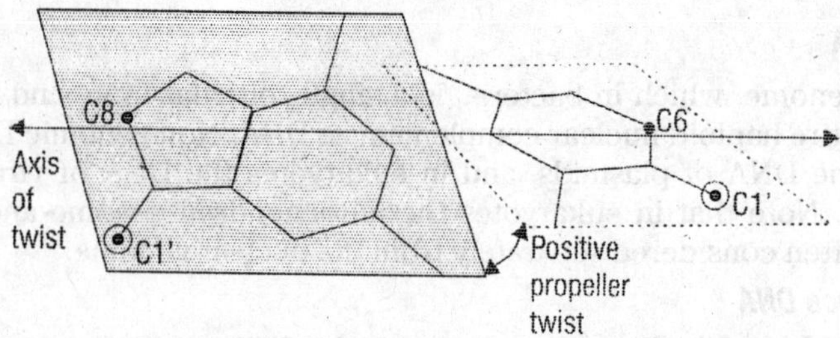

FIG. 9.4: Propeller twist is defined by a schematic drawing of a purine-pyrimidine base pair. A clockwise rotation of the nearer base as one sight down the long axis of the base pair in either direction is positive propeller twist.

The single crystal studies showed a significant deviation from the fiber studies predictions in a property of individual base pairs called propeller twist. Propeller twist is a rotation of the two bases of a base pair in opposite directions about their long axis. Fiber studies did not predict any propeller twist. The resulting models showed the flat purine and pyrimidine rings aligned relative to each other in the same plane. It is now known that propeller twist exists. It serves to stabilize the helix. The mean values for **propeller twist** are 15 degrees for A-DNA and 12 degrees for B-DNA. Individual propeller twist values for specific base pairs in the helix vary considerably to produce the most stable double helix structure. The range is 3 to 25 degrees.

The single crystal studies confirmed the fiber studies conclusion that the A-DNA double helix is short and broad with a deep major groove and a broad and shallow minor groove. The B-DNA double helix is thinner and taller for the same number of base pairs than A-DNA, with a wide major groove and a narrow minor groove; both grooves are of similar depths.

Guanine
Cytosine
B
syn
anti
O
C5'
O2
Z
O
O3'
C5'
O3'

FIG. 9.5: Two conformations of the C-N bond connecting eachbase to its sugar ring are shown. The anti conformation (right) appears in all A- and B-DNA and at the cytosines in Z-DNA. The syn conformation (left) appears at the quanines in Z-DNA. The three points labelled A, B, and Z indicates for each type of helix the location of the helix axis with respect to a base pair. The axis is in the major groove in A-DNA, passes through the base pair in B-DNA, and is in the minor groove in Z-DNA.

FEATURES OF Z-DNA

Z-DNA was discovered when single crystals of an alternating cytosine-guanine DNA oligomer were subjected to X-ray diffraction analysis. DNA oligomers containing A-T base pairs do not form Z-DNA. Z-DNA is a left-handed helix generated because of a different bond conformation between one of its bases and the deoxyribose sugar.

This bond is called a glycosidic bond. The syn conformation is found at the guanines in Z-DNA. The alternation of a purine (G) and a pyrimidine (c) along a Z-DNA double strand produces the anti-syn alternation that gives the backbone of Z-DNA a zigzag appearance.

In Z-DNA there are 12.0 base pairs per complete helical turn. The bases are inclined away from the perpendicular from the helix axis by 8.8 degrees. The base pairs have smaller mean propeller twist values than A- or B-DNA that is 4.4 degrees. The Z-DNA helix is thin and elongated with a deep, minor groove. The major groove is pushed to the surface of the helix so that it is not really evident as a distinct groove. The helix axis passes through the minor groove in Z-DNA, through the major groove in A-DNA, and through the base pairs in B-DNA.

Schematic molecular drawings of Z-DNA and B-DNA contrasting the zigzag backbone of the former with the regular helical backbone of the latter are presented in Fig. 9.5.

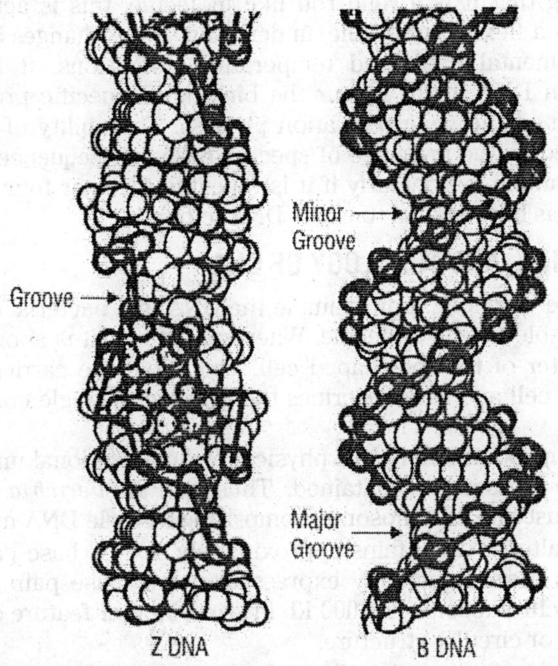

FIG. 9.6: Molecular models of Z-DNA and B-DNA. The irregularity of the Z-DNA backbone is illustrated by the heavy lines that go from phosphate to phosphate residue along the chain. The groove in Z-DNA is quite deep, extending to the axis of the double helix. In contrast, B-DNA has a smooth line connecting the phosphate groups and two grooves.

DNA IN THE CELL

In solution, DNA usually is found in the B form. Hence B-DNA is the form typically found in cells. Since A-DNA is found only under relatively low humidity conditions. It is unlikely that any lengthy sections of A-DNA exist within cells. There is evidence, however, that Z-DNA exists in at least some cells. This evidence has come from the use of antibodies made against Z-DNA molecules. That is, Z-DNA is much more immunogenic (capable of inducing antibodies against it) than B-DNA. Experiments have been done in which anti-Z-DNA antibodies have been added to fixed cells and chromosomes. These antibodies will bind to any region of the chromosomes that contain stretches of Z-DNA. The locations where these antibodies bound could not be seen directly. To do that, antibodies were prepared against the anti-Z-DNA antibodies and fluorescent molecules were attached to these new antibodies. When the new antibodies were added to the fixed cells they bound specifically to the anti-Z-DNA antibodies and the fluorescent tags they carried served to pinpoint the Z-DNA regions. The fluorescence is evidence for the presence of Z-DNA in the chromosomes.

It is possible that the Z-DNA has an important biological role, although at present this is uncertain. It has been postulated that Z-DNA may be involved in gene regulation. Although it has not been proved that any proteins bind specifically to Z-DNA, which presumably would be a facet of gene regulatory events.

BENDS IN DNA

Even though the DNA molecule is drawn in text books as a very straight, helical molecule, implying that it is a rigid rod like molecule, this is actually not so. In solution, B-DNA is a flexible molecule, undergoing shape changes as it responds to the local environmental ionic and temperature conditions. It is possible that particular bends in DNA might favour the binding of specific protein molecules. This could be part of the gene regulation process. The ability of B-DNA to bend slightly is enhanced by the presence of specific base-pair sequences. The sequence CAAAAAT, for example, particularly if it is repeated three or four times separated by 10 base pairs, has been shown to cause DNA to bend.

SIZE, ORGANIZATION, AND TOPOLOGY OF DNA

DNA molecules are very long. For example the DNA in a bacterial cell is contained within a single double-helical molecule. When spread out, it is about 1,000 times as long as the diameter of the rod-shaped cell. This molecule carries all the genetic information of the cell and thus describes the genome (a single complement of the genetic material).

The term chromosome refers to a physical or organizational unit. Within which part of or all the genome is contained. Thus, the *Escherichia coli* genome is contained within just one chromosome, comprising a single DNA molecule. It has a size of 2.5×10^9 daltons. It contains approximately 4×10^6 base pairs. The size of DNA molecules is more commonly expressed in kilo base pairs (Kb, 1000 base pairs). The *E. coli* chromosome is 4,000 Kb in size. Another feature of this particular molecule is closed or circular structure.

The DNA of bacteriophage ϕX 174 is single stranded. In this case the genome size given in Kb. It refers to the number of base pairs in an equivalent duplex form. In progressing from the simple viruses to eukaryotes, the amount of information in the genome increases through many orders of magnitude. Approximately the number of kilobases in the genome of viruses and bacteria can be considered equivalent to the number of genes. This equivalency of base number to genes does not exist in eukaryotes because of the presence of unexpressed sequences of bases. In viruses and bacteria, circular DNAs are common. In eukaryotes, each chromosome contains a single linear molecule of double helical DNA. In eukaryotes, DNA does not exist free. It is complexed with an approximately equal mass of basic proteins called histones. For a long time it was thought that bacterial DNA did not form such complexes. While histones are absent from bacteria. There is evidence for the presence of histone like proteins in them that enable condensation of their DNA into its compact nucleoid form.

There are five types of histones, designated H_1, H_2A, H_2B, H_3, and H_4. They are of low molecular weight (11,000 - 23,000). They contain a large portion of the basic amino acids arginine and lysine. In each case there is an unusual distribution of amino acids along the single polypeptide chain. The basic amino acids are being clusteredin one half, with the other half being relatively hydrophobic. In addition, the histones contain many modified amino acid side chains, *e.g.*, methylated arginines and acetylated lysines.

The nucleoprotein complex that is formed is called chromatin. Chromatin can be isolated as fibres from nuclei. When the fibers are spread and examined under the electron microscope, they appear as beads on a string. Most of the histone is contained in the beads, called nucleosomes. The nucleosome bead consists of a set of eight histones, two each of H_2A, H_2B, H_3 and H_4 around which approximately 200 base pairs of DNA are wrapped. Digestion of chromatin with nucleases yields a core particle still containing the eight histones but only 140 base pairs of DNA. The rest of the DNA, which is digested, functions as a linker between the cores, and histone H_1 is associated with this linker. The structure of the core particles has been examined by X- ray diffraction. To compact the DNA further within the nucleus. The nucleosoms are organized into condensed higher-order structures.

SUPERCOILED DNA

Supercoiling represents a twisting of the DNA double helix upon itself. It occurs in circular DNAs and in DNAs that are topologically constrained by being complexed to proteins. If a circular DNA molecule in a relaxed conformation (no super-coiling) is broken across both strands and one or more additional right handed helical turns are inserted before rejoining the ends, the molecule would twist on itself to form a positive super-coil. If, on the other hand, the helix is unwound (given left-handed turns) before rejoining the result is a negative super-coil. Each super-coil depicted here contains a single super-twist. The number of super-twists in a molecule can be very large. Super-coiled DNA is readily converted into its relaxed form by the introduction of a break (or nick) between adjacent nucleotides in one of the two strands this form is no longer topologically constrained.

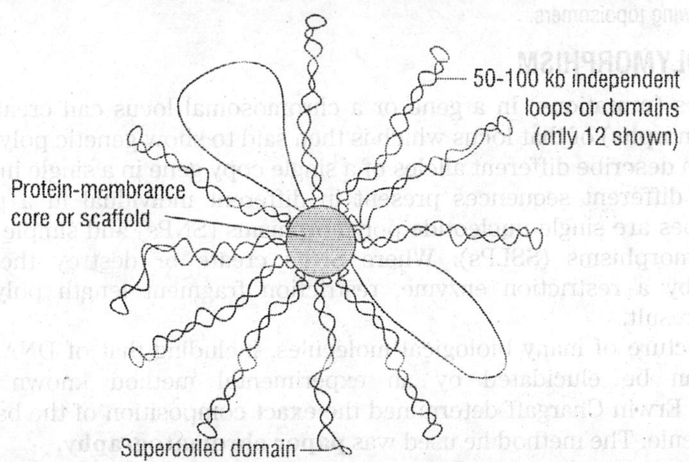

FIG. 9.7: Presentation of *E. coli* chromosome (4600 kb). The thin line present double halix.

When DNA is twisted the structure is known as **super-coiled**. This can only be achieved if the ends of the DNA are prevented from rotating, for example if the DNA is circular s in a plasmid. Enzymes known as **topoisomerases** can introduce or remove turns in the DNA double helix and, if this is done, the DNA twists up into a helical structure, that is the double helix as described is then further twisted into a helix.

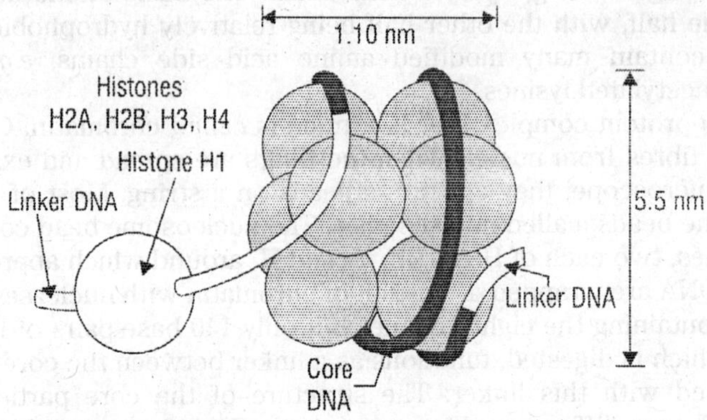

FIG. 9.8: The arrangement of histone and DNA in a nucleosome.

This super-coiling may be **positive super-coiling**, resulting from the addition of turns to the double helix, or alternatively **negative super-coiling** as a result of the removal of turns from the helix.

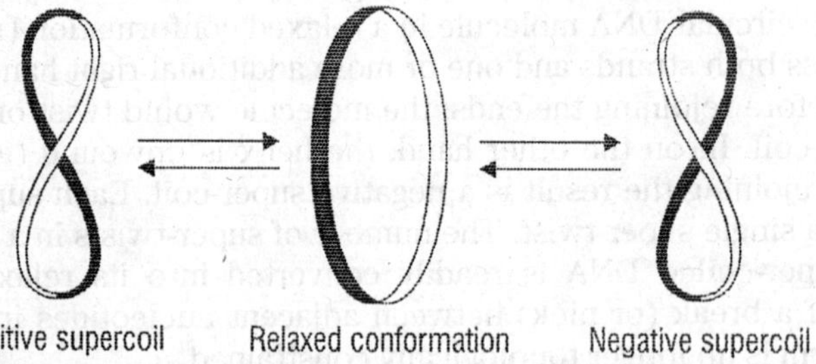

FIG. 9.9: Showing topoisomers.

GENETIC POLYMORPHISM

Base changes (mutations) in a gene or a chromosomal locus can create multiple forms (polymorphs) of that locus which is then said to show genetic polymorphism. The term can describe different alleles of a single copy gene in a single individual as well as the different sequences present in different individual in a population. Common types are single nucleotide polymorphisms (SNPs) and simple sequences length polymorphisms (SSLPs). Where SNPs create or destroy the sequence recognized by a restriction enzyme, restriction fragment length polymorphism (RFLP) will result.

The structure of many biological molecules, including that of DNA and many proteins, can be elucidated by an experimental method known as **X-ray diffraction**. Erwin Chargaff determined the exact composition of the bases within a DNA molecule. The method he used was **paper chromatography.**

DNA AS GENETIC MATERIAL

Transformation Experiment

One of the first studies to suggest that DNA is the genetic material involved the bacterium *Diplococcus pneumoniae* (also called pneumococcus). Two strains of pneumococcus are known. The smooth (S) strain is infectious (virulent) and results in the death of the infected animal. Each bacterial cell of the S type is surrounded by a polysaccharide coat. It gives the strain its infectious properties and results in the smooth, shiny appearance of S colonies. The rough (R) strain is non infectious (a virulent) since R cells lack the polysaccharide coat, colonies of this strain are non-shiny.

The S strains can be classified further into IIS and IIIS. These have clearly defined differences in the chemical composition of the polysaccharide coat. The differences between IIS and IIIS are genetically determined. In about one S cell out of 10^7 there will be a mutational change from S to R. The R colony rarely gives rise to an S cell by a second mutation. The S colony that develops has the same capsule type as the S strain from which the R mutant was originally derived. That is, if the original was IIS, then the reversion of the R strain that derives from it will also be IIS.

Griffith (1928), injected mice with R bacteria, produced by mutation of type II S bacteria. Genetically these bacteria lacked the ability to make the polysaccharide coat, and if a coat could be made at all, had the genes for making a type II polysaccharide coat. The R bacteria did not affect the mice, and after a while the bacteria disappeared from the animals blood streams. Griffith also injected mice with living type III S bacteria. Those animals died, and living type III S bacteria could be isolated from their blood. If the type III S bacteria were heat-killed before injecting them, however, the mice survived. These two experiments showed that the bacteria had to be alive and had to possess the polysaccharide coat for them to be infectious.

Lastly, Griffith infected mice with a mixture of living R bacteria (derived from II S) and heat-killed type III S bacteria. In this case the mice died. Moreover, living S bacteria were present in the blood. These living cells were all of type III S. Therefore, these could not have arisen by mutation of the R bacteria since that would have produced II S colonies. Grifith concluded that some R bacteria had somehow been transformed into smooth, infectious type III S cells by interaction with the dead type III S cells. The transformed III S cells retained their infectious properties and capsule type in successive generations indicating that the transformation was stable. Griffith believed the unknown agent responsible for the change in genetic material was a protein. He referred to the agent as the transforming principle.

Modern molecular perspective data indicate that the genetic material of the virulent bacteria was not destroyed by the heat treatment. It was released from the heat-killed cells and entered the living avirulent cells where recombination occurred to generate a virulent transformant. Subsequent research has verified this genetic hypothesis. That is, the S gene codes for a key enzyme needed for the synthesis of the carbohydrate- containing capsule. The R allele of the gene produces no active enzyme and the capsule is not made.

Avery, Macleod, and McCarty (1944), showed that the transforming principle was not protein but was instead a molecule of DNA. They used a system in which the transformation of bacteria from R to S type was done in the test tube. In their experiments they lysed type III S cells and separated the lysate into the various cellular macromolecular components such as lipids, polysaccharides, proteins and the nucleic acids RNA and DNA. They tested each component whether it was the transforming principle by checking whether or not it could transform living R bacteria derived from II S. The nucleic acids were the only components of the type III S cells that could transform the R cells into III S.

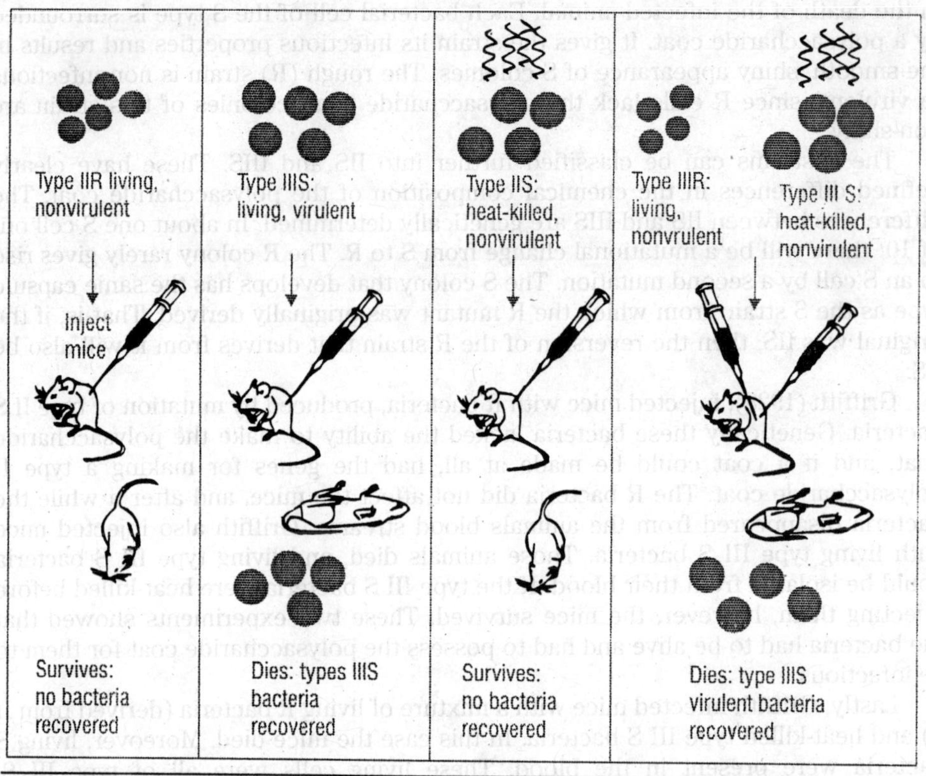

FIG. 9.10: Transformation experiment of Griffith. Mice injected with type III *S pneumococcus* died, whereas mice injected with either type II R or heat-killed type III S bacteria survived. An injected mixture of living type II R and heat-killed type III S bacteria, however, caused the mice to die.

Avery and his colleagues then used specific nucleases (enzymes that degrade nucleic acids) to determine whether DNA or RNA was the transforming principle. They treated the nucleic acids with ribonuclease (RNAase), which degrades RNA, the transforming activity was still present. However, when they used deoxyribonuclease (DNAase), which degrades DNA, no transformation resulted. Thus, according to.the results, DNA was the genetic material.

Support for the hypothesis that DNA was the genetic material came from the results of experiments that showed that DNA was located almost exclusively in the nucleus of eukaryotic cells. A Mirsky and H. Ris (1949) demonstrated by cytochemical analysis that the amount of DNA per diploid set of chromosomes was constant for a given organism. Moreover, the diploid DNA amount was shown to be equal to twice the amount of DNA found in the haploid sperm. Different organisms were shown to have different amounts of DNA. We now know that the DNA content of certain cells of an organism can vary depending on the tissue of origin. In general the DNA content is usually a multiple of the DNA content of a zygote cell. For example, the DNA content of human cells may be from two to four times that of diploid cell.

Three other aspects of DNA that strongly supported its role as the genetic material were:

1. DNA is metabolically stable, that is, it is not rapidly synthesised and degraded like many other cellular molecules.
2. The amount of DNA per cell is approximately related to the complexity of the

organism. Thus cells of higher organisms contain considerably more DNA than do bacterial cells.

3. The amount of DNA in gametes is approximately one-half that found in 2N cells of the same species.

The Hershey and Chase Bacteriophage Experiments

In 1953, Alfred D. Hershey and Martha Chase reported experimental results showing that DNA is genetic material. They were studying the replication of bacteriophage T2 to see which phage components were needed to complete the life cycle. The T2 infected *Escherichia coli* by first attaching itself with its tail fibers and then injecting into the bacterium some genetic material that controlled the production of progeny T2 particles. Ultimately, the bacterium lysed, releasing 100-200 phages that could infect other bacteria.

Hershey and Chase grew cells of *E. coli* in media containing either a radioactive isotope of phosphorus (P^{32}) or a radioactive isotope of Sulphur (S^{35}). They used these isotopes since DNA contains phosphorus but no sulphur, and protein contains sulphur but no phosphorus. They infected the bacteria with T2 and collected the progeny phages. At this point, Hershey and Chase had two batches of T2: one had the proteins radioactivity labelled with S^{35}, and the other had the DNA labelled with P^{32}.

Since they knew that each T2 phage consisted of approximately half DNA and half protein and one of these two classes of molecules must be the genetic material. To determine which class it was, they next infected unlabeled *E. coli* with the two types of radioactively labelled T2 and obtained the results.

When the infecting phage was P^{32} labelled, most of the radioactivity could be found within the bacteria soon after infection. Very little could be found in protein parts of the phage (the phage ghosts) released from the cell surface by mixing the cells in a kitchen blender. After lysis, some of the P^{32} was found in the progeny phages. In contrast, after infecting *E. coli* with S^{35} labelled T2, virtually none of the radio activity appeared within the cell, and none was found in the progeny phage particles.

Most of the radioactivity could be found in the phage ghosts released after treating the cultures in the kitchen blender. Since genes serve as the blue print for making the progeny virus particles, it was also presumed that the blue print must get into the bacterial cell for new phage particles to be built.

Therefore, it was DNA and not protein that entered the cell, as evidenced by the presence of P^{32}. Hershey and Chase reasoned that DNA must be the material responsible for the function and reproduction of phage T2. They hypothesized that the protein provided a structural framework to contain the DNA and the specialized structures required to inject the DNA into the bacterial cell.

Similarly, other experiments showed that the DNA of mouse polyoma virus is the genetic material. In these experiments DNA was purified from the virus particles (which consist of DNA and proteins) and used to infect mouse cells in culture. The DNA was able to direct the synthesis of thousands of new polyoma particles, thereby proving that DNA is the genetic material and proteins function as a protective framework for the DNA in its movement between cells.

Other evidences from transduction and conjugation in bacteria support that DNA is the genetic material of cells.

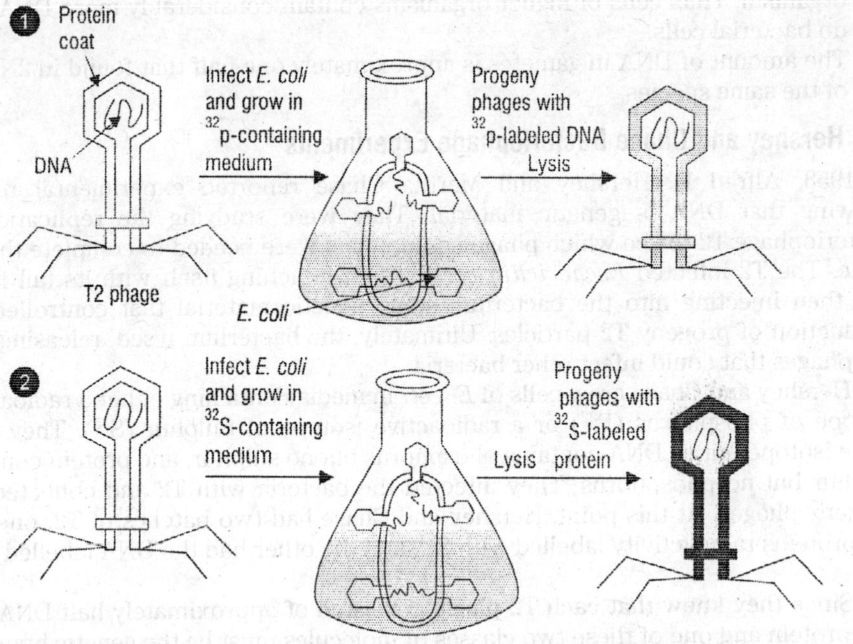

FIG. 9.11: Hershey-Chase experiment- the production of T2 phages either with (1) P^{32} labelled DNA or with (2) S^{35} labelled protein.

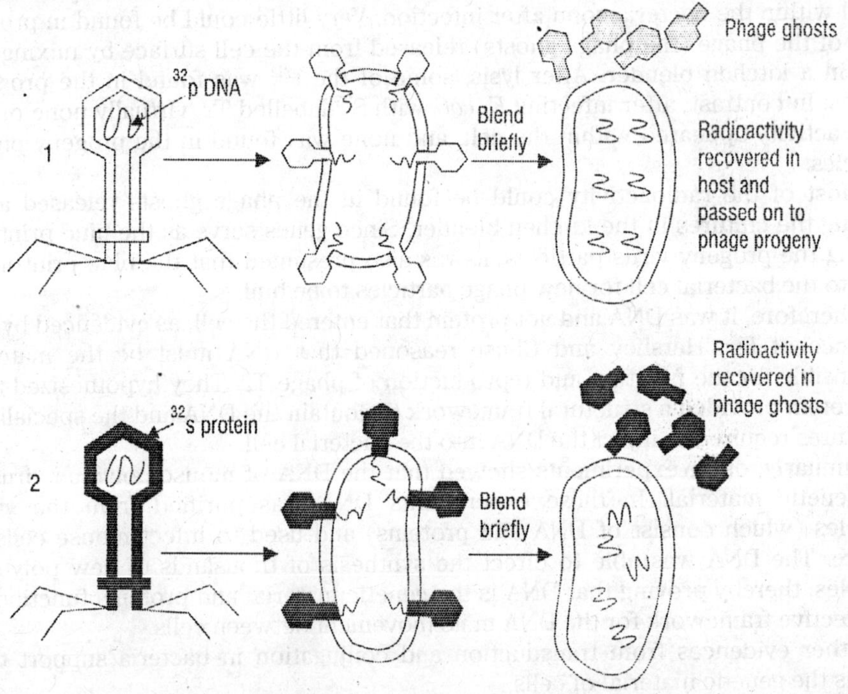

FIG. 9.12: Hershey-Chase experiment- the experimental evidence showing that DNA is the genetic material in T2: (1) the p^{32} is found within the bacteria and appears in progeny phages, (2) while the S^{35} is not found within the bacteria and is released with the phage ghosts.

DNA REPLICATION

Every cell in a body synthesises proteins and contains the appropriate DNA encoding for that information. All the cells are derived from a single cell in the very early stages of development, then DNA has to be replicated many millions of times in order for all the cells to receive their complement.

Replication is the preparation of DNA copies prior to reproduction of the cell or organism. In this process whereby a copy of a DNA molecule is made and thus the genetic information it contains is duplicated. The parental double-stranded DNA molecule is replicated semi-conservatively, that is each copy contains one of the original strands paired with a newly synthesised strand which is complementary in terms of AT and GC base pairing and of opposite chemical polarity. Though conceptually it is simple but mechanistically it is a complex process, in which many enzymes are involved. Because of specific base pairing, upon separation of the DNA double helix, each strand can reproduce (serving as a template for) the other. This process is called semiconservative replication.

Important Aspects of DNA Replication

1. Meselson and Stahl (1958) used heavy nitrogen (15N) to demonstrate that replication is semiconservative. This flash animation at "*DNA from the Beginning*" describes this experiment and some of Lederbergs experiments with DNA polymerase: A half DNA ladder is a template for copying the whole.
2. Taylor (1957) used 3H-thymidine label and colchicine on bean root-tip cells to also demonstrate semiconservative replication also shows that chromosome is probably one long DNA molecule.
3. Cairns (1963) used autoradiography to show that in *E. coli* replication proceeded from one point (he saw theta structures) called the origin of replication.
4. **One direction or two?:** Use of two labels demonstrated that replication was bidirectional in prokaryotes and eukaryotes (the later label showed up on both sides of the initial label, showing that new DNA was being made in two places, on either side of the starting point). There are some exceptions (Col El plasmid).
5. **Continuous, semidiscontinuous or discontinuous?:** Kornberg isolated a DNA polymerase and it only worked in the 5' to 3' direction. Okazaki suggested that on at least one strand DNA synthesis must be discontinuous. He used labelling experiments to show that DNA was initially synthesised in small fragments (Okazaki fragments). If no DNA ligase was present these would never get combined into larger DNA fragments. It is discontinuous only on the lagging strand - semidiscontinuous As DNA polymerase requires a free 3' end there must be a primer. Tuneko Okazaki finally demonstrated that an RNA primer of 10-12 bp was at the end of Okazaki fragments.
6. The existence of Okazaki fragments? Reiji Okazaki fed radioactive DNA precursors for brief periods (a "pulse") to growing *E. coli*, then isolated the DNA, "denatured" it to separate single strands, and measured the size of the radioactive DNA by centrifugation (today we would probably use gel electrophoresis). He found two peaks, very large DNA strands (from the leading strand) and small fragments of about 1000 nucleotides ("Okazaki" fragments) with little in between. The Okazaki fragments were shown to be true precursors of chromosomal DNA by following the pulse with a period of growth in medium containing non-radioactive ("cold") precursor, after which the label moved from the small fragment region into large DNA strands (this is called a pulse-

chase experiment). Later, other scientists, by studying the DNA replication of viruses in which one could separate the two DNA strands in intact form (due to their different base composition), were able to confirm that, as expected, the Okazaki fragments arose predominantly from only one of the two strands at the replication fork.

7. **Enzymology of DNA Replication:** Following enzymes are involved in replication:
   a) **Helicase:** separates the DNA strands.
   b) **SSB, single:** strand binding protein - keeps the strands apart.
   c) **Topoisomerases:** wind and unwind the DNA.
   d) **Gyrase nicks:** both strands and acts as a swivel.
   e) **Primosome:** a protein complex that makes the RNA primer.
   f) **Replisome**: - carries out elongation of the DNA.
   g) DNA polymerase III is the key enzyme.
      i) two DNA pol III molecules per replisome so probably does both leading and lagging strands at the same time.
      ii) 400 nt/second.
      iii) Here is an interactive tutorial on the structure of the beta subunit of DNA polymerase III, you'll need the chime plugin to see this.
8. DNA polymerase I replaces the RNA primer with DNA.
9. DNA ligase links the fragments together.

DNA replication begins with a partial unwinding of the double helix at an area known as the *replication fork.* This unwinding is accomplished by an enzyme known as *DNA helicase.* This unwound section appears under electron microscopes as a "*bubble*" and is thus known as a replication bubble. As the two DNA strands separate ("*unzip*") and the bases are exposed, the enzyme DNA polymerase moves into position at the point where synthesis will begin.

The Role of Accessory Proteins in Replication

Accessory proteins have been identified in two major ways. Geneticists have identified mutations in genes that block DNA replication (called *dnaA*, *dnaB*, etc.). Since these genes are usually essential for survival, the mutations isolated are temperature-sensitive mutations that block DNA replication and growth at high temperature (about 39 °C for *E. coli*) but not low temperature (about 25 °C). Subsequently, the genes can be isolated (as in the yeast cell cycle video) to examine what their function in replication is. Biochemists set up *in vitro* systems based on purified DNA polymerase and other pure proteins and then seek addition accessory proteins in cell extracts that enhance replication or make it more similar to *in vivo* replication. These proteins are purified using such *in vitro* assays. When a protein and its gene are identified as important by both biochemical and genetic analysis, this gives the strongest indication that it is really involved in *in vivo* replication.

DNA Replication is Semiconservative

Since each new strand is complementary to its old template strand, two **identical** new copies of the DNA double helix are produced during replication. In each new helix, one strand is the old template and the other is newly synthesised, a result described by saying that the replication is **semi-conservative**. Crick described the DNA replication process and the fitting together of two DNA strands as being like a hand in a glove. The hand and glove separate, a new hand forms inside the old glove, and a new glove forms around the old hand. As a result, two identical copies now exist. The whole molecule may be replicated into a new molecule in which the

old DNA remains the same and the **new** DNA is made of all new components. This is called **conservative replication**. The DNA may be broken up and new bits are added in a random pattern in which both strands of both resulting DNA molecules contain old and new components. This is called **dispersive replication**.

When the replication process is complete, two DNA molecules - identical to each other and identical to the original - have been produced. Each strand of the original molecule has remained intact as it served as the template for the synthesis of a complementary strand. This mode of replication is described as semi-conservative: one-half of each new molecule of DNA is old; one-half new. Watson and Crick had suggested that this was the way the DNA would turn out to be replicated. Proof of the model came from the experiments of Meselson and Stahl. DNA could be unzipped into two single strands and use each one of these strands to form the basis of the new DNA molecules. This type of replication is called **semiconservative replication.** The direction of replication is always in 5'-3' direction.

Dispersive Replication

Watson and Crick were completely wrong. Perhaps replication did not occur by one strand acting as a template for another. Perhaps enzymes created new DNA sequences at each generation, destroying the old sequences every time. This was called **dispersive** replication, and, if it were true, would essentially say that DNA was not really the genetic material.

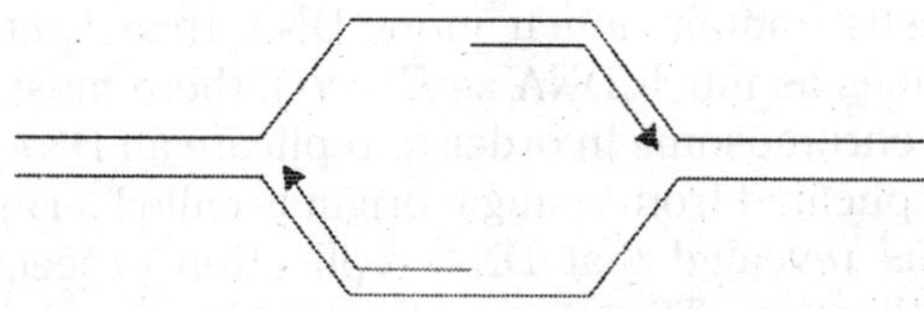

FIG. 9.13: Diagram illustrates that how each original DNA molecule branches, or forks, at the point where replication is occurring. These branch points are called replication forks. Because replication is bi-directional, two replication forks form at each origin of replication. (Some rare examples have been seen where replication is unidirectional from the origin.) The open area of the chromosome between the replication forks is called a replication bubble.

According to Meselson and Stahl, *How might one show which of these three models of DNA replication were correct*? Meselson and Stahl, working in the mid 1950's, realized that they had to be able to distinguish between DNA that remains intact and DNA that is newly synthesised. Radioactive isotopes had been tried, but were found wanting. Instead they took a new tack, and while doing so, invented an entirely new technique that is still widely used today.

Origins of Replication

1. Prokaryotes usually have a single origin on a circular chromosome.
2. Most viral DNAs have a single origin.
3. Eukaryotic chromosomes have multiple origins spaced along their length.

Although the semiconservative nature of DNA replication had been confirmed, many questions about replication remained. One of these questions was: is replication initiated at a specific site on the chromosome, or is it initiated at random sites, or even multiple sites?

The answer to this question depends somewhat on the organism being considered. Bacteria, for example, have a single specific origin of replication; in other words, bacterial replication begins at the same spot on the chromosome every time. In *E. coli*, this site is called *OriC*. *OriC* is a 9 base-pair (bp) sequence that is repeated four times within the region.

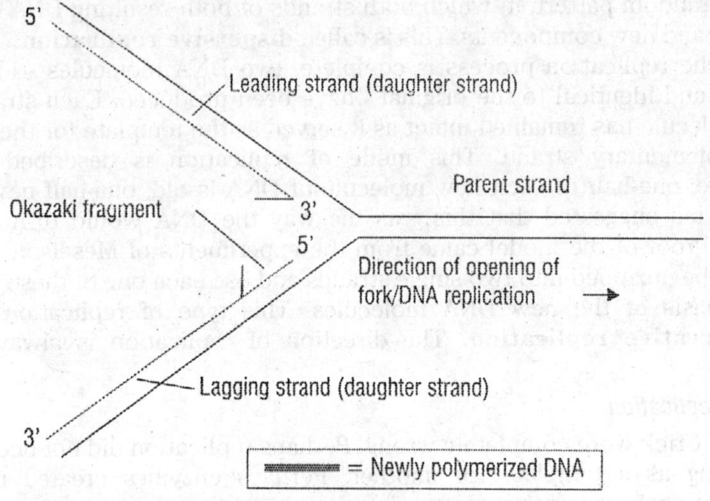

FIG. 9.14: Semiconservation DNA replication – replication fork.

Eukaryotes also have specific sites at which replication is originated. However, because eukaryotic cells contain much more DNA than bacteria (humans have approximately 1500 times as much DNA as *E. coli*), there must be multiple origins of replication on each chromosome in order to replicate all DNA in a timely fashion. The amount of DNA replicated from a single origin is called a **replicon.**

Other research has revealed that DNA replication proceeds **bidirectionally** from an origin of replication. This means that replication proceeds in opposite directions away from the origin.

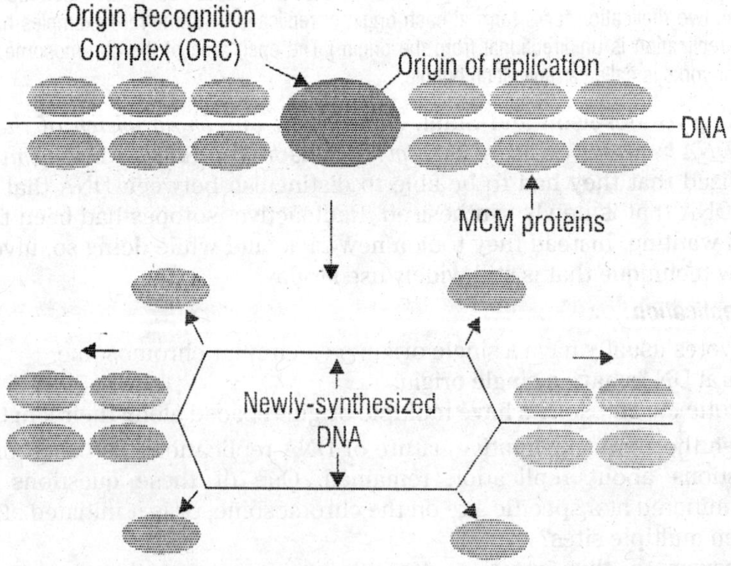

FIG. 9.15: Showing origin of replication in DNA.

Speed of Replication

Prokaryotes

The single molecule of DNA that is the *E. coli* genome contains 4.7×10^6 nucleotide pairs. DNA replication begins at a single, fixed location in this molecule, the **replication origin**, proceeds at about 1000 nucleotides per second, and thus is done in no more than 40 minutes. And thanks to the precision of the process (which includes a **proof-reading** function), the job is done with only about one incorrect nucleotide for every 10^9 nucleotides inserted. In other words, more often than not, the *E. coli* genome (4.7×10^6) is copied without error!

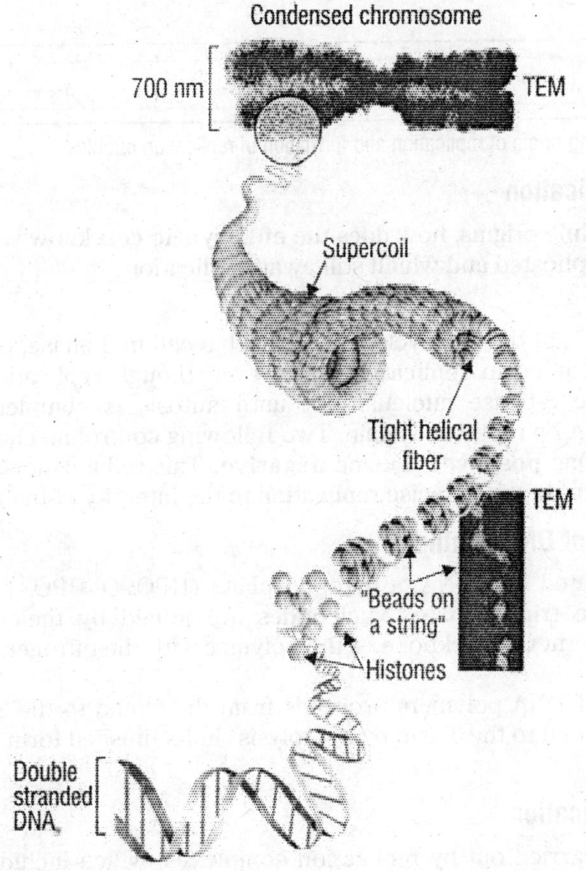

FIG. 9.16: Illustration of complex organization of DNA in chromosome. Note the process of DNA replication.

Eukaryotes

The average human chromosome contains 150×10^6 nucleotide pairs which are copied at about 50 base pairs per second. The process would take a month (rather than the hour it actually does) but for the fact that there are many places on the eukaryotic chromosome where replication can begin.

Replication begins at some replication origins earlier in S phase than at others, but the process is completed for all by the end of S phase. As replication nears completion, "*bubbles*" of newly replicated DNA meet and fuse, finally forming two new molecules.

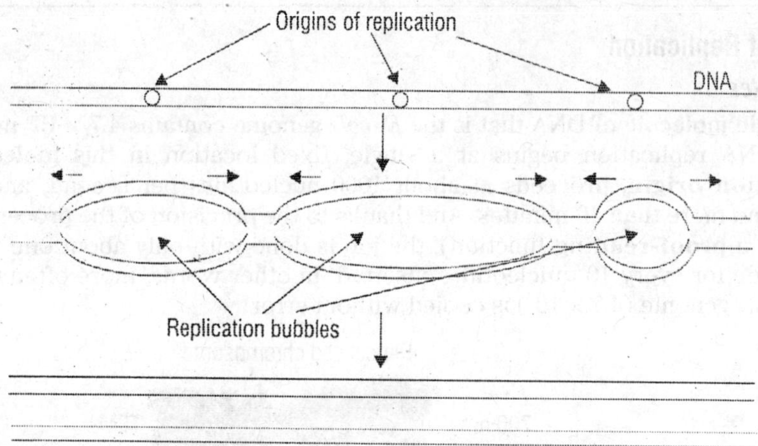

FIG. 9.17: Showing origin of replication and formation of replication bubbles.

Control of Replication

With their multiple origins, how does the **eukaryotic cell** know which origins have been already replicated and which still await replication?

An Observation

When a cell in G_2 of the cell cycle is fused with a cell in S phase, the DNA of the G_2 nucleus does not begin replicating again even though replication is proceeding normally in the S-phase nucleus. Not until mitosis is completed, can freshly-synthesised DNA be replicated again. Two following control mechanisms have been identified: (1) One **positive,** (2) One **negative.** This redundancy probably reflects the crucial importance of precise replication to the integrity of the genome.

The Chemistry of DNA Synthesis

When incorporated into DNA, a pyrophosphate (HPO_3-O~HPO_3-) is released from each nucleotide triphosphate. Nucleotides are joined by their phosphates and sugars, which form the backbone of the polymer with the nitrogen bases projecting off the side.

Synthesis of DNA polymers proceeds from the 5' end to the 3' end (each new nucleotide is added to the 3' end by hydrolysis) in its finished form, DNA is a double helix.

Origins of Replication

Replication is carried out by replication complexes, which include the unwinding enzyme helicase and the nucleotide adding enzyme DNA polymerase. The structure of helicase has now been revealed. The helix is unwound, the separated strands form replication forks, and new nucleotide triphosphates are added according to base pairing. The Replication complex untwists the helix in only one direction. The addition of bases to the leading end of a polymer (5' to 3') occurs smoothly, one base at a time. Replication on the complimentary strand also occurs in the 5'-to-3' direction but in short segments (Okazaki fragments) since the DNA polymerase must constantly wait for the original helix to unwind creating a large enough gap to get started. The Okazaki fragments are united by a special ligase enzyme. Because each new polymer is base-paired to an old one, DNA replication is called **semiconservative** (half is conserved).

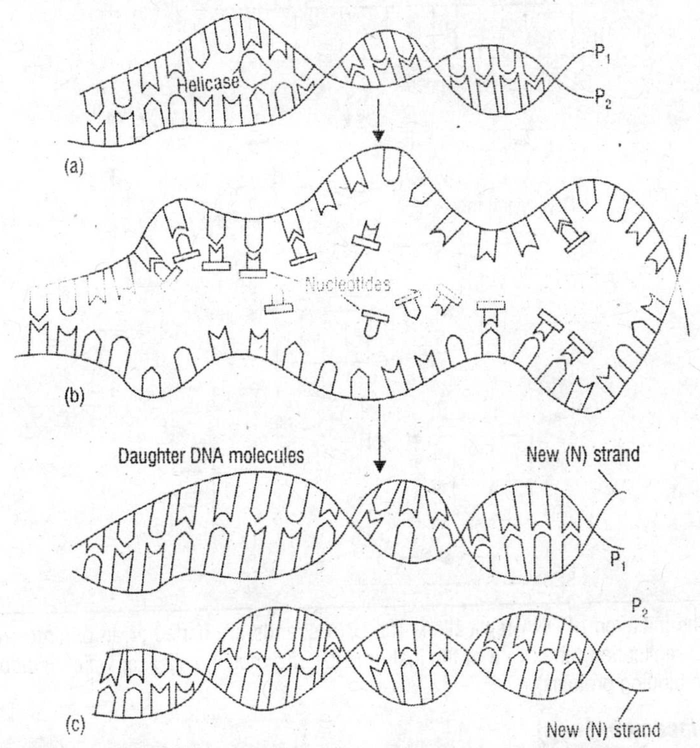

FIG. 9.18: Simplified steps in semiconservative replication of DNA. (*a*) A helicase unwinds the double helix into two parent strands (P_1 and P_2). (*b*) The replication of new complementary strands proceeds through the action of an enzyme that attaches nucleotides, using the exposed strands as templates. (*c*) Completed daughter molecules each contain one strand that is newly synthesised and one of the original parent strands.

Licensing: Positive Control of Replication

In order to be replicated, each origin of replication must be bound by:

1. an **O**rigin **R**ecognition **C**omplex of proteins (**ORC**). These remain on the DNA throughout the process.
2. accessory proteins called **licensing factors**. These accumulate in the nucleus during G_1 of the cell cycle. They include:

CDC-6 and CDT-1 bind to the **ORC** and are essential for coating the DNA with **MCM proteins**. Only DNA coated with MCM proteins (there are 6 of them) can be replicated. Once replication begins in S phase, CDC-6 and CDT-1 leave the ORCs (the latter by ubiquination and destruction in proteasomes). The MCM proteins leave in front of the advancing replication fork.

Geminin: Negative Control of Replication

G_2 nuclei also contain at least one protein called **geminin.** It prevents assembly of MCM proteins on freshly-synthesised DNA (probably by sequestering Cdt1). As the cell **completes mitosis**, geminin is degraded so the DNA of the two daughter cells will be able to respond to licensing factors and be able to replicate their DNA at the next S phase. Some cells deliberately cut the cell cycle short allowing repeated S phases without completing mitosis and/or cytokinesis. This is called **endoreplication**. How these cells regulate the factors that normally prevent DNA replication if mitosis has not occurred is still being studied.

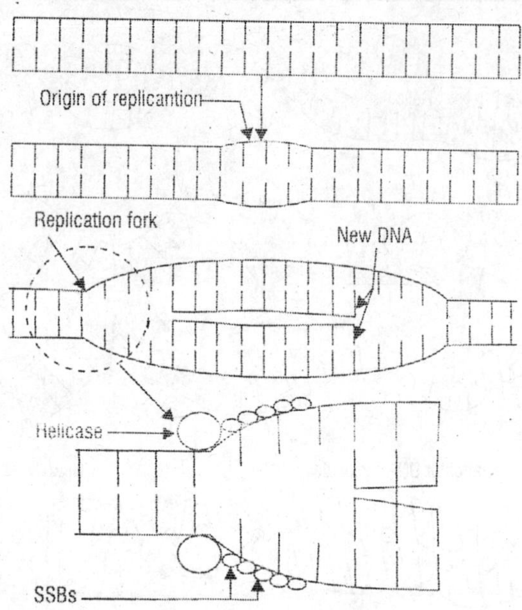

FIG. 9.19: The initiation of DNA replication. The DNA strands are parted at an origin of replication. It results in two replication forks. One of the instrumental enzymes is helicase, which is added by SSBs (single strand binding proteins).

D Loop (Displacement Loop)

Structure produced in replicating DNA when a *bubble* or *eye* of DNA is formed and can be visualized in the electron microscope. The production of the loop results from the DNA being double stranded on one side of the bubble and single-stranded on the other and it is exploited as a technique to demonstrate hybridization of an exogenous sequence to one of the complementary strands, leaving the other displaced strand looping out from the helix.

How Replication is Achieved?

Firstly the DNA is in a double helical structure and therefore, for each of the new DNA molecules to receive an old strand, this helical structure has to be **unzipped** (unwinding) and then the new DNA is synthesised. The unwinding of the DNA helix is carried out by a group of enzymes called **topoisomerases**. These enzymes fall into two groups, Type I and Type II. Type I enzymes. These enzymes (Type I) break one of the DNA strands, while Type II enzymes break both DNA strands. **DNA gyrase** is common examples is an enzyme.

Once the DNA has been unwound the two strands have to be separated. This starts at a point known as the **replication origin**. The DNA strands are separated by breakage of the **hydrogen bonding** between the bases. This process should remain continue to enable the replication. This process is catalysed by an enzyme called a **helicase** and the break points are refereed to as the **replication forks**. However, DNA would naturally reanneal back to the double helix with the reformation of the double bonds is allowed to do so. Therefore, proteins known as **single-strand binding proteins** (SSBs) bind to the single-stranded DNA formed by the helicase and stop the rennealing process from taking place.

Once the DNA opened up into a single-stranded structure, then the synthesis of new strand starts. However, how does this process start, as the first bases have to be there for strand extension to take place? The process starts by using a **primer**. This is a short length of RNA synthesised by an enzyme called **RNA polymerase,** or **primase**. The RNA primer may be of 6 to 30 bases length. **DNA polymerase**, catalyses synthesis of the DNA strand. In *E. coli* there are three DNA polymerases, but types I and III seem to be most essential for DNA replication, the function of type II remaining more obscure. Most of the DNA synthesis is carried out by polymerase III, with polymerase I removing the RNA primers and basically tidying up the missing bits.

The polymerases catalyse the formation of new **phosphodiester bonds**. The newly formed strand has a **free 3'** OH group on the ribose sugar, and to this the α phosphate of the next nucleotide can be attached, with the formation of the phosphodiester bond and the release of the β-γ phosphates as an orthophosphate unit. However, such a process means that the DNA synthesis always takes place in the 5'-3' direction. This is fine for synthesis along one strand of the DNA molecule, known as the **leading strand**, as continuous synthesis is obtained.

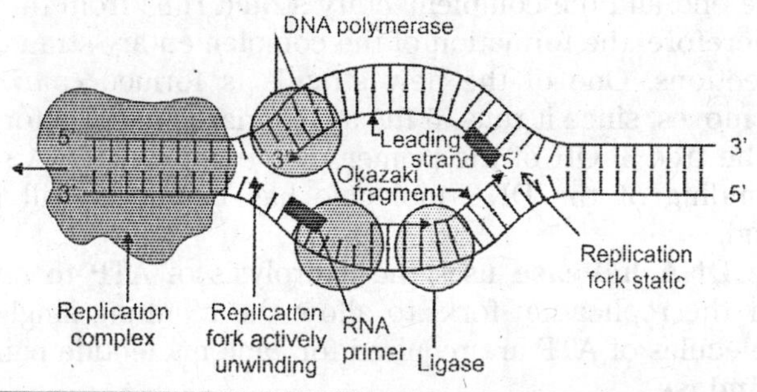

FIG. 9.20: Illustration of DNA replication and enzyme involved in the process.

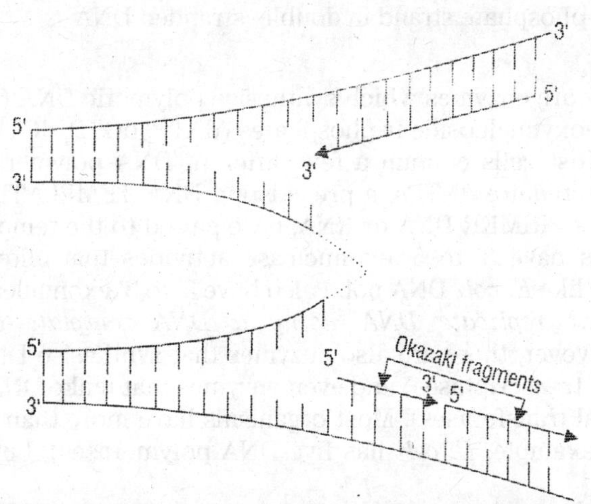

FIG. 9.21: Synthesis of DNA in the 5' to 3' direction. This results in continuous 5'-3' synthesis on one strand but discontinuous 5'-3' synthesis on the other strand with the formation of Okazaki fragments.

The other strand **(lagging strand)** is being effectively unteased backwards, synthesis along the second strand cannot follow the direction of the replication fork. Therefore, on the second strand the synthesis of the new DNA is **discontinuous.** This process was confirmed by the work of **Okazaki** (1968). Hence the short stretches of DNA are known as **Okazaki fragments**. Each fragment has to be started by its own RNA primer and so, along this strand, you would find short DNA stretches following short RNA stretches and gaps between. An enzyme called **DNA ligase** carries out the final covalent bonding, which joins all the resulting Okazaki fragments together.

Enzymes Involved in Replication

The initiation of the replication process requires a **primer**. It is a short RNA molecule, which is complementary to one of the DNA strands. Once the RNA molecule is formed by the enzyme RNA polymerase, DNA synthesis begins. Next, the RNA primer is removed and replaced by newly formed DNA. The replication process continues with the enzyme DNA polymerase adding nucleotides only to the 3' hydroxyl (OH) end of the replicating DNA strand or strands in a 5 to 3' direction. The fact that this polymerase can only add nucleotides to the 3' hydroxyl group of the strand raises the question of how DNA synthesis can proceed in proper order (sequentially) from the single starting point along *both* DNA strands. As mentioned earlier, the two strands of DNA are antiparallel, since on strand runs from the 5' OH to the 3' OH free end and the complementary strand runs from the 3' OH to the 5' OH free end. Therefore, the formation of the complementary strands must proceed in opposite directions. One of the new strands is formed *continuously* as the replication fork moves, since it runs in the appropriate direction for the addition of nucleotides to the free 3' OH of the primer. The other new DNA strand can start only after unwinding of the DNA molecule has begun, it will leg behind the continuous strand.

The enzyme **DNA helicase** uses the hydrolysis of ATP to unwind the DNA helix, usually at the replication fork, to allow the resulting single strands to be copied. Two molecules of ATP are required for each nucleotide pair of the duplex, also called **unwindase.**

The enzyme **DNA ligase** catalyses the covalent union of segments of interrupted sugar-phosphate strand in double- stranded DNA.

DNA Polymerase

DNA Polymerases are enzymes which synthesise polymeric DNA (double-stranded) starting with 5' deoxynucleoside triphosphates (dATP, dGTP, dCTP, dTTP, together called dNTPs). Most cells contain a few different DNA polymerase activities. All DNA polymerases require dNTPs, a pre-existing DNA TEMPLATE STRAND and a FREE 3' END on a PRIMER DNA or RNA, base-paired to the template strand. Most DNA polymerases have 3' to 5' exonuclease activities that allow polymerase to proofread. (Some like *E. coli* DNA pol. I also have 5' to 3' exonuclease activity.)

Enzymes that replicate DNA using a DNA template are called DNA polymerases. However, there are also enzymes that synthesise DNA using an RNA template (reverse transcriptases) and even enzymes that make DNA without using a template (terminal transferases). Most organisms have more than one type of DNA polymerase (for example, *E. coli* has five DNA polymerases), but all work by the same basic rules.

1. Polymerization occurs only 5' to 3'
2. Polymerization requires a **template** to copy: the complementary strand.
3. Polymerization requires 4 dNTPs: dATP, dGTP, dCTP, dTTP (TTP is sometimes not designated with a 'd' since there is no ribose containing equivalent).
4. Polymerization requires a pre-existing **primer** from which to extend. The primer is RNA in most organisms, but it can be DNA in some organisms; very rarely the primer is a protein in the case of certain viruses only.

TABLE 9.1: The various enzyme activities of the different polymerases

Enzyme Activity	DNA Polymerase I	DNA Polymerase II	DNA Polymerase III
5' to 3' polymerase	Yes	Yes	Yes
3' to 5' exonuclease	Yes	Yes	Yes
5' to 3' exonuclease	Yes	No	No

DNA Polymerase I

DNA replication is catalysed by a family of enzymes called DNA polymerases. The first of these enzymes to be discovered, DNA polymerase I, was isolated from bacteria (specifically, *E. coli*). Characterization of the activity of this enzyme *in vitro* revealed that it had certain requirements for activity. It needed 5'-triphosphate forms of the four nucleotides, and it required the presence of pre-existing DNA. The DNA serves two following purposes:

1. It serves as a template for the synthesis of the new DNA (the template determines the sequence of the new DNA strand, through the specificity of base pairing).
2. It serves as a primer for DNA synthesis. It turns out that DNA polymerase I cannot initiate DNA synthesis without having a free 3'-OH to add a new nucleotide to. DNA synthesis therefore needs a primer, a pre-existing piece of nucleic acid to serve as an initiator of DNA synthesis.

TABLE 9.2: Some examples of DNA polymerases

Enzyme	Template	Primer	Other activities	Other features
E. coli DNA pol I	DNA	DNA/RNA	3'-5' exo, 5'-3' exo	monomeric
E. coli DNA pol I (Klenow fragment)	DNA	DNA/RNA	3'-5' exo	C-terminal fragment
E. coli DNA pol III	DNA	DNA/RNA	3'-5' exo (on a separate subunit)	multimeric structure
Taq pol	DNA	DNA/RNA	Extendase (adds 3'-A overhangs)	thermostable, used in PCR
Reverse transcriptase	DNA/RNA	DNA/RNA	(ribonuclease H)	used to make cDNA
Terminal transferase	None required	DNA		will synthesise DNA in non-templated reaction

DNA polymerase I synthesises DNA by forming a bond between the 5' phosphate of the incoming nucleotide (the other two phosphate groups from the nucleotide triphosphate are lost) and the 3' OH group of the nucleotide at the end of the growing DNA chain. If you draw this out for yourselves, you'll realize that this means the DNA chain being synthesised grows in a 5' to 3' direction. This is an important rule to remember: DNA polymerase synthesises DNA only in a 5' to 3' direction.

In addition to its polymerase activity, DNA polymerase I has two other enzymatic activities, both of which are exonuclease activities. Exonucleases are enzymes that digest DNA (cleave phosphodiester bonds), chewing away at nucleotides from the end of the DNA chain. DNA polymerase I has 3' to 5' exonuclease activity, which degrades DNA in a direction opposite to that of synthesis. This provides the enzyme with a proofreading function: if a wrong nucleotide gets inserted into a growing chain, the enzyme can digest it out with the

3' to 5' exonuclease activity (almost like using the delete key while word processing), and insert the correct nucleotide.

DNA polymerase I also has a 5' to 3' exonuclease activity, which degrades nucleic acids in the same direction as synthesis. As we'll see, this activity is used to remove primers during DNA replication.

Why don't DNA Polymerases Elongate Chains in the 3' to 5' Direction?

If a DNA polymerase could synthesise DNA in the 3' to 5' direction, then nucleotides would add to the primer terminus which has a 5'-triphosphate: (a) The 3'-OH of each incoming deoxyribonucleoside triphosphate would attack the 5'-triphosphate of the growing chain. (b) The **editing** removal of an incorrect 5'-terminal nucleoside triphosphate would **prevent** the DNA chain from being further extended because the primer terminus would now be a 5'-monophosphate, not a 5'-triphosphate.

Interestingly, *E. coli* temperature-sensitive (conditional lethal) mutants lacking the 5'-to-3' exonuclease activity of DNA polymerase I at non-permissive temperatures (40 degrees C) are NOT viable (remember, the polymerizing activity of DNA polymerase I is dispensible!). This enzymatic activity plays an essential role by removing RNA primers during replication. This process is similar to nick translation (here translation means "*movement*" not "*protein synthesis*"), the simultaneous polymerization in the 5'-to-3' direction from a nick with concurrent removal of nucleotides ahead by the 5'-to-3' exonuclease activity.

Multiple Polymerases

DNA polymerase I, it turns out, is not the main enzyme involved in replicating the bacterial chromosome. When the gene encoding the enzyme was mutated in *E. coli*, the bacteria were still able to replicate their chromosomes. They were, however, deficient in DNA repair. This suggested that DNA polymerase I is primarily involved in DNA repair (although it does play a role in replication, as we will see), and that another yet-to-be-discovered enzyme would be responsible for replication.

Eventually, two other DNA polymerases were identified, and named DNA polymerases II and III. These both had 5' to 3' polymerase activity like DNA polymerase I, and 3' to 5' exonuclease activity. Neither of these enzymes had the 5' to 3' exonuclease activity found in DNA polymerase I. DNA polymerase III turns out to be the main enzyme involved in DNA replication. DNA polymerase II is a minor enzyme involved in DNA repair. DNA polymerase I is the main polymerase involved in DNA repair, and plays a specialized role in DNA replication, using its 5' to 3' exonuclease activity.

Any one of the systems of enzymes which catalyses the formation of DNA from deoxyribonucleotides, using one strand of DNA as a template is called **DNA Polymerase I, Holoenzyme.** It has 5' exonuclease activity in addition to DNA synthesis. This bifunctional activity enables the enzyme to use nicks or gaps in double stranded DNA as a starting point for DNA synthesis. The 5'-exonuclease activity degrades that DNA strand which is complementary to the template strand and thus forming nick. DNA synthesis begins at 3'-end of nick and produces a new strand of DNA complementary to the template. The net result is the movement of nick along the template strand (nick translation) until all the DNA complementary to the template strand (starting from the site of the origin of nick to the 5'-end of the template strand) is replaced.

Enzyme is used with radioactive or biotinylated nucleotides to prepare labelled DNA of high specific activity. It can also catalyse *de novo* DNA synthesis. **DNA Polymerase I, Klenow Fragment.** Treatment of DNA polymerase I holoenzyme with protease, results in production of two protein fragments. The larger fragment, also known as Klenow fragment does not exhibits 5'-exonuclease activity (5'-exonuclease activity is exhibited by intact enzyme). This Klenow fragment is used to synthesise DNA when there is no need of removing the DNA strand which is complementary to the template strand.

T_4 DNA Polymerase

T_4 DNA polymerase lacks 5' exonuclease activity but has a very active 3'-exonuclease action. This allows radiolabelling of DNA fragment by replacement synthesis. In this method it acts on both the 3'-end of double stranded DNA resulting in long 5'-single stranded extensions. DNA complementary to these single strands is synthesised with deoxynucleoside triphosphase enzyme in the presence of radiolabelled compounds. It is useful in generating 5'-single stranded ends. The enzyme can also be used in radiolabelling of DNA.

Taq DNA Polymerase

Taq DNA polymerase is isolated from the thermophilic bacterium *Thermus aquaticus*. This enzyme has the highest DNA polymerase activity at a pH of 9 and temperature around 75 °C. DNA polymerase activity is resistant to incubation at high temperatures like 95°C. This enzyme consists of a single polypeptide chain with a molecular weight of 95,000. It lacks 5'to 3' and 3'to 5' exonuclease activities. The highly thermostable DNA polymerase from *Thermus aquaticus* (Taq) is ideal for both manual and automated DNA sequencing because it is fast, highly progressive, has little or no 3'- exonuclease activity and is active over a broad range of temperatures. It can be used in DNA sequencing studies. (2) It is used in 'Polymerase chain reaction (PCR)' as it can withstand high temperatures.

E. coli DNA Polymerase I

DNA Polymerase I from *E. coli* was the first DNA polymerase characterized. There are approximately 400 molecules of the enzyme per cell. *E. coli* DNA polymerase I is abbreviated pol I. The enzyme is a single large protein with a molecular weight of approximately 103 kDa (103,000 grams per mole). The enzyme requires a divalent cation (Mg^{++}) for activity and has three enzymatic activities associated with it:

1. 5'-to-3' DNA Polymerase activity
2. 3'-to-5' exonuclease (Proofreading activity)
3. 5'-to-3' exonuclease (Nick translation activity)

The location of the three enzymatic activities within the protein is known, and it is possible to remove the 5'-to-3' exonuclease activity using an enzyme called a protease to cut DNA pol I into two protein fragments.

Like all known DNA polymerases, DNA polymerase I requires a primer from which to initiate replication and polymerizes deoxyribonucleotides into DNA in the 5' to 3' direction using the complementary strand as a template. Newly synthesised DNA is covalently attached to the primer, but only hydrogen-bonded to the template. The template provides the specificity according to Watson-Crick base pairing. Only the **alpha** phosphate of the dNTP is incorporated into newly synthesised DNA.

The rate of DNA synthesis by pol I is only 20 nucleotides/second, much slower than the rate of 1,000 nucleotides/second measured for the replication of *E. coli* DNA. In 1969, John Cairns isolated a viable mutant (pol A) lacking DNA polymerase I activity, an indication that pol I is **NOT** the main enzyme used to replicate DNA. Interestingly, pol A mutants are defective in repairing DNA damage.

DNA polymerase I also has two additional activities: a 3'-to-5' exonuclease activity and a 5'-to-3' exonuclease activity. The role of the 3'-to-5' exonuclease is to

EDIT DNA (remove incorrectly polymerized nucleotides;. This was first discovered in "*mutator*" mutants of phage T_4 DNA polymerase; these mutants had diminished 3'-to-5' exonuclease activity. "*Antimutator*" mutants had increased 3'-to-5' exonuclease activity. In general, the 3'-to-5' exonuclease increases the accuracy or fidelity of DNA synthesis by a factor of 10 to 1000. Thus, when the 5' to 3' polymerization of the enzyme accidentally puts the wrong base into DNA, the 3'- to 5' exonuclease proofreading activity immediately removes it. Thus errors due to incorporation of the wrong bases by the DNA polymerase are low because it must escape screening by two systems: the base pairing rules recognized by the 5' to 3' catalytic site as well as the 3' to 5' exonuclease proofreading site.

Applications of Klenow Fragment

(1) In DNA sequencing by dideoxy-method. (2) For production of second strand of cDNA. (3) Radiolabelling by filling in 5'-single stranded extension on double stranded DNA. (4) Mutagenesis of DNA with synthetic oligonucleotides. (5) In labelling the DNA by random primer method.

DNA Replication in Procaryotes

Viruses

Viral genomes contain information which: (1) ensures replication of viral genomes, (2) ensures packaging of genomes into virions, (3) alters the structure and/or function of the host cell to a greater or lesser degree.

DNA-Virus Replication Strategies: Viral strategy refers to the manner in which each virus carries out the above functions. Since a virus is an intracellular parasite, it has to operate within limits imposed by the host cell, or circumvent these limitations are as follows:

1. The virus needs to make mRNAs that can be translated into protein by the host cell translation machinery.
2. The virus needs to replicate its genome.
3. Host enzymes for mRNA synthesis and DNA replication are nuclear (except for those in mitochondrion) and so, if a virus is to avail itself of these enzymes, it needs to enter the nucleus.

Nuclear DNA Viruses

Parvovirus family: Parvoviruses are very small (18 to -25 nm diameter), single stranded DNA viruses (*parvum* = small). They have an icosahedral capsid, and are non-enveloped. DNA replication occurs in the nucleus. Human parvovirus B-19 replicates in dividing cells – primarily in erythrocyte progenitors in the bone marrow - and causes fifth disease (erythema infectiosum). This is usually a mild disease but the decreased production of red blood cells can be a problem in people with various types of severe hemolytic anemia.

Replication of the linear genome of Parvoviridae (eg. feline parvovirus; human parvovirus B19) involves use of a semi-stable genomic 3'-hairpin structure as a primer, synthesis of new DNA by host delta DNA pol using genomic DNA as template, covalent closure then re-opening of the circular DNA product, and re-initiation at the new 3'-terminus as primer after transient hairpin formation at the new 3' end after "*melting*" of the newly-formed duplex. A complicated series of steps follows, involving repeated single-strand cleavages and 5'-end binding by virus-coded NS1 protein, and possible topoisomerase action. The other circular genomes can replicate by a "rolling circle" model: nicking of one strand of the cccDNA RF allows use of an exposed 3'-terminus as a primer, which results in elongation of the "primer" strand, and displacement of the other. Ordinarily this results in greater-than- unit-length ss- and dsDNA forms being found in infected cells.

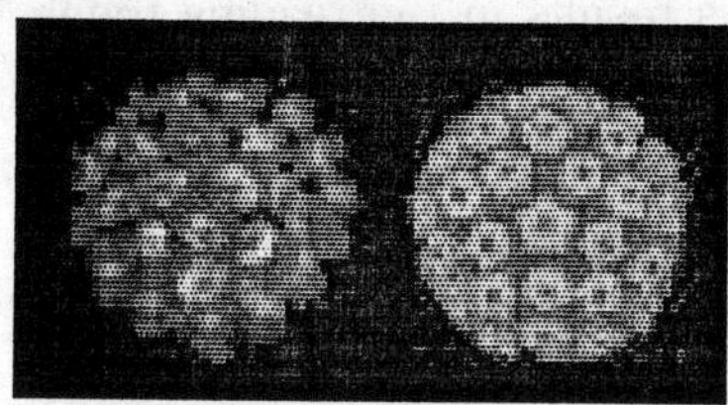

Papilloma virus

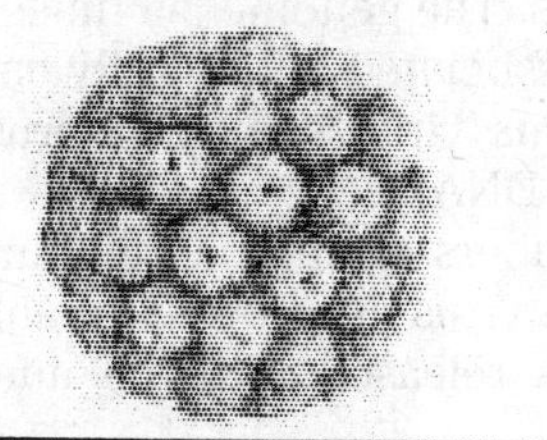

SV40 virus, a polyoma virus

Papillomavirus family: The Papillomavirus family was formerly grouped with the Polyomavirus family into the Papovavirus family (Papilloma, Polyoma, simian Vcuolating virus 40) because members of both families have a similar structure. However, it is now clear that the two families have a very different replication strategy and so each group has now been given its own family status. Papillomaviruses are small (40 – 60 nm), icosahedral: major capsid protein is VP1, with lesser amounts of VP2, VP3. They are non-enveloped and have circular, double-stranded DNA is associated with cell histones (nucleosomes). Papillomaviruses are difficult to grow in culture.

Polyomaviruses: These include SV40, BK, JC and polyoma viruses. All have a similar strategy for DNA replication. They are small (~40 nm diameter), icosahedral, non-enveloped viruses that replicate in the nucleus. Depending on the host cell, they can either transform the cell or replicate the virus and lyze the cell. Viral capsid proteins interact with cell surface receptors and penetration is probably via endocytosis. Virions are transported to the nucleus and uncoated. DNA (and associated histones) enters nucleus, probably through a nuclear pore. Gene expression is divided into early and late phases.

1. Early genes encode enzymes and regulatory proteins needed to start viral replication processes.
2. Late genes encode structural proteins, proteins needed for assembly of the mature virus.
3. Multiple use of the same DNA sequence (alternative splicing, overlapping reading frames)
4. Multifunctional protein (T antigen)
5. Small genome - so not surprising that virus codes for a very limited number of proteins.
6. Host cell provides RNA synthesis machinery, RNA modification machinery, DNA synthesis machinery, histones for packaging DNA.

SV40/polyoma DNA replication occurs in the nucleus. Large T antigen is needed for DNA replication. This binds to the origin of replication. Polyoma viruses use the host cell DNA polymerase, which recognizes the viral origin of replication if the T antigen is present. DNA replication is bidirectional (There are two replication forks per circular DNA genome and replication involves leading/lagging strands, Okazaki fragments, DNA ligase, etc.). This process of DNA replication is very similar to that which occurs in the host cell - which is not surprising as the virus is using mainly host machinery except for the involvement of the T antigen.

Host histones complex with the newly made DNA: Late mRNAs are made after DNA replication (a lot of newly made viral DNA is now available as template). Early mRNAs are still transcribed, but at a very much lower rate. T antigen is involved in controlling increased transcription from the late promoter and decreased transcription from early promoter. It also interacts with host proteins and changes the properties of the host cell, thus playing a role in cell transformation and tumor formation. VP1, 2, and 3 are made from same primary transcript which undergoes differential splicing. This results in the reading frame for VP1 being different from that for VP2 and VP3. Thus, one region of the DNA can code for two different amino acid sequences according to which reading frame is used. This is another way that viruses (and cells) can use a short stretch of DNA to code for more than one protein sequence. VP1, 2 and 3 mRNAs are translated in the cytoplasm, the proteins are transported to nucleus, and capsids assemble with DNA (and cell histones) inside the capsid. Large numbers of capsids accumulate in the nucleus and form inclusion bodies. Virions are released by cell lysis.

Adenoviruses: Adenoviruses are larger than papovaviruses (70 nm diameter). They are non-enveloped, icosahedral viruses with fibers. The genome 3-5 times size of papovavirus. The DNA is linear, double stranded, associated with virally coded, basic proteins in virion (unlike papilloma and polyoma viruses (papovaviruses), adenoviruses do not use cell histones to package virion DNA).

Adenoviruses usually infect epithelial cells. The fibers bind to a cell surface receptor and the virus is engulfed by endocytosis. The virus appears to be able to lyze endosomes. Un-coating occurs in steps. DNA is released into the nucleus (probably at a nuclear pore).

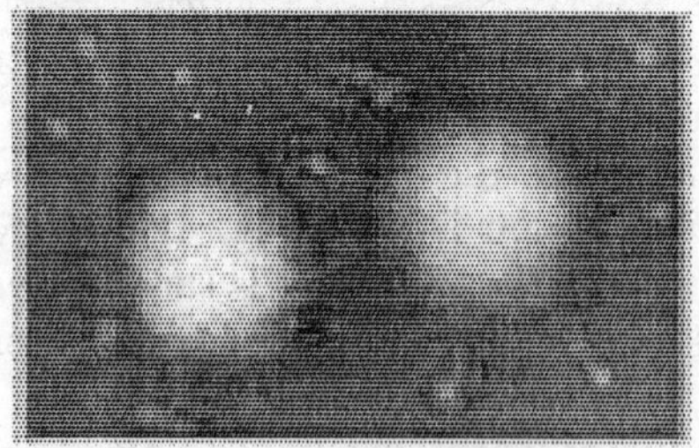

Adenoviruses

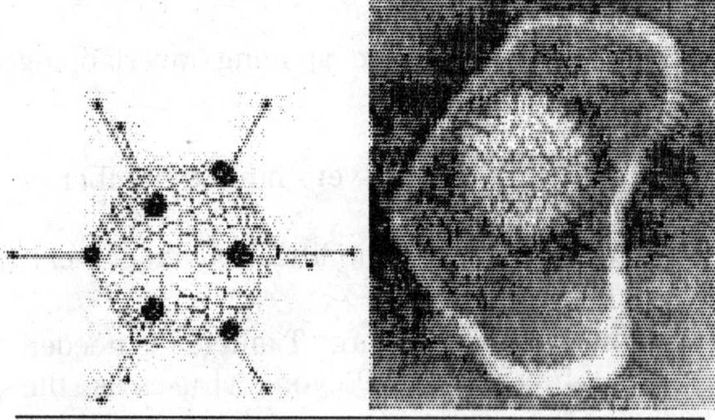

Herpes simplex virus

Adenovirus encodes its own DNA polymerase (which is one of the early proteins). The DNA is replicated by a strand displacement mechanism. There are no Okazaki fragments, both strands are synthesised in a continuous fashion. DNA polymerases cannot initiate synthesis *de novo*, they need a primer. In the case of adenovirus, the virally coded terminal protein (TP) acts as a primer. It is thus found covalently linked to the 5' end of all adenovirus DNA strands. Assembly of adenovirus particles occurs in the nucleus. DNA enters the particles after immature capsids are formed. The capsids then undergo a maturation process, after which the cells lyse and virions leak out.

More structural proteins are made than are needed and excess structural proteins accumulate in the nucleus where they form inclusion bodies.

Adenoviruses code for their own DNA polymerase and DNA packaging proteins. However, although adenoviruses code for their own DNA polymerase, they use host factors in addition to viral proteins for DNA replication, and they use host RNA polymerase and RNA modification systems and so nucleic acid synthesis needs to be in the nucleus.

Herpesviruses: Herpesviruses are larger virions (180 – 200 nm) than adenoviruses. It has linear, double-stranded DNA. Enveloped icosahedral virus (this means that lipid solvents readily inactivate these viruses).

Beta-mRNAs are transcribed the by host RNA polymerase. (Beta-genes are still "early" since they are transcribed prior to DNA synthesis. Sometimes therefore, alpha-genes are called "immediate early" and beta-genes are called "early"). Beta proteins are involved in gene expression regulation. They decrease alpha-gene expression and are needed for gamma gene expression. They are also involved in various aspects of DNA synthesis; for example, herpes beta -genes code for:

1. DNA polymerase.
2. DNA binding proteins.
3. Thymidine kinase.
4. Ribonucleotide reductase.

Since these beta proteins are virally-coded and not host-coded enzymes, they are potentially weak links in the virus life cycle and thus promising targets for viral chemotherapy. Herpesviruses code for several proteins, in addition to the DNA polymerase, that are needed for this DNA replication. The precise mechanism of DNA replication is not known. DNA replication is accompanied by a lot of recombination. The replicated DNA is present as long concatameric molecules (tandem repeats of the genome linked head-to-tail). These are cleaved to genome-size lengths when the DNA is packaged into the virion.

Some herpesviruses (*e.g.* herpes simplex virus) have a genome structure in which two parts of the genome can invert relative to each other, others do not. The significance of this is unclear. By definition, late transcription occurs after DNA replication. Gamma mRNAs are made and are translated in the cytoplasm. Gamma proteins are predominantly structural. There is decreased expression of beta genes in the late stage. This is probably due to down-regulation of transcription of beta genes by one of the gamma proteins. In herpesviruses there is no apparent organization of the genome into blocks for either early or late transcription.

Cytoplasmic DNA Viruses

Poxviruses: Certain poxviruses are of historic note, such as smallpox and vaccinia (cow pox, which was used in the smallpox vaccine.

Pox viruses may be possible agents of bioterrorism. They are used in new techniques of vaccine development (such as genetically-engineered vaccinia). Some members of this family infect man (molluscum contagium, monkey pox, cow pox).

It has complex morphology. It has large virions, large genome, double-stranded DNA 130 – 240 × 10^6 mol. Wt and enveloped. Poxviruses replicate in the cytoplasm. This means that they must provide their own mRNA and DNA synthetic machinery. Vaccinia is the most intensively studied member of the poxvirus family. Virus production "factories" are seen in the cytoplasm - sites of vaccinia virus replication. The early proteins are involved in DNA replication, RNA transcription, RNA modification and un-coating. They also include a few structural proteins. DNA synthesis occurs in "factories" and uses an unusual mechanism. This is a complex process. Some late proteins are made throughout the late phase, but others only at the beginning of late phase. Some early proteins are not synthesised once DNA replication commences while other early proteins are made during late as well as early phases. Thus, there is complex control of which proteins are made by vaccinia and when they are made. This means that there are controls other than just early/late controls. (This is a very large virus, thus the complexity is not surprising.)

Assembly occurs in *factories* in the cytoplasm. The new, immature virus particles acquire a membrane while in the cytoplasm - the exact mechanism is not fully understood but it seems that the virus gets "wrapped" by cellular membranes. The older idea that a membrane is formed directly from lipids rather than from a pre-existing membrane is not correct. There is a gradual maturation of enveloped particles. The virus is usually released by host cell disintegration, but some may get out by budding through membranes (in which case they have an extra membrane). Both forms appear to be infectious. The exact mechanism by which the virus gets out of infected cells may depend on host cell type.

Bacteria

The genetic material of a bacterium is contained within a single double-stranded molecule that is generally organized into one circular chromosome. Unlike eukaryotic chromosomes, the bacterial chromosome is not enclosed by a nuclear envelope and may be attached to one or several points along a bacterium cell membrane. While the replication process in bacteria is complex, and involves the enzymes described earlier, it is not difficult to visualize overall process.

The Mechanism of Prokaryotic DNA Replication: DNA replication in *E. coli* begins at *OriC*. It starts when the polypeptide products of the *dnaA* gene bind to the origin. These polypeptides cause localized strand separation.

This allows a complex of the protein products of the *dnaB* and *dnaC* genes to bind. This complex acts as a **helicase**, which functions to unwind the DNA further. This unwinding produces two **replication forks**. The unwound single-stranded region is kept single stranded through the action of **single-strand binding proteins**. There are other proteins that are found at the replication forks at this time, but their function is not well understood, so they will not be addressed.

At this point, a primer is needed so that DNA polymerase III can begin to act. As mentioned earlier, DNA synthesis needs a primer, so how is a primer produced? An enzyme called **primase** serves this purpose, by synthesizing a short stretch of RNA (generally from 5 to 15 nucleotides in length). RNA synthesis does not require a primer, so primase (which is a type of enzyme called an RNA polymerase) is able to synthesise a short primer where needed. Once this primer is made, DNA synthesis can begin, extending the polynucleotide chain originating with the RNA primer. Priming and synthesis occurs on both strands in the helicase complex moves along the parental DNA, shifting the replication fork, and allowing synthesis to continue.

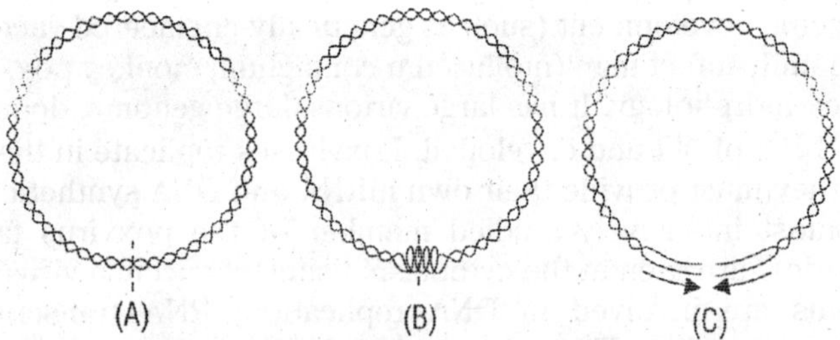

FIG. 9.22: Generation of supercoiled DNA. If a closed circle (a) of DNA is cut, rotated in the opposite direction to the twist of the helix (b), and then cut ends rejoined, (c), the DNA becomes negatively supercoiled.

This leads to another problem that has to be solved. As more DNA unwinds, and the replication fork moves along, synthesis of one strand (the lower strand in the diagram) can just continue, following the movement of the replication fork. The other strand being synthesised, however, cannot do this. As the replication fork moves along, it leaves a gap behind, as shown in panel B of the of below figure. To compensate for this, a second RNA primer must be synthesised a bit behind the first one, and DNA synthesised until it reaches the first primer. You can easily imagine that as the replication fork progresses a bit further, this process will have to be repeated. Therefore, at each replication fork, the synthesis of one new DNA strand (the lower one in the figure) is continuous, while synthesis of the other strand must be accomplished in small increments, short stretch after short stretch; this type of synthesis is termed discontinuous. The strand of DNA that is synthesised continuously is called the leading strand, and the strand that is synthesised discontinuously is called the lagging strand. The small fragments of DNA making up the lagging strand are named Okazaki fragments, after the researcher who discovered them. Okazaki fragments are typically about 1000 to 2000 nucleotides each.

Discontinuous replication solves one problem but still leaves one matter unsettled: the lagging strand will be composed of individual, unjoined fragments of DNA and RNA. This is where DNA polymerase I comes into play. DNA polymerase I uses its 5' to 3' exonuclease activity to digest away the primer RNA, and replaces the primer with DNA by extending the strand from the adjacent Okazaki fragment. At this point all that is left to be done is to physically join the Okazaki fragments. This is accomplished by an enzyme known as DNA ligase. DNA ligase is able to join the 5' end of one DNA strand to the 3' end of another DNA strand.

The DNA replication process begins at single points of origin, or nick site, and moves in two opposite directions (*bi-directional*) around the chromosome, resulting in the formation of two replicating forks. The newly created replication forks move away from the original starting site and around the circular chromosome to form a newly replicated DNA molecule. Since the chromosome is a circular structure, the forks eventually meet when the process is completed. DNA replication is relatively rapid process in bacteria at their respective optimal temperatures. Bacterial cells regulate the initiation of the process to conserve energy and limit the formation of unwanted products.

DNA gyrase (swivelase) is an enzyme that increases or introduces positive Supercoiling in covalently closed circular DNA molecules generated by DNA replication. This is achieved by breakage of at least one DNA strand followed by untwisting and subsequent resealing of the duplex. DNA gyrase is an alternative term for a 2 Topoisomerase.

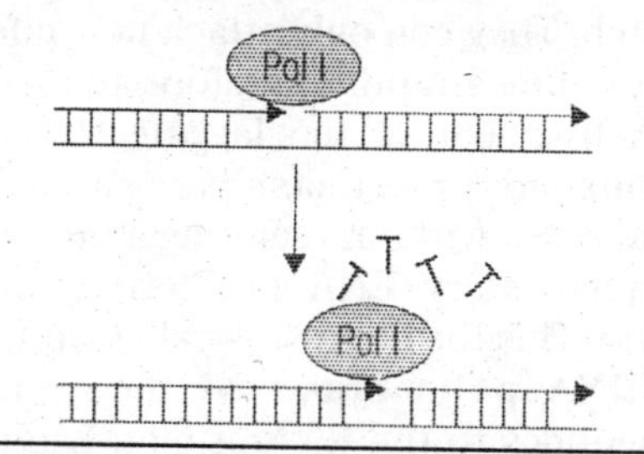

FIG. 9.23: Role of polymerase I in exonuclease activity.

In general, DNA is replicated by uncoiling of the helix, strand separation by breaking of the hydrogen bonds between the complementary strands, and synthesis of two new strands by complementary base pairing. Replication begins at a specific site in the DNA called the origin of replication.

DNA replication is **bidirectional** from the origin of replication. To begin DNA replication, unwinding enzymes called **DNA helicases** cause the two parent DNA strands to unwind and separate from one another at the origin of replication to form two "Y"-shaped **replication forks**. These replication forks are the actual site of DNA copying. **Helix destabilizing proteins** bind to the single-stranded regions so the two strands do not rejoin. Enzymes called **topoisomerases** produce breaks in the DNA and then rejoin them in order to relieve the stress in the helical molecule during replication.

As the strands continue to unwind and separate in both directions around the entire DNA molecule, new complementary strands are produced by the hydrogen bonding of free DNA nucleotides with those on each parent strand. As the new nucleotides line up opposite each parent strand by hydrogen bonding, enzymes called DNA polymerases join the nucleotides by way of phosphodiester bonds. Actually, the nucleotides lining up by complementary base pairing are deoxynucleoside triphosphates, composed of a nitrogenous base, deoxyribose, and three phosphates. As the phosphodiester bond forms between the 5' phosphate group of the new nucleotide and the 3' OH of the last nucleotide in the DNA strand, two of the phosphates are removed providing energy for bonding. In the end, each parent strand serves as a template to synthesise a complementary copy of itself, resulting in the formation of two identical DNA molecules. During this process, the DNA molecules may attach to the cytoplasmic membrane and, as the cell elongates, the two DNA molecules are physically separated.

In reality, DNA replication is more complicated than this because of the nature of the DNA polmerases. DNA polymerase enzymes are only able to join the phosphate group at the 5' carbon of a new nucleotide to the hydroxyl (OH) group of the 3' carbon of a nucleotide already in the chain. As a result, DNA can only be synthesised in a 5' to 3' direction while copying a parent strand running in a 3' to 5' direction. Remember as mentioned above, each DNA strand has two ends. The 5' end of the DNA is the one with the terminal phosphate group on the 5' carbon of the deoxyribose; the 3' end is the one with a terminal hydroxyl (OH) group on the deoxyribose of the 3' carbon of the deoxyribose. The two strands are antiparallel that is they run in opposite directions. Therefore, one parent strand, - the one running 3' to 5' and called the leading strand - can be copied directly down its entire length. However, the other parent strand - the one running 5' to 3' and called the lagging strand - must be copied discontinuously in short fragments (Okazaki fragments) of around 100-1000 nucleotides each as the DNA unwinds.

In addition, DNA polymerase enzymes cannot begin a new DNA chain from scratch. They can only attach new nucleotides onto 3' OH group of a nucleotide in a pre-existing strand. Therefore, to start the synthesis of the leading strand and each DNA fragment of the lagging strand, an **RNA polymerase complex called a primosome or primase** is required. The primase, which is capable of joining RNA nucleotides without requiring a pre-existing strand of nucleic acid, first adds several complementary RNA nucleotides opposite the DNA nucleotides on the parent strand. This forms what is called an **RNA primer**.

DNA polymerase III then replaces the primase and is able to add DNA nucleotides to the RNA primer. Later, **DNA polymerase II** digests away the RNA primer and replaces the RNA nucleotides of the primer with the proper DNA nucleotides to fill the gap. Finally, the DNA fragments themselves are hooked together by the enzyme **DNA ligase**. Yet even with this complicated procedure, a 1000 micrometer-long macromolecule of tightly-packed, supercoiled DNA can make an exact copy of itself in only about 10 minutes time under optimum conditions, inserting nucleotides at a rate of about 1000 nucleotides per second!

There is a great deal of genetic information in the bacterial chromosome. For example *Escherichia coli*, the most studied of all bacteria, has a genome containing 4,639,221 base pairs, which code for at least 4288 proteins.

Replication of DNA in Eukaryotes

Synthesis of DNA in eukaryotes is less well understood, but the process appears to be basically the same as in prokaryotes, with a few notable exceptions. For one thing, eukaryotic DNA is complexed with histones to form chromatin. Every round of replication, therefore, requires that the histones be removed, then replaced after replication is complete. This requirement understandably slows the whole replication process down.

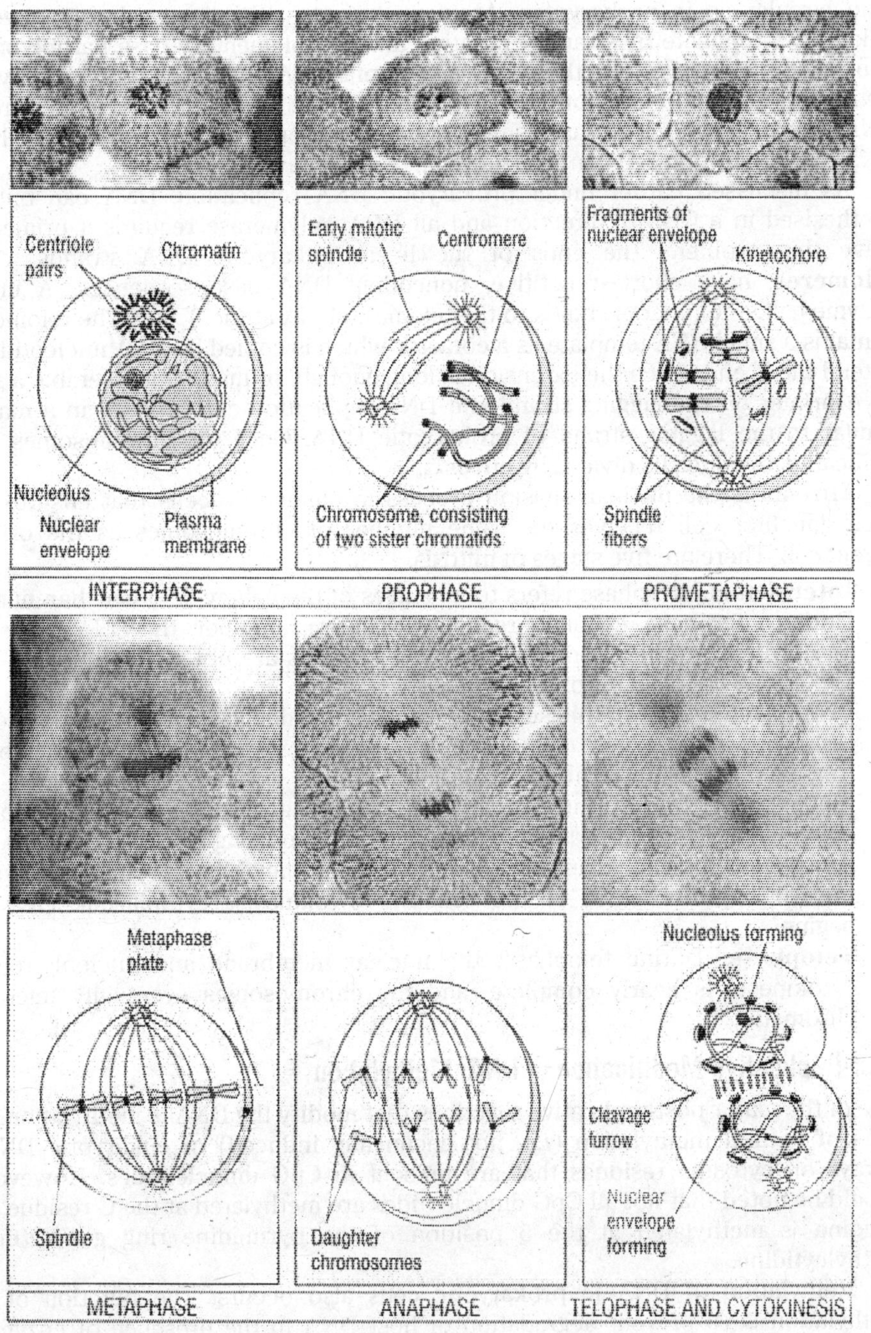

FIG. 9.24: A diagrammatic representation of the mitotic events and cytokinesis.

Eukaryotic cells are also much more complex than prokaryotes, because they contain organelles such as mitochondria and chloroplasts (in plants) that contain their own DNA, which must also be replicated. Eukaryotic cells therefore have more than three DNA polymerases; there have been five DNA polymerases identified so far.

Eukaryotes have another difference: eukaryotic chromosomes are linear, rather than circular as in prokaryotes. As in prokaryotes, the linear chromosomes of eukaryotes replicate by strand separation and complementary base pairing of free deoxyribonucleotides with those on each parent DNA strand. As with prokaryotes, DNA replication in eukaryotic cells is bidirectional. However, unlike the circular DNA in prokaryotic cells that usually has a single origin of replication, the linear DNA of a eukaryotic cell contains multiple origins of replication.

As discussed earlier under prokaryotic DNA replication, DNA can only be synthesised in a 5' to 3' direction and all DNA polymerase requires a primer. To solve this problem, the ends of the linear eukaryotic DNA strands, called **telomeres**, have short, repetitive, noncoding DNA base sequences. A unique enzyme called telomerase binds to the telomeric DNA at the 3' end. The telomerase contains a small RNA template as a cofactor which is copied by DNA nucleotides to extend the 3' end. Once the extension is long enough, primase can assemble a short RNA primer on the lagging strand and DNA replication can proceed in a manner similar to the lagging strand of prokaryotic DNA. Once the chromosomes have replicated, the nucleus divides by **mitosis**.

Mitosis is the nuclear division process in eukaryotic cells that ensures that each daughter cell receives the same number of chromosomes as the original parent cell. There are five stages of mitosis:

1. **Interphase:** Interphase refers to all stages of the cell cycle other than mitosis. During interphase, cellular organelles double in number, the DNA replicates, and protein synthesis occurs. The chromosomes are not visible and the DNA appears as uncoiled chromatin.
2. **Prophase:** During prophase, the chromatin condenses and the chromosomes become visible. Also the nucleolus disappears, the nuclear membrane fragments, and spindle fibers are assembled.
3. **Metaphase:** During metaphase, the nuclear membrane fragmentation is complete and the duplicated chromosomes line up along the cell's equator.
4. **Anaphase:** During anaphase, diploid sets of daughter chromosomes move toward opposite poles of the cell and cytokinesis (cytoplasmic cleavage) begins.
5. **Telophase:** During telophase, the nuclear membrane and nucleoli reform, cytokinesis is nearly complete, and the chromosomes eventually uncoil to chromatin.

Post-Replicative Modification of DNA: Methylation

One of the major post-replicative reactions that modify the DNA is methylation. The sites of natural methylation (*i.e.* not chemically induced) of eukaryotic DNA is always on cytosine residues that are present in CpG dinucleotides. However, it should be noted that not all CpG dinucleotides are methylated at the C residue. The cytidine is methylated at the 5 position of the pyrimidine ring generating 5-methylcytidine.

Methylation of DNA in prokaryotic cells also occurs. The function of this methylation is to prevent degradation of host DNA in the presence of enzymatic activities synthesised by bacteria called restriction endonucleases. These enzymes recognize specific nucleotide sequences of DNA. The role of this system in prokaryotic cells (called the restriction-modification system) is to degrade invading viral DNAs. Since the viral DNAs are not modified by methylation they are degraded by the host restriction enzymes. The methylated host genome is resistant to the action of these enzymes.

The precise role of methylation in eukaryotic DNA is unclear. It was originally thought that methylated DNA would be less transcriptionally active. Indeed, experiments have been carried out to demonstrate that this is true for certain genes. For example, under-methylation of the MyoD gene (a master control gene regulating the differentiation of muscle cells through the control of the expression of muscle-specific genes) results in the conversion of fibroblasts to myoblasts. The experiments were carried out by allowing replicating fibroblasts to incorporate 5-azacytidine into their newly synthesised DNA. This analog of cytidine prevents methylation. The net result is that the maternal pattern of methylation is lost and numerous genes become under methylated. However, lack of methylation nor the presence of methylation is a clear indicator of whether a gene will be transcriptionally active or silent.

The pattern of methylation is copied post-replicatively by the maintenance methylase system. This activity recognizes the pattern of methylated C residues in the maternal DNA strand following replication and methylates the C residue present in the corresponding CpG dinucleotide of the daughter strand.

The phenomenon of genomic imprinting refers to the fact that the expression of some genes depends on whether or not they are inherited maternally or paternally. Insulin-like growth factor-2 (Igf2) is a gene whose expression is required for normal fetal development and growth. Expression of Igf2 occurs exclusively from the paternal copy of the gene. Imprinted genes are "marked" by their state of methylation. In the case of Igf2 an element in the paternal locus, called an insulator element, is methylated blocking its function. The function of the un-methylated insulator is to bind a protein that when bound blocks activation of Igf2 expression. When methylated the protein cannot bind the insulator thus allowing a distant enhancer element to drive expression of the Igf2 gene. In the maternal genome, the insulator is not methylated, thus protein binds to it blocking the action of the distant enhancer element.

DNA, RNA, AND THE FLOW OF GENETIC INFORMATION

DNA and RNA are long linear polymers, called nucleic acids which carry information in a form that can be passed from one generation to the next. These macromolecules consist of a large number of linked nucleotides. Each nucleotide is composed of a sugar, a phosphate, and a base. Genetic information is stored in the sequence of bases along a nucleic acid chain. The bases have an additional special property: they form specific pairs with one another that are stabilized by hydrogen bonds. The base pairing results in the formation of a double helix, a helical structure consisting of two strands. These base pairs provide a mechanism for copying the genetic information in an existing nucleic acid chain to form a new chain.

RNA probably also function as the genetic material very early in evolutionary history, the genes of all modern cells and many viruses are made of DNA. DNA is replicated by the action of DNA polymerase enzymes.

The 'flow' of information from an organism's genome to the synthesis of its encoded proteins is referred to as the **central dogma** and emphasises the crucial roles that nucleic acids play within the cell. The synthesis of proteins (translation) is directed by the base order in mRNA, copied directly from that in the DNA of the genes by transcription.

DNA —Transcription→ RNA —Translation→ Protein

FIG. 9.25: The Central Dogma.

Genes specify the kinds of proteins that are made by cells, but DNA is not the direct template for protein synthesis. Rather, the templates for protein synthesis are RNA (ribonucleic acid) molecules. A class of RNA molecules called messenger RNA (mRNA). They carry information intermediates in protein synthesis. Other RNA molecules are transfer RNA (tRNA) and ribosomal RNA (rRNA). These are the part of the protein-synthesizing machinery. All forms of cellular RNA are synthesised by RNA polymerases that take instructions from DNA templates. This process of transcription is followed by translation, the synthesis of proteins according to instructions given by mRNA templates. Thus, the flow of genetic information, or gene expression, in normal cells is:

Transcription

Unlike prokaryotic RNA polymerase that initiates the transcription of all different types of RNA, RNA polymerase in eukaryotes (including humans) comes in three variations, each encoding a different type of gene.

1. RNA polymerase I is responsible for transcribing RNA that codes for genes that become structural components of the ribosome, a protein responsible for the translation of RNA into proteins.
2. RNA polymerase II transcribes protein-encoding genes, or messenger RNAs, which are the RNAs that get translated into proteins.
3. RNA polymerase III transcribes a different structural region of the ribosome, transfer RNAs, which are also involved the translation process, as well as non-protein encoding RNAs.

Translation

DNA translation is the process that converts an mRNA sequence into a string of amino acids that form a protein. This fundamental process is responsible for creating the proteins that make up most cells. It also marks the final step in the journey from DNA sequence to a functional protein; the last piece of the central dogma to molecular biology.

Translation involves RNAs in the form of the ribosome and tRNAs. In this unit we will be focusing on the relationship between the molecular components of nucleic acids, the structures they form in the cell and how they interact with proteins. We will also examine the approaches used to analyse these structures and discuss how the unique properties of nucleic acids have been exploited as molecular tools.

This flow of information is dependent on the genetic code, which defines the relation between the sequence of bases in DNA (or its mRNA transcript) and the sequence of amino acids in a protein. The code is nearly the same in all organisms: a sequence of three bases, called a *codon*, specifies an amino acid. Codons in mRNA are read sequentially by tRNA molecules, which serve as adaptors in protein synthesis. Protein synthesis takes place on ribosomes, which are complex assemblies of rRNAs and more than 50 kinds of proteins.

The last theme to be considered is the interrupted character of most eukaryotic genes, which are mosaics of nucleic acid sequences called *introns* and *exons*. Both are transcribed, but introns are cut out of newly synthesised RNA molecules, leaving mature RNA molecules with continuous exons. The existence of introns and exons has crucial implications for the evolution of proteins.

Similarities between DNA Replication and DNA Transcription

The processes that synthesise DNA and RNA are similar in that they use similar nucleotide building blocks. They also use the same chemical method of attack by a terminal -OH group of the growing chain on the triphosphate group of an incoming nucleotide. Both replication and transcription are fuelled by the hydrolysis of the pyrophosphate group that is released upon attack. There are, however, a number of important differences between these two distinct processes.

One major difference rests on the fact that while DNA replication copies an entire helix, DNA transcription only transcribes specific regions of **one** strand of the helix. During DNA transcription, only short stretches (about 60 base pairs) of the template DNA helix are unwound. As the RNA polymerase transcribes more of the DNA strand, this short stretch moves along with the transcription machinery. This process is different from that in DNA replication in which the parent helix remains separated until replication is done.

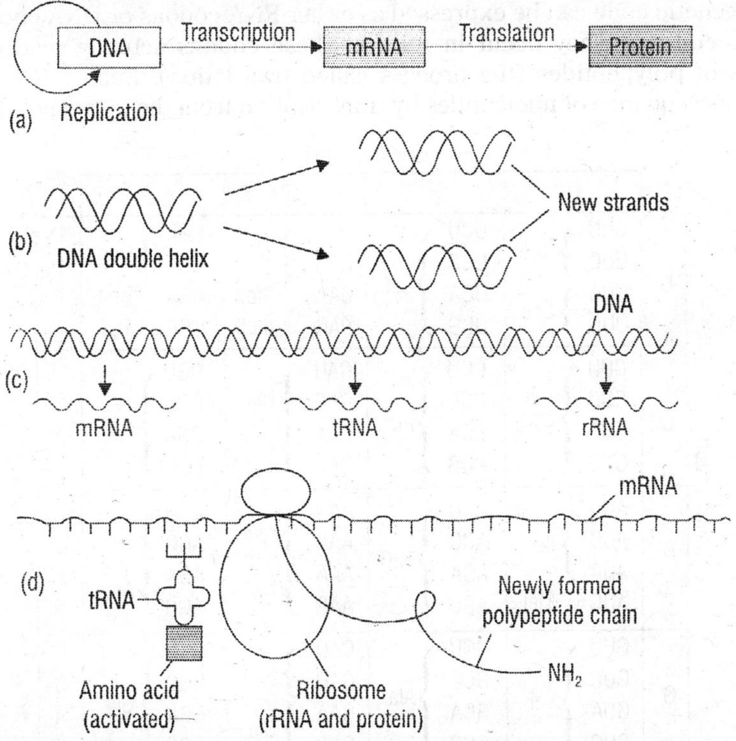

FIG. 9.26: The roles of nucleic acids in the cell. (a) The central dogma describes the flow of genetic information in the cell. (b) Replication of DNA. (c) Transcription of DNA to give mRNA, tRNAs and rRNAs. (d) Translation of mRNA in protein synthesis involves rRNAs in the ribosome and tRNAs.

The key problem in DNA replication lay in the initiation of the addition of nucleotides. RNA primers are needed to begin replication because DNA polymerase is unable to do it alone. DNA transcription does not have the same problem because RNA polymerase is capable of initiating RNA synthesis. The structure of the RNA polymerase is necessary for understanding all of the processes that underlie initiation, elongation, and termination and also explain some of its added capabilities.

GENETIC CODE

Genetic code is a sequence of nucleotides in DNA or RNA that determines the specific amino acid sequence in the synthesis of proteins. It is the biochemical basis of heredity and nearly universal in all organisms.

The genetic code consists of 64 triplets of nucleotides. These triplets are called **codons**. With three exceptions, each codon encodes for one of the 20 amino acids used in the synthesis of proteins. That produces some redundancy in the code: most of the amino acids being encoded by more than one codon. The code is triplet, unpunctuated and nonoverlapping. Three bases are required to specify each amino acid. There are no gaps between codons. Codons do not overlap.

One codon, **AUG** serves two related functions:

- It signals the start of translation.
- It codes for the incorporation of the amino acid methionine (Met) into the growing polypeptide chain.

The genetic code can be expressed as either RNA codons or DNA codons.

RNA codons: They occur in mRNA. These codons actually read during the synthesis of polypeptides (the process called translation). Each mRNA molecule acquires its sequence of nucleotides by transcription from the corresponding gene.

First Letter	Second Letter: U	C	A	G	Third Letter
U	UUU Phe	UCU Ser	UAU Tyr	UGU Cys	U
	UUC Phe	UCC Ser	UAC Tyr	UGC Cys	C
	UUA Leu	UCA Ser	UAA Stop	UGA Stop	A
	UUG Leu	UCG Ser	UAG Stop	UGG Trp	G
C	CUU Leu	CCU Pro	CAU His	CGU Arg	U
	CUC Leu	CCC Pro	CAC His	CGC Arg	C
	CUA Leu	CCA Pro	CAA Gln	CGA Arg	A
	CUG Leu	CCG Pro	CAG Gln	CGG Arg	G
A	AUU Ile	ACU Thr	AAU Asn	AGU Ser	U
	AUC Ile	ACC Thr	AAC Asn	AGC Ser	C
	AUA Ile	ACA Thr	AAA Lys	AGA Arg	A
	AUG Met	AGC Thr	AAG Lys	AGG Arg	G
G	GUU Val	GCU Ala	GAU Asp	GGU Gly	U
	GUC Val	GCC Ala	GAC Asp	GGC Gly	C
	GUA Val	GCA Ala	GAA Glu	GGA Gly	A
	GUG Val	GCG Ala	GAG Glu	GGG Gly	G

FIG. 9.27: The standard genetic code. Synthesis of a peptide always starts from methionine (Met), coded by **AUG**. The stop codon (UAA, UAG or UGA) signals the end of a peptide. This table applies to mRNA sequences. For DNA, U (uracil) should be replaced by T (thymine). In a DNA molecule, the sequence from an initiating codon (ATG) to a stop codon (TAA, TAG or TGA) is called an **open reading frame** (ORF), which is likely (but not always) to encode a protein or polypeptide.

DNA codons: These are the codons as they are read on the sense (5' to 3') strand of DNA. Except that the nucleotide thymidine (T) is found in place of uridine

(U). They read the same as RNA codons. However, mRNA is actually synthesised using the antisense strand of DNA (3' to 5') as the template.

The genetic code is **unambiguous**; no codon specifies more than one amino acid. The exceptions so far are AUG, UGA and UAG. In the first case, AUG specifies both Methionine and N-formyl-Methionine, which is used to initiate protein synthesis in bacteria. In the second case, UGA specifies the twenty-first amino-acid selenocysteine as well as being a stop codon and, in the last case, UAG specifies the twenty second amino acid (the most recent to be added to the list), pyrrolysine.

- There are 3 stop codons: UAA, UAG, and UGA.
- **There is one start codon:** AUG. However, note that GUG and UUG are occasionally found as start codons.
- The Genetic Code is Universal. Although there are a number of exceptions to this rule - particularly in organelle systems - the genetic code is remarkably the same in all organisms. The most common exception is the use of UGA as a codon for Tryptophan in mitochondria.

Codon Bias

Two of the amino acids (Met and Trp) can be encoded by from 2 to 6 different codons. However, the genome of most organisms reveals that certain codons are preferred over others. In humans, for example, alanine is encoded by GCC four times as often as by GCG. Why should this be? The answer is uncertain but probably reflects greater translation efficiency by the translation apparatus (e.g., ribosomes) for certain codons over their synonyms. Codon bias even extends to pairs of codons: wherever a human protein contains the amino acids Ala-Glu, the gene encoding those amino acids is seven times as likely to use the codons GCAGAG rather than the synonymous GCCGAA. Codon bias is exploited by the biotechnology industry to improve the yield of the desired product. The ability to manipulate codon bias may also usher in a era of safer vaccines.

Exceptions to the Code

The genetic code is **almost** universal. The same codons are assigned to the same amino acids and to the same **START** and **STOP** signals in the vast majority of genes in animals, plants, and microorganisms. However, some exceptions have been found. Most of these involve assigning one or two of the three STOP codons to an amino acid instead.

TRANSFER OF GENETIC MATERIAL

The transfer of genetic material is a *one-way gene transfer from donor to recipient.* Recombination in eucaryotes tends to be reciprocal- that is all of the DNA is conserved in the gametes that eventually arise from meiosis and recombination. There are three ways of movement of DNA from a donor bacterium to the recipient:

1. **Conjugation:** Direct transfer between two bacteria temporarily in physical contact,
2. **Transformation:** Transfer of a naked DNA fragment. Transformation is similar to tranduction but 'naked' DNA is transferred, and
3. **Transduction:** Transport of bacterial DNA by bacteriophages.

Whatever the mode of transfer, the **exogenote** has only four possible fates in the recipient.

- **First**, when the exogenote has a sequence homologous to that of the endogenote, integration may occur; that is, it may pair with the recipient DNA and be incorporated to yield a recombinant genome.

- **Second**, the foreign DNA sometimes persists outside the endogenote and replicates to produce a clone of partially diploid cells.
- **Third**, the exogenote may survive, but not replicate, so that only once cell is a partial diploid.
- **Finally**, host cell nucleases may degrade the exogenote, a process called **host restriction**.

Transformation

The gene transfer is usually set up between a donor strain that possesses the wild-type alleles and a recipient with the recessive alleles, transfer into the recipient strain being recessive alleles, transfer into the recipient strain being recessive alleles, transfer into the recipient strain being monitored by looking for acquisition of the biochemical.

F. Griffith (1928) initially conducted the experiments on transformation. He injected the mixture of two strains on *Pneumococcus (Diplococcus pneumoniae)* into mice. One of these strains SIII was virulent (causing infection) and another strain RII was non-virulent. When heat killed virulent strain (SIII) was injected to a mice, it did not cause death. This suggests that SIII strain became non-virulent after heating. If a mixture of RII (living) and SIII (heat killed) was injected to mice, it caused death. The virulent pneumococci were also isolated from killed mice.

Based on the Griffith's experiment O.T. Avery, CM. Macleod and M. Mc Carthy performed further experiment in an in vitro system. The identified the transforming substance in 1944 as the polysaccharide present in the capsule of virulent strain (SIII) of *Pneumococcus*, which was absent in non-virulent strain (RII). It is a simple type of recombination method in which the bacterial cells are changed into another recombinant form by absorbing a transforming principle from a filtrate of another strain of organisms. Transformation is entirely a laboratory procedure and never occurs in nature.

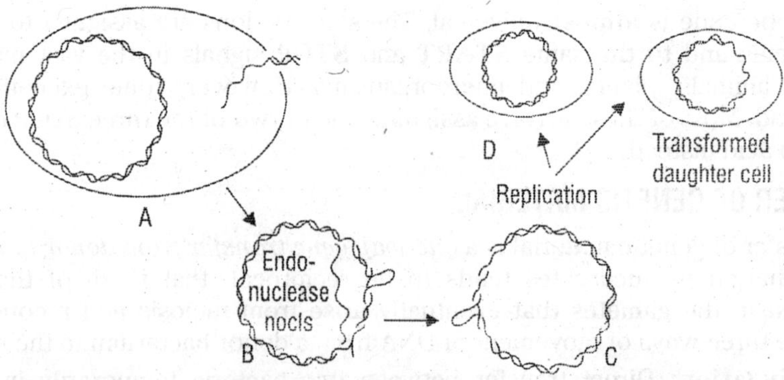

FIG. 9.28: Representation of transforming principle (DNA) during transformation.

In the laboratory, DNA of one strain of the bacterium is extracted and is prepared in the purified form. Thereafter, the pure DNA is mixed with the culture medium which contains entirely different strain. The mixture of extracted DNA and strain allows to grow for a considerable period. After this period, the progeny will bear the properties of that strain from which DNA was extracted. It shows that extracted DNA from the donor cell has incorporated into the chromosome of the receptor cell which was transformed thereby in a cell with entirely new properties.

The transformation of genetic material has also been demonstrated in several bacterial genera like *Bacillus, Haemophilus, Salmonella, Streptococcus, Rhizobium, Neisseria.* Such transformation does not normally occur in *E. coli.* Transformation may occur within species and very rarely between the two species.

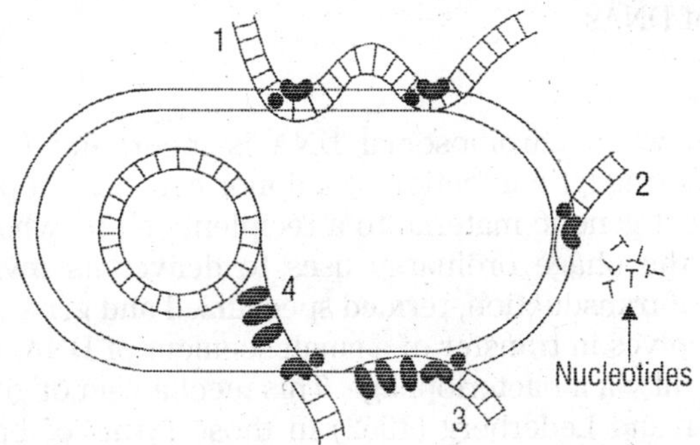

FIG. 9.29: The mechanism of transformation. (1) a long double-stranded DNA molecule binds to the surface with the aid of a DNA-binding protein (●) and is nicked by a nuclease (). (2) one strand is degraded by the nuclease. (3) the undegraded strand associates with competence-specific protein ( ). (4) the single strand of the homologous region of the host DNA.

Competence

The transformation in bacterial cell is dependent on the competence of recipient bacteria to accept DNA. This character of a bacterial cell is not a permanent feature but can be observed only at certain times in the life cycle. For example in *Pneumococcus* the ability of competence develops 90-100 minutes after beginning of the culture and lasts for 7-15 minutes. Probably, competence has no relation with cell division. The physical contact of cells is also not required for transfer of competence. The structure of cell and synthesis of specific receptor sites has important note in development of competence.

Binding

During transformation, DNA comes into contact with outer surface of the bacterial cell. This occurs due to random collision. The first step of the binding of donor DNA is temoprary and lasts for 4-5 seconds only. Then donor DNA binds permanently to the cell surface at several receptor sites and probably associated themselves with the mesosomes. The DNA which is transformed, is double strand (dsDNA) while the non-transforming DNA (ssDNA) is single strand and it is less effective (Miao and Guild, 1970).

Penetration

The transformed dsDNA penetrates the bacterial cell and bound permanently. It has been observed that dsDNA is converted into ssDNA upon the entering the *Pneomococcus* cell. The minimum length of penetrating DNA should be of about 750 base pairs. In *Bacillus subtilis,* the donor DNA reaches from outer surface to the chromosome DNA within 15-60 minutes.

Synapsis

The chromosome DNA of the recipient cell then unwinds and initiates the pairing with ssDNA (donor). This is called synapsis. The donor ssDNA and recipient ssDNA are homologous but not identical. This character of ssDNA is found to be essential for transformation.

Integration

The endonuclease cut the unsynapsed portion of recipient of DNA at two places and endonuclease then degrades the ssDNA. There is an involvement of **repair synthesis** during the **integration** of the donor DNA to close the gaps present between two types of DNAs.

Transduction

Transduction occurs when chromosomal DNA is incorporated, by error, into a bacteriophage coat during lytic infection of a donor cell. Such a phage then acts as a vector to deliver host genetic material to a recipient cell by whatever mechanism (e.g. injection) that the phage ordinarily uses to deliver its own infection DNA. There are two types of transduction, termed specialized and generalized.

Transduction involves in transfer of a small segment of DNA, up to 50 kb or so, from donor to recipient via a bacteriophage. This mechanism of gene exchange was discovered by Zinder and Lederberg (1952) in those forms of bacteria which are responsible for mouse typhoid (*Salmonella typhimurium*). Transduction has now been verified in other bacteria like *E. coli* and *Micrococcus*. In this process, the particle of genetic material is transferred by the means of a bacteriophage.

In the first stage of transduction, bacteriophage are allowed to develop on one strain of bacterium of known genetic constitution- for example, the bacterial strain, which is unable to synthesis amino acid- histidine (hist$^-$). Bacteriophages then allowed to infect a different strain of bacterium which has the property to synthesis histidine (hist$^+$). Those bacteria which survive even after the infection of bacteriophages are than isolated and observed.

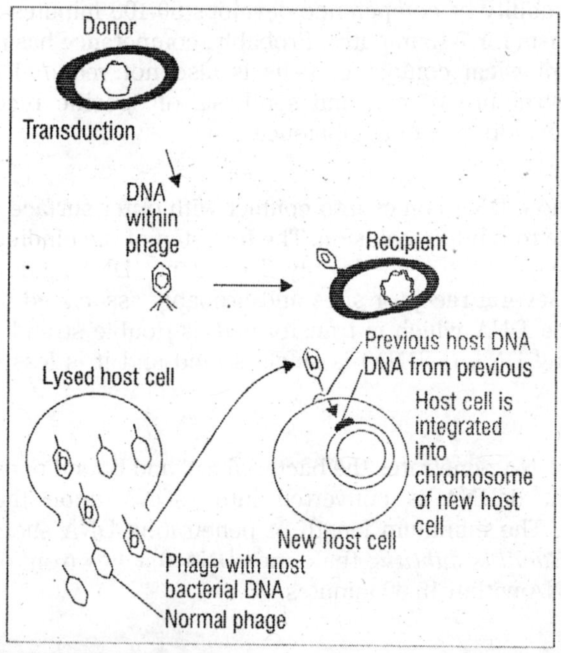

FIG. 9.30: Mechanism of transduction.

Transduction is of two types:

1. Generalized transduction.
2. Specialized transduction.

The **specialized (restricted) transduction** occurs only in some temperature viruses where DNA from a specific region of the host chromosome is integrated into the virus genome usually replacing some of the genes. It was first discovered in *Escherichia coli* bringing transduction of the galactose genes by temperate phage lambda. The region in which lambda integrates is immediately adjacent to the cluster of host genes which control the enzymes involved in galactose utilization, and the DNA of lambda is inserted into the host DNA at that site.

The **abortive transduction** is the fate of transduced DNA in the recipient cell (not called a transduction) depends on various factors. If DNA is a complete recipient (e.g., a plasmid) it may be stably inherited by the transductant. But is the DNA is a fragment of a chromosome or plasmid it may meet one of the three possible fates.

1. It may be completely degraded by the recipient cell's restriction endonuclease system.
2. It may undergo recombination with a homologous region of the parents chromosome (or plasmid) such that at least some of the gene it carriers can be stably inherited also called complete transduction.

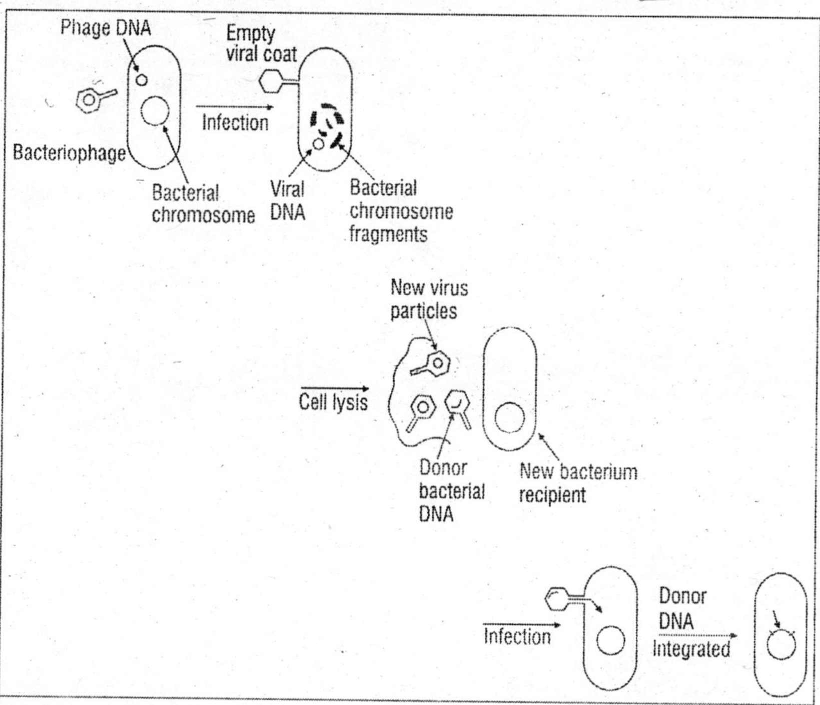

FIG. 9.31: Bacteriophage infects bacteria by injecting contents of the virus head (viral nucleic acid-DNA) through a puncture hole in cell wall. The viral nucleic acid then takes control of the cell metabolism and directs the synthesis of more viral nucleic acid and other materials needed for making complete virus. In a short time the newly formed virus is released by a sudden rapture of the cell wall-*lysis*, and the virus particles are free to infect other susceptible bacteria.

It may persist in the cell in a stable but not replicating form which I referred to as abortive transduction. The transduced DNA in an abortive transductant may exist as a circular DNA protein complex.

These cells acquired one of the properties of the first strain i.e. the cell does not synthesis histidine or will be of hist$^-$ type. This change in the property of bacteria is due to the transference of genetic material of one strain (hist-) into another strain (hist$^+$) by bacteriophage in bacterial cell.

During this process a small part of genetic material from the first type (hist$^-$) attached with the second cell. In fine, transduction can be referred as a specific type of genetic exchange which its possible with the help of bacteriophages. Here, only very small pieces are transferred from one host to another and leads to the formation of new recombinant.

CHAPTER 10

Organic Component Nucleic Acids: RNA

The primary structure of RNA is similar to that of DNA, except for the presence of ribose and uracil in place of deoxyribose and thymine. The base composition of RNA does not follow Chargaff's rules, since RNA molecules are made of only one chain.

RNA is synthesised within the nucleus by using only one strand of DNA as template. RNA is a faithful copy of the information contained in the DNA. There are three major classes of ribonucleic acid: ribosomal RNA (rRNA), transfer RNA (tRNA), and messenger RNA (mRNA). All of them are involved in protein synthesis.

Although each RNA molecule has only a single poly nucleotide chain, RNA is not a simple, smooth, linear structure. RNA molecules have extensive regions of complementary in which hydrogen bonds between AU and GC pairs are formed between different regions of the same molecule. As a result the molecule folds upon itself, forming structures called hairpin loops. In the base-paired regions the RNA molecule adopts a helical structure which is comparable to that of DNA. The compact structure of RNA molecules folded upon themselves has important biological consequences.

Viroids are made of an RNA molecule only that is not covered with protein. These are infectious agents that attack plant cells. Viroid that attacks potatoes has a circular RNA only 359 nucleotides long. It has a compact base paired structure. This probably protects the molecule from ribonucleases.

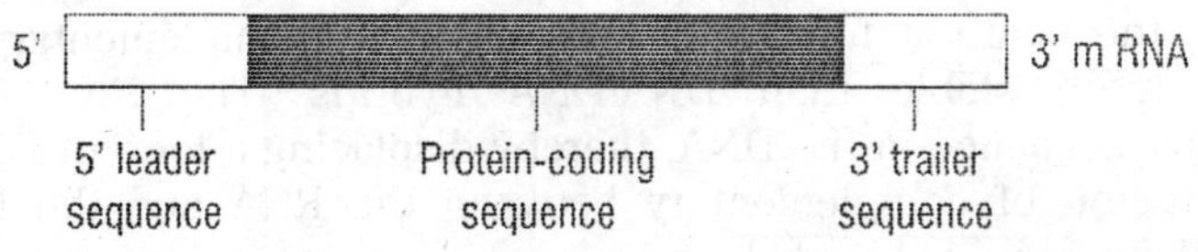

FIG. 10.1: General structure of mature prokaryotic and eukaryotic mRNA molecules.

In prokaryotes and eukaryotes, when the process of transcription is completed, three major products result: messenger RNA, transfer RNA, and ribosomal RNA. In eukaryotes only, a fourth end product, small nuclear RNA, results. The initial RNA transcripts of most genes must be modified before the biological active (mature) RNA molecules are produced. These initial transcripts are called precursor RNA molecules (pre-RNAs). Their processing may involve the addition and/or removal of bases, the chemical modification of some bases, or the cleavage of sequences from the precursor. Processing of precursor-RNA molecules is similar in both prokaryotes and eukaryotes for tRNA and rRNA precursors. Prokaryotic mRNA is not processed while many eukaryotic mRNAs are generated by processing of precursor molecules.

THE STRUCTURE AND FUNCTION OF MESSENGER RNA

All mRNA molecules are transcribed from structural genes. The mRNA molecule has three main parts. At the 5' end is a leader sequence whose length varies from mRNA to mRNA. This leader sequence has the coded information. The ribosome reads to tell it where to begin the synthesis of the protein. The bases of the leader sequence are not translated into amino acids. Following the 5' leader sequence is the actual coding sequence of the mRNA. This sequence determines the amino acid sequence of a protein during translation. Since the number of amino acids varies from proteins to protein, the mRNAs vary in length. Following the coding sequence is a trailer sequence. This is also not translated and varies in length from mRNA to mRNA. An mRNA molecule contains more nucleotides than are necessary for the protein coding. The trailer sequence indicates that the coding sequence has been completely translated, and that all specified amino acids have assembled into the specified protein.

The production of functioning mRNA is different in prokaryotes and eukaryotes. In prokaryotes the RNA transcript functions directly as the mRNA molecule for translation. In eukaryotes the primary RNA transcript must be modified in the nucleus by RNA processing to produce the mature mRNA. In prokaryotes an mRNA begins to be translated on ribosomes before it has been completely transcribed. This process is called the coupling of transcription and translation. In eukaryotes the mRNA must migrate from the nucleus to the cytoplasm before it can be translated by the ribosomes. Therefore a eukaryotic mRNA is always completely transcribed and processed before it is translated.

In prokaryotes the initial transcript of a structural gene is processed very little, if at all, to produce the mature mRNA molecule. An exact point-by-point relationship exists between genes and mRNA. The relationship exists between the orders of base pairs ion the structural gene and the order of the corresponding bases in both the initial transcript of the gene and the mature mRNA.

The evidence for this relationship came from experiments using a technique called R looping. This procedure was developed by M. Thomas, R. White, and R. Davis. The molecules of double-stranded DNA are incubated at temperatures below their denaturing temperature.

This opens up short stretches of the DNA double helix and single-stranded RNA molecules in the solution form RNA/DNA hybrids (complementary). RNA/DNA hybrids are more stable than DNA/DNA hybrids. The RNA binds to its complementary sequence in the DNA, thereby displacing a loop of DNA. The DNA outside the region of complementary between the RNA and the DNA remains double-stranded.

Eukaryotic mRNAs are usually modified at both the 5' and 3' ends. The modifications that occur after the molecule has been transcribed are catalysed by specific enzymes.

At the 5' end the mRNA is modified by the addition of a cap. It consists of the addition of a guinine nucleotide (7-methylquanosine) to the terminal 5' nucleotide by an unusual 5' to 5' linkage and the addition of two methyl groups (CH_3) to the first two nucleotides of the RNA chain.

This 5' capping structure is found in most eukaryotes, although slight variations are seen in the structure of the cap itself. There is no DNA template for the 5' cap. The cap is essential for mRNA to attach to the ribosome. A cap-binding protein is responsible for the capped end of the mRNA to bind to the small ribosomal subunit. Once bound, the large subunit binds with the complex and translation can begin. Thus, capping is necessary for the initiation of protein synthesis.

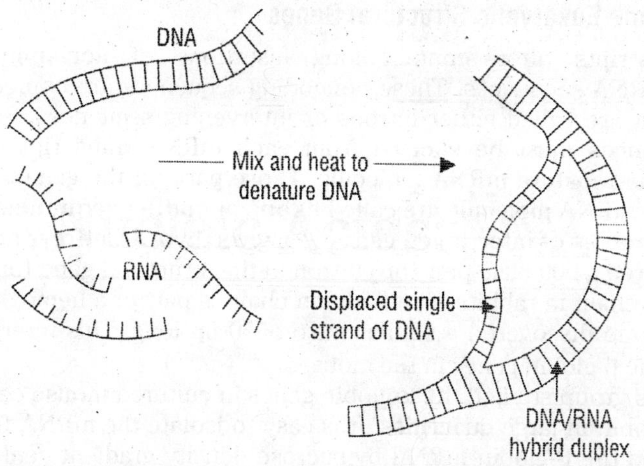

FIG. 10.2: The R- looping procedure.

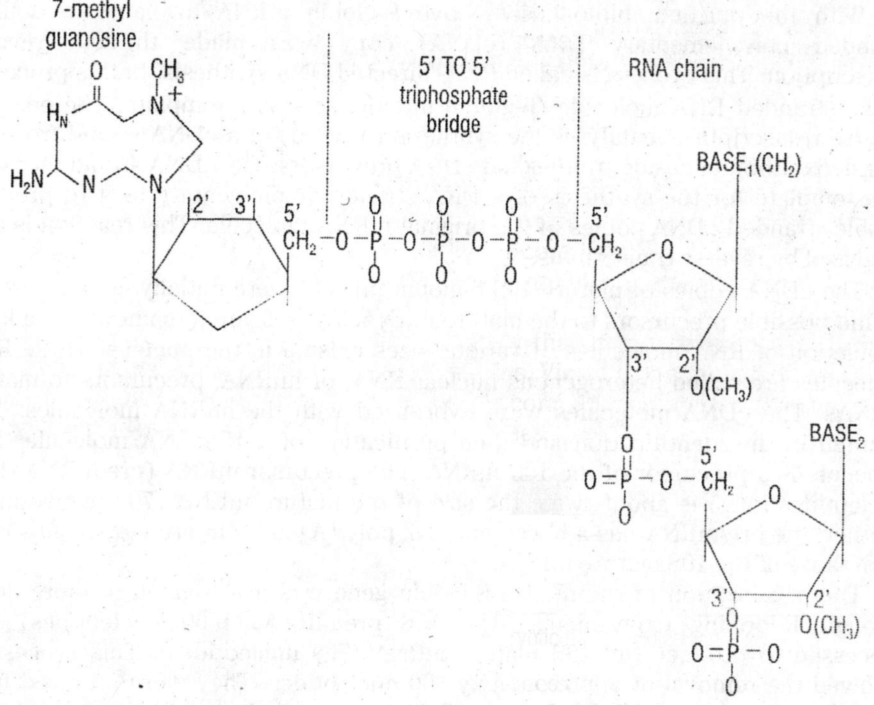

FIG. 10.3: Structure of the 5' cap of mRNA.

The post transcriptional modification at the 3' ends of most eukaryotic mRNAs is an addition of a sequence of about fifty to two hundred and fifty adenine nucleotides. This sequence is called a poly (A) tail. Its production is catalysed by the enzyme polyadenylate or poly (A) polymerase. The energy for the reaction is supplied by ATP.

$$\text{mRNA} + \text{nATP} \xrightarrow{\text{Poly (A) polymerase}} \text{mRNA} - \text{(A)n} + \text{nPP}_1$$

Introns in Some Eukaryotic Structural Genes

mRNA transcripts often contain long insertions of non-amino acid-coding (noncoding) RNA sequences. These noncoding sequences are copied from regions of a gene that are called either introns or intervening sequences. Such noncoding mRNA sequences must be excised from each mRNA transcript to convert the transcript into a mature mRNA molecule. Those parts of the gene that correspond to the mature mRNA molecule are called **exons** or **coding sequences**.

The first report of interrupted genes was given by A.J. Jeffreys and R.A. Flavell. They discovered a 600 base-pair (bp) intron in the structural gene for the 146 amino acid ß-globin chain in rabbits. The ß-globin chain is part of a hemoglobin molecule. P. Leder's group discovered a similar intron 550 bp long. This interrupted each of two non-allelic β-globin genes in the mouse.

P. Leder's group studied the ß-globin genes in cultured mouse cells. These cells produce ß-globin in large quantities. It is easy to isolate the mRNA for this protein. They analysed the ß-globin mRNA by sucrose density gradient centrifugation. The β-globin mRNA has a size of about 10S, which is about the size expected if all the mRNA codes for the β-globin protein.

With the purified, biologically active ß-globin mRNA available, a double-stranded complementary DNA (cDNA) copy was made through reverse transcription. This process is called RNA- directed DNA synthesis. In this process a single stranded RNA molecule (β-globin mRNA) acts as a template. The enzyme, reverse transcriptase, catalyses the synthesis (in vitro) of a cDNA strand from the four deoxyribonucleoside triphosphate DNA precursors. The cDNA strand then acts as a template for the synthesis of a DNA strand complementary to it to produce double-stranded cDNA copies of the original mRNA molecule. This reaction is also catalysed by reverse transcriptase.

The cDNA copies of mature 10S β-globin mRNAs were initially used as probes to find possible precursors to the mature mRNA. The researchers knew that a large population of RNA molecules of various sizes existed in the nucleus. These RNA molecules are called heterogenous nuclear RNA, or hnRNA, precursors to mature mRNAs. The cDNA molecules were hybridized with the hnRNA molecules. This resulted in the identification and then purification of a 15S RNA molecule. This molecule is a precursor of the 10S mRNA. The precursor mRNA (pre-mRNA, 1500 nucleotides long) is about twice the size of the mature mRNA (700 nucleotides). Further, the pre-mRNA has a 5' cap and a 3' poly (A) tail that are indistinguishable from those of the 10S mature mRNA.

The organization of the mouse ß-globin gene was investigated in more detail through R-looping experiments. The 15S pre-mRNA (1500 nucleotides) was processed to produce the 10S mature mRNA (700 nucleotides). This processing involved the removal of approximately 800 nucleotides. They were removed from the middle of the pre-mRNA. The pre-mRNA is a base-by-base copy of the base-pair sequence of the DNA, where as the mature mRNA is not. The evidence was that the gene contained an intervening sequence of base pairs that is transcribed as part of the pre-mRNAs then removed in subsequent processing. The intron is spliced out to produce the mature mRNA molecule.

Experiments have established that a large number of eukaryotic protein-coding genes are interrupted by introns. All mammalian β-globin genes contain two introns. The structure of the mammalian β-globin gene, its transcription and the removal of introns generate the mature mRNA.

$Exon_1$ is the 5' non-translated sequence and the sequence for amino acids 1-30.

Exon2 codes for amino acids 31-104 and Exon3 contains sequences for the remaining amino acids and the 3' non-translated sequence.

Recently, an intron was found in a prokaryotic system in the bacteriophage T4 thymidylate synthetase gene. In that gene, a single intron split the coding sequence into two parts, one coding for 183 amino acids and the other coding for the other 103 amino acids. The intron is 1017 base pairs long.

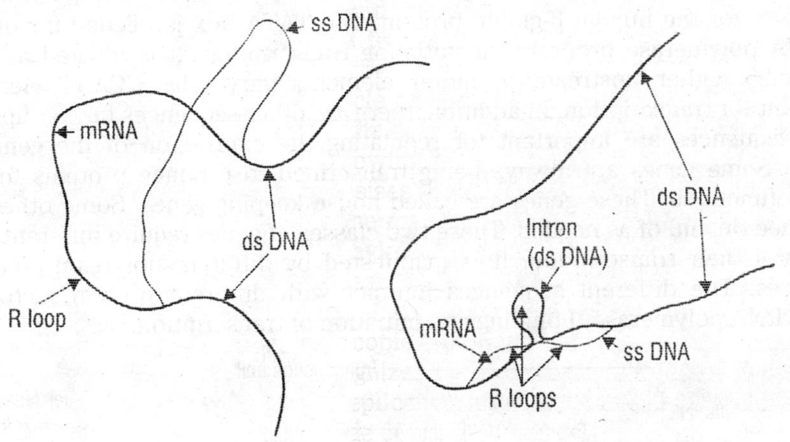

FIG. 10.4: Diagrams showing R loops: (a) formed between 15S β-globin pre-mRNA and the β-globin gene (b) formed between 10S mature β-globin mRNA and the β-globin gene.

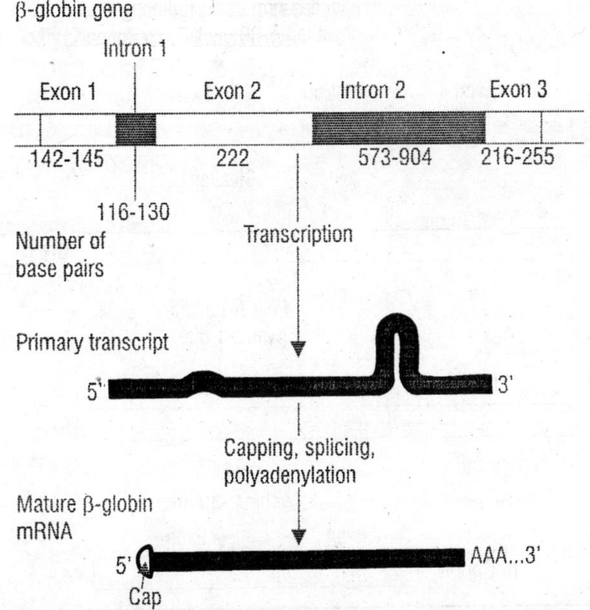

FIG. 10.5: Structure of the mammalian β-globin gene and model for the removal of introns as mature mRNA is produced.

Production of mature mRNA in Eukaryotes: The steps of mRNA production from genes with introns are the initiation of transcription by RNA polymerase II, the addition of the methylated 5'-cap, the addition of the poly (A) tail, and the splicing of the primary transcripts, if necessary, in the nucleus to remove the introns and produce the mature mRNAs. The sequence of steps is the same for genes without introns, except that the step involving the removal of introns is not needed.

Initiation of Transcription by RNA Polymerase II

RNA polymerase II is responsible for transcribing a large number of different RNA molecules including all the protein-coding genes and the genes for small nuclear RNAs. Fig. 10.6 presents the sequence of the mammalian RNA polymerase II promoters for the human β-globin protein. The TATA box is needed for orienting the RNA polymerase properly for initiating transcription. It is located at around position-25. Other upstream promoter elements vary. The CCAAT element is important for transcription. In addition, there are other sequences farther upstream. These sequences are important for regulating the expression of the genes they control. Some genes are always being transcribed to produce proteins for basic cellular functions. These genes are called house keeping genes. Some other genes are turned on and of as needed. These two classes of genes require different signals to control their transcription. It is manifested by different upstream controlling sequences. The different sequences interact with different protein factors that control RNA polymerase II binding and initiation of transcription.

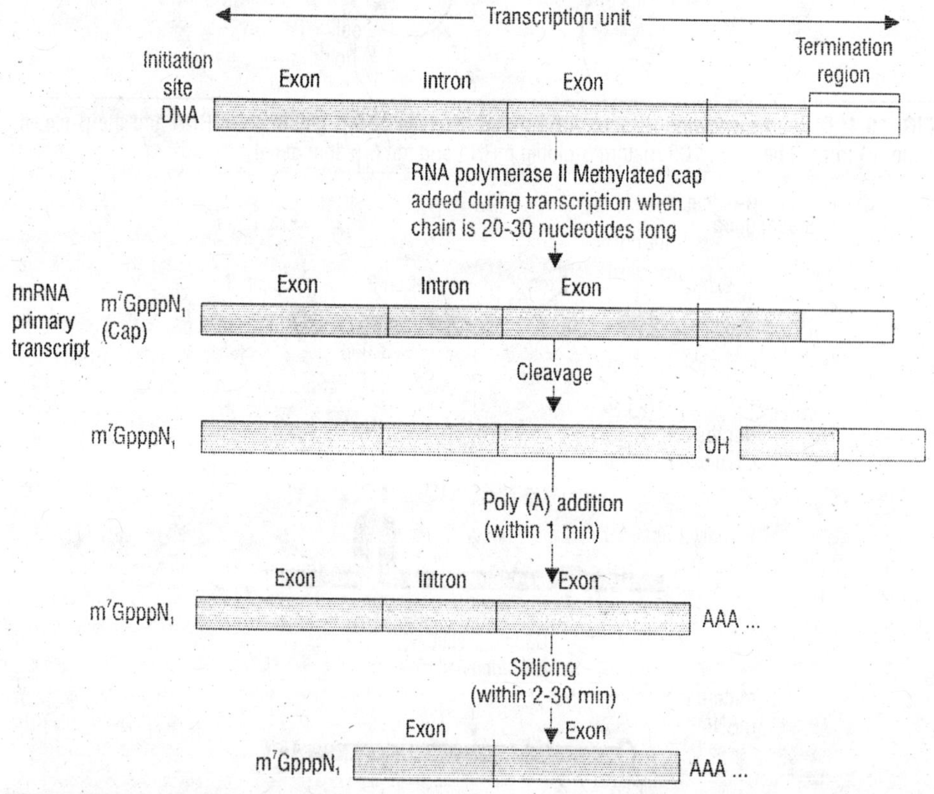

FIG. 10.6: General sequence of steps in the formation of eukaryotic mRNA. All steps are not necessary for all mRNAs.

The addition of a methylated cap occurs at the 5'end of each primary transcript when it is about 20 to 30 nucleotides long. The RNA polymerase II starts transcription at the base to which the cap is added. That site in the DNA is called the cap site. There is no DNA template for the 5'cap.

Poly (A) tails are added to the 3' ends of most mRNA molecules. There is no DNA template for the poly (A) tails of mRNA molecules. There are no specific transcription termination sequences near the end of an mRNA molecule.

mRNA transcription catalysed by RNA polymerase II proceeds for hundreds or thousands of nucleotides transcription past the so-called poly (A) addition site. It is cleaved by an RNA-specific endonuclease generates a 3' OH end to which the poly (A) tail is added. The sequences for controlling poly (A) addition have been identified. About 10 to 30 nucleotides upstream from the poly (A) site there is the strongly conserved sequence AAUAAA. This AAUAAA sequence and the sequences down stream from the poly (A) site function together to signal the location of the poly (A) site. This leads to cleavage of the primary transcript and addition of the poly (A).

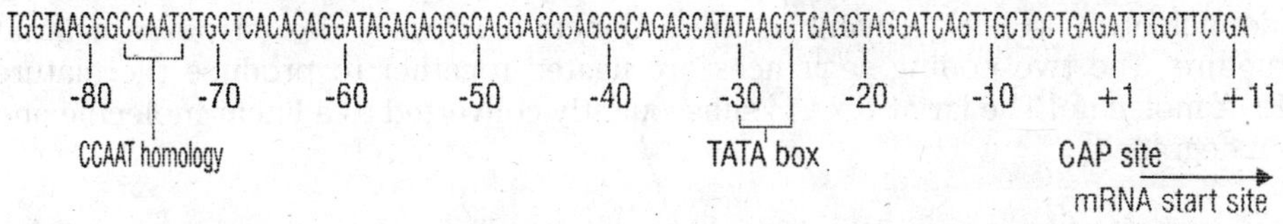

FIG. 10.7: Essential sequences in one of the best characterized mammalian RNA polymerase II promoters, the human β-globin promoters.

The cleavage of the primary transcript at the poly (A) site and the addition of the poly (A) tail occur almost simultaneously. They are actually two separate events.

In the nucleus, the length of the poly (A) tail remains constant in a given species. Once the poly (A) containing mRNA reaches the cytoplasm, the poly (A) tail is shortened to produce irregularly sized tails. In mammals, for example, the tail in nuclei is about 250 As long where as in the cytoplasm it varies from 30 to 250 As. The function of the tail is not clear. mRNAs with poly (A) tails have longer lives in cells than mRNAs without the tails. Splicing of the primary transcript produces mature mRNA.

1. Pre-mRNA molecule containing two coding sequences, 1 and 2, separated by an intron.
2. An endonucleolytic cleavage event separates coding sequence 1 from the rest of the RNA molecule. The G at the 5' end of the intron pairs covalently with an A residue located about 20 to 40 bases before the 3' splice junction to produce a lariat structure.
3. An endonucleolytic cleavage event at the 3' junction releases the intron as a lariat structure, and the two coding sequence are ligated together to produce the mature mRNA molecule.

The primary transcript is a precursor-mRNA molecule since it contains noncoding as well as coding sequences. The pre-mRNA molecules are processed to remove the introns. The introns are looped out, and the loop is removed by nuclease cleavage. The now adjacent coding sequences (the exons) are then ligated together to generate a contiguous molecule.

These events are called mRNA splicing. There are specific nucleotide signals that indicate exon-intron junctions. At the RNA level the introns begin with a 5' GU and end with a 3' AG. It is called the GU-AG rule. The 5' splice junction probably involves at least seven nucleotides and the 3' splice junction at least ten nucleotides.

The sequence of events involved in splicing two coding sequences together with the elimination of an intron. Pre-mRNA molecule with two coding sequences 1, and 2, are separated by an intron.

The first step in splicing is a cleavage at the 5' splice junction. Thus, results in the separation of coding sequence 1 from an RNA molecule that contains the intron and coding sequence 2. The G at the 5' end of the intron joined to an A that is part of a sequence about 18 to 40 nucleotides upstream of the 3' splice junction to produce a lariat structure (the looped back structure). In mammalian systems the branch-point consensus sequence is Py X Py Pu A Py, where Py is a pyrimidine, Pu is a purine, X is any base, and within the consensus sequence.

The branch point in the RNA, that gives the lariat structure. It involves an unusual 2'-5' phosphodiester bond formation between the 2' OH of the adenine nucleotide and the 5' phosphate of the guanine nucleotide at the end of the intron. The A remains in normal 3'-5' linkage with its adjacent nucleotides of the intron. An endonucleolytic cleavage event at the 3' junction releases the intron as a lariat structure. The two coding sequences are ligated together to produce the mature mRNA molecule. The lariat RNA is subsequently converted to a linear molecule and then degraded.

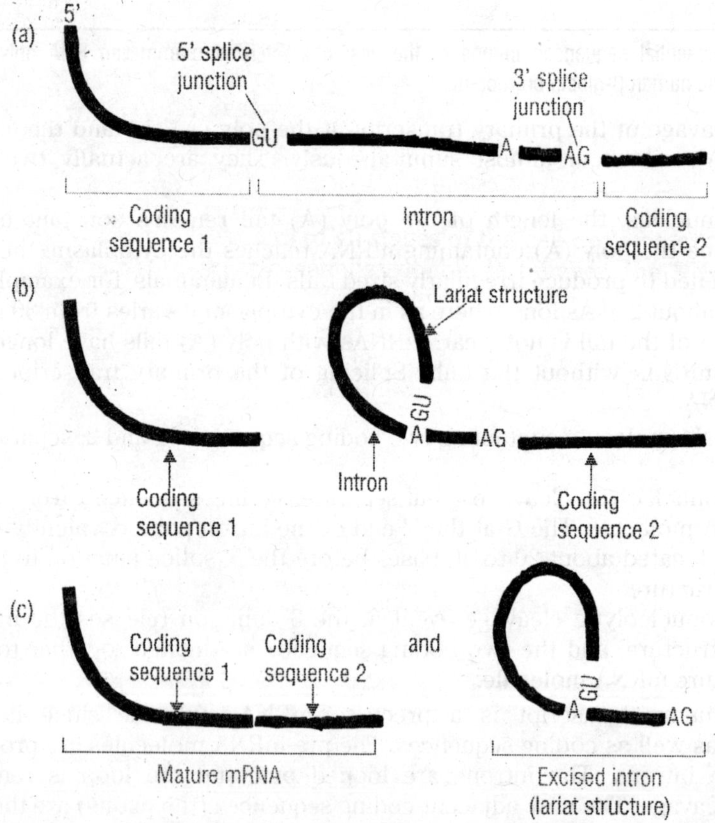

FIG. 10.8: Model for splicing of introns out of mRNA precursor molecules to produce a mature mRNA molecule.

The processing of pre-mRNA molecules do not occur with naked RNA molecules. It occurs in a large structure called a spliceosome, which contains several protein and RNA molecules. Spliceosomes have been described in both yeast and mammalian systems. In yeast they are 40S and in mammalian systems they are 60S. Different ribonucleoprotein particles called small nuclear ribonucleoprotein particles (SnRNPs) are essential components of the spliceosome. The nucleoplasm of eukaryotic cells contains a number of different SnRNP particles. They contain small nuclear RNA (SnRNA) molecules and five to nine proteins. In animal cells there are five distinct types of SuRNA called U_1, U_2, U_4, U_5, and U_6. U_1, U_2, and U_5 SnRNAs are found individually in their corresponding SnRNPs. The U_4 and U_6 SnRNAs coexist within a single SnRNP particle.

During the splicing reaction, U_2 SnRNPs are associated with the intron branch point while U_1 SnRNPs interact with the 5' splice site. U_5 SnRNP binds the 3' splice junction in the pre-mRNA. Not all of their roles have been described in splicing events. Probably the whole spliceosome structure is required to hold the pre-mRNA molecule in a stable configuration for the splicing events.

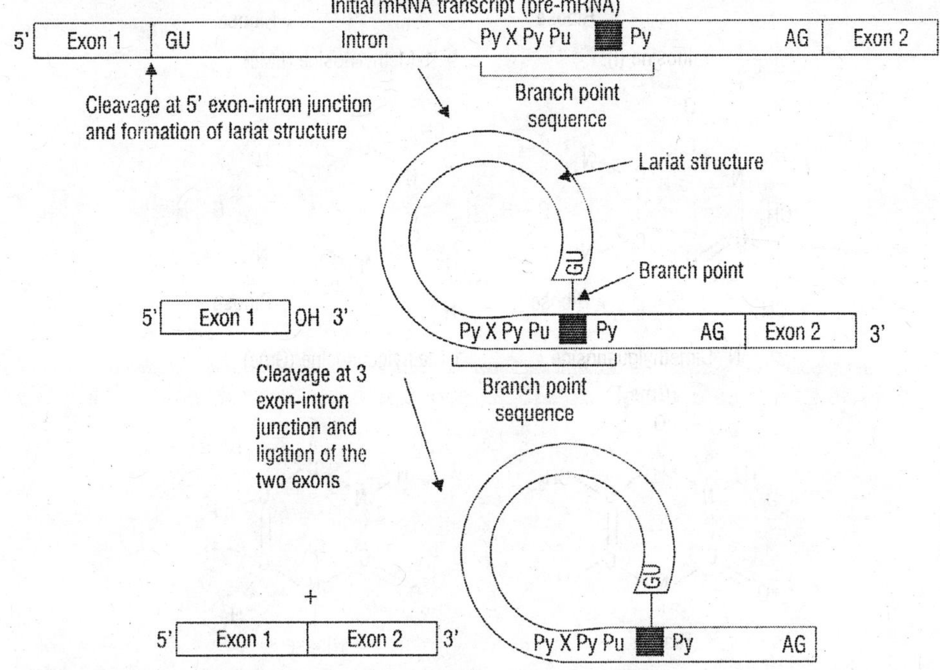

FIG. 10.9: Details of intron removal from a pre-mRNA molecule.

THE STRUCTURE AND FUNCTION OF TRANSFER RNA

In a cell, transfer RNAs bring amino acids to the ribosome-mRNA complex. Here the amino acids are polymerised into protein chains in the translation process. The sequence of amino acids in the protein is directed by the sequence of codons in the mRNA molecule.

Molecular Structure of tRNA

Transfer RNA molecules constitute 10 to 15 per cent of the total cellular RNA in both prokaryotes and eukaryotes. They have a size of about 4S. They consist of a single chain of 75 to 90 nucleotides. Their sequence varies from one tRNA molecule to another. The differences in nucleotide sequences of a particular tRNA molecule have the ability to bind a particular amino acid.

The nucleotide sequences found in tRNA molecules are different from those in mRNAs and rRNAs. Primary transcript of tRNA genes in prokaryotes and eukaryotes the precursor tRNAs (pre-tRNAs), are extensively modified as they mature.

Fig. 10.10 (left) gives examples of the modified bases that are found in tRNAs. These modifications are brought about by specific enzyme action. The modifications themselves cause the particular two-and three-dimensional configurations of tRNAs. This in turn determines the function of those molecules.

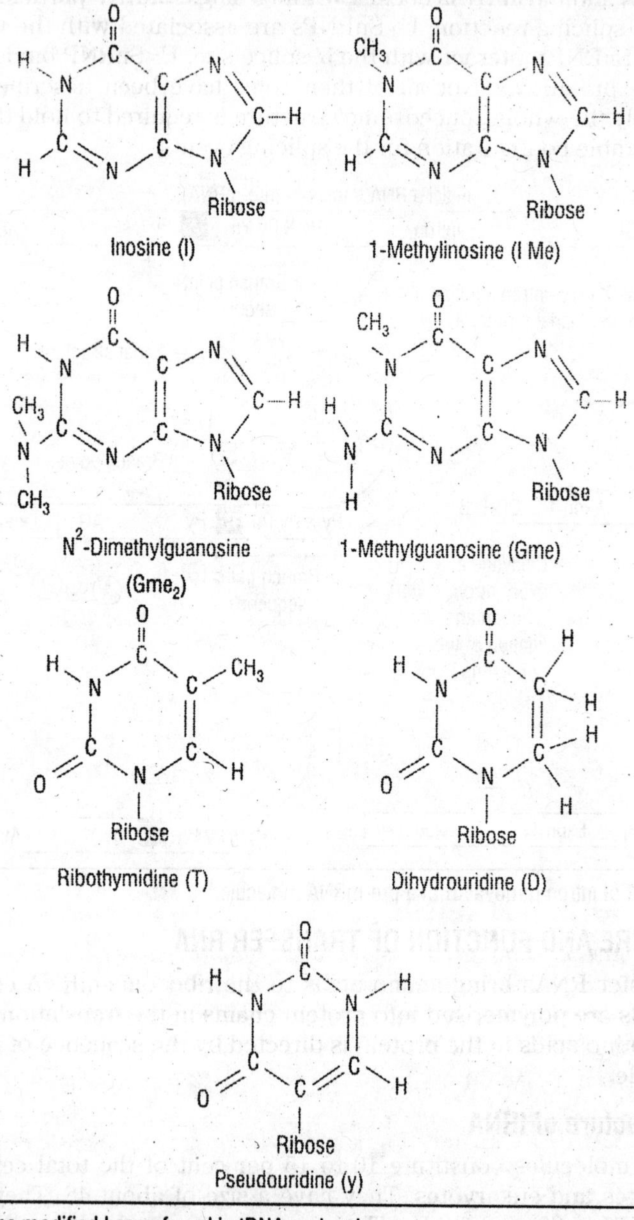

FIG. 10.10: Some modified bases found in tRNA molecule.

Two types of modification occur in pre-tRNAs. The addition of a 5' C-C-A 3' sequence to the 3' end of the molecule and the extensive chemical modification of a number of nucleotides at many places with in the chain. The type and extent of the modifications vary from tRNA to tRNA. The typical modifications include the addition of methyl groups to the base, the reduction of certain uridines to dihydrouridine, or the rearrangement of some uridines to produce pseudouridine (ψ).

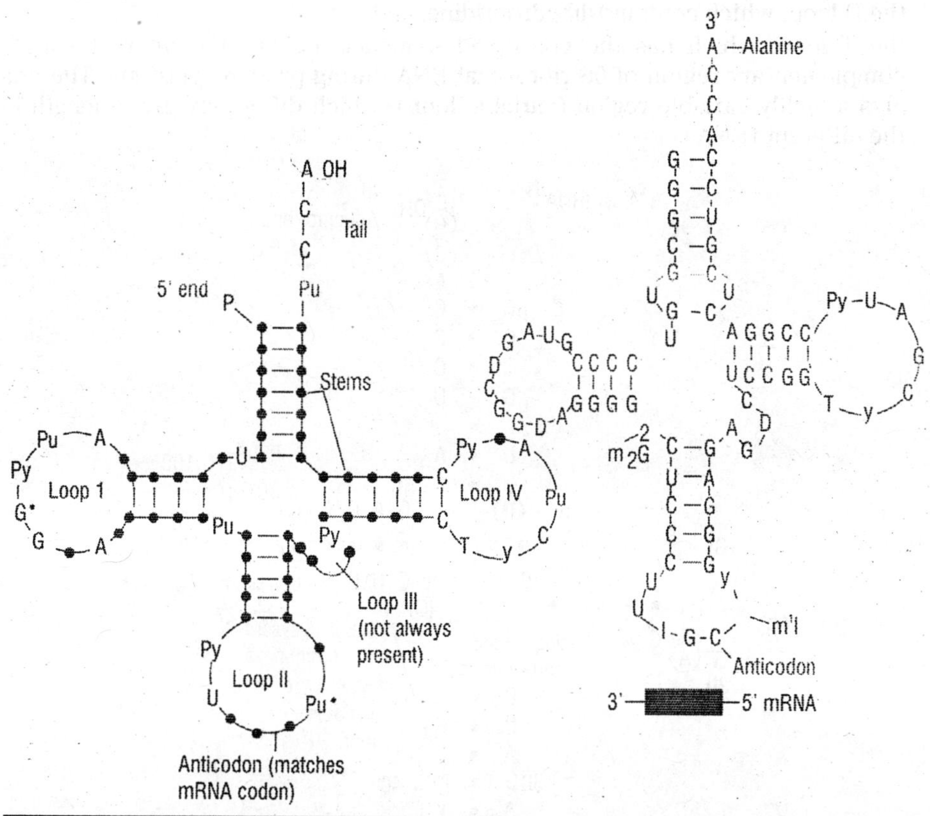

FIG. 10.11: (a) Schematic diagram of a tRNA molecule shown in the two-dimensional clover-leaf configuration (b) The complete nucleo-tide sequence of yeast alanine tRNA showing the unusual bases and anticodon position.

Clover Leaf Model for the Structure of tRNA

Many tRNAs have been sequenced from a number of organisms. All the sequences can be arranged into a clover leaf model for tRNA. This model is a secondary structure in that only two dimensions are used to display the sequence. Although in reality, a tRNA has a well-defined three-dimensional structure.

The clover leaf it self results from complementary-base pairing between different sections of the molecule. Three base-paired **stems** appear in the molecule, with the number of base pairs in each stem varying from tRNA to tRNA.

All tRNAs share common features of which the most notable is the folding into cloverleaf structure. Other common features include the sequence CCA at the 3' end, the conserved nucleotides indicted by circles and unusual bases such as pseudouridine, inosinic acid, methylcytosine, methylguanosine, ribothymidine, and others. These unusual bases are modified post transcriptionally by the action of methylating enzymes.

The cloverleaf structure consists of a series of loops separated by short stems of helical double stranded regions between four and seven complementary bases.

Fig. 10.12 shows the principal elements of this structure as follows:

1. The acceptor end (CCA), at which the specific amino acid becomes attached,
2. the anticodon loop, directly opposite from the acceptor end and having the three bases that recognize and form hydrogen bonds with the mRNA codon,
3. the D loop, which contains dihydrouridine, and
4. the T loop, which has the conserved sequence Tψ CG and interact with a complementary region of 5S ribosomal RNA during protein synthesis. There is also a highly variable region (variable loop), which differs greatly in length in the different tRNAs.

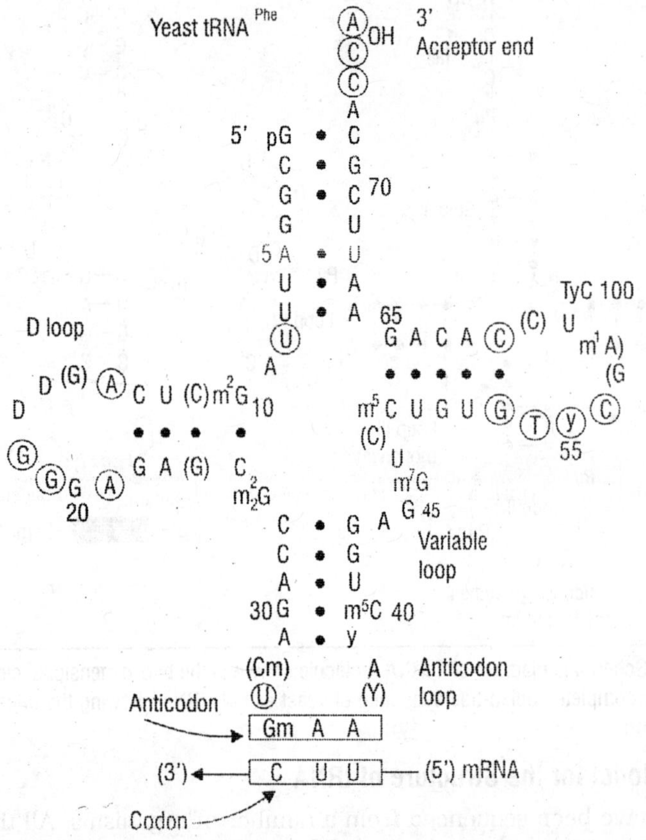

FIG. 10.12: Diagram of the clover leaf model of tRNA showing the amino acid acceptor end, the anticodon loop, the D loop and the TψC loop.

All tRNAs have three unpaired loops, I, II, and IV. Loop I usually contains 9 bases. Loop II has 7 bases and contains with in it the three nucleotide sequence called the anticodon. It pairs with a codon in mRNA by complementary base pairing. This codon-anticodon pairing is crucial for adding the correct amino acid to the growing polypeptide chain. Some, but not all, tRNAs have a small loop of variable length, called loop III, or the lump. Loop IV has 7 bases, and within this loop the sequence 5'T-ψ-C3' is universal among tRNAs.

The cloverleaf model shows four stems and three (sometimes four) loops containing unpaired nucleotides. The particular nucleotides present in the stems vary greatly from tRNA to tRNA, but the number of nucleotides in a particular stem is fairly constant.

THREE DIMENSIONAL STRUCTURE OF TRNA

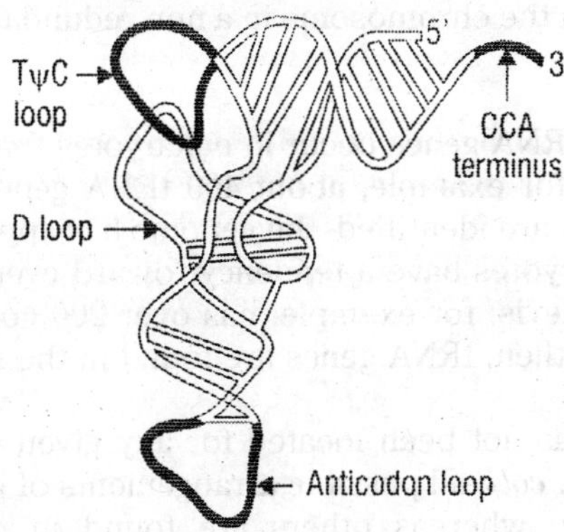

FIG. 10.13: Tertiary structure of yeast phenyl-alanine tRNA showing a compact configuration with L form.

The two- dimensional clover leaf model is a schematic drawing based on an analysis of the primary nucleotide sequences of tRNAs. However, like all molecules, the tRNA has a three dimensional structure. Because the tRNA molecules are so small, they can be crystallized, and X- ray crystallography can be used to develop a three-dimensional model. Fig. 11.13 shows the tertiary- structure model for yeast tRNA. Phe was the first tRNA to be analysed in this way. All other tRNAs show similar three-dimensional structures.

The tertiary structure of tRNA has been determined by X- ray diffraction studies. The tRNA has a compact configuration in which the side arms are folded together, rather than the open configuration. The molecule has the shape of an L, with the anticodon at one end of the L and the acceptor at the other. The D and T loops are hydrogen- bonded together at the corner of the structure, and the nucleotides involved in these crucial hydrogen bonds are highly conserved in different tRNAs (circled nucleotides). All tRNAs share these characteristics of folding, with the exception of mitochondrial tRNAs, which sometimes do not follow these rules.

TRANSFER RNA GENES

Three base sequences (codon) in an mRNA, specifies each amino acid to be added to a polypeptide chain. Only 20 different amino acids can be used to make a protein.

Actually 61 different codons can be used in an mRNA to specify amino acids, and 3 additional codons do not specify amino acids but instead are used as termination signals for protein synthesis. Each amino acid- specifying codon must be matched by an appropriate anticodon on a tRNA molecule. Theoretically any cell must have at least 61 different tRNA types. Eukaryotes commonly have a number of different tRNA molecules with the same anticodon. Therefore, the instructions for the manufacture of such large numbers of tRNAs can be coded, the genome has many tRNA genes. The number and the distribution of tRNA genes in prokaryotes and eukaryotes have been extensively studied in a number of organisms.

The tRNA Genes of *E. coli*

Examples of virtually all possible sorts of arrangements of tRNA genes occur in *E. coli*. Some tRNA genes occur only once on the *E. coli* chromosome, while others are present two or more times, a situation called gene redundancy. In some cases redundant genes are not linked on the chromosome, but in others they are linked, side by side, in what is called a redundant gene cluster. Unique tRNA genes can also be linked side by side on the chromosome in a non redundant tanden gene cluster.

Eukaryotic tRNA Genes

In general, many more tRNA genes occur in eukaryotes than in prokaryotes. In the lower eukaryote yeast, for example, about 400 tRNA genes occur in the genome. Since 61 different tRNAs are identified, the genome has approximately six copies of each gene. Higher eukaryotes have a tendency toward even greater redundancy of these genes. Xenopus laevis, for example, has over 200 copies of each tRNA gene per genome. As a class, then, tRNA genes are found in the middle repetitive kinetic class of DNA.

Every tRNA gene has not been located for any given eukaryote. Experiments have shown that, as in *E. coli*, all possible arrangements of genes occur. Thus, some genes are found singly, whereas others are found in clusters, which can be redundant or non-redundant tandem clusters. There is also evidence of tandem clusters, in which some genes are a single copy while others are redundant.

Some (nuclear) tRNA genes contain introns. Depending on the tRNA, the intron is 14 to 60 base pairs long. It is always located between the first and second nucleotides 3' to the anticodon. The anticodon itself often pairs with intron sequences in the tRNA precursor. Introns are found only in tRNAs for tyrosine, phenylalanine, tryptophan, lysine, proline, serine, leucine, and isoleucine. The introns present in any given family of tRNA genes (*e.g.*, the eight yeast tyrosine tRNAs) are identical.

The first tRNA gene in which an intron was found was the yeast gene for tRNA Tyr. This is the gene for the tRNA that carries the amino acid tyrosine to the ribosome. The 14-base intron is located in the pre-tRNA to the 3' side of the anticodon. The presence of the transcript of the intron sequence in the pre-tRNA results in a significant change in the anticodon loop. Abelson's j group has identical an enzyme, RNA ligase, that splices together the RNA pieces, once the intron is removed from the pre-tRNA.

Removal of the intron involves cleavage with a splicing endonuclease, associated with the nuclear membrane. About 10 per cent of the 400 tRNA genes in yeast have intervening sequences, and all are spliced in the same way.

Deletion studies on cloned tRNA genes have shown that nucleotide sequences within the genes are essential for transcription by RNA polymerase III. They include two separate regions (A and B boxes) of about 10 nucleotides each that contain many of the nucleotides that are constant in all tRNAs. There are two transcription factors that must bind to these sequences before a tRNA gene can be recognized by RNA polymerase III. So far only one of these protein factors, the one that binds to box B, has been isolated.

The RNA polymerase III is a large molecule. Its one part recognizes the promoter sequence (called an internal control region) within the gene sequence. Its other part initiates transcription at the appropriate point upstream from the promoter. Internal control regions are also found in the eukaryotic 5S rRNA genes. These genes are also transcribed by RNA polymerase III. Two regions of conserved sequences (A and B blocks) are important for RNA polymerase III to transcribe the tRNA and 5S rRNA genes. The distance between A and B blocks can vary considerably, especially in tRNA genes with introns.

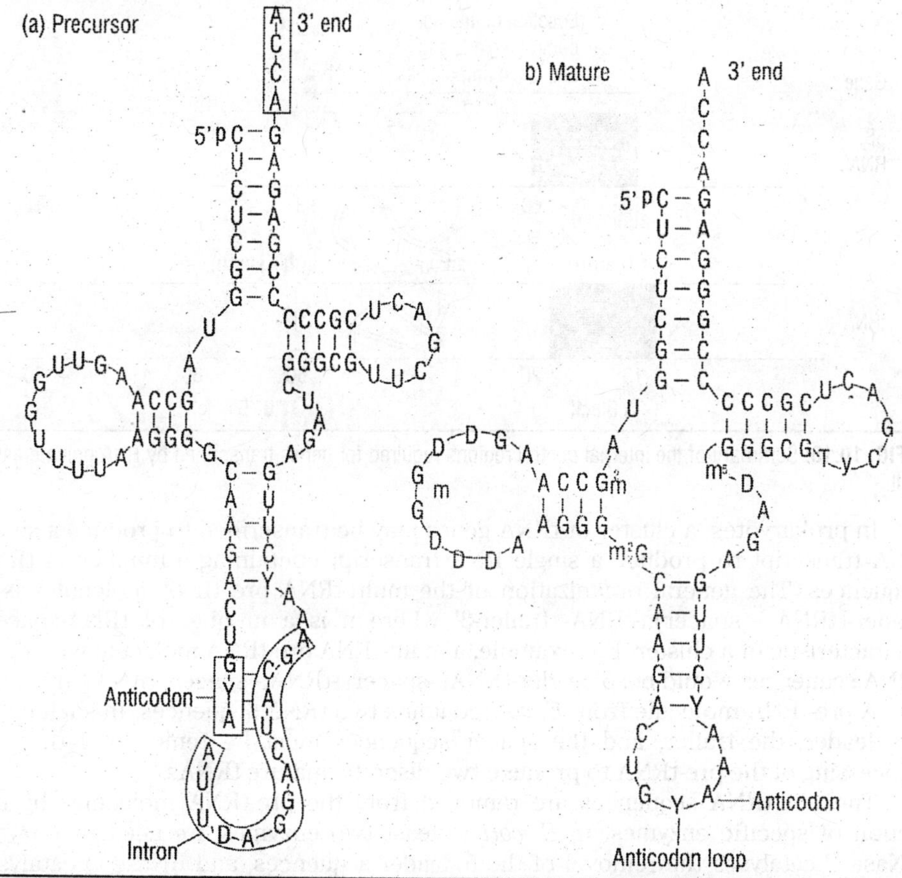

FIG. 10.14: Clover leaf models for precursor yeast tRNA. Tyr and mature tRNA. Tyr (a) the 14-base intervening sequence in the precursor, located adjacent to the anticodon-coding region, (b) the intervening sequence removed from the precursor to produce a mature tRNA. Tyr molecule.

Internal control regions are essential sequences for RNA polymerase III to initiate transcription. There are other specific upstream and downstream sequences that are needed for the expression of some RNA polymerase III-transcribed genes. These sequences are the sites for specific DNA binding proteins whose functions are not yet clearly defined.

tRNA PROCESSING

tRNA is not transcribed directly into its final form but into a longer precursor. *E. coli* tyrosine tRNA is synthesised as a precursor having 39 extra nucleotides at the 5' end and 3 at the 3' end. These nucleotides must be removed before an active tRNA molecule is obtained. Other *E. coli* tRNAs are synthesised as very long precursors. These contain several tRNAs (5 to 7) in a single RNA chain. These precursor tRNAs are cleaved by a specific enzyme, ribonuclease P (P for processing). Ribonuclease P has two components or subunits. One is a protein, the other RNA of about 300 nucleotides in *E. coli*. The RNA moiety is the catalytic subunit of the enzyme. It can cleave tRNA precursors even without the protein subunit, although slower. This finding constituted the first identification of an RNA enzyme.

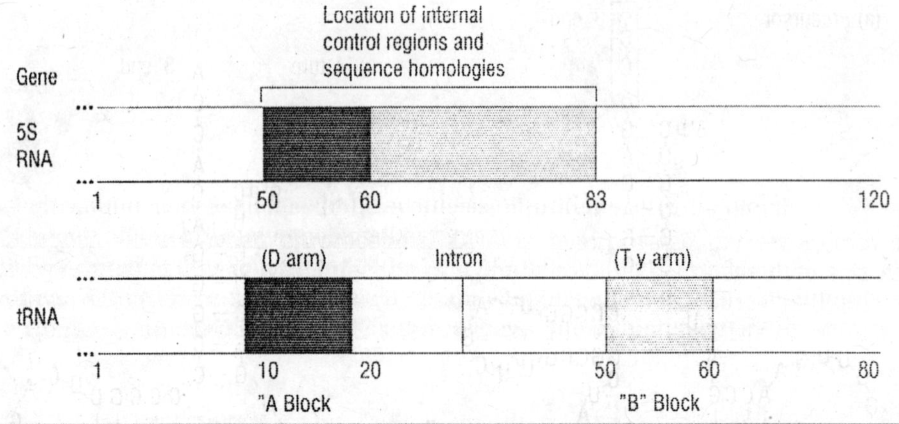

FIG. 10.15: Summary of the internal control regions required for genes transcribed by RNA polymerase III.

In prokaryotes, a cluster of tRNA genes may be transcribed to produce a single RNA transcript to produce a single RNA transcript containing a number of tRNA sequences. The general organization of the multi-tRNA pre tRNA molecules is 5' leader-(tRNA - spacer)n-tRNA- trailer-3' where n is a number of tRNA-spacers characteristic of a cluster. For example, a multi-tRNA pre-tRNA molecule with three tRNA sequences would be: 5'-leader-$tRNA_1$ -$spacer_1$ -$tRNA_2$- $spacer_2$-$tRNA_3$- trailer-$_3$'

A pre-tRNA molecule from *E. coli* contains two tRNA sequences. In such cases the leader, the trailer, and the spacer sequences must be removed during the processing of the pre-tRNA to produce two discrete, mature tRNAs.

The non-tRNA sequences are removed from the pre-tRNA molecules by the action of specific enzymes. In *E. coli* at least two enzymes are needed. Enzyme RNase P catalyses the removal of the 5' leader sequences, and RNase Q catalyses the removal of the 3' trailer sequence.

Figure 10.16 shows the precursor molecules transcribed from a cloned eukaryotic tRNA gene microinjected into frog oocytes. The initial transcript has a 5' leader, a 3' trailer, and an intervening sequence in addition to the coding region. The ends of the molecule are trimmed first, and then the molecule is spliced and religated. The tRNA precursors are present in the nucleus, while the mature tRNA exists into the cytoplasm. Intervening sequences have been found in several eukaryotic tRNA genes. They are always inserted just after the anticodon. The tRNA intervening sequences, which vary from 10 to 60 nucleotides in length, are removed by two enzymes, a splicing endonuclease and a ligase. tRNA introns do not have conserved sequences at their ends. The splicing enzymes recognize the conserved general shape of the folded tRNA molecule.

Other post transcriptional modifications important in tRNA function are the base modifications (such as methylations). These are unusual bases found in tRNA, and the terminal addition of CCA. The 3' end of tRNA always terminates with CCA. Its integrity is essential because it is at this end that the amino acid is attached. The 3' end can be degraded in vivo. Cells have an enzyme capable of terminal addition of CCA, which regenerates the active end. In eukaryotes most tRNA genes do not contain in their DNA the information for the CCA end, which is added post transcriptionally.

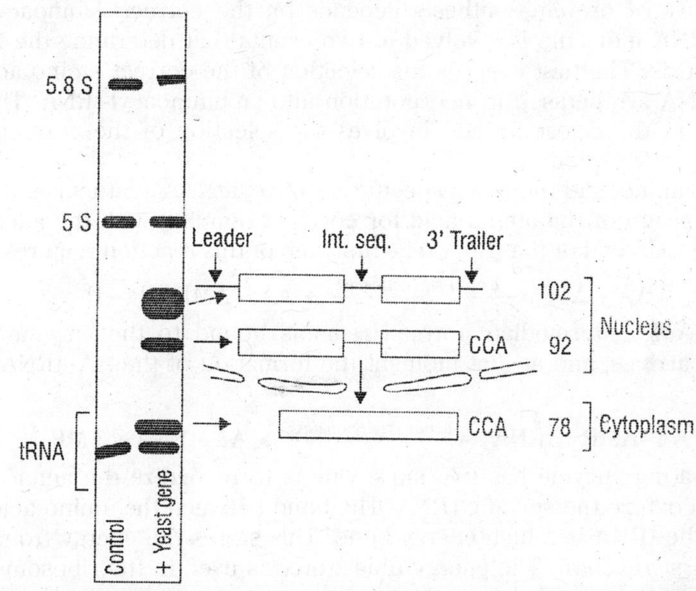

FIG. 10.16: RNAs synthesised in frog oocytes microinjected into the nucleus with a cloned tRNA gene. The structure and intracellular location of the different precursor bands are indicted.

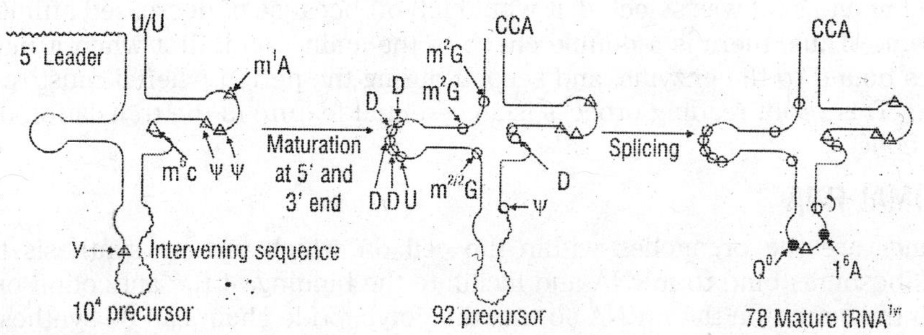

FIG. 10.17: Diagram summarizing the sequential addition of base modifications to tRNA. Tyr precursors in oocytes.

A possible function for the 5' leader sequence became apparent when the order in which the various base modifications appear. The RNA precursors having a leader sequence have particular base modifications added, while other modifications are not present and can only be added after the leader sequence is removed. The precursor RNA with a 5' leader sequence is a substrate for some base modification enzymes, and the RNA without it is a substrate for others.

1. Modified bases added at the early precursor stage;
2. Modified bases detected at 92-nucleotide stage;
3. Modifications added after splicing of the intervening sequence.

Intervening sequences are also important in controlling base modifications. When the intervening sequence was deleted from the gene using genetic engineering methods, it was observed that the 16 A modification can not be introduced into the tRNA molecules and the resulting tRNA had an impaired function. Thus, an intervening sequence is required to obtain a fully functional tRNA. The extra segments of RNA present in tRNA initial transcripts may introduce a precise order in the intricate processing events involved in tRNA biosynthesis.

The fidelity of protein synthesis depends on the correct Aminoacylation of tRNA- The tRNA molecule is involved in two events that determine the fidelity of protein synthesis. The first event is the selection of the correct amino acid by the aminoacyl-tRNA synthetase for incorporation into an aminoacyl-tRNA. The second event occurs in the ribosome and involves the selection of the correct tRNA in response to an mRNA codon.

For each amino acid there is a specific aminoacyl-tRNA synthetase. It activates the carboxyl group of the amino acid for covalent bonding with the adenylic acid residue of the CCA end of the tRNA. The first step of this reaction requires ATP:

$$\text{AA} + \text{ATP} \xleftrightarrow{a\min oacylsynthetase} \text{AA} \sim \text{AMP} + \text{P} \sim \text{P}$$

The AA~AMP intermediate formed remains bound to the enzyme until the proper tRNA arrives, and at that moment the formation of the AA~tRNA complex occurs.

$$\text{AA}\sim\text{AMP} + \text{tRNA} \xleftrightarrow{a\min oacylsynthetase} \text{AA}\sim\text{tRNA} + \text{AMP}$$

The activating enzyme has two sites. One is to recognize the amino acid and another to recognize the specific tRNA. The bond between the amino acid and the CCA end of the tRNA is a high energy bond. This stores the energy from the ATP used in the first reaction. The energy thus stored is used in the ribosome to drive the formation of the peptide bond.

The accuracy of the tRNA synthetases depends on this two-step reaction. The AA ~AMP must remain attached to the enzyme until the tRNA arrives. If an incorrect amino acid was selected, it would fall off because of decreased affinity for the enzyme. Thus, there is a double check of the amino acid, first when it initially becomes bound to the enzyme, and second during the period when it must remain attached. This proof reading process is an essential feature in the translation of the genetic code.

RIBOSOMAL RNA

Ribosomes are the organelles within the cell on which protein synthesis takes place. Ribosomes bind to mRNA and facilitate the binding of the anticodon on the tRNA to the codon on the mRNA so that the polypeptide chain can be synthesised. Ribosomes are complex structures, and one has yet to understand fully how they function. In both prokaryotes and eukaryotes the ribosomes consist of two unequally sized subunits, the large and small ribosomal subunits, each of which consists of a complex between RNA molecules and proteins. Each subunit contains at least one ribosomal RNA (rRNA) molecule and a large number of ribosomal proteins. Ribosomal subunits, mRNA and tRNAs are required for protein synthesis.

Structure of the Prokaryotic Ribosome in *E. coli*

The *E. coli* ribosome has been the subject of intensive study, and because the features of *E. coli* are generally found in other bacterial ribosomes. So it is used as a model of a prokaryotic ribosome. Fig. 10.18 shows the general structure and composition of the *E. coli* ribosome.

The *E. coli* ribosome has a size of 70S. The sizes of the two subunits are 50S (large subunit) and 30S (small subunit). The shape of the *E. coli* ribosomes and its two subunits has been discovered through electron microscopic examination of purified components. The models of *E. coli* ribosomes and ribosomal subunits have been constructed from the electron micrographic studies. In interpreting these models, however, one must exercise caution. Material studied under the electron microscope must be placed in a vacuum, which causes dehydration. Therefore, the organelle in the living cell may be more plumped up. Nonetheless, the two subunits have distinct three-dimensional shapes, unlike the two-dimensional circles or ellipses of text book drawings.

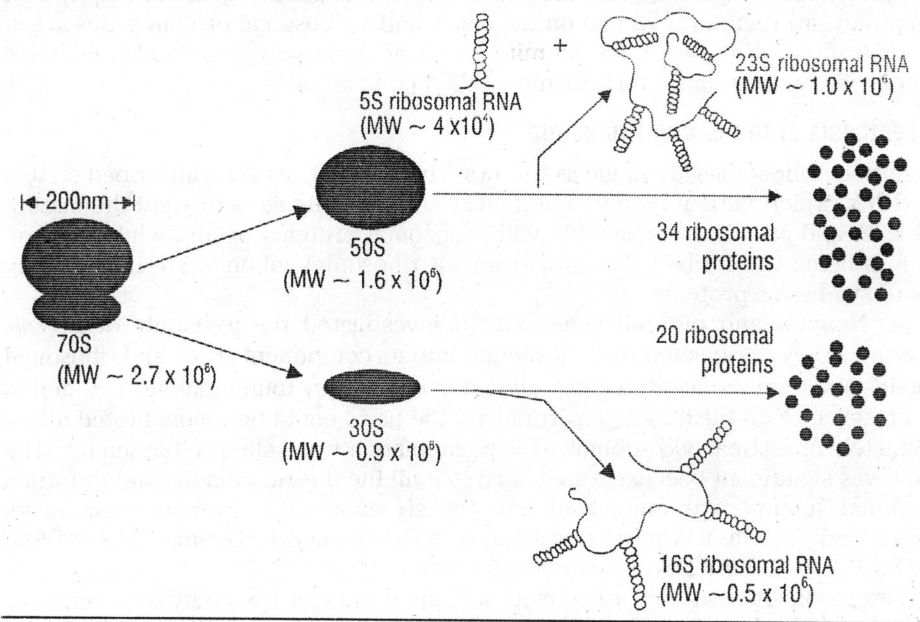

FIG. 10.18: Structure of the *E. coli* 70S ribosome. Each of the two subunits (50S and 30S) contains rRNA and ribosomal protein molecules.

Approximately two-thirds of the *E. coli* ribosome consists of ribosomal RNA, the rest of the mass consisting of ribosomal proteins. The large 50S subunit has 34 different proteins and two rRNA molecules, one with a size of 23S (2904 nucleotides) and the other with a size of 5S (120 nucleotides). The small 30S ribosomal subunit has 20 different proteins and one rRNA molecule has a size of 16S (1542 nucleotides).

The rRNA Genes of *E. coli*

In prokaryotes and eukaryotes the regions of DNA that contain the genes for the three types of rRNA are called ribosomal DNA (rDNA). In *E. coli* production of equal amounts of the three rRNAs is transcribed by three adjacent genes in rDNA into a single precursor rRNA (pre-rRNA) molecule. One transcription unit consists of one gene each for the three rRNAs. *E. coli* has seven such transcription units (rRNA regions).

In each transcription unit, the three rRNA genes are arranged in the order 16S-23S-5S. So the arrangement of the rRNAs on the pre-rRNA transcript is 5'16S-23S-5S 3'. Transfer RNA genes are associated with the rRNA transcription units. In all seven transcription units, one or two tRNA genes are found in the spacer between the 16S and 23S rRNA coding sequences. Another one or two tRNA genes are found between the 5S rRNA gene and the 3' end of the transcription unit in three of the seven transcription units. During RNA synthesis from the transcription units, the tRNA genes are transcribed as part of the pre-rRNA molecule. The tRNAs are then removed from the precursor by the action of specific processing enzymes.

The Ribosomal Protein Genes of *E. coli*

The *E. coli* ribosome has one gene for each of 54 ribosomal proteins. The genes are located at many places on the *E. coli* genetic map. Some of the ribosomal genes are found clustered on the map, for example, 27 ribosomal protein genes are in the strA (streptomycin) region (at 72 min on the map), and 4 ribosomal protein genes are in the *rif* (*rifampicin*) region (at 88 min). Each of these two regions also contains other genes, such as genes for subunits of RNA polymerase.

Biosynthesis of the *E. coli* Ribosome

Prokaryotic ribosomes are made as follows: The rRNA genes are transcribed on to a pre-rRNA, which is then modified and cleaved at specific sites to yield the mature 16S, 23S, and 5S rRNAs. Assembly with ribosomal proteins occurs while the pre-rRNA is being transcribed. The two resultant ribosomal subunits are immediately able to synthesise proteins.

M. Nomura and his colleagues (1969) investigated the assembly of *E. coli* ribosome. They dissociated the 30S subunit into its component rRNA and ribosomal proteins and then reconstituted the subunit in vitro. They found that at 37 °C and in the presence of an ideal ionic environment, the parts could be reconstituted into a completely functional 30S subunit. This phenomenon was called self-assembly. The result was significant because it indicated that all the information needed to form a functional subunit was embedded into the structure of individual components (mRNA and proteins). A few years later, K. Nierhans and F. Dohme reconstituted the 50S *E. coli* subunit from its component parts.

The reconstitution of a ribosomal subunit occurs in a step-by-step series of reactions, each of which adds specific ribosomal proteins to the rRNAs. Many of the steps have been elucidated by omission experiments in which one ribosomal protein is left out of the reactions and the resulting particle is analysed. In this way geneticists have shown definite pathways for the assembly of the subunits. In addition, even the omission of a protein results in an essentially complete subunit, it is not functional in protein synthesis. Thus, all proteins must be available for the ribosomal subunit to function.

The P30S molecule is cleaved by RNase III to produce three precursor molecules in one each of the P16S, P23S, and P5S molecules. In normal cells, RNase III cleaves the transcript of the rRNA transcription unit to produce the three rRNA precursors while transcription is still occurring. It is a rapid process, so the P30S is not seen in normal cells.

For correct ribosome assembly, the 16S and 23S rRNAs must be methylated at a few places. The methylation is catalysed by specific methylases. The most of the methyl groups are added to the bases, a few added to the 2'-OH of the ribose group. Methylation of the 23S sequence occurs on the P30S molecule, where as methylation of the 16S rRNA occurs when it is in its mature state.

As the rRNA genes are being transcribed, the transcript rapidly associated with ribosomal proteins. The cleavage of the transcript takes place within a complex formed between the rRNA transcript and ribosomal proteins. It is most probably directed by the conformational state of the rRNA- ribosomal protein particle at the time of the cleavage.

Structure of the Eukaryotic Ribosome

Eukaryotic ribosomes are larger and more complex than the prokaryotic ribosomes. The size of the ribosome and the molecular weights of the rRNA molecules differ from organism to organism.

Lower eukaryotes have the smallest ribosomes, while mammalian ribosomes are the largest. All eukaryotic ribosomes have many common structural and chemical features. Fig. 10.19 shows the general composition of mammalian ribosomes. They have a size of 80S and have two subunits, a large one with a size of 60S and a small one with a size of 40S. The 40S subunit is quite similar in size and mass in all the eukaryotes, but the 60S subunit varies greatly.

The 80S mammalian ribosome consists of about equal weights of rRNA and ribosomal proteins. There are four rRNA types, with sizes of 18S (2300 nucleotides), 28S (4200 nucleotides), 5.8S (156 nucleotides) and 5S (120 nucleotides). The ribosomes of higher eukaryotes have 70-80 different ribosomal proteins. They are about 30 in the small subunit and about 50 in the large subunits in mammalian cells.

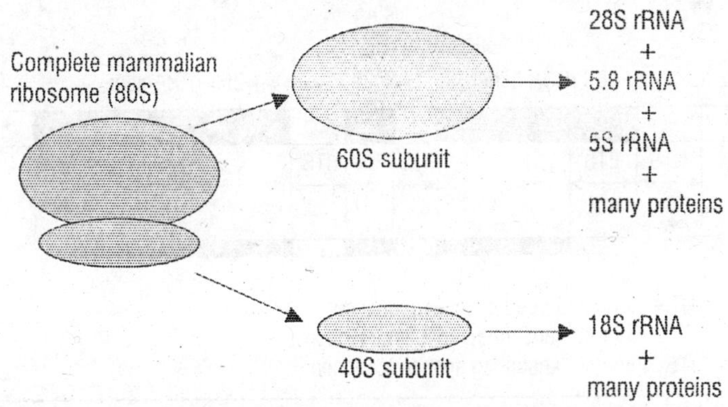

FIG. 10.19: Composition of whole ribosomes and of ribosomal subunits in mammalian cells.

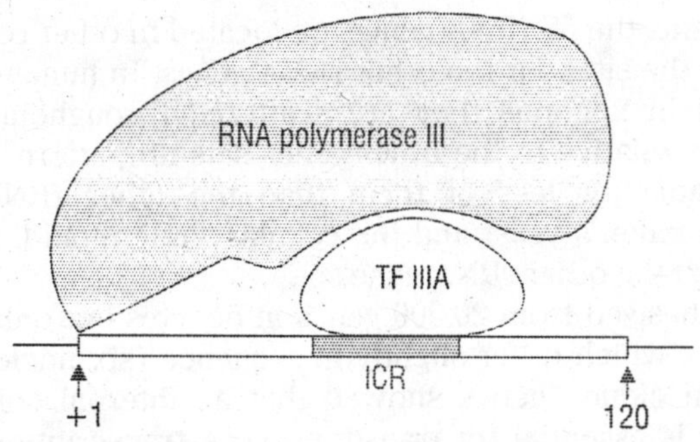

FIG. 10.20: Transcription of a 5S gene requires a transcription factor bound to the middle of the gene, which is then recognized by RNA polymerase III.

In all eukaryotes the 40S ribosomal subunit contains only one rRNA type, 18S rRNA. The large 60S subunit contains the other three types of rRNA, the 28S, 5.8S, and 5S rRNAs. The 5.8S rRNA is hydrogen-bonded to the 28S rRNA in the functional ribosome.

Location and Organization of rRNA Genes

Within the nucleus, the rRNA genes are located in regions of DNA called ribosomal DNA (rDNA). rDNA is present in the chromatin around the nucleolus. The nucleus has one or more nucleoli within which rRNAs are transcribed and the rRNAs associate with ribosomal proteins to form the two ribosomal subunits. The ribosomal proteins are synthesised in the cytoplasm. These proteins migrate into the nucleolus.

Once the ribosomal subunits are made they are transported to the cytoplasm to take part in protein synthesis. Most eukaryotes have a large number of copies of the genes for each of the four rRNA species. The genes for 18S, 5.8S, and 28S rRNAs are found in one or more clusters in the genome, and around each cluster a nucleolus is formed. The extent of gene repetition varies from eukaryote to eukaryote and is in the range of 100 to 1000, with a few exceptions. Thus, the rRNA genes are representative of the middle- repetitive DNA in the genome. For example, yeast has 140 rRNA gene sets, *Neurospora* has 200, *Drosophila* has 260, and HeLa (human) cells have 1250.

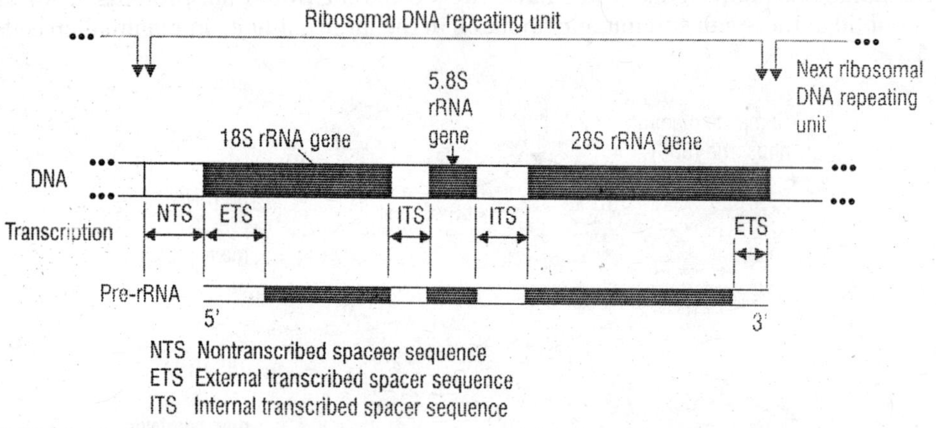

FIG. 10.21: Generalized diagram of a eukaryotic, rDNA repeating unit. The coding sequences for 18S, 5.8S, and 28S rRNAs are indicated.

In most organisms the 5S rRNA genes are located in other regions of genome at a site distinct from the sites for the other rRNA genes. In humans, the 5S genes are clustered, whereas in xenopus they are scattered throughout the genome. The number of 5S genes relative to the other genes sets no pattern. The organism may have more, the same, or less of them than the other rRNA genes. In some organisms, such as baker's yeast and the cellular slime mould, the 5S rRNA genes are inter-spread with the other rRNA gene set.

5S RNA is synthesised from 20,000 genes in oocytes but only from 400 somatic genes in other cells, which differ slightly in sequence (six nucleotides out of 120). Deletion studies on cloned genes showed that an internal control region in the middle of the gene is essential for transcription. A transcription factor called (TF IIIA) binds to this region and enables RNA polymerase III to start transcription at an upstream site. TF IIIA has nine **fingers** or small domains held together by zinc ions, which bind to DNA every half turn of the DNA helix. TF IIIA, in addition to binding to 5S genes, can also bind to their product, 5S RNA. Young oocytes have massive amounts of TF IIIA (15% of total protein), most of it bound to 5S RNA, forming a 7S storage particle. The complex between TF IIIA and a 5S gene is stable, and the gene can be transcribed hundreds of times without the protein being displaced. It is because of TF IIIA binds to the non-coding strand of 5S RNA, which has a sequence similar to that of 5S RNA. The protein might recognize analogous structural features in 5S DNA and RNA.

In oocytes all 20,000 5S genes are switched on because the massive amounts of transcription factor available can saturate every gene. It is not yet known why oocyte-type genes are switched off in somatic cells. At the chromosome level the

18S, 5.8S and 28S rRNA genes are usually found adjacent to one another. They are in the order 18S-5.8S-28S, with each set of three genes repeated many times to form repeated tandem arrays of rRNA genes called rDNA repeat units. Each set of three rRNA genes is transcribed on to a single pre-rRNA molecule. The rDNA repeat units of four eukaryotes are diagrammed in Fig. 10.22. In general, variation in transcribed spaces lengths is largely responsible for variation in length of the transcribed region (the rRNA transcription unit) of the rDNA repeat unit.

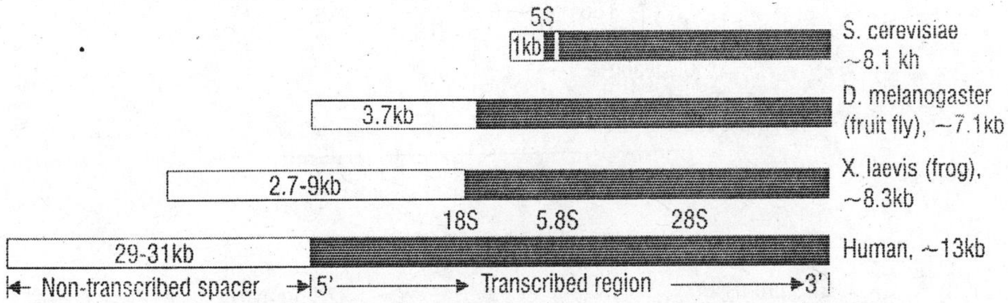

FIG. 10.22: The ribosomal transcription units of four eukaryotes. The 18S, 5.8S, and 28S regions code for the rRNAs found in all ribosomes. The transcribed spacer regions vary in length and result in large differences in the lengths of the transcription units. Variation in the length of the nontranscribed spacer regions is also considerable.

At the gene level several types of noncoding sequences have been discovered. The pre-rRNA is longer than the three mature rRNAs together. Two types of transcribed spaces sequences in the DNA have been identified. The external transcribed spacers (ETS) are DNA sequences. These are transcribed into the pre-rRNA. These are located immediately adjacent to the 5' end of the 18S sequence and to the 3' end of the 28S sequence in the pre-rRNA molecule. The internal transcribed spacer (ITS) is transcribed into the pre-rRNA on either side of the 5.8S sequence. Finally, another class of rDNA spacer sequences, nontranscribed spacer (NTS) sequences are found at the extreme 5' and 3' ends of each tandom rRNA repeating unit in the rDNA. These regions are not transcribed into the pre-rRNA molecule. NTS sequences vary in length among eukaryotes, from 2 kb in frogs to 30 kb in humans.

Introns are also present in the rRNA genes of eukaryotes. However, they have been found in only a few systems, *e.g.*, *Drosophila* and *Tetrahymena*. So they are not as widespread as the introns in mRNA genes. The introns are spliced out of the pre-rRNA transcript as it is processed to produce the mature rRNAs. The splicing reactions are distinct from those involved in removing introns from pre-mRNA and pre-tRNA.

BIOSYNTHESIS OF EUKARYOTIC RIBOSOMES

Eukaryotic ribosomes are assembled in a process similar to that found in prokaryotes. Ribosomal proteins are made in the cytoplasm from mRNAs that are transcribed by RNA polymerase II. The 5S rRNA is transcribed by RNA polymerase III in the nucleus directly from individual rRNA genes. Ribosomal proteins and the 5S rRNA molecule both migrate to the nucleolus.

In the nucleolus the 18S, 5.8S, and 28S rRNA genes are transcribed by RNA polymerase I into a large pre-rRNA molecule (45S pre-rRNA, 13,700 nucleotides) found in human HeLa cells or the 40S pre-rRNA of the frog *xenopus laevis*, rRNA sequences are highly conserved in the evolutionary sense. RNA polymerase I only synthesises one type of product. Different organisms have different sequences around their initiation sites for pre rRNA synthesis. Strong homologies are seen only for closely related species such as mouse, rat, and human.

The bases and ribose sugars of eukaryotic pre-rRNA are methylated after transcription. The methylation occurs primarily on the 2'-OH group of the ribose.

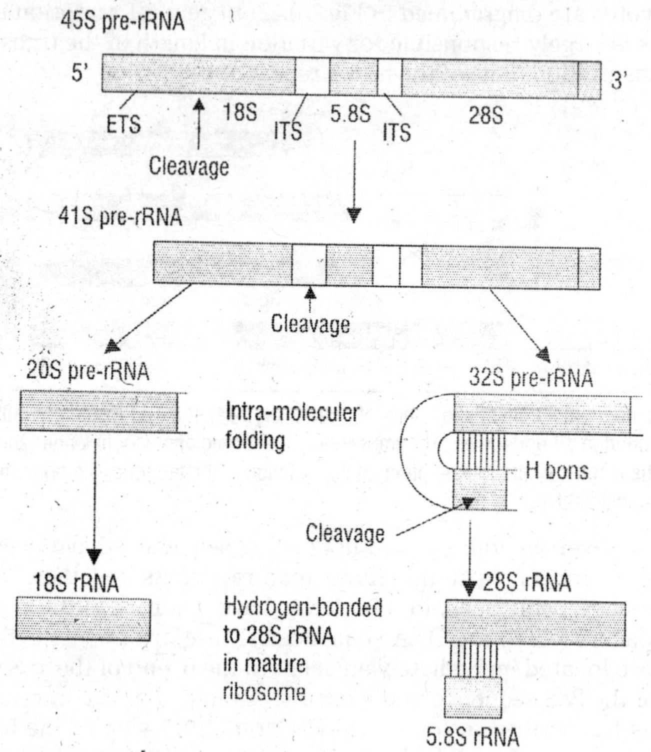

FIG. 10.23: Processing of 45S pre-rRNA in HeLa cells to produce the mature 18S, 5.8S, and 28S rRNAs.

After the pre-rRNA is modified, it is cleaved at specific sites by specific RNases to remove ITS and ETS. The processing of the pre-rRNA does not occur until the molecule is completely synthesised. These discrete rRNAs are finally processed into mature 18S, 5.8S, and 28S rRNAs. Fig. 10.23 shows the pre-rRNA -processing pathway in HeLa cells. The pathway is similar in all eukaryotes. The first cleavage removes the 5'ETS sequence and produces a precursor molecule that still contains all three rRNA sequences. In the next step an internal cleavage occurs to produce the precursor to 18S rRNA and the precursor to 28S and 5.8S rRNAs. For the former the generation of 18S rRNA involves the removal of the ITS sequence to the 3' side of the sequence. For the latter the molecule folds so that the 5.8 S sequence becomes hydrogen-bonded to the 28S sequence. After that the ITS sequence is removed from between those two rRNAs, and the mature rRNAs are produced.

All the pre-rRNA processing events takes place in complexes formed between the pre rRNA, 5S rRNA, and the ribosomal proteins. As the pre-rRNA is cleaved, conformational changes take place in the complexes, so that when the processing is finished, the 60S and 40S ribosomal subunits are produced. These have the appropriate mature rRNAs and ribosomal proteins. Once assembled the two ribosomal subunits then migrate out of the nucleolus and into the cytoplasm, where they associate with mRNAs and tRNAs and begin the process of protein synthesis.

BIOLOGICAL SIGNIFICANCE OF RNA

1. RNAs play an important role in protein synthesis.
2. RNA also acts as a genetic material.

RNA as Genetic Material

Some bacterial viruses and some animal and plant viruses have RNA as their genetic material. The following classical experiment showed that RNA is the genetic material of the tobacco mosaic virus (TMV).

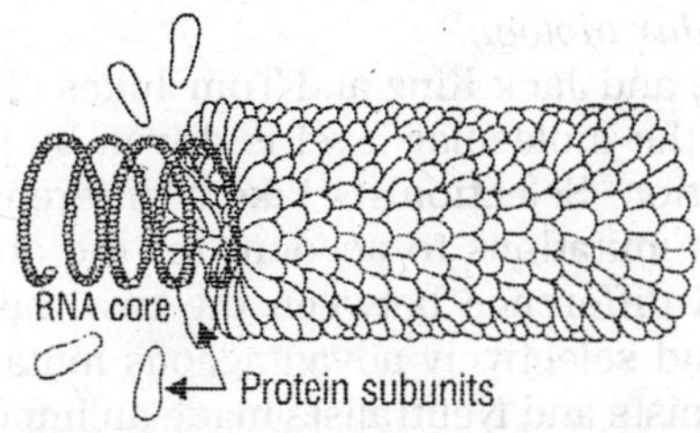

FIG. 10.24: Typical tobacco mosaic virus (TMV) particle. The helical RNA core is surrounded by a helical arrangement of protein subunits.

TMV contains two chemical components, RNA and protein in a spiral (helical) configuration. The protein surrounds the RNA core, protects it from attack by nucleases, and along with the RNA core, functions in the infection of the plant cells. Many varieties of TMV are known. They differ in the plants they infect and by the extent to which they affect these plants.

A. Giorer and G. Schramm showed that when tobacco plants were inoculated with the purified RNA of TMV (*i.e.*, without the protein coat), they developed typical virus-induced lesions. This result indicated strongly that RNA was the genetic material of TMV. This was supported by the observation that no lesions were produced when the RNA had been degraded by treatment with ribonuclease and then injected into the plant.

In an experiment H. Fraenkel- Conrat and B. Singer (1957), confirmed Gierer and Schramm's conclusions. They isolated the RNA and protein components of two distinct TMV strains. They reconstituted the RNA of one type with the protein of the other type, and vice versa. They then infected the tobacco leaves with the two hybrid viruses. The progeny viruses isolated from the resulting lesions were of the type specified by the RNA and not by the protein. Thus, they concluded that RNA is the genetic material of TMV.

DNA MAKES RNA AND RNA MAKES PROTEIN

The precise replication of the hereditary information in the DNA leads to the question of how this information actually influences the activities of the cell. A major step forward in understanding this was the discovery that the DNA is copied, in a process called *transcription*, into a single-stranded molecule of the related ribonucleic acid (RNA). As in the replication of DNA, the information in the bases of DNA is precisely copied by base pairing to produce the RNA. After further processing, the so-called messenger RNA (mRNA) moves to sub-cellular particles called ribosomes where it is translated into protein. This translation is governed by the *genetic code* in which each combination of three bases, or triplet, directs the addition of a particular amino acid on to the protein chain: ACC directs addition of threonine, CCC of proline, and so on. Hence the genetic information contained in the linear array of bases in the DNA directs the production of a linear array of amino acids within a protein. Thus genetic changes in the bases in the DNA result in changes in the protein produced. For example, an A to C change in an ACC triplet would lead to the addition of a proline instead of a threonine. As specific proteins have particular biological effects, changes which affect the function of the protein will lead to an alteration in the appearance or function of an organism. In this way differences in the information in the DNA are observed as inherited differences between individuals, such as eye colour, or genetic diseases, such as haemophilia. The conclusion that DNA makes RNA makes protein has been referred to as the *central dogma of molecular biology*.

Motto Kimura (1968) and Jack King and Tom Jukes (1969) suggested that most of the variation seen at the molecular level is driven by random drift and did not result from neural selection. Selectionists take a different view: there are enough selectively advantageous mutations to account for the observed polymorphism in populations. The crucial difference between the two theories lies in the relative frequencies of neutral and selectively advantageous mutations (Ridley, 1993). The debate between Selectionists and Neutralists made an important contribution to the comparison of gene sequences between species. This gives rise to a molecular clock and enables an estimate to be made of the time of species' divergence from a simple comparison of the DNA sequences of two species. It seems likely that both natural selection and genetic drift determine the fate of DNA sequence changes. Changes in non-coding DNA probably have no measurable effect on fitness, but it is likely that changes to the coding regions will do so.

Sequencing DNA developed the techniques to infer molecular phylogenies. The theory of evolution suggests that all of the life has a common ancestry. Phylogenetic tree building methods using DNA sequences allow the evolutionary pas to be reconstructed.

CHAPTER 11

Enzymes

The existence and power of enzymes were first recognized in the nineteenth century when a reaction that has been considered to occur only in the presence of cells and were found also to be mediated by cell-free extracts. It was the Buchner brothers who showed that yeast, a living organism, yielded a non-living cell-free extract capable of fermenting sugars. Subsequent investigators of this field showed that this was due to the presence in the extract of many powerful catalysts, each one is highly specific for one of the steps in the breakdown of glucose. Similar catalysts, or enzymes, are found in all living things. These are organic catalyst found within a living organism which alter the fate of a chemical change, but itself remains unchanged at the end of the reaction; and it is produced by the living organisms but is not itself alive.

Enzymes are catalysts, ubiquitous in nature, responsible for thousands of chemical reactions involved in biological processes of living systems. Most are proteins. A few ribonucleoprotein enzymes have been discovered and, for some of these, the catalytic activity is in the RNA part rather than the protein part. These are thermolabile proteins produced by living cells catalysing specific chemical reactions. Enzymes bind temporarily to one or more of the reactants of the reaction they catalyse. In doing so, they lower the amount of **activation energy** needed and thus speed up the reaction. They are highly specific in their action, promoting only one biochemical reaction with their substrate ensuring synthesis of a bimolecular product under the conditions of mild pressure and temperature and in aqueous solutions. Enzymes are complex organic catalysts, produced by living cells but capable of acting independently of the cells that produce them.

ENZYMES AND LIFE PROCESSES

The living cell is the site of tremendous biochemical activity called metabolism. This is the process of chemical and physical change which goes on continually in the living organism. Build up of new tissue, replacement of old tissue, conversion of food to energy, disposal of waste materials, reproduction - all the activities that we characterize as *life*.

In a broad sense, there are two types of enzyme. Some that help *join specific molecules together to form new molecules* and others that *help break specific molecules apart into separate molecules.* Enzymes play many important roles outside the cell as well. One of the best examples of this is the **digestive system**. For instance, it is enzymes in your digestive system that break food down in our digestive system break food down into small molecules that can be absorbed by the body. Some enzymes in our digestive system break down starch, some proteins and others break down fats.

Building up and tearing down takes place in the face of an apparent paradox. The greatest majority of these biochemical reactions do not take place spontaneously. The phenomenon of catalysis makes possible biochemical reactions necessary for all life processes. **Catalysis** is defined as the acceleration of a chemical reaction by some substance which itself undergoes no permanent chemical change. The catalysts of biochemical reactions are enzymes and are responsible for bringing about almost all of the chemical reactions in living organisms. Without enzymes, these reactions take place at a rate far too slow for the pace of metabolism.

The oxidation of a fatty acid to carbon dioxide and water is not a gentle process in a test tube - extremes of pH, high temperatures and corrosive chemicals are required. Yet in the body, such a reaction takes place smoothly and rapidly within a narrow range of pH and temperature. In the laboratory, the average protein must be boiled for about 24 hours in a 20% HCl solution to achieve a complete breakdown. In the body, the breakdown takes place in four hours or less under conditions of mild physiological temperature and pH.

Microorganisms are Natural Enzyme Factories

The researchers wanted to find an organism that was easier and less expensive to raise than a calf. An organism that reproduced quickly does require a lot of space and food. So they began looking among the smallest and simplest organisms that is microorganisms. Microorganisms are very small living organisms. Some examples are **bacteria, fungi and yeast.** They live in soil or water in every corner of the earth. Because they are so small, they are obviously a lot less complex than a calf. Even so, a simple microorganism can produce many different types of enzyme. In fact, it doesn't take a complex organism to create a complex enzyme - as was proved by a simple **fungus** named *Mucor.* This fungus produced a chymosin-like enzyme that was almost the same as that produced by calves. But finding the right enzyme is only half the battle.

Compounds and Specificity of Enzymes

Specific enzymes work on specific compound. We need the right type of enzyme for the foods you want it to break down. Think of the foods you have problems with and then choose a product that contains at least those types of enzymes. Here is a list of the common enzyme types and foods they act on.

In order to do its work, an enzyme must unite - *even if ever so briefly* - with at least one of the reactants. In most cases, the forces that hold the enzyme and its substrate are **noncovalent**, an assortment of: hydrogen bonds, ionic interactions and hydrophobic interactions.

Catalase. It catalyses the decomposition of hydrogen peroxide into water and oxygen. One molecule of catalase can break 40 million molecules of hydrogen peroxide each second.

Carbonic anhydrase. It is found in red blood cells where it catalyses the reaction. It enables red blood cells to transport carbon dioxide from the tissues to the lungs. One molecule of carbonic anhydrase can process one million molecules of CO_2 each second.

Acetylcholinesterase. It catalyses the breakdown of the neurotransmitter **acetylcholine** at several types of synapses as well as at the neuromuscular junction – the specialized synapse that triggers the contraction of skeletal muscle. One molecule of acetylcholinesterase breaks down 25,000 molecules of acetylcholine each second. This speed makes possible the rapid "resetting" of the synapse for transmission of another nerve impulse.

ENZYMES: NOMENCLATURE

A commission has been established for deciding the name of enzyme and is known as **Enzyme nomenclature commission.** The terminology generally agreed for use in relation to enzymes. As in the example of urease, enzyme and classes of enzymes usually are named by appending the suffix *-ase* to the name or abbreviated name of a compound on which the enzyme acts. Enzymes that cause the cleavage of the polypeptide chain by reaction with water are referred to, as a class, as proteinase. Naming of enzymes is known as nomenclature. There are various ways of naming enzymes. The principles of nomenclature are as follows:

1. The enzymes are named by adding the *suffix-ase* to the name of the substrate on which they act *e.g.* Proteinase, sucrase, nucleases, which break up proteins, sucrose and nucleic acids respectively.
2. The enzymes can be named according to the type of *function* they are performing *e.g.* dehydrogenases remove hydrogen, carboxylase help in adding CO_2, decarboxylases help in removal of CO_2, oxidases helping in oxidation and so on.
3. The enzymes are sometimes given *double name*, one after the nature of substrate upon which they act and second according to their function *e.g.* pyruvic decarboxylases catalyses the removal of CO_2 from pyruvic acid, alcoholic dehydrogenase catalyses the removal of hydrogen from alcohol.

TERMS RELATED TO ENZYMES

Enzyme (Gr. *En* = *in zyma* = yeast): Any substance, protein in whole or in part, that regulates the rate of a specific biochemical reaction in living organisms. It is capable of catalysing a reaction in which substrate(s) are converted to product(s) through the formation of an intermediate enzyme - substrate complex. As with other catalysts, enzymes are responsible for accelerating the rate of a chemical reaction.

- **Endoenzyme:.** The enzymes, which act within the cell in which they are synthesised inner core of or substrate molecule, are known as endoenzyme.
- **Exoenzyme:** The enzymes, which act outside the cell in which they are synthesised or at the outer ends of substrate molecule, are termed exoenzyme.

The chemical nature of enzymes remained in dispute for a long time. Willstatter, working with peroxidase, had chosen an enzyme of such high catalytic efficiency that he believed that his active preparations were protein-free. By contrast, Sumner's crystalline urease was of such relatively modest activity that his critics attributed the catalysis to a highly active trace contaminant rather than the purified protein itself. Improvements in fractionation procedures have since allowed the purification of many hundreds of enzymes from diverse sources, leading to the realization that all enzymes are proteins. On the basis of chemical composition enzymes are of two types:

1. **Purely proteinaceous enzyme:** These are made up of only proteins *e.g.*, proteases that split protein and amylase that split starch.
2. **Conjugate enzymes:** These are made up of protein molecule, to which a non-protein group (*prosthetic group* or cofactor) is also attached *e.g.*, oxidizing enzymes.

It is essential to understand the difference between the following important terms - substrates, prosthetic group, apoenzyme, coenzymes and cofactor.

Substrate: The substance upon which an enzyme act is known as substrate. They are produced by the living protoplast of a cell.

Prosthetic group: It is the non-protein group of enzyme. The prosthetic group is firmly bound to protein component of the enzyme by chemical bonds and is not removed by dialysis. Haem, biotin and pyridoxal phosphate, like flavin, usually function as prosthetic groups.

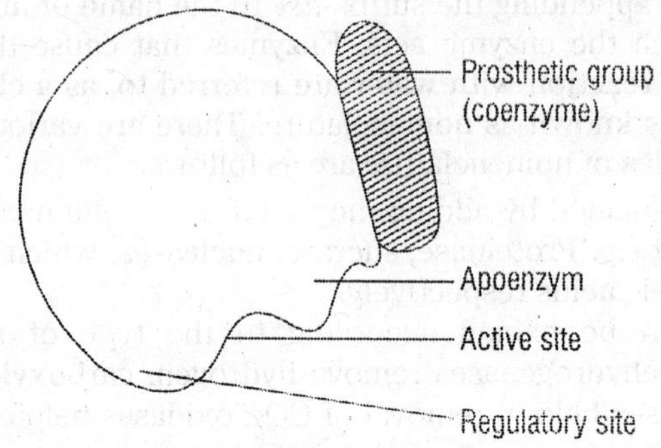

FIG. 11.1: Hypothetical morphology of an enzyme.

Apoenzyme: If, however, the prosthetic group is removed, the remaining protein part of the enzyme is called as apoenzyme and it is inert.

Coenzyme: It is the co-worker of prosthetic group and apoenzyme because neither the apoenzyme nor the prosthetic group alone is enzymatically active. The prosthetic group, the coenzymes, as the name specifics, act as co-worker of an apoenzyme. They combine with the enzyme and leave it again in the course of a single catalytic cycle. The coenzymes are vitamins, DPN, NAD, tetrahydrofolate, Coenzyme A and adenosine phosphates, like $NAD(P)^+$.

Cofactor: When a prosthetic group consists of single atom of some metal like Mg^{++}, Fe^{++}, Cu^{++}, Mo^{+++}, Zn^{++}, then it is known as cofactor and can be easily separated from rest of the protein part. Many enzymes employ either metal ion cofactors or organic cofactors. Thus many oxidoreductases utilize haem, nicotinamide or flavin; the transferases use folate, coenzyme A, pyridoxal phosphate, thiamine and adenosine phosphates; some of the ligases use biotin etc. Among these cofactors, some may be regarded as integral parts of their enzymes; they are known as *prosthetic* groups.

ENZYME CLASSIFICATION

A system of rules by which enzymes are classified on the basis of the substrate they react with and the type of reaction catalysed. This system provides both a systematic name and a four-part number code. The first number of the code places the enzyme into one of six groups indicating the type of reaction involved. The next two numbers indicate the groups involved in the reaction and the fourth number provides the absolute identification of the enzyme. The enzymes have been classified into following types:

1. Hydrolysing enzymes, 2. Oxidation-reduction enzymes, 3. Ligases, 4. Group transfer enzymes, 5. Desmolases, 6. Isomerizing enzymes, 7. Carboxylation enzymes.

Hydrolysing Enzymes

They catalyse the hydrolysis of complex big molecules into simple, smaller molecules. Depending upon the nature of food substance upon which the enzyme act, they can be further classified into following types.

1. **Carbohydrases:** They hydrolyse the complex polysaccharides into simpler

monosaccharides *e.g.* sucrases, maltases, lactase, cellulase etc.

2. **Esterases:** They hydrolyse the substances containing ester linkage *e.g.* lipase, phosphatase.
3. **Proteolytic enzymes:** They hydrolyse the proteins into peptones, and amino acids *e.g.* pepsin, peptidases.
4. **Amidases:** They hydrolyse amides into ammonia and acids *e.g.* urease.
5. **Oxidation-Reduction Enzyme:** They catalyse oxidation and reduction reactions *e.g.* Oxidases, reductases, catalases, dehydrogenases.
6. **Ligases:** They bring linkage between two molecules with the simultaneous breakdown of ATP molecule which supplies energy *e.g.* synthetases which join amino acids to RNA.

Oxidoreductases	$A^- + B \rightleftharpoons A + B^-$
Transferases	$A - B + C \rightleftharpoons A + B - C$
Transferases	$A - B + H_2O \rightleftharpoons A - H + B - OH$
Lyases	$\underset{A - B}{\overset{X \quad Y}{\mid \quad \mid}} \rightleftharpoons A = B + X - Y$
Isomerases	$\underset{A - B}{\overset{X \quad Y}{\mid \quad \mid}} \rightleftharpoons \underset{A - B}{\overset{Y \quad X}{\mid \quad \mid}}$
Ligases	$A + B \rightleftharpoons A - B$

FIG. 11.2: Main classes of enzymes with their reactions.

7. **Group transfer enzymes:** They catalyse the transformation of a group from one kind of molecule to another or from one position of molecule to another position in the same molecule and are called as transferases.
8. **Desmolases:** They catalyse the reactions in which long carbon chain is broken or lengthened *e.g.* aldolase.
9. **Isomerizing enzymes:** They catalyse reactions in which an organic molecule is transformed into its isomeric form and are known as isomerases.
10. **Carboxylation enzymes:** They catalyse the reactions in which CO_2 is added or removed and are known as carboxylases or decarboxylases.

TABLE 11.1: Digestive enzymes are enzymes that break down food into usable material

Enzyme type	Functions
Amylase	Breaks down carbohydrates, starches, and sugars which are prevalent in potatoes, fruits, vegetables, and many snack foods
Lactase	Breaks down lactose (milk sugars)
Diastase	Digests vegetable starch
Sucrase	Digests complex sugars and starches
Maltase	Digests disaccharides to monosaccharides (malt sugars)
Invertase	Breaks down sucrose (table sugar)
Glucoamylase	Breaks down starch to glucose
Alpha-Glactosidase	Facilitates digestion of beans, legumes, seeds, roots, soy products, and underground stems
Protease	Breaks down proteins found in meats, nuts, eggs, and cheese
Pepsin	Breaks down proteins into peptides
Peptidase	Breaks down small peptide proteins to amino acids

Enzyme type	Functions
Trypsin	Derived from animal pancreas, breaks down proteins
Alpha	Chymotrypsin, an animal-derived enzyme, breaks down proteins
Bromelain	Derived from pineapple, breaks down a broad spectrum of proteins, has anti-inflammatory properties, effective over very wide pH range
Papain	Derived from raw papaya, broad range of substrates and pH, works well breaking down small and large proteins
Lipase	Breaks down fats found in most dairy products, nuts, oils, and meat
Cellulose	Breaks down cellulose, plant fiber; not found in humans
Other Stuff	
Betaine HCL	Increases the hydrochloric acid content of the upper digestive system; activates the protein digesting enzyme pepsin in the stomach (does not influence plant- or fungal-derived enzymes)
Cerecalase™	A unique cellulase complex from National Enzyme Company that maximizes fiber and cereal digestion and absorption of essential minerals; an exclusive blend of synergistic phytase, hemicellulase, and beta-glucanase
Endoprotease	Cleaves peptide bonds from the interior of peptide chains
Exoprotease	Cleaves off amino acids from the ends of peptide chains
Extract Of Ox Bile	An Animal-Derived Enzyme, Stimulates The Intestine To Move
Fructooligosaccharides (FOS)	Helps support the growth of friendly intestinal microbes, also inhibits the growth of harmful species
L-Glutamic Acid	Activates the protein digesting enzyme pepsin in the stomach
Lysozyme	An animal-derived enzyme, and a component of every lung cell; lysozyme is very important in the control of infections, attacks invading bacterial and viruses
Papayotin	From papaya
Pancreatin	An animal-derived enzyme, breaks down protein and fats
Pancrelipase	An animal-derived enzyme, breaks down protein, fats, and carbohydrates
Pectinase	Breaks down the pectin in fruit
Phytase	Digests phytic acid, allows minerals such as calcium, zinc, copper, manganese, etc. To be more available by the body, but does not break down any food proteins
Xylanase	Breaks down xylan sugars, works well with grains such as corn

PROPERTIES OF ENZYMES

Enzymes are characterized by many biophysical properties. Some common properties are given below:

Enzymes are Natural

Enzymes are proteins. Like all other proteins, enzymes are organic. Once they have done their job, enzymes break down swiftly and can be absorbed back into nature. They may be either purely made up of protein or associated with a non-protein part.

Enzymes are Catalysts

While it is true that an enzyme can only perform one specific job, it is important to know that one enzyme can perform that same job over and over again, millions of times, without being consumed in the process. And enzymes do their job best in the **mild pH and temperature conditions** found in nature. Enzymes act as catalysts and influence the speed of chemical reaction but themselves remain unchanged. A small quantity of enzyme can catalyse the transformation of a very large quantity of the substrate into end product *e.g.* the sucrase can hydrolyse 100,000 times of sucrose as compared with its own weight.

Enzymes are Efficient

Not only do enzymes work hard, they also work with blinding speed. For instance, there is an enzyme in the liver that helps hydrogen peroxide break down into water and oxygen. What's amazing is that one enzyme can process 5 million hydrogen peroxide molecules in one minute.

Enzymes are Specific

An enzyme that is able to break fat down would not be able to dissolve protein or starch. Enzymes perform only one specific job. That means an enzyme can do its job with very few side effects. It also explains why there are so many different types of enzyme. To date, 3,000 different types have been identified and there are many more waiting to be discovered. The ability of an enzyme to catalyse one specific reaction and essentially no other is perhaps its most significant property. Close examination reveals that most enzymes can catalyse the same type of reaction.

Reversibility of Enzyme Action

Most of the reactions catalysed by enzymes are reversible and the enzyme can catalyse the reactions in both directions *e.g.* the lipase can catalyse not only the hydrolysis of fats into fatty acids and glycerol but also can synthesise fats from fatty acids and glycerol as shown below.

$$\text{Fats} \xleftrightarrow{\textit{Lipase}} \text{Fatty acids} + \text{Glycerol}$$

Sensitivity of the Enzymes

The enzymes are sensitive to rise of temperature and are destroyed at higher temperature of 60-70 °C. However, most of the enzymes act best between 20° C to 35° C. At 0° C or below, they are not destroyed but become inactivated.

ENZYMES AS CHEMO-THERMO REGULATORS

The chemical reactions that take place within the living cell must be precisely controlled and coordinated. These reactions interlock somewhat as a multitude of assembly processes interlock on a modern day production line. For example glucose and other fuel molecules release energy in cells by burning. The cells could not tolerate and survive in such high temperatures nor effectively utilize energy that is released in short sudden bursts. Therefore, cells need a slow, steady release of energy that they must be able to regulate to meet metabolic energy requirements. In cellular respiration, fuel molecules are slowly oxidized and energy is extracted in small amounts, which includes sequences of 30 or more reactions (as many as 20 or 30 chemical transformations before it reaches some final state). Such chemical transformations require a system of flexible chemical control. The key elements of this control system are the remarkable enzymes.

MECHANISM OF ENZYME ACTION

An enzyme increases the speed of a chemical reaction without being consumed itself. An enzyme affects the rate of a reaction by lowering the energy needed to activate the reactions. Even a strongly exergonic reaction that releases more than enough energy as it proceeds is prevented from beginning by an energy barrier. This is because before new chemical bonds can be formed, existing ones must be broken. The energy required to overcome this barrier and get the reaction going is called **activation energy.** An enzyme greatly reduces the activation energy necessary to initiate a chemical reaction. Enzyme promotes a chemical reaction. It does not influence the *direction of a chemical reaction* or the final concentrations of the molecules involved. They simply *speed up reactions.* For example carbonic anhydrase is an enzyme, which promotes the interconversion of carbonic acid with carbon dioxide and water:

$$CO_2 + H_2O \underset{\text{Carbonic acid}}{\overset{\text{Carbonic anhydrase}}{\rightleftharpoons}} H_2CO_3$$

This reaction occurs in animal cells and tissues (kidneys and red blood cells). Carbonic acid ionizes, forming bicarbonate ions, the forms in which most of the carbon dioxide in the blood is transported. The reaction shown above could proceed without the enzyme but would be very slow. However, in presence of carbonic anhydrase, the reaction proceeds about 10 million times faster. This reveals that a single molecule of this enzyme can promote the conversion of an estimated 600,000 molecules of carbon dioxide into carbonic acid each second.

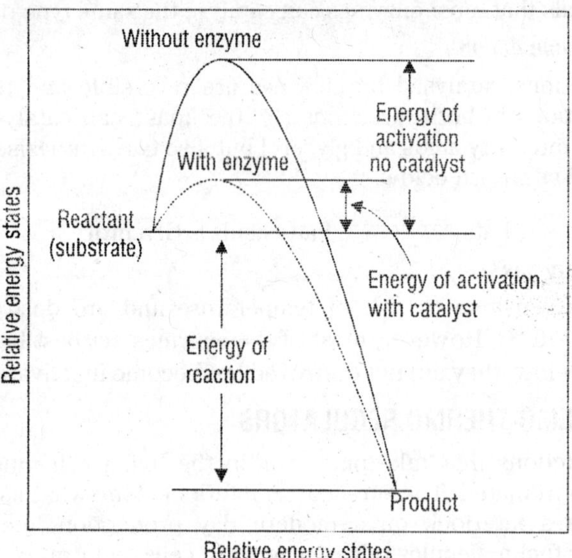

FIG. 11.3: An enzyme accelerates the rate of reaction through lowering activation energy.

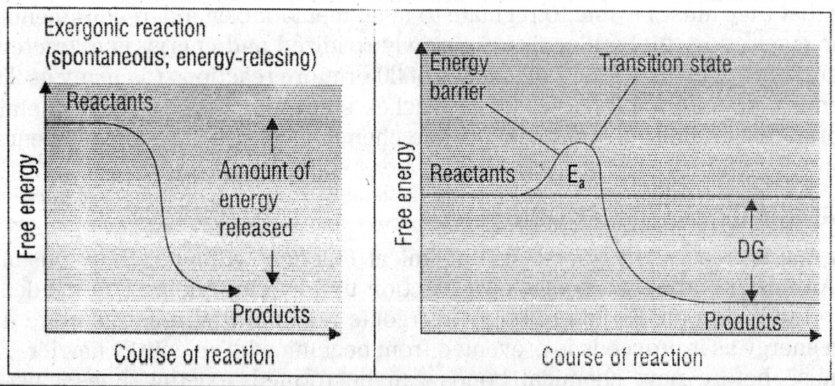

FIG. 11.4: Time-energy graphs of an exergonic reaction (left) and endergonic reaction (right).

Endergonic and Exergonic

Energy releasing processes, ones that generate energy, are termed exergonic reactions. Reactions that require energy to initiate the reaction are known as endergonic reactions. All natural processes tend to proceed in such a direction that the disorder or randomness of the universe increases (the second law of thermodynamics).

Oxidation/Reduction

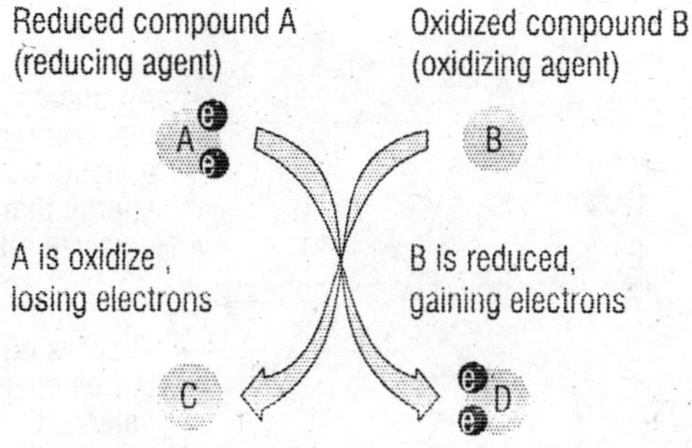

FIG. 11.5: Reduction and oxidation of compounds.

Biochemical reactions in living organisms are essentially energy transfers. Often they occur together, "linked", in what are referred to as oxidation/reduction reactions. Reduction is the gain of an electron. Sometimes we also have H ions along for the ride, so reduction also becomes the gain of H.

Oxidation is the loss of an electron (or hydrogen). In oxidation/reduction reactions, one chemical is oxidized, and its electrons are passed (like a hot potato) to another (reduced, then) chemical. Such coupled reactions are referred to as redox reactions. The metabolic processes glycolysis, Kreb's Cycle, and Electron Transport Phosphorylation involve the transfer of electrons (at varying energy states) by redox reactions.

Passage of electrons from compound A to compound B, when A loses its electrons it is oxidized; when B gains the electrons it is reduced.

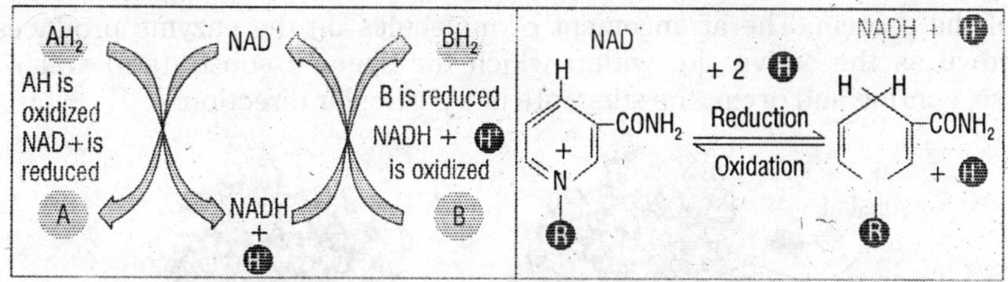

FIG. 11.6: Oxidation/reduction via an intermediary (energy carrier) compound, in this case NAD^+.

Catabolism and Anabolism

Anabolism is the total series of chemical reactions involved in synthesis of organic compounds. Autotrophs must be able to manufacture (synthesise) all the organic compounds they need. Heterotrophs can obtain some of their compounds in their diet (along with their energy). For example humans can synthesise 12 of the 20 amino acids, we must obtain the other 8 in our diet. Catabolism is the series of chemical reactions that breakdown larger molecules. Energy is released this way, some of it can be utilized for anabolism. Products of catabolism can be reassembled by anabolic processes into new anabolic molecules.

Enzymes: Organic Catalysts

Enzymes allow many chemical reactions to occur within the homeostasis constraints of a living system. Enzymes function as organic catalysts. A catalyst is a chemical involved in, but not changed by, a chemical reaction. Many enzymes function by lowering the activation energy of reactions. By bringing the reactants closer together, chemical bonds may be weakened and reactions will proceed faster than without the catalyst.

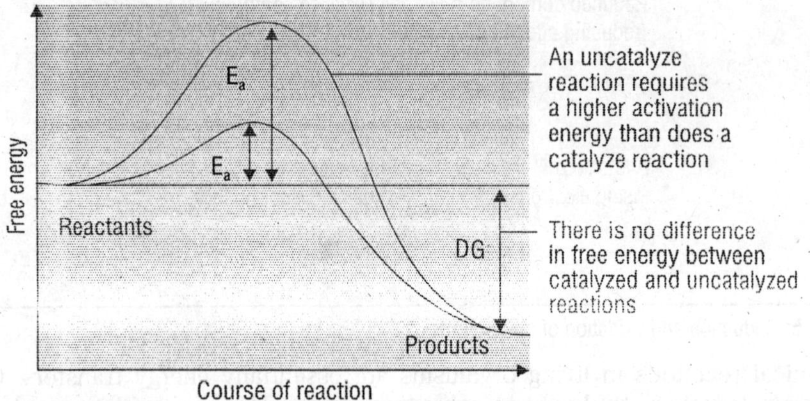

FIG. 11.7: The use of enzymes can lower the activation energy of a reaction (Ea).

Enzymes can act rapidly, as in the case of carbonic anhydrase (enzymes typically end in the -ase suffix), which causes the chemicals to react 107 times faster than without the enzyme present. Carbonic anhydrase speeds up the transfer of carbon dioxide from cells to the blood. There are over 2000 known enzymes, each of which is involved with one specific chemical reaction. Enzymes are substrate specific. The enzyme peptidase (which breaks peptide bonds in proteins) will not work on starch (which is broken down by human-produced amylase in the mouth).

Enzymes are proteins. The functioning of the enzyme is determined by the shape of the protein. The arrangement of molecules on the enzyme produces an area known as the active site within which the specific substrate(s) will *fit*. It recognise, confine and orient the substrate in a particular direction.

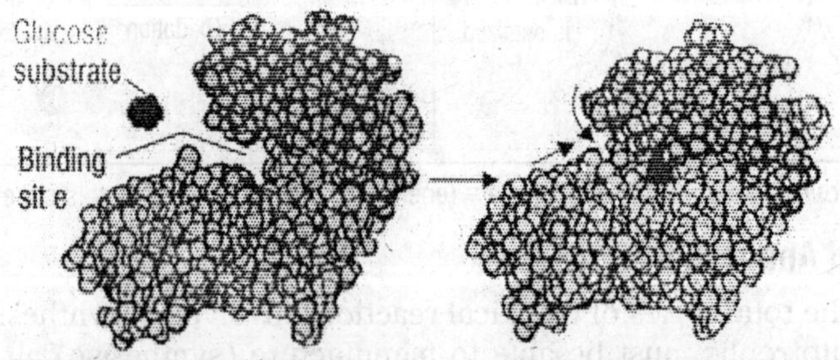

FIG. 11.8: Space filling model of an enzyme working on glucose. Note the shape change in the enzyme (indicated by the red arrows) after glucose has fit into the binding or active site.

The induced fit hypothesis suggests that the binding of the substrate to the enzyme alters the structure of the enzyme, placing some strain on the substrate and further facilitating the reaction. Cofactors are non-proteins essential for enzyme activity. Ions such as K^+ and Ca^{+2} are cofactors. Coenzymes are non-protein organic molecules bound to enzymes near the active site (NAD - nicotinamide adenine dinucleotide).

Enzymatic pathways form as a result of the common occurrence of a series of dependent chemical reactions. In one example, the end product depends on the successful completion of five reactions, each mediated by a specific enzyme. The enzymes in a series can be located adjacent to each other (in an organelle or in the membrane of an organelle), thus speeding the reaction process. Also, intermediate products tend not to accumulate, making the process more efficient.

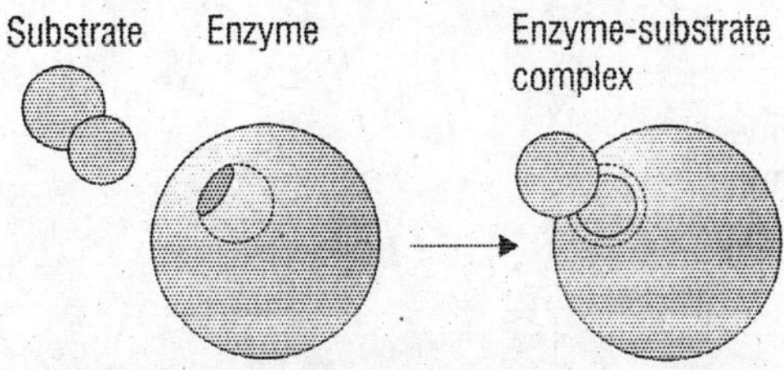

FIG. 11.9: The formation of an enzyme-substrate complex.

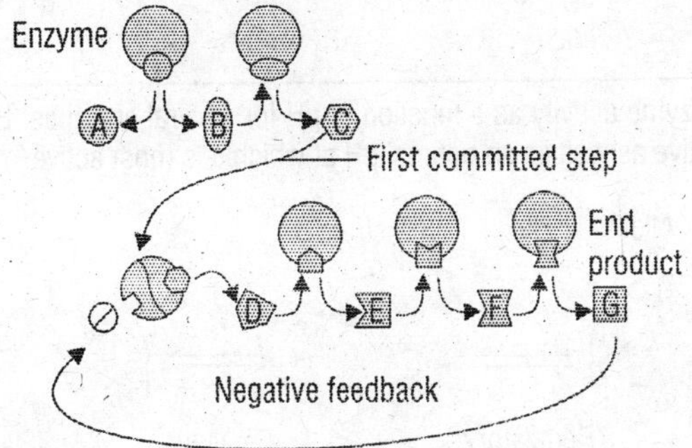

FIG. 11.10: Negative feedback and a metabolic pathway. The production of the end product (G) in sufficient quantity to fill the square feedback slot in the enzyme will turn off this pathway between step C and D.

By removing intermediates (and by inference end products) from the reactive pathway, equilibrium (the tendency of reactions to reverse when concentrations of the products build up to a certain level) effects are minimized, since equilibrium is not attained, and so the reactions will proceed in the *preferred* direction.

Temperature: Increases in temperature will speed up the rate of non-enzyme mediated reactions, and so temperature increase speeds up enzyme mediated reactions, but only to a point. When heated too much, enzymes (since they are proteins dependent on their shape) become denatured. When the temperature drops, the enzyme regains its shape. Thermolabile enzymes, such as those responsible for the colour distribution in Siamese cats and colour camouflage of the Arctic fox, work better (or work at all) at lower temperatures.

Concentration of substrate and product also control the rate of reaction, providing a biofeedback mechanism.

Activation, as in the case of chymotrypsin, protects a cell from the hazards or damage the enzyme might cause.

Changes in pH will also denature the enzyme by changing the shape of the enzyme. Enzymes are also adapted to operate at a specific pH or pH range.

Allosteric Interactions may allow an enzyme to be temporarily inactivated. Binding of an allosteric effector changes the shape of the enzyme, inactivates it while the effector is still bound. Such a mechanism is commonly employed in feedback inhibition. Often one of the products, either an end or near-end product act as an allosteric effector, blocking or shunting the pathway.

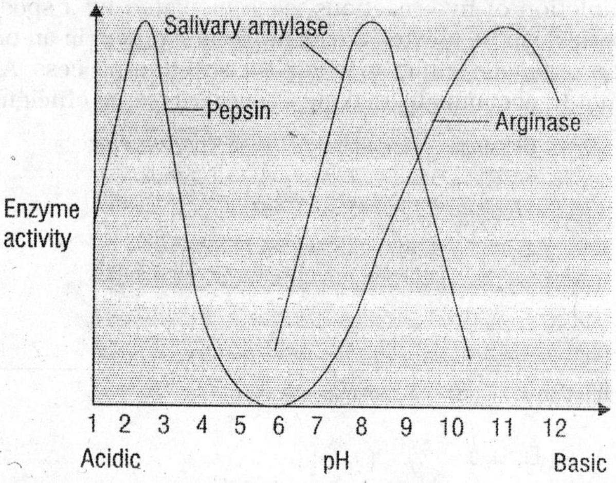

FIG. 11.11: Plot of enzyme activity as a function of pH for several enzymes. Each enzyme has a range of pH at which it is active as well as an optimal pH at which it is most active.

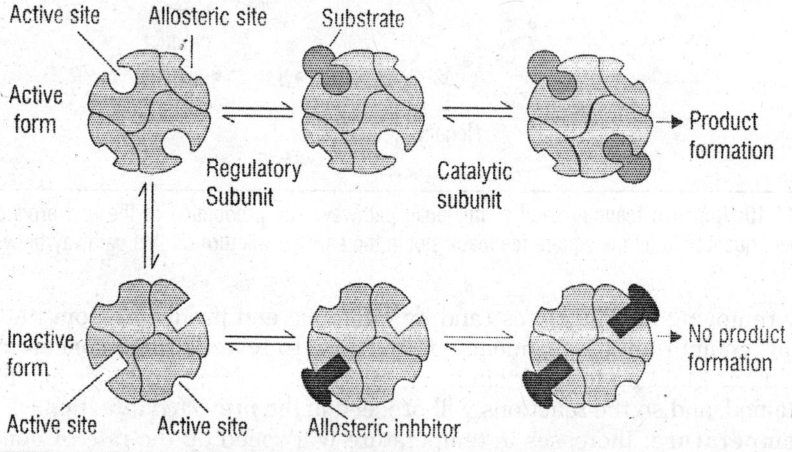

FIG. 11.12: Action of an allosteric inhibitor as a negative control on the action of an enzyme.

Competitive Inhibition works by the competition of the regulatory compound and substrate for the binding site. If enough regulatory compound molecules bind to enough enzymes, the pathway is shut down or at least slowed down. PABA, a chemical essential to bacteria that infects animals, resembles a drug, sulphanilamide that competes with PABA, shutting down an essential bacterial (but not animal) pathway.

Noncompetitive Inhibition occurs when the inhibitory chemical, which does not have to resemble the substrate, binds to the enzyme other than at the active site. Lead binds to SH groups in this fashion. Irreversible Inhibition occurs when the chemical either permanently binds to or massively denatures the enzyme so that the tertiary structure cannot be restored. Nerve gas permanently blocks pathways involved in nerve message transmission, resulting in death. Penicillin, the first of the "wonder drug" antibiotics, permanently blocks the pathways certain bacteria use to assemble their cell wall components.

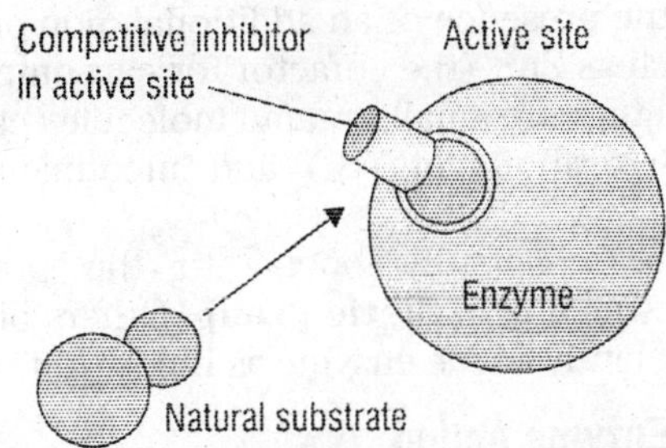

FIG. 11.13: General diagram showing competitor in the active site normally occupied by the natural substrate.

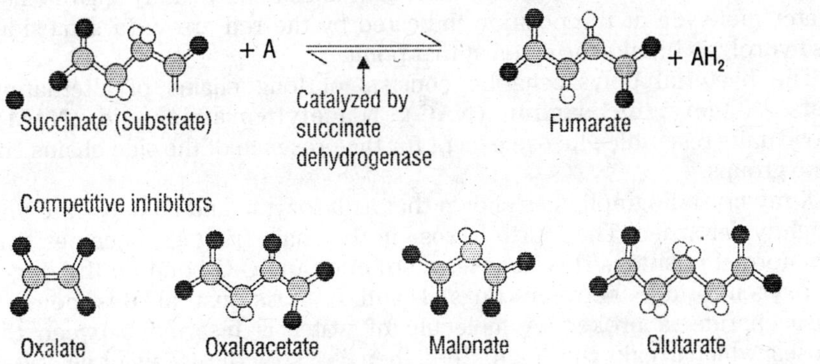

FIG. 11.14: Specific case of succinate dehydrogenase and its natural substrate (succinate) and competitors (oxalate).

Multi-Substrate Enzymes

When enzymes catalyse reactions of two or more substrates, they work in a varied ways and can be differentiated by kinetic analysis. Certain enzymes bind their substrates in a fixed order; while some bind in random order. Sometimes, bindings of one substrate gives a partial reaction before the second substrate binds.

The only true one-substrate enzymes are the isomerases, which catalyse reactions of the type A = B, and the lyases, which catalyse reactions of the type A - B + C, and are, therefore, one-substrate enzymes in one direction only. Hydrolases, catalysing reactions of the general type A-B + H_2O = A-OH + BH may be regarded as honorary one-substrate enzymes, since one substrate, water, is normally present at constant concentration.

The remaining groups are:

$$\text{Oxidant + Reductant} \xrightarrow{\text{oxidoreductases}} \text{Reduced product + Oxidized product}$$

$$\text{A + BX} \xrightarrow{\textit{transferases}} \text{AX + B}$$

This large group includes methyl transferases, transaldolase, transketolase, acyl transferases, glycosyl transferases, transaminases, kinases etc.

$$\text{X + Y + ATP} \xrightarrow{\textit{ligases}} \text{XY + ADP + P}_i \text{ (or AMP + PP}_i\text{)}$$

These three groups of enzymes by definition catalyse reactions involving more than one substrate.

ENZYME COFACTORS

Many enzymes require the presence of an additional, non-protein cofactor. Some of these are metal ions such as Zn^{2+} (the cofactor for carbonic anhydrase), Cu^{2+}, Mn^{2+}, K^+, and Na^+. Some cofactors are small organic molecules called **coenzymes**. The B vitamins (thiamine (B_1), riboflavin (B_2) and nicotinamide are precursors of coenzymes.

Coenzymes may be covalently bound to the protein part (called the **apoenzyme**) of enzymes as a **prosthetic group**. Others bind more loosely and, in fact, may bind only transiently to the enzyme as it performs its catalytic act.

Lysozyme: A Model of Enzyme Action

A number of lysozymes are found in nature; in human tears and egg white, for examples. The enzyme is antibacterial because it degrades the polysaccharide that is found in the cell walls of many bacteria. It does this by catalysing the insertion of a water molecule at the position indicated by the red arrow (a glycosidic bond). This hydrolysis breaks the chain at that point.

The bacterial polysaccharide consists of long chains of alternating amino sugars: **N-acetylglucosamine** (**NAG**), **N-acetylmuramic acid** (**NAM**). These hexose units resemble glucose except for the presence of the side chains containing amino groups.

X-ray crystallography has shown that as lysozyme and its substrate unite, each is slightly deformed. The fourth hexose in the chain (ring #4) becomes twisted out of its normal position. This imposes a strain on the C-O bond on the ring-4 side of the oxygen bridge between rings 4 and 5. It is just at this point that the polysaccharide is broken. A molecule of water is inserted between these two hexoses, which breaks the chain. Here, then, is a structural view of what it means to lower activation energy. The energy needed to break this covalent bond is lower now that the atoms connected by the bond have been distorted from their normal position.

As for lysozyme itself, binding of the substrate induces a small (~0.75 Å) movement of certain amino acid residues so the cleft closes slightly over its substrate. Therefore, *lock* as well as the *key* changes shape as the two are brought together. (This is sometimes called **induced fit**.)

The amino acid residues in the vicinity of rings 4 and 5 provide a plausible mechanism for completing the catalytic act. Residue 35; glutamic acid (**Glu-35**), is about 3 Å from the -O- bridge that is to be broken. The free carboxyl group of glutamic acid is a hydrogen ion donor and available to transfer H^+ to the oxygen atom. This would break the already-strained bond between the oxygen atom and the carbon atom of ring 4.

Now having lost an electron, the carbon atom acquires a positive charge. Ionized carbon is normally very unstable, but the attraction of the negatively-charged carboxyl ion of **Asp-52** could stabilize it long enough for an -OH ion (from a spontaneously dissociated water molecule) to unite with the carbon. Even at pH 7, water spontaneously dissociates to produce H^+ and OH^- ions. The hydrogen ion (H^+) left over can replace that lost by **Glu-35**.

In either case, the chain is broken, the two fragments separate from the enzyme, and the enzyme is free to attach to a new location on the bacterial cell wall and continue its work of digesting it.

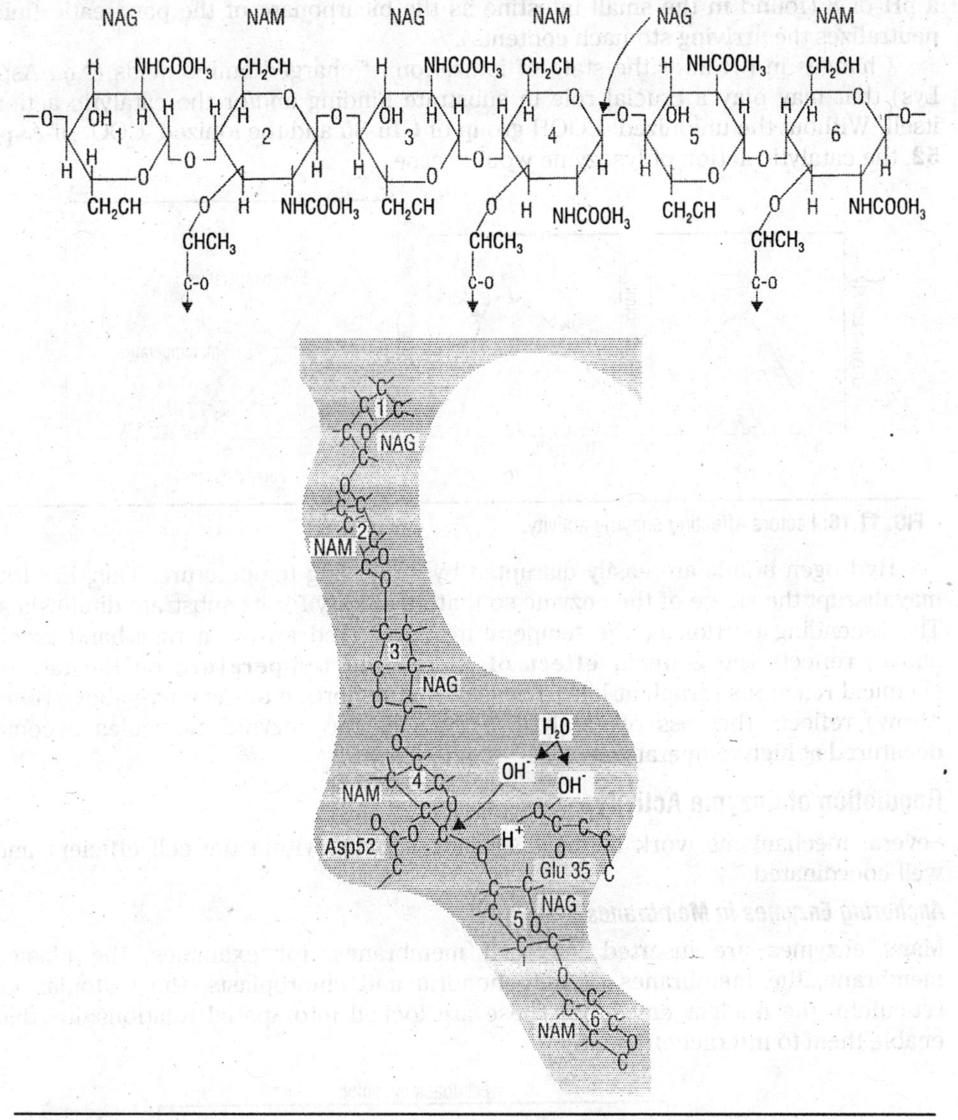

FIG. 11.15: Lysozyme is a globular protein with a deep cleft across part of its surface. Six hexoses of the substrate fit into this cleft. With so many oxygen atoms in sugars, as many as 14 hydrogen bonds form between the six amino sugars and certain amino acid R groups such as Arg-114, Asn-37, Asn-44, Trp-62, Trp-63, and Asp-101. Some hydrogen bonds also form with the C = O groups of several peptide bonds. In addition, hydrophobic interactions may help hold the substrate in position.

FACTORS AFFECTING ENZYME ACTION

The activity of enzymes is strongly affected by changes in **pH** and **temperature**. Each enzyme works best at a certain pH (left graph) and temperature (right graph), its activity decreasing at values above and below that point. This is not surprising considering the importance of tertiary structure (*i.e.* shape) in enzyme function and noncovalent forces, *e.g.*, ionic interactions and hydrogen bonds, in determining that shape. For examples: the protease **pepsin** works best as a pH of 1–2 (found in the stomach) while the protease **trypsin** is inactive at such a low pH but very active at a pH of 8 (found in the small intestine as the bicarbonate of the pancreatic fluid neutralizes the arriving stomach contents).

Changes in pH alter the state of ionization of charged amino acids (*e.g.*, Asp, Lys) that may play a crucial role in substrate binding and/or the catalytic action itself. Without the unionized -COOH group of **Glu-35** and the ionized -COO^- of **Asp-52**, the catalytic action of lysozyme would cease.

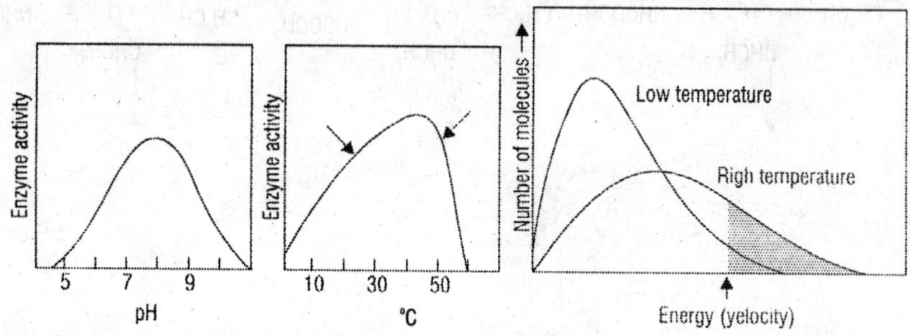

FIG. 11.16: Factors affecting enzyme activity.

Hydrogen bonds are easily disrupted by increasing temperature. This, in turn, may disrupt the shape of the enzyme so that its affinity for its substrate diminishes. The ascending portion of the temperature curve (red arrow in right-hand graph above) reflects the **general effect of increasing temperature** on the rate of chemical reactions (graph at left). The descending portion of the curve above (blue arrow) reflects the loss of catalytic activity as the enzyme molecules become denatured at high temperatures.

Regulation of Enzyme Activity

Several mechanisms work to make enzyme activity within the cell efficient and well-coordinated.

Anchoring Enzymes in Membranes

Many enzymes are inserted into cell membranes, for examples, the plasma membrane, the membranes of mitochondria and chloroplasts, the endoplasmic reticulum, the nuclear envelope. These are locked into spatial relationships that enable them to interact efficiently.

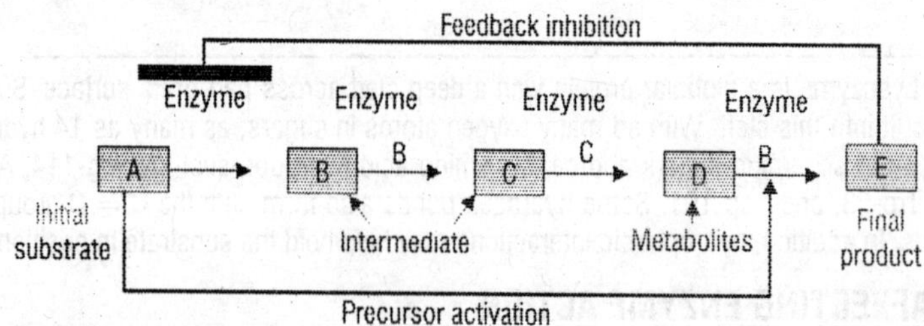

FIG. 11.17: Enzyme feedback inhibition.

Inactive Precursors

Enzymes, such as proteases, that can attack the cell itself are inhibited while within the cell that synthesises them. For example, pepsin is synthesised within the chief cells (in gastric glands) as an inactive precursor, **pepsinogen**. Only when exposed to the low pH outside the cell is the inhibiting portion of the molecule removed and active pepsin produced.

Feedback Inhibition

If the product of a series of enzymatic reactions, *e.g.*, an amino acid, begins to accumulate within the cell, it may specifically inhibit the action of the first enzyme involved in its synthesis. Thus further production of the enzyme is halted.

Precursor Activation

The accumulation of a substance within a cell may specifically activate (blue arrow) an enzyme that sets in motion a sequence of reactions for which that substance is the initial substrate. This reduces the concentration of the initial substrate.

In the case if feedback inhibition and precursor activation, the activity of the enzyme is being regulated by a molecule which is **not** its substrate. In these cases, the regulator molecule binds to the enzyme at a different site than the one to which the substrate binds. When the regulator binds to its site, it alters the shape of the enzyme so that its activity is changed. This is called an **allosteric** effect.

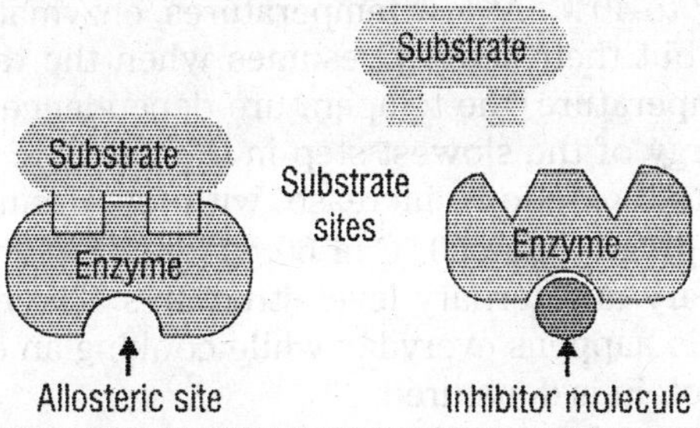

FIG. 11.18: Enzyme-substrate relationship.

1. In feedback inhibition, the allosteric effect lowers the affinity of the enzyme for its substrate.
2. In precursor activation, the regulator molecule increases the affinity of the enzyme in the series for its substrate.

REGULATION OF ENZYME SYNTHESIS

The four mechanisms described above regulate the activity of enzymes already present within the cell.

What about enzymes which are not needed or are needed but not present?

Here, too, control mechanisms are at work, which regulate the rate at which new enzymes are synthesised. Most of these controls work by *turning on – or off –* the **transcription of genes.**

If, for example, ample quantities of an amino acid are already available to the cell from its extracellular fluid, synthesis of the enzymes that would enable the cell to produce that amino acid for itself is shut down.

Conversely, if a new substrate is made available to the cell, it may induce the synthesis of the enzymes needed to cope with it. Yeast cells, for example, do not ordinarily metabolize lactose and no **lactase** can be detected in them. However, if grown in a medium containing lactose, they soon begin synthesizing lactase – by transcribing and translating the necessary gene(s) – and so can begin to metabolize the sugar.

E. coli also has a mechanism which regulates enzyme synthesis by controlling **translation** of a needed messenger RNA.

ENZYME OPTIMA

Enzymes act better under certain narrowly defined conditions known as *optima*. These conditions are optimum at appropriate temperature, pH, and salt concentration, etc. Enzyme pepsin (protein-digesting enzyme of the stomach) works best at the strongly acid pH of 2 whereas amylase (starch-digesting enzyme in saliva and pancreatic juice) works better at pH 8.5 (slightly alkaline). Strong acids or bases irreversibly *inactivate* most enzymes. pH permanently changes the molecular conformation of enzymes.

The Michaelis constant Km is affected by ionizations of both free and bound species, and tends to behave in a more complicated way. Enzymes like other proteins, are stable over a limited range of pH. Outside this range, changes in the charges on ionisable amino acid residue results in modifications and eventually lead to denaturation. But within the range of its pH is a characteristic more of the enzyme than of the substrate.

Temperature dependence of the catalytic process can only be meaningful when studied within a range in which thermal denaturation is not significant. It has a narrow range from 5° to 40°C. At low temperatures, enzymatic reactions occur very slowly or not at all, but their activity resumes when the temperature is raised to normal. At lower temperature, the temperature dependence of Kcat can be related to the activation energy of the slowest step in the catalytic pathways. The rates of most enzyme-regulated reactions increase with increasing temperature, within limits. Temperatures greater than 50 °C or 60 °C rapidly inactivate most enzymes by altering their secondary and tertiary level structures. The enzyme is denatured at high temperature. This happens everyday while cooking an egg white; it changes in consistency as the protein is denatured.

ENZYME KINETICS AND THE MICHAELIS-MENTEN EQUATION

Study of the impact made on the rate of an enzyme-catalysed reaction by changes in experimental conditions is known as **enzyme kinetics**. Knowledge of kinetics can be a very useful tool in understanding the **mechanism** by which an enzyme carries out its catalytic activity.

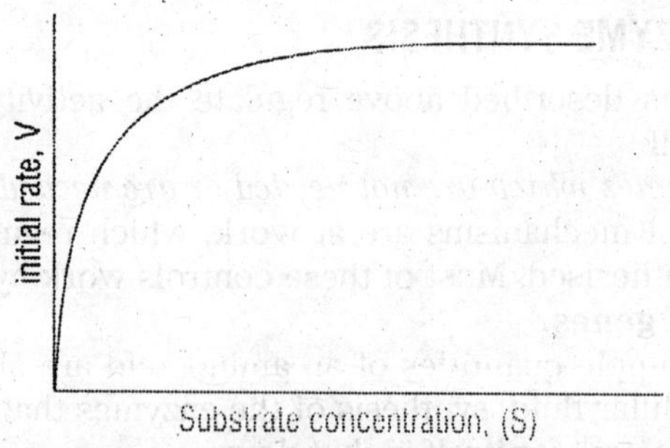

Fig. 11.19: The relation between substrate concentration (s) and initial rate of reaction (v).

The effect of substrate concentration on the initial rate of an enzyme-catalysed reaction is a central concept in enzyme kinetics. When data are generated from experiments of this type and the results plotted as a graph of initial rate (**v, y-axis**) against substrate concentration (**[S]*, x-axis**), many enzymes exhibit a rectangular hyperbolic curve like the one shown in the diagram below.

Observations of this type set Leonor Michaelis and Maud Menten thinking about the underlying reasons why a curve should follow this shape and led them to derive an algebraic equation that now bears their names. There are several modern ways to explain the way in which the Michaelis-Menten equation is derived, and one is spelt out below.

Deriving the Michaelis-Menten Equation

Start with the generalised scheme for enzyme-catalysed production of a product (P) from substrate (S). The enzyme (E) does not magically convert S into P, it must first come into physical contact with it, i.e. E binds S to form an enzyme-substrate complex (ES). Michaelis and Menten therefore set out the following scheme:

$$E + S \underset{k_{-1}}{\overset{k_1}{\rightleftharpoons}} ES \xrightarrow{k_2} E + P$$

The terms k_1, k_{-1} and k_2 are rate constants for, respectively, the association of substrate and enzyme, the dissociation of unaltered substrate from the enzyme and the dissociation of product (= altered substrate) from the enzyme. Note that there is the theoretical possibility of a reverse reaction, with ES complex forming from E and P, but this can be ignored because we are considering *initial* rates of reaction, i.e. when the enzyme is first provided with substrate, so there should not be any product available to combine with enzyme.

The overall rate of the reaction (v) is limited by the step ES to E + P, and this will depend on two factors - the rate of that step (i.e. k_2) and the concentration of enzyme that has substrate bound, i.e. [ES]. This can be written as:

$$v = k_2 [ES] \quad \text{.......(i)}$$

At this point it is important to draw your attention to two assumptions that are made in this scheme. The first is the availability of a vast excess of substrate, so that [S]>>[E]. Secondly, it is assumed that the system is in steady-state, i.e. that the ES complex is being formed and broken down at the same rate, so that overall [ES] is constant. The formation of ES will depend on the rate constant K_1 and the availability of enzyme and substrate, i.e. [E] and [S]. The breakdown of [ES] can occur in two ways, either the conversion of substrate to product or the non-reactive dissociation of substrate from the complex. In both instances the [ES] will be significant. Thus, at steady state we can write:

$$k_1[E][S] = k_{-1} [ES] + k_2[ES]$$

The next couple of steps are rearrangements of this equation. First of all we can collect together the rate constants on the right-hand side because they are both multiplied by [ES], this gives us:

$$k_1[E][S] = (k_{-1} + k_2) [ES]$$

Then dividing both sides by ($k_{-1} + k_2$), this becomes:

$$\frac{k_1[E][S]}{(k_{-1} + k_2)} = [ES]$$

Notice that the three rate constants are now on the same side of the equation. As the name implies, these terms are constants, so we can actually combine them into one term. This new constant is termed the Michaelis constant and is written K_M.

$$\frac{(k_{-1} + k_2)}{k_1} = K_M$$

Notice that the three rate constants in the definition of K_M are actually inverted (the other way up) compared with our previous equation. This is a 'trick' that makes for easier calculation at a later stage. Substituting this definition of K_M into our previous equation now gives us:

$$\frac{[E][S]}{K_M} = [ES] \quad \text{.........(ii)}$$

The total amount of enzyme in the system must be the same throughout the experiment, but it can either be free (unbound) E or in complex with substrate, ES. If we term the total enzyme E_0, this relationship can be written out:

$$[E_0] = [E] + [ES]$$

This can be rearranged (by subtracting [ES] from each side) to give:

$$[E] = [E_0] - [ES]$$

So, the [E] free in solution is equal to the total amount of enzyme minus the amount that has substrate bound. Substituting this definition of [E] back into equation 2 gives us:

$$\frac{([E_0] - [ES])[S]}{K_M} = [ES]$$

This can now be rearranged in several steps. First of all, open the bracket so that the terms $[E_0]$ and [ES] are separately multiplied by [S]

$$\frac{[E_0][S] - [ES][S]}{K_M} = [ES]$$

Next, multiply each side by K_M, this gives us:

$$[E_0][S] - [ES]\,[S] = K_M\,[ES]$$

Then collect the two [ES] terms together on the same side (you can either think of this as adding [ES][S] to both sides or as 'carry over and change the sign' - your preference will probably be an indication of how long ago you went to school). This gives:

$$[E_0][S] = K_M[ES] + [ES][S]$$

Then because both terms on the right-hand side are multiplied by [ES] we can collect them together into a bracket:

$$[E_0][S] = (K_M + [S])\,[ES]$$

Dividing both sides by $(K_M + [S])$ now gives us:

$$\frac{[E_0][S]}{[S] + K_M} = [ES]$$

Substituting this left-hand side into Equation 1 in place of [ES] results in:

$$v = k_2 \frac{[E_0][S]}{[S] + K_M}$$

The maximum rate, which we can call V_{max}, would be achieved when all of the enzyme molecules have substrate bound. Under conditions when [S] is much greater than [E], it is fair to assume that all E will be in the form ES. Therefore $[E_0]$ = [ES]. Thinking again about Equation 1, we could substitute the term V_{max} for v and $[E_0]$ for [ES]. This would give us:

$$V_{max} = k_2[E_0]$$

Notice that $k_2[E_0]$ was present in our previous equation, so we can replace this with Vmax, giving a final equation:

$$v = \frac{V_{max}[S]}{K_M + [S]}$$

This final equation is actually called the Michaelis-Menten equation.

Perhaps this derivation still leaves you puzzled about the importance of the Michaelis-Menten equation. The significance becomes clearer when you consider the case when the rate of reaction (v) is exactly half of the maximal reaction rate (V_{max}). Under those circumstances, the Michaelis-Menten equation could be written:

$$\frac{V_{max}}{2} = \frac{V_{max}[S]}{K_M + [S]}$$

On dividing both sides by V_{max} this becomes:

$$\frac{1}{2} = \frac{[S]}{K_M + [S]}$$

Multiplying both sides by (KM + [S]) gives:

$$\frac{K_M + [S]}{2} = [S]$$

And then multiplying both sides by 2 further resolves the equation to:

$$K_M + [S] = 2[S]$$

2[S] on the right-hand side is the same as [S] + [S], so we can take away one [S] from each side. Thus when the rate of the reaction is half of the maximum rate:

$$K_M = [S]$$

The K_M of an enzyme is therefore the substrate concentration at which the reaction occurs at half of the maximum rate. If we now reconsider the graph that came at the start of this tutorial it could be written:

What does this all mean in physical terms? K_M is an indicator of the affinity that an enzyme has for a given substrate, and hence the stability of the enzyme-substrate complex.

Look at the shape of the graph. At low [S], it is the availability of substrate that is the limiting factor. Therefore as more substrate is added there is a rapid increase in the initial rate of the reaction - any substrate is rapidly mopped up and converted to product. At the K_M, 50% of active sites have substrate bound. At higher [S] a point is reached (at least theoretically) where all of the enzyme has substrate bound and is working flat out. Adding more substrate will not increase the rate of the reaction, hence the levelling out observed in the graph.

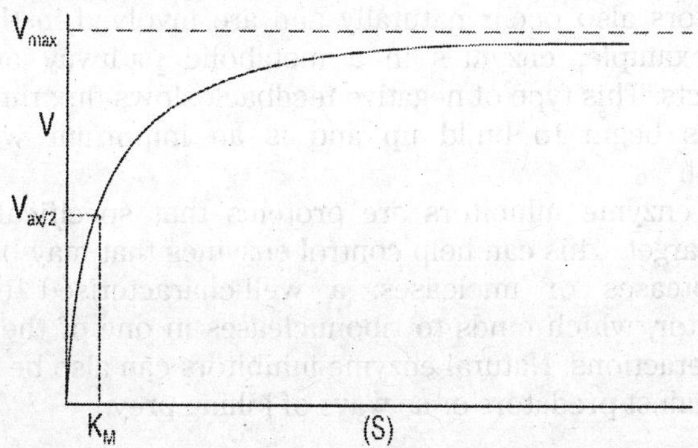

There are limitations in the quantitative (i.e. numerical) interpretation of this type of graph, known as a Michaelis plot. The V_{max} is never really reached and therefore V_{max} and hence K_M values calculated from this graph are somewhat approximate. A more accurate way to determine V_{max} and K_M (though still not perfect) is to convert the data into a linear Lineweaver-Burk plot. See separate page:

"Experimental calculation of V_{max} and K_M"

If, incidentally, we consider that the rate constant k_2 for the conversion of ES to E + P in the initial scheme is the step determining the overall rate of production of P (as we have in deriving the Michaelis-Menten equation) then k_2 is actually the same term as k_{cat}, the turnover number.

$$E + S \underset{k_{-1}}{\overset{k_1}{\rightleftharpoons}} ES \xrightarrow{k_2} E + P$$

EFFECTS OF INHIBITORS ON ENZYME ACTIVITY

Enzyme inhibitors are substances which alter the catalytic action of the enzyme and consequently slow down, or in some cases, stop catalysis. There are three common types of enzyme inhibition - competitive, non-competitive and substrate inhibition.

Enzyme Inhibitor

Enzyme inhibitors are molecules that bind to enzymes and decrease their activity. Since blocking an enzyme's activity can kill a pathogen or correct a metabolic imbalance, many drugs are enzyme inhibitors. They are also used as herbicides and pesticides. However, not all molecules that bind to enzymes are inhibitors; *enzyme activators* bind to enzymes and increase their enzymatic activity.

The binding of an inhibitor can stop a substrate from entering the enzyme's active site and/or hinder the enzyme from catalysing its reaction. Inhibitor binding is either reversible or irreversible. Irreversible inhibitors usually react with the enzyme and change it chemically. These inhibitors modify key amino acid residues needed for enzymatic activity. In contrast, reversible inhibitors bind non-covalently and different types of inhibition are produced depending on whether these inhibitors bind the enzyme, the enzyme-substrate complex, or both.

Many drug molecules are enzyme inhibitors, so their discovery and improvement is an active area of research in biochemistry and pharmacology. A medicinal enzyme inhibitor is often judged by its specificity (its lack of binding to other proteins) and its potency (its dissociation constant, which indicates the concentration needed to inhibit the enzyme). A high specificity and potency ensure that a drug will have few side effects and thus low toxicity.

Enzyme inhibitors also occur naturally and are involved in the regulation of metabolism. For example, enzymes in a metabolic pathway are inhibited by downstream products. This type of negative feedback slows flux through a pathway when the products begin to build up and is an important way to maintain homeostasis in a cell.

Other cellular enzyme inhibitors are proteins that specifically bind to and inhibit an enzyme target. This can help control enzymes that may be damaging to a cell, such as proteases or nucleases; a well-characterised example is the ribonuclease inhibitor, which binds to ribonucleases in one of the tightest known protein–protein interactions. Natural enzyme inhibitors can also be poisons and are used as defenses against predators or as ways of killing prey.

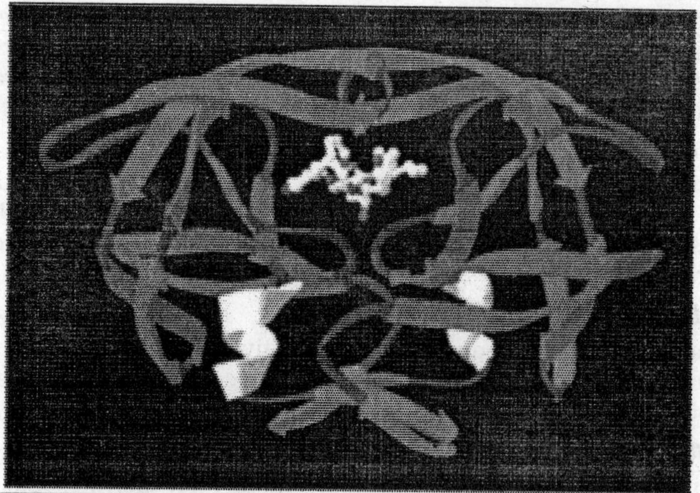

FIG. 11.20: HIV protease in a complex with the protease inhibitor Ritonavir. The structure of the protease is shown by shaded ribbons. The inhibitor is shown as the smaller ball-and-stick structure near the centre.

Reversible Inhibitors

Reversible inhibitors bind to enzymes with non-covalent interactions such as hydrogen bonds, hydrophobic interactions and ionic bonds. Multiple weak bonds between the inhibitor and the active site combine to produce strong and specific binding. In contrast to substrates and irreversible inhibitors, reversible inhibitors generally do not undergo chemical reactions when bound to the enzyme and easily be removed by dilution or dialysis.

There are three kinds of reversible enzyme inhibitors. They are classified according to the effect of varying the concentration of the enzyme's substrate on the inhibitor.

Competitive Inhibition

In competitive inhibition, the substrate and inhibitor cannot bind to the enzyme at the same time, as shown in the Fig. 11.21 on the left. This usually results from the inhibitor having an affinity for the active site of an enzyme where the substrate also binds; the substrate and inhibitor *compete* for access to the enzyme's active site. This type of inhibition can be overcome by sufficiently high concentrations of substrate, *i.e.*, by out-competing the inhibitor. Competitive inhibitors are often similar in structure to the real substrate.

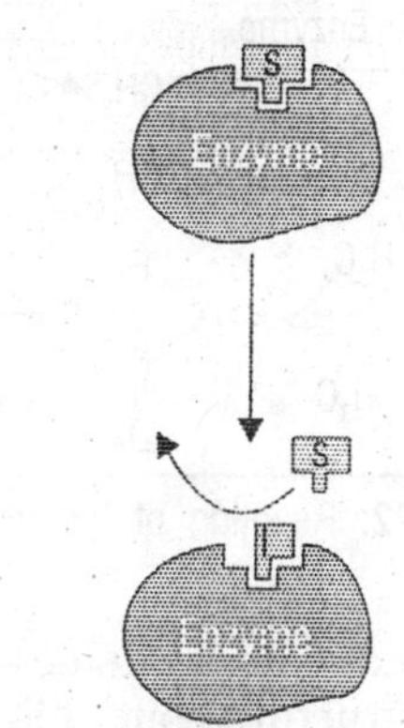

FIG. 11.21: Competitive inhibition: substrate (S) and inhibitor (I) compete for the active site.

Mixed Inhibition

In mixed inhibition, the inhibitor can bind to the enzyme at the same time as the enzyme's substrate. However, the binding of the inhibitor affects the binding of the substrate, and vice versa. This type of inhibition can be reduced, but not overcome by increasing concentrations of substrate. Although it is possible for mixed-type inhibitors to bind in the active site, this type of inhibition generally results from an allosteric effect where the inhibitor binds to a different site on an enzyme. Inhibitor binding to this allosteric site changes the conformation (*i.e.*, tertiary structure or three-dimensional shape) of the enzyme so that the affinity of the substrate for the active site is reduced.

Non-competitive Inhibition

Non-competitive inhibition is a form of mixed inhibition where the binding of the inhibitor to the enzyme reduces its activity but does not affect the binding of substrate. As a result, the extent of inhibition depends only on the concentration of the inhibitor.

Reversible Inhibition

Reversible inhibition can be described quantitatively in terms of the inhibitor's binding to the enzyme and to the enzyme–substrate complex, and its effects on the kinetic constants of the enzyme.

Irreversible Inhibitors

Irreversible inhibitors usually covalently modify an enzyme, and inhibition cannot therefore be reversed. Irreversible inhibitors often contain reactive functional groups such as nitrogen mustards, aldehydes, haloalkanes or alkenes. These electrophilic groups react with amino acid side chains to form covalent adducts. The residues modified are those with side chains containing nucleophiles such as hydroxyl or sulphahydryl groups; these include the amino acids serine (as in DFP, right), cysteine, threonine or tyrosine.

Irreversible inhibition is different from irreversible enzyme inactivation. Irreversible inhibitors are generally specific for one class of enzyme and do not inactivate all proteins; they do not function by destroying protein structure but by specifically altering the active site of their target.

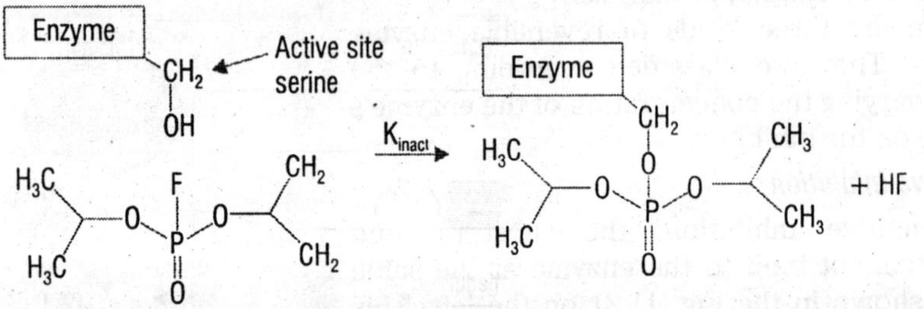

FIG. 11.22: Reaction of the irreversible inhibitor diisopropylfluorophosphate (DFP) with a serine protease.

For example, extremes of pH or temperature usually cause denaturation of all protein structure, but this is a non-specific effect. Similarly, some non-specific chemical treatments destroy protein structure: for example, heating in concentrated hydrochloric acid will hydrolyse the peptide bonds holding proteins together, releasing free amino acids.

As shown in the Fig. 11.23 to the left, irreversible inhibitors form a reversible non-covalent complex with the enzyme (EI or ESI) and this then reacts to produce the covalently modified "dead-end complex" EI*. The rate at which EI* is formed is called the inactivation rate or k_{inact}. Since formation of EI may compete with ES, binding of irreversible inhibitors can be prevented by competition either with substrate or with a second, reversible inhibitor. This protection effect is good evidence of a specific reaction of the irreversible inhibitor with the active site.

Suicide inhibition: Suicide inhibitionis a unusual type of irreversible inhibition where the enzyme converts the inhibitor into a reactive form in its active site. An example is the inhibitor of polyamine biosynthesis, α-difluoromethylornithine or DFMO, which is an analogue of the amino acid ornithine, and is used to treat African trypanosomiasis (sleeping sickness).

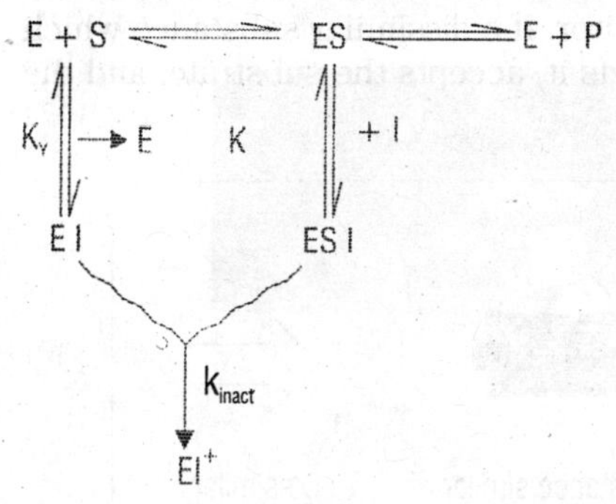

FIG. 11.23: Kinetic scheme for irreversible inhibitors.

Uses of Inhibitors

Enzyme inhibitors are found in nature and are also designed and produced as part of pharmacology and biochemistry. Natural poisons are often enzyme inhibitors that have evolved to defend a plant or animal against predators. These natural toxins include some of the most poisonous compounds known today. Artificial inhibitors are often used as drugs, but can also be insecticides such as malathion, herbicides such as glyphosate, or disinfectants such as triclosan.

The most common uses for enzyme inhibitors are as drugs to treat disease. Many of these inhibitors target a human enzyme and aim to correct a pathological condition. However, not all drugs are enzyme inhibitors. Some, such as anti-epileptic drugs, alter enzyme activity by causing more or less of the enzyme to be produced. These effects are called enzyme induction and inhibition and are alterations in gene expression, which is unrelated to the type of enzyme inhibition discussed here. Other drugs interact with cellular targets that are not enzymes, such as ion channels or membrane receptors.

Enzyme inhibitors are also important in metabolic control. Many metabolic pathways in the cell are inhibited by metabolites that control enzyme activity through allosteric regulation or substrate inhibition. A good example is the allosteric regulation of the glycolytic pathway.

Physiological enzyme inhibition can also be produced by specific protein inhibitors. This mechanism occurs in the pancreas, which synthesises many digestive precursor enzymes known as zymogens. Many of these are activated by the trypsin protease, so it is important to inhibit the activity of trypsin in the pancreas to prevent the organ from digesting itself.

Acetylcholinesterase inhibitors: Acetylcholinesterase (AChE) is an enzyme found in animals from insects to humans. It is essential to nerve cell function through its mechanism of breaking down the neurotransmitter acetylcholine into its constituents, acetate and choline.

Mechanism of Enzyme Inhibition

Most theories concerning inhibition mechanisms are based on the existence of the **enzyme-substrate complex** (ES). As mentioned earlier, the existence of temporary ES structures has been verified in the laboratory.

Competitive inhibition occurs when the substrate and a substance resembling the substrate are both added to the enzyme. A theory called the *lock-key theory* of enzyme catalysts can be used to explain why inhibition occurs.

The lock and key theory utilizes the concept of an *active site.* The concept holds that one particular portion of the enzyme surface has a strong affinity for the substrate. The substrate is held in such a way that its conversion to the reaction products is more favourable. If we consider the enzyme as the lock and the substrate the key - the key is inserted in the lock, is turned, and the door is opened and the reaction proceeds. However, when an inhibitor which resembles the substrate is present, it will compete with the substrate for the position in the enzyme lock. When the inhibitor wins, it gains the lock position but is unable to open the lock. Hence, the observed reaction is slowed down because some of the available enzyme sites are occupied by the inhibitor. If a dissimilar substance which does not fit the site is present, the enzyme rejects it, accepts the substrate, and the reaction proceeds normally.

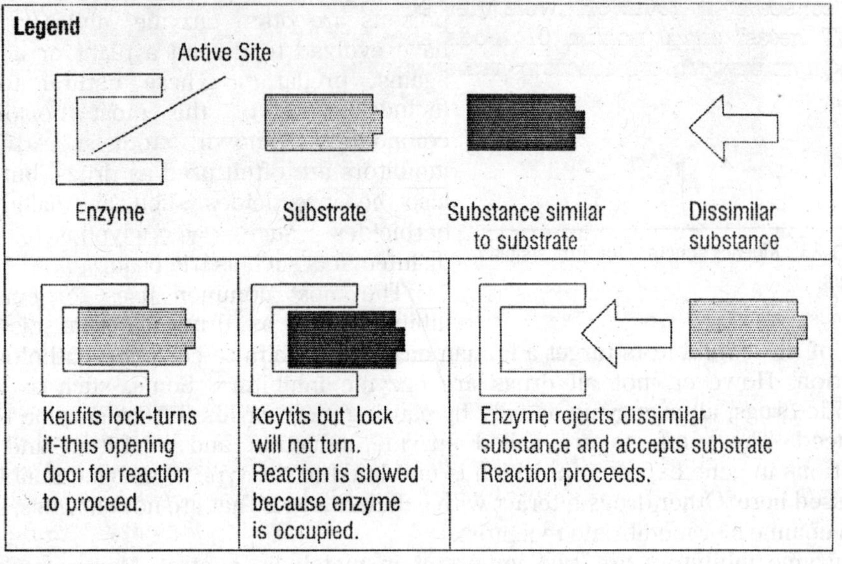

FIG. 11.24: Lock and key theory - competitive inhibition.

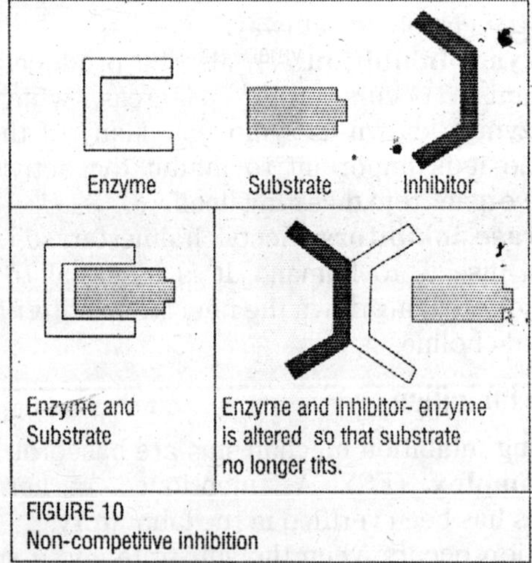

Fig. 11.25: Enzyme, substrate and inhibitor relationship.

Non-competitive inhibitors are considered to be substances which when added to the enzyme alter the enzyme in a way that it cannot accept the substrate.

Substrate inhibition will sometimes occur when excessive amounts of substrate are present.

Additional amounts of substrate added to the reaction mixture after this point actually decreases the reaction rate. This is thought to be due to the fact that there are so many substrate molecules competing for the active sites on the enzyme surfaces that they block the sites and prevent any other substrate molecules from occupying them.

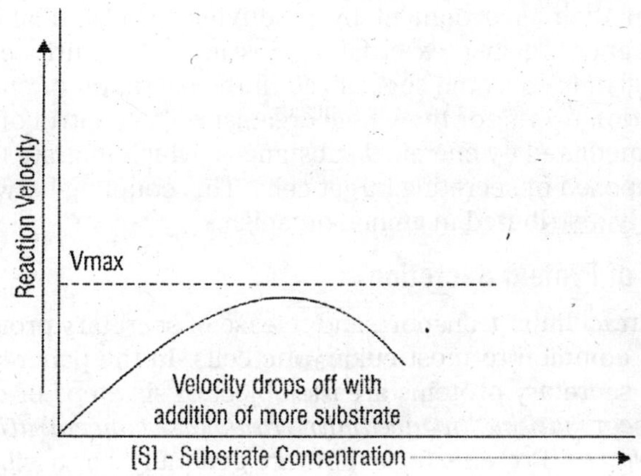

Fig. 11.26: Substrate become rate- inhibiting.

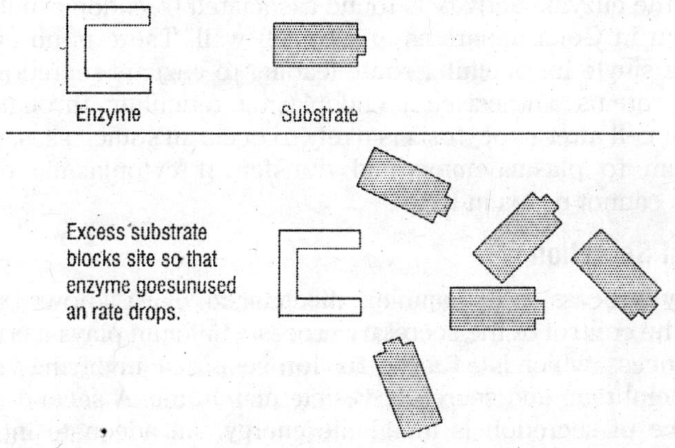

FIG. 11.27: Substrate inhibition.

This causes the reaction rate to drop since all of the enzyme present is not being used.

Evolutionary Dynamics of Enzymes

The large diversity in reaction rates and substrate specificity of enzymes in terms of a model which postulates that the kinetic properties of present-day enzymes are the consequence of the evolutionary force of mutation and selection acting on a class of primordial enzymes with poor catalytic activity and broad substrate specificity. Enzymes are classified in terms of their thermodynamic parameters, activation enthalpy ΔH^* and activation entropy ΔS^*, in their kinetically significant transition states as follows: type 1, $\Delta H^* > 0, \Delta S^* < 0$; type 2, $\Delta H^* \leq 0, \Delta S^* \leq 0$; type 3, $\Delta H^* > 0$, $\Delta S^* > 0$. Lloyd Demetrius of Department of Organismic and Evolutionary Biology,

Harvard University Cambridge, (USA) studied the evolutionary dynamics of these three classes of enzymes subject to mutation, which acts at the level of the gene which codes for the enzyme and selection, which acts on the organism that contains the enzyme.

ENZYME SECRETION BY PLANT CELLS

Enzyme secretion is a common feature of most of the living cells. By this process, cells may act on their environment by modifying its chemical composition and ensure their defence against external aggressions. In pluri cellular organisms, groups of cells (glands) become specialized in the secretion of enzymes, which are utilized in a function useful for the whole organism. The control of such a release of enzyme is often mediated by chemical substances, which migrate from one tissue to another one composed of secreting target cells. This coupling between stimulus and secretion is widely distributed in animal organisms.

The Mechanism of Protein Secretion

The synthesis, intracellular transport, and release of secretary proteins are the basic cellular function common to most eukaryotic cells. In the pancreatic exocrine cell for example, the secretary proteins are the object of six steps or operations, which are: *synthesis*, *segregation*, *intracellular transport*, *concentration*, *intracellular storage* and *discharge*. Proteins for export are generally synthesised on polysomes, attached to the membrane of the rough endoplasmic reticulum. In etiolated radish seedlings submitted to far red light, which is known to induce the synthesis of β-fructosidase, the enzyme activity is found associated to endoplasmic reticulum in a first time, then in Golgi apparatus and in cell wall. There is no evidence for the existence of a single intracellular route leading to enzyme secretion. Although the migration of proteins synthesised in endoplasmic reticulum, through Golgi towards the exterior of cell after exocytosis is likely to occur in some cases, direct transport from reticulum to plasmalemma and transfer of cytoplasmic enzymes across plasmalemma cannot be excluded.

The Control of Secretion

The secretary process is a complex mechanism. It is known that calcium is essential for the control of the secretary process. Calcium plays a critical role in the exocytosis process, which is a **fusion-fission response** involving the interaction of the plasma membrane and secretary vesicle membrane. A second requirement for the occurrence of secretion is metabolic energy. An adequate intracellular ionic equilibrium is also necessary. Ca^{2+} is considered as a second messenger in animal cells, able to trigger the discharge of secretary products by regulated cells. In addition, non-regulated cells often requires Ca^{2+} for their secretion. It would be, therefore, necessary to explore the knowledge concerning the regulation of Ca^{2+} in plant cells. It is known that Ca^{2+} is involved in the secretion of some plant enzymes, including peroxidase, α-amylase, and phosphatase. Recently, several articles have been published, which described modes of Ca^{2+} transport across pant membranes. An ATP-dependent Ca^{2+} uptake by isolated membranes vesicles was reported.

Some elements exist for substantiating the hypothesis that phytohormones regulate protein secretion through the mediation of second messenger such as c-AMP or Ca^{2+}/calmodulin. However, the existence of an exchange mechanism between H^+ and Ca^{2+} suggests that plant growth regulators especially auxin could modify secretary process by a modification of the distribution of protons which indirectly affect Ca^{2+} compartmentation.

SOURCES OF ENZYMES

Industrial enzymes can be obtained from plants, animals or microorganisms. Various factors such as specificity, pH, thermostability, activation or inhibition, availability and cost, determines the selection from any of the sources.

The important enzymes are:

Table **11.2**: Sources of animal enzymes

Type of enzyme	Source
α-amylase	Pancreatic gland of bovine and porcine
catalase	Liver
chymotrypsin	Pancreatic gland
esterase	Pancreatic gland
glucose oxidase	Liver
lipase	Pancreatic gland
pepsin	Fundus portion of hog stomach
rennet	Calf stomach
trypsin	Pancreatic gland of bovine and porcine

1. **Glucose Oxidase:** This converts glucose to gluconic acid, which is an industrial chemical and is also used extensively in diagnostic test for measuring blood sugar.
2. **Rennin**, used in cheese production.
3. **Papain**, an enzyme derived from papaya plant and added to animal feeds as a meat tenderizer.
4. **Glucose isomerase** converts glucose to fructose.
5. **Pepsin**, an animal protease used in the manufacturing industry.
6. **Urokinase** is used as an anti-thrombotic agent in treatment or pulmonary embolism.
7. **Asperaginase** used in cancer chemotherapy.
8. **Tyrosine hydroxylase,** which converts tyrosine into DOPA, which is specific against Parkinson's disease.

Animal gland *e.g.*, the pancreas, is source for hydrolysing enzymes used as digestive aids. The pancreas, a very rich source contain about 23% trypsinogen and 10-14% chymotrypsinogen. Pancreatin contains several enzymes such as amylase, lipase and protease.

TABLE 11.3: Sources of animal enzymes

Type of enzyme	Source
Proteases (bromelain and papain)	Pineapple and papaya
Amylolytic	Tissues of barley, wheat, rye, potatoes, sweet potatoes, beans and soybeans.
α-amylase and β-amylase	Seeds

The extraction from plants and animals sources is much simpler and less costly. The toxicity and contamination is not a problem.

MICROBIAL ENZYMES

The enzymes used in the industry are obtained from bacteria, fungi and yeasts. Efforts have been made to replace animal or plant enzymes with enzymes from microbes *e.g.*, amylases and endo-β-glucanases from *Bacillus* and *Aspergillus*, plant and animal proteases by *Aspergillus* and *Thermoactinomyces*, calf chymosin by

Mucor.

TABLE 11.4: Enzymes produced by bacteria

Type of enzyme	Bacteria
α-Amylase	*Bacillus* spp.
β-Amylase	*Bacillus cereus*
Dextran sucrose	*Leuconostoc mesenteroides*
β-1,3-Gluconase	*Bacillus subtilis*
Glucose isomerase	*Arthrobactor globiformis Bacillus coagulans. Streptomycin* spp.
Penicillin acylase	*Bacillius megaterium*
Proteinase	*Escherichia coli, Bacillus subtilis*
Pullulanase	*Klebsiella aerogens*
Streptokinase	*Streptococcus* spp.
Urokinase	*Bacillus fastidions*

TABLE 11.5: Enzymes produced by fungi

Types of enzyme	Fungi
α-Amylase	*Aspergillus oryzae*
Amyloglucosidase	*Aspergillus niger*
Catalase	*Aspergillus niger*
Cellulase	*Trichoderma reesi, Penicillium funiculsosum*
Dextranse	*Penicillium* spp.
Glucoamylase	*Aspergillus niger, A. awamori and Rhizopus* spp.
Glucose oxidase	*Aspergillus niger*
Invertase	*Saccharomyces* spp.
Lactose	*Kluveromyces fragilis, Aspergillus niger*
Lipase	*Candida lipoytica, Rhizopus delemar*
Pectinase	*Aspergillus niger*
Proteinase	*Aspergills oryzae*
Raffinase	*Mortierella vinacea*
Renin	*Mucor miehes*

The application of genetic engineering gained enormous potential in the use of microbes to produce animal or plant enzyme. The strains constructed by gene cloning procedure will be able to produce abnormal amounts of foreign proteins *i.e.*, synthesis of calf prochymosin by *E. coli/Saccharomyces cerevisiae* and urokinase normally occurring in human urine by *E. coli* cells. The various enzymes and their microbial sources are given in following.

Microorganisms have biochemical diversity. The enzyme concentration in these microorganisms is dependent on environmental and genetic manipulations. Microbial enzymes may have varied stability and different pH and temperature optima. Microbial cells require short fermentation times, inexpensive media and simple screening procedures. Microbes can be genetically manipulated.

MECHANISM OF ENZYME BIOSYNTHESIS IN MICROBIAL CELLS

Metabolism is mainly regulated by change in the rate of enzyme reaction. The enzyme concentration can be varied by controlled protein synthesis and controlled protein degradation.

Enzymes synthesis by microbial cells may be influenced by various mechanisms. These include:

Induction

Biosynthesis requires the presence of substrates in the medium. The structural genes coding for the production of many catabolic enzymes are repressed in the absence of substrates of the enzymes. By the addition of substrate, enzyme synthesis is induced or derepressed. In the case of lac operon, the inducer (lactose) binds the repressor and inactivates it. Transcription thus begins and enzyme synthesis occurs. This can be seen in case of the synthesis of sugar or amino acid degrading enzymes.

Repression

The repressed enzymes can be of the constitutive or of the inducible types. Enzymes are often repressed by the presence of intermediates or end products of the enzyme reactions. This type of regulation is known as end product repression or feedback repression.

Catabolic Repression

The biosynthesis of the most of extra-cellular depolymerases is influenced by the level of catabolites present in the growth medium. It occurs when cells grow rapidly on readily metabolisable carbon sources *i.e.*, repression of β-galactosidase in *E. coli* by glucose.

Transcription Control during the Growth Cycle

The different behaviour of enzyme biosynthetic in relation to the growth is caused by different regulatory mechanisms. In case of *B. subtilis* transition of cells from vegetative growth to sporulation is accompanied by changes in the predominant forms of RNA polymerase. Four forms of RNA polymerase have been identified in vegetative and early stationary phase cells. Sporulation forms of RNA polymerase recognize the specific promoter sequence that control expression of genes during growth cycle and sporulation. Genes coding for extracellular proteases indicates overlapping promoters, ensuring gene expression during both growth and stationary phase. Therefore, altered promoter specificity is one of the cellular tools to influence gene expression on the level of transcription during growth. Further, translation efficiency of a gene is influenced by several factors. Rate of translation efficiency of a gene is influenced by several factors. Rate of translation efficiency is affected by the ribosomal binding site, stability of mRNA, the codon usage by the translation machinery, occurrence of repression by attenuation, effectiveness of the secretion etc.

APPLICATIONS OF ENZYMES: THE TAILOR'S SCISSORS

Enzymes help industry and nature. Today enzymes are helping industry make the products used by society in a way that is less harmful to the environment. In the cases mentioned above, enzymatic processes are, in general, safer and more environment-friendly than the traditional processes they replace. That is why enzymes in particular and biotechnology in general hold such promise.

The applications of enzymes (biological catalysts) can change plants and animals in precise and often remarkably in dramatic fashion. In the hands of knowledgeable biotechnologists, *enzymes become the tailor's scissors and the surgeon's scalpel.* There is a strong dependence of advances in enzymology on agriculture. Discovery of enzymes and elucidation of many of their properties were by agricultural chemists. The earliest enzymes studied were primarily from agriculturally important animals, plants and microorganisms. The fast developing wide fields of biochemical engineering/biotechnology, and its tremendous promise for the future, have brought basic and applied scientists together in private sector companies, at universities, and in government and industrial laboratories. There are many reasons for the use of microbes in enzyme production. The important characteristics that have made microbes useful in enzyme industries are:

1. Their ability to grow on easily available and cheap raw materials.
2. Their ability to maintain a physiological constancy.
3. Their ability to bring about biochemical transformations under simple culture conditions. Microbes have tremendous capacity of carrying out a variety of reactions (especially secondary metabolism resulting in an inexhaustible supply of secondary metabolites available for commercial exploitation.
4. High ratio of surface area to volume, which facilitates the rapid uptake of nutrients required to support high rates of metabolism and biosynthesis.

TABLE 11.6: Industrial enzymes

Carbohydrases	Regular Liquefying and Saccharifying Alpha-Amylases, Alkaline Alpha-Amylase, Beta-Amylase, Cellulase, Dextranase, Alpha-Glucosidase, Alpha-Galactosidase, Glucoamylase, Hemicellulase/Pentosanase/Xylanase, Invertase, Lactase, Naringanase, Pectinase, Pullulanase
Proteases	Acid protease, alkaline protease, bromelain, ficin, neutral proteases, papain, pepsin, peptidases, rennet/rennin/chymosin, subtilisin, thermolysin, trypsin
Lipases and esterases	Triglyceridases, phospholipases, pregastric esterases, phosphatases, phytase, amidases, aminoacylase, glutaminase, lysozyme, penicillin acylase
Isomerases	Glucose isomerase
Oxidoreductases	Amino acid oxidase, catalase, chloroperoxidase, glucose oxidase, hydroxysteroid dehydrogenase, peroxidase
Lyases	Acetolactate decarboxylase, aspartic beta-decarboxylase, fumarase, histadase
Transferases	Cyclodextrin glycosyltranferase.

These properties of microorganisms are highly preferable over synthetic processes. Presently, industrial microbiology has established as a strong arm of applied microbiology and it is unlikely that, at least in some industries, the microbes could be replaced by synthetic techniques in the near future.

There are about 3000 enzymes known and only few are manufactured at large scale. These enzymes are mainly extracellular hydrolytic enzymes which have the property to degrade naturally occurring polymers for example starch, proteins, pectins, and cellulose. Enzyme glucose isomerase is, however, an exception in this regard.

1. Mutants that produce the enzyme constitutively, that is without an inductor;
2. The production of the desired enzyme in the mutants is not inhibited by repressors; and
3. The yield of enzyme is considerably increased by multiplication of the gene copies, and/or changes in the single sequences preceding the structural gene to bring about high expression.

TABLE 11.7: Areas of industrial use of microorganisms, and of application of major enzymes from microbes (main microbial species)

Industrial uses	**Microorganisms used**
Alcoholic Fermentation	Bear (*Saccharomyces cerevisiae*, *S. carlsbergensis*, *S. uvarum*), cider (*S. cidri*), wine (*S. cerevisiae*), sake (*Aspergillus oryzae*, Lactobacillus and *Leuconostoc* spp., *S. cerevisiae*) and other fermentations of fruit juices, distilled sprits, etc.
Amino acids,	*Corynebacterium glutamicum*.

Industrial uses	Microorganisms used
mononucleotides	
Baker's yeast	*Saccharomyces cerevisiae; Candida milleri.*
Carotenoids	β-carotene: *Blakeslea trispora*; astaxanthin: *Phaffia rhodozyma*.
Dextrans	*Leuconostoc mesenteroides*, levans, mannans; xanthans -*Xanthomonas campestris*.
Ethanol	*Kluyveromyces fragilis, S. cerevisiae, Zymomonas mobilis*, n-butanol and acetone (*Clostridium acetobutylicum, C. saccharoacetobutylicum*); acetic acid (*Acetobacterium woodii, Clostridium aceticum*) citric acid (*Aspergillus niger, Saccharomycopsis lipolytica*).
Fodder yeast	*Candida utilis, Saccharomycopsis lipolytica*
Fumaric and lactic acids	*Lactobacillus delbrueckii*
Gibberellins	*Gibberella fujikuroi* and other plant growth hormones
Production of antibiotics	Penicillins (*Penicillium chrysogenum*), cephalosporins (*Cephalosporium acremonium*), streptomycin, kanamycins, neomycins, tetracyclines, etc. (*Strptomyces* spp.), gramicidin-S (*Bacillus brevis*), polymyxin-B (*Bacillus polymyxa*), bacitracin (*Bacillus subtilis*)
Production of Fermented Beverages and Foods	Dried bonito, soyasauce (*Aspergillus oryzae, Pediococcus soyae, Saccharomyces rouxii, Torulopsis* spp.), vinegar (*Gluconobacter suboxidans*), pickled vegetables, cheeses (*Penicillium camembertii, P. roqueforti, Propionibacterium shermanii, Streptococcus* spp.), yoghurt (*Lactobacillus bulgaricus, Streptococcus thermophilus*), lactic acid (sour) drinks.
Single-cell proteins	Spirulinas, Chlorellas, and other unicellular algae; (*Methylophilus methylotrophus, Candida tropicalis, C. utilis, Saccharomycopsis lipolytica*).
Steroids	*Arthrobacter simplex, Mycobacterium sp., Rhizopus arrhizus, R. nigricans*.
Vaccine and microbial bioinsecticides	*Bacillus popilliae, B. thuringiensis*
Vitamins	Riboflavin: *Eremothecium ashbyi*; Vitamin B_{12}: *Propionibacterium* sp., Pseudomonas denitrificans).
Abatement of pollution	Treatment of industrial effluents, waste waters, disposal of wastes and garbage; recovery and reuse of biodegradable wastes. Microbial leaching of ores and recovery of mining and colliery wastes.

GENETICS OF ENZYME FORMATION

Modern biotechnology was born because scientists learned to manipulate (engineer) DNA. This capability was derived largely from the field of nucleic acid enzymology. Prior to 1970, there was simply no method available for cutting a duplex DNA molecule into discrete fragments. Discovery of DNA metabolizing enzymes allowed scientists to contemplate and initiate genetic engineering. With the advent of genetic engineering, barriers to the exchange of genetic material were crossed to improve the formation of desired enzyme in various organisms.

Genes coding for intra- or extra- cellular enzymes in Gram negative bacteria are organized in operons while in gram-positive bacteria, however, the occurrence of genes for exoenzyme synthesis in clusters in uncertain. Linkage of regulating genes with related structural genes are known to be occurring *i.e.*, amylase (amy R, amy E), levansucrase (sac Q, sac B). Besides regulatory genes, there are genes enhancing the level of simultaneous secretion of different genes (pap, amy B, sac U, in *B. subtilis*.

Classical genetic methods have also been used for the improvement of enzyme formation, *e.g.*, mutation. Through mutations overproduction of enzymes can be obtained by altering mechanisms of regulation and by increasing copies of responsible genes for the production of enzymes. The method of site directed mutagenesis of cloned gene is also being used for genetic alteration of enzyme. Chemical mutation is also helpful for the isolation of natural enzyme and modification to its structure.

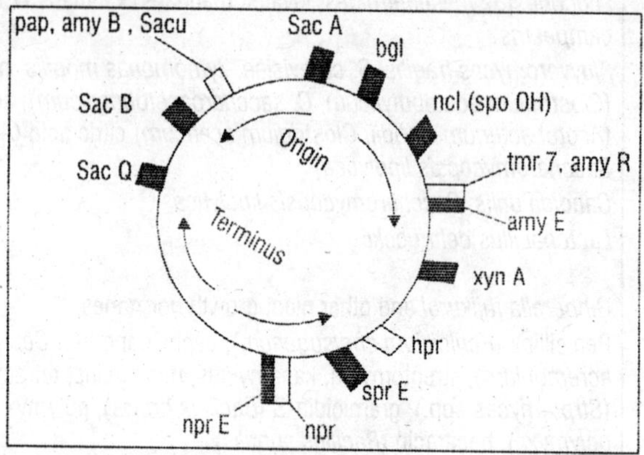

FIG. 11.28: Presentation of location of genes coding for extra cellular enzymes on the linkage map of *B. subtilis.*

'In vitro' methods is also being used to overcome the barriers to genetic exchange between unrelated organisms. Genes encoding enzymes with industrial importance have been cloned. The potential hosts expressing cloned genes are *E. coli, B. sublitis* and *S. cereviiae.* Cloning of the structural genes for α-amylase from *B. amyloliquefaciens* in plasmid pUB 110, and transformation of this hybrid plasmid into *B. subtilis* has enhanced the productivity of α-amylase to about 2000 fold. Thus, it becomes possible to produce industrial enzymes of very high quality and purity. Enzymes can be modified by evaluation of their performance. The main objective is to enhance the activity, stability change in pH or temperature optima, change the specificity to catalyse the conversion of different substrate and to perform function in a changed environment.

ISOENZYMES

Isoenzymes were discovered about 30 years ago and were at first regarded as interesting but of rare occurrences. Since then a wealth of information on enzyme heterogeneity has accrued and it now seems likely that at least half of all enzymes exist as isoenzymes. This is important in many areas of biological and medical science. Thus isoenzyme studies have provided the main experimental substance for the neutral drift controversy in genetics and evolution. Isoenzymes have greatly extended our understanding of metabolic regulation not only in animals but also in bacteria and plants; their existence has made available a multitude of highly sensitive markers for the study of differentiation and development, as well as providing indices of aberrant gene expression in carcinogenesis and other pathological processes. Isoenzymes are also being used increasingly in diagnostic clinical biochemistry. Isoenzymology crosses the traditional boundaries of the biological disciplines and it seems that advances and new applications in this field, are likely to occur at a rapid pace.

The Cause of Enzyme Multiplicity

There are various causes of enzyme multiplicity and they may be divided into two categories. These are (a) **genetic** or **primary causes**, whereby the organism carries multiple genes each one encoding a different type of enzyme subunit; and (b) post-translational or secondary causes, whereby homogeneous enzyme subunits are modified differentially so as to produce a range of subunits from a single gene. There are, in turn, two types of genetic multiplicity; firstly, multiple alleles at a single genetic locus and secondly, multiple genetic loci.

Primary or Genetic Isoenzymes

Isoenzymes due to Multiple Alleles at a Single Genetic Locus

In the diploid genome, each genetic locus is represented twice. For each locus, the individual will either be homozygous, possessing two identical alleles, or heterozygous, possessing two different alleles. Where the genetic locus encodes an enzyme subunit, the homozygous individual can only produce one type of subunit. However, the heterozygous individual with two different allelic variants will produce two different types of enzyme subunits. Within the individual the degree of enzyme multiplicity produced by multiple alleles is limited, as two different alleles per diploid locus is the maximum possible genetic variation of this type. However, from one individual to another, there may be considerable variation in the range of enzyme subunit types produced since there may be a variety of different alleles for the locus in the gene pool of the species.

The enzyme subunit types produced as a result of multiple alleles are likely to differ from each other only in minor ways, such as by individual amino acid substitutions caused by point mutations in the DNA nucleotide sequences. Where a genetic locus is active, both alleles will usually be expressed. Thus in the same individual, although the total activity of the enzyme may vary considerably between different kinds of cell, the isoenzyme profile will be constant throughout, homozygous individuals displaying only one subunit type, and heterozygotes possessing two different subunit types.

Isoenzymes due to Multiple Genetic Loci

Many enzymes are encoded at more than one genetic locus and, where the case, each locus will produce a different type of enzyme subunit. As the expression of each genetic locus can be controlled independently, the organism may synthesise one type of enzyme subunit in a particular cell, and another enzyme subunit elsewhere. Furthermore, the expression of genetic loci may alter during the course of development, and therefore the type of enzyme subunit produced in a tissue may change. Thus multiple genetic loci permit **differences in isoenzyme profile** both from one tissue to another, and from one developmental stage to another even within the same tissue. Such variation in isoenzyme profile are not possible where the enzyme multiplicity is due to multiple alleles operating at only a single locus. However, as all members of the same species will possess the same genetic loci, multiple enzyme-encoding loci in the absence of multiple alleles cannot account for isoenzyme differences between one member of the species and the next. The isoenzyme subunits produced by multiple loci are likely to differ extensively as a result of numerous amino acid substitutions, and also deletions and additions of residues may have occurred resulting in small size differences between the subunits.

Secondary or Post-Translational Isoenzymes

Proteins may be modified in numerous ways following their synthesis. Such possible modifications include the addition of carbohydrate, limited proteolysis, and the covalent modification of amino acid side chains. Post-synthetic alterations affecting only part of the enzyme subunit population, so that modified and unmodified subunits are found in the same organism, will result in isoenzymes. Where the post-translational modification process is very active in some tissues but not in others, the result will be a tissue-specific distribution pattern of the secondary isoenzymes, mimicking the effect of multiple genetic loci. An example of this is one particular pyruvate kinase subunit type. There are three major types of pyruvate kinase subunit in mammals, each coded for by an independent genetic locus.

TABLE 11.8: Enzymes, coenzymes, and mineral activating substances

Enzyme	Coenzyme or co-factor In processes of glycolysis	Activating mineral elements
Phosphatase	Phosphate	Mg^{2+},
Phosphoglutamase		Mg^{2+}
Aldolase		Fe^{2+}, Zn^{2+}, Co^{2+}
Enolase		Mg^{2+}, Mn^{2+}, Zn^{2+}
Pyruvate kinase	ADP In the tricarboxylic cycle	Mg^{2+}, K^{+}, NH_4^{+}
Decarboxylase and pyruvate dehydrase	Thiamine pyrophosphyte fatty acids, CoA	Mg^{2+}, Mn^{2+}
Isocitrate dehydrogenase	NADP	Mn^{2+}, Mg^{2+}
Succinate dehydrogenase	FAD	Ca^{2+}, Cr^{3+}, Al^{3+}
Aconitase		Fe^{2+}, -SH
Dehydrogenase		Mn^{2+}, K^{+}, Rb^{+}, NH_4^{+}
Pyruvic oxidase		Mg^{2+}, Mn^{2+}

Among the three most important functions namely, the mechanical, osmotic and chemical, the later depends solely upon the bio acceleration of metabolic reactions caused by enzymes.

CHAPTER 12

Metabolism: General

BIOENERGETICS AND ATP

Bioenergetics is also referred to as biochemical thermodynamics. It is the study of the energy changes accompanying biochemical reaction. It provides the underlying principles to explain why some reactions may occur while others do not. The biologic systems are essentially isothermic and use chemical energy to power the living processes, whereas, non-biologic systems may utilize heat energy to perform work.

Biologic systems conform to the general laws of **Thermodynamics** and the free energy is the useful energy in a system. Change in free energy (G) is that portion of the total energy change in a system that is available for doing work; *i.e.*, it is the useful energy, also known in chemical system as the **chemical potential.**

Laws of Thermodynamics

Like all other kinds of changes in the universe, chemical changes obey and are governed by the laws of **thermodynamics**. These most fundamental rules of nature deal with the energy relations of a system and its surroundings.

The first law of thermodynamics, often also called the **law of conservation of energy**, states that, in any process, the sum of all energy changes must be zero, *i.e.*, energy can be neither created nor destroyed. Thus, if a given chemical system gains energy, that amount of energy must be lost by the environment of the system, or if a chemical system loses energy, that amount must be gained by the environment. This law implies that energy can only be redistributed, changed in form or both. Indeed, chemical, thermal, radiant, mechanical, or other energy forms are mutually interconvertible. Moreover, energy may also be transformed into mass or vice versa.

The second law of thermodynamics states that, if left to itself, any system tends toward a state of greatest stability. The actual stability of chemical systems depends on two factors: **Enthalpy** and **Entropy**.

Ethalpy (H) denotes the total energy content of a chemical system. Compounds are most stable if 'H' is at a minimum. Chemical reactions tend to proceed in such a way that, end, the total energy content of all participants will be least. Consider the following reaction.

$$A + B \rightarrow C + D$$

The total energy of all the bonds in A and B together is greater than the total energy of all the bonds in C and D together. In other words, more potential energy is available in the bonds of the starting materials than is needed to form the bonds of the end products. C + D containing less total energy is more stable than A + B. The energy difference is lost in the environmental, usually in the form of heat. The energy difference (H) or total energy change passes from the reaction system to the environment, it is given a negative sign (energy loss from the system). It can be written as:

$$A + B \rightarrow C + D - \Delta H$$

Energy yielding reactions of this sort are called **exergonic or exothermic** reactions.

Assume, in above reaction, the total energy of A + B is less than that of C + D, then A + B will be more stable, and according to the second law no process proceeds from a more stable to a less stable state. However, the reaction could be made to occur if energy were supplied from the environment to the reaction system. Then the total energy of A + B, together with H from the environment, would actually suffice to form C + D. The reaction would be written as:

$$A + B \rightarrow C + D + \Delta H$$

(The "+" sign denotes energy gain by the system). Energy requiring reactions are called **endergonic or endothermic** reactions.

Entrpy(s) is the second factor which determines the relative stability of a chemical system. Entropy denotes the energy distribution in a chemical system. It also refers to the extent of disorder of randomness of the system. If 'S' is at a maximum, a system is more stable, i.e., if energy is distributed as uniformity or randomly as possible.

In a reaction A + B → C + D, assume that H is zero, that is, the total energy content is the same before and after the reaction. Assume, however, that A is very rich in energy, that B is very poor in energy. Under such conditions, the energy distribution in the A + B state is more uneven, or less random, than in the C + D state; the latter will therefore be the more stable. If other conditions are suitable, the reaction may consequently occur and it will increase the entropy of the system. If we symbolize the net entropy change as S, we may write:

$$A + B \rightarrow C + D + \Delta S$$

Under conditions of constant temperature and pressure, the relationship between the free energy change (G) of a reacting system and the change in entropy (S) is given by the following equation which combines the I and II laws of thermodynamics:

$$G = H\text{-}T\ S$$

Where, H = change in enthalpy (heat), T = absolute temperature, S = change in entropy.

Under the conditions of biochemical reactions, because H is approximately equal to E, the total change in internal energy of the reaction, the above relationship may be expressed in the following way:

$$G = E - T\ S$$

If G is negative in sign, the reaction proceeds spontaneously with loss of energy (exergonic reaction). If in addition, G is of great magnitude, the reaction goes virtually to completion and is essentially, irreversible. The negative G is largely characteristic of **decomposition reactions.**

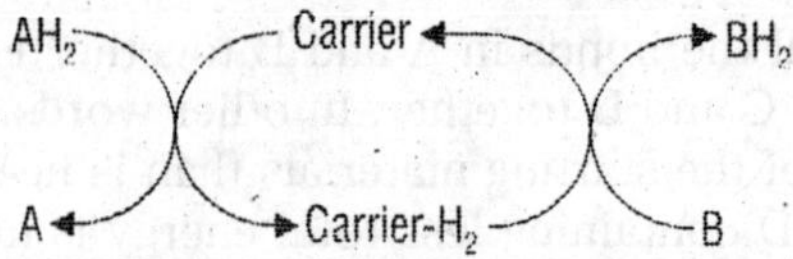

FIG. 12.1: Coupling of dehydrogenation and hydrogenation reactions by an intermediate carrier.

FIG. 12.2: Adenosine triphosphate (ATP).

If G is positive, the reaction proceeds only if free energy can be gained (endergonic reaction). If, in addition, the magnitude of G is great, the system is stable with little or no tendency for a reaction to occur. If G is zero, the system is at equilibrium and no net change takes place. The positive G is characteristic of synthesis reactions.

In living system, actually, decomposition reactions (catabolism) are coupled intimately to synthesis reactions (anabolism). Reactions with negative G yield the energy needed to sustain those with positive G. The total of all of the catabolic and anabolic processes is metabolism.

The coupling of exergonic and endergonic reactions are provided by:

1. An intermediate carrier: It is illustrated in the following Fig. 12.1.

FIG. 12.3: The magnesium complex of ATP.

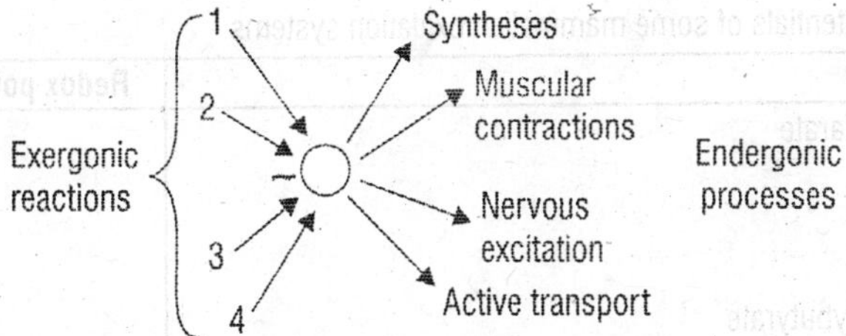

FIG. 12.4: Transduction of energy through a common high energy compound to energy requiring (energonic biologic) processes.

2. Synthesising a compound of high-energy potential in the exergonic reaction and to incorporate this new compound into the endergonic reaction, thus, effecting a transference of free energy from an exergonic to the endergonic pathway. In the living cells, the principal high-energy intermediate is adenosine triphosphate (ATP). The high energy phosphates are designated by the symbol ~P. In its reactions in the cell, ATP functions as the Mg^{+2} complexes.

ATP serves as a transducer of energy from a wide range of exergonic reactions to an equally wide range of endergonic reactions or processes, as shown in (Fig.

12.4).

An ATP/ADP (Adenosine di-phosphate) cycle connects those processes which generate ~P to those processes that utilize ~P. Thus ATP is continuously consumed and regenerated. There are three major sources of ~P taking part in energy conservation or energy capture:

1. **Oxidative phosphorylation:** This is the greatest quantitative source of ~P in aerobic organisms. The free energy to drive this process comes from respiratory chain, oxidation within mitochondria.
2. **Glycolysis:** A net formation of 2 ~P results from the formation of lactate from one molecule of glucose generated in 2 reactions catalysed by phosphoglycerate kinase and pyruvate kinase, respectively.
3. **The citric acid cycle:** One ~P is generated directly in the cycle at the succinyl thiokinase step.

Phosphagens, are another group of compounds which act as storage forms of high-energy phosphate. These include creatine phosphate (occurring in vertebrate muscle and brain), and arginine phosphate (occurring in invertebrate muscle). When ATP concentration is plentiful, it transfers one of its high energy phosphate group to creatine to produce creatine phosphate. During muscle contraction creatine phosphate transfers ~P to ADP to produce ATP.

$$\text{Creatine phosphate} \xleftrightarrow[\text{ADP} \quad \text{ATP}]{\text{Creatine kinase}} \text{Creatine}$$

FIG. 12.5: Transfer of high-energy phosphate between ATP and creatine.

The high-energy ~P from ATP can also be transferred in following forms:

$$\text{ATP + AMP} \xleftrightarrow{\text{Adenylate kinase (myokinase)}} \text{2ADP}$$

$$\text{ATP + UDP} \xleftrightarrow{\text{Uridine diphosphate kinase}} \text{ADP + UTP (Uridine triphosphate)}$$

$$\text{ATP + GDP} \xleftrightarrow{\text{Guanosine diphosphate kinase}} \text{ADP + GTP (Guanosine triphosphate)}$$

$$\text{ATP + CDP} \xleftrightarrow{\text{Cytidine diphoshpate kinase}} \text{ADP + CTP (Cytidine triphosphate)}$$

BIOLOGICAL OXIDATION

TABLE 12.1: Redox potentials of some mammalian oxidation systems

System	Redox potential (Eo', Volts)
Succinate / α-keto glutarate	-0.62
H^+/H_2	-0.42
NAD^+/NADH	-0.32
Lipoate; ox/red	-0.29
Acetoacetate/β-hydroxybutyrate	-0.27
Pyruvate/Lactate	-0.19
Oxaloacetate/Maltate	-0.17
Flavoprotein-old yellow enzyme; ox/red	-0.12
Fumarate/Succinate	+0.03
Cytochrome b; Fe^{3+}/Fe^{2+}	+0.08
Ubiquinoen; ox/red	+0.10
Cytochrome C; Fe^{3+}/Fe^{2+}	+0.22
Cytochrome a; Fe^{3+}/Fe^{2+}	+0.29
Oxygen/Water	+0.82

Chemically, **oxidation** is defined as the removal of electrons and reduction is the gain of electrons, as illustrated by the oxidation of ferrous to ferric ion.

$$Fe^{2+} \xrightarrow{e^- \text{ (electron)} \uparrow} Fe^{3+}$$

It follows that oxidation is always accompanied by reduction of an electron acceptor. This principle of oxidation-reduction applies equally to biochemical systems and is an important concept underlying understanding of the nature of biologic oxidation. Oxidation brought about by the loss of electron (or hydrogen) are usually directly or indirectly dependent upon involvement of oxygen at one or the other stage.

The oxidation-reduction reactions are termed as **redox reactions**. In redox systems free energy changes can be expressed in terms of **redox potential** (Eo') and the free energy exchange is proportionate to the tendency of reactants to donate or accept electrons. For biologic systems it is normal to express the redox potential (Eo') at pH 7.0 at which pH the electrode potential of the hydrogen electrode is -0.42 volts. The redox potentials of some redox systems in mammalian biochemistry are shown in Table 12.1. The list of redox potentials shown in the table allows prediction of the direction of flow of electrons from one redox couple to another.

It should be noted that a system having relatively more positive redox potential can oxidize another system having lower positive redox-potential. In the mitochondria, the hydrogen/electrons pass through different carriers in the sequence of increasing order of their positive potential, releasing energy in the form of packets.

If redox-potential of a oxygen-reduction system is known, the corresponding free-energy change which might occur in the system on oxidation or reduction may be calculated from the following equation:

$$\Delta F = -(n \times F \times \Delta Eo') \text{ Coulomb Joules}$$

Where, F = Standard free energy in the reactions; n = number of electrons or hydrogen atoms involved in oxidation; F = Faraday (965,000 Coulombs) constant; Eo' = differenece between redox-potentials of the systems.

ENZYMES

The enzymes involved in oxidation and reduction reactions are designated **oxidoreductases.** The oxidoreductases are classified into four groups:

1. Oxidases.
2. Dehydrogenases.
3. Hydroperoxidases.
4. Oxygenases.

Oxidases: The oxidases use oxygen as a hydrogen acceptor. These catalyse the removal of hydrogen from a substrate and transfer it to oxygen. They form water or hydrogen peroxide as a reaction product.

$$AH_2 \text{ (Red.)} \xrightarrow[\text{Oxidase}]{1/2\,O_2} A(ox) + H_2O$$

$$AH_2 \text{ (Red.)} \xrightarrow[\text{Oxidase}]{1/2\,O_2} A(ox) + H_2O_2$$

Oxidation of a metabolite catalysed by an oxidase (A) forming H_2O and (B) forming H_2O_2

Copper Containing Oxidases

Some oxidases contain Copper as a part of their molecular structure.

1. **Cytochrome Oxidase (cytochrome a₃):** It is a hemeprotein, widely distributed in many plant and animal tissues. It is the terminal component of the chain of respiratory carriers found in mitochondria and is therefore responsible for the reaction, whereby electrons resulting from the oxidation of substrate molecules by dehydrogenases are transferred to their final acceptor, oxygen. The enzyme is poisoned by carbon monoxide, cyanide, and hydrogen sulphide.
2. **Phenolase** (tyrosinase, polyphenol oxidase, caetchol oxidase) is a Copper-containing enzyme of broad specificity. It is able to convert monophenols or O-diphenols to O-quinones.

Flavoproteins as Oxidases

Flavoprotein enzymes contain flavin mononucleotide (FMN) or flavin adenine dinucleotide (FAD) as prosthetic groups. FMN and FAD are formed in the body from riboflavin vitamin. Many flavoprotein enzymes contain one or more metals as essential co-factors and are known as **metalloflavoproteins.**

1. **L-amino acid oxidase:** It is an FMN linked enzyme found in kidney for the oxidative deamination of naturally occurring L-amino acids.

$$\text{Amino acid} + H_2O + O_2 \xrightarrow{\text{L-Amino acid oxidase}} \alpha\text{-Keto acid} + NH_3 + H_2O_2$$

2. **Xanthine oxidase:** It is widely distributed and found in milk, small intestine, kidney and live. It contains molybdenum and plays an important role in the conversion of purine bases to uric acid. It is of particular significance in the liver and kidney of birds, which excrete uric acid as the main nitrogenous end product, not only of purine metabolism but also of protein and amino acid catabolism.

$$\text{Hypoxanthine} + H_2O + O_2 \xrightarrow{\text{Xanthine oxidase}} \text{Xanthine} + H_2O_2$$

$$\text{Xanthine} + H_2O + O_2 \xrightarrow{\text{Xanthine oxidase}} \text{Uric acid} + H_2O_2$$

3. **Ascorbic oxidase:** It oxidizes ascorbic acid.

$$\text{Ascorbic acid} + O \xrightarrow{\text{Ascorbic oxidase}} \text{Dehydroascorbic acid} + H_2O$$

4. **Aldehyde dehydrogenase:** It is an FAD linked enzyme present in mammalian liver. It is a metalloflavoprotein containing molybdenum and nonheme iron and acts upon aldehydes and N-heterocyclic substrates.
5. **Glucose Oxidase:** It is an FAD- specific enzyme prepared from fungi. It is used in glucose estimation.

Dehydrogenases: The dehydrogenases do not use oxygen as a hydrogen acceptor. These fall into two categories:

1. Dehydrogenases which transfer hydrogen from one substrate to another in a coupled oxidation-reduction reaction. These dehydrogenases are specific for their substrates but often utilize same co-enzyme of hydrogen carrier as other dehydrogenases.
2. Dehydrogenases which are components of a respiratory chain of electron transfer from substrate to oxygen.

Nicotinamide Co-Enzymes Dependent Dehydrogenases

Many dehydrogenases fall into this category. They are specific for either nicotinamide adenine dinucleotide (NAD^+) or nicotinamide adenine dinucleotide phosphate ($NADP^+$) as coenzymes. These are produced in the body from the vitamin niacin. The coenzymes are reduced by the specific substrate of the dehydrogenase and reoxidised by a suitable electron acceptor. They may freely and reversibly dissociate from their respective apoenzymes.

$$NAD^+ + AH_2 \underset{+H^+ + A}{\longleftrightarrow} NADH$$

(AH_2 = Substrate)

NAD-linked dehdyrogenases catalyse oxidoreduction reactions in the oxidative pathways of metabolism, particularly in glycolysis citric acid cycle and respiratory chain of mitochdondria. NADP-linked dehydrogenases are found characteristically in reductive syntheses, as in the extra-mitochondrial pathway of fatty acid synthesis and steroid synthesis and pentose phosphate pathway.

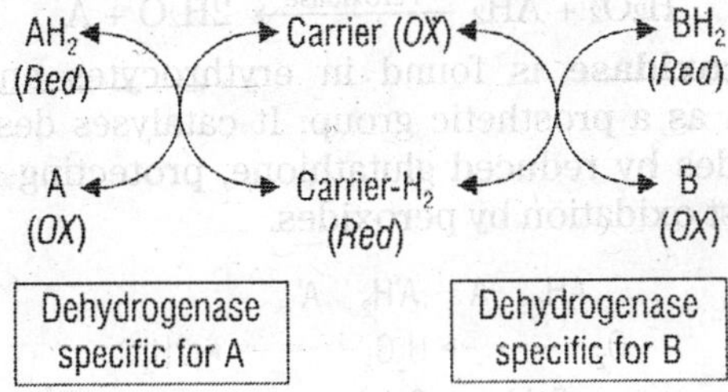

FIG. 12.6: Oxidation of a metabolite catalysed by dehydrogenases, not involving a respiratory chain.

Riboflavin Dependent Dehydrogenases

Most of the riboflavin-linked dehydrogenases are concerned with electron transport in (or to) the respiratory chain. NADH dehydrogenase is a member of the respiratory chain acting as a carrier of electrons between NADH and the more electro-positive components. Other dehydrogenases such as succinate dehydrogenase, acyl-CoA dehydrogenase, and mitochondrial glycerol-3 phosphate dehydrogenase transfer reducing equivalents directly from the substrate to the respiratory chain. Flavin-dependent dehydrogenase also plays role in the dehydrogenation of reduced lipoate, an intermediate in the oxidative decarboxylation of pyruvate and α-ketoglutarate.

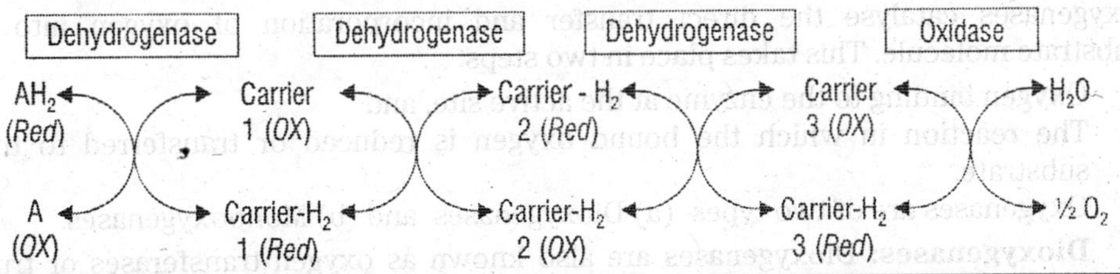

FIG. 12.7: Oxidation of a metabolite by dehydrogenases and finally by an oxidase in a respiratory chain.

Cytochromes as Dehydrogenases

The cytochromes, except cytochrome oxidase act as dehydrogenases in respiratory chain, they are involved as carriers of electrons from flavoproteins on one hand to cytochrome oxidase on the other. These are iron-containing hemoproteins in which the iron atom oscillates between Fe^{3+} and Fe^{2+} during oxidation and reduction. These may be cytochromes b, c1 and c. Cytochromes may also be found in other locations, *e.g.*, the endoplasmic reticulum (cytochrome P-450 and b5), plant cells, bacteria, and yeasts.

Hydroperoxidases

Hydroperoxidases protect the body against harmful peroxides. These use hydrogen peroxide or an organic peroxide as substrate. Accumulation of peroxidase can lead to generation of free radicals, which in turn can disrupt membranes etc. and possibly cause cancer and artherosclerosis. Hydroperoxidases include- **peroxidases and catalase.**

1. **Peroxidases:** Peroxidases occur in milk, leucocytes, platelets etc. These contain protoheme prosthetic group, which is loosely bound to the apoprotein. In peroxidase catalysed reactions, hydrogen peroxide is reduced at the expense of several substances that will act as electron acceptor, such as ascorbate, quinines, and cytochrome C.

$$H_2O_2 + AH_2 \xrightarrow{\text{Peroxidase}} 2H_2O + A$$

Glutathione peroxidase is found in erythrocytes and other tissues and contains selenium as a prosthetic group. It catalyses destruction of H_2O_2 and lipid hydroperoxides by reduced glutathione, protecting membrane lipids and hemoglobin against oxidation by peroxides.

$$O_2 \xrightarrow[\text{Oxidase}]{AH_2 \quad A} H_2O_2 \xrightarrow[\text{Catalase}]{A'H_2 \quad A'} 2H_2O$$

$$H_2O_2 \quad O_2$$

FIG. 12.8: Role of catalase in the destruction of H_2O_2.

2. **Catalse:** Catalase uses hydrogen peroxide as electron donor and electron acceptor (oxidant). It is a hemoprotein containing 4-hemegroups. Catalase is found in blood, bone marrow, mucous membranes, kidney and liver. It destroys H_2O_2 formed by the action of oxidases. Microbodies or peroxisomes are found in many tissues including liver. They are rich in oxidases and in catalase, which suggests that there may be a biologic advantage in grouping the enzymes that produce H_2O_2 with the enzyme that destroys it.

Oxygenases

Oxygenases catalyse the direct transfer and incorporation of oxygen into a substrate molecule. This takes place in two steps:

1. Oxygen binding to the enzyme at the active site, and
2. The reaction in which the bound oxygen is reduced or transferred to the substrate.
3. Oxygenases are of two types- (a) Dioxygenases, and (b) Monooxygenases.

Dioxygenases: Dioxygenases are also known as oxygen transferases or true oxygenases. These incorporate both oxygen atoms into the substrate. The basic reaction is as follows:

$$A + O_2 \rightarrow AO_2$$

Dioxygenases include enzymes that contain iron such as **homogentisate dioxygenase** and **3-hydroxyanthranilate dioxygenase** from the supernatant fraction of the liver, and enzymes utilizing heme such as **L-tryptophan dioxygenase.**

Monooxygenases: Monooxygenases are also known as mixed function oxidases or hydroxylases. These incorporate only are atom of oxygen into the substrate. The other oxygen atom is reduced to water, an additional electron donor or co-substrate being necessary for this purpose.

$$A - H + O_2 + ZH_2 \rightarrow A - OH + H_2O + Z$$

Certain examples of monooxygenases may be as follows:

Microsomal cytochrome P-450 monooxygenases bring about hydroxylation of many drugs. These monooxygenases are found in the microsomes of the liver together with cytochrome P-450 and cytochrome b5. Both NADH and NADPH donate reducing equivalents for the reduction of these cytochromes, which in turn are oxidized by substrates in a series of enzymatic reactions collectively known as the **hydroxylase cycle.**

$$\text{Drug - H} + O_2 + \underset{(P-450)}{2Fe^{2+}} + \underset{(P-450)}{2H^+} \xrightarrow{\text{Hydroxylase}} \text{Drug - OH} + H_2O + 2Fe^{3+}$$

This system metabolizes drugs like benzypyrene, aminopyrene, aniline, morphine, and benzphetamine.

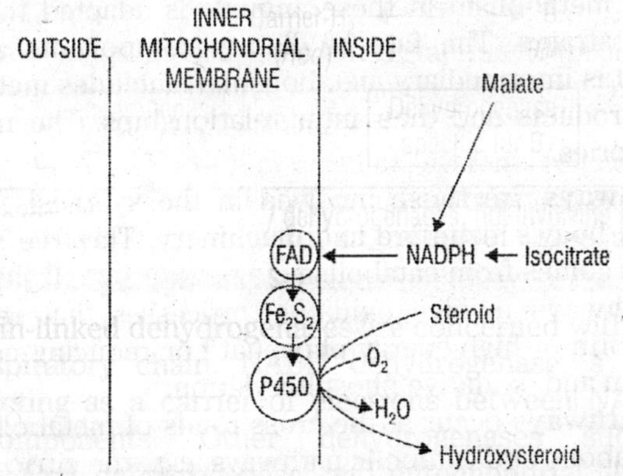

FIG. 12.9: Mitochondrial cytochrome P-450 monooxygenase system. Because NADP (H) cannot penetrate the mitochondrial membrane, sources of reducing equivalents are confined to substrates such as malate and isocitrate for which there are intramitochondrial NADP- specific dehydrogenases.

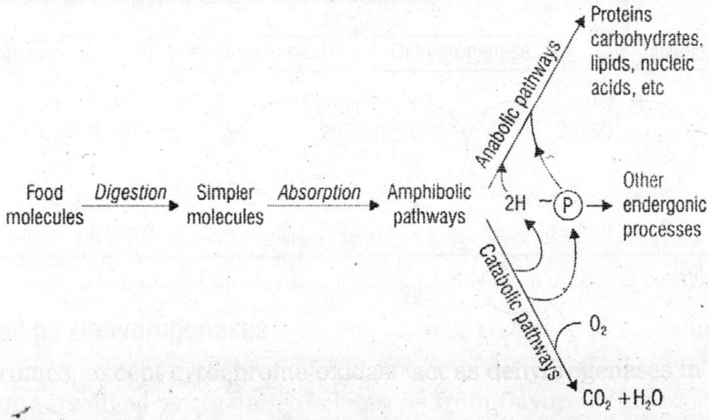

FIG. 12.10: The 3 major categories of metabolic pathways. Catabolic release free energy in the form of reducing equivalents (2H) or high-energy phosphate (~P) to power the anabolic pathways act as links between the other 2 categories of pathways.

Mitochondrial cytochrome P-450 monooxygenases catalyse steroidal hydroxylation. These systems are found in sterioidogenic tissues such as adrenal cortex, testis, ovary, and placenta and are concerned with the biosynthesis of steroid hormones from cholesterol (hydroxylation at C22 and C20 inside chain cleavage and at the 11B and 18 positions). Renal systems catalyse 1 α- and 24-hydroxylation of 25-hydroxycholecalciferol and the liver catalyses 26-hydroxylation in bile acid biosynthesis.

The monooxygenase system consists of three components situated on the inside of the inner mitochondrial membrane; and NADP- specific FAD containing flavoprotein, and Fe_2S_2 iron sulphur protein (edrenodoxin), and cytochrome P-450.

OVERVIEW OF INTERMEDIARY METABOLISM

The nature of the diet sets the basic pattern of metabolism in the tissues. Animals need to process the absorbed products of dietary carbohydrates, lipids and proteins. These are mainly glucose, fatty acids and glycerol, and amino acids respectively. In ruminants (and to lesser extent other herbivores), cellulose in the diet is digested by symbiotic microorganisms to lower fatty acids (acetic, propionic, butyric) and tissue metabolism in these animals is adapted to utilize lower fatty acids as major substrates. The fate of dietary components after digestion and absorption constitutes intermediary metabolism. It includes metabolic pathways of various digestion products and their inter relationships. The metabolic pathways fall into three categories.

1. **Anabolic pathways** are those involved in the synthesis of the compounds constituting the body's structure and machinery. The free energy required for these processes comes from catabolism.
2. **Catabolic pathways** involve oxidative processes that release free energy, usually in the form of high-energy phosphate or reducing equivalents, *e.g.*, the respiratory chain and oxidative phosphorylation.
3. **Amphibolic pathways** occur at the cross roads of metabolism, acting as links between the anabolic and catabolic pathways, *e.g.*, the citric acid cycle.

CHAPTER 13

Metabolism: Carbohydrates

The carbohydrates are digested in the alimentary canal by the action of salivary amylase in the buccopharyngeal cavity and pancreatic amylase, oligo-1,6-glucosidase and disaccharases in the intestine. The digestive products are monosaccharides, which are absorbed into the blood and taken to liver and other tissues. The major metabolic pathways for carbohydrates may be as follows:

1. Anabolic pathway: Glycogenesis.
2. Catabolic pathways: Glycogenolysis, Glycolysis and the oxidation of pyruvate, Citric acid cycle and pentase phosphate pathway.
3. Gluconeogenesis.

The goal of glycolysis, glycogenolysis, and the citric acid cycle is to conserve energy as ATP from the catabolism of carbohydrates. If the cells have sufficient supplies of ATP, then these pathways and cycles are inhibited. Under these conditions of excess ATP, the liver will attempt to convert a variety of excess molecules into glucose and/or glycogen.

TABLE 13.1: Differences between Glycogenesis, Glycogenolysis and Gluconeogenesis

Glycogenesis	**Glycogenolysis**	**Gluconeogenesis**
Glycogenesis is the formation of glycogen from glucose. Glycogen is synthesised depending on the demand for glucose and ATP (energy). If both are present in relatively high amounts, then the excess of insulin promotes the glucose conversion into glycogen for storage in liver and muscle cells. In the synthesis of glycogen, one ATP is required per glucose incorporated into the polymeric branched structure of glycogen. actually, glucose-6-phosphate is the cross-roads compound. Glucose-6-phosphate is synthesised directly from glucose or as the end product of gluconeogenesis.	In glycogenolysis, glycogen stored in the liver and muscles, is converted first to glucose-1-phosphate and then into glucose-6-phosphate. Two hormones which control glycogenolysis are a peptide, glucagon from the pancreas and epinephrine from the adrenal glands. Glucagon is released from the pancreas in response to low blood glucose and epinephrine is released in response to a threat or stress. Both hormones act upon enzymes to stimulate glycogen phosphorylase to begin glycogenolysis and inhibit glycogen synthetase (to stop glycogenesis).	Gluconeogenesis is the biosynthesis of glucose from non-carbohydrate sources. The starting point of gluconeogenesis is pyruvic acid, although oxaloacetic acid and dihydroxyacetone phosphate also provide entry points. Lactic acid, some amino acids from protein and glycerol from fat can be converted into glucose. Gluconeogenesis is similar but not the exact reverse of glycolysis, some of the steps are the identical in reverse direction and three of them are new ones. Without going into detail, the general gluconeogenesis sequence is given in the graphic on the left.
	Glycogen is a highly branched polymeric structure containing glucose as the basic monomer. First individual glucose molecules are hydrolyzed from the chain,	The oxaloacetic acid is synthesised from pyruvic acid in the first step. Oxaloacetic acid is also the first compound to react with acetyl CoA in the citric acid

Glycogenesis	Glycogenolysis	Gluconeogenesis
	followed by the addition of a phosphate group at C-1. In the next step the phosphate is moved to the C-6 position to give glucose 6-phosphate, a cross road compound. Glucose-6-phosphate is the first step of the glycolysis pathway if glycogen is the carbohydrate source and further energy is needed. If energy is not immediately needed, the glucose-6-phosphate is converted to glucose for distribution in the blood to various cells such as brain cells.	cycle. The concentration of acetyl CoA and ATP determines the fate of oxaloacetic acid. If the concentration of acetyl CoA is low and concentration of ATP is high then gluconeogenesis proceeds. Also notice that ATP is required for a biosynthesis sequence of gluconeogenesis. Gluconeogenesis occurs mainly in the liver with a small amount also occurring in the cortex of the kidney. Very little gluconeogenesis occurs in the brain, skeletal muscles, heart muscles or other body tissue. In fact, these organs have a high demand for glucose. Therefore, gluconeogenesis is constantly occurring in the liver to maintain the glucose level in the blood to meet these demands.

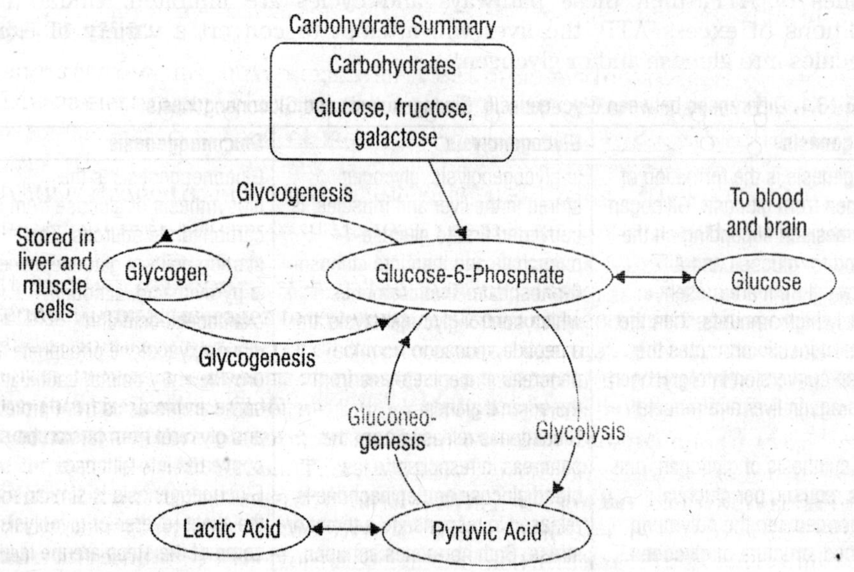

FIG. 13.1: Presentation of Glycogenesis, Glycogenolysis and Gluconeogenesis.

GLYCOGENESIS

Glycogenesis refers to the process of biosynthesis of glycogen from glucose. Glycogen is synthesised in practically all the tissues of the body but the major sites are liver and muscles. It forms about 6% and 1% of liver and muscle weight respectively. The various steps of glycogenesis are as follows:

Glucose is phosphorylated to glucose 6-phosphate. This reaction is catalysed by **hexokinase** in muscle and **glucokinase** in liver, in the presence of ATP and Mg^{++}.

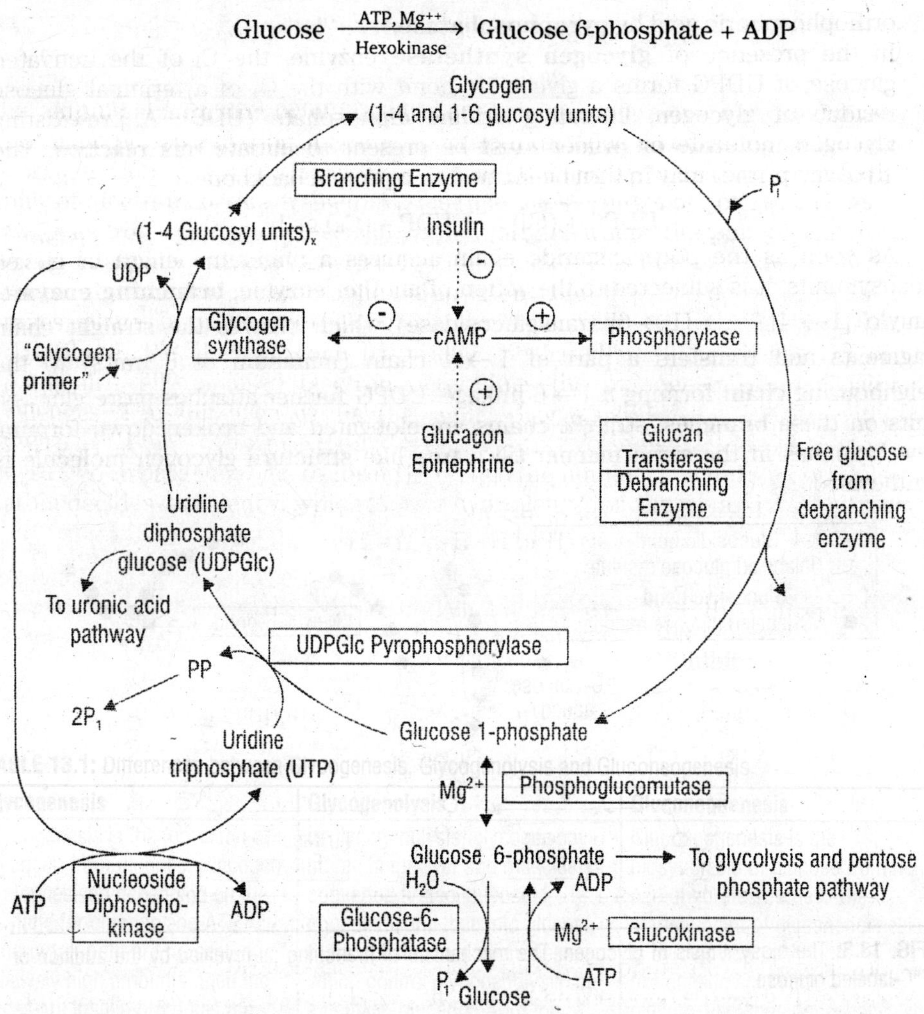

FIG. 13.2: Pathway of glycogenesis and of glycogenolysis in the liver. Two high-energy phosphates are used in the incorporation of 1 mol of glucose into glycogen. + stimulation, -, inhibition. Insulin decreases the level of cAMP only after it has been raised by the glucagen or epinephrine, *i.e.*, it antagonizes their action. Glucagen is active in heart muscle but not in skeletal muscle. Glucan transferase and debranching enzyme appear to be two separate activities of the same.

Glucose 6-phosphate is converted to glucose 1-phosphate in a reaction catalysed by the enzyme **phosphoglucomutase.**

The enzyme itself is phosphorylated, and the phospho-group takes part in a reversible reaction in which glucose 1,6-diphosphate is an intermediate.

$$\text{Enz} - \text{P} + \text{Glucose 6-phosphate} \longleftrightarrow \text{Enz} + \text{Glucose 1,6-diphosphate} \longleftrightarrow \text{Enz-P} + \text{Glucose 1-phosphate.}$$

1. Glucose 1-phosphate reacts with uridine triphosphate (UTP) to form the active nucleotide uridine diphosphate glucose (UDPG) in the presence of an enzyme, **UDPG pyrophosphorylase.**

UDPG + PPi

In this reaction pyrophosphate is liberated which is converted to orthophosphoric acid by **pyrophosphatase.**

2. In the presence of **glycogen synthetase** enzyme, the C_1 of the activated glucose of UDPG forms a glycosidic bond with the C_4 of a terminal glucose residue of glycogen, liberating uridine diphosphate (UDP). A pre-existing glycogen molecule or primer must be present to initiate this reaction. The glycogen primer may in turn be formed on a protein backbone.

$$UDPG + (C_6)\,n \rightarrow UDP + (C_6)\,n{+}1$$

As soon as the polysaccharide chain acquires a minimum length of eleven glucosyl-units, it is subjected to the action of another enzyme, **branching enzyme,** (amylo [1→ 4] ——→ [1→ 6] transglucosidase) which cleaves the straight chain fragments and transfers a part of 1→ 4 chain (minimum of 6 units) to the neighbouring chain forming a 1→ 6 linkage. UDPG further attaches more glucosyl units on these branches, straight chains are elongated and broken down forming new branches in the same manner till a tree like structure glycogen molecule is synthesised.

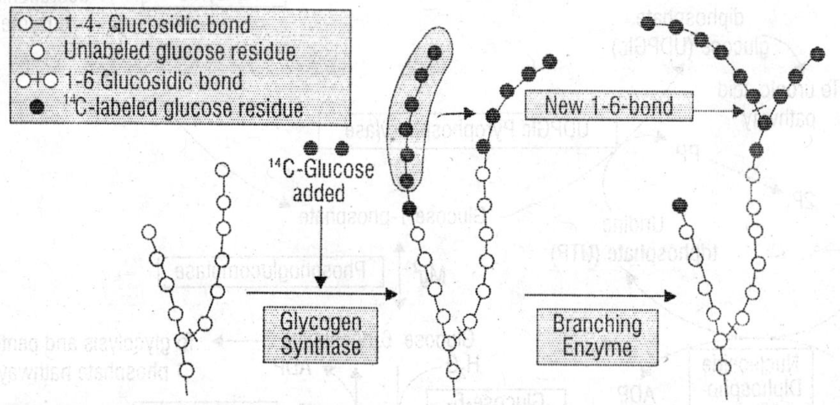

FIG. 13.3: The biosynthesis of glycogen. The mechanism of branching as revealed by the addition of ^{14}C-labeled glucose.

Other sugars such as galactase or fructose may also be converted into glycogen by way of first forming glucose, glucose 6-phosphate or glucose 1-phosphate.

GLYCOGENOLYSIS

Glycogenolysis refers to the breakdown of glycogen into glucose (in liver and kidney) or glucose 6-phosphate (in muscles). The various steps of glycogenolysis are as follows:

1. Glycogen is acted upon by **phosphorylase** enzyme in the presence of inorganic phosphate (Pi). This enzyme is specific for the phosphorolytic breaking (phosphorolysis) of the 1→ 4 linkages of glycogen to yield glucose 1-phosphate.

$$\underset{\text{glycogen}}{(C_6)n} + Pi \rightarrow \underset{\text{glycogen}}{(C_6)n - 1} + \text{Glucose 1 - phospate}$$

2. Glucosyl residues from the outermost chains of the glycogen molecule are removed until approximately 4-glucose residues remain on either side of a 1→ 6 branch.

3. Another enzyme (a-[1 → 4] → a-[1 → 4] glucan transferase) transfers a

trisaccharide unit from one branch to the other, exposing the 1→ 6- branch points.

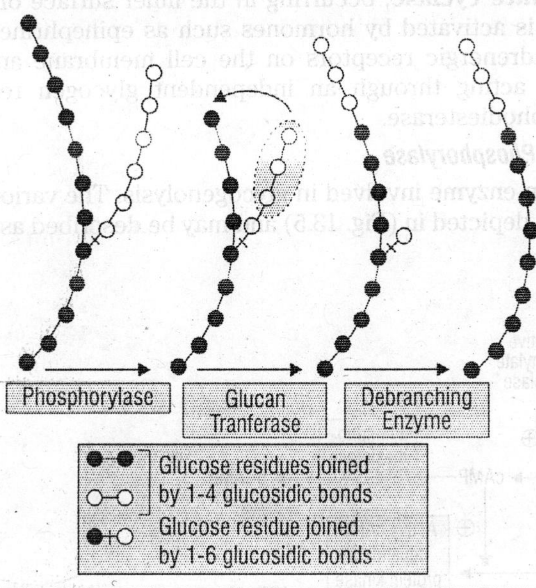

FIG. 13.4: Steps in glycogenolysis.

4. **Debranching enzyme** (amylo- [1→ 6]- glucosidase) brings about hydrolytic splitting of the 1→ 6 linkages.
5. With the removal of the branch, further action of phosphorylase can proceed. Combined action of various enzymes mentioned above, leads to the complete breakdown of glycogen into glucose 1-phosphate.
6. Glucose 1-phosphate is converted into glucose 6-phosphate in the presence of **phosphoglucomutase** and Mg^{++} ions.
7. In liver and kidney (not in muscles) **glucose-6-phosphatase enzyme** removes phosphate from glucose 6-phosphate, enabling glucose to diffuse from the cell into the blood.

FIG. 13.5: 3' 5' Adenylic acid (cyclic AMP; cAMP).

ROLE OF CYCLIC AMP IN GLYCOGEN METABOLISM

The activities of two principal enzymes involved in glycogen metabolism, *i.e.* glycogen **phosphorylase** and glycogen synthetase, are regulated by cyclic AMP (cAMP, 3',5'- cyclic adelylic acid). cAMP is the intracellular intermediate compound or second messenger through which many hormones act. It is formed from ATP by an enzyme, **adenylate cyclase**, occurring in the inner surface of cell membranes. Adenylate cyclase is activated by hormones such as epinephrine and epinephrine acting through ß-adrenergic receptors on the cell membrane and additionally in liver by glucagon acting through an independent glycogen receptor. cAMP is destroyed by phosphodiesterase.

Activation of Muscle Phosphorylase

Phosphorylase is an enzyme involved in glycogenolysis. The various steps involved in its activation are depicted in (Fig. 13.5) and may be described as follows:

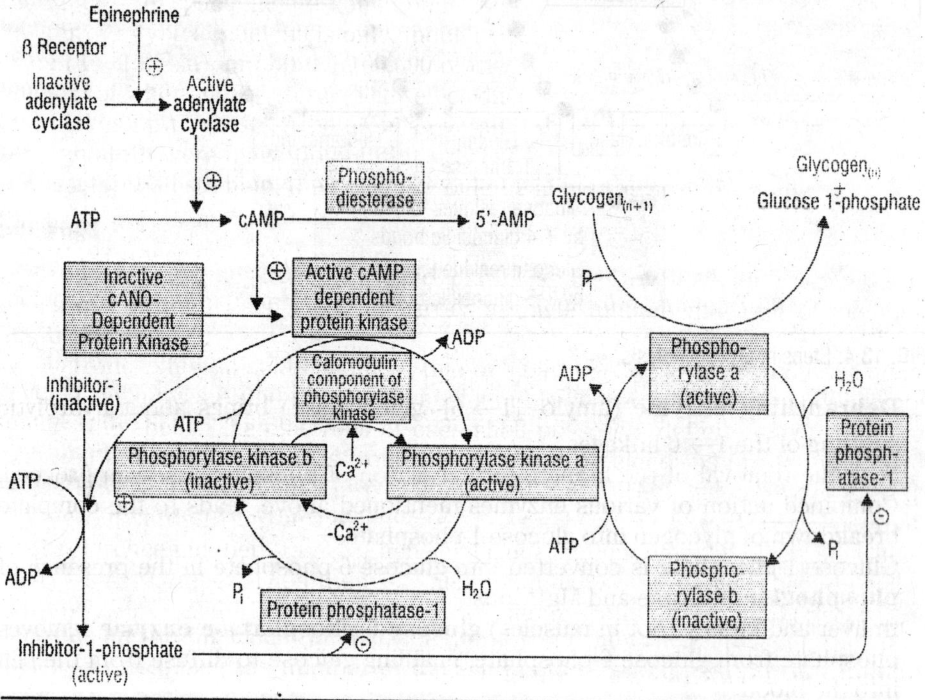

FIG. 13.6: Control of phosphorylase in muscle (n =. number of glucose residues). The sequence of reactions arranges as a cascade amplification of the hormonal signal at each step.

1. Epinephrine activates adenylate cyclase, which in turn converts ATP into cAMP.
2. cAMP activates **cAMP-dependent protein kinase** (an enzyme of wide specificity), which catalyses the phosphorylation by ATP of inactive **phosphorylase kinase** b to active **phosphorylase kinase** a, which in turn, by means of a further phosphorylation, activates phosphorylase b to phorphorylase a.
3. During muscle contraction, rapid activation of phosphorylase takes place owing to activation of phosphorylase kinase by Ca^{+2}, the same signal that initiates contraction. Thus, Ca^{+} synchronizes the activation of phosphorylase with muscle contraction.

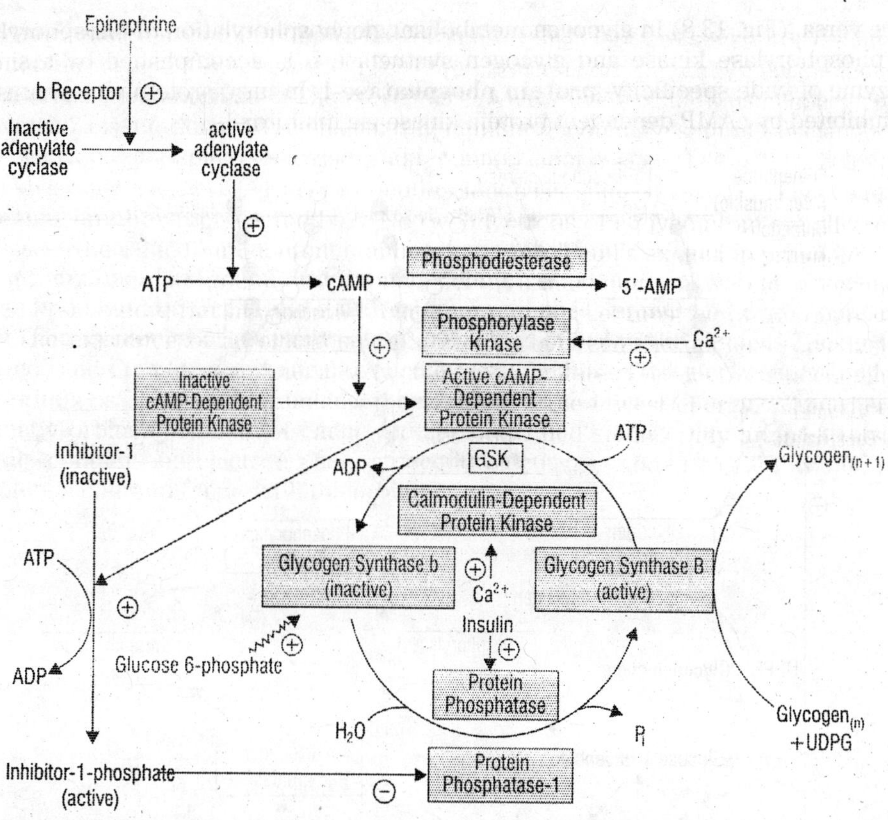

FIG. 13.7: Control of glycogen synthase in muscle (n= number of glucose residues). The sequence of reactions arranged in a cascade causes amplification at each step, allowing only nanomole quantities of hormone to cause major changes in glycogen concentration. GSK, glycogen synthase kinase-3, -4, and -5; wavy arrow, allosteric activation.

4. Both phosphorylase a and phosphorylase kinase a are dephosphorylated and inactivated by **protein phosphatase**-1. Protein phosphatase-1 is inhibited by a protein called **inhibitor-1**, which becomes active after it has been phosphorylated by cAMP- dependent protein kinase. Thus, cAMP controls both the activation and inactivation of phorphorylase.

Activation of Glycogen Synthetase

Activation of glycogen synthetase has been depicted in Fig. 13.7. Unlike phosphorylase, the active form of glycogen synthetase is dephosphorylated (glycogen synthetase a) and inactive form is phosphorylated (glycogen synthetase b). Protein phosphatase converts inactive form of glycogen synthetase b to active form a.

Glucose 6-phosphate acts as an allosteric activator of glycogen synthetase a, causing a decrease in Km. for UDP-glucose and allowing glycogen synthesis by the phosphorylated enzyme.

Glycogen synthetase and phosphorylase are under substrate control (through allostery) as well as hormonal control. Increase in cAMP concentration (via phosphorylase kinase), phosphorylase is activated and glycogen synthetase converted to the inactive form. Both these activities are mediated via cAMP dependent protein base. Inhibition of glycogenolysis enhances net glycogenesis and vice versa. (Fig. 13.8) In glycogen metabolism, dephosphorylation of phosphorylase a, phosphorylase kinase and glycogen synthetase b is accomplished by a single enzyme of wide specificity- **protein phosphatase**-1. In turn, protein phosphatase-1 is inhibited by cAMP dependent protein kinase via inhibitor-1.

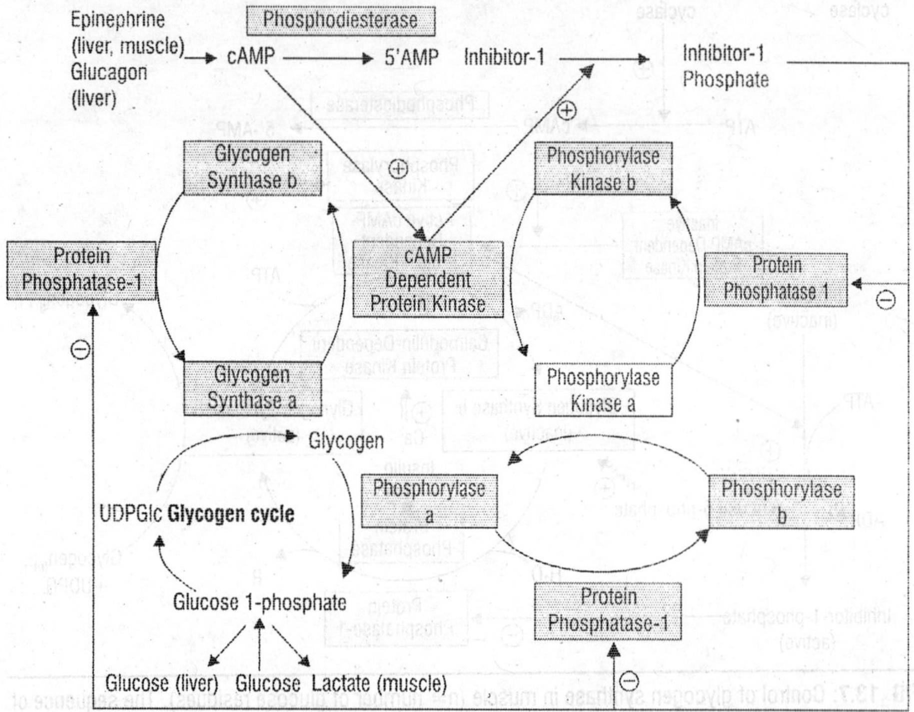

FIG. 13.8: Coordinated control of glycogenolysis and glycogenesis by cAMP-dependent protein kinase. The reactions that lead to glycogenolysis as a result of an increase in cAMP concentrations are shown with bold arrows, and those that are inhibited are shown as broken arrows. The reverse occurs when cAMP concentrations decrease as a result of phosphodiesterase activity, leading to glycogenesis.

Glycolysis and the Oxidation of Pyruvate

Glycolysis is a biochemical process in which glucose is converted into pyruvate (pyruvic acid). It is also known as **Embden-Meyerhof pathway**. Glycosis can occur in anaerobic as well as aerobic conditions. Under aerobic conditions pyruvate breaks down into CO_2 and H_2O and in absence of O_2 pyruvate is changed into lactic acid. The enzymes involved in glycolysis are found in the extra mitochondrial soluble fraction of the cell, the cytosol. The sequences of reactions occurring in glycolysis are as follows (Fig. 13.9).

Phosphorylation

Glucose undergoes phosphorylation to produce glucose 6-phosphate in the presence of an enzyme **hexokinase** and in liver parenchymal cells by **glucokinase**. ATP is the phosphate group donor and reacts as the Mg-ATP complex. It donates one high energy phosphate group and converted into ADP. The reaction is accompanied by considerable loss of free energy as heat and therefore, under physiologic conditions, may be regarded as irreversible.

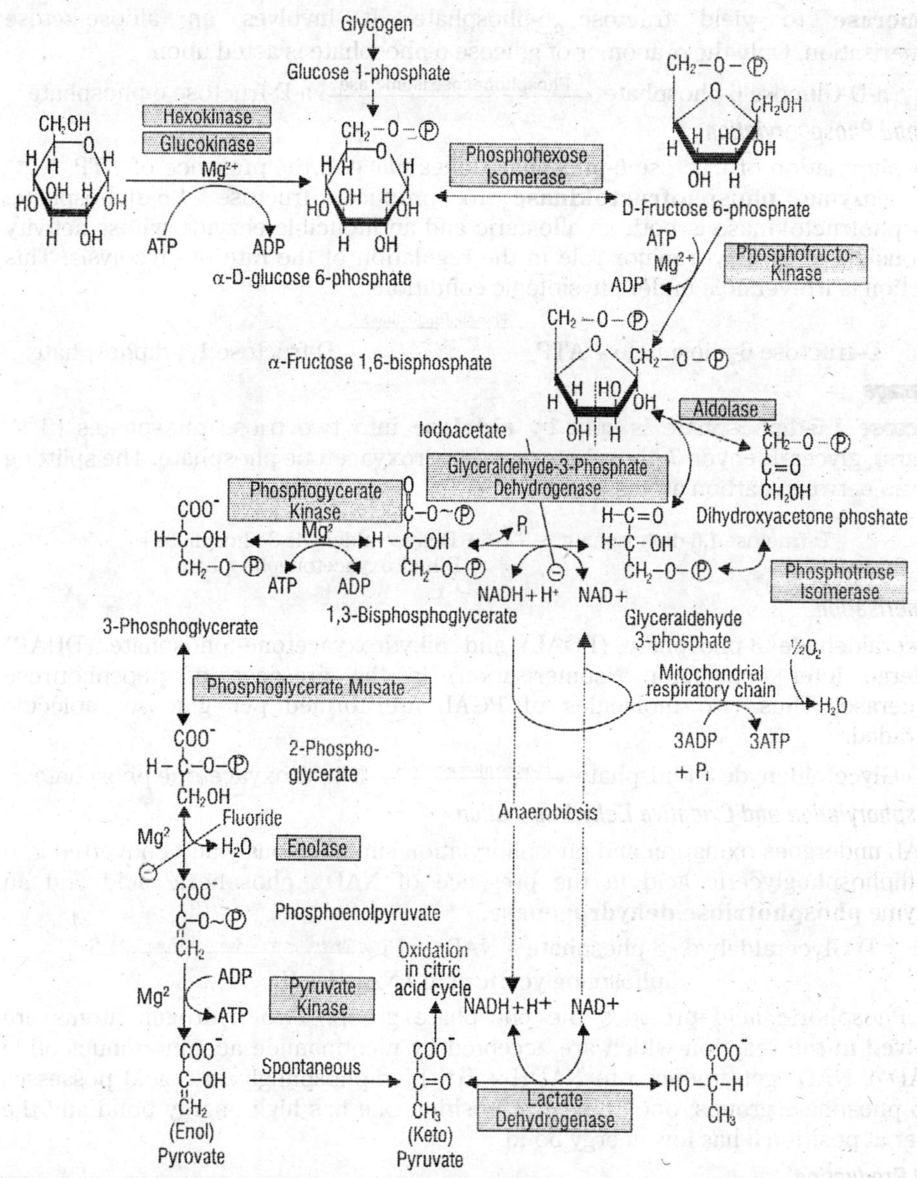

FIG. 13.9. The pathway of glycolysis. P,-PO_3^{2-}; P1, HPO_3^{2-}; -, inhibition. Carbon atoms 1-3 of fructose bisphosphate form dihydroxyacetone phosphate, whereas carbons 4-6 form glyceraldehyde 3-phosphate. The term bis-as in bisphosphate indicates that the phosphate groups are separated, whereas diphosphate, as in adenosine diphosphate, indicates they are joined.

Hexokinase is inhibited in an allosteric manner by the product, glucose 6-phosphate.

Hexokinase/Glucokinase

$$\text{Glucose} + \text{ATP} \xrightarrow[Mg^{++}]{} \text{Glucose 6-phosphate} + \text{ADP}$$

Isomerisation

Glucose 6-phosphate undergoes isomerisation in the presence of **phosphohexose**

isomerase to yield fructose 6-phosphate. It involves an aldose-ketose isomerisation. Only the α-anomer of glucose 6-phosphate is acted upon.

$$\text{a-D-Glucose 6-phosphate} \xleftrightarrow{\text{Phosphohexose isomerase}} \text{a-D-fructose 6-phosphate}$$

Second Phosphorylation

Phosphorylation of fructose 6-phosphate takes place in the presence of ATP, Mg^{++} and enzyme **phosphofructokinase** to produce, fructose 1,6-diphosphate. Phosphofructokinase is both an allosteric and an inducible enzyme whose activity is considered to play a major role in the regulation of the rate of glycolysis. This reaction is irreversible under physiologic conditions.

$$\text{D-fructose 6-phosphate + ATP} \xrightarrow[Mg^{+-}]{\text{Phosphofructokinase}} \text{D-fructose 1,6-diphosphate}$$

Cleavage

Fructose 1,6-diphosphate is split by **aldolase** into two triose phosphates (3 C-sugars), glyceraldehyde 3-phosphate and dihydroxyacetone phosphate. The splitting occurs between carbon atoms 3 and 4.

$$\text{D-fructose 1,6-diphosphate} \xleftrightarrow{\text{Aldolase}} \text{D-glyceraldehyde 3-phosphate + Dihydroxyacetonephosphate}$$

Isomerisation

Glyceraldehyde 3-phosphate (PGAL) and dihydroxyacetone phosphate (DHAP) undergo inter conversion (isomerisation) in the presence of phosphotirose isomerase. Thus two molecules of PGAL are formed per glucose molecule degraded.

$$\text{D-Glyceraldehyde 3-phosphate} \xleftrightarrow{\text{Phospho triose isomerase}} \text{Dihydroxyacetone phosphate}$$

Phosphorylation and Oxidative Dehydrogenation

PGAL undergoes oxidation and phosphorylation simultaneously and converted into 1,3-diphosphoglyceric acid in the presence of NAD^+, phosphoric acid and an enzyme **phosphotriose dehydrogenase**.

$$\text{D-Glyceraldehyde 3-phosphate} + NAD^+ + Pi \xleftrightarrow{\text{Glyceraldehyde 3-dehydrogenase}} \text{1,3-diphosphoglyceric acid} + NADH + H^+$$

Phosphoric acid provides the phosphate group. Two hydrogen atoms are evolved in the reaction which are accepted by nicotinamide adenine dinucleotide (NAD^+), NAD^+ gets reduced to $NADH + H^+$. 1,3-diphosphoglyceric acid possesses two phosphate groups; one present at position one has high energy bond and the other at position 3 has low energy bond.

ATP Production

1,3-diphosphoglyceric acid transfers its phosphate group with high-energy bond to ADP, forming 3-phosphoglyceric acid (PGA) and ATP. The reaction is catalysed by **phosphoglycerate kinase**.

$$\text{1,3-diphosphog lyceric acid + ADP} \xleftrightarrow{\text{Phosphoglycerate kinase}} \text{3-phospho-glyceric acid + ATP}$$

Since 2-molecules of PGA are formed per molecule of glucose undergoing glycolysis, 2-molecules of ATP are generated.

Isomerisation

The 3-phosphoglyceric acid arising from the above reaction is converted to 2-phosphoglyceric acid in the presence of enzyme, **phosphoglycerate mutase**.

$$\text{3-phosphoglyceric acid} \xleftrightarrow{\text{Phosphoglycerate mutase}} \text{2-phosphoglyceric acid}$$

Dehydration

The 2-phosphoglyceric acid undergoes dehydration and redistribution of energy within the molecule, raising the phosphate on position 2 to the high energy state, thus, forming phosphoenol pyruvic acid. The reaction is catalysed by enolase, whose activity can be inhibited by fluoride. Enolase requires Mg^{++} or Mn^{++} for the activity.

$$\text{2-phosphoglyceric acid} \xleftrightarrow{\text{Enolase}} \text{Phosphoenolpyruvic acid} + H_2O$$

ATP Generation

The high energy phosphate of phosphoenol pyruvic acid is transferred to ADP by the enzyme **pyruvate kinase** to generate, at this stage, 2 molecules of ATP per molecule of glucose oxidized. Pyruvic acid (enol) is formed in this reaction which is converted spontaneously to the keto form. This step is accompanied by considerable loss of free energy as heat and must be regarded as physiologically irreversible.

$$\text{Phosphoenolypyruvic acid} + \text{ADP} \xrightarrow{\text{Pyruvate kinase}} \text{Pyruvic acid} + \text{ATP}$$

At the end of glycolysis 2-molecules of pyruvic acid are produced per glucose molecule.

Under anaerobic conditions, the reoxidation of NADH by transfer of reducing equivalents through the respiratory chain to oxygen is prevented. Pyruvic acid is reduced by the NADH to lactic acid in the presence of enzyme, **lactate dehydrogenase**.

$$\text{Pyruvic acid} + \text{NADH} + H^+ \xleftrightarrow{\text{Lactate dehydrogenase}} \text{Lactic acid} + \text{NAD}^+$$

Oxidation of cytosol NADH: It is mediated by substrate shuttles because NADH can not penetrate the mitochondrial membrane. The shuttles used are (Fig. 13.10):

1. **Glycerophosphate shuttle**: As shown in the Fig. 13.10a, since the mitochondrial enzyme is linked to the respiratory chain via FAD rather than NAD, only 2 molecules of ATPase are produced.
2. **Malate shuttle**: As depicted in the Fig. 13.10b, 3 ATP molecules will be produced per $2H^+$ transferred into mitochondria. The hydrogen atoms are transferred to NADH in mitochondria.

Oxidation of Pyruvic acid: The pyruvic acid is first oxidized to acetyl CoA, which in turn enters the Krebs cycle. The conversion of pyruvic acid to acetyl-CoA is an irreversible reaction. It takes place in mitochondria. Pyruvic acid from the cytosol is transported into the mitochondrion via a special **pyruvate transporter** that aids its passage across the inner mitochondrial membrane. This involves a symport mechanism whereby one proton is transported.

Within the mitochondrion, pyruvic acid is oxidatively decarboxylated to acetyl-CoA. This reaction is catalysed by a multi enzyme complex, known as **pyruvate dehydrogenase complex**. The various steps may be as follows (Fig. 13.11).

1. Pyruvic acid reacts with thiamin pyrophosphate or thiamine diphosphate (TPP) in the presence of Mg^{++} and **pyruvate dehydrogenase** to produce an **addition compound.**
2. The addition compound undergoes decarboxylation to produce hydroxyethyl thiamine diphosphate in the presence of pyruvate dehydrogenase.
3. Hydroxy ethyl thiamine diphosphate reacts with oxidized lipoamide in the presence of enzyme, **dihydrolipoyl transacetylase** to produce acetyl

lipoamide.

4. Acetyl lipoamide reacts with coenzyme A to form acetyl- CoA and reduced lipoamide in the presence of dihydrolipoyl transacetylase. Acetyl-CoA contains one high energy bond.

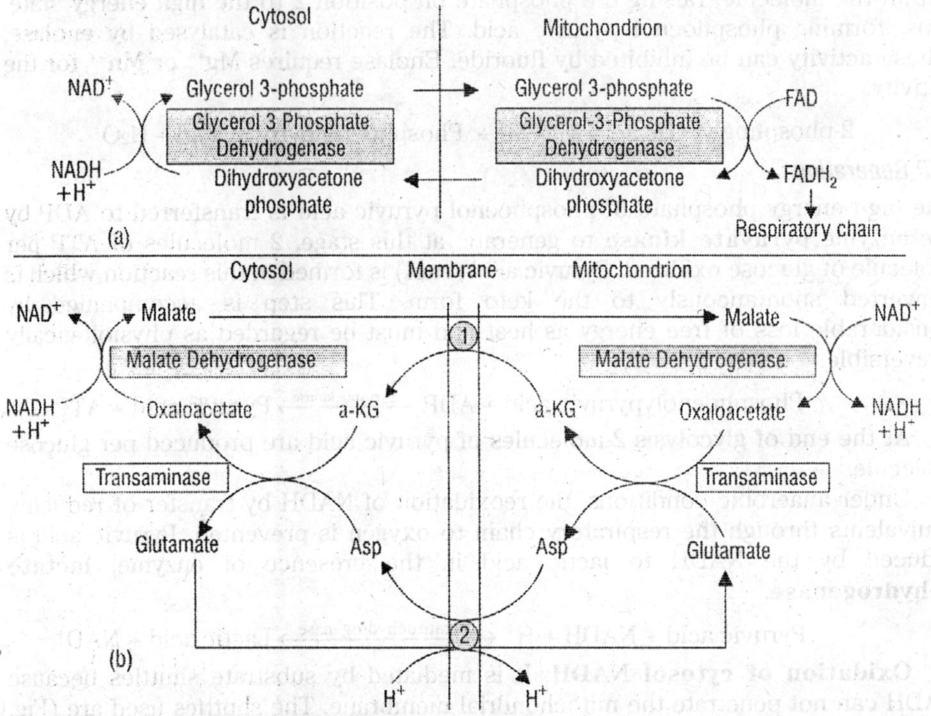

FIG. 13.10: (a) Glycerophosphate shuttle for transfer of reducing equivalents from the cytosol into the mitochondrion. (b) Malate shuttle for transfer of reducing equivalents from the cytosol into the mitochondrion. 1. ketoglutrate transporter. 2. Glutamate-aspartate transporter (note the proton symport).

Reduced lipoamide is reoxidised by a flavoprotein in the presence of dihydrolipoyl dehydrogenase. Finally, the reduced flavoprotein is oxidized by NAD, which in turn transfers reducing equivalents to the respiratory chain.

The oxidation of pyruvic acid to acetyl-CoA can be summarized as follows:

Pyruvic acid + NAD^+ + CoA $\longrightarrow$ Acetyl-CoA + NADH + H^+ + CO_2

2-acetyl-CoA molecules are produced per glucose molecule oxidized.

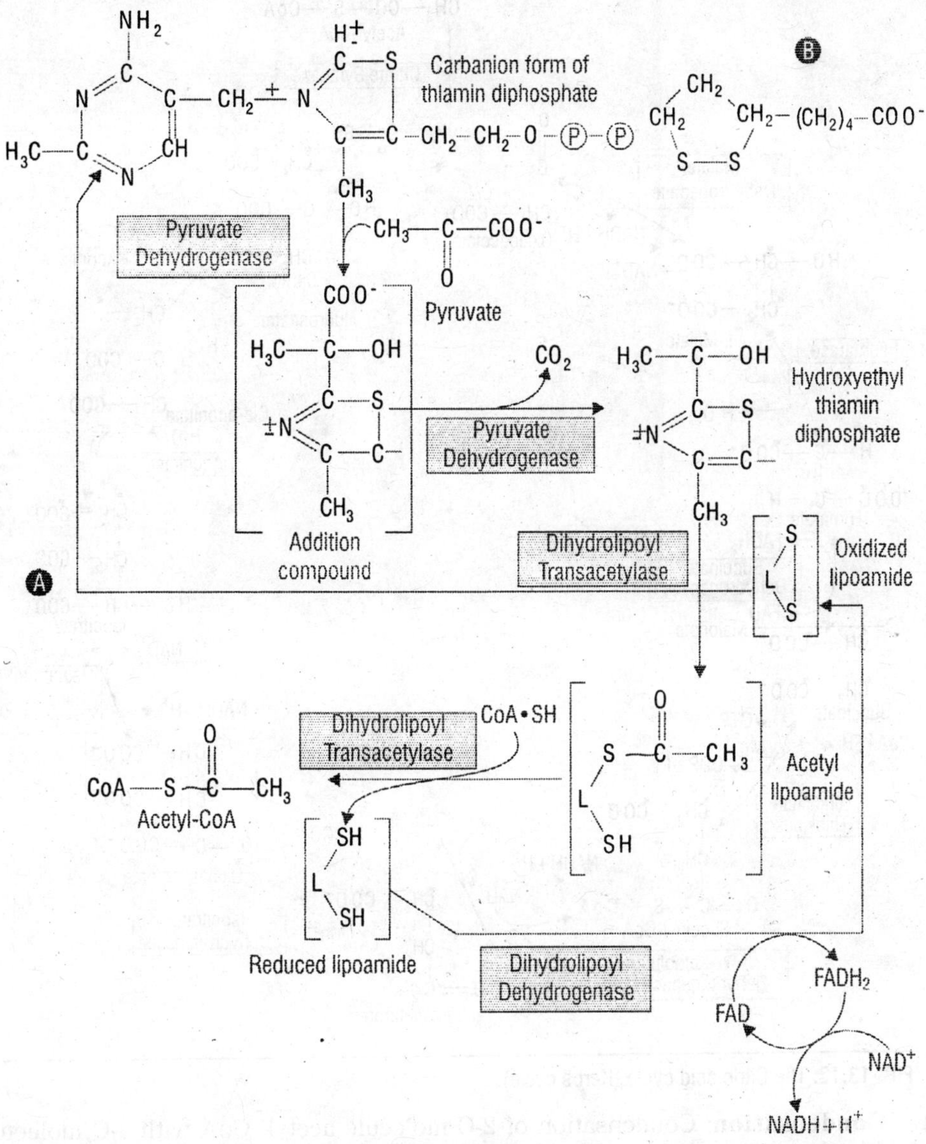

FIG. 13.11: (A) Oxidative decarboxylation of pyruvate by the pyruvate dehydrogenase complex. (B) Lipoic acid. Lipoic acid is joined by an amide link to a lysine residue of the transacetylase component of the enzyme complex.

CITRIC ACID CYCLE

The citric acid cycle (Krebs cycle, tricarboxylic acid cycle) is a series of reactions in mitochondria that bring about the catabolism of acetyl residues which are in the form of acetyl-CoA (CH_3-Co~S-CoA, active acetate), an ester of coenzyme A. CoA contains the vitamin pantothenic acid. The various steps of citric acid cycle are as follows (Fig. 13.12).

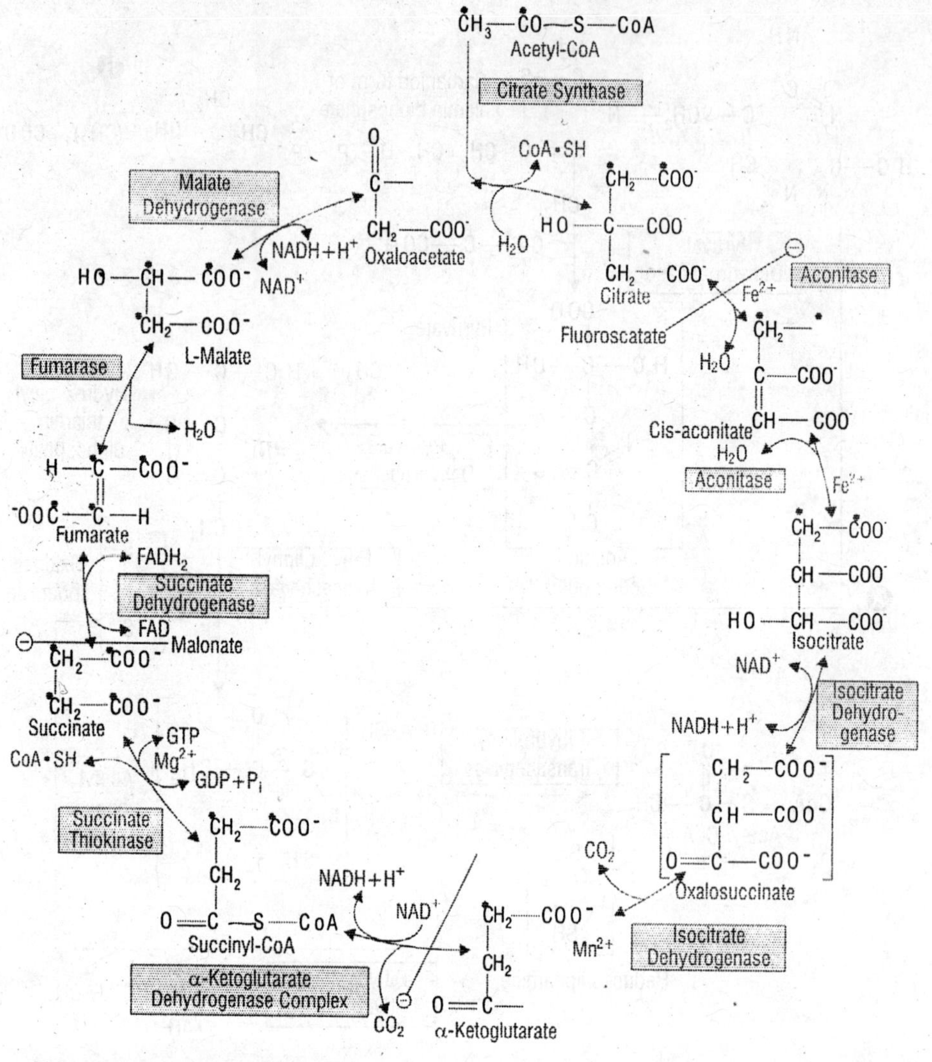

FIG. 13.12: The Citric acid cycle (Kerbs cycle).

1. **Condensation**: Condensation of 2-C molecule acetyl- CoA with 4-C molecule oxalo acetic acid takes place in the presence of a condensing enzyme, **citrate synthetase** to produce 6-C molecule citric acid and CoA.

$$\text{Acetyl-CoA} + \text{Oxaloacetate} + H_2O \longrightarrow \text{Citric acid} + \text{CoA}$$

2. **Dehydration**: Under the action of enzyme **aconitase**, citric acid undergoes dehydration to form cis-aconitic acid.

$$\text{Citric acid} \xleftrightarrow{\text{Aconitase}} \text{Cis-aconitic acid} + H_2O$$

3. **Hydration-I**: Cis-aconitic acid undergoes hydration to form isocitric acid under the influence of aconitase:

$$\text{Cis – aconitic acid} + H_2O \xleftrightarrow{\text{Aconitase}} \text{Isocitric acid}$$

Dehydrogenation I: Isocitric acid is dehydrogenated in the presence of isocitrate dehydrogenase to yield oxalosuccinic acid. The oxalosuccinic acid is an enzyme bound molecule. The hydrogen atoms produced in the reaction are accepted by NAD^+.

$$\text{Isocitric acid} + NAD^+ \xleftrightarrow{\text{Isocitrate dehydrogenase}} \text{Oxalosuccinic acid} + NADH + H^+ \quad \text{(Enzyme bound)}$$

Decarboxylation I: Enzyme bound oxalosuccinic acid undergoes decarboxylation to produce α- ketoglutaric acid.

$$\underset{\text{(Enzyme bound)}}{\text{Oxalosuccinic acid}} \leftrightarrow \text{a - Ketoglutaric acid} + CO_2$$

Dehydrogenation II and decarboxylation II: α -Ketoglutaric acid undergoes oxidative decarboxylation.

$$\alpha\text{-Ketoglutaric acid} + NAD^+ + \text{CoA} \longrightarrow \text{Succinyl-CoA} + CO_2 + NADH + H^+$$

This reaction is catalysed by an **α-ketoglutarate dehydrogenase** complex which requires cofactors thiamine diphosphate, lipoate, NAD^+, FAD and CoA and results in the formation of succinyl-CoA, a thioester containing a high-energy bond.

Formation of Succinic acid: Succinyl - CoA is converted to succinic acid by the enzyme succinate thiokinase (succinyl-CoA synthetase). The reaction requires GDP (Guanosine di-phosphate) or IDP (inosine diphosphate) which undergoes phosphorylation in the presence of inorganic phosphate to produce GTP or ITP. GTP or ITP possesses a high energy phosphate bond, the energy is released. In the citric acid cycle, this is the only example where energy is generated at the substrate level. By means of nucleoside diphosphate kinase, ATP may be formed from either GTP or ITP.

$$\text{Succinyl-CoA} + Pi + GDP \longleftrightarrow \text{Succinic acid} + GTP + \text{CoA}$$

$$GTP + ADP \longleftrightarrow GDP + ATP$$

Dehydrogenation III: Succinic acid undergoes dehydrogenation which is catalysed by succinate dehydrogenate, which is bound to the inner surface of the inner mitochondrial membrane unlike the other enzymes of the cycle, which are found in the matrix. The hydrogen atoms are accepted directly by FAD.

$$\text{Succinic acid} + FAD \longleftrightarrow \text{Fumaric acid} + FADH_2$$

Hydration II: Fumaric acid reacts with water to form malic acid. The reaction is catalysed by fumarase (fumarate hydrolase).

$$\text{Fumaric acid} + H_2O \longleftrightarrow \text{Malic acid}$$

Dehydration IV: Malic acid is converted to oxaloacetic acid by malate dehydrogenase. The hydrogen atoms released in the reaction are accepted by NAD^+.

$$\text{Malic acid} + NAD^+ \longleftrightarrow \text{Oxaloacetic acid} + NADH + H^+$$

PENTOSE PHOSPHATE PATHWAY

Pentose Phosphate Pathway is also known as **alternative pathway** or **hexose monophosphate shunt** (HMP-shunt). This operates in liver, lactating mammary glands, adrenal cortex, adipose tissues, testis, thyroid, RBC and certain other tissues and is an alternative route for the oxidation of glucose.

HMP shunt performs two major functions:

1. The generation of NADPH for reductive syntheses such as fatty acids and steroid biosynthesis, and
2. The provision of ribose for nucleotide and nucleic acid biosynthesis.

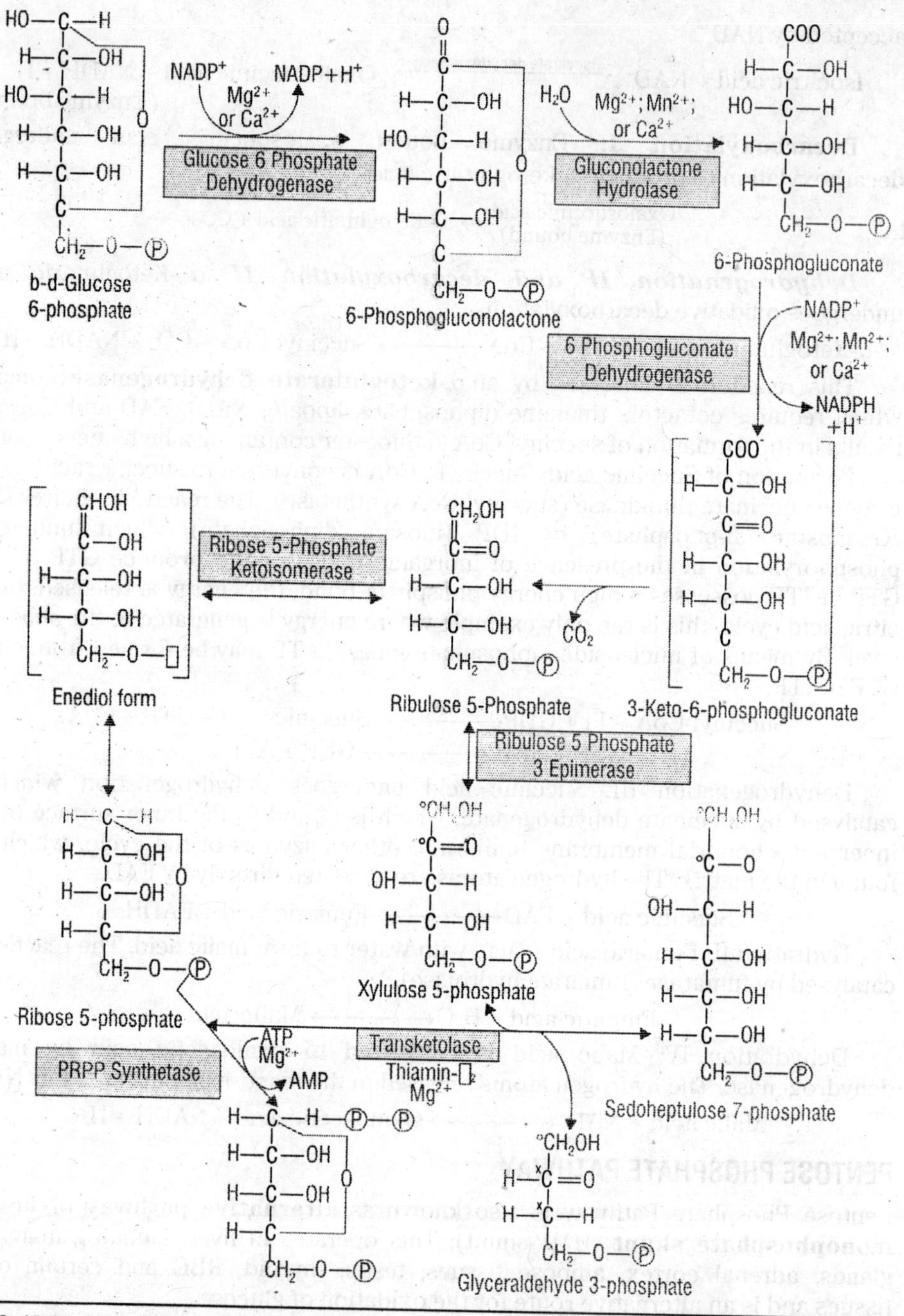

FIG. 13.13: The pentose phosphate pathway, ~P, -PO^{32-}, PRPP, 5-phosphoribosyl-1-pyrophosphate.

The HMP-shunt reactions occur in the cytosol. The sequences of this shunt may be grouped into two phase (Fig. 13.13).

1. The oxidative phase,
2. The non-oxidative phase.

The oxidative phase: During this phase NADPH are generated. The sequences of reactions are as follows:

Phosphorylation: Glucose is converted into glucose 6-phosphate in the presence of ATP and an enzyme hexokinase.

$$\text{Glucose + ATP} \xrightarrow{\text{Hexokinase}} \text{Glucose 6-phosphate + ADP}$$

Dehydrogenation: Glucose 6-phosphate is oxidised into 6-phosphogluconolactone in the presence of glucose 6-phosphate dehydrogenase. NADP accepts the evolved hydrogen atoms and Mg^{+2} act as cofactors.

$$\text{Glucose 6-phosphate} \xrightarrow[\text{Glucose 6-phosphate dehydrogenase}]{NADP^+ \quad Mg^{+2} \quad NADPH + H^+} \text{6-phosphogluconolactone}$$

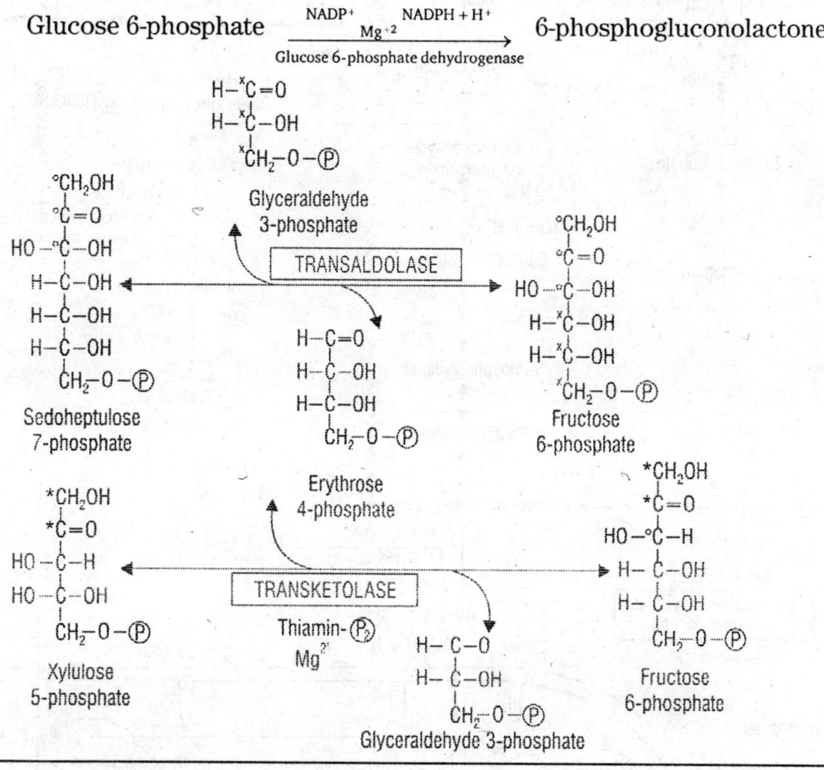

FIG. 13.14: The pentose phosphate pathway (contu.).

Hydrolysis: 6-phosphogluconolactone undergoes hydrolysis to produce 6-phosphogluconic acid. This reaction is catalysed by Gluconolactone hydrolase in the presence of water and Mg^{+2} and Mn^{+2} ions.

$$\text{6-phosphogluconolactone} \xrightarrow[\text{Gluconolactone hydrolase}]{H_2O \quad Mg^{+2}.Mn^{+2}} \text{6-phosphogluconic acid}$$

Dehydrogenation: 6-phosphogluconic acid is oxidized into 3-keto-6-phosphogluconic acid in the presence of an enzyme 6-phosphogluconate dehydrogenase and cofactors Mg^{+2} and Mn^{+2} ions. NADP acts as hydrogen acceptor.

$$\text{6-phosphogluconic acid} \xrightarrow[\text{6-phosphogluconate dehydrogenase}]{NADP^+ \quad Mg^{+2}, Mn^{+2} \quad NADPH + H^+} \text{3-keto-6-phospho-gluconic acid}$$

Decarboxylation: Dehydrogenation of step-4 is followed by decarboxylation to produce ketopentose, ribulose 5-phosphate. 3-keto-6-phosphogluconic acid of step-4 acts as in intermediate. Thus a hexose (glucose) is converted into a pentose (Ribulose 5-phosphate).

$$\text{3-keto-6-phosphogluconic acid} \longrightarrow \text{Ribulose 5-phosphate} + CO_2$$

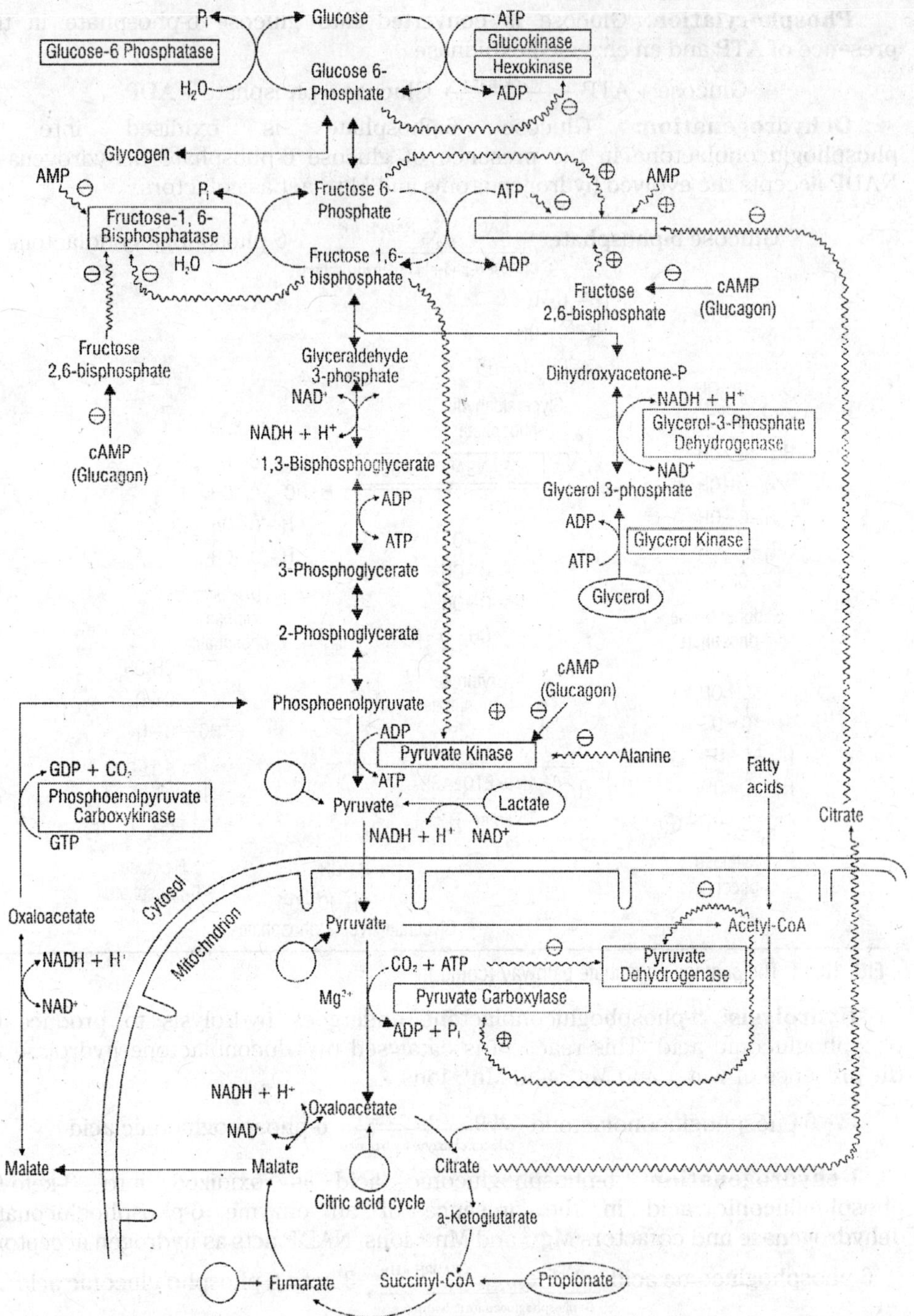

FIG. 13.15: Major pathways and regulation of gluconeogensis and glycolysis in the liver.

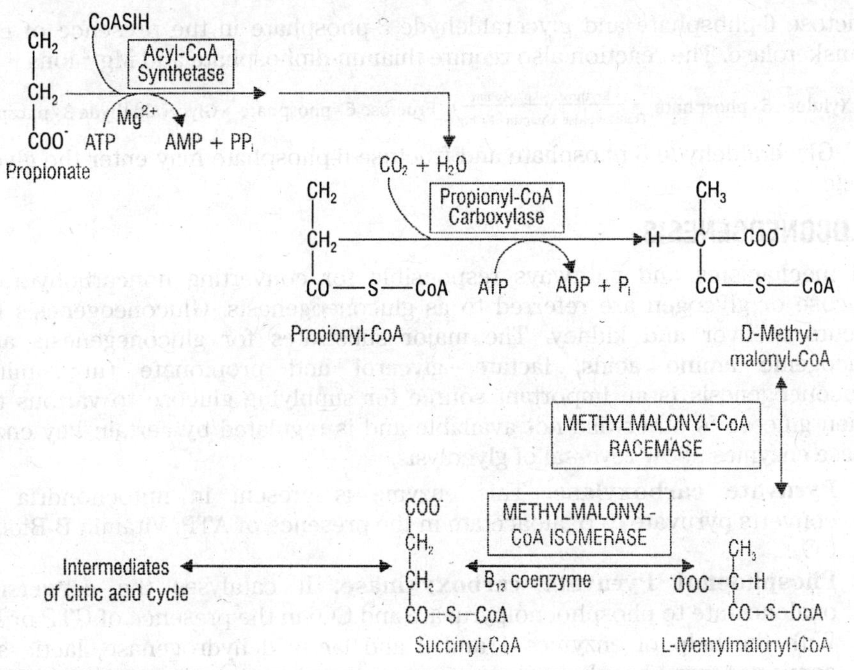

FIG. 13.16: Metabolism of propionate.

The non-oxidative phase: During this phase ribose precursors are generated. The various sequences of reactions are as follows:

Ribulose 5-phosphate generated during oxidative phase is acted upon by 2-different enzymes:

Ribose 5-phosphate ketoisomerase converts Ribulose 5-phosphate into an aldopentose, Ribose 5-phosphate, and

Ribulose 5-phosphate 3-epimerase converts ribulose 5-phosphate into xylulose 5-phosphate.

$$\text{Ribulose 5-phosphate} \xleftrightarrow{\text{Ribose 5-phosphate keto isomerase}} \text{Enediolform} \leftrightarrow \text{Ribose 5-phosphate}$$

$$\text{Ribulose 5-phosphate} \xleftrightarrow{\text{Ribulose 5-phosphate 3-epimerase}} \text{xylulose 5-phosphate}$$

Transketolase catalyses the transfer of the 2-carbon unit from xylulose 5-phosphate to ribose 5-phosphate, producing the 7-carbon ketose sedoheptulose 7-phosphate and the aldose glyceraldehyde 3-phosphate. The reaction requires a B-vitamin, thiamin, as the coenzyme thiamine diphosphate and Mg^{+2} ions. This step is referred to as transketolation.

$$\text{Ribose 5-phosphate} \xleftrightarrow[\text{Transketolase thiamin-}P_2\ Mg^{+2}]{\text{+ xylulose 5-phosphate}} \text{Sedoheptulose 5-phosphate + glyceraldehyde 3-phosphate}$$

Transaldolation: Transaldolase allows the transfer of a 3-carbon moiety, i.e. carbon 1-3 (active dihydroxyacetone) from the ketose sedoheptulose 7-phosphate to the aldose glyceraldehyde 3-phosphate to form the ketose fructose 6-phosphate, and the 4-carbon aldose erythrose 4-phosphate.

$$\text{Sedoheptubse 7-phosphate} \xleftrightarrow{\text{+ Glyceraldehyde 3-phosphate}} \text{Fructose 6-phosphate + Erythrose 4-phosphate}$$

Xylulose 5-phosphate may also react with erythrose 4-phosphate to produce fructose 6-phosphate and glyceraldehyde 3-phosphate in the presence of enzyme transketolase. The reaction also require thiamin-diphosphate and Mg^{+2} ions.

$$\text{Xylulose 5-phosphate} \xrightleftharpoons[\text{Transketalose Thiamin - P}_2\ \text{Mg}^{+2}]{+\ \text{Erythrose 4-phosphate}} \text{Fructose 6-phosphate + Glyceraldehyde 3-phosphate}$$

Glyceraldehyde 3-phosphate and fructose 6-phosphate may enter the glycolytic cycle.

GLUCONEOGENESIS

All mechanisms and pathways responsible for converting noncarbohydrates to glucose or glycogen are referred to as gluconeogenesis. Gluconeogenesis mainly occurs in liver and kidney. The major substrates for gluconegenesis are the glucogenic amino acids, lactate, glycerol and propionate (in ruminants). Gluconeogenesis is an important source for supplying glucose to various tissues when glucose is otherwise not available and is regulated by certain key enzymes. These enzymes allow reversal of glycolysis.

1. **Pyruvate carboxylase.** This enzyme is present in mitochondria which converts pyruvate to oxaloacetate in the presence of ATP, Vitamin B-Biotin and CO_2.
2. **Phosphoenol Pyruvate carboxykinase.** It catalyses the conversion of oxaloacetate to phosphoenolpyruvate and CO_2 in the presence of GTP or ITP.
3. With the help of enzymes 1 and 2 and lactic dehydrogenase, lactic acid is converted into phosphoenol pyruvate.
4. **Fructose-1,6-diphosphatase.** Fructose-1,6-diphosphatase converts fructose-1,6-diphosphate into fructose 6-phosphate, necessary to achieve a reversal of glycolysis.
5. **Glucose 6-phosphatase.** The conversion of glucose 6-phosphate to glucose is catalysed by glucose 6-phosphatase.
   a) Glucogenic amino acids (glutamic acid, glycine, serine, aspartic acid, arginine and ornithine), after transamination or deamination, form either pyruvate or members of the citric acid cycle which ultimately produce glucose or glycogen.
   b) Lactic acid (an oxidation product of glucose in skeletal muscles) undergoes dehydrogenation and produces pyruvate which enters the mitochondria before conversion to oxaloacelate and ultimate conversion to glucose in liver. Glucose again enters muscles for oxidation. This is known as **lactic acid cycle or cori cycle**.
   c) Propionate is a major source of glucose in ruminants. It enters the main gluconeogenic pathway via the citric acid cycle after conversion to succinyl-CoA.
6. **Glycerol kinase.** This enzyme catalyses conversion of glycerol (which is a metabolic product of adipose tissues) into glycerol 3-phosphate in the presence of ATP. Glycerol 3-phosphate is oxidized to dihydroxyacetone phosphate, which is a member of glycolytic cycle. This reaction is catalysed by **glycerol-3-phosphate dehydrogenase**. NAD^+ act as hydrogen acceptor.

TABLE 13.2: Generation of ATP during aerobic oxidation of one molecule of glucose to carbon dioxide and water

Pathway	Stage of Metabolism	Enzyme Catalysing the Stage of Reaction	Method of ATP Production	No. of ATP Formed
Glycolysis	1,3-diphosphoglyceric acid to 3-phosphoglyceric acid	Phosphoglycerate kinase	Oxidation at substrate level (1 X 2 = 2)	2
	Phosphoenol pyruvic acid to pyruvic acid	Pyruvate kinase	Oxidation at substrate level (1 X 2 = 2)	2
	PGAL to 1,3-diphosphoglyceric acid	Glyceraldehyde 3-phosphate dehydrogenase	Respiratory chain oxidation of 2 NADH (3 X 2 = 6)	6
			Total =	10
	Glucose to glucose 6-phosphate	Hexokinase	1-ATP consumed	
	Fructose 6-phosphate to fructose 1,6- diphosphate	Phosphofructokinase	1-ATP consumed	-2
			Net gain =	8
Oxidation of Pyruvic acid	Pyruvic acid to Acetyl CoA	Pyruvate dehydrogenase	Respiratory chain oxidation of 2NADH (3 X 2 = 6)	6
Citric acid cycle	Isocitric acid to oxalosuccinic acid	Isocitrate dehydrogenase	Respiratory chain oxidation of 2 NADH (3 X 2 = 6)	6
	α-Ketoglutaric acid to succinyl CoA	-αKetoglutarate dehydrogenase	Respiratory chain oxidation of 2 NADH (3 X 2 = 6)	6
	Succinyl Co-A to succinic acid	Succinate thiokinase	Oxidation at substrate level (1 X 2 = 2)	2
	Succinic acid to fumaric acid	Succinate dehydrogenase	Respiratory chain oxidation of 2FADH₂ (2 X 2 = 4)	4
	Malic acid to oxaloacetic acid	Malate dehydrogenase	Respiratory chain oxidation of 2 NADH (3 X 2 = 6)	6
			Total =	24

Comments

Total net ATP produced under aerobic conditions:

(a)	**Glycolysis**	**8**
(b)	Pyruvic acid oxidation	6
(c)	+ Citric acid cycle	24
Total		**38**

Total net ATP produced under Anaerobic conditions = 2

If NADH produced during glycolysis is transported into mitochondria via glycerophosphate shuttle, only two ATP are produced hence total production of ATP would be 36 only instead of 38. If Malate shuttle is used 38 ATP are produced. (refer to glycolysis).

CHAPTER 14

Metabolism: Protein

The proteins of the food are digested in stomach and intestine and are hydrolysed into amino acids. The protein digestive enzymes are endopeptidases (pepsin, trypsin, chymotrypsin and exopeptidases (carboxipeptidases, aminopeptidases, dipeptidases). The dietary proteins consist of 20-amino acids which exhibit L-configuration. These amino acids can be grouped into two:

1. **Indispensable (Nutritionally essential) amino acids:** Indispensable amino acids are those which can not be synthesised in the body but must be supplied through food. These amino acids are: arginine, histidine, isoloucine, leucine, methionine, phenylalanine, threonine, tryptophan, valine. (Arginine and histidine are synthesised at rates inadequate to support growth of children, hence also termed as semi essential amino acids).
2. **Dispensable (Nutritionally non-essential) amino acids:** Dispensable amino acids are those which can be synthesised at the desired rates in the body. These amino acids are: alanine, asparagine, aspartate, cysteine, glutamate, glutamine, glycine, proline, serine, tyrosine.

After digestion in the alimentary canal, the amino acids are absorbed through the intestinal epithelium into the blood. The sources of amino acids which constitute the "amino acid pool" are:

1. Dietary proteins,
2. Intra-cellular synthesis and
3. Tissue protein breakdown.

The amino acids from the pool undergo various metabolic pathways which may be as follows:

1. Catabolic pathways:
   a) Deamination.
   b) Transamination.
   c) Decarboxylation.
2. Anabolic pathway.
   a) Protein synthesis.

DEAMINATION

During deamination process an amino acid is converted into an a-Keto acid and ammonia. This may occur in the following ways:

Amino acid oxidases: The amino acid oxidases are flavoproteins and oxidize amino acids to the corresponding keto acids after generating a molecule of ammonia. Oxygen takes part in this oxidation process.

$$\text{Amino-acid} + 1/2\ O_2 \longrightarrow \alpha\text{-Keto acid} + NH_3$$

Amino acid dehydrogenases: Amino acid dehydrogenases are NAD or NADP linked enzymes. This enzyme catalyses the removal of hydrogen atoms from an a-amino acid. The hydrogen atoms are accepted by NAD or NADP. The amino acid is hydrolysed to produce α-keto acid and ammonia.

$$\alpha\text{-Amino acid} \xrightarrow[\text{Amino acid dehydrogenase}]{NAD^+} \alpha\text{-Imino acid} + NADH + H^+$$

$$\alpha\text{-imino acid} + H_2O \longrightarrow \alpha\text{-Keto acid} + NH_3$$

Non-oxidative deamination: Serine and threonine dehydratases catalyse a non-oxidative deamination reaction resulting from a primary dehydration of the substrate and yield pyruvate or α-ketobutyrate.

$$\text{Serine} \xrightarrow[H_2O]{} \alpha\text{ - imino acid} \xrightarrow{H_2O} \text{Pyruvate} + NH_3$$

$$\text{Theronine} \rightarrow \alpha\text{ - imino acid} \xrightarrow{H_2O} \text{a - ketobutyrate} + NH_3$$

Non-oxidative deamination reactions also occur in microorganisms. Some of the reactions are as follows:

$$\text{Aspartate} \xrightarrow{\text{Aspartase}} \text{Fumarate} + NH_3$$

$$\text{Histidine} \xrightarrow{\text{Histidinase}} \text{Uroeonic acid} + NH_3$$

Glucogenic and Ketogenic Amino Acids

Glucogenic amino acids are those amino acids which form pyruvic acid, oxaloacetic acid, α-ketoglutaric acid and succinyl-CoA on deamination. These citric acid cycle intermediates and pyruvic acid can produce phosphoenol pyruvic acid and then glucose via the gluconeogenic pathway.

Ketogenic amino acids yield acetyl-CoA or acetoacetic acid on deamination, which leads to **ketone body** formation. Certain amino acids can give rise to both glucose and ketone bodies. Various groups of these amino acids are shown in Table 14.1.

TABLE 14.1: Glucogenic and ketogenic amino acids

Glucogenic		Ketogenic	Glucogenic and Ketogenic
Alanine	Hydroxyproline	Leucine	Isoleucine
Arginine	Methionine		Lysine
Asparagine	Proline		Phenylalanine
Aspartic acid	Serine		Tyrosine
Cystine	Threonine		
Glutamic acid	Tryptophan		
Glycine			
Histidine	Valine		

Fate of Ammonia: Ammonia produced by deamination is highly toxic. It is converted into urea in liver. The formation of urea is known as **urea cycle** or **ornithine cycle** or Krebs-Henseleit cycle. The various steps of this cycle are as follows:

- **Step 1: Synthesis of Carbamyl phosphate:** One molecule of NH_3 reacts with CO_2 to produce carbamyl phosphate. ATP donates phosphate group. **Carbamyl phosphate synthetase** catalyses the reaction in the presence of Mg^{+2} and N-acetyl glutamic acid.

$$CO_2 + NH_3 + 2ATP \xrightarrow[Mg^{+2}]{\text{Carbamyl phosphate Synthetase}} \text{Carbamyl phosphate} + 2ADP + Pi$$

- **Step 2: Synthesis of Citrulline:** Carbamyl phosphate reacts with ornithine in the presence of ornithine carbamyl transferase to produce citrulline.

$$\text{Ornithine + Carbamyl phosphate} \xrightarrow{\text{Ornithine carbamyl transferase}} \text{Citrulline + PI}$$

- **Step 3: Synthesis of arginosuccinic acid:** Citrulline reacts with aspartic acid (which injects one more NH_3) to produce arginosuccinic acid. The reaction is catalysed by **arginosuccinate synthetase** in the presence of Mg^{+2}. ATP provides energy.

  $$\text{Citrulline + Aspartic acid + ATP} \xrightarrow{\text{Arginesuccinate synthetase}} \text{Arginosuccinic acid + AMP + PPi}$$

- **Step 4: Cleavage of arginosuccinic acid: Arginosuccinase** catalyses the cleavage of arginosuccinic acid into arginine and fumaric acid. The fumaric acid enters the Krebs cycle.

  $$\text{Arginosuccinic acid} \xrightarrow{\text{Arginosuccinase}} \text{Arginine + Fumaric acid}$$

- **Step 5: Hydrolysis of Arginine:** Arginine undergoes hydrolysis releasing urea and regenerating ornithine. This reaction is catalysed by the enzyme **arginase**.

  $$\text{Arginine} + H_2O \xrightarrow{\text{Arginase}} \text{Urea + arnithine}$$

Urea is the main excretory products of ureotelic animals. It is less toxic than that of NH_3.

TRANSAMINATION

In transamination reactions α-amino group is transferred to α-carbon atom of an α-keto acid. Transamination is catalysed by transaminases or aminotransferases. Pyridoxal phosphate acts as a cofactor.

$$\text{Amino acid 1 + Keto acid-1} \xleftrightarrow{\text{Transaminases}} \text{Keto acid 2 + Amino acid - 2}$$

The transaminases are specific in their reactions. These are widely distributed in the tissues, however, their concentration in heart and liver is relatively high. Some specific examples of transaminases are as follows:

1. **GOT (Glutamic- oxaloacetate transaminase):** GOT activity is very high in heart tissue. It catalyses the reaction between glutamic acid and oxaloacetic acid to produce aspartic acid and α-ketoglutaric acid.

   $$\text{Glutamic acid + Oxaloacetic acid} \xleftrightarrow{\text{GOT}} \text{Aspartic acid} + \alpha\text{-ketoglutaric acid}$$

   In a state of cardiac infarction, GOT is released in blood and its level in serum (SGOT) is abnormally raised.

2. **GPT (Glutamic-pyruvate transaminase):** GPT concentration is very high in liver. Glutamic acid and pyruvic acid react in the presence of GPT to produce α-ketoglutaric acid and alanine.

   $$\text{Glutamic acid + Pyruvic acid} \xleftrightarrow{\text{GPT}} \alpha\text{-ketoglutaric acid + Alanine}$$

   If liver cells are damaged, SGPT activity is abnormally raised.

DECARBOXYLATION

Amino acids undergo decarboxylation to produces CO_2 and an amine in the presence of **amino acid decarboxylases**. These enzymes require pyridoxal phosphate for their activity. The amines produced are called biogenic amines and have important pharmacological effects.

$$\text{Amino acids} \xrightarrow[\text{Pyridoxal phosphate}]{\text{Decarboxylase}} \text{Amine} + CO_2$$

Some metabolically significant amines formed as a result of decarboxylation are as follows:

1. **γ -Aminobutyric Acid (GABA):** GABA is formed in brain by the decarboxylation of glutamic acid. It stimulates neural activity and inhibits synaptic transmission in the central nervous system.

$$\text{Glutamic acid} \xrightarrow{\text{Glutamic acid decarboxylase}} \gamma\text{-Aminobutyric acid} + CO_2$$

2. **Tyramine:** Tyramine is obtained as a result of decarboxylation of tyrosine. It increases blood pressure.

$$\text{Tyrosine} \xrightarrow{\text{Tyrosine decarbosylase}} \text{Tyramine} + CO_2$$

3. **Histamine:** L-histidine is decarboxylated to produce histamine.

$$\text{Histidine} \xrightarrow{\text{Histidine decarboxylase}} \text{Histamine} + CO_2$$

It's concentration is high in animal tissues such as liver, kidney, duodenum, large intestine and lungs. Histamine is a powerful vasodilator.

4. **Serotonin (5-hydroxytryptamine):** Serotonin is produced by the decarboxylation of 5-hydroxytryptophan.

$$\text{5-hydroxytry ptophan} \xrightarrow{\text{5-hydroxytry ptophan decarboxyl ase}} \text{5-hydroxytry ptamine} + CO_2$$

(Serotonin)

Serotonin acts as a potent vasoconstrictor substance. It is mainly found in brain, blood platelets and intestinal tissues.

5. **Dopamine (3,4-dihydroxyphenyl ethylamine):** Dopamine is produced by the decarboxylation of 3,4-dihydroxyphenylalanine (Dopa).

$$\text{Dopa} \xrightarrow{\text{Dopa decarboxylase}} \text{Dopamine} + CO_2$$

Dopamine is present in kidney and adrenal tissues as well as sympathetic ganglia and nerves. It acts as an inhibitory neurotransmitter.

PROTEIN SYNTHESIS

The protein synthesis involves two major steps- (1) Transcription, (2) Translation.

Transcription

Transcription refers to the transfer of genetic information from DNA to mRNA. This process takes place in nucleus and involves an enzyme, **RNA polymerase**. In prokaryotes (bacteria), a single type of RNA polymerase occurs which is responsible for the synthesis of all kinds of RNAs (mRNA, tRNA, rRNA). In eukaryotes atleast three types of RNA polymerases occur. The prokaryotic RNA polymerase consists of two parts:

1. **Core enzyme:** Consisting two alpha (α), one beta (β) and one beta dash (β') polypeptide chains.
2. **Sigma (d) polypeptide chain**. It attaches with core enzyme during transcription.

TABLE 14.2: Components of RNA polymerase

Polypeptide Chains			**Molecular weight**
RNA polymerase (Holoenzyme)	Core enzyme	α_2	40,000
		β	155,000
		β'	165,000
	Sigma (δ) factor	δ	95,000

The transcription process takes place as follows:

1. RNA polymerase binds at the **promoter site** on the DNA molecules forming closed promoter complex. It then diffuses to a binding site consisting of a sequence of seven bases (TATUATG). mRNA synthesis starts at the initiation site.
2. The unwinding of DNA helix takes place to form open (binary) promoter complex. A short segment of DNA molecule opens up and mRNA synthesis begins on one of the two strands of DNA.

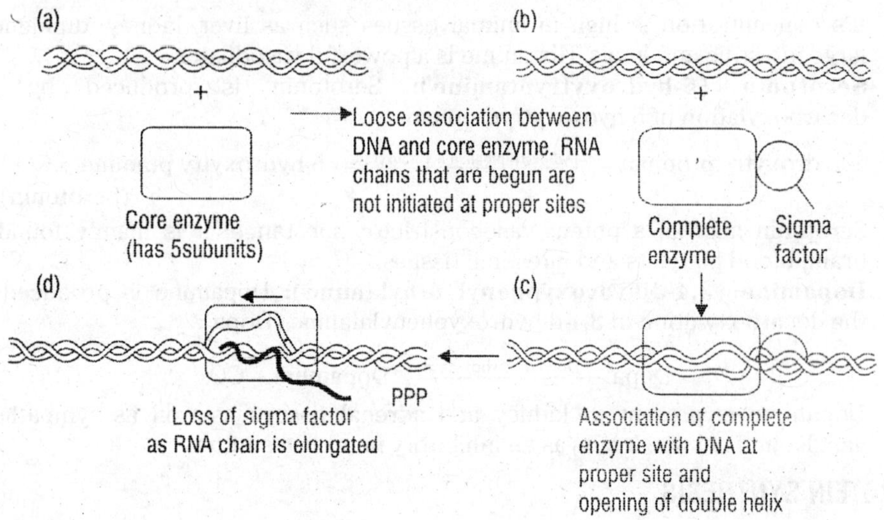

FIG. 14.1: The enzymatic synthesis of RNA in *E. coli* (a) The core enzyme in the absence of the sigma factor cannot interact with the DNA at specific initiation sites. (*b-d*) When the core enzyme is associated with the sigma factor, the enzyme initiates RNA syn thesis at the proper site and then dissociates from the core enzyme, which is capable of elongation.

3. The first base which is incorporated is always a purine molecule, either pppA or pppG.
4. Activated ribonucleoside triphosphates (ATP, UTP, GTP and CTP) are added to the DNA template. One pyrophosphate (PPi) is produced per nucleotide added to the growing chain. Mg^{++} are required for transcription.
5. Sigma factor dissociates from the core enzyme after the commencement of the chain elongation. Now the core enzyme moves along the DNA template and elongation of the chain takes place.
6. The chain growth takes place in the 5' ⟶ 3' direction.
7. The region of DNA that has been transcribed regains its' helical configuration and the next region of DNA unwinds.
8. The termination of transcription can be brought about by the presence of certain termination signals on DNA molecule or the involvement of **rho (p)**

factor. The rho factor binds to RNA polymerase which then releases RNA chain.

Translation

The process of translation involves translation of the language of mRNA into the sequence of amino-acids of the proteins. The mRNA possesses 64 types of triplet codons (Fig. 14.2). Translation occurs in following steps:

Activation of amino acids: Activation of amino acids is brought about by the binding of an amino acid with ATP in the presence of activating enzymes known as **amino acyl RNA synthetases.** This reaction results in the formation of amino acyl AMP-enzyme complex.

$$\text{Amino acid } (AA_1) + ATP + EnZ_1 \longrightarrow (AA_1\text{-}AMP)\ EnZ_1 + PP$$

The amino-acyl RNA synthetases are amino acid specific, *i.e.* for different amino acids, different enzymes are required. These enzymes can recognize their respective tRNAs also.

The transfer of amino acid to tRNA: The complex (amino-acyl AMP-enzyme complex) formed in the preceding step reacts with a particular tRNA and transfers the amino acid to the tRNA to form amino acyl tRNA complex.

$$(AA_1\text{-}AMP)\ EnZ_1 + tRNA_1 \longrightarrow AA_1\text{-}tRNA_1 + AMP + EnZ_1$$

The AMP and enzyme are liberated.

The tRNAs are also amino acid specific, it means atleast 20 types of tRNAs are required. It is now well known that one type of amino acid can be carried by more than one type of tRNA, but one type of tRNA can not carry more than one type of amino acid.

A high-energy ester bond is formed between the carboxyl group of the amino acid and the 3'- hydroxy group of the terminal adenosine of tRNA.

Initiation of Polypeptide Synthesis: For the initiation of polypeptide synthesis certain initiation factors (IF) are required. In prokaryotes these factors are of three types: IF_1, IF_2 and IF_3, and are loosely associated with the 30S subunits. IF_3 is required for the binding of 30S subunit of ribosomes to the mRNA starting site, it also functions as a ribosome dissociation factor. IF_2 binds the initiator fMet-tRNA and carries it to the ribosome, together with GTP in response to the first AUG codon. IF_1 participates in the interaction between IF_2 and the initiator tRNA. During protein synthesis, once the large ribosomal subunit binds to the small subunit, the initiation factors are released and can be reutilized.

In eukaryotes at least 10 soluble protein factors have been purified.

TABLE 14.3: Eukaryotic Initiation factors (Jagus, R., Anderson, W. F and Safer, B. Prog. Nucleic acid Res., 25, 134, 1981)

Factors	Molecular Wt.	Functions
eIF_1	15,000	Stabilization of 48S preinitiation complex
eIF_2	125,000	Activated by GTP; necessary for Met-$tRNA_f$ binding to $43S_N$ subunit
eIF_2A	65,000	Not known
eIF_3	700,000	Formation of $43S_N$ binding to 43S preinitiation complex.
eIF_4A	50,000	mRNA binding
eIF_4B	80,000	mRNA binding by cap recognition.
eIF_4C	17,000	Formation of $43S_N$, stimulates subunit joining.
eIF_4D	15,000	Not known
eIF_4E	24,000	Cap binding protein.
eIF_5	125,000	Initiation factor release, 60S subunit joining

Note: (1) eIF_4E is a contaminant of eIF_3 and eIF_4B and may be a new factor.
(2) The 43 S_N complex contains 40S ribosomal subunit and two factors, eIF_4C and eIF_3.

First Letter	Second Letter: U		C		A		G		Third Letter
U	UUU	Phe	UCU	Ser	UAU	Tyr	UGU	Cys	U
	UUC		UCC		UAC		UGC		C
	UUA	Leu	UCA		UAA	Stop	UGA	Stop	A
	UUG		UCG		UAG	Stop	UGG	Trp	G
C	CUU	Leu	CCU	Pro	CAU	His	CGU	Arg	U
	CUC		CCC		CAC		CGC		C
	CUA		CCA		CAA	Gln	CGA		A
	CUG		CCG		CAG		CGG		G
A	AUU	Ile	ACU	Thr	AAU	Asn	AGU	Ser	U
	AUC		ACC		AAC		AGC		C
	AUA		ACA		AAA	Lys	AGA	Arg	A
	AUG	Met	AGC		AAG		AGG		G
G	GUU	Val	GCU	Ala	GAU	Asp	GGU	Gly	U
	GUC		GCC		GAC		GGC		C
	GUA		GCA		GAA	Glu	GGA		A
	GUG		GCG		GAG		GGG		G

FIG. 14.2: Genetic code directory: *Chain initiation codons; **Chain termination codons.

Initiation process in prokaryotes takes place as follows (Fig. 14.3):

1. In *E. coli*, there are two tRNAs for methionine, $tRNA_m^{met}$ (non-formylatable) and $tRNA_f^{met}$ (formylatable). The later one deposits methionine at first place, whereas the former deposits methionine at the intercalary positions only.
2. Methionine binds with $tRNA_f^{met}$ to form met- $tRNA_f^{met}$ complex.
3. met-$tRNA_f^{met}$ complex undergoes formylation in the presence of transformylase enzyme and a formate source.

$$\text{met} \quad tRNA_f\text{met} \quad \xrightarrow[\text{formyl tetrahydrofolic acid}]{\text{formate}} \quad \text{transformylase} \quad \text{f met} \quad tRNA_f\text{met}$$

4. IF_3 factor binds to 30S subunit.

$$30S \quad IF_3 \qquad IF_3 \quad 30S$$

5. IF_3 - 30S complex binds to mRNA.

$$IF_3 \quad 30S \quad mRNA \qquad IF_3 \quad 30S \quad mRNA$$

The 30S subunit binds with the first AUG codon (**initiation codon**) of the message on mRNA. The recognition of this particular AUG message results from the interaction between complementary nucleotides of the 3' end of the 16S ribosomal RNA of the 30S subunit with a region of the message just before the initiation codon. The mRNA is read in a 5' to 3' direction.

It has been indicated that just to the 5' side of the initiation codon (AUG) is a stretch of nucleotides very rich in purines, both A and G. For example, the phages that infect *E. coli* have the sequence 5'-AGGAGGU-3' in a similar position just to the 5' side off the initiation codon. The 3'-terminus of the 16S rRNA of *E. coli* has the complementary sequence of 3'-UCCUCCA-5' near its terminus. The binding of the

30S subunit to the mRNA requires the factor IF_3, and the interaction seems to be stabilized by IF_1.

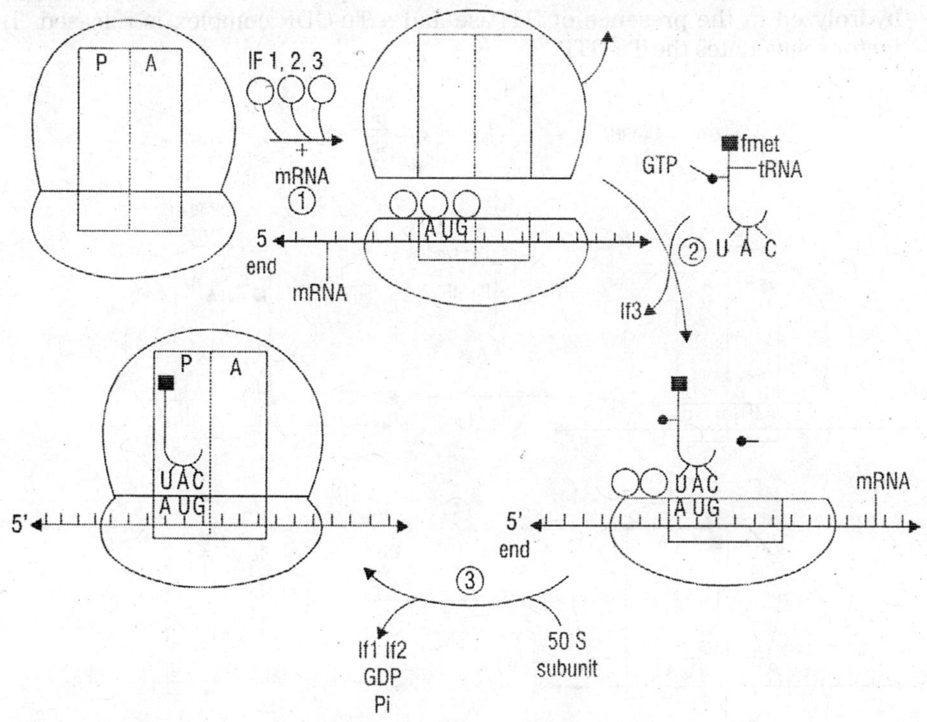

FIG. 14.3: Proposed steps in the initiation of protein synthesis as described in the text.

1. The ribosomes have two sites for association with tRNA:
   a) Amino-acyl site (A),
   b) Peptidyl site (P).

   The `A' site is generally the one in which aminoacyl- tRNAs enter the ribosome- mRNA complex, while `P' site is the one from which the tRNA leaves after having donated its amino acid. In the case of first amino-acyl-tRNA, that of f-met-$tRNA_f^{met}$., the entry appears to be directly into the `P' site of the 30S subunit. Recognition and binding of the f-met $tRNA_f^{met}$ molecule to the mRNA- 30S complex is mediated by the factor IF_2.
2. GTP is also required for the successful initiation. GTP binds to the f-met-tRNAfmet before it attaches to the mRNA-30S complex. Once the initiator tRNA becomes attached, IF_3 is released and the 50S ribosomal subunit attaches to the 30S subunit to form the complete initiation complex (70S-mRNA-f-met-tRNAfmet). At this point GTP is hydrolyzed and the GDP, IF_1 and IF_2 are released.

Elongation of Polypeptide

After the formation of 70S-mRNA-f-met-$tRNA_f^{met}$ complex, elongation of polypeptide chain is brought about by the regular addition of amino acids.

1. The elongation process requires the presence of factors **Tu**, **Ts**, and **G**.
2. The first step in elongation involves the entry of the second aminoacyl-tRNA to the `A' site of the ribosome. This and the subsequent amino acyl- tRNA combine with Tu factor and GTP before binding to the A site to form **Tu-GTP-AA-**

tRNA complex. Tu factor probably helps in codon recognition and for orientation of the incoming aminoacyl- tRNA.

3. After the attachment of Tu-GTP-AA-tRNA complex to the ribosome, GTP is hydrolyzed in the presence of GTPase and a Tu-GDP complex is released. Ts factor rejuvenates the Tu-GTP.

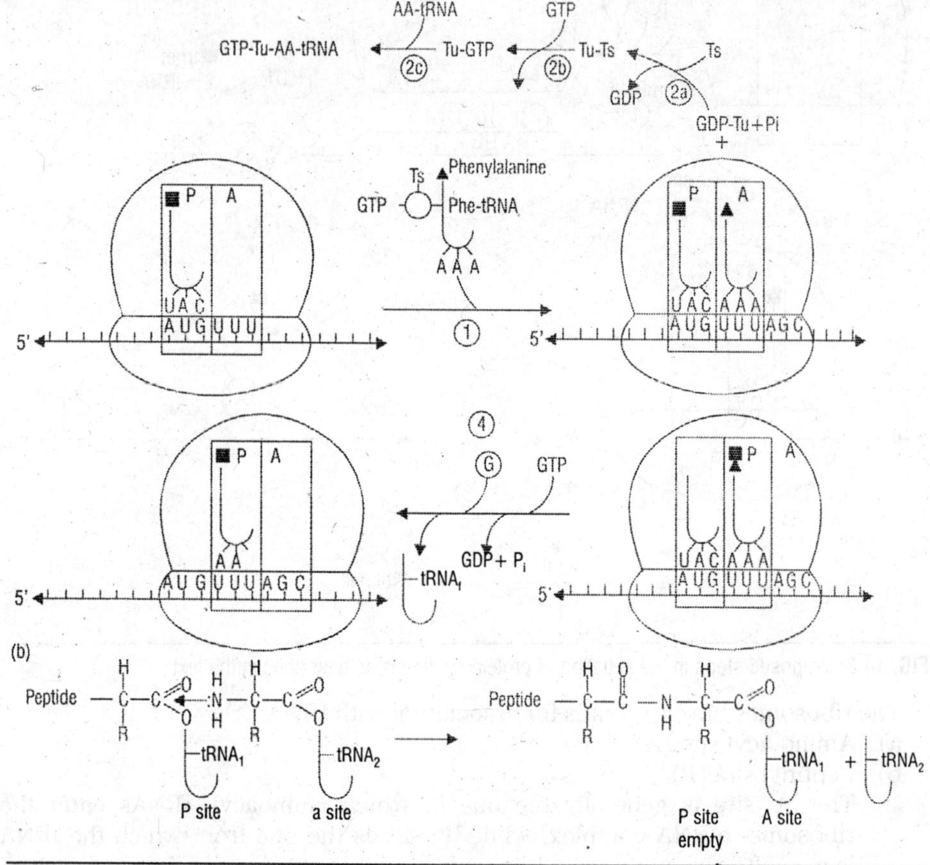

FIG. 14.4: (A) Steps in the elongation of the polypeptide chain as described in the text. (B) Peptide bond formation and the subsequent displacement of the deacylated tRNA of the P site.

4. A peptide bond is formed between first and the second amino acids. This bond is formed between the free carboxyl group (-COOH) of the peptidyl tRNA at the `P' site and the free amino group (-$NH_2$) present with amino acyl tRNA at the `A' site. The enzyme involved in this reaction is **peptidyl transferase** and is an integral part of 50S ribosomal unit. After the formation of peptide bond, the tRNA at `P' site is deacylated and the tRNA at `A' site now carries the dipeptide/polypeptide. The tRNA from `P' site is dislodged.

5. The ribosome moves one codon along the mRNA in the 3' direction, it is termed as **translocation**, it is accompanied by the movement of the tRNA- dipeptide to the `P' site of the ribosome, still hydrogen-bonded to the second codon of the mRNA. The translocation requires G factor (also termed as **translocase**) and hydrolysis of GTP.

6. When the tRNA with its attached dipeptide moves to the `P' site, the `A' site is open to the entry of another aminoacyl-tRNA, in this case one whose anticodon

is complementary to the third codon. A peptide bond is formed between second and third amino acid and the cycle is repeated as discussed in preceding paras.

7. As the ribosome continues to move along the mRNA, the growing polypeptide chain, termed the **nascent polypeptide**, increases in length. The direction of elongation is from amino to carboxylic group.

Termination of Polypeptide

The termination of the polypeptide chain is brought about by the presence of any of the three termination codons, namely UAA, UAG, UGA. There are not tRNAs with anticodons complementary to these triplets. When the ribosome reaches one of these termination codons, the signal is received to stop further elongation and terminate the polypeptide.

The termination involves release factors (RF-1 RF-2 and RF-3) and GTP. The RF-1 and RF-2 help in the recognition of the termination codons, whereas, RF-3 (or S) is involved in the release of the completed polypeptide from its attachment at its C-terminal end to the last tRNA that had entered. After this process, the ribosome separates from mRNA and dissociates once again into its subunits in the presence of IF-3.

It has been revealed that more than one ribosome is involved at one time in the translation of a single message. During translation a number of ribosomes attach along the length of mRNA. This complex of ribosomes and mRNA is termed a **polyribosome**, or **polysome**. Each one of the ribosomes initially attaches to the mRNA at the initiation codon and then moves from that point toward the 3' end. As soon as a recently attached ribosome moves a sufficient distance along mRNA, the next ribosome becomes associated with the mRNA and begins to translate the message simultaneously with its translation by previously attached ribosomes. The simultaneous translation of the same mRNA by several ribosomes greatly increases the potential for protein synthesis.

Modification of Released Polypeptide

The released polypeptide is deformylated in the presence of an enzyme **deformylase**, which removes the formyl group of first amino acid methionine. Due to action of some exopeptidases, amino acids are removed from either the N-terminal end or the C-terminal end or both. The polypeptide chains singly or in association with other chains become folded and form tertiary structures.

CHAPTER 15

Metabolism Lipids

Bloor included fats and fat-like substances in lipids. The term lipid includes fats, oils, waxes, phospholipids, cerebrosides, steroids, carotenes, vitamins A, D, E, K. Fats are hydrolyzed (digested) in the alimentary canal in the presence of lipase enzyme; the main products are glycerol and fatty acids. The digestive products are absorbed (mainly) into the lymph through intestinal epithelium. In the body fats are stored mainly in adipose tissues like, (i) Subcutaneous connective tissue, (ii) abdominal cavity, (iii) inter-muscular connective tissue.

The metabolic pathways of lipids can be discussed as follows:

1. **Catabolic pathways:** (a) Oxidation of glycerol, (b) Oxidation of fatty acids.
2. **Anabolic pathways:** (a) Synthesis of fatty acids, (b) Synthesis of triglycerides.

OXIDATION OF GLYCEROL

Glycerol is derived from the dietary fats and oils and by hydrolysis of circulatory or tissue triglycerides and phospholipids. The metabolism of glycerol takes place as follows:

1. **Phosphorylation:** Glycerol undergoes phosphorylation to produce glycerol 3-phosphate. ATP donates one phosphate group to glycerol. The reaction is catalysed by **glycerol kinase**. This enzyme has been found in significant amounts in liver, kidney, intestine, brown adipose tissue, and lactating mammary gland.

   Glycerol ATP $\xrightarrow{\text{Glycerol kinase}}$ Glycerol 3 phosphate ADP

2. **Dehydrogenation:** Glycerol 3-phosphate becomes dehydrogenated to produce dihydroxyacetone phosphate in the presence of glycerol **3-phosphate dehydrogenase**. NAD$^+$ accepts the hydrogen atoms released.

Glycerol 3 - phosphate NAD $\xrightarrow[\text{dehydrogenase}]{\text{Glycerol 3-phosphate}}$ Dihydroxyacetone phosphate NADH H

Dihydroxyacetone phosphate may later on follow the following paths:

1. It may enter the forward reactions of glycolysis and finally oxidized to CO_2 and water by Krebs cycle reactions, or
2. It may be isomerised into glyceraldehyde 3-phosphate in the presence of **triosephosphate isomerase (TPI)**.

   Dihydroxyacetone phosphate $\xrightarrow{\text{TPI}}$ Glyceraldehyde 3-phosphate

3. Dihydroxyacetone phosphate and glyceraldehyde 3-phosphate may condense in the presence of Aldolase to produce fructose 1,6-diphosphate, which ultimately forms glycogen.

Glycerol 3-phosphate may also form triglycerides and phospholipids in different tissues.

OXIDATION OF FATTY ACIDS

Fatty acid oxidation takes place in mitochondria (except activation step of long chain fatty acids which takes place outside mitochondria). In fatty acid oxidation, 2 carbons are cleaved at a time, starting at the carboxyl end. The chain is broken between the α-(2) and ß-(3) carbon atoms, hence the name **ß-oxidation**. The various steps of ß-oxidation are as follows (Fig. 15.1).

1. **Activation**: Fatty acid, in the presence of ATP, coenzyme A and enzyme **acyl-CoA synthetase** is converted into an active fatty acid or acyl-CoA. ATP is converted into AMP, thus in the activation process two high energy phosphate bonds, are utilized. Acyl CoA synthetases are found in the endoplasmic reticulum and inside and on the outer membrane of mitochondria.
   Long chain fatty acids can penetrate the inner mitochondrial membrane only in combination with **carnitine** (ß-hydroxy-Y-trimethylammonium butyrate) (Fig. 15.2). Carnitine is synthesised from lysine and methionine in liver and kidney.
   Active fatty acid (Acyl-CoA) reacts with carnitine in the presence of **Carnitine palmitoyl-transferase I**, present on the inner side of the outer mitochondrial membrane, to produce **Acyl carnitine** and CoA. Activation of lower fatty acids and their oxidation may occur within the mitochondria. Acyl carnitine penetrates the mitochondria. **Carnitine-acyl carnitine translocase** acts as a membrane carnitine exchange transporter. Acyl carnitine is transported in, coupled with the transport out of one molecule of carmitine. Acyl carnitine then reacts with CoA, catalysed by **carnitine palmitoy transferase II**, to produce Acyl-CoA again in the matrix and carnitine is liberated.
2. **Dehydrogenation**: Acyl-CoA undergoes dehydrogenation in the presence of enzyme **Acyl-CoA dehydrogenase** to produce Δ^2-trans-enoyl-CoA. The two hydrogen atoms are removed from 2(a) and 3 (ß) carbon atoms creating a double bond. A flavoprotein containing FAD as a prosthetic group acts as the coenzyme for the dehydrogenase and accepts hydrogen atoms.
3. **Hydration:** A molecule of water is added to Δ^2-trans-enoyl-CoA, producing an alcoholic group at β-carbon thus, saturating the double bond. The product is 3-hydroxy-acyl-CoA and the reaction is catalysed by Δ^2-Enoyl-**CoA hydratase**.
4. **Dehydrogenation:** The 3-Hydroxyacyl-CoA undergoes dehydrogenation in the presence of **3-hydroacyl-CoA dehydrogenase** to produce 3-Ketoacyl-CoA. Hydrogen atoms are removed from β-carbon to produce keto-group and are accepted by NAD^+.
5. **Cleavage:** The 3-ketoacyl-CoA combines with a molecule of CoA and undergoes thiolytic cleavage at the 2,3 position to produce a molecule of acetyl-CoA and an acyl-CoA derivative containing 2-carbons less than the original acyl-CoA molecule that underwent oxidation. The reaction is catalysed by thiolase or 3-ketoacylthiolase or acetyl-CoA acyltransferase.

The acyl-CoA produced, renters the oxidative pathway at reaction **step-2**, the process will be repeated till complete degradation occurs. The acetyl-CoA enters the Krebs cycle and oxidized to CO_2 and H_2O.

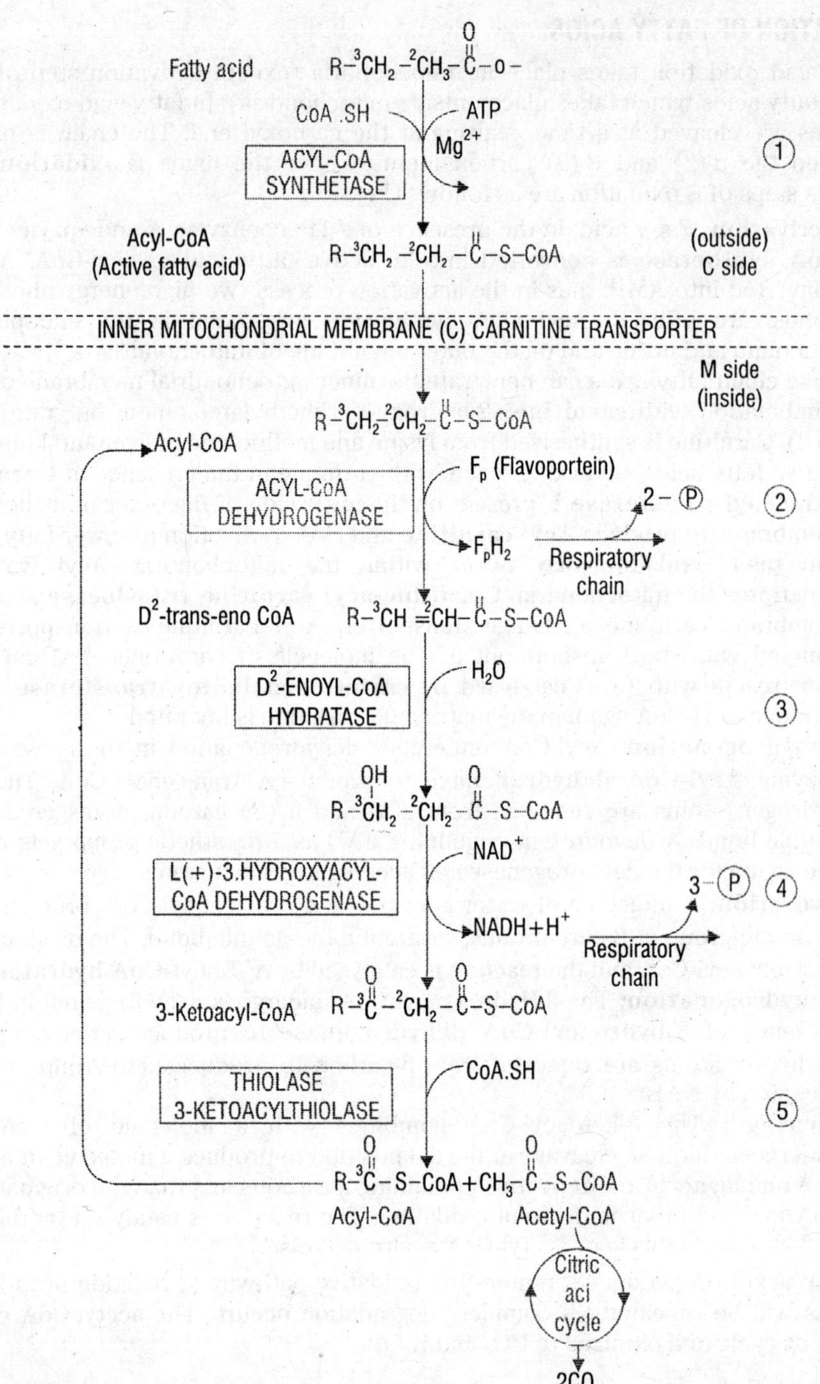

FIG. 15.1: ß-Oxidation of fatty acids. Long-chain acyl-CoA is cycled through reactions 2-5, acetyl-CoA being split off each cycle by thiolase (reaction 5). When the acyl radical is only 4 carbon atoms in length, 2 acetyl-CoA molecules are formed in reaction 5.

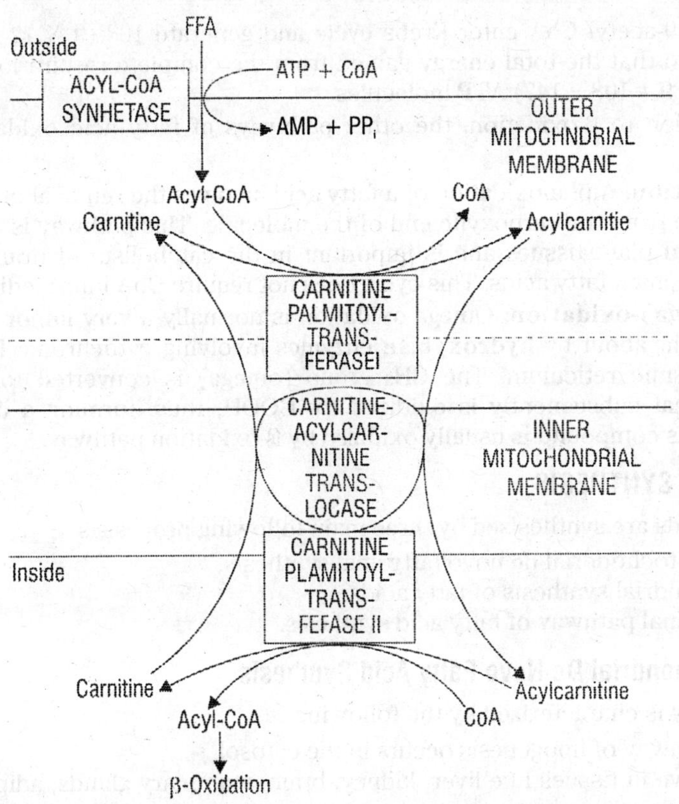

FIG. 15.2: Role of carnitine in the transport of long-chain fatty acids through the inner mitochondrial membrane. Long-chain acyl-CoA cannot pass through the inner mitochondrial membrane, but its metabolic product, acylcarnitine, can.

Fatty acids with an odd number of carbon atoms are oxidized by the pathway of ß-oxidation producing acetyl-CoA until a 3-carbon (propionyl-CoA) residue remains. The propionyl-CoA is converted to succinyl-CoA, a constituent of the citric acid cycle.

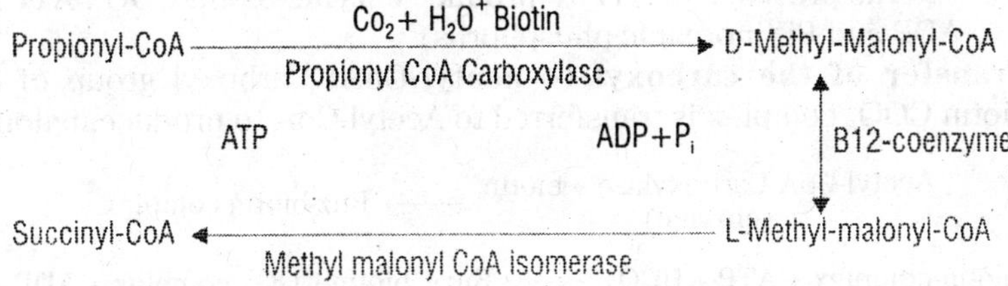

Energetics of Fatty Oxidation

In β-oxidation at steps 2 and 4 dehydrogenation occurs, the hydrogen acceptors being FAD and NAD^+ respectively. Thus, producing 5 ATP molecules per Acetyl-CoA molecule formed. If say, fatty acid is stearic acid (C_{18} molecule), then β-oxidation will occur successively eight times, yielding acetyl-CoA each time and leaving a ninth acetyl-CoA as a remainder. At 5 ATP per β-oxidation cycle, the yield is therefore 5 × 8 = 40 ATP, minus 1-ATP expended for the original activation of stearic acid at step -1. Therefore, one C_{18} fatty acid yields a net of 39 ATP and 9-

acetyl CoA. 9-acetyl CoA enter Krebs cycle and generate 108 (9 X 12 = 108) ATP molecules, so that the total energy gained from the complete respiration of stearic acid is 147 (39 + 108 = 147) ATP molecules.

In addition to β-oxidation, the other pathways of fatty acid oxidation are as follows:

1. **β-oxidation:** Alpha-oxidation of a fatty acid involves the removal of one carbon at a time from the carboxylic end of the molecule. This pathway is operative in brain and plant tissues and is important in the catabolism of branched chain and odd chain fatty acids. This cycle does not require CoA intermediates.
2. **Ω-(omega)-oxidation:** Omega-oxidation is normally a very minor pathway. It is brought about by **hydroxylase** enzymes involving cytochrome P-450 in the endoplasmic reticulum. The CH_3 group (omega) is converted to a (CH_2OH group that subsequently is oxidized to -COOH, thus, forming a dicarboxylic acid. This compound is usually oxidized by ß-oxidation pathway.

FATTY ACID SYNTHESIS

The fatty acids are synthesised by three main following processes:

1. Extra mitochondrial de novo fatty acid synthesis.
2. Mitochondrial synthesis of fatty acids.
3. Microsomal pathway of fatty acid synthesis.

Extra Mitochondrial De Novo Fatty Acid Synthesis

This pathway is characterized by the following:

1. This pathway of lipogenesis occurs in the cytosol.
2. It is active in tissues like liver, kidney, brain, mammary glands, adipose tissues and lungs.
3. Acetyl-CoA is the immediate substrate and free polmitic acid is the end product.
4. NADPH, ATP, Mn^{+2}, biotin, and HCO_3^- (as a source of CO_2) act is cofactors.

This pathway is completed in following steps.

1. **Carboxylation of Acetyl-CoA:** This occurs in two sub-stages.
   a) **Carboxylation of biotin:** Acetyl-CoA carboxylase enzyme combines with a co-factor biotin to form enzyme-biotin complex. This complex reacts with HCO_3 in the presence of ATP to produce enzyme-biotin-COO^-. ATP changes into ADP and Pi (inorganic phosphorus).
   b) **Transfer of the carboxyl to acetyl-CoA:** Carboxyl group of enzyme-biotin-COO^- complex is transferred to Acetyl-CoA to produce malonyl-CoA.

Acetyl-CoA-Carboxylase + biotin (enzyme) ⟶ Enz-biotin complex

Enz-biotin-complex + ATP + HCO_3 ⟶ Enzy. biotin-COO^- complex + ADP + Pi

Enzy-biotin-COO-complex Enz-biotin

Acetyl CoA Malonyl CoA

2. **Condensation of acetyl-CoA and malonyl-CoA:** After the formation of malonyl-CoA, the further reactions are catalysed by an enzyme complex known as **fatty acid synthase complex**. It is a dimer. In mammals, each monomer is identical (Fig. 15.3) and consists of one polypeptide chain containing all the 6 enzymes of fatty acid synthase (Transacylase, ketoacyl synthase, ketoacyl reductase, hydratase, enol-reductase and thioesterase) and an ACP (acyl carrier protein) with a 4'-phosphopantetheine-SH group. In close proximity is another

thiol of a cysteine residue attached to 3-ketoacetyl synthetase (condensing enzyme) of the other monomer. Since both thiols participate in the synthetase activity, only the dimer is active not the monomer.

The various sub stages are as follows (Fig. 15.4).

1. A priming molecule of acetyl-CoA combines with the cysteine-SH group of synthase. This is catalysed by **transacylase**.
2. **Molanyl-CoA** combines with the adjacent-SH on the 4'-phosphopantetheine of ACP of the other monomer. This is also catalysed by **transacylase**.

The result of sub stages (i) and (ii) is to form acetyl (acyl)- malonyl enzyme.

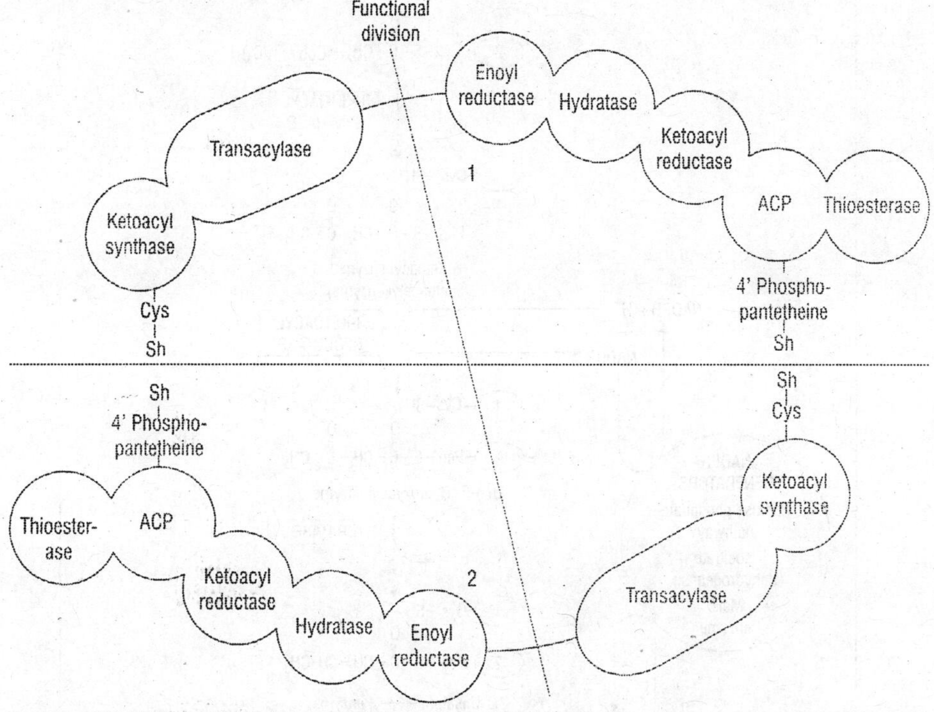

FIG. 15.3: Fatty acid synthase multienzyme complex. The complex is a dimer of 2 identical polypeptide monomers. 1 and 2 each consisting of 6 separate enzyme activities and the acyl carrier protein (ACP).

1. **Decarboxylation:** Acyl-malonyl enzyme is decarboxylated to produce 3' ketoacyl enzyme (acetoacyl enzyme) catalysed by **3-keto acyl synthase**. This frees cysteine-SH group, which was occupied by the acetyl group.
2. **Hydrogenation:** 3-ketoacyl enzyme is reduced to produce 3-hydroxyacyl enzyme. The reaction is catalysed by **3-ketoacyl reductase**. $NADPH_2$ is the hydrogen atom donor.
3. **Dehydration:** The enzyme **hydratase** catalyses the dehydration of 3-hydroxyacyl enzyme to produce 3,3-unsaturated acyl enzyme.
4. **Hydrogenation:** In the presence of **enoyl reductase**, 2,3- unsaturated acyl enzyme accepts hydrogen atoms from $NADPH_2$ to produce saturated acyl enzyme.
5. **Saturated acyl-enzyme reenters the cycle.** A new molecule of malonyl-CoA combines with the-SH of 4'-phospho pantetheine, displacing the saturated acyl residue onto the free cysteine-SH group. The sequence of reactions is repeated 6 more times, a new malonyl residue being incorporated during each sequence,

until a saturated 16-carbon acyl radical (palmityl) has been assembled.

6. **Acyl palmityl enzyme** liberates palmitic acid, which is catalysed by **thioesterase** (deacylase), a sixth enzyme in the complex.

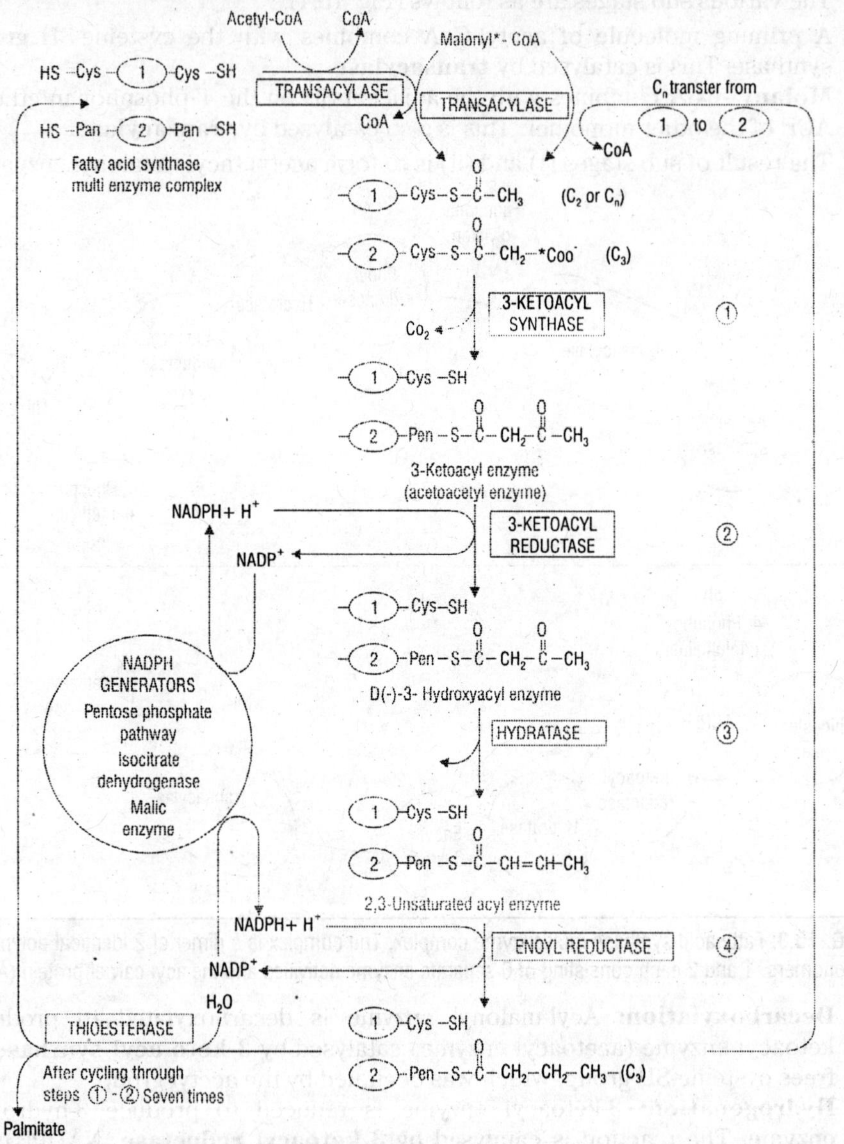

FIG. 15.4: Biosynthesis of long-chain fatty acids. Details of how addition of a malonyl residue causes the acyl chain to grow by 2 carbon atoms.

One dimer of enzyme produces two chains of palmitic acid simultaneously, as there are two centers of activity. Palmitic acid is usually esterified into acylglycerols.

Mitochondrial Synthesis of Fatty Acids

This is a chain elongation pathway where the addition of acetyl CoA to the existing long chain fatty acids takes place. Palmitic acid is the precursor of long chain saturated and unsaturated fatty acids. Elongation of palmitic acid to stearic acid is more abundant.

Microsomal Pathway of Fatty Acid Synthesis

Microsomal pathway also provides a means of elongation of both saturated and unsaturated fatty acids utilizing malonyl-CoA as the acetyl donor instead of acetyl-CoA.

SYNTHESIS OF TRIGLYCERIDES

The triglycerides are mainly synthesised in liver, kidney, heart, intestinal mucosa, adipose tissues, lactating mammary glands. These are principally synthesised in the microsomal fraction. The various steps of triglyceride synthesis are as follows:

Activation of Glycerol: Glycerol is phosphorylated (activated) in the presence of an enzyme **glycerokinase** and ATP and α-glycerophosphate is produced. Glycerokinase is found in almost all the tissues except adipose tissues and muscles.

$$\text{Glycerola} \xrightarrow[\text{ATP}]{\text{Glycerokinase}} \alpha - \text{glycerophosphate}$$

In muscles and adipose tissues α-glycerophosphate is produced by reduction of dihydroxyacetone phosphate (an intermediate of glycolysis) in the presence of α-**glycerophosphate dehydrogenase** and $NADH_2$.

Activation of fatty acids: The fatty acids are activated by ATP and CoA-SH to produce fatty acyl-CoA. The reaction is catalysed by **thiokinase (acyl CoA synthetase)**; Mg^{+2} act as co-factors.

$$\text{Fatty acids + ATP + CoA-SH} \quad \overset{\text{thiokinase}}{\underset{Mg^{+2}}{}} \quad \text{Fatty Acyl-CoA + AMP + PPi}$$

Fatty Acyl group transfer: Two molecules of fatty acyl CoA react with α-glycerophosphate to produce 1,2-diglyceride phosphate (phosphatidate) in the presence of **glycerophosphate acyl transferase.**

$$\text{-glycerophosphate + 2-Fatty acyl CoA} \quad \overset{\text{Glycerophosphate acyl transferase}}{} \quad \text{1,2-diglyceride phosphate} \quad \text{phosphatidate}$$

Dephosphorylation: 1,2-diglyceride phosphate undergoes dephosphorylation to produce 1,2-diglyceride. **Phosphatidate Phosphohydrolase** catalyses the reaction. α-hydroxyl group of glycerol is set free.

$$\text{1,2-diglyceride phosphate} + H_2O \quad \overset{\text{Phosphatidate phosphohydrolase}}{} \quad \text{1,2-diglyceride}$$

Fatty acyl group transfer: The free hydroxyl group present in 1,2-diglycerol is esterified by the third molecule of acyl CoA in the presence of **acyl transferase** resulting in the formation of a neutral triglyceride molecule.

$$\text{1,2-diglyceride + Fatty acyl-CoA} \quad \overset{\text{Acyl transferase}}{} \quad \text{Triglyceride}$$

Ketone Bodies

Ketone bodies include **acetoacetic acid**, **ß-hydroxybutyric acid** and **acetone**. These are normally produced in the liver tissue during fatty acid metabolism. Under normal conditions the level of the ketone bodies in blood is approximately 1 mg per 100 ml. of blood. During starvation or due to intake of fatty diets, the level of ketone bodies increases in blood, the condition is known as **ketonemia**, and their higher levels in urine is referred to as **ketonuria**. The condition when ketonemia and ketonuria exist, is called as **ketosis**. Ketosis may also occur in pathological conditions such as **diabetes mellitus** or **toxaemia of pregnancy**. The ketone bodies are formed in the body as follows:

Formation of Acetoacetic Acid

1. Two molecules of acetyl-CoA undergo condensation to produce acetoacetyl-CoA in the presence of **thiolase**.

   2Acetyl-CoA $\xrightarrow{\text{thiolase}}$ Acetoacetyl-CoA

2. Acetoacetyl-CoA is deacylated in the presence of **deacylase** to produce acetoacetic acid.

   Acetoacetyl-CoA $\xrightarrow{\text{Deacylase}}$ Acetoacetic acid + CoASH

3. Acetoacetic acid may also be formed from acetoacetyl-CoA through a different pathway.
   a) One molecule of acetoacetyl-CoA and acetyl-CoA condense to form ß-hydroxy-ß-methylglutaryl-CoA (HMG-CoA).
   b) HMG-CoA undergoes cleavage to produce acetoacetic acid and acetyl CoA. The reaction is catalysed by **HMG-CoA cleavage enzyme**.

Acetoacetyl – CoA + Acetyl CoA $\xrightarrow{\text{Condensing enzyme}}$ HMG-CoA + CoA - SH

HMG – CoA $\xrightarrow{\text{HMG-CoA cleavage enzyme}}$ Acetoacetic acid + Acetyl - CoA

Formation of β-Hydroxy Butyric Acid

Acetoacetyl-CoA is reduced to produce β-hydroxybutyryl-CoA. NADH is the hydrogen donor and the reaction is catalysed by β-hydroxybutyryl-CoA dehydrogenase.

Acetoacetyl – CoA + NADH $\xrightarrow{\text{ß-hydroxybutyryl-CoA dehydrogenase}}$ ß-hydroxybutyrul CoA + NAD

Acetoacetyl-CoA may be formed, either:

- by condensation of 2-molecules of acetyl-CoA, or
- in the extra hepatic tissues, aceto acetic acid in the presence of ATP and an enzyme **acetoacetic thiokinase** is converted to acetoacetyl-CoA.
- β-hydroxy butyryl-CoA is deacylated in the presence of deacylase enzyme to yield ß-hydroxybutyric acid.

ß – hydroxybutyrul – CoA $\xrightarrow{\text{Deacylase}}$ ß hydroxybutyric acid CoA SH

Formation of Acetone

Acetoacetic acid is decarboxylated to produce acetone.

Acetoacetic acid $\xrightarrow{\text{Spontaneous}}$ Acetone + CO_2

Acetoacetic acid and β-hydroxybutyric acid are fairly strong acids and are buffered when present in blood or tissues. Their excretion in the urine results in the loss of buffer cations. Since ketone bodies are negatively charged, their excretion from the body is accompanied by excretion of positively charged ions usually sodium as sodium salts, which depletes the alkali reserve of the body leading to fall in plasma bicarbonate level, giving rise to fall in pH. This leads to **ketoacidosis**, which may be the cause of danger in uncontrolled diabetes mellitus.

CHAPTER 16

Metabolism: Nucleic Acid

All animal and plant cells contain nucleic acids in the form of nucleoproteins. Nucleoproteins are hydrolyzed in stomach and give rise to nucleic acids and proteins. In duodenum nucleic acids are broken down into nucleotides in the presence of nucleases. Nucleotidases further breakdown nucleotides into nucleosides and phosphoric acid. Nucleosides are hydrolyzed into organic bases and sugar in the presence of nucleosidases. Organic bases and nucleosides are absorbed through the intestinal epithelium into the blood stream and circulated.

PURINE METABOLISM

Catabolism

The purine bases, adenine and guanine are absorbed as such or in the form of their nucleosides, adenosine and guanosine.

The catabolism of adenine/adenosine takes place as follows.

1. Adenine is converted into hypoxanthine and ammonia in the presence of **adenase**. Adenase is found in low quantities in mammalian tissues.
2. Adenosine is deaminated to inosine in the presence of an enzyme, **adenosine deaminase**.
3. Inosine splits into hypoxanthine by phosphorolysis, the reaction is catalysed by **nucleoside phosphorylase**.
4. Hypoxanthine undergoes oxidation in the presence of **xanthine oxidase** and oxygen to produce xanthine.
5. Xanthine oxidase further oxidises xanthine into uric acid. In most of the mammals uric acid is the main end-product of purine catabolism.
6. In non-primate mammals uric acid further oxidises into allantoin in the presence of **uricase**.
7. In non-mammalian animals such as teleost fishes, amphibians, bivalves allantoin in the presence of **allantoinase** is converted into allantoic acid.
8. Amphibians and bivalves can further degrade allantoic acid into urea and glyoxylic acid. The reaction is catalysed by **allantoicase**.

Guanosine catabolism takes place as follows (Fig. 16.2):

1. Guanosine splits into guanine and sugar phosphate by a phosphorolytic reaction catalysed by the enzyme **nucleoside phosphorylase**.
2. The enzyme guanase deaminates guanine into xanthine and ammonia. Xanthene is further metabolised as has already been described.

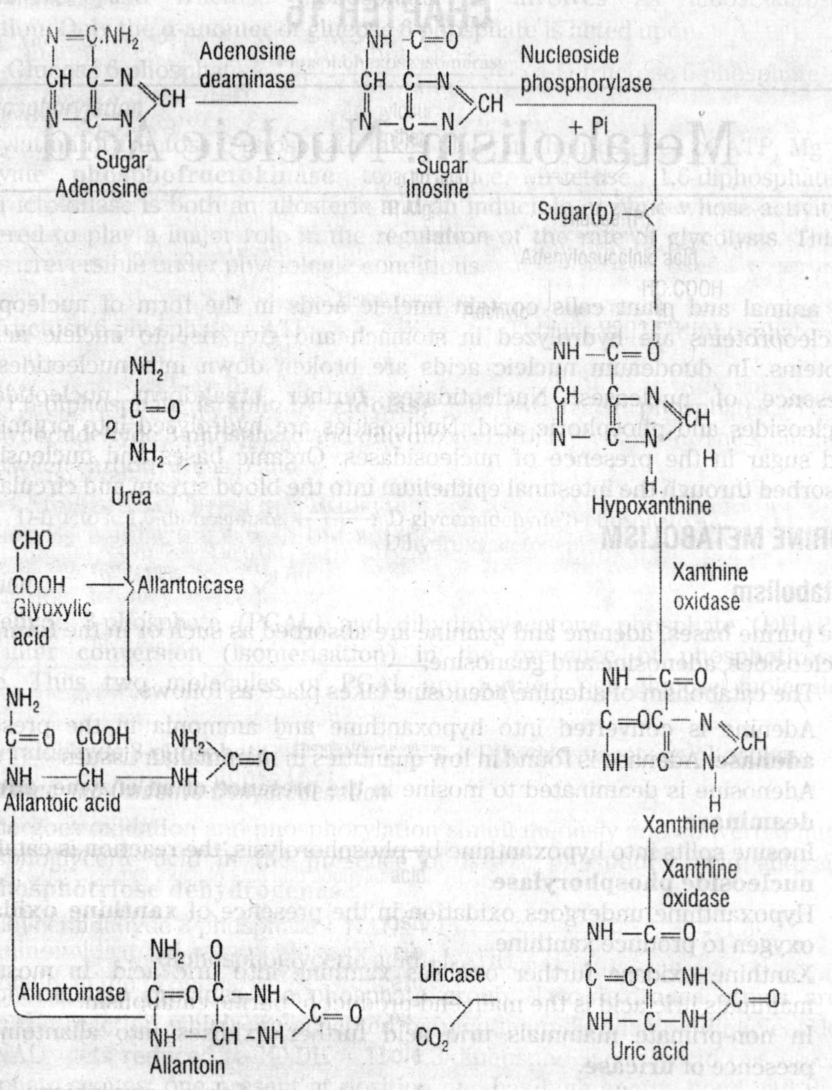

Adenosine → (Adenosine deaminase) → Inosine → (Nucleoside phosphorylase, + Pi) → Sugar(p) + Hypoxanthine → (Xanthine oxidase) → Xanthine → (Xanthine oxidase) → Uric acid → (Uricase) → CO_2 + Allantoin → (Allontoinase) → Allantoic acid → (Allantoicase) → Glyoxylic acid ($CHO-COOH$) + 2 Urea ($NH_2-C=O-NH_2$)

FIG. 16.1: Metabolism of adenine nucleoside in the animal tissues.

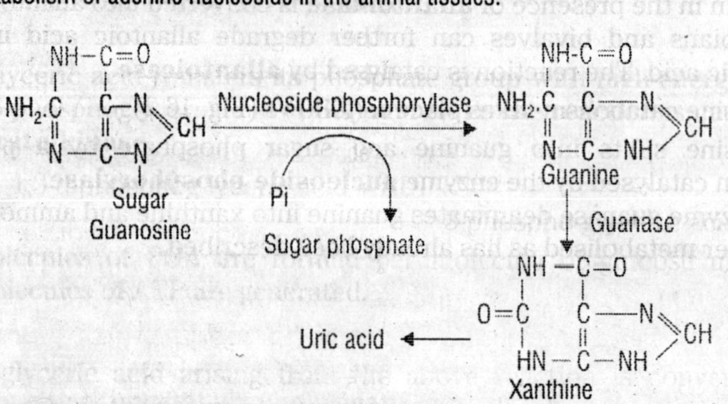

Guanosine → (Nucleoside phosphorylase, Pi → Sugar phosphate) → Guanine → (Guanase) → Xanthine → Uric acid

FIG. 16.2: Catabolism of guanosine in animal tissues.

Purine Biosynthesis

Biosynthesis of purines takes place from precursors available in the body. Origin of different atoms in the purine ring is being depicted in Fig. 16.3. It has been confirmed by radioisotope labelling technique. The various steps in purine biosynthesis are as follows:

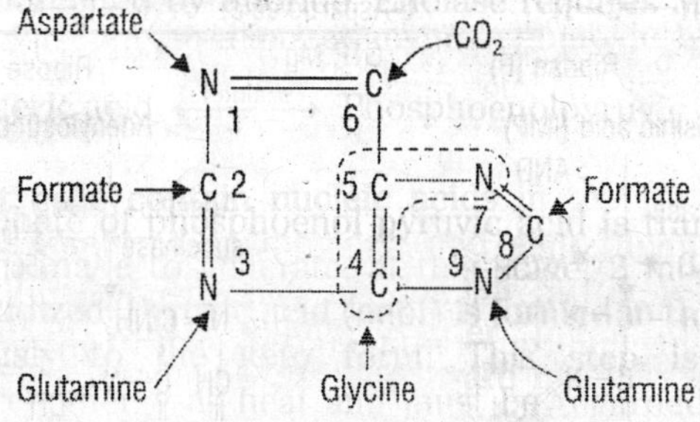

FIG. 16.3: Origin of different atoms in the Purine ring.

1. In the first step ribose-5 phosphate is activated in the presence of ATP and Mg^{++} ions and forms 5-phosphoribosyl-1-pyrophosphate (PRPP). The reaction is catalysed by kinase enzyme.

   $$\text{Ribose} - 5 - \text{phosphate} + \text{ATP} \xrightarrow[\text{Kinase}]{\text{Mg}} \text{PRPP} + \text{AMP}$$

2. PRPP is converted into 5-phosphoribosyl-1-amine in the presence of glutamine and Mg^{++} ions. The reaction is catalysed by glutamine phosp**horibosyl pyrophosphate amidotransferase** enzyme. Pyrophosphate and glutamic acid are released in the reaction.

   $$\text{PRPP} \xrightarrow[\text{Mg}^{++}]{\text{Glutamine}} \text{5-Phosphoriboxyl-1-amine} + \text{glutamic acid} + \text{pyrophosphate}$$

3. One molecule of glycine is conjugated with 5-phosphoribosylamine in the presence of ATP and Mg^{++} with the formation of glycinamide ribonucleotide.

   $$\text{5-phosphoribosyl} - 1 - \text{amine} \xrightarrow{\text{ATP + Glycine}} \text{Glycinamide ribonucleotide} + \text{ADP} + \text{Pi}$$

4. Glycinamide ribonucleotide undergoes formylation by formylfolic acid to produce formylglycinamide ribonucleotide (FGR) in the presence of an enzyme, **glycinamide ribonucleotide transformylase.**

   Glycinamide ribonucleotide → FGR

5. FGR receives an amino group from glutamine to produce N-formylglycinamide ribonucleotide in the presence of ATP and Mg^{++} ions.

   FGR + Glutamine + ATP → N-formylglycinamide-ribonucleotide + Glutamic acid + ADP + Pi

6. N-formylglycinamide ribonucleotide is converted into an imidazole nucleus, 5-Aminoimidazole ribonucleotide in the presence of ATP and K^+ ions.

   $$\text{N-formylglycinamide ribonucleotide} \xrightarrow{\text{ATP + K}} \text{5 – Aminoimida zole ribonucleotide} + \text{ADP} + \text{Pi}$$

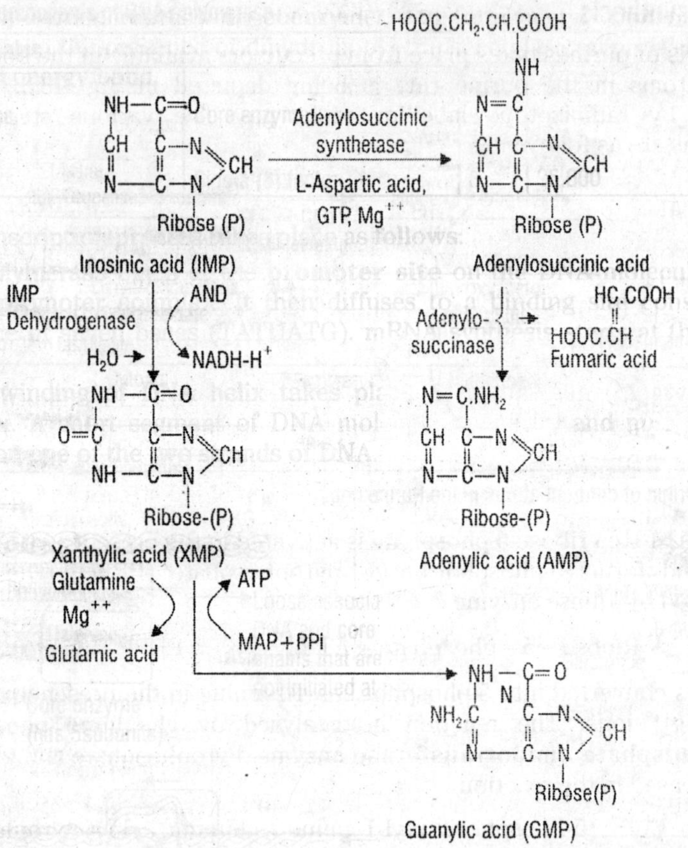

FIG. 16.4: Conversion of Inosinic acid into adenylic acid, Xanthylic acid and guanylic acid.

7. A molecule of CO_2 gets attached at position-4 of the imidazole ring to produce 5-Aminoimidazole-4-carboxylic acid ribonucleotide. This reaction proceeds optimally in the presence of high concentration of HCO_3^- ions.

5 – Aminoimidazole ribonucleotide $\xrightarrow{HCO_3}$ 5 – Aminoimidazole – 4 – Carboxylic acid ribonucleotide

8. 5-Aminoimidazole-4-carboxylic acid reacts with aspartic acid in the presence of ATP and Mg^{++} ions to produce 5-aminoimidazole-4- succinocarboxamide ribonucleotide is subsequently cleaved into 5-aminoimidazole-4-Carboxamine ribonucleotide and fumaric acid in the presence of enzyme, adenylosuccinase.

5-Aminoimidazole-4-Carboxylic acid ribonucleotide $\xrightarrow[\text{ATP}]{\text{Aspartic acid + Mg}}$ 5-Aminoimidazole-4-succinocarboxamide ibonucleotide + ADP + Pi $\xrightarrow{\text{Adenylosuccinase}}$ 5-Aminoimidazole 4-carboxamie ribonucleotide + Fumaric acid

9. 5-Aminoimidazole-4-carboxamide ribonucleotide undergoes formylation in the presence of K^+ ions, formyl-tetrahydrofolic acid (formyl THFA) and a **transformylase** enzyme to produce 5-formamidoimidazole-4-Carboxamide ribonucleotide.

5-Aminoimidazole-4-Carboxamide ribonucleotide $\xrightarrow[\text{Transformylase}]{\text{Formyl-THFA + K}^+}$ 5-Formamidoimidazole-4-Carboxyamide ribonucleotide

10. In the presence of inosinicase enzyme, 5-Formamidoimidazole-4 carboxamide ribonucleotide is converted into inosinic acid.

5-Formamidoimidazole-4-carboxamide ribonucleotide $\xrightarrow{\text{Inosinicase}}$ Inosinic acid + H_2O

11. Inosinic acid may be converted into adenylic or guanylic acid (Fig. 16.4).

PYRIMIDINE METABOLISM

Catabolism

The pyrimidine bases found in nucleic acids are uracil, thymine and cytosine. The steps of catabolic pathway of cytosine and uracil are as follows:

1. Cytosine undergoes hydrolytic deamination to produce uracil.

Cytosine $\xrightarrow{H_2O}$ Uracil + NH_3

2. Uracil is reduced reversibly to dihydrouracil in the presence of NADPH.

Uracil + NADPH ⇌ Dihydrouracil + NADP

3. Dihydrouracil is converted into ß-Ureidopropionic acid. One molecule of water is added and the pyrimidine ring is opened.

Dihydrouracil + H_2O → ß- Ureidopropionic acid

4. ß-ureidopropionic acid undergoes hydrolysis to produce ß-alanine, CO_2 and NH_3

ß-ureidopropionic acid + H_2O → ß-alanine + CO_2 + NH_3

Uracil has also been shown to undergo hydrolysis to produce formyloxaluric acid which then is cleaved into urea, oxalic acid and CO_2.

Thymine is catabolised as follows:

1. Thymine is reversibly reduced to dihydrothymine in the presence of NADPH.

Thymine + NADPH ⇌ Dihydrothymine + NADP

2. In the second step, the pyrimidine ring is ruptured hydrolytically and dihydrothymine is converted into ß-ureidoisobutyric acid.

Dihydrothymine → ß-Ureidoisobutyric acid

3. ß-Ureidoisobutyric acid splits hydrolytically into CO_2, NH_3 and ß-aminoisobutyric acid.

ß-Ureidoisobutyric acid + H_2O → ß-aminoisobutyric acid + CO_2 + NH_3

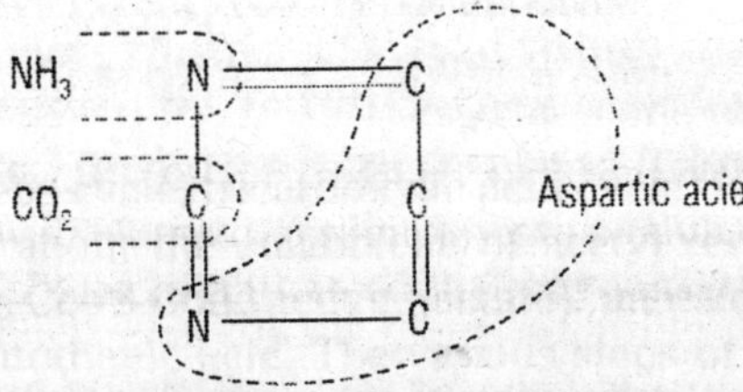

FIG. 16.5: Origin of different atoms in the Pyrimidine Nucleus.

Pyrimidine Biosynthesis

Pyrimidine biosynthesis occurs from precursors, namely ammonia, carbon dioxide, aspartic acid and 5-phosphoribosyl pyrophosphate. Sources of different atoms in pyrimidine nucleus have been depicted in Fig. 16.5, based on radioisotopic investigations.

The various steps in pyrimidine biosynthesis are as follows:

1. NH_3 and CO_2 react in the presence of ATP to produce carbonyl phosphate. The reaction is catalysed by carbamyl phosphate synthetase.

$$NH_3 + CO_2 + ATP \xrightarrow{\text{Carbamyl phosphate synthetase}} \text{Carbamyl phosphate}$$

2. Carbamyl phosphate reacts with aspartic acid forming N-carbamyl aspartate (ureidosuccinic acid), the reaction is catalysed by aspartate transcarbamylase.
3. N-carbamyl aspartate undergoes ring closure to produce dihydro-orotic acid in the presence of an enzyme dihydro-orotase.

$$\text{N – Carbamyl aspartate} \xrightarrow{\text{Dihydro-orotase}} \text{Dihydro – orotic acid} + H_2O$$

4. Dihydro-orotic acid is dehydrogenated to produce orotic acid. NAD accepts hydrogen atoms and the reaction is catalysed by **dihydro-orotic acid dehydrogenase**.

$$\text{Dihydro – Orotic acid + NAD} \xrightarrow{\text{Dihydro-orotic acid dehydrogenase}} \text{Orotic acid + NADH}$$

Orotic acid reacts with 5-phosphoribosyl-1-pyrophosphate (PRPP) giving rise to orotidine-5'-phosphate (orotidylic acid)

$$\text{Orotic acid + PRPP} \longrightarrow \text{Orotidine-5'-phosphate + PPi}$$

5. Orotidine-5'- Phosphate is decarboxylated to uridine-5'-phosphatic (Uridylic acid, UMP) in the presence of decarboxylase enzyme.

$$\text{Oratidine – 5' phosphate} \xrightarrow{\text{Decarboxylase}} \text{UMP} \quad CO_2$$

6. UMP is converted into uridine diphosphate (UDP) and uridine triphosphate (UTP) by reacting with ATP.

$$\text{UMP} \xrightarrow{\text{ATP}} \text{UDP} \xrightarrow{\text{ATP}} \text{UTP}$$

Cytidine triphosphate (CTP) is produced by the amination of UTP.

$$\text{UTP} \xrightarrow{NH_3 \text{ ATP and Mg}} \text{CTP}$$

Dephosphorylation of CTP results in the formation of CDP and CMP.

CHAPTER 17

Pheromones

The term *pheromone* was introduced by Peter Karlson and Martin Lüscher (1959) and is based on the Greek *pherein* (to transport) and *hormon* (to stimulate) is a chemical that triggers a natural behavioral response in another member of the same species. There are alarm pheromones, food trail pheromones, sex pheromones, and many others that affect behavior or physiology. Their use among insects has been particularly well documented, although many vertebrates and plants also communicate using pheromones.

Pheromones are chemicals released by an organism into its environment enabling it to communicate with other members of its own species.

Karlson and Martin Lüscher proposed the term to describe **chemical signals** from conspecifics which elicit innate behaviours soon after Butenandt characterized the first such chemical.

TYPES OF PHEROMONES

Insect Pheromones

Alarm Pheromones

Some species release a volatile substance when attacked by a predator that can trigger flight (in aphids) or aggression (in bees) in members of the same species.

When an ant is disturbed, it releases a pheromone that can be detected by other ants several centimeters away. They are attracted by low concentrations of the pheromone and begin to move toward the region of increasing concentration. As they get nearer to their disturbed nestmate, their response changes to one of alarm. The higher concentration causes them to run about as they work to remedy the disturbance.

Unless additional amounts of the alarm pheromone are released, it soon dissipates. This ensures that once the emergency is over, the ants return quietly to their former occupations.

Honeybees also have an alarm pheromone (which is a good thing not to elicit around a colony of "Africanized" bees).

Queen Mandibular Pheromone

Honeybee queens spend their lives literally surrounded by a retinue of worker bees. The workers are attracted to her by a pheromone that she releases from her mandibular glands. The pheromone is a mixture of alcohols and organic acids. Among its effects:

1. inducing the workers to feed and groom her;
2. inhibiting the workers from building queen cells and rearing new queens;
3. inhibiting ovary development in the workers.

Epideictic Pheromones

These pheromones were deducted in insects. These pheromones are different from territory pheromones. According to Fabre, those *females who lay their eggs in these fruits deposit these mysterious substances in the vicinity of their clutch to signal to other females of the same species so that they will clutch elsewhere.*

Sex Attractants

Hundreds of pheromones are known with which one sex (usually the female) of an insect species attracts its mates. Many of these sex attractants - or their close chemical relatives - are available commercially. They have proved useful weapons against insect pests in two ways:

Male Confusion: Distributing a sex attractant throughout an area masks the insect's own attractant and thus may prevent the sexes getting together. This *communication disruption* has been used successfully against a wide variety of important pests. For example, the sex attractant of the cotton boll weevil has reduced the need for conventional chemical insecticides by more than half in some cotton-growing areas.

Insect Monitoring: Insect sex attractants are also valuable in monitoring pest populations. By baiting traps with the appropriate pheromone, a build-up of the pest population can be spotted early. Even if a conventional insecticide is the weapon chosen, its early use reduces :

1. The amount needed.
2. damage to the crop.
3. Cost to the grower.
4. Possible damage to the environment.

Early detection of pest build-up is a key ingredient in the system known as **integrated pest management (IPM)**.

Aggregation Pheromones

They are produced by one or the other sexs, these pheromones attract individuals of both sexes.

Releaser Pheromones

These are powerful attractant molecules and some organisms may use to attract mates from a distance of 2 miles or more. This type of pheromone generally elicites rapid response but is quickly degraded. In contrast, a primer pheromone would have a slower onset but a longer duration.

Primer Pheromones

These pheromones trigger a change of developmental events.

Territorial Pheromones

These pheromones are laiddown in the environment and mark the boundaries of an organism's territory. In dogs, these hormones are present in the urine, which they deposit on landmarks serving to mark the perimeter of the claimed territory.

Trail Pheromones

These pheromones are common in social insects. For example, ants mark their paths with these pheromones, which are non-volatile hydrocarbons.

Some of the ants lay down an initial trail of pheromones as they return to the nest with food. This trail attracts other ants and serves as a guide. As long as the food source remains, the pheromone trail will be continually renewed. The pheromone should be continually renewed because it evaporates quickly. When the supply begins to dwindle, the trail making ceases. In at least one species of ant, trails that no longer lead to food are also marked with a repellent pheromone.

Sex Pheromones

In animals, sex pheromones indicate the availability of the female for breeding. Male animals may also emit pheromones that convey information about their species and genotype. Many insect species release sex pheromones to attract a mate and many lepidopterans can detect a potential mate from as far away as 10 km (6.2 miles). Pheromones are used in gametes to trail the opposite sex's gametes for fertilization. Pheromones are also used in the detection of oestrus in sows. Boar pheromones are sprayed into the sty, and those sows which exhibit sexual arousal are known to be currently available for breeding.

Other Pheromones (Unclassified)

This classification, based on the effects on behavior, remains artificial. Pheromones fill many additional functions.

There are four classes of pheromones: territorial markers, mother-infant, menstrual synchrony, and the fourth class that is the area of expertise for Dr. Cutler and Athena Institute as human sex-attractant pheromones.

Plant Pheromones: Pheromones also exist in plants: certain plants emit alarm pheromones when grazed upon, resulting in tannin production in neighboring plants. These tannins make the plants less appetizing for the herbivore.

Human pheromones

In women during synchronization of menstrual cycles there is an unconscious odor cues (the McClintock effect, named after the primary investigator, Martha McClintock, of the University of Chicago). The studied proposed that there are two types of pheromone involved:

1. Produced prior to ovulation, shortens the ovarian cycle.
2. Produced just at ovulation, lengthens the cycle". This is analogous to the Whitten effect.

Other studies have suggested that people might be using odor cues associated with the immune system to select mates who are not closely related to themselves. Using a brain imaging technique, Swedish researchers have shown that homosexual and heterosexual males' brains respond differently to two odors that may be involved in sexual arousal, and that the homosexual men respond in the same way as heterosexual women. The possible role for human pheromones is in the biological basis of sexual orientation.

Another study shows that the smell of androstadienone, a chemical component of male sweat, maintains higher levels of cortisol in females. The scientists suggest that the ability of this compound to influence the endocrine balance of the opposite sex makes it a human pheromonal chemosignal.

In 2006 it was shown that a second mouse receptor sub-class is found in the olfactory epithelium. Called the trace amine-associated receptors (TAAR), some are activated by volatile amines found in mouse urine, including one putative mouse pheromone. Orthologous receptors exist in humans providing, the authors propose, evidence for a mechanism of human pheromone detection.

Some body spray advertisers claim that their products contain human sexual pheromones which act as an aphrodisiac. In the 1970's "copulins" were patented as products which release human pheromones, based on research on rhesus monkeys. Subsequent to that androstenone, axillary sweat, and vomodors have been claimed to act as human pheromones. Despite these claims, no pheromonal substance has ever been demonstrated to directly influence human behavior in a peer reviewed study.

BIOCHEMISTRY

During the past 40 years, pheromones of hundreds of insect species have been chemically elucidated, including the sex pheromone of the codling moth. Its main component is (E,E)-8,10-dodecadien-1-ol, a primary alcohol containing a straight chain of 12 carbons and two conjugated double bonds. Other moth pheromones are hydrocarbons, epoxides, acetates or aldehydes. These molecules all vaguely resemble fatty acids, from which they are indeed biogenetically derived.

FIG. 17.1: (E,E)-8,10-Dodecadien-1-ol, Codlemone.

Most pheromones consist of blends of two or more chemicals which need to be emitted at exactly the right proportions to be biologically active. The female effluvia or sex gland can contain additional compounds which are related to the pheromone components and whose biological function is often unclear. On the other hand, many attractants of male moths have been discovered simply by field screening. In several cases it could later be shown that the attractant found with this technique was identical to the natural pheromone produced by the female. In most others, the composition of the true pheromone is still unknown. You can click the structural formula above to see a list of all the species for which (E,E)-8,10-dodecadien-1-ol has been reported as a pheromone or attractant component.

CHAPTER 18

Animal Hormone

GENERAL INTRODUCTION

A **hormone** (from Greek ὁρμη - impetus) is a chemical messenger that carries a signal from one cell (or group of cells) to another, in most instances via the blood. All multicellular organisms produce hormones (including plants and algae – phytohormone).

Organic compound (often a steroid or peptide) produced in one part of a multicellular organism and travels to another part to exert its action. Animal hormones are produced in specifie glands.

In animals, William Bayliss and Ernest Starling (1902) showed that a substance taken from the lining of the intestine could be injected into a dog to stimulate the pancreas to secrete fluid. They called the substance secretin which set in motion. Today more than 100 hormones have been identified.

In general, hormones regulate the function of their target cells, i.e., cells that express a receptor for the hormone. The action or net effect of hormones is determined by a number of factors including its pattern of secretion and the response of the receiving tissue - the signal transduction response.

Therefore, hormone is a chemical that transfers information and instructions between cells in animals and plants. Often described as the body's chemical messengers, hormones regulate growth and development, control the function of various tissues, support reproductive functions, and regulate metabolism (the process used to break down food to create energy). Unlike information sent by the nervous system, which is transmitted via electronic impulses that travel quickly and have an almost immediate and short-term effect, hormones act more slowly, and their effects typically are maintained over a longer period of time.

In animals, hormones are made by specialized glands or tissues that manufacture and secrete these chemicals as the body needs them. The majority of hormones are produced by the glands of the endocrine system, such as the pituitary, thyroid, adrenal glands, and the ovaries or testes. These endocrine glands produce and secrete hormones directly into the bloodstream. However, not all hormones are produced by endocrine glands. The mucous membranes of the small intestine secrete hormones that stimulate secretion of digestive juices from the pancreas. Other hormones are produced in the placenta, an organ formed during pregnancy, to regulate some aspects of fetal development.

Types of Hormones (Chemical basis)

Hormones are classified into two basic types based on their chemical makeup.

1. **Peptides Hormones:** The majority of hormones are peptides, or amino acid derivatives that include the hormones produced by the anterior pituitary, thyroid, parathyroid, placenta, and pancreas. Peptide hormones are typically produced as larger proteins. When they are called into action, these peptides are broken down into biologically active hormones and secreted into the blood

to be circulated throughout the body.

2. **Steroid hormones:** Steroid hormones include those hormones which secreted by the adrenal glands and ovaries or testes. They are synthesised from cholesterol (a fatty substance produced by the body) and modified by a series of chemical reactions to form a hormone ready for immediate action.

Hormones and Life History Evolution (LHE)

Hormones are also major players in LHE in a broad array of animal groups. This proposal seems eminently sensible: evolutionary alterations in organismal life histories must involve a radical reorganization of the mechanisms underlying the LHTs themselves. Hormones coordinate and orchestrate LHTs in a wide variety of taxa; therefore, LHE likely involves modifications in the production, regulation and/or tissue-specific response of these very same hormones.

Evolutionary alterations in organismal life histories have been profitably described in terms of **heterochronies**: changes in the relative timing of developmental events. For example, neoteny, which has evolved multiple times independently in various salamanders from metamorphic ancestors, can be described in terms of alterations in the relative timing of sexual and somatic differentiation. Specifically, metamorphosis is blocked, and sexual maturation proceeds within an otherwise larval morphology. This metamorphic block has different physiological causes in different taxa.

A careful interspecific comparison of reasonably closely related species allows that shifts in timing of hormonal release and/or the cellular response to hormones in the target tissues are intimately associated with the evolution of derived life history strategies.

Types of Hormones (Animal Basis)

Hormones exist in mammals, including invertebrates, as well as in humans and plants. The hormones of vertebrates, and other humans, mammals are nearly identical in chemical structure and function in the body. They are generally characterized by their effect on specific tissues.

Vertebrate Hormones

Amine-derived hormones: They are derivatives of the amino acids tyrosine and tryptophan. Examples are catecholamines and thyroxine.

Peptide hormones: Consist of chains of amino acids. Examples of small peptide hormones are TRH and vasopressin. Peptides composed of scores or hundreds of amino acids are referred to as proteins. Examples of protein hormones include insulin and growth hormone. More complex protein hormones bear carbohydrate side chains and are called glycoprotein hormones. Luteinizing hormone, follicle-stimulating hormone and thyroid-stimulating hormone are glycoprotein hormones.

Lipid and phospholipid-derived hormones: They derive from lipids such as linoleic acid and arachidonic acid and phospholipids. The main classes are the steroid hormones that derive from cholesterol and the eicosanoids. Examples of steroid hormones are testosterone and cortisol. Sterol hormones such as calcitriol are a homologous system. The adrenal cortex and the gonads are primary sources of steroid hormones. Examples of eicosanoids are the widely studied prostaglandins.

Invertebrate Hormones

Invertebrate hormones are less satisfactory in the acoelomate and pseudocoelomate animals. The researchers are more evident on the functions of endocrine of the coelomate vertebrates and mainly confined to the arthropods, annelids, mollusks and echinoderms. Squids, octopuses, and other mollusks, for example, have hormonally controlled pigment cells that enable the animals to change colour to blend in with their surroundings.

Hormones from Colenterata: The neurosecretory cells are in abundance in hypostome (region of new growth) of an adult Hydra. If the buds are detached from the parent body prior to the secretion of neurosecretory cells in the hypostome they fail to develop. During sexual maturity, the neurosecretory cells disappear therefore, at this stage the growth and reproduction processes become antagonistic processes. The neurosecretory growth hormone then activates cell proliferation and causes interstitial cells to develop into somatic structures such as nematocysts. When the hormone is not present the growth ceases and interstitial cells form gametes.

Hormones from Annelida: Neurosecretions in annelids regulates various processes like seasonal swarming of distinct sexual forms, hermaphroditism and viviparity. For example in Nereis (polychaete worm) various neurosecretory cells are found in the brain and other ganglia. These cells control processes like growth, metamorphosis, sexual development and reproductive behaviour.

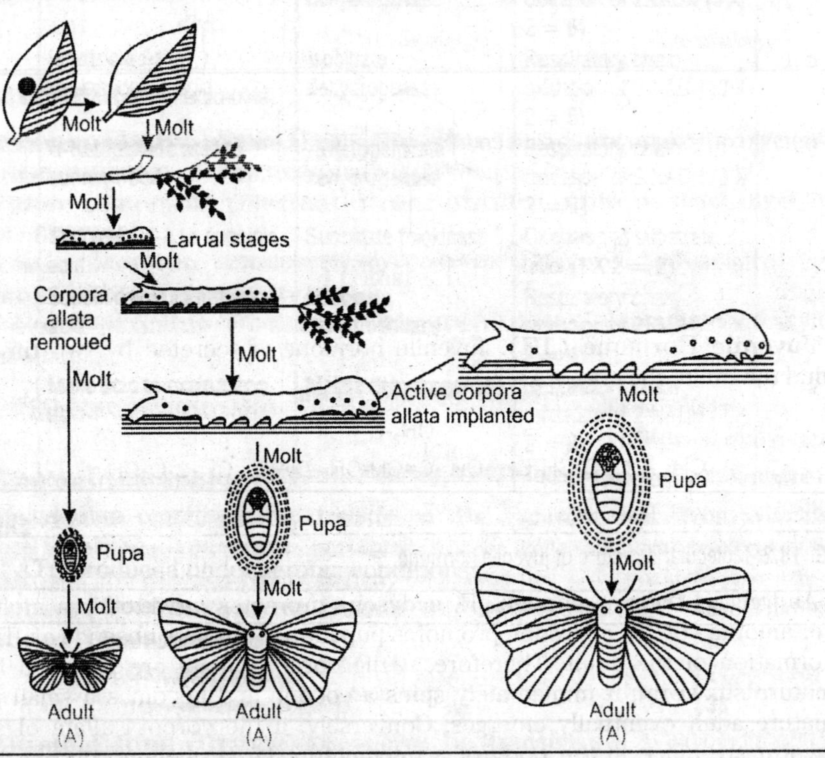

FIG. 18.1: Effect of molting hormone on insect.

Hormones from Arthropoda: Insects and crustaceans are the best examples of hormone secretion which controls various metabolic processes including growth, development reproduction, colour adaptation etc.

Insect Hormones: Insects have rigid exoskeleton and can grow only by periodically shedding their exoskeleton - called molting. Molting occurs repeatedly during larval development. At the final molt, the adult emerges. In several insect orders, the adult looks entirely different from the larva that preceded it. This marked transformation is called metamorphosis. Metamorphosis takes place during a dormant stage called the pupa. The sequence ending in the center panel (B) shows the larval, pupal, and adult stages during normal development of the domestic silkworm moth, Bombyx mori. Knowledge of insect hormones has provided a number of opportunities to enlist them – or molecules related to them – in the battle against insect pests.

Prothoracicotropic Hormone (PTTH): Molting and pupation require the hormone, PTTH, secreted by a two pairs of cells in the brain of the larva. If these cells are cut out of the brain of a full-grown larva, pupation does not occur. This is not because of the trauma of surgery; if transplanted somewhere else in the caterpillar's body, pupation occurs normally. PTTH is a homodimer of two polypeptides of 109 amino acids. PTTH does not drive pupation directly but, as its name suggests, acts on the prothoracic glands.

Ecdysone: There are two prothoracic glands located in the thorax. Under the influence of PTTH, they secrete the steroid hormone ecdysone. Acting together, PTTH and ecdysone trigger every molt: larva-to-larva as well as pupa-to-adult. It accounts for the dramatic changes of metamorphosis.

Juvenile Hormone (JH): Juvenile hormone is secreted by two tiny glands behind the brain, the corpora allata.

$$CH_3\text{-}CH_2\text{-}\overset{O}{\widehat{\underset{CH_3}{C}\text{-}CH}}\text{-}CH_2\text{-}CH_2\text{-}\overset{CH_2\text{-}CH_3}{C}{=}CH\text{-}CH_2\text{-}CH_2\text{-}\overset{CH_3}{C}{=}CH\text{-}\overset{O}{\overset{\|}{C}}\text{-}O\text{-}CH_3$$

FIG. 18.2: Molecular structure of juvenile hormone.

As long as there is enough JH, ecdysone promotes larva-to-larva molts. With lower amounts of JH, ecdysone promotes pupation. Complete absence of JH results in formation of the adult. Therefore, if the corpora allata are removed from an immature silkworm, it immediately spins a cocoon and becomes a small pupa. A miniature adult eventually emerges. Conversely, if the corpora allata of a young silkworm are place in the body of a fully-mature larva, metamorphosis does not occur. The next molt produces an extra-large caterpillar.

JH affects gene expression: Adult insects do not normally molt, but if extra amounts of **PTTH** are given to an adult Rhodnius (the *kissing bug*), it is forced into an extra molt. The English insect physiologist V.B. Wigglesworth showed that if juvenile hormone is first applied to the insect's exoskeleton, the regions affected by it revert to larval type after this extra molt. Presumably, JH interacts with hormone receptors in the cells to produce a new set of transcription factors.

The swallowtail butterfly, *Papilio xuthus*, passes through 5 larval stages, growing larger after each molt.

Recently, R. Futahashi and H. Fujiwara reported in the 22 February 2008 of *Science* that the dramatic transition from the 4th to the 5th instar is accompanied by a drop in the level of JH. When they applied synthetic JH soon after the molt from the 3rd to the 4th instar, when the final molt occurred, the 5th instar continued to resemble a bird dropping. They also found that treatment with synthetic JH altered the expression of several genes associated with the transition from the bird-dropping phenotype to the camouflaged phenotype.

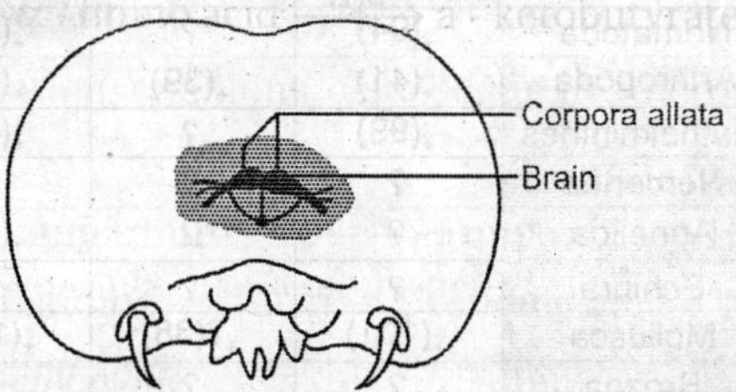

FIG. 18.3: Insect – showing corpora allata and brain.

Hormones from Mollusca: In Octopus (cephalopod) the subpeduncular lobes of brain controls a pair of optic glands through the action of inhibitory nerves. The optic glands are found on the optic stalks, one on either side of the central part of the supraesophageal brain. The brain gets environmental hints through the eyes. The optic glands dominate the reproductive endocrinology (comparable to vertebrate pituitary). Optic gland secretion promotes vitellogenesis in females and spermatogenesis in males. The secretion also governs differentiation of the female reproductive tract. In male the sex differentiation is directly controlled by testicular hormones.

In *Lymnaea stagnalis* (gastropods) well-established endocrine organs are dorsal bodies are present. The small discrete bodies are present on the cerebral ganglia and develop from glial cells and the perineurium of the cerebral ganglia. The secretions of these stimulate vitellogenesis and differentiation of the female genital tract. The dorsal bodies perform varied functions than those conducted by the cephalopod optic glands.

Hormones from Enchinodermata: In starfishes hormones regulates sexual maturation and spawning these hormones are:

Gonad-stimulating substance (GSS): Gonad-stimulating substance, a hormonal peptide released from the nervous system of starfishes, acts on the ovary to produce 1-methyladenine, an inducer of oocyte maturation. It is a low molecular weight (2,000) protein which is synthesised in the radial nerves. Addition of methionine to the incubation mixture of ovarian fragments and gonad-stimulating substance enhanced the production of 1-methyladenine, whereas ethionine inhibited it. Incubation of ovarian tissue with methionine alone failed to produce 1-methyladenine. Use of a radioactive label showed that methionine is a methyl donor in the biosynthesis of 1-methyladenine, suggesting that gonad-stimulating substance is involved in the methylation process.

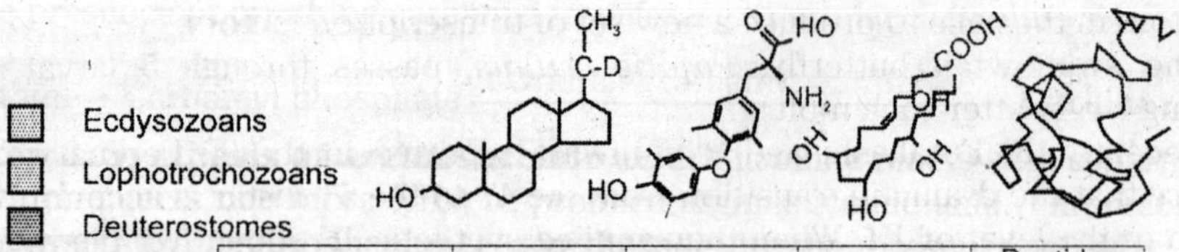

Ecdysozoans
Lophotrochozoans
Deuterostomes

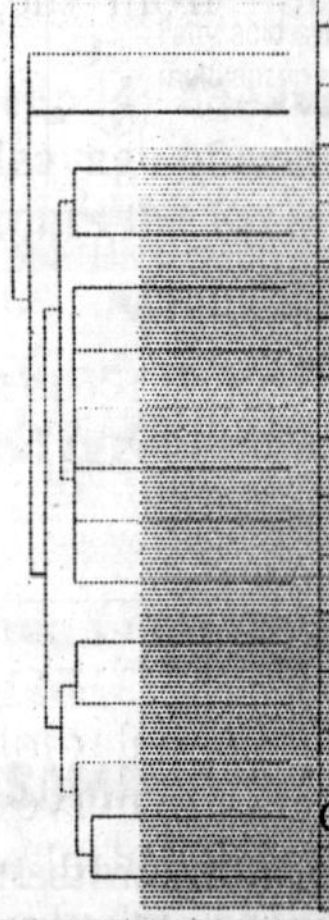

	Steroids	Amino acid derivatives	Eicosanoids	Peptides & proteins
Porifera	?	?	?	?
Cnidaria	?	+(92)	?	+(40)
Ctenophora	?	?	?	?
Nematoda	+(94)	?	+(95)	+(96)
Arthropoda	+(41)	+(39)	+(97)	+(98)
Plathelminthes	+(99)	?	+(93)	?
Nemertea	?	?	?	+(3)
Annelida	?	?	+(44)	+(100)
Echiura	?	?	?	?
Mollusca	+(101)	+?(38)	+(101)	+(101)
Brozoa	?	?	?	?
Echinodermata	+(86)	+(37)	+(93)	+(86)
Hemichordata	?	?	?	?
Urochordata	+?(102)	+(102)	?	+?(302)
Cephalochordata	?	?	?	+?(102)
Vertebrata	+(105)	+(32)	+(104)	+(32)

FIG. 18.4: Wide-spread involvement of hormones in life history transitions (LHTs). "þ" indicates that this class of hormones has been demonstrated to be involved in a LHT [hatching, larval transitions, resting stages (such as dauer and diapause), metamorphosis, settlement, migration, attainment of reproductive maturity and ovulation/oviposition]. "?" indicates that such a role has not been demonstrated for that phylum. "þ?" indicates preliminary evidence for such a role. Note that the four large hormone families considered here do not encompass all animal hormones involved in LHTs; certain hormones were omitted for simplicity. Also, hormones belonging to the same family can be synthesised by vastly different biochemical pathways. Furthermore, the original sources of animal hormones might have been plants/algae in many cases, and could well have been independently derived from plants/algae in distinct lineages. Finally, different taxa use similar hormones to regulate totally different LHTs (such as steroids in molluscan reproduction and arthropod molting) and, in some cases, the same hormone has been coopted multiple times to regulate independently evolved metamorphoses (TH is the prime example; see the text). For these reasons, the reader should not interpret the presence of a similar class of hormones in related animal taxa to be sufficient evidence for common ancestry of particular LHT functions and their hormonal control.

Maturation-inducing substance (MIS): It is 1-methyl adenine and is produced by the ovarian follicles. 1-Methyladenine(1-MeA) has been identified as the oocyte maturation-inducing substance (MIS) in starfish, but little is known about its biosynthesis. The spawning reactions in starfishes are synchronized. The presence of eggs or sperms in the water stimulates other individuals to spawn. The endocrinological explanation of this mass spawning is simple. A neurosecretion released by the radial nerves acts on the follicle cells to induce the synthesis of MIS, which triggers ovulation, release of gametes and reproductive behaviour.

Arthropoda: In invertebrates, hormones regulate metamorphosis (the process in which many insects, crustaceans, and mollusks transform from eggs, to larva, to pupa, and finally to mature adults). A hormone called ecdysone triggers the insect molting process, in which these animals periodically shed their outer covering, or exoskeletons, and grow new ones. The molting process is delayed by juvenile hormone, which inhibits secretion of ecdysone. As an insect larva grows, secretion of juvenile hormone declines steadily until its concentrations are too low to prevent the secretion of ecdysone. When this happens, ecdysone concentrations increase until they are high enough to trigger the metamorphic molt.

In insects that migrate long distances, such as the locust, a hormone called octopamine increases the efficiency of glucose utilization by the muscles, while adipokinetic hormone increases the burning of fat as an energy source. In these insects, octopamine levels build up in the first five minutes of flight and then level off as adipokinetic hormone takes over, triggering the metabolism of fat reserves during long distance flights.

Hormones also trigger colour changes in invertebrates.

HUMAN HORMONES

Human hormones significantly affect the activity of every cell in the body. They influence mental acuity, physical agility, and body build and stature. Growth hormone is a hormone produced by the pituitary gland. It regulates growth by stimulating the formation of bone and the uptake of amino acids, molecules vital to building muscle and other tissue.

Sex hormone

Sex hormones regulate the development of sexual organs, sexual behavior, reproduction, and pregnancy. For example, gonadotropins, also secreted by the pituitary gland, are sex hormones that stimulate egg and sperm production. The gonadotropin that stimulates production of sperm in men and formation of ovary follicles in women is called a follicle-stimulating hormone. When a follicle-stimulating hormone binds to an ovary cell, it stimulates the enzymes needed for the synthesis of estradiol, a female sex hormone. Another gonadotropin called luteinizing hormone regulates the production of eggs in women and the production of the male sex hormone testosterone. Produced in the male gonads, or testes, testosterone regulates changes to the male body during puberty, influences sexual behavior, and plays a role in growth. The female sex hormones, called estrogens, regulate female sexual development and behavior as well as some aspects of pregnancy. Progesterone, a female hormone secreted in the ovaries, regulates menstruation and stimulates lactation in humans and other mammals.

Hormone which regulates metabolism

Other hormones regulate metabolism. For example, thyroxine, a hormone secreted by the thyroid gland, regulates rates of body metabolism. Glucagon and insulin, which are secreted in the pancreas, control levels of glucose in the blood and the availability of energy for the muscles. A number of hormones, including insulin, glucagon, cortisol, growth hormone, epinephrine, and norepinephrine, maintain glucose levels in the blood. While insulin lowers the blood glucose, all the other hormones raise it. In addition, several other hormones participate indirectly in the regulation. A protein called somatostatin blocks the release of insulin, glucagon, and growth hormone, while another hormone, gastric inhibitory polypeptide, enhances insulin release in response to glucose absorption. This complex system permits blood glucose concentration to remain within a very narrow range, despite external conditions that may vary to extremes.

Hormones also regulate blood pressure and other involuntary body functions. Epinephrine, also called adrenaline, is a hormone secreted in the adrenal gland. During periods of stress, epinephrine prepares the body for physical exertion by increasing the heart rate, raising the blood pressure, and releasing sugar stored in the liver for quick energy.

Hormones used for medical treatment

Certain hormones are particularly used for treatment of diseases of the endocrine system. In people with diabetes mellitus type 1, for example, the pancreas secretes little or no insulin. Regular injections of insulin help maintain normal blood glucose levels. Sometimes, an illness or injury not directly related to the endocrine system can be helped by a dose of a particular hormone. Steroid hormones are often used as anti-inflammatory agents to treat the symptoms of various diseases, including cancer, asthma, or rheumatoid arthritis. Oral contraceptives, or birth control pills, use small, regular doses of female sex hormones to prevent pregnancy.

Initially, hormones used in medicine were collected from extracts of glands taken from humans or animals. For example, pituitary growth hormone was collected from the pituitary glands of dead human bodies, or cadavers, and insulin was extracted from cattle and hogs. As technology advanced, insulin molecules collected from animals were altered to produce the human form of insulin.

SYNTHETIC HORMONES

With improvements in biochemical technology, many hormones are now made in laboratories from basic chemical compounds. This eliminates the risk of transferring contaminating agents sometimes found in the human and animal sources. Advances in genetic engineering even enable scientists to introduce a gene of a specific protein hormone into a living cell, such as a bacterium, which causes the cell to secrete excess amounts of a desired hormone. This technique, known as **recombinant DNA technology**, has vastly improved the availability of hormones.

Recombinant DNA has been especially useful in producing growth hormone, once only available in limited supply from the pituitary glands of human cadavers. Treatments using the hormone were far from ideal because the cadaver hormone was often in short supply. Moreover, some of the pituitary glands used to make growth hormone were contaminated with particles called prions, which could cause diseases such as Creutzfeldt-Jakob disease, a fatal brain disorder. The advent of recombinant technology made growth hormone widely available for safe and effective therapy.

MODE OF ACTION

Most hormones are released directly into the bloodstream, where they circulate throughout the body in very low concentrations. The important characteristics of hormone action are:

1. Some hormones travel intact in the bloodstream but others require a carrier substance, such as a protein molecule, to keep them dissolved in the blood. These carriers also serve as a hormone reservoir, keeping hormone concentrations constant and protecting the bound hormone from chemical breakdown over time.
2. Hormones travel in the bloodstream until they reach their target tissue, where they activate a series of chemical changes.
3. To achieve its intended result, a hormone must be recognized by a specialized protein in the cells of the target tissue called a **receptor**.
4. Typically, hormones that are water-soluble use a receptor located on the cell membrane surface of the target tissues.
5. A series of special molecules within the cell, known as **second messengers**,

transport the hormone's information into the cell.

6. Fat-soluble hormones (e.g., steroid hormones) pass through the cell membrane and bind to receptors found in the cytoplasm. When a receptor and a hormone bind together, both the receptor and hormone molecules undergo structural changes that activate mechanisms within the cell. These mechanisms produce the special effects induced by the hormone.
7. Receptors on the cell membrane surface are in constant turnover. New receptors are produced by the cell and inserted into the cell wall, and receptors that have reacted with hormones are broken down or recycled. The cell can respond, if necessary, to irregular hormone concentrations in the blood by decreasing or increasing the number of receptors on its surface.
8. If the concentration of a hormone in the blood increases, the number of receptors in the cell wall may go down to maintain the same level of hormonal interaction in the cell. This is known as **downregulation**.
9. If concentrations of hormones in the blood decrease, upregulation increases the number of receptors in the cell wall.
10. Some hormones are delivered directly to the target tissues instead of circulating throughout the entire bloodstream. For example, hormones from the hypothalamus, a portion of the brain that controls the endocrine system, are delivered directly to the adjacent pituitary gland, where their concentrations are several hundred times higher than in the circulatory system.

EFFECTS OF HORMONE

Hormonal effects are complex, but their functions can be divided into three broad following categories:

1. Some hormones change the permeability of the cell membrane.
2. Other hormones can alter enzyme activity.
3. Some hormones stimulate the release of other hormones.

Hormone is also responsible for activation of specific genes. For example, when a steroid hormone enters a cell, it binds to a receptor in the cell's cytoplasm. The receptor becomes activated and enters the cell's nucleus, where it binds to specific sites in the deoxyribonucleic acid (DNA), the long molecules that contain individual genes. This activates some genes and inactivates others, altering the cell's activity. Hormones have also been shown to regulate ribonucleic acids (RNA) in protein synthesises.

A single hormone may affect one tissue in a different way than it affects another tissue, because tissue cells are programmed to respond differently to the same hormone. A single hormone may also have different effects on the same tissue at different times in life. To add to this complexity, some hormone-induced effects require the action of more than one hormone. This complex control system provides safety controls so that if one hormone is deficient, others will compensate.

PEPTIDE HORMOMES

Peptides

Peptides are like steak. They arise from proteins. And like any protein, when taken orally, they are broken down into amino acids by digestive enzymes for absorption. This does not bode well for peptides as orally available drugs.

Synthesis and Metabolism of Peptides

Biologically, peptides come from proteins. The protein precursors of peptide hormones are called **prohormones**. Two examples of prohormones are listed below.

1. The formation of insulin (6,000 kDa) from proinsulin occurs in several stages. Preproinsulin (11,500 kDa) is the precursor for proinsulin.
2. In a more complex example, pro-opiomelanocortin (260 amino acids) is a common precursor for several peptide hormones, including ACTH (39 aa), β-endorphin (33 aa), met-enkephalin (5 aa), and others.

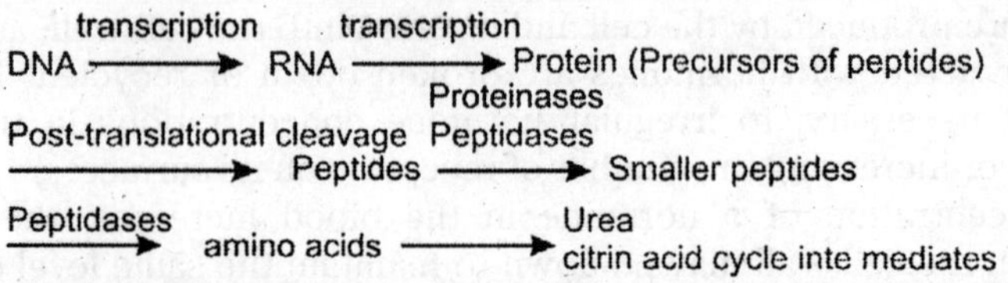

DNA —transcription→ RNA —transcription→ Protein (Precursors of peptides)

—Post-translational cleavage→ Peptides —Proteinases Peptidases→ Smaller peptides

—Peptidases→ amino acids —→ Urea, citrin acid cycle inte mediates

FIG. 18.5: Biosynthesis and fate of peptide hormones (Note: translation of proteins begins at N-terminus and proceeds to C-terminus).

There are three sources of proteins which are used in therapeutic products.

1. Natural: e.g., growth hormone isolated from both human and animal pituitary glands.
2. Recombinant DNA methods: e.g., insulin (and now growth hormone, tissue plasminogen activator).
3. Chemical synthesis: e.g., lysine vasopressin (9 aa). Chemical synthesis of peptides is practical only for small peptides.

Peptide Synthesis

Chemically, peptides are synthesised from their C-terminal amino acids to their N-terminal amino acids. The chemical peptide synthesis is less efficient than protein synthesis. Consequently, peptides longer than 30-35 amino acids are too long to be synthesised for use as drugs. Peptides are used in making antigens for antibody production and for studying enzyme and receptor substrates.

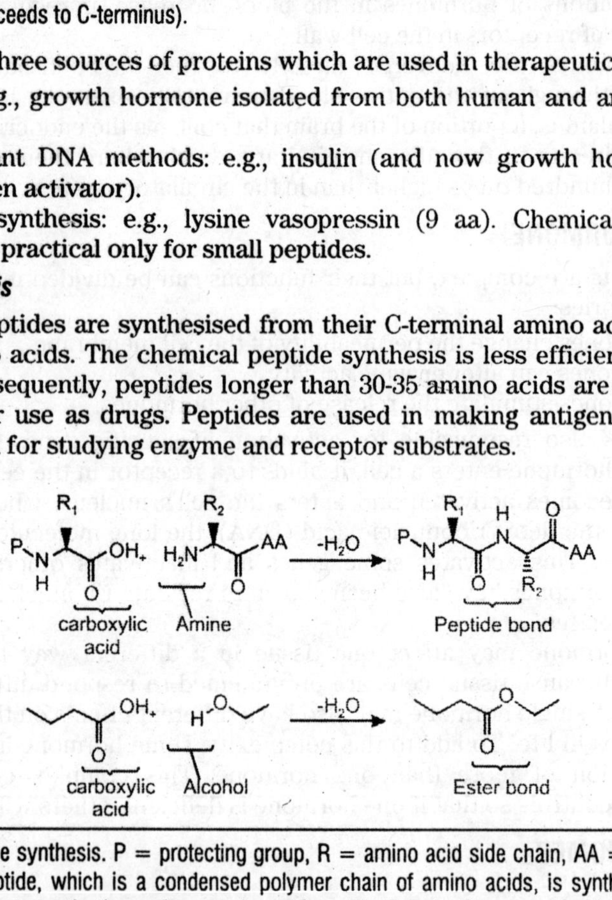

FIG. 18.6: Peptide synthesis. P = protecting group, R = amino acid side chain, AA = amino acid(s). Note that the peptide, which is a condensed polymer chain of amino acids, is synthesised from the carboxy terminus towards the amino terminus (C to N). This is the opposite direction to that of biological formation of protein chains, which form from N to C.

Above is a general example of peptide bond formation as. A peptide bond is formed when a carboxylic acid condenses (loses water) with an amine. This is analogous to ester bond formation, which results from condensation of a carboxylic acid with an alcohol.

In more specific peptide formation the protecting groups on the side chains, protecting group on the lysine side chain amine is different from that of the amino acid amine. The differences in protecting groups permit selective removal of one group in the presence of another.

Chemical Synthesis of Peptides

Chemical synthesis of peptides is usually carried out in an automated synthesiser using solid supports. Again, peptides are synthesised chemically from the C-terminus to the N-terminus.

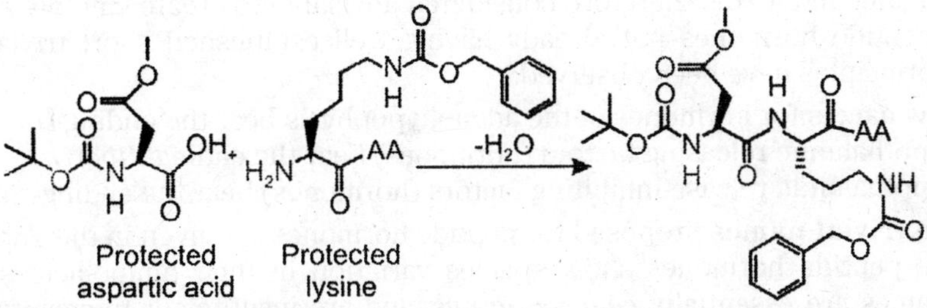

Protected aspartic acid, Protected lysine, $-H_2O$

FIG. 18.7: Specific example of peptide formation.

One key reaction of peptide synthesis is the removal of water to form the peptide bond. A coupling agent is used and a peptide bond is a dehydrated amine and acid. Water needs to be removed. This is done chemically by allowing water to react with a carbodimide to form an urea. One such coupling reagent is dicyclohexylcarbodiimide (DCC). There are several other types, but the principles are very similar. Learn DCC as the prototypical coupling agent. In all likelihood, you will need to know what DCC does and what it forms on the test.

1. Peptide coupling agents help form peptide bonds by making high energy intermediates that eventually bind up water in the form of urea. One example of such a reagent is DCC.
2. Peptide bonds are formed by removal of water between a carboxylic acid and an amine.
3. Functional groups on the side chains of amino acids used in a peptide synthesis must be protected from unwanted chemical reactions. To remove these protecting groups requires an extra step at the end of chemical peptide synthesis which contributes to the practical limits to the length a peptide can be made synthetically. 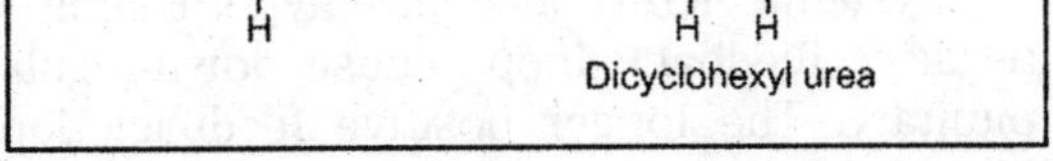
4. To avoid degradation, hormones contain special structural features, are part of larger polypeptides (peptides or proteins), or are bound to protecting carrier proteins.

DCC: Dicyclohexylcarbodiimide; Remove H_2O; Gan H_2O; Dicyclohexyl urea

Nomenclature of Peptide Hormones

Many amino acid and peptide hormones are elaborated by neural tissue, with ultimate impact on the entire system. When their composition was still unknown, hypothalamic secretory products were known as releasing factors, since their effect was to release endocrine hormones from the pituitary. More recently the releasing factors have been renamed releasing hormones. Currently, both names are in common use.

In the last two decades, the structures of many peptide hormones have been elucidated and other peptide hormones have been obtained in pure form. However, there are presently no accepted guidelines for nomenclature in this field.

The peptide hormones have well established trivial names indicating either natural source (e.g. insulin) or physiological action (e.g. relaxin, prolactin). However, some of the trivial names are so long that these hormones are known mainly by abbreviations (e.g. FSH for follicle-stimulating hormone). This is unfortunate, and it was therefore considered advisable to create suitable names for those peptide hormones not already having well established short trivial names. Three principles have been observed:

1. New names for hormones of the adenohypophysis bear the ending-*tropin*.
2. Hypothalamic releasing factors (hormones) bear the ending- *liberin*.
3. Hypothalamic release-inhibiting factors (hormones) bear the ending-*statin*.

The **trivial names** proposed for peptide hormones are given in the Table 18.2.

The peptide hormones show species variation in their amino-acid sequence, their names are essentially *generic names*, and are insufficient to specify a single chemical compound.

HORMONE SECRETION

The secretion of hypothalamic, pituitary, and target tissue hormones is under tight regulatory control by a series of feedback and feed- forward loops. This complexity can be demonstrated using the growth hormone (GH) regulatory system as an example.

The stimulatory substance growth hormone releasing hormone (GRH), and the inhibitory substance somatostatin (SS), both are products of the hypothalamus, control pituitary GH secretion. Under the influence of GRH, growth hormone is released into the systemic circulation, causing the target tissue to secrete IGF-1. (Growth hormone also has other more direct metabolic effects; it is both hyperglycemic and lipolytic). The principal source of systemic IGF-1 is the liver, although most other tissues secrete and contribute to systemic IGF-1. Liver IGF-1 is considered to be the principal regulator of tissue growth. In particular, the IGF-1 secreted by the liver is believed to synchronize growth throughout the body, resulting in a homeostatic balance of tissue size and mass. IGF-1 secreted by peripheral tissues is generally considered to be autocrine or paracrine in its biological action.

Systemic IGF-1 also has hypothalamic and pituitary regulatory targets. The negative feedback loops cause down-regulation of GH secretion directly at the pituitary. The longer positive feedback loop, involving IGF-1 regulation at the hypothalamus, stimulates the secretion of somatostatin (SS, also called growth hormone-inhibiting hormone, GIH), which in turn inhibits the secretion of growth hormone by the pituitary. The latter is a relatively unusual negative feed-forward regulatory process. In addition, a shorter negative feedback loop is shown that involves direct IGF-1 action on the pituitary, leading to down-regulation of GH secretion. Similar feedback loops exist for all the major endocrine hormones, and many subtle nuances modulate each regulatory loop.

Special Groups of Hormones

Hypothalamic Factors (Hormones): The hypothalamic **releasing factors** or **releasing hormones** have no well established trivial names. They are based on the ending- *liberin* added to the prefix of the pituitary hormone released by the factor.

Thus, *thyroliberin* indicates the hypothalamic peptide stimulating the release (and perhaps also the biosynthesis) of thyrotropin, the corresponding tropic hormone, from the pituitary gland. (Note that the ending "-tropin" is no longer retained in the name; it is implied in the definition of- *liberin.*) The names of those factors inhibiting the release (and perhaps the synthesis) of pituitary hormones are formed in a similar way with the suffix- *statin.*

The central nervous system regulates body function in two fundamentally different ways.

1. Direct innervation of organs by the sympathetic and parasympathetic nervous system. Responses are essentially instantaneous - within fractions of a second.
2. Indirect control by products of the endocrine system. Endocrine control is not instantaneous but instead requires seconds or longer to produce physiological response.

The general Scheme for endocrine control is as follows:

1. **Hypothalamus:** produces stimulatory and inhibitory releasing factors which act upon the anterior pituitary gland. Also produces the hormones oxytoxin and vasopressin.
2. **Pituitary gland:** produces peptide hormones which have endocrine and non-endocine targets.
3. **Endocrine organs:** Thyroid, adrenals, gonads, pancreas.
4. **Other organs:** Kidney, heart, arteries also secrete hormones.

The overall picture is that signals from direct innervation and from the endocrine system are coordinated to generate the most appropriate response to an external stimuli.

The Endocrine System

The endocrine system is composed of a number of tissues that secrete their products, called **endocrine hormones**, into the circulatory system; from there they are disseminated throughout the body, regulating the function of distant tissues and maintaining homeostasis. In a separate but related system, exocrine tissues secrete their products into ducts and then to the outside of the body or to the intestinal tract.

Endocrine hormones are considered to be derived from amino acids, peptides, or sterols and to act at sites distant from their tissue of origin. However, the latter definition has begun to blur as it is found that some secreted substances act at a distance (classical endocrines), close to the cells that secrete them (paracrines), or directly on the cell that secreted them (autocrines).

Insulin-like growth factor-I (IGF-I), which behaves as an endocrine, paracrine, and autocrine, provides a prime example of this difficulty.

Hormones are normally present in the plasma and interstitial tissue at concentrations in the range of 10^{-7}M to 10^{-10}M. Because of these very low physiological concentrations, sensitive protein receptors have evolved in target tissues to sense the presence of very weak signals. In addition, systemic feedback mechanisms have evolved to regulate the production of endocrine hormones.

Once a hormone is secreted by an endocrine tissue, it generally binds to a specific plasma protein carrier, with the complex being disseminated to distant tissues. Plasma carrier proteins exist for all classes of endocrine hormones. Carrier proteins for peptide hormones prevent hormone destruction by plasma proteases. Carriers for steroid and thyroid hormones allow these very hydrophobic substances to be present in the plasma at concentrations several hundred-fold greater than their solubility in water would permit. Carriers for small, hydrophilic amino acid–

derived hormones prevent their filtration through the renal glomerulus, greatly prolonging their circulating half-life.

Tissues capable of responding to endocrines have 2 properties in common: they posses a receptor having very high affinity for hormone, and the receptor is coupled to a process that regulates metabolism of the target cells. Receptors for most amino acid--derived hormones and all peptide hormones are located on the plasma membrane. Activation of these receptors by hormones (the first messenger) leads to the intracellular production of a second messenger, such as cAMP, which is responsible for initiating the intracellular biological response. Steroid and thyroid hormones are hydrophobic and diffuse from their binding proteins in the plasma, across the plasma membrane to intracellularly localized receptors. The resultant complex of steroid and receptor bind to response elements of nuclear DNA, regulating the production of mRNA for specific proteins.

The endocrine system is replete (abundantly supplied with) feedback mechanisms of both a positive and negative nature.

The endocrine system is composed of a number of tissues that secrete their products, called endocrine hormones, into the circulatory system; from there they are disseminated throughout the body, regulating the function of distant tissues and maintaining homeostasis. In a separate but related system, exocrine tissues secrete their products into ducts and then to the outside of the body or to the intestinal tract. Classically, endocrine hormones are considered to be derived from amino acids, peptides, or sterols and to act at sites distant from their tissue of origin. However, the latterdefinition has begun to blur as it is found that some secreted substances act at a distance (classical endocrines), close to the cells that secrete them (paracrines), or directly on the cell that secreted them (autocrines). Insulin-like growth factor-I (IGF-I), which behaves as an endocrine, paracrine, and autocrine, provides a prime example of this difficulty. Hormones are normally present in the plasma and interstitial tissue at concentrations in the range of 10^{-7}M to 10^{-10}M.

Mode of Action: These hormones has very low physiological concentrations, therefore, sensitive protein receptors have evolved in target tissues to sense the presence of very weak signals. In addition, systemic feedback mechanisms have evolved to regulate the production of endocrine hormones. Once a hormone is secreted by an endocrine tissue. It generally binds to a specific **plasma protein carrier**, with the complex being disseminated to distant tissues.

- **Plasma carrier proteins** exist for all classes of endocrine hormones. Carrier proteins for peptide hormones prevent hormone destruction by plasma proteases. Carriers for steroid and thyroid hormones allow these very hydrophobic substances to be present in the plasma at concentrations several hundred-fold greater than their solubility in water would permit. Carriers for small, hydrophilic amino acid--derived hormones prevent their filtration through the renal glomerulus, greatly prolonging their circulating half-life. Tissues capable of responding to endocrines have 2 properties in common: they posses a receptor having very high affinity for hormone, and the receptor is coupled to a process that regulates metabolism of the target cells.
- **Receptors** for most amino acid- derived hormones and all peptide hormones are located on the plasma membrane. Activation of these receptors by hormones (the first messenger) leads to the intracellular production of a **second messenger.** The main second messengers are cAMP, Ca^{2+}, inositol triphosphate (IP_3), and diacylglycerol (DAG). Proteins are phosphorylated on serine and threonine by cAMP-dependent protein kinase (PKA) and DAG-

activated protein kinase C (PKC). Additionally a series of membrane-associated and intracellular tyrosine kinases phosphorylate specific tyrosine residues on target enzymes and other regulatory proteins. Steroid and thyroid hormones are hydrophobic and diffuse from their binding proteins in the plasma, across the plasma membrane to intracellularly localized receptors. The resultant complex of steroid and receptor bind to response elements of nuclear DNA, regulating the production of mRNA for specific proteins.

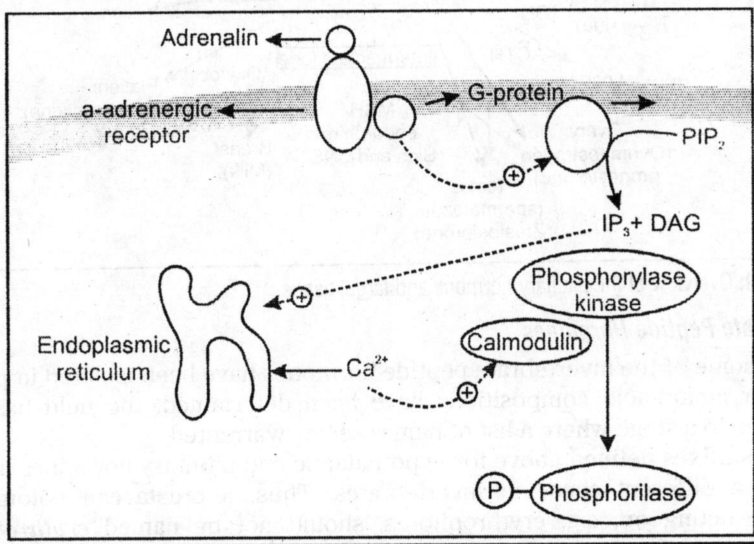

FIG. 18.8: α-receptor mediated responses on phosphorylase. Pathways involved in the regulation of glycogen phosphorylase by epinephrine activation of α-adrenergic receptors.

The endocrine system consider the systems controlled by the hypothalamus and pituitary glands, which are found at the base of the brain. From their sheltered position beneath the brain, they control organs scattered throughout the body, without the use of direct innervation.

TABLE 18.1: Types of pituitary hormones

Anterior Pituitary (adenohypophysis)	Growth Hormone (GH), Thyrotropin, thyroid stimulating hormone (TSH), Corticotropin, adrenocorticotropic hormone (ACTH), Follicle stimulating hormone (FSH), Luteinizing hormone (LH), Prolactin
Posterior Pituitary (neurohypophysis)	Vasopressin, antidiuretic hormone (ADH), oxytocin.
Intermediate Pituitary	Melanocyte stimulating hormone (MSH), Corticotropin-like intermediate lobe peptide (CLIP), ACTH 18-39 AA, Enkephalins, Endorphins

Pituitary Hormones: Most of the hormones of the adenohypophysis have acceptable trivial names ending in- *tropin*. This suffix should be restricted to pituitary and similar hormones and should not be used for, e.g. crustacean hormones acting on pigment cells.

Some placental hormones are physiologically very similar to pituitary hormones. They are named accordingly with the prefix- *chorio*, e.g. choriogonadotropin for chorionic gonadotropin.

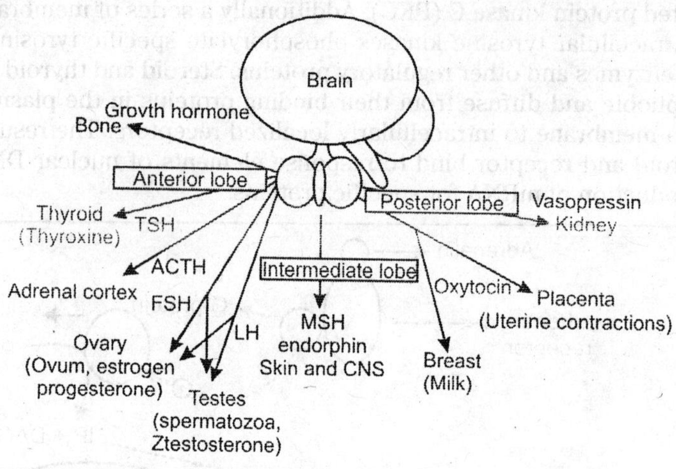

FIG. 18.9: Overview of the pituitary hormone and target tissue.

Invertebrate Peptide Hormones

Though some of the invertebrate peptide hormones have been isolated in pure form and their amino-acid compositions have been determined, the field has not yet developed to a stage where a list of names seems warranted.

The suffixes defined above for hypothalamic and pituitary hormones are not be used in a different sense in invertebrates. Thus, a crustacean colour change hormone acting on, e.g. erythrophores, should not be named *erythrotropin*, a hormone causing release of eggs and/or sperm in sea urchins should not be called *gametoliberin*.

TABLE 18.2: List of Peptide Hormones

Trivial name	Other names	Current abbreviation
Hypothalamic Factors		
Corticoliberin	Corticotropin-releasing factor	CRF
Folliberin	Follicle-stimulating-hormone-releasing factor	FSH-RF
Gonadoliberin	Gonadotropin-releasing factor	(LH/FSH-RF)
Luliberin	Lutemizing hormone-releasing factor	LH-RF (LRF)
Melanoliberin	Melanotropin-releasing factor	MFR
Melanostatin	Melanotropin release-inhibiting factor	MIF
Prolactoliberin	Prolactin-releasing factor	PRF
Prolactostatin	Prolactin release-inhibiting factor	PIF
Somatoliberin	Somatotropin-releasing factor; growth hormone-releasing factor	SRF GH-RF
Somatostatin	Somatotropin release-inhibiting factor	
Thyroliberin	Thyrotropin-releasing factor	TRF
Pituitary and Related Hormones		
Choriogonadotropin	Chorionic gonadotropin	CG
Choriomammotropin	Chorionic somatomammotropin	CS
Corticotropin	Adrenocorticotropic hormone	
Follitropin	Follicle-stimulating hormone	FSH
Gonadotropin	Gonadotropin hormone	
Glumitocin	[Ser^4, Gln^8]Ocytocin[j]	
Isotocin	[Ser^4, Ile^8]Ocytocin[j]	

Trivial name	Other names	Current abbreviation
Lipotropin	Lipotropic hormone	LPH
Lutropin	Luteinizing hormone; (Interstitial cell-stimulating hormone)	LH (ICSH)
Melanotropin	Melanocyte-stimulating hormone	MSH
Mesotocin	[Ile^8]Ocytocin[i]	
Ocytocin (Oxytocin)		OXT
Prolactin	Mammatropic hormone; mammatropin; lactotropic hormone; lactotropin	PRL
Somatotropin	Somatropic hormone; growth hormone	STH GH
Thyrotropin	Thyrotropic hormone	TSH
Urogonadotropin	(Human) Menopausal gonadotropin	HMG
Vasopressin	Adiuretin; antidiuretic hormone	VP, ADH
Vasotocin	[Arg^8]Ocytocin[i]	
Other Peptide Hormones		
Angiotensin	Angiotensin II	
Bradykinin	Kinin-9	
Calcitonin	Thyrocalcitonin	
Erythropoietin		
Gastrin		
Gastrin sulphate	Gastrin II	
Glucagon	Hyperglycemic factor	(HGF)
Insulin		
Kallidin	Kinin-10	
Pancreozymin	Cholecystokinin	
Parathyrin	Parathyroid hormone; Parathormone	
Proangiotensin	Angiotensin I	
Relaxin		
Secretin		
Somatomedin	Sulphation factor	
Thymopoietin	Thymin	

Types

Prolactin (PRL)

Prolactin is produced by acidophilic pituitary lactotropes. Prolactin is the lone tropic hormone of the pituitary that is routinely under negative control by prolactin inhibiting hormone (PIH), which is now known to be dopamine. Decreased hypophyseal dopamine production, or damage to the hypophyseal stalk, leads to rapid up-regulation of PRL secretion. A number of other hypothalamic releasing hormones induce increased prolactin secretion; as a result, it is unclear whether a specific PRH exists for up-regulating PRL secretion. PRL initiates and maintains lactation in mammals, but normally only in mammary tissue that has been primed with estrogenic sex hormones.

Human Placental Lactogen (HPL)

Human placental lactogen is produced by the placenta late in gestation. At its height it is secreted at a rate of about 1 g/day, the highest secretory rate of any known human hormone. However, little hPL reaches the fetal circulation, and hPL has only about 1% the activity of PRL or GH in producing biological effects, leading some to question its functional importance in humans.

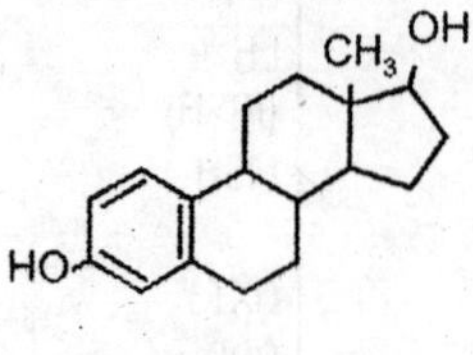

Estradiol 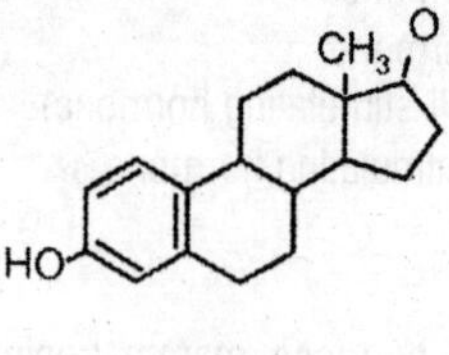 Estrone 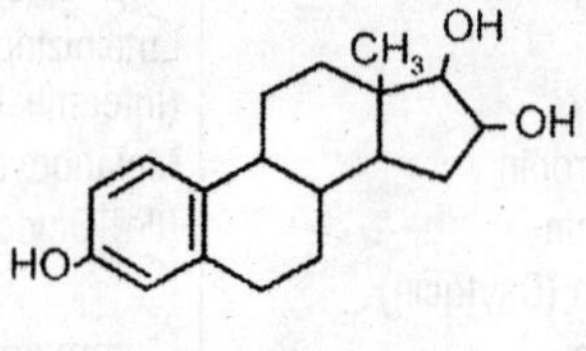 Estriol

The Gonadotropins

The glycoprotein hormones are the most chemically complex family of the peptide hormones. All members of the family are highly glycosylated. Each of the glycoprotein hormones is an (α:β) heterodimer, with a subunit being identical in all members of the family. The biological activity of the hormone is determined by the β-subunit, which is not active in the absence of a subunit. The molecular weight of the gonadotropins FSH, LH, and CG is about 25,000, whereas that of the thyroid tropic hormone TSH is about 30,000. All members of the glycoprotein family transduce their intracellular effects via the receptor, G-protein, adenylate cyclase, second-messenger system.

The gonadotropins (LH, FSH and CG) bind to cells in the ovaries and testes, stimulating the production of the steroid sex hormones estrogen, testosterone (T), and dihydrotestosterone (DHT). In males, luteinizing hormone (LH) binds to Leydig cells of the testes to induce the secretion of T, while follicle stimulating hormone (FSH) binds to Sertoli cells and induces the secretion of T and DHT. In females, LH induces thecal cells to secrete estradiol, and FSH stimulates estrogen synthesis by granulosa cells.

Human chorionic gonadotropin (HCG) is a placental hormone. The production of hCG increases markedly after implantation; its appearance in the plasma and urine is one of the earliest signals of pregnancy and the basis of many pregnancy tests.

Thyroid Stimulating Hormone (TSH)

Secretion of TSH, the final member of the glycoprotein hormone family, is stimulated by TRH from the hypothalamus. While cAMP causes increased secretion of TSH by thyrotropes, it is not yet certain that cAMP is the physiological signal that regulates TSH production. Circulating TSH binds to receptors on the basal membrane of thyroid follicles.

Thyroglobulin produced on rough endoplasmic reticulum has a molecular weight of 660,000. It is glycosylated and contains more than 100 tyrosine residues, which become iodinated and are used to synthesise T3 and T4. Thyroglobulin is exoctosed through the apical membrane into the closed lumen of thyroid follicles, where it accumulates as the major protein of the thyroid and where maturation takes place.

Thyroid hormones act by binding to cytosolic receptors very similar to steroid hormone receptors, and for this reason T3 and T4 are often classified along with the hydrophobic steroid hormones. The principal role of thyroid hormones is also like that of steroid hormones. In adults, the ligand receptor combination binds to thyroid hormone response elements in nuclear DNA and is responsible for up-regulating general protein synthesis and inducing a state of positive nitrogen balance. In the embryo, thyroid hormone is necessary for normal development.

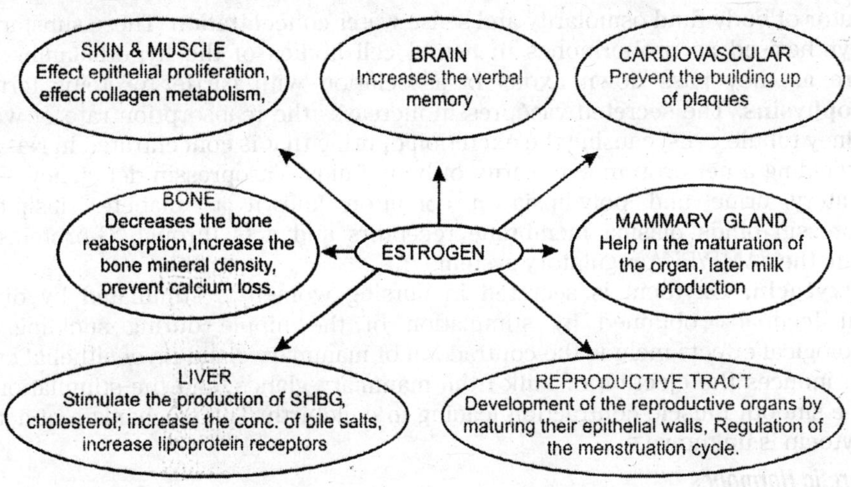

FIG. 18.10: Functions of estrogen.

The Pro-Opiomelanocortin Family

The POMC gene is expressed in both the anterior and intermediate lobes of the pituitary gland. The primary protein product of the POMC gene is a 285 amino acid precursor that can undergo differential processing to yield at least 8 peptides, dependent upon the location of synthesis and the stimulus leading to their production. The important are:

1. **Adrenocorticotropic hormone** (ACTH) and β-lipotropin are products generated in the corticotrophic cells of the anterior pituitary under the control of corticotropin releasing hormone (CRH). The biological role of ACTH is to stimulate the production of adrenal cortex steroids, principally cortisol and costicosterone. The mechanism of action of ACTH involves activation of adenylate cyclase, elevation of cAMP, and increased PKA activity of adrenal cortex tissue.
2. **Alpha-melanocyte stimulating hormone** (α-MSH), corticotropin-like intermediate lobe peptide (CLIP), γ-lipotropin and β-endorphin are products generated in the intermediate lobe of the pituitary under the control of dopamine. α–, β- and γ-MSH are collectively referred to as melanotropin or intermedin.
3. **Corticotropin releasing hormone** (CRH) induces rapid secretion of adrenocorticotropic hormone (ACTH, also called corticotropin) and a variety of other peptides from corticotropes of the anterior pituitary. The rapid CRH-stimulated secretion of ACTH is associated with induction of adenylate cyclase activity and an increase in cAMP and PKA in corticotropes. Longer-term responses of corticotropes to CRH include a marked increase in POMC mRNA.

Many of the other POMC products are synthesised in other neural tissue that contains proteases with appropriate specificity. In human embryos and in pregnant women, the intermediate lobe is active and leads to the production of endorphins and enkephalins.

Vasopressin and Oxytocin

The principal hormones of the posterior pituitary are the nonapeptides **oxytocin** and **vasopressin**.

Vasopressin: It is also known as or antidiuretic hormone [(ADH). It is the main regulator of body fluid osmolarity and sense water concentration. These substances are synthesised as prohormones in neural cell bodies of the hypothalamus and mature as they pass down axons in association with carrier proteins termed **neurophysins**. The secreted vasopressin increases the reabsorption rate of water in kidney tubule cells, causing the excretion of urine that is concentrated in Na^+ and thus yielding a net drop in osmolarity of body fluids. Vasopressin deficiency leads to watery urine and polydipsia, a condition known as diabetes insipidus. Vasopressin binds plasma membrane receptors and acts through G-proteins to activate the cAMP/PKA regulatory system.

Oxytocin: Oxytocin is secreted in nursing women is stimulated by direct neural feedback obtained by stimulation of the nipple during suckling. Its physiological effects include the contraction of mammary gland myoepithelial cells, which induces the ejection of milk from mammary glands, and the stimulation of uterine smooth muscle contraction leading to childbirth. The mechanism of action of oxytocin is unknown.

Natriuretic Hormones

Natriuresis refers to enhanced urinary excretion of sodium. This can occur in certain disease states and through the action of diuretic drugs. Following natriuretic hormones have been identified:

Atrial natiuretic peptide (ANP): It was the first cardiac natriuretic hormone identified. The hormone is secreted by cardiac muscle when sodium chloride intake is increased and when the volume of the extracellular fluid expands. Active ANP is a 28-amino acid peptide containing a 17-amino acid ring formed by intrachain disulphide bonding. Two smaller forms of ANP have also been isolated from the brain.

Brain natriuretic peptide (BNP): It was first isolated from porcine brain and is found in human heart and blood (but not human brain). BNP has various amino acids in its 17-amino acid ring and is encode for by a different gene. In humans, a third natriuretic peptide (CNP) is present in the brain but not in the heart. ANP causes natriuesis presumably by increasing glomerular filtration rate (its exact mechanism of action remains unclear). ANP induces relaxation of the mesangial cells of the glomeruli and thus may increase the surface area of these cells so that filtration is increased. ANP may act on tubule cells and increases sodium excretion. It reduces blood pressure, decreasing the responsiveness of adrenal glomerulosa cells to stimuli that result in aldostreone production and secretion. It inhibits secretion of vasopressin and decreasing vascular smooth muscle cell responses to vasoconstrictive agents. These latter actions of ANP are counter to the effects of angiotensin II. In fact, ANP also lowers renin secretion by the kidneys thus lowering circulating angiotensin II levels.

ANPR-A, ANPR-B and ANPR-C are different ANP receptors. These receptors when bind to receptor, an increase in **guanylate cyclase** activity results leading to production of cyclic GMP (cGMP).

Renin-Angiotensin System

The renin-angiotensin system is responsible for regulation of blood pressure. A rise in pressure results in the release of renin from the juxtaglomerular cells of the kidneys. The only function for renin is to cleave a 10-amino acid peptide from the N-terminal end of angiotensinogen. This decapeptide is called angiotensin I.

Angiotensin II was also refered to as hypertensin and angiotonin. It is one of the most potent naturally occurring vasoconstrictors. Two distinct types of angiotensin II receptors have been identified (AT_1 and AT_2).

Parathyroid Hormone (PTH)

Parathyroid hormone (molecular weight 9,500) is synthesised and secreted by chief cells of the parathyroid in response to systemic Ca^{2+} levels. The Ca^{2+} receptor of the parathyroid gland responds to Ca^{2+} by increasing intracellular levels of PKC, Ca^{2+} and IP_3; this stage is followed, after a period of protein synthesis, by PTH secretion. PTH acts by binding to cAMP-coupled plasma membrane receptors, initiating a cascade of reactions that culminates in the biological response.

Calcitonin (CT)

Calcitonin (CT) is a 32-amino acid peptide secreted by C cells of the thyroid gland. Calcitonin is employed therapeutically to relieve the symptoms of osteoporosis, although details of its mechanism of action remain unclear. However, it has been observed that CT induces the synthesis of PTH in isolated cells, which leads *in vivo* to increased plasma Ca^{2+} levels.

Insulin, Glucagon and Somatostatin

The primary function of the pancreatic hormones is the regulation of whole-body energy metabolism, principally by regulating the concentration and activity of numerous enzymes involved in catabolism and anabolism of the major cell energy supplies.

Insulin: Insulin is a hormone. Insulin is secreted by the pancreas is directly infused via the portal vein to the liver, where it exerts profound metabolic effects. It is a member of a family of structurally and functionally similar molecules that include the insulin-like growth factors (IGF-1 and IGF-2), and relaxin. Its major function is to counter the concerted action of a number of hyperglycemia-generating hormones and to maintain low blood glucose levels. Insulin generally leads to severe hyperglycemia and shortened life span. It is synthesised as a preprohormone in the β cells of the islets of Langerhans. Insulin secretion from β cells is principally regulated by plasma glucose levels, but the precise mechanism by which the glucose signal is transduced remains unclear. Its signal peptide is removed in the cisternae of the endoplasmic reticulum and it is packaged into secretory vesicles in the Golgi, folded to its native structure, and locked in this conformation by the formation of 2 disulphide bonds. In addition to its role in regulating glucose metabolism, insulin stimulates lipogenesis, diminishes lipolysis, and increases amino acid transport into cells. Insulin also modulates transcription, altering the cell content of numerous mRNAs. It stimulates growth, DNA synthesis, and cell replication, effects that it holds in common with the IGFs and relaxin.

One possibility is that the increased uptake of glucose by pancreatic β-cells leads to a concommitant increase in metabolism. The increase in metabolism leads to an elevation in the ATP/ADP ratio. This in turn leads to an inhibition of an ATP-sensitive K^+ channel. The net result is a depolarization of the cell leading to Ca^{2+} influx and insulin secretion.

Glucagon: It is a 29-amino acid hormone synthesised by the α cells of the islets of Langerhans as a very much larger proglucagon molecule. Like insulin, glucagon lacks a plasma carrier protein, and like insulin its circulating half life is also about 5 minutes. The principal effect of glucagon is on the liver, which is the first tissue perfused by blood containing pancreatic secretions. The role of glucagon is well established. It binds to plasma membrane receptors and is coupled through G-proteins to *adenylate cyclase*. The resultant increases in cAMP and PKA reverse all of the effects described above that insulin has on liver. The increases also lead to a marked elevation of circulating glucose, with the glucose being derived from liver gluconeogenesis and liver glycogenolysis.

Somatostatin: It is secreted by δ cells of the pancreas, is a 14-amino acid peptide, identical to somatostatin secreted by the hypothalamus. In neural tissue somatostatin inhibits GH secretion and thus has systemic effects. In the pancreas, somatostatin acts a paracrine inhibitor of other pancreatic hormones and thus also has systemic effects. It has been speculated that somatostatin secretion responds principally to blood glucose levels, increasing as blood glucose levels rise and thus leading to down-regulation of glucagon secretion.

Gastrointestinal Hormones

Although the many gastrointestinal hormones each have significant physiological roles, the focus here will be a discussion of the glucagon-like peptides (principally glucagon-like peptide-1, GLP-1) and glucose-dependent insulinotropic peptide (GIP). Both of these gut hormones constitute the class of molecules referred to as the **incretins**. Incretins are molecules associated with food intake-stimulation of insulin secretion from the pancreas.

GLP-1 is derived from the product of the proglucagon gene. This gene encodes a preproprotein that is differentially cleaved dependent upon the tissue in which it is synthesised. For example, in pancreatic α-cells prohormone convertase 2 action leads to the release of glucagon. In the gut prohormone convertase 1/3 action leads to release of several peptides including GLP-1. The primary physiological responses to GLP-1 are glucose-dependent insulin secretion, inhibition of glucagon secretion and inhibition of gastric acid secretion and gastric emptying. It leads to increased satiety with reduced food intake along with a reduced desire to ingest food. This activity has significance in the context of diabetes.

Glucose-dependent insulinotropic peptide (GIP): It is derived from a 153-amino acid proprotein encoded by the GIP gene and circulates as a biologically active 42-amino acid peptide. GIP is synthesised by enteroendocrine K-cells whose locations are primarily in the duodenum and proximal jejunum. The GIP inhibits gastric acid secretion hence called **gastric inhibitory peptide**. It stimulates lipoprotein lipase activity leading to increased uptake and incorporation of fatty acids by adipocytes. The GIP receptor (GIPR) is a seven-transmembrane G-protein coupled protein found on pancreatic β-cells.

TABLE 18.3: Major peptide hormones, their locations and action

Hormone	Location	Major action
Glucagon-like peptide 1 (GLP-1)	Enteroendocrine L cells predominantly in the ileum and colon	Potentiates glucose-dependent insulin secretion, inhibits glucagon secretion, inhibits gastric emptying
Glucose-dependent insulinotropic polypeptide (GIP) originally called gastric inhibitory polypeptide	Enteroendocrine K cells of the duodenum and proximal jejunum	Inhibits secretion of gastric acid, enhances insulin secretion
Gastrin	Gastric antrum, duodenum	Gastric acid and pepsin secretion
Cholecystokinin (CCK)	Duodenum, jejunum	Pancreatic amylase secretion
Secretin	Duodenum, jejunum	Pancreatic bicarbonate secretion
Vasoactive intestinal peptide (VIP)	Pancreas	Smooth muscle relaxation; stimulates pancreatic bicarbonate secretion
Motilin	Small bowel	Initiates interdigestive intestinal motility
Pancreatic polypeptide (PP)	Pancreas	Inhibits pancreatic bicarbonate and protein secretion
Enkephalins	Stomach, duodenum,	Opiate-like actions

Hormone	Location	Major action
	gallbladder	
Substance P	Entire gastrointestinal tract	Physiological actions uncertain
Bombesin-like immunoreactivity (BLI)	Stomach, duodenum	Stimulates release of gastrin and CCK
Neurotensin	Ileum	Physiological actions unknown
Enteroglucagon	Pancreas, small intestine	Physiological actions unknown

STEROID HORMONES

Steroid hormones are crucial substances for the proper function of the body. They mediate a wide variety of vital physiological functions ranging from anti-inflammatory agents to regulating events during pregnancy. They are synthesised and secreted into the bloodstream by endocrine glands such as the adrenal cortex and the gonads (**ovary and testis**). Steroid hormones are all characterized by the steroid nucleus which is composed of three six member rings and one five member ring, ingeniously labelled A, B, C, and D respectively.

The steroid hormones are all derived from cholesterol (except retinoic acid). Except vitamin D, these hormones contain the same cyclopentano-phenanthrene ring and atomic numbering system as cholesterol. The conversion of C_{27} cholesterol to the 18-, 19-, and 21-carbon steroid hormones (designated by the nomenclature C with a subscript number indicating the number of carbon atoms, e.g. C_{19} for androstanes) involves the rate-limiting, irreversible cleavage of a 6-carbon residue from cholesterol, producing pregnenolone (C_{21}) plus isocaproaldehyde. Common names of the steroid hormones are widely recognized, but systematic nomenclature is gaining acceptance and familiarity with both nomenclatures is increasingly important. Steroids with 21 carbon atoms are known systematically as pregnanes, whereas those containing 19 and 18 carbon atoms are known as androstanes and estranes, respectively. The important mammalian steroid hormones are shown below along with the structure of the precursor, pregneolone. Retinoic acid and vitamin D are not derived from pregnenolone, but from vitamin A and cholesterol respectively.

Classes of Steroid Hormones

The following steroid hormones (glucocorticoids, mineralocorticoids, estrogens, androgens, progestines, and vitamin D) will be defined according to their origin and their major effects.

Glucocorticoids: Glucocorticoids originate in the adrenal cortex and affect mainly metabolism in diverse ways; decrease inflammation and increase resistance to stress.

Mineralocorticoids: Mineralocorticoids originate in adrenal cortex and maintain salt and water balance.

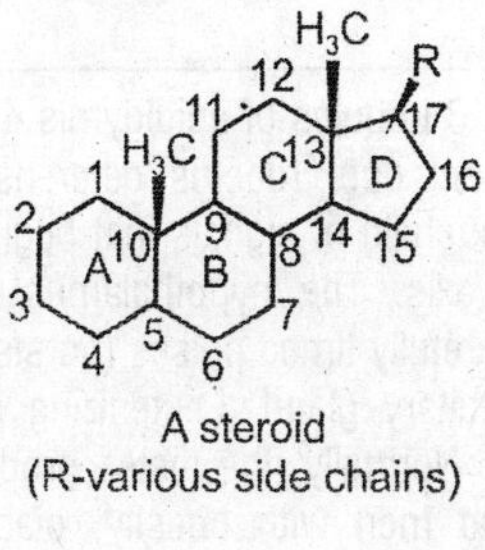

H_3C R 12 11 17 H_3C C 13 D 16 1 2 9 14 15 A 10 B 8 3 5 7 4 6

A steroid
(R-various side chains)

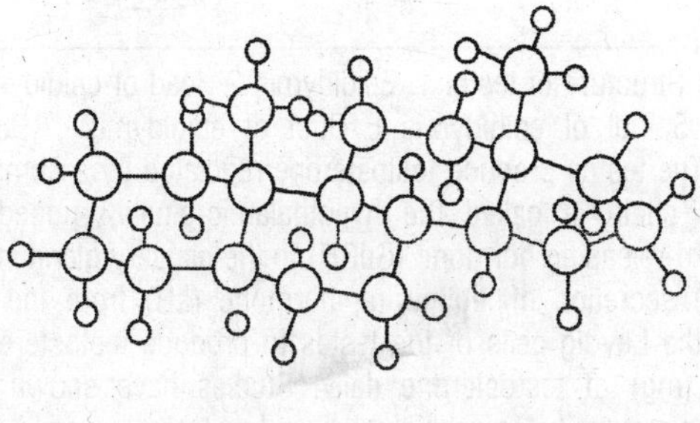

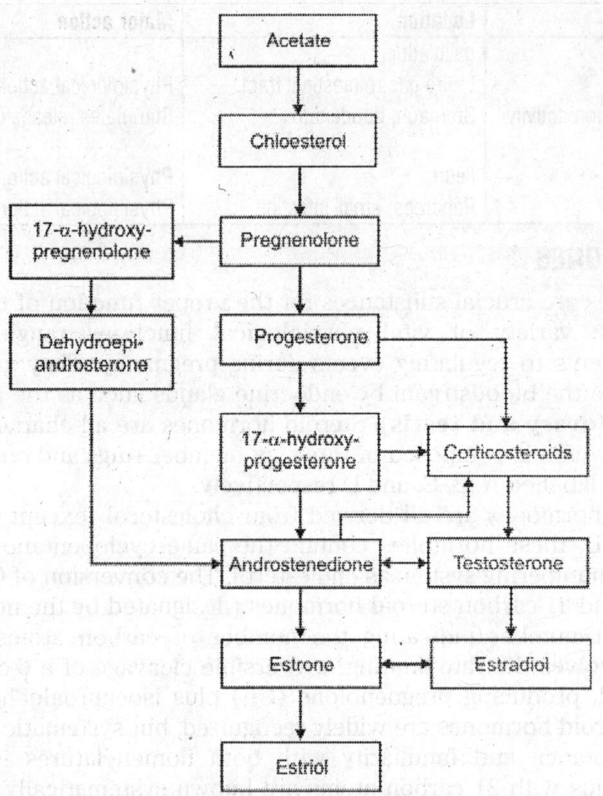

FIG. 18.11: Biosynthetic pathway of steroid synthesis.

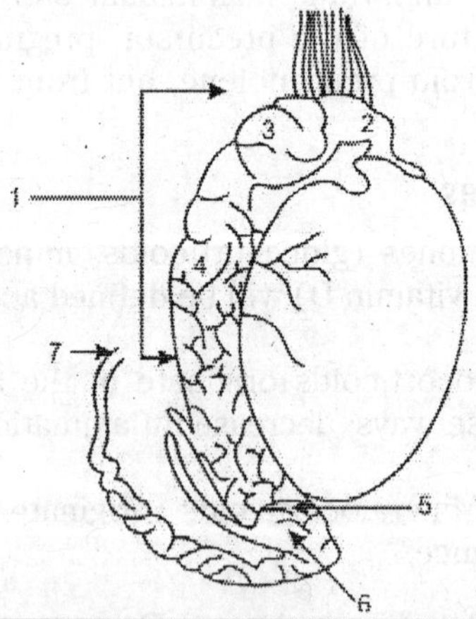

FIG. 18.12: Structure of teste. 1. Epididymis 2 Head of epididymis 3 Lobules of epididymis 4 Body of epididymis 5 Tail of epididymis 6 Duct of epididymis 7 Deferent duct (ductus deferens or vas deferens). The testes produce testosterone regulated by a complex chain of signals that begins in the brain. This chain is called the hypothalamic-pituitary-gonadal axis. The hypothalamus secretes gonadotropin-releasing hormone (GnRH) to the pituitary gland in carefully timed pulses (bursts), which triggers the secretion of leutenizing hormone (LH) from the pituitary gland. Leutenizing hormone stimulates the Leydig cells of the testes to produce testosterone. Normally, the testes produce 4–7 milligrams (mg) of testosterone daily. Studies have shown that men with obesity, diabetes, or hypertension may be twice as likely to have low testosterone levels.

Estrogens: Estrogens originate in the adrenal cortex and gonads and primarily affect maturation and function of secondary sex organs (female sexual determination). Estrogens, or estrone and estradiol are forms of testosterone synthesised in the ovaries, which control female secondary characteristics and regulation of the menstrual cycle. Another sex hormone is needed for preparing the uterus for implantation of the ovum, this hormone is progesterone.

Androgens: Androgens originate in the adrenal cortex and gonads and primarily affect maturation and function of secondary sex organs (male sexual determination). Androgens and estrogens play a major role in the development of both sexes secondary characteristics. Androgens, or testosterone and androsterone give the male its sex characteristics during puberty and for promoting tissue and muscle growth.

Progestins: Progestins originate from both ovaries and placenta, and mediate menstrual cycle and maintain pregnancy.

Steroids of the Adrenal Cortex

The adrenal cortex is responsible for production of three following major classes of steroid hormones:

1. **Glucocorticoids:** They regulate carbohydrate metabolism.
2. **Mineralocorticoids:** They regulate the body levels of sodium and potassium.
3. **Androgens:** Their actions are similar to that of steroids produced by the male gonads.

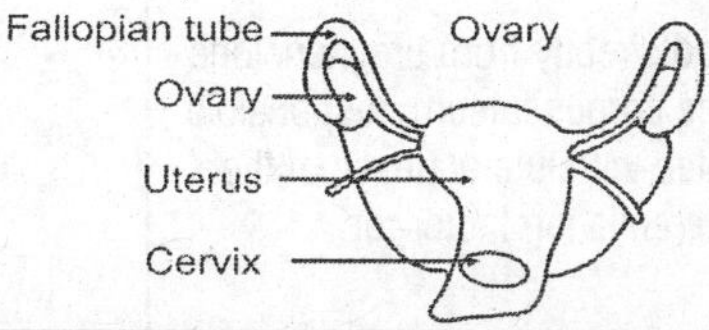

FIG. 18.13: (Left) Position of an ovary in a female. (Right) Section of an ovary.

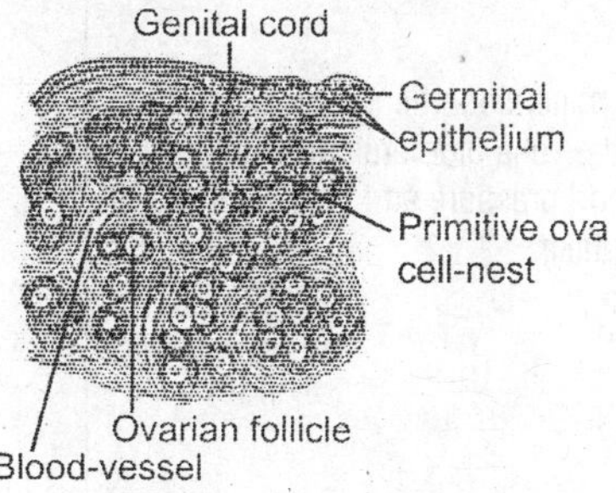

Adrenal insufficiency is known as Addison disease. The disease falls into the same category as pernicious anemia, systemic lupus erythematosus, myasthenia gravis, and Hashimoto thyroiditis, in which an autoimmune basis is suggested by some evidence and in which familial aggregation occurs. In the absence of steroid hormone replacement therapy can rapidly cause death (in 1 to 2 weeks).

Although the pathway to pregnenolone synthesis is the same in all zones of the cortex, the zones are histologically and enzymatically distinct, with the exact steroid hormone product dependent on the enzymes present in the cells of each zone.

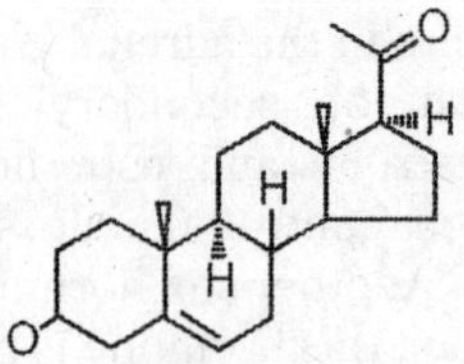

Both androgens and estrogens affect sexual development and function. They regulate sexual differentiation, the secondary sex characteristics, and sexual behavior patterns. A second class of ovarian steroids, C-21 compounds called progestins. Progestins help mediate the menstrual cycle and pregnancy. The mineralocorticoids largely function to regulate the excretion of salt and water by kidney. The glucocorticoids affect carbohydrate, protein, and lipid metabolism in manner nearly opposite to that of insulin, and influence a wide variety of other vital functions including inflammatory reactions and the capacity to cope with stress.

TABLE 18.4: Types of steroid hormones

Pregnenolone	It is produced directly from cholesterol, the precursor molecule for all C18, C19 and C21 steroids	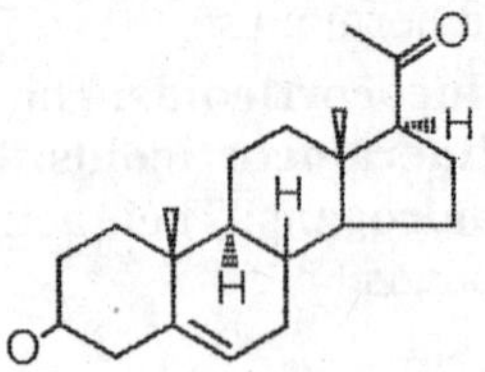
Progesterone	A progestin, produced directly from pregnenolone and secreted from the corpus luteum, responsible for changes associated with luteral phase of the menstrual cycle, differentiation factor for mammary glands.	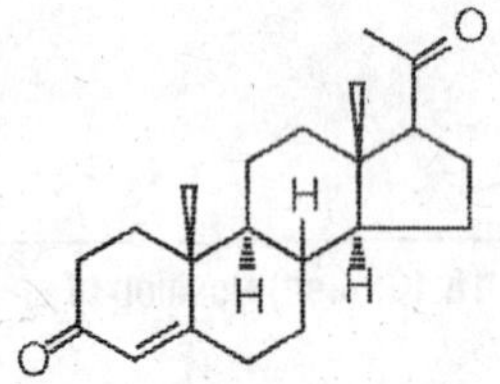
Aldosterone	It is the principal mineralocorticoid, produced from progesterone in the zona glomerulosa of adrenal cortex, raises blood pressure and fluid volume, increases Na^+ uptake.	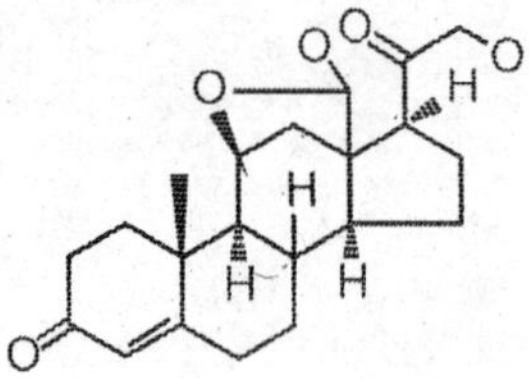
Testosterone	An androgen, male sex hormone synthesised in the testes, responsible for secondary male sex characteristics, produced from progesterone.	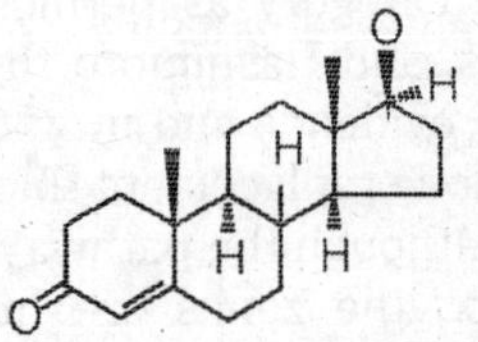

Estradiol	An estrogen, principal female sex hormone, produced in the ovary, responsible for secondary female sex characteristics.	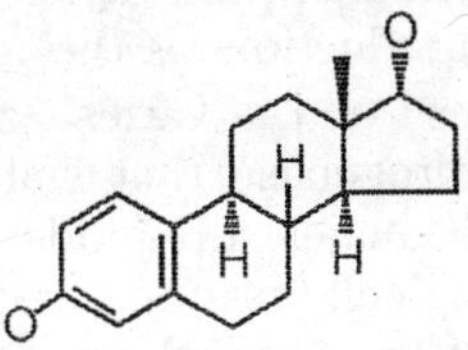
Cortisol	Dominant glucocorticoid in humans, synthesised from progesterone in the zona fasciculata of the adrenal cortex, involved in stress adaptation, elevates blood pressure and Na^+ uptake, numerous effects on the immune system.	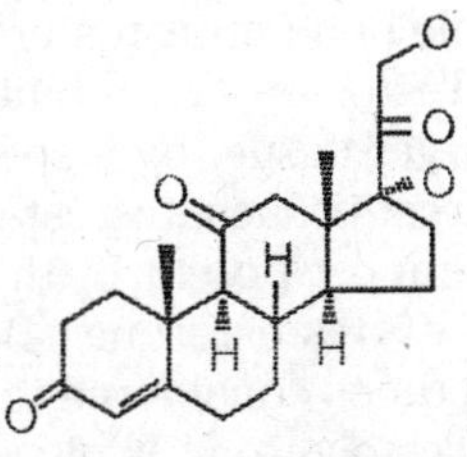

Gonadal Steroid Hormones

Steroids are also produced by the testes and the ovaries, the two most important are **testosterone** and **estradiol**.

Testosterone

Testosterone is a steroid hormone from the androgen group. In mammals, testosterone is primarily secreted in the testes of males and the ovaries of females, although small amounts are also secreted by the adrenal glands. It is the principal male sex hormone and an anabolic steroid.

In both men and women, testosterone plays a key role in health and well-being as well as in sexual functioning.

Testosterone production declines naturally with age. Low testosterone, or testosterone deficiency (TD), may result from disease or damage to the hypothalamus, pituitary gland, or testicles that inhibits hormone secretion and testosterone production, and is also known as **hypogonadism**. Depending on age, insufficient testosterone production can lead to abnormalities in muscle and bone development, underdeveloped genitalia, and diminished virility.

Examples include enhanced libido, increased energy, increased production of red blood cells and protection against osteoporosis. Libido in its common usage means sexual desire. Osteoporosis is a disease of bone that leads to an increased risk of fracture.

These compounds are under tight biosynthetic control, with short and long negative feedback loops that regulate the secretion of **follicle stimulating hormone (FSH)** and **leutinizing hormone (LH)** by the pituitary and **gonadotropin releasing hormone (GnRH)** by the hypothalamus.

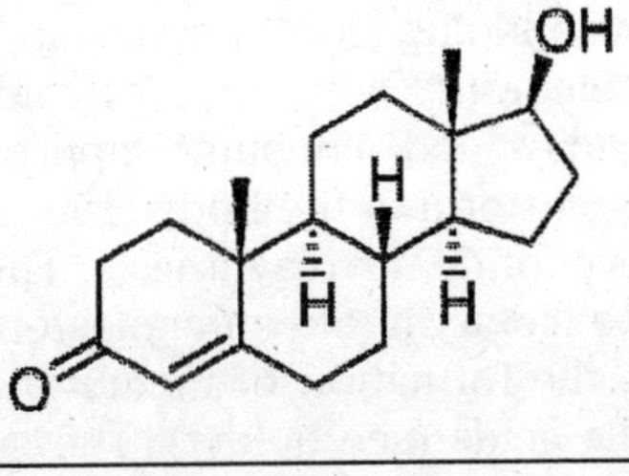

Low levels of circulating sex hormone reduce feedback inhibition on GnRH synthesis (the long loop), leading to elevated FSH and LH. The latter peptide hormones bind to gonadal tissue and stimulate P450ssc activity, resulting in sex hormone production via cAMP and PKA mediated pathways.

The roles of cAMP and PKA in gonadal tissue are the same as that described for glucocorticoid production in the adrenals, but in this case *adenylate cyclase* activation is coupled to the binding of LH to plasma membrane receptors. The biosynthetic pathway to sex hormones in male and female gonadal tissue includes the production of the androgens- androstenedione and dehydroepiandrosterone. Testes and ovaries contain an additional enzyme, a 17β-hydroxysteroid dehydrogenase, that enables androgens to be converted to testosterone.

In males, LH binds to **Leydig cells**, stimulating production of the principal Leydig cell hormone, testosterone. Testosterone is secreted to the plasma and also carried to **Sertoli cells** by androgen binding protein (ABP). In sertoli cells the Δ-4 double bond of testosterone is reduced, producing dihydrotestosterone.

Testosterone and dihydrotestosterone are carried in the plasma, and delivered to target tissue, by a specific gonadal-steroid binding globulin (GBG). In a number of target tissues, testosterone can be converted to dihydrotestosterone (DHT). DHT is the most potent of the male steroid hormones, with an activity that is 10 times that of testosterone. Because of its relatively lower potency, testosterone is sometimes considered to be a prohormone.

Testosterone is also produced by Sertoli cells but in these cells it is regulated by FSH, again acting through a cAMP- and PKA-regulatory pathway. In addition, FSH stimulates Sertoli cells to secrete androgen-binding protein (ABP), which transports testosterone and DHT from Leydig cells to sites of spermatogenesis. There, testosterone acts to stimulate protein synthesis and sperm development. In females, LH binds to thecal cells of the ovary, where it stimulates the synthesis of androstenedione and testosterone by the usual cAMP- and PKA-regulated pathway. An additional enzyme complex known as *aromatase* is responsible for the final conversion of the latter 2 molecules into the estrogens. *Aromatase* is a complex endoplasmic reticulum enzyme found in the ovary and in numerous other tissues in both males and females. Its action involves hydroxylations and dehydrations that culminate in aromatization of the A ring of the androgens.

Aromatase activity is also found in granulosa cells, but in these cells the activity is stimulated by FSH. Normally, thecal cell androgens produced in response to LH diffuse to granulosa cells, where granulosa cell aromatase converts these androgens to estrogens.

Biosynthesis

The synthesis of steroid hormone in a cell depends upon its complement of peptide hormone receptors and its response to peptide hormone stimulation and its genetically expressed complement of enzymes. The following indicates which peptide hormone is responsible for stimulating the synthesis of which steroid hormone:

The first reaction occurs by conversion of cholesterol to C_{18}, C_{19} and C_{21}. Steroids involve in the cleavage of a 6-carbon group from cholesterol which is a sterol which is a natural product derived from the steroid nucleus. Besides being the building block for steroid hormones, cholesterol is also a component of the cell membrane. Normal healthy adults synthesise cholesterol at a rate of approximately 1g/day and consume approximately 0.3g/day. A relatively constant level of cholesterol in the body (150 - 200 mg/dL) is maintained primarily by controlling the level of de novo synthesis. The level of cholesterol synthesis is regulated in part by the dietary intake of cholesterol. Cholesterol from both diet and synthesis is utilized in the formation of membranes and in the synthesis of the steroid hormones and bile acids (see below). The greatest proportion of cholesterol is used in bile acid synthesis. Therefore, cholesterol is the principal committing, regulated, and rate-limiting step in steroid biosynthesis.

Biosynthesis of Progesterone

Progesterone is manufactured in the corpus luteum of the human ovary through the conversion of pregnenolone to progesterone. The theca internal cells of the corpus luteum have all the enzymes necessary to convert cholesterol to estradiol, whereas the granulosa cells- which acquire a rich blood supply after ovulation and follicular rupture–convert pregnenolone to progesterone. Thus, a physiologic increase in progesterone occurs during the luteal phase of the menstrual cycle, probably stimulated by the luteinizing hormone (LH) surge at this time. It is thought that LH stimulates the uptake of acetate into the cholesterol molecule, which forms pregnenolone.

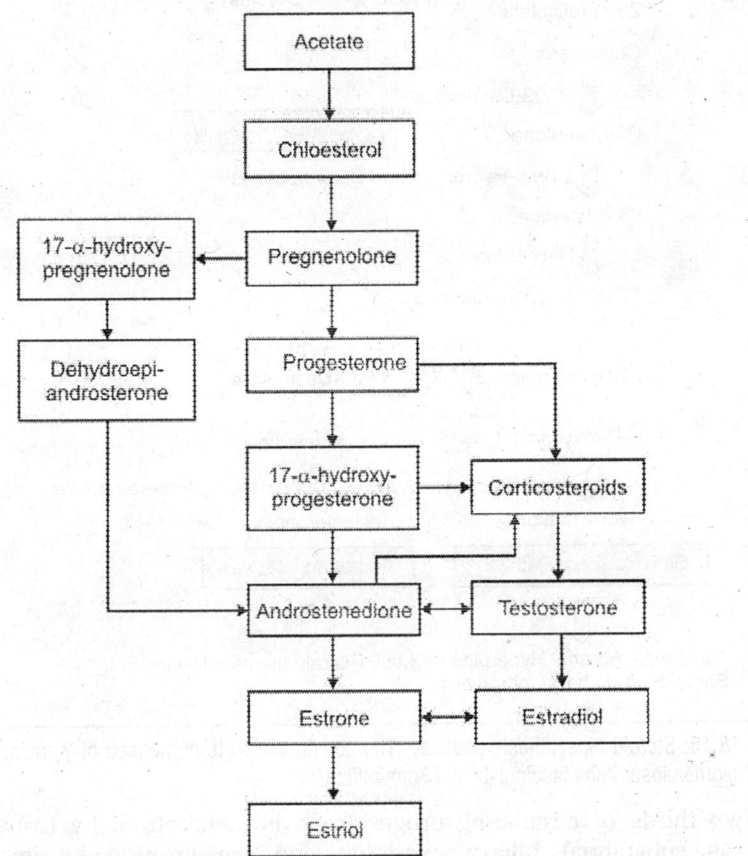

FIG. 18.14: Biosynthetic pathway of steroid synthesis.

Progesterone is also synthesised by the placenta, mainly by hydroxylation of the low-density lipoprotein fraction of cholesterol to pregnenolone, then to progesterone. It is also found in the adrenals and even the uterus in many mammals, including humans. A substantial portion of progesterone is stored in adipose tissue.

Plasma concentrations of progesterone in women vary with the menstrual cycle, with levels during the follicular phase being low, generally under 2 ng/mL. During the luteal phase, levels rise to 2 to 20 ng/mL in a surging pattern after ovulation. In the first trimester of pregnancy, blood levels are about 10 to 40 ng/mL, and they rise to 100 to 200 ng/mL near term. Progesterone levels also vary throughout the day;a decline in plasma levels of as much as 15% occurs 1 hour after a meal and in the early morning hours.

Progesterone is a key precursor to the biosynthesis of cortisol and the C-18 and C-19 steroids, such as androstanedione, estrone, and estradiol. When progesterone circulates in blood, 90% is bound to an albumin fraction, but a specific binding protein has not been identified. Only 3% circulates unbound. Very little progesterone is present in erythrocytes.

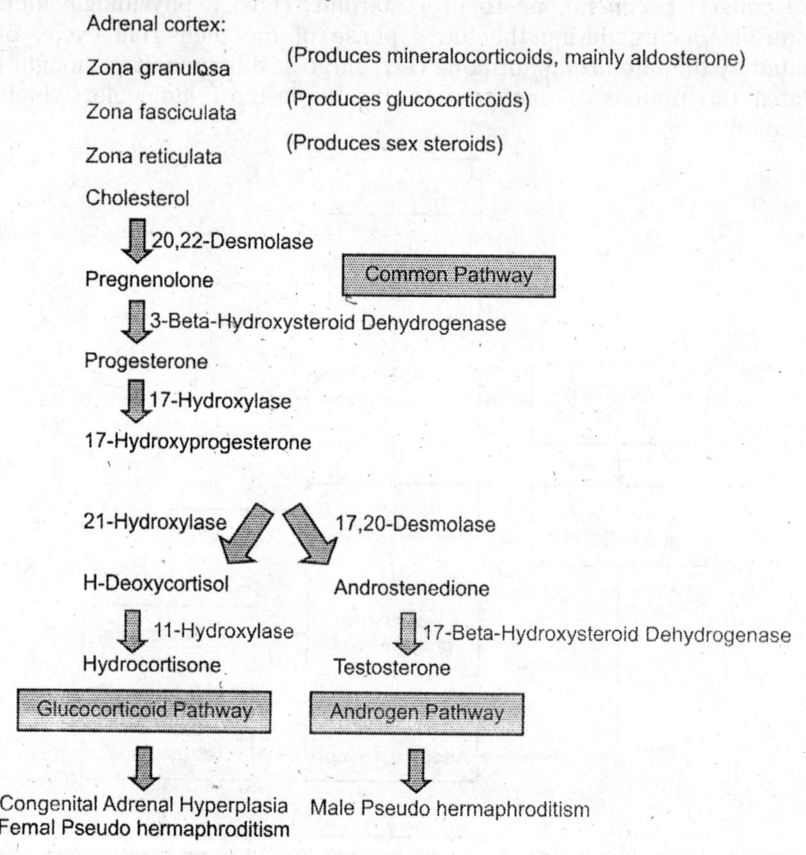

FIG. 18.15: Steroid biosynthesis pathway, The adrenal cortex is composed of 3 main tissue regions: Zona glomerulosa, Zona fasciculate and Zona reticularis.

Two thirds of circulating progesterone is metabolized by liver conjugation; there is substantial biliary excretion and reabsorption of its metabolites. Progesterone is excreted primarily by the kidney. Pregnanediol is one of the major specific urinary metabolites of progesterone; its measurement can be used as an index of endogenous progesterone secretion.

Progesterone formation appears to be regulated by both the stimulatory and inhibitory effects of other steroids and neurochemicals. Stimulatory modulators include ß-adrenergic signals, human chorionic gonadotropin, dehydroepiandrosterone (DHEA), and estrogens. Progesterone has been measured in the adrenal vein of postmenopausal women after corticotropin (ACTH) stimulation. Inhibitory signals come from DHEA-sulphate and androgens.

The enzyme catalyses in reaction the system is P450-linked side chain cleaving enzyme (P450ssc), or desmolase. This is present in the mitochondria of steroid-producing cells, but not in significant quantities in other cells. **Mitochondrial**

desmolase is a complex enzyme system consisting of cytochrome P450, and adrenadoxin (a P450 reductant).

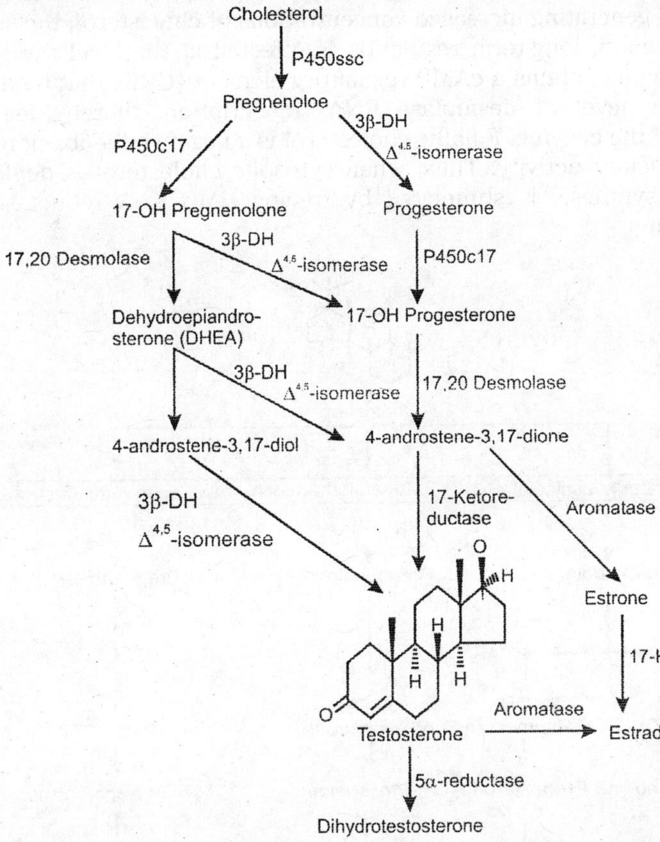

Cholesterol
P450ssc
Pregnenoloe
P450c17
3β-DH
Δ4,5-isomerase
17-OH Pregnenolone
Progesterone
3β-DH
17,20 Desmolase
Δ4,5-isomerase
P450c17
Dehydroepiandro-sterone (DHEA)
17-OH Progesterone
3β-DH
Δ4,5-isomerase
17,20 Desmolase
4-androstene-3,17-diol
4-androstene-3,17-dione
3β-DH
Δ4,5-isomerase
17-Ketore-ductase
Aromatase
Estrone
17-Ketoreductase
Testosterone
Aromatase
Estradiol
5α-reductase
Dihydrotestosterone

FIG. 18.16: Synthesis of the male sex hormones in Leydig cells of the testis. P450SSC, 3β-DH, and P450c17 are the same enzymes as those needed for adrenal steroid hormone synthesis. 17,20-desmolase is the same as 17,20-lyase of adrenal hormone synthesis.

Testosterone
17-Ketoreductase
Androstenedione
Aromatase
Aromatase
17-Ketoreductase
Estradiol
Estrone

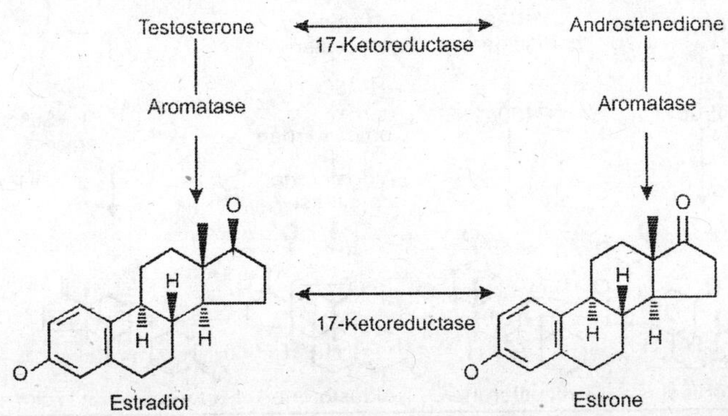

FIG. 18.17: Synthesis of the major female sex hormones in the ovary- testosterone and androstenedione from cholesterol occurs by the same pathways as indicated for synthesis of the male sex hormones.

The activity of each of these components is increased by 2 principal cAMP- and PKA-dependent processes.

First, cAMP stimulates PKA, leading to the phosphorylation of a cholesteryl-ester esterase and generating increased concentrations of cholesterol, the substrate for desmolase. Second, long-term regulation is effected at the level the gene for desmolase. This gene contains a cAMP regulatory element (CRE) that binds cAMP and increases the level of desmolase RNA transcription, thereby leading to increased levels of the enzyme. Finally, cholesterol is a negative feedback regulator of HMG CoA reductase activity. Thus, when cytosolic cholesterol is depleted, *de novo* cholesterol synthesis is stimulated by freeing HMG CoA reductase of its feedback constraints.

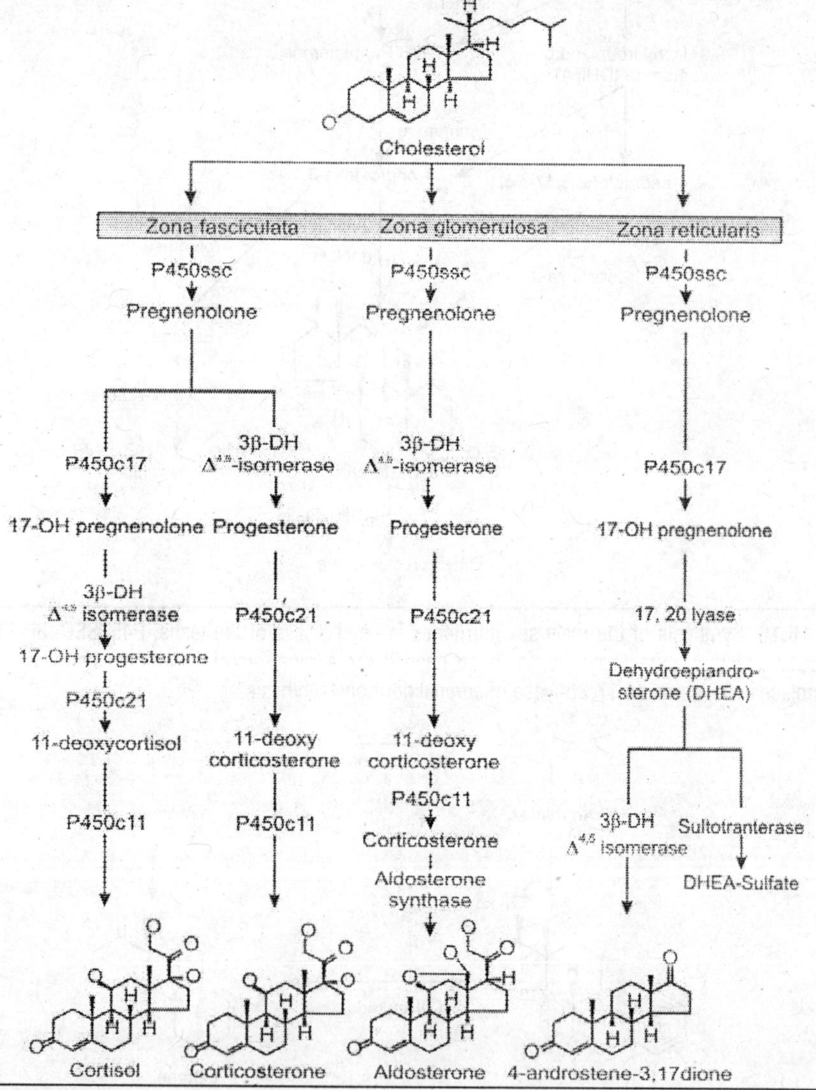

FIG. 18.18: Synthesis of the various adrenal steroid hormones from cholesterol. Only the terminal hormone structures are included. 3β-DH is 3β-dehydrogenase, P450c11 is 11β-hydroxylase, P450c17 is 17α-hydroxylase, P450c21 is 21β-hydroxylase.

Subsequent to desmolase activity, pregnenolone moves to the cytosol, where further processing depends on the cell (tissue) under consideration. The various hydroxylases involved in the synthesis of the steroid hormones have a nomenclature that indicates the site of hydroxylation (e.g. 17α-hydroxylase introduces a hydroxyl group to carbon 17). These hydroxylase enzymes are members of the cytochrome P450 class of enzymes and as such also have a nomenclature indicative of the site of hydroxylation in addition to being identified as P450 class enzymes (e.g. the 17α-hydroxylase is also identified as P450c17).

Zona glomerulosa

The cells of Zona glomerulosa are devoid of *P450c17* which is responsible to convert pregnenolone and progesterone to their C_{17} hydroxylated analogs. Therefore, the pathways to the glucocorticoids (deoxycortisol and cortisol) and the androgens [dehydroepiandosterone (DHEA) and androstenedione] are blocked in these cells. Zona glomerulosa cells are unique in the adrenal cortex in containing the enzyme *P450c18* (or *18α-hydroxylase*) or *aldosterone synthase* responsible for converting corticosterone to aldosterone, the principal and most potent mineralocorticoid.

Zona glomerulosa is mainly responsible for the conversion of cholesterol to the weak mineralocorticoid, corticosteron and the principal mineralocorticoid, aldosterone.

Zona fasciculata and Zona reticularis

The cells of the zona fasciculata and zona reticularis lack *aldosterone synthase* (*P450c18*) that converts corticosterone to aldosterone, and thus these tissues produce only the weak mineralocorticoid corticosterone. However, both these zones do contain the *P450c17* missing in zona glomerulosa and thus produce the major glucocorticoid, cortisol.

The cells of zona fasciculata and zona reticularis also contain the *C17,20 lyase.* The enzyme is responsible for producing the androgens, dehydroepiandosterone (DHEA) and androstenedione. Thus, fasciculata and reticularis cells can make corticosteroids and the adrenal androgens, but not aldosterone.

P450ssc is a mitochondrial activity and its product is pregnenolone. It moves to the cytosol and is converted either to androgens or to 11-deoxycortisol and 11-deoxycorticosterone by enzymes of the endoplasmic reticulum. The latter 2 compounds then re-enter the mitochondrion, where the enzymes are located for tissue-specific conversion to glucocorticoids or mineralocorticoids, respectively.

Regulation of Adrenal Steroid Synthesis

Adrenocorticotropic hormone (ACTH) of the hypothalamus regulates the hormone production of the zona fasciculata and zona reticularis. ACTH receptors in the plasma membrane activate adenylate cyclase with production of the second messenger, cAMP. The effect of ACTH on the production of cortisol is particularly important, with the result that a classic feedback loop is prominent in regulating the circulating levels of corticotropin releasing hormone, (CRH), ACTH, and cortisol.

The gene is located on chromosome 2 at locus 2p23. Elucidation of the organization of the 5α-reductase type 2 gene revealed 5 coding exons. The enzyme has a pH optimum at pH 5.5 in broken cell preparations, but may function optimally in the native state in intact cells at neutral pH and has an apparent M. HighμKm for testosterone of 4-50 nM. The apparent Km for NADPH is 3-10 expression of 5α - reductase type 2 is observed in prostate tissue. Mutations and deletions in the 5α-reductase type 2 gene have been found to be the molecular basis for the syndrome of 5α-Reductase type 2 deficiency.

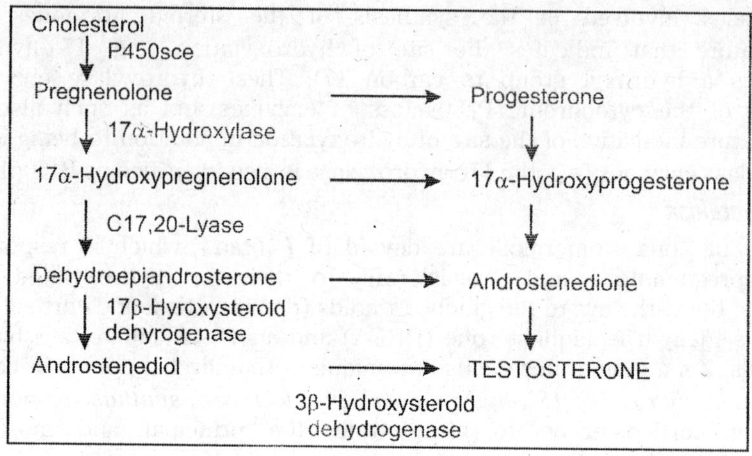

FIG. 18.19: Biosynthetic pathway of testicular Testosterone synthesis.

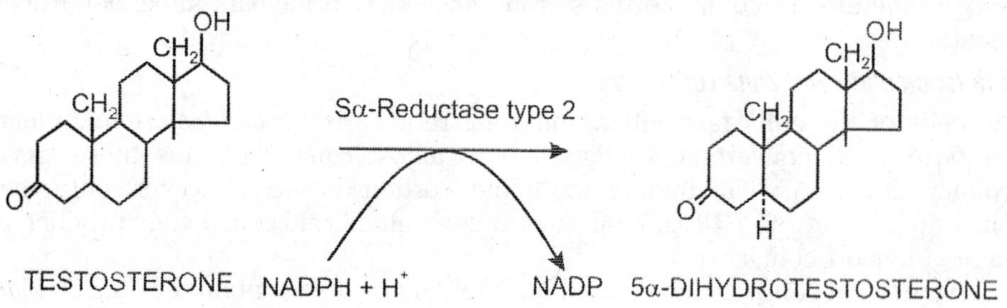

PH optimum:	acidic
apparent Km for Testosterone:	4-50nM
apparent Km for NADPH:	3-10 μM

FIG. 18.20: Metabolism of Testosterone to 5α-Dihydrotestosterone by the enzyme 5α-Reductase type 2 (SDR5A2).

The 5α-reductase type 1 (SRD5A1) isozyme, also catalyzing the conversion of testosterone to 5α-dihydrotestosterone, differs from the type 2 isozyme in its amino acid composition, kinetics, biochemical properties, substrate specificity, tissue distribution and pH optimum. The type 1 enzyme is not involved in androgen dependent male sexual differentiation. The 5α-reductase type 1 cDNA codes for a protein of 259 amino acid residues with a predicted molecular mass of 29,462 Dalton. The gene is located on chromosome 5 at locus 5p15 (11). The apparent Km for testosterone is 1-5 mM and for NADPH 3-10 mM. The expression of 5α-reductase type 1 in prostate tissue is relatively low. The difference in pH optima between 5α -reductase type 1 and type 2 enzymes can be used diagnostically for the differential assessment of the individual isoenzymes in genital skin fibroblasts of patients with the syndrome of 5α -reductase deficiency.

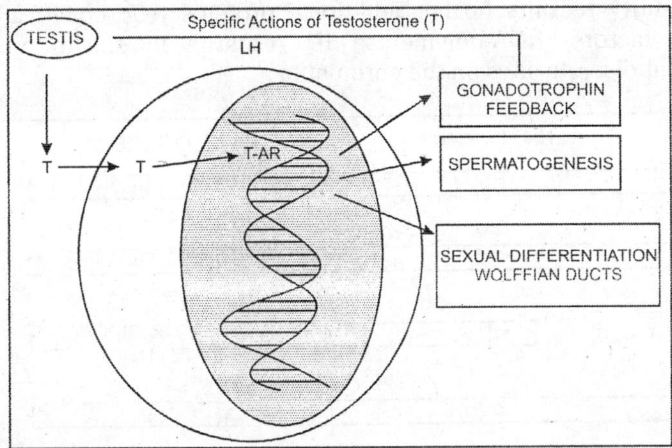

FIG. 18.21: Specific actions of Testosterone (T). T is synthesised in the testis under control of luteinizing hormone (LH) from the pituitary. After entering the target cells (in the hypothalamus, pituitary, testis and wolffian duct) T is directly bound to the androgen receptor (AR) and the complex T-AR binds to specific DNA sequences and regulates gene transcription, which can result in negative feedback regulation of gonadotrophin synthesis and secretion, in initiation and regulation of spermatogenesis and in differentiation and development of the wolffian duct.

Mode of Action

Steroid hormones act by passing through the plasma `membrane and bind with intracellular receptors. The mechanism of action of the thyroid hormones is similar; they interact with intracellular receptors. Both the steroid and thyroid hormone-receptor complexes exert their action by binding to specific nucleotide sequences in the DNA of responsive genes. These DNA sequences are identified as hormone response elements, HREs. The interaction of steroid-receptor complexes with DNA leads to altered rates of transcription of the associated genes.

The Androgen Receptor and the Nuclear Receptor Family

Androgens actions are mediated by the androgen receptor (NR_3C_4; NCIB). This ligand dependent transcription factor belongs to the super family of nuclear receptors. This family includes receptors for steroid hormones, thyroid hormones, all-trans and 9-cis retinoic acid, 1,25 dihydroxy-vitamin D, ecdysone and peroxisome proliferator-activated receptors. In addition an increasing number of nuclear proteins have been identified with a protein structure homologous with that of nuclear receptors, but without a known ligand. These so-called "orphan" receptors form an important subfamily of transcription factors acting either in the absence of any ligand or with as yet unknown endogenous ligands. Comparative structural and functional analysis of nuclear hormone receptors has revealed a common structural organization in 4 different functional domains: a NH_2-Terminal Domain, a DNA-Binding Domain, a Hinge Region and a Ligand Binding Domain.

The current model for androgen action involves a multi step mechanism. Testosterone after entry into the androgen target cell, binds to the androgen receptor either directly or after its conversion to 5α-dihydrotestosterone, followed by dissociation of heat shock proteins in the cytoplasm, simultaneously accompanied by a conformational change of the receptor protein resulting in a transformation and a translocation to the nucleus. After binding in the nucleus to specific DNA-sequences the receptor dimerizes with a second molecule and the homodimer entity recruits further additional proteins (e.g. coactivators, general transcription factors, RNA-polymerase II) resulting in specific activation of transcription at discrete sites on the chromatin.

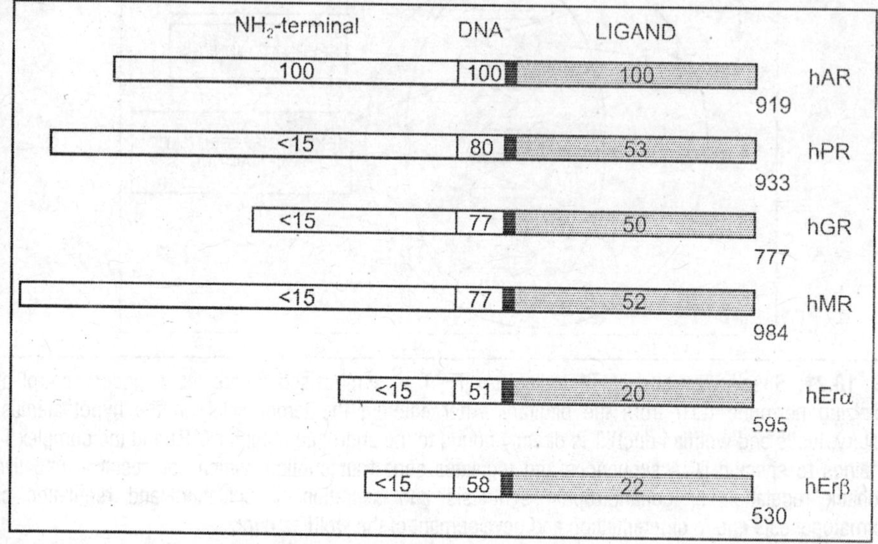

FIG. 18.22: Sequence homologies between the human Androgen Receptor (hAR), human Progesterone Receptor (hPR), human Glucocorticoid Receptor (hGR), human Mineralocorticoid Receptor (hMR) and the human Estrogen Receptors α (hERα) and β (hERβ).

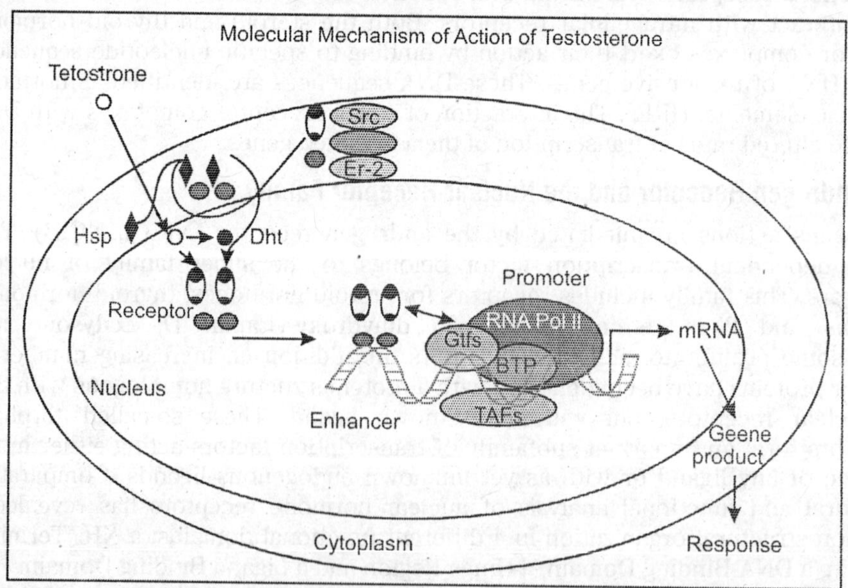

FIG. 18.23: Simplified model of androgen action in an androgen target cell.

Androgen receptor is the key protein and binds directly with testosterone or its active metabolite 5α-dihydrotestosterone (DHT). After dissociation of heat shock proteins (hsp) the receptor enters the nucleus via an intrinsic nuclear localization signal. After binding in the cytoplasm or in the nucleus, the androgen receptor binds as homodimer to specific DNA elements which are present as enhancers in upstream promoter sequences of androgen target genes. The next step is recruitment of co-activators and can form the communication bridge between receptor and several components of the transcription machinery. The direct and indirect communication of the androgen receptor complex with several components of the transcription machinery (e.g. RNA-polymerase II [RNA-Pol II], TATA box binding protein [TBP], TBP associating factors [TAF's], general transcription factors [GTF's]) are key events in nuclear signaling. This communication triggers subsequently mRNA synthesis and consequently protein synthesis, which finally results in an androgen response. A non-genomic pathway involving the classical androgen receptor via cross-talk with the Src/Raf-1/Erk-2 pathway is also known.

Interestingly androgen signaling via the androgen receptor can also occur in a non-genomic, rapid and sex-nonspecific way by crosstalk with the Scr, Raf-1, Erk-2 pathway (19, 20). The classical androgen receptor is also involved in androgen-mediated induction of Xenopus oocyte maturation via the (MAPK)-signalling cascade in a transcription independent way (21).

Cloning and structural organization of androgen receptor gene: Since cloning of the human androgen receptor cDNA our insights into the mechanism of androgen action has been increased tremendously. Only one androgen receptor cDNA hve been identified and cloned. Despite of the two different ligands. The androgen receptor gene is located on the X-chromosome at Xq11.2 -12 [approx. 186 kb] and codes for a protein with a molecular mass of approx. 110 kDa. The gene consists of 8 coding exons and the structural organization of the coding exons is essentially identical to those of the genes coding for the other steroid hormone receptors (e.g. exon/intron boundaries are highly conserved). Androgen receptor polymorphisms have also been identified.

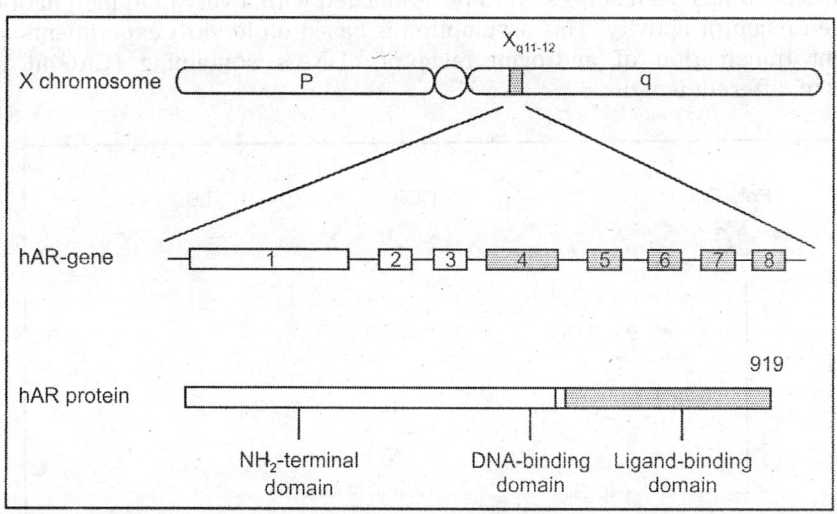

FIG. 18.24: Human androgen receptor gene on long arm of the X-chromosome. The human androgen receptor protein (919 amino acid residues) is encoded by 8 exons.

Steroid Hormone Receptors

Ligand-activated proteins are the receptors of steroid hormones. They bind with the receptor and regulate transcription of selected genes. Similar to peptide hormone receptors, that span the plasma membrane and bind ligand outside the cell, steroid hormone receptors are found in the cytosol and the nucleus. The steroid hormone receptors belong to the steroid and thyroid hormone receptor super-family of proteins. These include receptors for steroid hormones, thyroid hormones, vitamin D and vitamin A (retinoic acid).

When these receptors bind ligand they undergo a conformational change and bind to specific nucleotide sequences in DNA called **hormone-response elements** (HREs). The ligand-receptor complexes interact with DNA and alter the transcriptional level (responses can be either activating or repressing) of the associated gene. Thus, the steroid-thyroid family of receptors has following three distinct domains:

1. A ligand-binding domain.
2. A DNA-binding domain.
3. A transcriptional regulatory domain.

Binding of thyroid hormone to its receptor results in release the of receptor from DNA.

Androgen Receptor Polymorphisms

The androgen receptor DNA - and ligand-binding domains have a high homology with the corresponding domains of the other members of the steroid receptor subfamily.

There is a remarkably low homology between the androgen receptor NH_2^- terminal domain and that of the other steroid receptors. A poly-glutamine stretch, encoded by a polymorphic (CAG)nCAA - repeat is present in the NH_2^- terminal domain. The length of the repeat has been used for identification of X-chromosomes for carrier detection in pedigree analyses.

Variation in length (9 - 38 glutamine residues) is observed in the normal population and has been suggested to be associated with a very mild modulation of androgen receptor activity. This assumption is based on in vitro experiments after transient transfection of androgen receptor cDNA's containing (CAG)nCAA - repeats of different lengths.

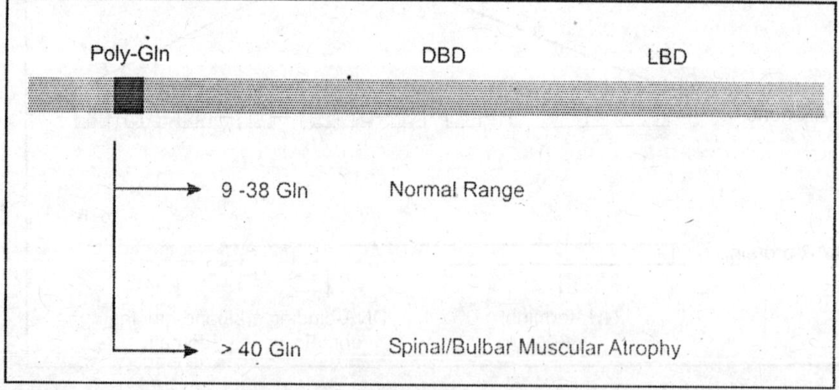

FIG. 18.25: Variation in the polyglutamine stretch in human androgen receptors, encoded by the (CAG)nCAA repeat in exon 1. The normal range of this stretch is 9 - 38 glutamine residues, while in the motor neuron disease spinal/bulbar muscular atrophy (also called Kennedy's disease) the (CAG)nCAA repeat in exon 1 is expanded and codes for more than 40 Glutamine residues.

The functional domain structure androgen receptors are as follows:

The NH_2^- terminal domain: The androgen receptor NH_2^- terminal domain harbors the major transcription activation functions. Within its 537 amino acids, two independent activation domains have been identified: activation function 1 (AF-1) (located between residues 101 and 370) that is essential for transactivity of full length AR, and activation function 5 (AF-5) (located between residues 360-485) that is required for transactivity of a constitutively active androgen receptor, which lacks its LBD.

Another function of the androgen receptor NH_2^- terminal domain is its binding to the COOH-terminal LBD (55, 56). The NH_2^-terminal regions required for the binding of the LBD have been mapped to two essential units: the first 36 amino acids and residues 370-494.

The DNA-binding domain: The DNA-binding domain is the best conserved among the members of the receptor super family. It is characterised by a high content of basic amino acids and by nine conserved cysteine residues. More recently also 3D-information became available for AR-DNA interaction at different types of response elements.

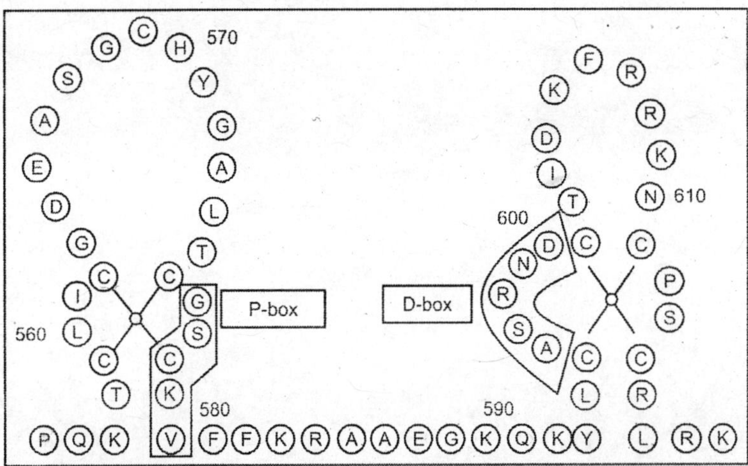

FIG. 18.26: Structure of the DNA binding domain of the human androgen receptor. The protein structure is represented in the one-letter code. The domain consists of two zinc cluster modules, which are stabilized by the coordination binding of a zinc atom (red dot) by 4 cysteine residues. The first zinc cluster contains the P-box (proximal box), of which three residues determine androgen response element recognition. The second zinc cluster contains the D-box (distal box) in which amino acids are located that are involved in protein-protein interactions with a second receptor molecule in the homodimer complex.

Briefly, the DNA-binding domain has a compact, globular structure in which two substructures can be distinguished. Both substructures contain centrally one zinc atom, which interacts via coordination bonds with four cysteine residues.

The hinge region: Between the DNA-binding domain and the ligand-binding domain a non-conserved hinge region is located, which is also variable in size in different steroid receptors. The hinge region can be considered as a flexible linker between the ligand-binding domain and the rest of the receptor molecule.

The ligand-binding domain: This domain is encoded by approximately 250 amino acids, residue in the C-terminal end of the molecule. The crystal structure of the human androgen receptor ligand binding in complex with the synthetic ligand methyltrienolone (R1881) and 5α-dihydrotestosterone, respectively, have been determined.

Anti-Androgens

Androgen receptor antagonists are compounds that interfere in some way in the biological effects of androgens and are frequently used in the treatment of androgen-based pathologies. The steroidal anti-androgens, cyproterone acetate (CPA) and RU38486 (RU486; mifepristone), have partial agonistic and antagonistic actions.

CHAPTER 19

Plant Hormones / Growth Regulators

STATUS AND DEFINITION

The conventional definition of plant hormone originally proposed by Fitting (1909) has met with several difficulties of explaining the different responses and reactions of several compounds, both synthetic and natural. Pincus and Thimann (1948) proposed definitions for these two categories of substances. Plant hormone has been defined as *organic substance produced naturally in the higher plants, controlling growth or other physiological functions at a site remote from its place of production, and active in minute amounts.* To distinguish it from animal hormone, it was termed as phytohormone. However, Trewavas (1982) has started a discussion about abolishing the name of plant growth substances as hormone. This idea was hotly debated at the XI International Conference of Plant Growth Substances (1982). The crisis of the phytohormone theory is caused mainly by the following problem: a poor correlation was found between growth rate and endogenous hormone levels in developing plants. Further, unexpected results were frequently found with exogenously applied hormone, i.e., in some cases, promoters cause an inhibition and inhibitors a promotion of growth. Plant growth regulators are defined as *organic compounds other than nutrients, which in small amount promotes/inhibit or otherwise modify any physiological response in plants.* The definitions adopted by the American Society of Plant Physiology in 1954 are as follows:

Plant Hormones: *Plant hormones are regulators produced by plants which in low concentration regulate a physiological plant processes. Hormones usually move within plant from a site of production to site of action.* There are five classes of phytohormones: auxins, gibberellins, cytokinins, abscisic acid and ethylene.

The Plant Hormone Concept: Terminology and the concept of the plant hormones are largely derived from the mammalian concept of a hormone. This involves a localized site of synthesis, transport in the bloodstream to a target tissue and the control of a physiological response in the target tissue via the concentration of the hormone (Trewavas, 1981). Auxin was similarly thought to produce a growth response at a distance from its site of synthesis and thus to fit the definition of a transported chemical messenger. It is now clear that plant hormones do not fulfill the requirements of a hormone in the mammalian sense. There is lack of direct rigid parallels between animal hormones he has argued that we should abandon the term **plant hormone**. In a rebuttal Cleland (1983) has characterized this argument as **Semantic Quibbling**. Trewavas suggests we should replace the term hormones with **plant growth substances**. The important plant growth substances are as follows:

1. **Auxins:** These are organic substances, which at low concentration (less than 0.001 M) promote growth along the longitudinal axis, when applied to shoot of plants freed as far as placing from their own inherent growth promoting substances.
2. **Gibberellins:** These substances are having gibbane ring skeleton capable of producing the same physiological responses as gibberellic acid that it must be active in specific gibberellin bioassay. The gibberellins are phytohormones, which are active in regulating dormancy, flowering, fruit setting, and stimulating germination of seeds and extending growth of shoots.
3. **Cytokinins:** These are substances composed of hydrophilic group of higher specificity (adenine) and one lipophilic group without specificity. The CKs form a group of plant hormones having similar effects as those of GA in breaking the dormancy of a wide range of seeds and in increased fruit set. These hormones mainly stimulate cell divisions and prevent chlorophyll degradation.
4. **Abscisic acid (ABA):** ABA is a naturally occurring sesquiterpene, which regulate plant growth and metabolism in various ways and have been detected in nearly all plants. It is involved in the abscission of plant organs, induction and vegetative buds, in regulation of fruits repining and generally in reduction of growth.
5. **Ethylene:** It is the only gaseous hydrocarbon hormone, which plays an important role in the ripening of fruits, inhibition of root growth, abscission and other growth processes.

Unlike the other hormones, ABA and ethylene are not discovered through any interaction with fungi.

Brassinosteroids: In 1979 a novel plant growth-promoting steroidal lactone, termed brassinolide was found in the pollen of *Brassica napus.*

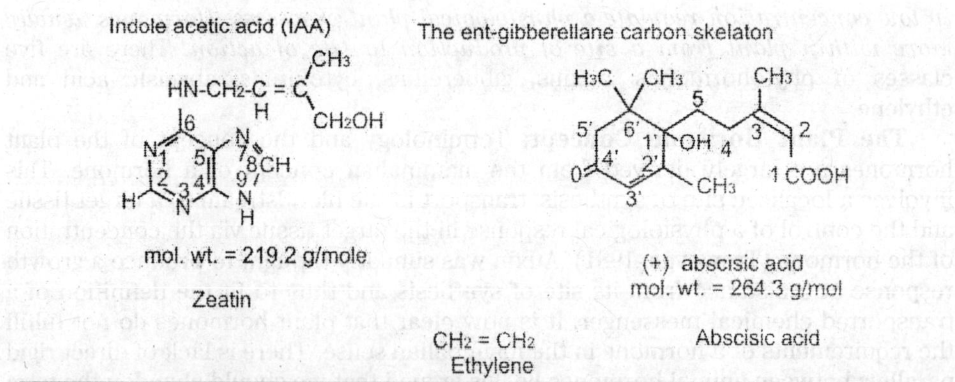

indoleacetic acid (IAA)
mol. wt. = 175.2 g/mole

Indole acetic acid (IAA)

The ent-gibberellane carbon skelaton

mol. wt. = 219.2 g/mole

Zeatin

(+) abscisic acid
mol. wt. = 264.3 g/mol

Abscisic acid

$CH_2 = CH_2$
Ethylene

FIG. 19.1: Structures of plant hormones.

Other Plant Compounds which Affect Growth

There are several other plant products, which have been shown to modify plant growth in different bioassays. Most of these compounds are inhibitors though their natural role remains to be established *in vivo*. In some cases their interaction is well characterized as growth hormones has also been postulated. In this category phenolic compounds are also included. We shall discuss the significance of these compounds in interacting with IAA oxidase and regulating the IAA levels in later discussion. Of the 550 secondary plant constituents nearly more than 200 are shown to be inhibitory in nature. For details the reader may consult a treatise by MacMillan (1980). We shall briefly mention some of these constituents:

Aromatic compounds: These compounds include simple phenols and benzoic acid to the polymeric tannins. It also includes cinnamic acids, the coumarins and the flavonoids. In fact next to carbohydrates these compounds are met with in abundance in plants. In plants these compounds occur as glycosides.

Nitrogen containing compounds: Hydrogen cyanide is liberated enzymatically from cyanogenic glucosides and inhibits growth. Amygdalin is also highly inhibitory and is believed to be responsible for the 'peach re-plant' problem. Several of the alkaloids are reported to affect plant growth processes e.g. colchicine, delcosine, lycorine, cocaine, cinchonin, quinone, caffeine, codeine, berberine, etc. Several of the non-protein amino acids are also shown to have variable effects on plant growth.

Terpenoid compounds: Several naturally occurring terpenoids from plants are shown to affect plant growth. These include monoterpene-essential oils (growth inhibitors), α-methylene lactones (both growth inhibition and root stimulation), and unsaturated lactones.

Aliphatic alcohols: Several of the naturally occurring aliphatic alcohols, aldehydes, ketones, acids and esters are reported to affect plant growth. Besides the compounds mentioned so far, extracts of leaves of *Eucalyptus grandis* have yielded rooting inhibitors; ascorbic acid is shown to have variable effects on plant growth and has been suggested as plant hormone.

Triazoles: New types of plant growth retardants developed (as fungicides) in recent years include triazoles, a group of compounds known as ergosterol biosynthesis inhibitors. These compounds include paclobutrazol, Lab-117-682, uniconazol, triapentenol, Bas-111, fluruprimidol, etc. These compounds have greater broad spectrum retarding ability than earlier retardants-CCC, Mepiquat chloride and diaminozide, etc. They inhibit endogenous production of gibberellic acid; control excessive shoot growth.

Polyamines: Polyamines have a wide range of effects in plants and seem to be essential for cell division, normal morphology and hence plant development. It appears that polyamines are ubiquitous and universally present in all cells rather than having a specific site of synthesis.

With increasing cost of energy, PGRs were sure to play an increasingly important role in energy conservation as a consequence of enhanced yield following their use. The dire need to double world's food supply by the end of the 21st century is also another significant incentive to improve agricultural production through novel approaches e.g. PGRs technology. Plant growth regulators are known to bring about diversity of effects, which include effects on plant functions e.g., induction of rooting, control of flowering, regulation of sex and control of ageing. Additionally PGRs are also shown to control the shape, size or growth rate. Several of the biochemical processes and suitable metabolic changes are affected by PGRs leading to alterations in both quality and quantity of economic products. In the near future it is very likely that novel PGRs will be developed and rationally used for crop improvement. In fact these chemicals open new opportunities to improve agriculture and horticulture by circumventing several barriers imposed by genetics and environment.

Other Growth Regulators

New other growth regulators have been identified in different plant species and these are briefly described below:

- **Lunaric acid:** In liverworts, lunaric acid is present in gemmae formed in gemma cups and prevents germination of these gemmae unless the acid is leached out. Lunaric acid is reported to be present in many species of lower plants excluding algae.
- **Batasins:** These chemicals cause dormancy in bulbils of *Dioscorea batatas* and arise from swelling of the aerial lateral buds. The precise mechanism of its action and transport is not well understood.
- **Jasmonic acid:** Jasmonic acid and its methyl ester are widely distributed in plant species and are isolated from oil of jasmine. They are biosynthesised from free linoeic acid through the action of lipoxygenase. These compounds promote senescence and inhibit growth of some plant species.
- **Turgorins:** Some German chemists have isolated and identified compounds that activate pulvini in *Mimosa* and *Acacia* leaves. Compounds like ß-glucosides of gallic acid with the glucoside bonding at its para-hydroxyl group, ß-D-glucoside-6-sulphate, ß-D-glucoside-3, 6-disulphite of gallic acid, etc., are included in this category. These compounds are called turgorins since they act on the turgor of pulvinus cells and are active at very low concentrations (10-5 to 10-7 M).
- **Tuberonic acid:** Tuberonic acid is closely related to jasmonic acid and has strong potato tuber-inducing activity. It has been isolated from the leaves of the potato plant and identified as 3-oxo-2-(5-ß-D-glucopyranosyloxy-2-in-pentenyl) cyclopentane-1-acetic acid. The substance isolated is a glucoside and aglycone of the substance is called **tub onic acid**.
- **Cucurbic acid:** This is an inl ...r isolated from cucumber seedlings.

QUANTIFICATION OF PLANT HORMONES

The ability to identify and accurately measure plant hormones has become increasing important in studies on the physiology and biochemistry of plant hormones. The analysis of plant hormones in plant extracts poses many technical problems, primarily because only minute amounts are present. Their concentration in fresh material rarely exceeds 1 part in 10^6 but is often as low as 1 part in 10^{10} and therefore they require analytical procedures, which are able to measuring or rather pg quantities of hormone present in the extracts. In addition, plant extracts contain many more low molecular compounds than animal tissue and these are difficult to separate from the low molecular plant hormones. Thus, plant hormone analysis is not yet a routine procedure and at the present time not one particular method can be singled out as the best way to analyse plant hormones.

Extraction: During the extraction of plant hormones, plant material is homogenized to ensure maximum cellular disruption. Selection of an appropriate solvent for extraction is one of the more confusing aspects of plant hormone analysis. Although 80% aqueous methanol is most commonly selected, there are numerous reports of other solvent systems. In particular, the use of aqueous methanol to extract cytokinins has been questioned because phosphatases are active in this solvent. The use of Bieleski's mixture (methanol, chloroform and formic acid) overcomes this problem, thus minimizing degradation of cytokinin ribotides to ribosides. However, care must be taken when adopting complex solvent mixtures because they can solubilize compounds, which may interfere with analysis methods such as immunoassay.

Ethylene poses different extraction problems from the other plant hormones. It readily diffuses out of plant tissues and can be collected by incubating material in a gas–tight container. Samples can then be withdrawn for analysis. The beauty of this method is that it is non–destructive and by sequential sampling, determinations of the emanation of ethylene can be made during the course of a developmental event.

Purification: solvent partitioning and chromatography purify crude plant hormone extracts. The solvents commonly employed include chloroform, diethyl ether and ethyl acetate. Purification procedures are designed to reduce contamination by interfering molecules, but by their very nature reduce plant hormone recovery. To correct this loss, the extract is spiked prior to purification with a known quantity of an internal standard. Two types of internal standard have been employed. The first is purified plant hormone labelled with ^3H or ^{14}C to allow independent quantification at the end of the extraction procedure. The reliability of this approach is dictated by the degree of degradation of the standard, which takes place during the clean–up process. Ideally, this should be the same as the plant hormone being purified.

Identification and Quantification

Bioassays: Historically, bioassays have played a central role in botanical research, since they have been instrumental in the discovery and isolation of every major group of plant hormone. Many bioassays have been developed but for acceptance into general use they must be reproducible, sensitive and exhibit a high degree of specificity. In practice these specifications are often difficult to achieve, since extracts may contain compounds, which interfere synergistically or antagonistically with the assay. For this reason, a variety of assays based on different biological responses have been developed. Parameters such as cell elongation, cell division, and pigment accumulation or enzyme activity have all been employed to assay for different groups of plant hormones.

Immunological methods: These methods rely on the generation of antibodies to plant hormones, which have been coupled to carrier molecules such as bovine serum albumin (BSA) to enhance their antigenecity. In most cases, between 5 and 15 mol of hormone per mol of BSA have been used successfully to immunize rabbits or mice. From the collected blood, serum is produced by standard procedures and this is kept frozen at –20°C or lower temperature. Serum fractions from individual bloods and from different animals must be kept separately until screened for titer and gross selectivity. Only then should appropriate serum fractions be pooled. A single animal usually yields sufficient antiserum from successive bleedings to last for 10^5 to 10^7 immunoassay samples.

By coupling the carrier to plant hormone at different sites, it is possible to generate antibodies exhibiting different selectivity. This approach has been successfully used to produce antibodies capable of distinguishing between free and conjugated forms of ABA. Quantification using antibodies can be achieved by either radioimmunoassay (RIA) or enzyme–linked immunoassay (ELISA). In addition to offering a method of quantification, antibodies to plant hormones can be used in other ways. For example, immunoaffinity columns are used as a powerful tool for plant hormone purification. In this technique, antigenic molecules are encouraged to bind to antibodies immobilized to a column support and can therefore be separated from co–extracted compounds.

Gas Chromatography: Separation of volatile compound can be readily achieved using gas chromatography techniques. With the aid of suitable detectors, the compound may be identified and quantified. By simple derivatizations all plant hormones can be made amenable to GC analysis. In general, the stable volatiles that can be generated are the product of methylation or trimethylation reactions. These have been recent advances in the use of GC detectors for plant hormones.

Gas Chromatography–Mass Spectrometry: The selectivity problem of GC analysis may be resolved by linking a mass spectrometer to the system as a functional detector. The mass spectrometer analyses the fragments released from ionized compounds on the basis of their mass to charge ratio. Every compound produces a characteristic mass spectrum from which it can be identified by comparison with spectra from known standards. This technique has been used very effectively for the identification of trace amounts of hormones.

High Performance Liquid Chromatography: The great advantage of HPLC in the analysis of plant hormones lies in its powerful separation capabilities. HPLC is comparatively rapid and normally does not require the rigorous pre–purification procedures necessary for GC–MS analysis.

TABLE 19.1: Analytical methods for plant growth regulators

Plant growth regulator	Method of analysis	Problems associated with analysis
Auxins	GC–MS HPLC–F Immunoassay	Chemical and enzymic oxidation
Gibberellins	GC–MS Immunoassay	Ring arrangements oxidation
Cytokinins	GC–MS HPLC–UV Immunoassay	Isomerization to *trans*, *trans*–configuration
Ethylene	GC–FID GC–PID	Contamination by other volatiles from enclosing vessels

Detectors: EC, electron capture; FID, flame ionization; F, fluorescence; MS, mass spectrometer; PID, photoionization; UV, ultraviolet.

HOW HORMONE ACTS?

When a hormone acts upon a responsive plant system, it of course, enters into some direct and molecular interactions, which results, eventually in manifestation of measurable effects (biochemical or physiological responses). However, there are two aspects of hormone action:

1. **Mechanism of action:** The direct and specific molecular interaction, and
2. **Mode of action:** The succeeding series of steps (events and processes) which results in the measurable biochemical or physiological responses.

According to Paleg (1960) the two terms *Mechanisms* and *Mode* of action should be taken as different terms rather than synonyms. In the last years, numerous physiological and biochemical responses to plant hormones have been observed which have a bearing on the mechanism of hormone action. Some of the responses have been enumerated by Gordon (1952), they are as follows:

1. Increased plasticity of the shoot and elasticity of the root cell wall.
2. Increased permeability to water.
3. The enhanced capacity to retain water taken up.
4. The *active* uptake of water and solute apparently independent of osmotic forces and occurring even against an osmotic gradient.
5. A decrease in protoplasmic viscosity.

6. An accelerated rate of respiration and cyclosis.
7. More rapid synthesis of protein, with lowered levels of free amino acids.
8. The increase of monosaccharide at the expense of reserve polysaccharide.
9. An increment of (intermicellar) wall pectin and of cellulose and a stimulation of alcohol and malic dehydrogenase, catalase, phosphatase and ascorbic acid oxidase.
10. These large varieties of effects which hormone evokes depict the multiplicity of sites.

TABLE 19.2: List of hormones and their physiological responses*

Effects	Auxin	GA	CK	ABA	Ethylene
Cell elongation	+	+	+	-	-
Cell division	+		+		-
Apical dominance	+		-		
Leaf growth					
In length		+			
In breadth			+		
Chlorophyll formation			+		
Vein elongation	+				
Leaflet movement	+				
Root formation	+				
Fibre growth in cotton seed	+	+	-		
Abscission	-	-		+	+
Sex expression					
Maleness	-	+		-	
Femaleness	+		+		
Fruit growth and division	+	+	+	+	+
Xylem formation	+	+			
Flower	+	+	+	-	-
Dormancy break					
Seed dormancy		+	+	+	-
Bud dormancy		+	+	+	-
Senescence		-	-	+	+
Tropism	+	+		+	

+ promote, - inhibit, * After Sen, S.P. (1984).

An understanding of hormone action is dependent on location of sites of action (Sen, 1984). According to Overbeck (1966) *we are beginning to learn that the site of action of hormones is close to gene.* Usually two sites of plant hormone action at the subcellular/molecular level are discussed: an early (*primary*) effect at the cell membrane and a subsequent (*secondary*) effect at genetic level (Parthier, 1985).

SIGNAL TRANSDUCTION BY PLANT HORMONES

For optimal growth and development plant requires the ability to coordinate complex developmental processes and at the same time sense and respond to endogenous physiological factors and environmental cues. The responses can be divided into three main events: stimulus perception, generation and transmission of a signal and subsequent changes in downstream biochemical processes. For this process to be successfully executed specialized receptor molecules, which perceive the endogenous incoming cue, are required to generate chemical and/or protein effector molecules to transmit the signal in order to trigger subsequent downstream responses. A long–standing objective of plant hormone biology has been to elucidate the molecular basis of signaling systems that transduce hormone messages into the wide range of responses that they regulate. The last few years have witnessed substantial progress in our understanding of plant intracellular signaling mechanism and some of these have a direct bearing on hormone signal transduction.

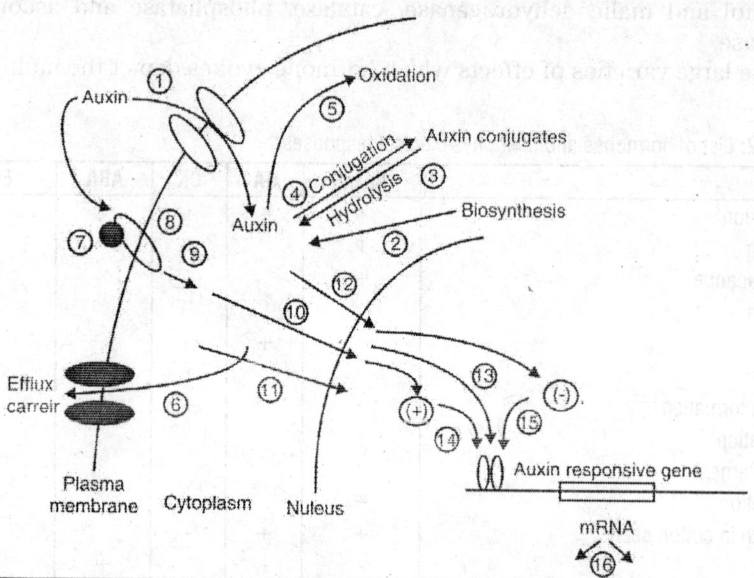

FIG. 19.2: Scheme showing auxin signal transduction. This scheme, which is a working model, aims to draw together some of the potential elements of auxin signal transduction pathways. Auxin external to the cell can enter the cell directly, or indirectly via a channel or permease. (1) Once within the cell, free auxin could form part of a pool whose amount is controlled by *de novo* synthesis (2), hydrolysis of conjugates (3), conjugation (4), breakdown by oxidation (5), or transport via the naphthylphthalamic acid/2,3,5–triiodobenzoic acid–inhibited auxin efflux carrier (6), Alternatively, external auxin could bind with a receptor that transduces a signal indirectly (7), or directly (8), across the plasma membrane, triggering a signal transduction cascade (9), This signal cascade may converge with a signal response triggered by the internal pool of auxin (10), or run parallel with it (11), These cascades culminate in changes in gene expression. Changes in gene expression could be triggered directly by auxin itself (12), or by elements of the putative signal transduction cascade (10,11). Changes in gene expression might be caused by direct interaction with transcription complexes (13), activation of transcriptional activators (14), or inactivation of repressors (15), Triggering gene expression will then, in turn, activate genes mediating the effects of auxin action (16).

Auxin

Using *Arabidopsis*, several groups have adopted genetic approaches to identify a collection of auxin–response mutants (Hobbie et al., 1994). One of these mutants, auxin–resistant 1 **(axr1)** was isolated as an auxin–resistant mutant that has a pleiotropic phenotype associated with a decreased response to auxin. In addition, **axr1** plants are deficient in auxin–regulated gene expression, suggesting that **AXR1** is required for auxin signal transduction. The **AXR1** gene encodes a protein that participates in activation of the ubiquitin–related protein **RUB1**. The auxin transport inhibitor resistant 1 **(tir1)** mutant also exhibits reduced auxin response. **TIR1** is a member of the F–box family of proteins. Studies in yeast and animals indicate that F–box proteins are part of an **E3** ubiquitin–ligase complex called the **SKP1 CDC53 F–box (SCF)**.

The cloning and characterization of these–response gene suggest that auxin signaling is mediated through the ubiquitin pathway. Recent molecular data on **AXR1** and **TIR1** function in *Arabidopsis* are providing new insight into auxin response. The first step in the ubiquitin pathway is the activation of ubiquitin or ubiquitin like protein by ubiquitin–activating enzyme **E1**. *Arabidopsis* **AXR1** was found to encode a protein with amino acid sequence homology to the N–terminus of **E1**.

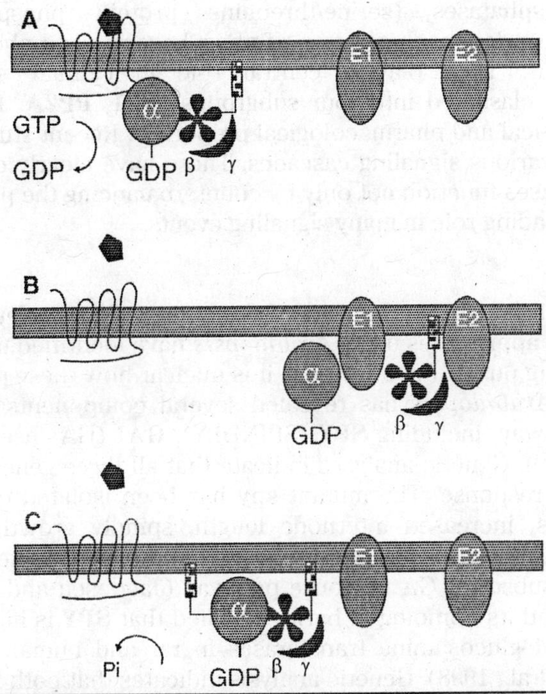

FIG. 19.3: Diagram depicting stages in the G protein cycle. **A**, Ligand activated GPCR interacts with G protein stimulating GDP/GTP exchange; **B**, activated G protein dissociates into G_α subunit and G_β/G_γ subunit complex. These can each interact with downstream effector molecules (E1 and E2). **C**, Inherent GTPase activity of the G_α subunit hydrolyses the bound GTP and the heterotrimeric form is re–established ready to enter another round of the cycle.

Earliest evidence suggests a possible involvement of **G–protein** in auxin signaling. One emerging signaling pathway in plants is the **G–protein-signaling cascade**. The **G**–protein signaling pathway is one of the most highly conserved membranes–located signaling mechanisms and is responsible for transducing the action of a wide variety of extracellular signaling molecules. **G**–protein–coupled receptors **(GPCRs)** are a primary component of this pathway and constitute a large and functionally diverse superfamily of integral membrane proteins. The DNA and deduced amino acid sequence of more than 700 **GPCRs** is known and all have seven stretches of between 20 and 28 hydrophobic amino acids capable of forming a transmembrane ∝–helix. **GPCRs** transverse the membrane seven times forming a helical cluster with associated extracellular and intracellular loops. **GPCRs** specifically recognize a huge range of ligands including plant hormone. When **GPCRs** bind their cognate ligand they undergo conformational changes that enable them to interact with heterotrimeric **GTP**–binding proteins **(G–proteins)**. These **G–proteins** consist of ∝, β and γ subunits which are associated with the cytoplasmic face of the plasma membrane. **G–protein** activation by **GPCRs** stimulates the exchange of **GDP** for **GTP** in the guanine nucleotide-binding site of the G_α subunit. This causes conformational changes in the G_α subunit that promotes dissociation from the $G\beta/\gamma$ complex. Both the **GTP–G_α** subunit and the **$G\beta/\gamma$** complex are capable of signaling to effectors. The inherent **GTPase** activity of the **G_α** subunit restores the inactive heterotrimer with bound **GDP** ready for another cycle of stimulation by the **GPCR**.

Protein phosphatases (serine/threonine protein phosphatases: PPase) specifically catalyse dephosphorylation of phosphoserine and phosphothreonine in protein substrates. They play a central role in cellular signaling. Protein phosphatases are classified into four subgroups (**PP1**, **PP2A**, **PP2B** and **PP2C**) based on biochemical and pharmacological properties. Recent studies place protein phosphatases in various signaling cascades. These have clearly demonstrated that protein phosphatases function not only by counterbalancing the protein kinases but also by taking a leading role in many signaling events.

Gibberellins

Little is known about the mechanism of GA action. Cereal aleurone system and molecular genetic approaches using *Arabidopsis* have identified many of the pieces of the GA–signaling puzzle but at present it is unclear how these pieces fit together. Recent work in *Arabidopsis* has revealed several components of the GA signal transduction pathway, including **SPY** (SPINDLY), **GAI** (GA–insensitive) and **RGA** (repressor of $ga1^{-3}$). Genetic analyses indicate that all three genes encode negative regulators of GA response. The mutant **spy** has been isolated on the basis of its light–green leaves, increased internode length, spindly growth habit and poor fertility. As this mutant is still responsive to GA, the **SPY** protein most probably only regulates a subset of GA response pathway (Jacobsen and Olszewski, 1993). Cloning of **SPY** and its homolog in barley revelled that **SPY** is highly similar to the O–linked N–acetyl–glucosamine transferases in rat and human (Jacobsen et al., 1996, Robertson et al., 1998). Genetic analysis indicates that both N–and C–terminal domains of O–linked N–acetylglucosamine transferases **(OGTs)** participate in GA signal transduction and **TPR** (tetratricopeptide repeat) domains participate in protein–protein interactions (Das et al., 1998). The mutational inactivation of the **SPY** protein would cause constitutive activation of a subset of the GA pathways.

The mutant, **gai**, shows reduced sensitivity to GA (Peng et al., 1997). The **gai** gene encodes a protein of 532 amino acid residues with a high degree of sequence homology to **SCARECROW** (SCR) at the C–terminal region and N–terminus of **GAI** contains a putative nuclear localization signal, which is absent in **SCR**. The **gai**–1 allele suffers a 51 bp in–frame deletion generating a mutant protein with an internal deletion of 27 amino acid residues at the N–terminus. This **gai** mutant protein acts most likely as a dominant negative mutant, thus explaining the semi–dominate nature of the mutation. A null allele of **gai** confers a weak **spy**–like phenotype hence it was proposed that **GAI** acts as a repressor of GA signaling. The repressor function of the **GAI** protein is inactivated by GA treatment but the **gai** mutant protein is refractory. Strong alleles of **spy** are able suppress the **gai** mutant phenotype, suggesting that **SPY** and **GAI** participate in the same pathway with **SPY** acting upstream of **GAI**.

Changes in phospholipid and choline metabolism are among the earliest effects of GA on wheat aleurone cells and provide support for lipid based GA signaling pathway. Within 2h of exposure to GA_3, the activities of two enzymes involved in phospholipid biosynthesis, phosphorylcholine–cytidyl transferase and phosphorylcholine–glyceride transferase increase substantially. During the lag phase between GA_3 addition and the induction of ∝–amylase there is a marked change in the rate of turnover of N–methyl and methylene carbons of the phosphotidyl choline head groups. In non–GA_3–treated controls, N–methyl carbons turn over more rapidly than methylene carbons. GA_3 promotes phospholipid breakdown and changes the pattern of phosphotidyl choline metabolism such that the whole choline head group rather than N–methyl carbon turn over. It has been suggested that this changes in phosphotidyl choline turn over may be a component of a lipid–based GA signal transduction pathway in aleurone.

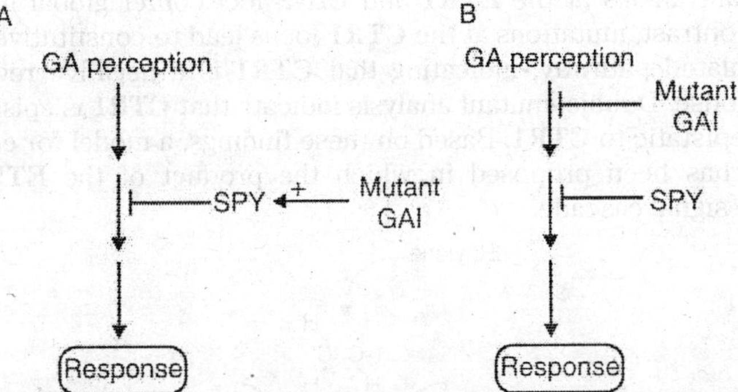

FIG. 19.4: Models of the role of SPY and the mutant GAI protein in the GA signal–transduction pathway that can explain the epistasis of spy–4–over *gai*. A, The mutant, GAI protein positively regulates (directly or indirectly) SPY activity, which in turn negatively regulates GA signal transduction. B, The mutant GAI protein and SPY independently negatively regulate GA signal transduction, and SPY acts downstream. Arrows may represent more than one biochemical event in the transduction pathway.

Cytokinin

Despite its importance in cell division and various differentiation processes, the pathway through which cytokinin transmits its signal is poorly understood. Using activation T–DNA tagging, Kakimoto (1998) isolated a gene **(CKI 1)** whose overexpression in plant calli confers a cytokinin–independence phenotype. The **CKI 1** gene encodes a protein with sequence homology to a histidine kinase domain as well as a receiver domain suggesting that two–component signal system may be involved in cytokinin responses. Several *Arabidopsis* cDNA clones encoding putative responses regulators have been isolated but their roles in cytokinin signaling are not clear at present.

Ethylene

Ethylene–dependent processes in plants are determined both by the production of the hormone and by the competence of cells of respond. Production occurs in response to developmental and environmental cues such as pathogenesis, wounding or pollination. Although the biosynthetic pathway leading to ethylene production is now known, the molecular events that stimulate basal and stress levels of ethylene production are much less clear. By contrast, as a result of simple genetic screens, much has been learned about the competence of cells of respond to ethylene.

During the past decade, considerable progress has been made in the genetic and molecular dissection of the ethylene–response pathway. A number of loci involved in ethylene signaling have been identified largely on the basis of the isolation of ethylene response mutants in *Arabidopsis* (Johnson and Ecker, 1998; Kieber, 1997; Bleecker et al., 1998). Many of these genes have been cloned, defining a pathway from ethylene perception to changes in gene expression. The seedling bioassay has been used to isolate mutants in *Arabidopsis* that are either insensitive to applied ethylene such as **etr1** (ethylene triple response mutant) or **ein3** (ethylene–insensitive mutant) type or show a constitutive triple response in the absence of added ethylene such as **ctr1** (constitutive triple response mutant). Several classes of mutants obtained using the seedling bioassay as a screen affect a broad range of ethylene responses throughout the life cycle of the plant indicating that these responses share a primary signal transduction pathway. In particular, multiple mutant alleles at the **ETR1** and **EIN2** loci confer global insensitivity to ethylene. In contrast, mutations at the **CTR1** locus lead to constitutive activation of ethylene regulated pathway, indicating that **CTR1** is a negative regulator of the ethylene response. Double–mutant analysis indicate that **CTR1** is epistatic to **ETR1** and **EIN2** is epistatic to **CTR1**. Based on these findings, a model for ethylene signal transduction has been proposed in which the product of the **ETR1** gene acts earliest in the signal cascade.

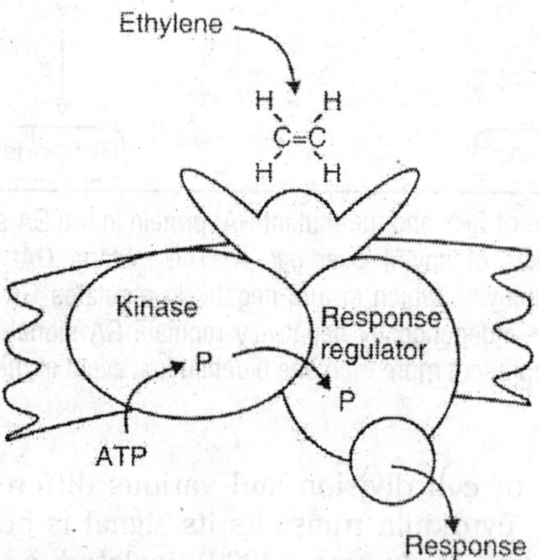

FIG. 19.5: Hypothetical model for two–component regulation of ethylene action. Binding of ethylene to the receptor activates the kinase domain, which autophosphorylates, then transfers the phosphate to the response regulator protein, resulting in release of a ligand, or activation of a transcription factor.

CTR1 has been isolated by epistatic relationships to be downstream from **ETR1** and **EIN4** and upstream from other **ein–type** mutations. The gene was isolated with the help of T–DNA tagged mutants. The C–terminus of **CTR1** contains domains typical of protein kinases of the **Raf** protein kinase family (Kieber et al., 1993). **Raf** is the retroviral protein that transduces signals from cell–surface receptors to transcriptional factors that mediate changes in cell growth and differentiation (Gupta and Roberts, 1994). Inspection of all mutant alleles of **CTR1** shows that they would yield truncated or non–functional kinases. This indicates that loss of function at this juncture results in constitutive activation of the pathway, suggesting that functional phosphorylating activity of **CTR1** represses downstream ethylene response. The N–terminal portion of **Raf**–type kinases has been hypothesized to regulate the kinase activity and in the case of **CTR1**, its N–terminal portion might be the direct or downstream target for **ETR1** activity.

In the absence of ethylene, the ethylene receptors (**ETR1** and **ERS**) interact and activate the **CTR1** protein kinase. Once activated, **CTR1** negatively regulates, possibly through a **MAPK** cascade, downstream events by repressing the positive regulator **EIN2**. Binding of ethylene to the receptors disrupts the interaction of the latter to **CTR1**. The released CTR1 is presumably inactive and is unable to repress

EIN2. Active **EIN2** positively regulates the nuclear elements **EIN3** and **EIL 1–3**, which are proposed in turn to activate EREBP transcription factors which may then regulate target gene expression.

Double mutants of **ein 3/ctr1** show an ethylene–insensitive phenotype thus positioning **EIN3** downstream from **CTR1**. **EIN3** was isolated by T–DNA tagging and found to be part of a multigene family called **EIL** (EIN–like) (Chao et al., 1997). It is likely that **EIN3** acts as a positive regulator of the ethylene pathway because over-expression of wild–type **EIN3** conferred a constitutive ethylene phenotype that is insensitive to the addition of inhibitors of ethylene sensing. **EIL** genes can also complement **ein2** mutants when regulated by the **CaMV–35 S** viral promoter. **EIN3** and **EIL1** show short stretches of positively charged amino acids that serve as nuclear–localization signals. **EIN3–GUS** reporter gene fusions are localized to the nucleus. Additional stretches of acidic, glutamine and proline–rich regions in **EIN3** might serve as transcriptional activation domain. The fusions of **EIN3** to the **GAL4 DNA** binding domain activate transcription in yeast.

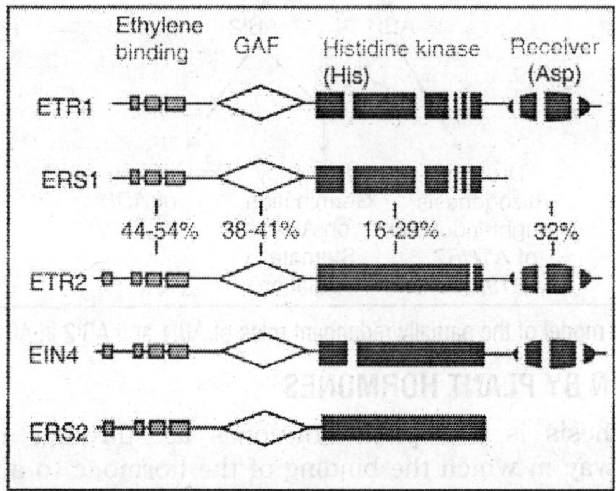

FIG. 19.6: The *Arabidopsis* ethylene receptor family. The five receptors fall into two subfamilies (ETR1/ERS1 and ETR2/EIN4/ERS2) on the basis of their gene (not shown) and protein structures. The percentage ranges of amino acid identities between domains of the two subfamilies are shown. (ETR1 and ERS1 domains have 64–78% identity to each other, and domains of the ETR2 subfamily have about 48–58% identity to each other). All members have an amino–terminal ethylene-binding domain containing three transmembrane subdomains indicated by shaded bars. The putative cytoplasmic portion is comprised of a 'GAF'–related domain of unknown function, a histidine protein kinase domain and a receiver domain (which is lacking in ERS1 and ERS2). The conserved sequence motifs of histidine kinase and receiver domains, including the His and Asp phophorylation sites, are indicated by white bars. The ETR2 subfamily contains an additional hydrophobic sequence at the amino–terminus, and lacks many of the histidine kinase sequence motifs.

A subset of pathogenesis–related proteins accumulates to high levels in the presence of stress–induced ethylene. Comparison of cis–regulatory sequences of the ethylene–sensitive genes revealed an 11 bp sequence referred to as the **GCC** box that was essential for ethylene–dependent transgene expression.

Ethylene signaling also modulates rhizobium–induced infection and nodulation in legumes. Exogenously applied ethylene inhibits formation of nodules in *Pisum sativum* and the effect can be reversed by addition of silver ions. Conversely, the ethylene–insensitive sickle mutant of *Medicago truncatula* becomes hyper-infected by the symbiotic nitrogen–fixing bacterium. In both organisms, ethylene does not affect the number of infections that initiate rather, the hormone governs progression of nodule formation.

Abscisic Acid

Advances in determining ABA signal transduction pathway has been hampered by the paucity of mutants. Two dominate *Arabidopsis* mutants, **abi1** and **abi2**, are insensitive to the ABA–mediated growth arrest. They were identified by their ability to germinate on concentrations of ABA that normally inhibit wild–type. Both **ABI 1** and **ABI 2** genes encode protein phosphatases of the **PP2C** family which indicates that protein phosphorylation and dephosphorylation are involved in ABA signaling. Molecular analysis revealed that both mutant proteins contain a mutation of glycine to aspartate, which renders the phosphatases inactive.

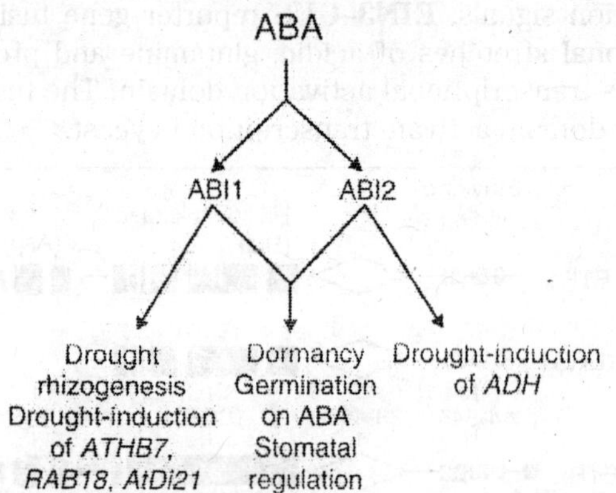

FIG. 19.7: Schematic model of the partially redundant roles of ABI1 and ABI2 in ABA signaling.

GENE EXPRESSION BY PLANT HORMONES

A working hypothesis is that plant hormones act through a standard signal transduction pathway in which the binding of the hormone to a receptor elicits a transduction cascade and expression of specific genes or gene families represent the end point of the signaling pathway. Numerous studies of hormonal action have identified abundance of specific mRNAs. Such results could be due to increased rates of gene transcription, mRNA processing, mRNA transport from the nucleus to the cytoplasm, increased mRNA stability or combination of these. Although there is increasing interest in mechanisms controlling mRNA stability and the possibility that plant hormones may be able to influence this process, so far there is little convincing evidence that any plant hormone regulates mRNA stability. On the other hand, there is good evidence that hormones regulate gene transcription.

A number of hormonally regulated genes have been isolated and sequenced including the ethylene regulated **E4** and **E8** genes of tomato, the auxin regulated **GH** and **JCW** series of genes from soybean, GA–regulated α–amylase and other genes from cereals and the ABA–regulated **Em** and **rab** genes from cereals.

There is a considerable body of evidence, which indicates that regulation of gene transcription is controlled by the interaction of **cis**–acting promoter sequences with **trans**–acting DNA–binding protcins. DNA–binding proteins are usually detected by retardation of the mobility of DNA fragments in electrophoretic gels (gel retardation assay) when protein/DNA complexes are formed. DNA–binding proteins have been detected for rice and barley α–amylase promoters fragments, for the **Em** gene, the **rab** gene and for the ethylene–regulated genes **E4** and **E8** of tomato (Cordes et al., 1989), the extensin gene of carrot (Holdsworth and Laties,

1989) and the **chitinase** gene of bean (Broglie et al., 1989).

Auxin

Recently, molecular biologists have attempted to gain insight into auxin action by analysing auxin–responsive gene expression and characterizing the gene and gene products induced or repressed by auxin application. Guilfoyle et al. (1990) have identified, sequenced and characterized a number of auxin–responsive genes and cDNAs from soybean. These include the genes that encode the mRNAs for **GH1**, **GH2/4**, **GH3** and **SAURs**. Nuclear run–on *in vitro* transcription assays have been used to show that all of these genes are transcriptionally activated within minutes after exogenous auxin are applied to excised, incubated plant organs or intact seedlings.

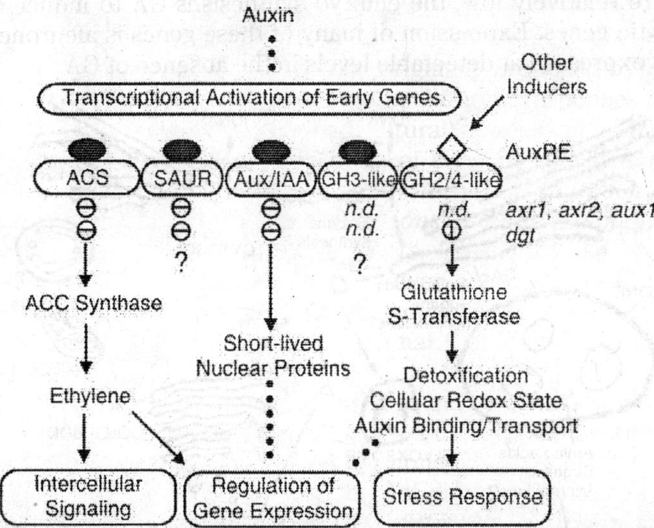

FIG. 19.8: A model of early–auxin–regulated gene expression and gene function. Expression of early auxin genes in several auxin–insensitive mutants of *Arabidopsis* (*axr1, axr2,* and *aux1*) and *Lycopersicon* (*dgt*) is either unaffected (open valve), inhibited to varying degrees (closed valve), or not determined (n.d.). A common set of functionally identified *AuxREs* used by at least three major classes of auxin–specific early genes is indicated by a shaded ellipsoid. An open diamond indicates the ocs–like element of GH2/4–like genes.

The most extensively studied early auxin–responsive plant gene promoters are those from the pea **PS–IAA 4/5** gene, the soybean **GH3** gene and the soybean **SAUR15A** gene.

The **Aux/IAA** gene family represents one class of early auxin responsive genes that has been identified in a variety of plants. Further studies showed that **ARF1** represents only one of the **ARF** families of transcription factors, which consist of so far of nine members. The **ARF** proteins range in size form 67 to 129 KDa. All contain the conserved N–terminal DNA–binding domain however, **ARF3** lacks the C–terminal domain implicated in dimerization. Transient expression assays in carrot protoplasts showed that **ARF1** acts as a repressor whereas **ARF 5,6,7** and **8,** which contain glutamine–rich activation domains can function as transcriptional activators of **AuxRE**–linked promoters. These results suggest that when **ARFs** bind to **AuxRE** directly they can either activate or repress transcription. The relationship between **ARFs** and **Aux/IAA** protein is not clear at present. **Aux/IAA** proteins do not contain DNA–binding or activation domain. They may act as auxin response repressor since they can dimerize with **ARFs**, thereby inhibiting their activation function.

Gibberellin

During germination, under the influence of GA, aleurone cells of the grain endosperm are induced to synthesise and secrete very large amounts of hydrolytic enzymes that digest the contents of starchy endosperm. Although α–amylase is made in large quantities, other hydrolases such as proteases, nucleases and cell wall degrading enzymes are also produced. This protein synthesis capacity is tightly controlled. During seed development, presumably mediated by ABA, expression of the genes for these hydrolytic enzymes is strongly suppressed such that essentially none of their gene products are produced. Conversely, during germination, when ABA levels are relatively low, the embryo synthesises GA to induce expression of these hydrolytic genes. Expression of many of these genes is aleurone–specific and many are not expressed at detectable levels in the absence of GA.

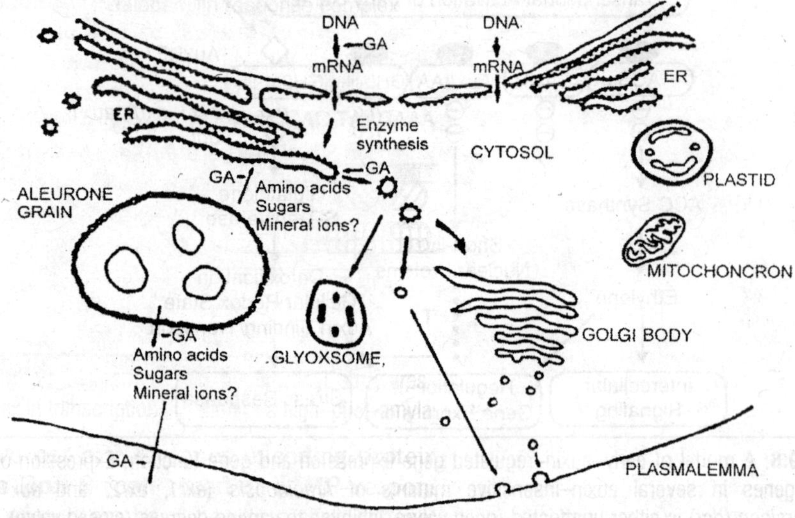

FIG. 19.9: Suggested sites of response to gibberellin in cereal aleurone cell.

Aleurone cells are an excellent system for studying GA and ABA regulation of gene expression. Aleurone cells and protoplasts synthesise α–amylase in response to GA and this synthesis is accompanied by an increase in the steady–state–levels of α–amylase mRNA. Run–on transcription with nuclei isolated from barley and wild-oat aleurone protoplasts have demonstrated that the GA–induced increase in α–amylase mRNA is due primarily to increased transcription of α–amylase gene which can be overcome by ABA. The expression of α–amylase increases only after several hours of exposure of aleurone cells to GA.

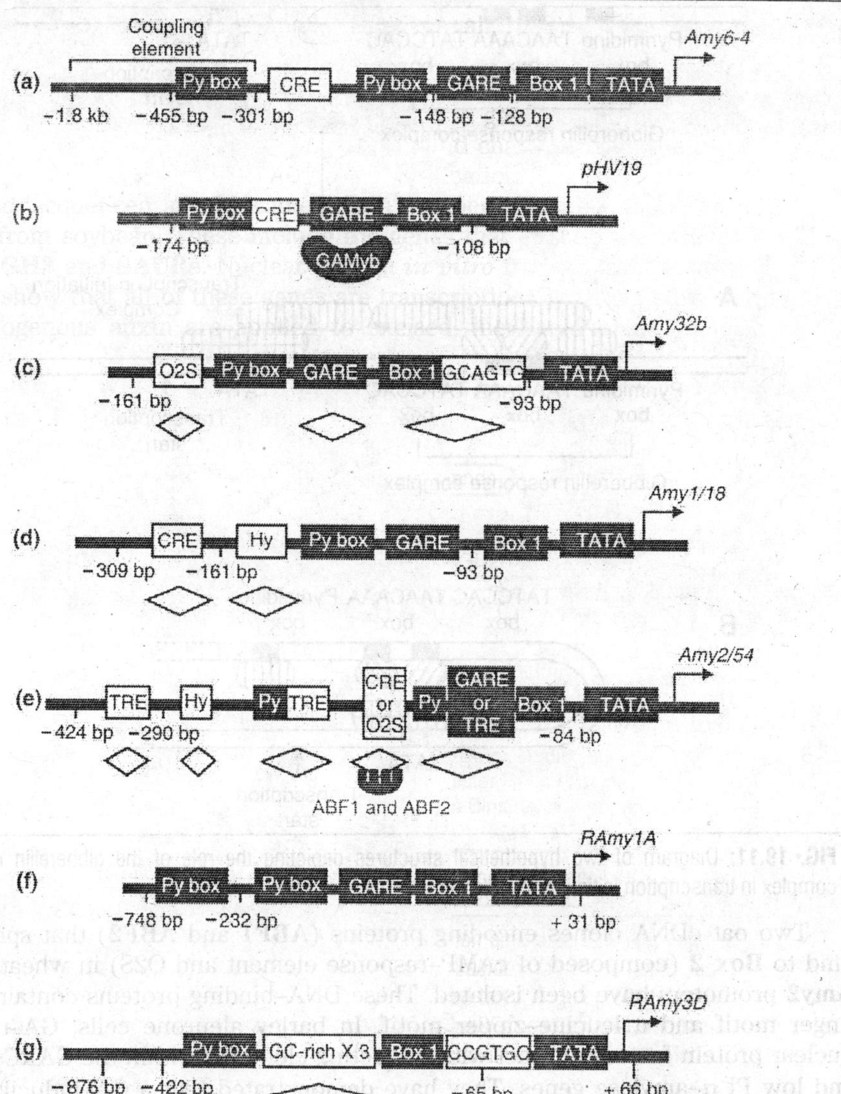

FIG. 19.10: Structural comparison of the promoter regions of α–amylase genes: (a) barley *Amy6–4*; (b) barley *pHV19*; (c) barley *Amy32b*; (d) wheat *Amy1/18*; (e) wheat *Amy2/54*; (f) rice *RAmy1A*; and (g) rice *RAmy3D*. In each diagram, the arrows indicates the transcription initiation site, highly conserved sequence elements are shown in bold boxes and the sites of DNA–protein interactions observed by foot–print experiments are indicated by diamonds. The sequence element GCAGTG, which is positioned immediately in the 3' direction of a Box 1 sequence, is highly conserved in barley low pl α–amylase promoters (*Amy32b*). The sequence CCGTGC of *RAmy3D* is common to the promoters of metabolically repressed plant genes. Cloned sequence–specific DNA–binding proteins–GAMyb, ABF1 and ABF2 – are also indicated. Abbreviations: GARE, gibberellic acid–response element; O2S, Opaque–2–binding sequence; CRE, cAMP–response element; TRE, phorbol ester–response element; Hy, hyphenated palindrome sequence; Py, pyrimidine.

GA induction of α–amylase mRNA is sensitive to cycloheximide and aminoethyl–L–cysteine (amino acid analog), suggesting that protein synthesis is required before α–amylase gene transcription can be activated.

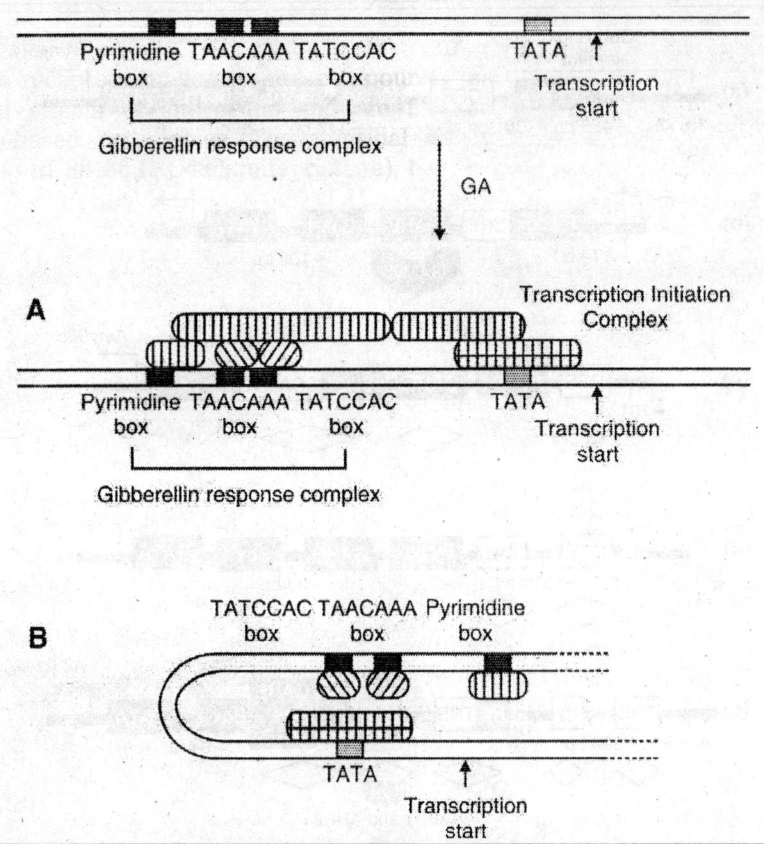

FIG. 19.11: Diagram of two hypothetical structures depicting the role of the gibberellin response complex in transcription initiation of a high–pI α–amylase gene.

Two oat cDNA clones encoding proteins (**ABF1** and **ABF2**) that specifically bind to **Box 2** (composed of cAMP–response element and O2S) in wheat and oat **Amy2** promoters have been isolated. These DNA–binding proteins contain a zinc-finger motif and a leucine–zipper motif. In barley aleurone cells, GA–regulated nuclear protein factors interact with cis–acting elements within the GARCs of high and low PI α–amylase genes. They have demonstrated that a GA–inducible Myb-related protein **(GAMyb)** from barley could bind to the **TAACAAA** box *in vitro*. This protein is capable of activating the transcription of a high PI α–amylase promoter fused to a reporter gene in the absence of GA in the transient expression system. Thus, the activation of **GAMyb** is an early step in the expression of GA-induced α–amylases in barley aleurone.

Many α–amylase genes have been isolated from different plant species and their nucleotide sequence, chromosomal locations and phylogenetic relationships determined. Barley α–amylase isoforms are divided into high and low PI groups based on their different isoelectric points. Peptide sequencing and DNA blotting experiments analysed the polymorphism. The high and low PI groups are located on chromosome 6 and 1 respectively. In contrast, in wheat, α–amylase genes fall into three subfamilies–**Amy1**, **Amy2** and **Amy3** which are located on chromosomes 6,7 and 5 respectively.

In rice seeds as well as calli derived from the embryo, 11 α–amylase isoforms were identified and at least 11 genes isolated and sequenced. The rice α–amylase genes have been divided into **RAmy1**, **RAmy2**, and **RAmy3** subfamilies and members of these families are located on chromosome 1, 2, 6, 8 and 9. In addition to **TATA** box, three conserved sequences were found in the 5'flaking regions of cereal α–amylase genes. One of these, **TATCAAT** is conserved in most of the α–amylase genes and is related to the **CATC** box found in many other plant genes. The second motif **TAACA [A/G]** A (termed box1) is specifically found in α–amylase genes. It is generally known as the amylase box or GA–responsive element **(GARE)**. This motif is not present in the **Amy3** subfamily. The third motif designated as the pyrimidine box, **[C/T] CTTT[C/T]** generally appears in other GA–inducible genes such as β–glucanase, carboxypeptidase and aleurain. All **Amy1** and **Amy2** genes and several members of the **Amy3** gene subfamily contain this motif. It is apparent that the promoter regions of **Amy3** genes differ considerably from other α–amylase genes.

GA–regulated α–amylase biosynthesis results from enhanced transcriptional activity. How cis– and trans–acting factors interact to regulate gene transcription is unknown.

Cytokinin

Cytokinins induce cell division in concert with auxin in non–dividing plant cells. Until recently, the molecular targets of this hormone in the cell cycle regulatory machinery have remained elusive. Cloning genes coding for plant cyclin–dependent kinase (Cdks eg. **Cdc2**), cyclins (**CycA, CycB** and **CycD** families) and cyclin–dependent protein kinase inhibitors (**ICKs**) has made the cell cycle accessible to molecular studies. The cytokinin–dependent cell division activity in *Arabidopsis* is mediated by the transcriptional regulation of the cyclin **D3** (Cyc D3) gene at the G1/S transition.

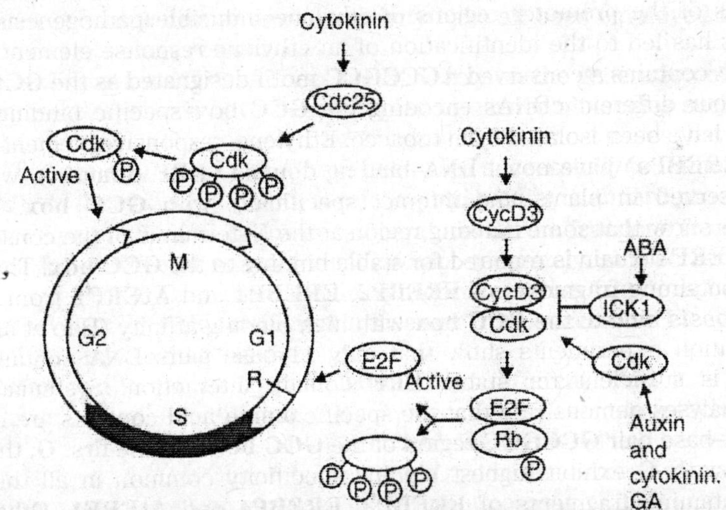

FIG. 19.12: Hormonal regulation of the plant cell cycle. The main activities of cytokinins, through the activation of CycD3 and Cdc25, are indicated. Cytokinins together with auxin and GA act through activation of Cdc2a, a Cdk. ABA induces ICKI, a kinase inhibitor protein that interacts with both Cdk and CycD3. abbreviations: ABA, abscisic acid; Cdc25, a cell division cycle phosphatase; Cdk, cyclin–dependent kinase; Cyc, cyclin; E2F, elongation factor; GA, gibberellin; ICK, cyclin–dependent protein kinase inhibitor; P, phosphorylation; R, restriction point; Rb, anti–proliferative tumour suppressor retinoblastoma protein.

Three plant Cyclin D gene homologues, **CycD1**, **CycD2** and **CycD3**, have been identified by complementation of yeast G1 cyclin mutants. They are similar to mammalian **D–type** cyclins whole expression is dependent on mitogenic stimuli such serum growth factor. Mammalian **D–type** cyclins are required during middle to late G1 phase and are considered to be the link between proliferating signals and emergence from the quiescent state. In *Arabidopsis* cell culture it has been found that **CycD3** is produced in the G1 phase by cytokinin and sucrose. The cytokinin overproducing **amp1** mutants show two to three–fold higher steady–state level of **CycD3** mRNA compared to the wild type.

Ethylene

Ethylene regulates many complex processes of stress responses and developmental adaptations in higher plants. Stimulation of expression of pathogenesis–related genes such as class 1 basic **chitinase** and β–1–3–**glucanases** is among the consequences of ethylene involvement (Ohme–Takagi and Shinshi, 1990).

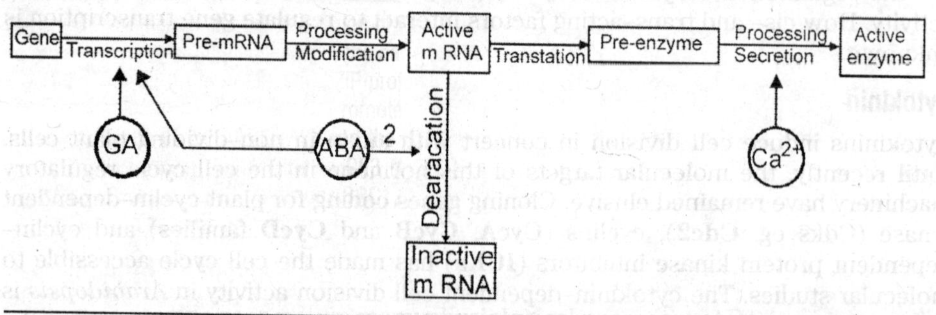

FIG. 19.13: Possible site of action of GA and ABA in α-amylase gene expression.

Analysis of the promoter regions of ethylene–inducible pathogenesis–related **(PR)** genes has led to the identification of an ethylene response element **(ERE)**. The element contains a conserved **AGCCGCC** motif designated as the **GCC** box or **PR box**. Four different cDNAs encoding the **GCC** box–specific binding factors **(EREBPs)** have been isolated from tobacco. Ethylene–responsive element–binding proteins **(EREBPs)** have novel DNA–binding domain **(ERF domain)**, which are highly conserved in plants and interact specifically with **GCC box**. Deletion experiments show that some flanking region at the N–terminus of the cónserved 59 amino acid **ERF** domain is required for stable binding to the **GCC box**. Three **ERF** domains–containing fragments of **EREBP2**, **EREBP4** and **AtERF1** from tobacco and *Arabidopsis* bind to the **GCC box** with high binding affinity (Hao et al., 1998). DNA truncation experiments show that only 11–base pairs DNA containing the **GCC box** is sufficient for stable **ERF** domain interaction. Systematic DNA mutation analyses demonstrate that the specific amino acid contacts are confined within the 6–base pair **GCCGCC** region of the **GCC box** and the first **G**, the fourth **G** and the sixth–**G** exhibit highest binding specificity common in all three ERF domain–containing fragments of **EREBP2**, **EREBP4** and **AtERF1**. Other bases within the **GCC box** are important for differential binding by different **EREBPs**. The conserved N–terminal half is responsible for formation of a stable complex with the **GCC box** and the divergent C–terminal half for modulating the specificity.

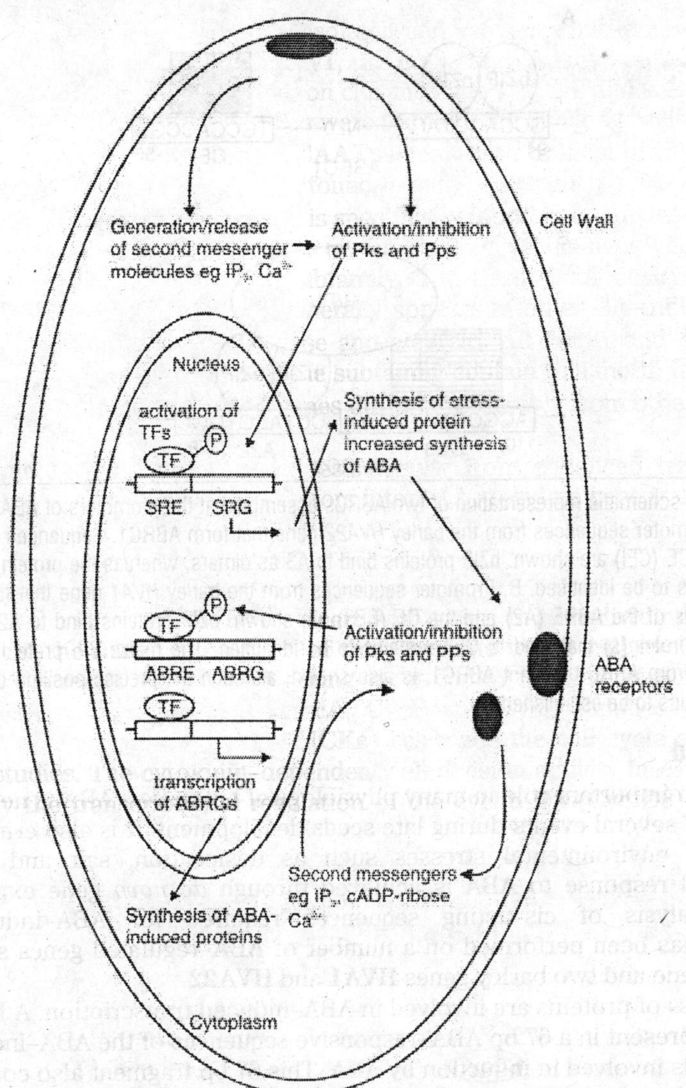

FIG. 19.14: Induction of gene expression by water deficit stress. In essence, transcriptional activation in plant cells involves the interaction of transcription factors (nuclear proteins that interact with other nuclear proteins and DNA) with regulatory DNA sequences (*cis*–elements) to effect conformational changes that modulate transcriptional activity. Intracellular signaling cascades involve the generation and release of short–lived second messenger molecules and changes in protein conformation and activity via ligand binding and alteration of phosphorylation status. ABRE: = ABA response element, ABRG = ABA responsive gene, ECM = extracellular matrix. PK = protein kinase, PP = protein phosphatase, SRE = stress–response element, SRG = stress–responsive gene, TF = transcription factor.

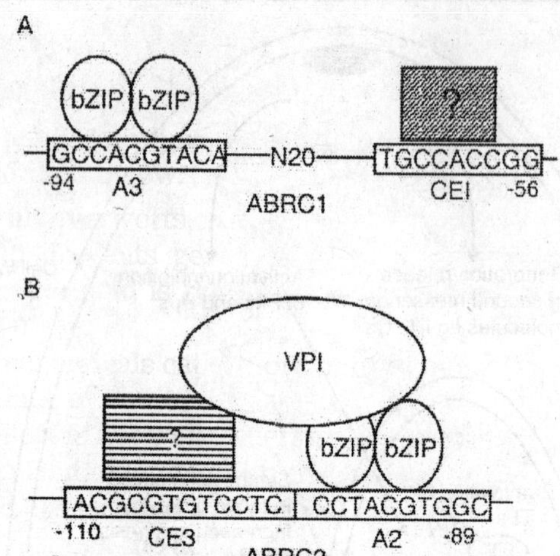

FIG. 19.15: A schematic representation of two ABRCs assembled at the promoters of ABA–responsive genes. A, Promoter sequences from the barley *HVA22* gene that form ABRC1. Sequences of the ABRE (A3) and the CE (CEI) are shown. bZIP proteins bind to A3 as dimers, whereas the protein(s) that bind to CEI remains to be identified. B, Promoter sequences from the barley *HVA1* gene that form ABRC3. The sequences of the ABRE (A2) and the CE (CE3) are shown. bZIP proteins bind to A2 as dimers, whereas the protein(s) that bind to CE3 remains to be identified. The maize VPI protein, which can transactivate from ABRC3 but not ABRC1, is also shown, although the precise position of VPI in the complex remains to be established.

Abscisic Acid

ABA plays an important role in many physiological processes. ABA is necessary for regulation of several events during late seeds development. It is also crucial for the response to environmental stresses such as desiccation, salt and cold. The physiological response to ABA is achieved through *de novo* gene expression. A detailed analysis of cis–acting sequences required for ABA–induced gene expression has been performed on a number of ABA–regulated genes such as the wheat **EM** gene and two barley genes **HVA1** and **HVA22**.

Myb class of proteins are involved in ABA–induced transcription. A **Myb** biding sequence is present in a 67 bp ABA–responsive sequences of the ABA–inducible rad 22 gene and is involved in induction by ABA. This 67 bp fragment also contains **Myc** binding site that participates in ABA–induction and binds a drought–inducible **Myc** homologue. Because the **Myc** and **Myb** factors are induced by water stress and seem to regulate the protein synthesis–dependent induction of **rd22**, it was suggested that **Myc** and **Myb** may regulate gene whose induction by ABA depends on protein synthesis while regulation via **ABRE** does not require protein synthesis (Abe et al., 1997).

In a different promoter context, **VP1** acts as repressor of α–amylase expression. **VP1** also plays a role in anthocyanin gene expression by transactivating the **C1** promoter through a DNA sequence called the **Sph** element. The **Sph** element is quite distinct from the **ABRE**. Only 120–amino acid, B_3 domain located at the C–terminus of **VP1** binds the **Sph** element with high specificity.

TABLE 19.3: Plant growth substances interacting in physiological responses and in gene expression (transcription)

Physiological Process	Plant Growth Substance
Cell Enlargement	Auxin
α-amylase synthesis in aleurone cells	Gibberellin
Plastid Development	Cytokinin
Embryogenesis and seed development	ABA
Fruit ripening	Ethylene

SECONDARY MESSENGER HYPOTHESIS

According to this hypothesis, auxin does not directly act on the cell wall. Addition of auxin to isolated cell walls or to cells with non-functional protoplasts has no effect on the wall extensibility. This has led to the conclusion that the initial action of the hormone is at a site on or within the protoplast, an action, which leads indirectly to wall loosening. Thus, there must be a signal or *Secondary messenger* of some type, which travels from the protoplast to cell wall, initiating wall loosening.

HORMONE RECEPTOR CONCEPT

This concept finds its origin in work on animal hormones, which can not elicit physiological responses unless they are recognized by their specific receptor sites. Each animal hormone has a specific site (gland) of synthesis and a target organ having affinity receptor (s) located in the plasma membrane and cytosol of the cell. These receptor molecules have been shown to be proteinaceous in nature.

Based on analogy with animal systems, it is currently thought that binding of hormones also inhibits most of the physiological and biochemical responses elicited by plant hormones to their respective receptors. A hormone must be recognized by the cell in order to evoke its specific physiological and biological response. The binding sites often referred to as receptors have been characterized as protein moieties, with a dissociation constant ranging from 10^{-8} M to 10^{-5} M.

Thus, according to the basic scheme, the mechanism by which plant hormones exert their effect(s) is by first binding to specific proteins or receptors, commonly envisaged to be an allosteric protein. As a consequence of binding to the hormone, forms a hormone-receptor complex, which initiates the biochemical events leading to a response. Binding to such a receptor would, therefore, represent the primary level of interaction of a hormone within a cell. The hormone-receptor complex may bind to the template DNA or to RNA polymerase or both and results in improved transcriptional activity. Such a participation of hormone receptor complex in the transcriptional process has not been fully established. Thus, in summary following could be various alternatives based on receptor concept in explaining the hormone regulation of gene expression of enzyme proteins.

1. Hormone receptor complex interacts with chromatin at specific initiation sites and thereby enhances the transcriptional activity.
2. Hormone receptor complex may bind to RNA polymerase and thereby result in enzyme activation; the modified RNA polymerase could then accelerate transcription activity.
3. An authentic receptor system should satisfy certain minimum criteria (Venis, 1980), namely:
   a) binding should be reversible, of high affinity and of finite capacity;
   b) the saturation range of binding should be consistent with the concentration range over which the physiological response saturates;

c) binding specificity for different analogues should reflect their relative biological activity, and binding should lead to a hormone specific response.

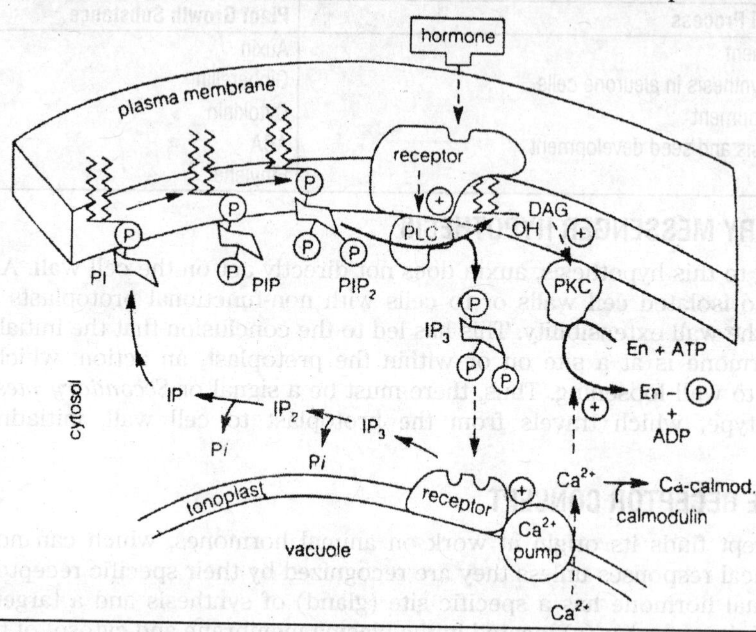

FIG. 19.16: Model for initial hormone transduction at the plasma membrane. Binding of a hormone to its receptor cause activation (+) of nearby phospholipase c (PLC). PLC hydrolyses a membrane lipid, phosphatidylinositol-4,5-bisphosphate (PIP_2) to release inositol-1,4,5-trisphosphate (IP_3) and diacylglycerol (DAG). IP_3 moves to the tonoplast in plant cells, where it combines with a receptor that activates (+) a $Ca^2{}_+$ pump or transporter that moves Ca^{2+} from the vacuole to the cytosol. DAG, which remains membrane-bound, activates protein kinase c (PKC). Ca^{2+} released from the vacuole, so various enzymes, when free or bound with calmodulin also activates PKC. IP_3 loses phosphates by hydrolysis to form IP_2. and IP, which is then converted back to phosphatidylinositol (PI) and other phosphoinositide lipids (PIP and PIP_2.) in the plasma membrane.

Until recently, it is difficult to obtain a plant system that approximated to even remotely accepted concept of hormone receptors. However, based on research during the two decades and especially because of the availability of labelled hormones and high affinity chromatography, receptor molecules all of proteinaceous nature have been isolated, and identified for their molecular weights and dissociation constants of the several hormone-receptor complexes.

PLANT HORMONE CONJUGATES

Plant hormone conjugates or conjugated plant hormones are defined as *those plant hormones which are metabolically coupled to other low molecular weight compounds by covalent binding* (Sembdner, 1974; Sembdner et al., 1986). This definition excludes non-covalent associations between plant hormones and high molecular weight particles like protein, cell particles and membranes. These complexes should be called *bound* plant hormones (Schneider, 1983). Non-identified polar fractions, which possibly belong to the conjugates, are also often termed *bound hormones*. The plant hormone conjugates are widely distributed in plants and occasionally present in high concentration. Investigations on the isolation and identification of plant hormones have led to the detection of these conjugates from which free hormones can be liberated by hydrolysis.

Conjugated plant hormones are those hormones, which are metabolically coupled to other low molecular weight compounds by covalent binding. Plant hormone conjugates are widely distributed in plants and generally occur in high concentrations. Plant hormones or free hormones are released from conjugates through hydrolysis. In fact conjugation is a part of metabolism of plant hormone but unlike biosynthesis, they occur in irreversible state. Conjugates are different from the corresponding free hormone physically, chemically and in their biological properties. Thus, conjugates may be concerned with the regulation of the level of free and active hormone. Plant hormones could form conjugates with lipids, carbohydrates, myoinositol, amino acids, etc. Membrane associated steroids may form hydrogen bond with the carboxyl or a negative charge center of the hormone molecule with the phosphate hydroxyl: the lecithin in the other side.

CHAPTER 20

Plant Hormone: Auxins

HISTORICAL BACKGROUND

One of the unique and timeless contributions from twenty century was the discovery of the plant hormones. In 1897, Charles Darwin and his son, Francis published a monograph on tropic responses of plant showing that the growth of canary grass seedlings toward light was controlled by the coleoptile tip. They also noticed that although the photoreceptor was located at the tip, the bending took place at a distance from the active part of the tip. They reasoned that some sort of signal must travel from the tip to the growing region. A few year earlier, the prominent physiologist Julius Sachs had hypothesized without the burden of evidence that there were special substances in plants for forming roots, others for forming leaves, others for forming stems and so on (Thimann, 1935). Thus, the idea of small molecules with biological effects was in the air.

In 1910, P. Boysen–Jensen showed that if the tip of an etiolated coleoptile was isolated from the lower coleoptile by the insertion of a small slab of gelatin, the phototropic stimulus produced in the tip traversed the gelatin and caused bending below the gelatin slab. Sixteen years later Fritz Went made the discovery that excised oat coleoptile tips secreted a substance, auxin, into gelatin blocks such that when applied to the sides of coleoptiles, the gelatin blocks could induce bending of decapitated coleoptile tips that were not otherwise stimulated. Went later recounted that he made the discovery at 3 a.m. on April 17, 1926 – an indication of the excitement of the experiments (Went, 1974). The gelatin block technique also provided for experimental verification of a theoretical proposal by Cholodny that phototropism was due to asymmetric distribution of a growth substance. Similarly, in 1930 Herman Dolk showed that gravitropism was also associated with an asymmetric distribution of auxin (Thimann, 1935). The discovery of polar transport of auxin by van der Weij in 1932 provided the first hint as to the possible identity of a factor that might be involved in controlling polarity in morphogenesis. Went and Thimann discovered the root inducing activity of auxin and the role in apical dominance, in 1930. Work on auxin also led to the discovery of the first widely useful herbicide, 2,4 – dichlorophenoxyacetic acid, a synthetic auxin that is still in use today.

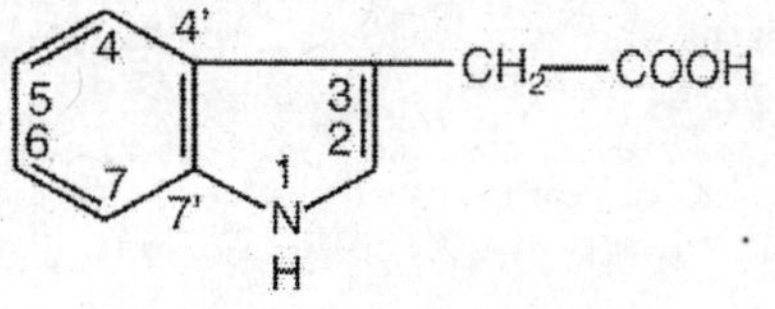

Indole 3-acetic acid

The primary auxin in plants is indole–3–acetic acid. Although other compounds with auxin activity such as indole–3–butyric acid, phenyl acetic acid and 4–chloro–IAA are also present in plants (Normanly et al., 1995) but little is known about their physiological role.

Free auxins

These are diffusible auxin, which move out of the tissue quite readily (e.g., auxin diffusing out of coleoptile tips into agar). They are readily extractable in various solvents like diethyl ether at 0° to 5 °C.

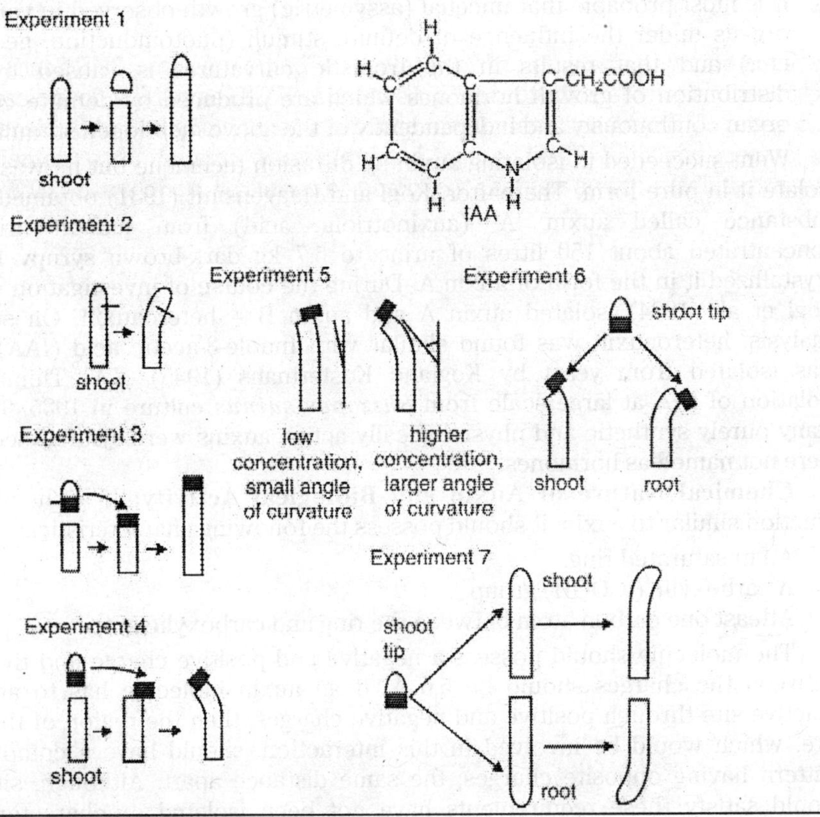

FIG. 20.1: Early experiments with *Avena* coleoptile.

Free auxins: These are diffusible auxin, which move out of the tissue quite readily (e.g., auxin diffusing out of coleoptile tips into agar). They are readily extractable in various solvents like diethyl ether at 0° to 5 °C.

Bound auxins: These auxins are released from the plant tissues only after they are subjected to hydrolysis, autolysis or enzymolysis. Bound auxins include *reserve*, or *storage*, forms and *detoxification* forms. Detoxification products are not active *per se*. They are often formed from excess IAA or from high levels of synthetic auxins that may be introduced into plant tissues.

A young Dutch Botany student of Utrecht named F.W. Went in 1926 made the definitive discovery of auxin. He designed very simple experiment with *Avena* coleoptile and named as *Avena* test. According to Went, if intact tips of *Avena* coleoptile were allowed to stand for an hour or two on small blocks of agar, then growth hormone activity could readily be detected afterwards in the Agar.

He adapted and improved Stark's procedure of applying the test material to one side of decaptitated coleoptile, in darkness (source of irradiation was weak red light, which had not been found to be phototropically active) then the resulting curvature could be easily measured and thus the quantitative concept was demonstrated experimentally.

This experiment enabled the Cholodny theory of tropisms (in 1927, Growth hormone and tropism in plants) to be directly verified:

1. Different tropisms in plant can be traced to growth regulation by so called growth hormones.
2. It is most probable that unequal (assymetric) growth observed in various plant organs under the influence of definite stimuli (photoinduction, geoinduction etc.) and that results in the tropistic curvatures is caused by unequal distribution of growth hormones which are produced by definite cells in the organ continuously and independently of the above mentioned stimuli.

Went succeeded in isolating auxin by diffusion technique but he was unable to isolate it in pure form. Thereafter, Kögl and Haagensmit (1931) obtained an active substance called auxin 'A' (auxinotriolic acid) from human urine. They concentrated about 150 litres of urine to 5.7 kg dark-brown syrupy liquid and crystallized it in the form of auxin A. During the course of investigation with urine Kögl et al. (1934) isolated auxin A and auxin B + heteroauxin. On subsequent analysis. heteroauxin was found similar with Indole-3-acetic acid (IAA). Later, it was isolated from yeast by Kogland Kostermans (1935). K.V. Thimann made isolation of IAA at large scale from *Rhizopus suinus* culture in 1935. Thereafter, many purely synthetic and physiologically active auxins were synthesised but they were not named as hormones.

Chemical Nature of Auxin and Biological Activity: If a chemical has a function similar to auxin, it should possess the following characteristics:

1. An unsaturated ring.
2. A carboxylic (-COOH) group.
3. Atleast one carbon atom between the ring and carboxylic group.

The molecule should possess a negative and positive charge and the distance between the charges should be 5.5 Å. If an auxin molecule has to attach to a reactive site through positive and negative charges, then the region of the reactive site, which would be involved in this interaction, should have a complementary pattern having opposite charges, the same distance apart. Although, sites which would satisfy these requirements have not been isolated or characterized; and charge distribution data of such auxin molecules as *ortho*-napthalene acetic acid (NAA) and several other auxins are not available, these studies provide a base to formulate the nature of the reactive sites. Depending on pH, the major cellular polymers- proteins and nucleic acids- would be differentially protonated providing a point of attachment for the negative charge on the carboxyl group of the auxin molecule; the small negative charge density due to the lone pair of electrons on the *O*-atom of the carboxyl or hydroxyl group or the phosphate residues may attract the functional positive charge centers of the auxin molecule. This would provide several regions of the macromolecules for selective associates with the auxins. If the auxin molecule interacts with a receptor protein, which then undergoes allosteric modification, a larger region of the chromatin will be involved in the interaction.

Antiauxins: Those compounds when applied to the plant and nullify the effects of auxin are known as antiauxins, e.g., 2,4-dichlororanizole (DCA). These compounds compete with auxin for the same site of action in the cell. Therefore, the inhibition of auxin activity by antiauxins is competitive in nature.

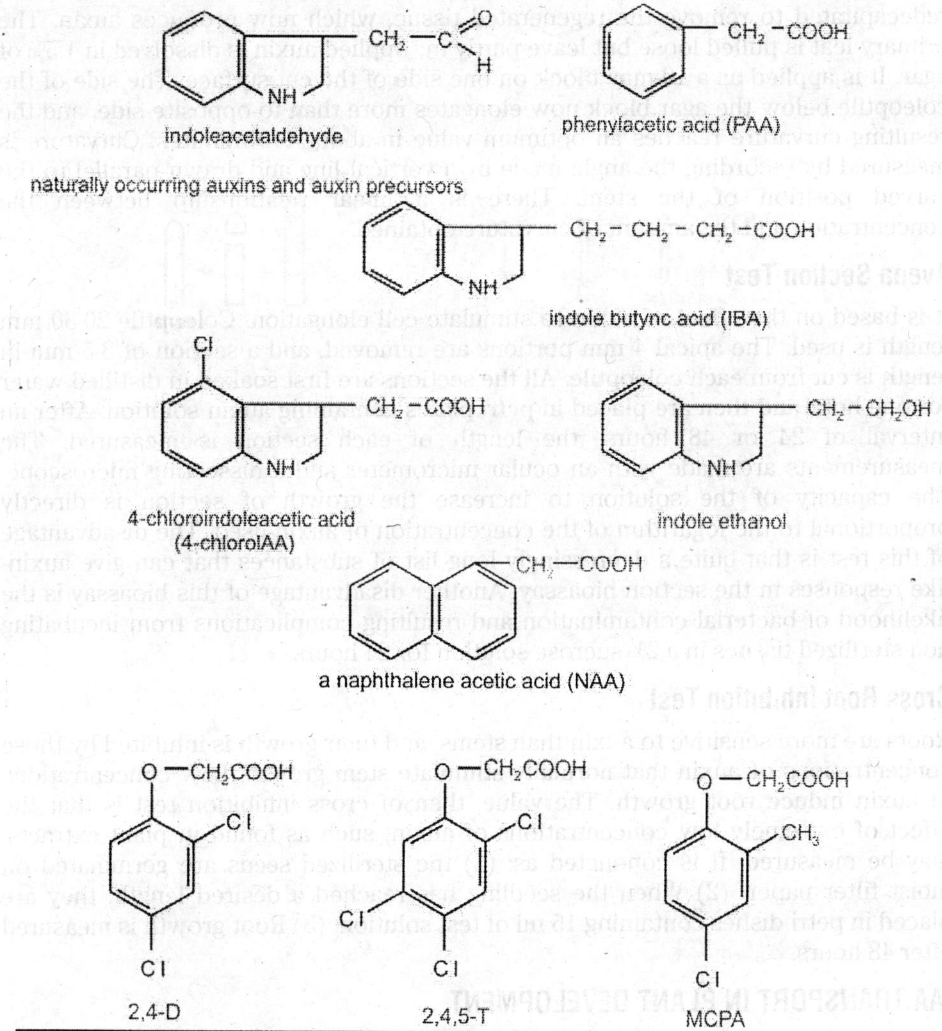

indoleacetaldehyde

phenylacetic acid (PAA)

naturally occurring auxins and auxin precursors

indole butync acid (IBA)

4-chloroindoleacetic acid
(4-chloroIAA)

indole ethanol

a naphthalene acetic acid (NAA)

2,4-D

2,4,5-T

MCPA

FIG. 20.2: Various types of auxins.

SITE OF SYNTHESIS

Auxin is characteristically synthesised in the stem tip and in young tissues and moves mainly down stem. It thus, seems to form a gradient from shoot tip to the root.

TECHNIQUES OF AUXIN BIOASSAY

A bioassay is a method of determining the presence or (more often) the amount of substance by measuring the substance's effect on a designated biological process. Hormone and vitamins are particularly likely to be measured by bioassay because they occur at very low concentration to be detected by the usual physical or chemical method.

Avena Curvature Test

Avena seedlings are grown in dark. After a height of 14-30 mm, the plants are decapitated to remove the auxin source in the tip, leave them for 3 hours and redecapitated to remove the regenerated tissue, which now produces auxin. The primary leaf is pulled loose but leave partly in. Applied auxin is dissolved in 1.5% of agar. It is applied as a 10-mm^3 block on one side of the cut surface. The side of the coleoptile below the agar block now elongates more than to opposite side, and the resulting curvature reaches an optimum value in about 110 minutes. Curvature is measured by recording the angle made by a vertical line and drawn parallel to the curved position of the stem. There is a linear relationship between the concentration and the amount of curvature obtained.

Avena Section Test

It is based on the effect of auxin to stimulate cell elongation. Coleoptile 20-30 mm length is used. The apical 4 mm portions are removed, and a section of 3-5 mm in length is cut from each coleoptile. All the sections are first soaked in distilled water for one hour and then are placed in petri plates containing auxin solution. After an interval of 24 or 48 hours, the length of each section is measured. The measurements are made with an ocular micrometer and a dissecting microscope. The capacity of the solution to increase the growth of section is directly proportional to the logarithm of the concentration of auxin used. The disadvantage of this test is that quite a depressingly long list of substances that can give auxin-like responses in the section bioassay. Another disadvantage of this bioassay is the likelihood of bacterial contamination and resulting complications from incubating non-sterilized tissues in a 2% sucrose solution for 24 hours.

Cross Root Inhibition Test

Roots are more sensitive to auxin than stems, and their growth is inhibited by those concentrations of auxin that normally stimulate stem growth. Low concentrations of auxin induce root growth. The value, then of cross inhibition test is that the effect of extremely low concentrations of auxin, such as found in plant extracts, may be measured. It is conducted as: (1) the sterilized seeds are germinated on moist filter paper. (2) When the seedling has reached a desired length, they are placed in petri dishes containing 15 ml of test solution. (3) Root growth is measured after 48 hours.

IAA TRANSPORT IN PLANT DEVELOPMENT

IAA is synthesised in apical growing regions of plants, the shoot apex and the young leaves. From these IAA-producing tissues, IAA is unidirectionally transported away towards basally located plant tissues and organs. The transport of IAA through the vascular system of leaves and stem is polar, occurs basipetally with a velocity of 5–20 cm/h, consumes energy and is specific for biologically active auxins (Lomax et al., 1995). Pharmacologically active drugs, phytotropins, inhibit auxin transport as well as gravitropic and phototropic responses in intact plants. The precise mode of action of these compounds is not yet known, they seem to interact with plasma membrane binding site thereby blocking auxin efflux.

The basipetal directionality of auxin transport from shoot apices and young leaves to root is thought to result from the polar distribution of specialized carrier molecules in the plasma membranes. According to the chemiosmotic hypothesis, auxin is assumed to leave transport competent cells only through the activity of specific carrier proteins that are localized at the basal end of these cells. According to the chemiosmotic theory, metabolic energy is expended in order to maintain the pH and electrical gradient and the polar permeability. Biochemical and physiological experiments suggested that at least three polypeptides are associated with the auxin efflux machinery. First, a plasma membrane localized carrier protein: second, an NPA binding protein: and third, a labile probably cytosolic component (Lomax et al., 1995). Several auxin transport inhibitors exist including NPA (Naphthylphthalamic acid) and TIBA (2,3,5–triidobenzoic acid). They bind to a protein at the plasma membrane and directly or indirectly inhibit the auxin efflux carrier. The subsequent accumulation of auxin blocks tropic responses and cell elongation.

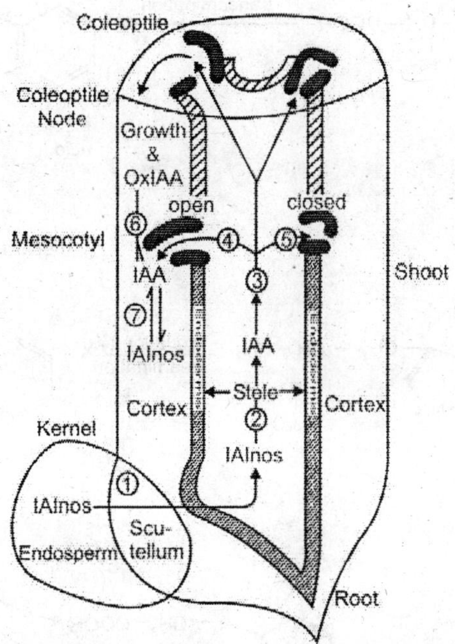

FIG. 20.3: A diagrammatic representation of the Potential Gating Theory showing the 2-step regulation of free IAA amount. First, regulation by gating IAA movement from stele to cortex, and second the enzyme–catalysed regulation of free to ester IAA.

ACCUMULATION OF AUXIN

The concentration of auxin is highest in stem tip and gradually decreases downwards. The basal portion of the stem contains lowest concentration of auxin. The production/biosynthesis of auxin by some tissue of plant is a continuous process. However, they do not accumulate it in large amount. This means that some process or processes of inactivation or destruction must take place. Inactivation of auxin is one of the important processes of the plant system by which growth is controlled. Auxin concentration at a given site is proportional not only to its rate delivery but also to its rate of destruction.

BIOSYNTHESIS OF IAA

Naturally auxin is produced by *Rhizopus suinus* when grown in peptone-containing medium. It synthesises auxin by oxidation of peptone. It has also been demonstrated that this mould could convert tryptophan into auxin. Ferri and Bonner indicated that enzymes, which are involved in conversion of tryptophan to IAA in *Avena* coleoptile, are similar in distribution of IAA.

There are several pathways for synthesis of IAA for example; it is possible to begin with tryptophan pathway for derivation of metabolic indole ring. Tryptophan is the precursor of IAA biosynthesis, which is converted into IAA by the removal of amino acid and carboxylic acid. Tryptophan is converted into tryptomena, that is - COOH group of tryptophan is replaced by NH_2. However, tryptophan may also be converted into indole 3 pyruvic acid. Both the compounds are independently converted into Indole 3- aldehyde, which may be further, converted into Indol-3-ethanol. The indole 3-aldehyde and indole-3- ethanol is interconvertable. In the last, the indole 3- aldehyde is converted into Indole-3-acetic acid. The biosynthesis of IAA is well established.

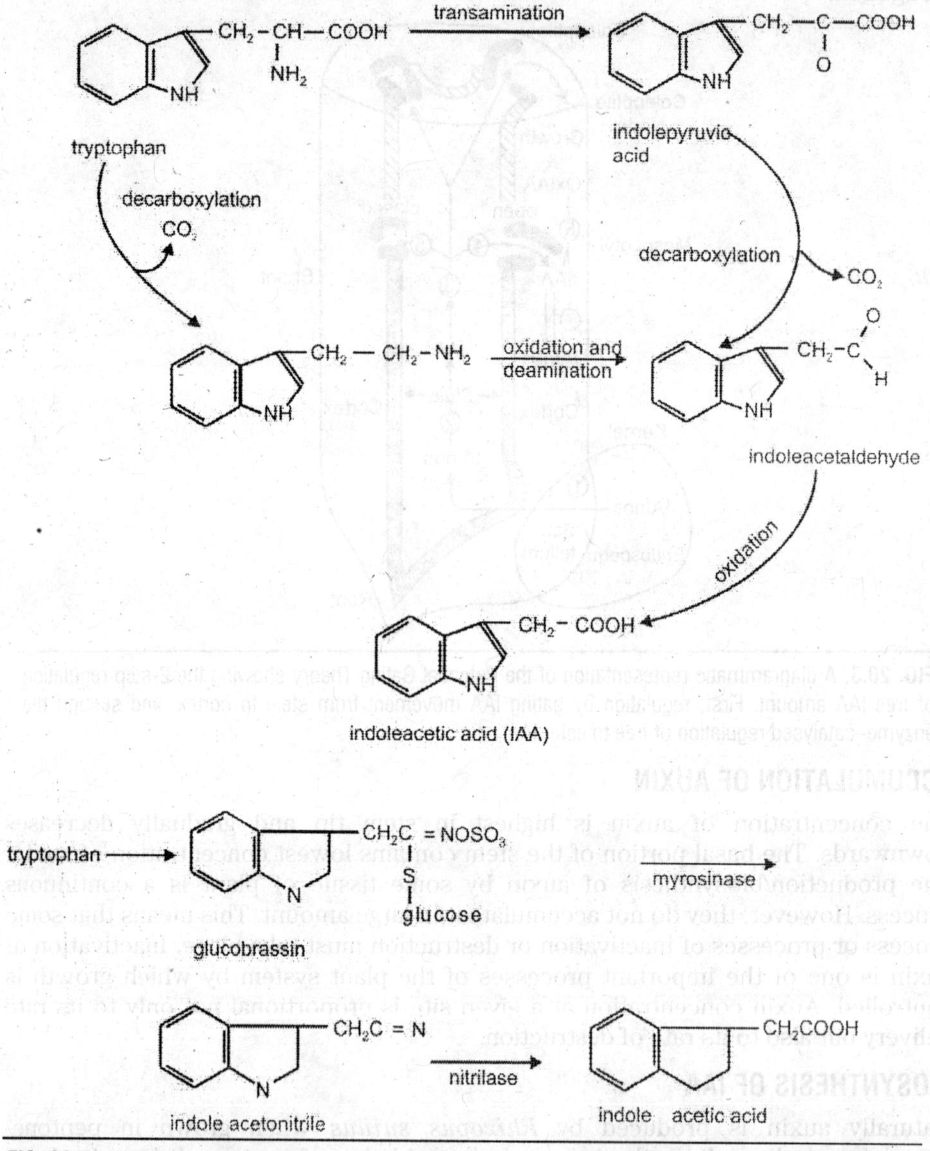

tryptophan — transamination → indolepyruvic acid

tryptophan — decarboxylation (CO_2) → tryptamine ($-CH_2-CH_2-NH_2$)

indolepyruvic acid — decarboxylation (CO_2) → indoleacetaldehyde

tryptamine — oxidation and deamination → indoleacetaldehyde

indoleacetaldehyde — oxidation → indoleacetic acid (IAA)

tryptophan → glucobrassin ($CH_2C=NOSO_3$, S, glucose) — myrosinase →

indole acetonitrile ($CH_2C\equiv N$) — nitrilase → indole acetic acid (CH_2COOH)

FIG. 20.4: (A) Pathway of auxin biosynthesis from tryptophan. (B) An alternate pathway of tryptophan biosynthesis in some plants.

Metabolism of IAA: In plants, IAA is metabolized by conjugation and oxidation via different pathways, which vary with species and probably also with the organ and stage of growth. The synthesis, conjugation and hydrolysis, oxidation and utilization during growth are determinants of IAA concentrations in plants.

Limited evidences show that metabolism of IAA can be effectively regulated by environmental factors, such as light, or by chemical applications, such as auxin, cytokinin, glyphosphate or phenol.

Conjugates of IAA

Plant hormones are sometimes available as bound, water soluble, or polar hormones from which free hormones can be liberated by hydrolysis. These compounds are widely distributed in plants and occasionally present in high concentrations. Some of them have been characterized as *conjugated plant hormones*. This term includes plant hormones, which are metabolically coupled to other low molecular weight compounds by covalent binding. This definition excludes noncovalent association between plant hormones and high molecular weight particles like protein, cell particles and membrane. These complexes are known as *bound* plant hormones.

Another important way in which IAA can be rapidly inactivated is by its conjugation with a variety of chemical normally available in the cell. IAA is metabolized in plants by oxidation and by conjugation with amino acids, sugars and the sugar alcohol *myo*-inositol. The existence of *bound IAA* was recognized long before chemical characterization. The discovery of the formation of IAA-aspartate from applied IAA by Andreae and Good (1955) open new direction in the field of IAA-conjugates. Since then, IAA-aspartate has been reported in several plant species, and conjugates of IAA with other amino acids have been found. Conjugation of IAA by ester linkage with glucose was first reported by Zenk (1961). Subsequently, Bandurski et al made extensive studies of ester conjugates of IAA with *myo*-inositol. Good et al. (1956) demonstrated the formation of indole-3-acetamide in tissues of several plants after incubation with IAA. In all endogenous auxin conjugates isolated from plant material up to now, the low molecular weight groups are covalently bound to the carboxylic group of the plant hormone, linkage to N-atom of the indole ring, e.g., by formation of N-glucosides has been observed. The endogenous auxin conjugates can be classified as either peptidyl conjugates, or sugar and *myo*-inositol derivatives, or alkyl esters.

DESTRUCTION OF AUXIN

There are two major means of inactivation or destruction of auxin in plants. (i) Enzymatic oxidation and (ii) Photooxidation.

Enzymatic Oxidation of Auxin

The enzyme involved in destruction of IAA in vivo is known as *IAA-oxidase*. It is ubiquitously isolated from plants. The oxidation of IAA in pea epicotyls appears to be catalysed by a peroxidase in which one molecule of oxygen is consumed (hence the termed as oxidase) and CO_2 is released for one molecule of IAA oxidized. A flavin protein is also found closely associated with the peroxidase, which possibly generate H_2O_2 needed for reaction. There is a basic difference between peroxidase and oxidase activity. During peroxidase reaction there is *no* need of oxygen, however, in the destruction of IAA, the enzyme possessing peroxidase activity also operates as an oxidase, and oxygen is consumed in the reaction. The end products of destruction of IAA are 3-methyleneoxindole or indolealdehyde. Mn^{++} ions and phenol (2,4-dinitrophenol) are required during the reaction. The chlorogenic acid, caffeic acid, scopoletin, and ferulic acid are the inhibitor of IAA-oxidase activity. The activity of first two compounds on IAA oxidase can be reversed by the addition of H_2O_2.

Photooxidation of Auxin

IAA is inactivated by ionizing radiation. Pure IAA is inactivated by X- and gamma-radiation due to oxidation by peroxidases formed during irradiation. The radiation directly affects the enzyme system responsible for the conversion of tryptophan to IAA. Ultraviolet radiations also inactivate IAA.

MECHANISM OF ACTION

The initial event of IAA action is the absorption of IAA to the hormone specific binding site. The IAA binding site complex then initiates the reaction. This is evidenced as follows:

1. Membrane phenomena leading to media acidification.
2. Nucleic acid related phenomena leading to longer changes involved in growth and morphogenesis - one of the results of (a) and (b) would be changes in the plasticity of the wall so that growth could occur. This increases in plasticity requires following changes: (1) In protein matrix of the wall. (2) In cellulose matrix. (3) Hemicellulose matrix.

   The working hypotheses is given below:

Nature of Wall Loosening Signals

Davis (1973) and Evans (1974) presented reviews to describe the nature of the wall loosening signal moving from protoplast to cell wall. They presented the following suggestions:

1. Auxin accumulates synthesis of pectic fraction of the cell wall or enhances methylation of carboxyl group.
2. Auxin acts to stimulate the de novo synthesis of wall loosening enzyme by the induction of transcription (the *gene expression hypothesis*) or translation.
3. Auxin stimulates total wall synthesis perhaps by stimulation of glucan synthetase activity.
4. Auxin stimulates wall turnover, especially the xyloglucan component.
5. Auxin stimulates the outward pumping of hydrogen ions leading to wall acidification (the *acid growth hypothesis*).

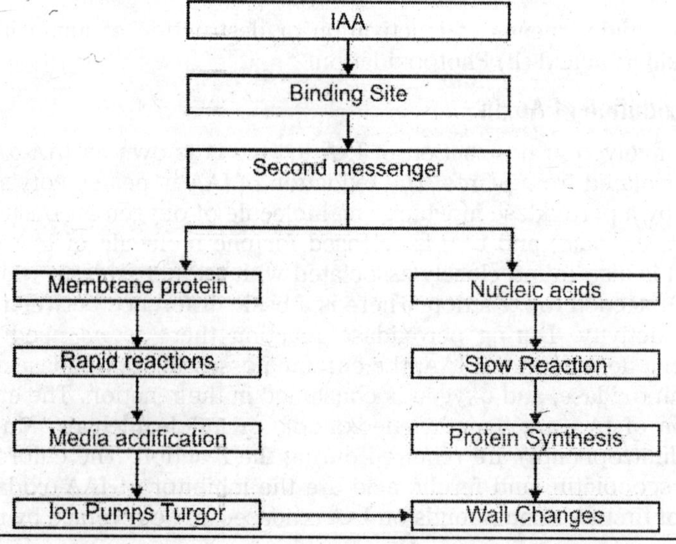

FIG. 20.5: Events in cellular enlargement and possible relationships of auxin action.

In order for an auxin induced effect to be a candidate for the cause of growth promotion it must be shown, of course, that the effect occurs before the growth promotion begins. Among the most rapidly occurring effects of auxin are:

1. Stimulation of protoplasmic streaming (almost immediate),
2. Apparent increase, in membrane permeability to water (about 6 minutes),
3. Stimulation of beta-glucan synthesis activity (about 15 minutes),
4. Stimulation of wall xyloglucan turnover (15 to 30 minutes), and
5. Enhancement of H^+ efflux (0 to 20 minutes depending on material and conditions).

The binding of IAA with cell wall/cell components is however reported by several workers, yet no plant hormone receptor has been isolated with certainly that the binding proline is involved in hormone action.

Membrane and Ion Movement Leading to Media Acidification

When auxin sensitive tissue is incubated in the solution of low pH, it promotes growth. The pH optimum for acid-induced growth is 4.5 to 5.0 for those tissues whose cuticle is removed. Hager, Melzel and Krauss (1971) proposed acid growth hypothesis for this low pH action. According to them IAA activates plasma membrane-ATP which stimulates active proton efflux from the cell. The low pH activates enzymes capable of hydrolysing wall polysaccharides to soften the wall and allow cell extension. The uncoupler like carboxyl cyanamide m-chlorophenyl hydrozone (CCCP) prevents both cells extension and media acidification. Similarly KCN, 2,4-dinitrophenol, valnomycine, abssisic acid and cyclohexamide also act similar to CCCP.

The finding supports the acid growth hypothesis that neutral buffers prevent IAA induced growth. The above discussion reveals that media acidification is some how related to IAA induced growth.

Mechanism Media Acidification

The exact mechanism of media acidification is still obscure. However calcium ions are required for this process. The ATP: ADP ratio decreased while ADP: ATP ratio is increased. Respiration rate also induced acidification.

Nucleic Acid and Protein Changes

The growth of a plant by IAA may be stimulated by an increase in RNA synthesis which may depress certain genes, thus develops change the rate of RNA and protein synthesis. It has been demonstrated that inhibitors of RNA and protein synthesis also inhibit auxin-induced growth. It now seems within reach to determine, what enzymes are encoded in sequences of messenger RNA, which becomes active after the plant is treated with IAA. There are several aspects like changes in RNA and protein, changes in cell wall plasticity, which are triggered after IAA treatment.

Changes in Wall Plasticity

Plant cell is enveloped by a cell wall, which stretches or grows as the protoplast enlarges. It may increase in size, about 50-fold, with a little or no increase in dry weight. The cellulosic microfibrils of cell wall are almost at right angles to the long axis of the plant prior to cell expansion, however, after expansion these fibrils become more nearly parallel to the plants axis. How this action occurs is still a matter of active research. It is also established that four polymeric substances determine the wall rigidity: wall protein, the cellulosic beta-1-4 glucane, beta-1-3 cellulose type glucane and lichemine type glucane. They also determine the wall plasticity.

It may, therefore, concluded that there is no single determinant of the structural rigidity of the wall when IAA-induced growth occurs; the solution bathing the cell wall becomes more acidic and thus wall is easily deformed. This change in the plasticity is not only due to low pH but also by enzyme-induced changes in the wall, however, the exact mechanism of cell wall plasticity needs further research.

Secondary Messenger Hypothesis

According to this hypothesis, auxin does not directly act on the cell wall. The addition of auxin to isolated cell walls or to the cells with non-functional protoplasts has no effect on wall extensibility. This has led to the conclusion that the initial action of the hormone is at a site on or within the protoplast, an action that leads indirectly to wall loosening. Thus, there must be a signal or *secondary messenger* of some type to enrich travels from the protoplast to cell wall, initiating wall loosening.

Composite Model? There are certain difficulties in proposing a composite model for auxin action, because of a lot of variabilities in different systems have been studied so far. Biswas and Roy (1978) proposed a working hypothesis based on their studies with IAA using coconut endosperm cells. According to them auxin may first form complex with receptor protein in the membrane and may activate membrane bound ATPasc initiating a proton pump, thus loosening of cell wall occurs. Auxin is then detached from this complex and transferred to cytoplasmic receptor. Lipids might play a role in this transfer. This complex than translocates to the nucleus and then to chromatin. It is proposed that this nucleoplasmic receptor-hormone complex is associated with the chromatin. Non-histone proteins in the chromatin may observe this binding. After transfer to the auxin to the chromosomal acceptor site, the genome would be switched on to be transcribed. However, this model needs further experimental evidences.

Theologis Model: In an attempt to elucidate the relationship between early auxin-induced mRNA accumulation, H^+ secretion, and cell elongation, a different model was proposed by Theologis (1986). According to it, auxin action involves a complex metabolic pathway, with auxin regulating at head of the pathway (*gene transcription*) and the secretory processes at the tail. The model involves H^+ extrusion systems. One system, activated indirectly by auxin, and is linked to secretory system. The second system, activated directly by fusicoccin, is isolated on cell membranes and is independent of secretory processes.

BIOLOGICAL EFFECTS

Coleoptile and Stem Growth

Auxin causes growth of the coleoptile of *Avena*, wheat, maize and other cereals. Such elongation is only possible when the tip of the coleoptile is removed before its treatment. This is because of when the tip is intact on the coleoptile, the tip itself produces the auxin, and hence the intact coleoptile shows rapid growth. After the removal of tip, the growth of the coleoptile stump ceases in the absence of endogenous supply of auxin from the tip. Externally applied auxin can promote the growth of detopped coleoptile. When auxin causes growth of a coleoptile, it is exclusively due to the elongation of pre-existing cells. Stem growth is also promoted by auxin-treatment through cell elongation.

Cell Elongation

Auxin causes cell elongation by loosening of cell wall. The rigid cell wall becomes softened and its plasticity (irreversible capacity to stretch) increases. The loosening of cell wall begins due to: (1) (i) dissolution of cell wall material. (2) Breaking of chemical bonds between cellulose and other wall materials, and (3) fresh synthesis of new cell wall material and its incorporation into existing cell wall. The loosening of cell wall decreases the wall pressure (WP), which is equal to turgor pressure (TP). This action causes greater uptake of water and increase in the size of vacuole; and cell stretches.

Secondary Growth

When the tip of the sunflower seedling is removed (source of auxin) it unable to show any symptoms of secondary growth. However, when it is treated with auxin, the formation of secondary vascular cambium resumes. Promotion in cambium activity by auxin continues secondary growth. It also controls total thickness of the stem.

Rooting of Cuttings

When the shoot-tip is removed it reduces the shoot's growth. In contrast, removal of the root-tip does not result in decreased root growth. The removal of less than 1 mm of the root tip results in a very small but a significant stimulation of growth rate. When the tip is replaced, the root growth is also retarded. The action of auxin in root is similar to that of stem, but the difference is dependent on the concentration of auxin which is stimulatory to stem and inhibitory to root growth. This also reflects that roots are more sensitive to auxins than that of stems. When high concentration of auxin is applied to root, it not only retards root elongation but also causes a remarkable increase in the number of branch roots. Application of IAA + linolin paste to the severed end of a young stem initiates the rate of root formation and develops large number of roots. This aspect has its commercial value.

Auxin (or IAA) plays a key role in a wide variety of growth and development processes. At the cellular level, auxin acts as a signal for division, extension and differentiation during the course of the plant life cycle. At the whole plant level, auxin plays an important role in root formation, apical dominance, tropism and senescence. In spite of the importance of this hormone, the molecular basis of auxin response is poorly understood. During the past decade, identifying mutants with altered responses and characterizing the corresponding genes and gene product have made progress toward the mode of action of auxin.

Initiation of Lateral Roots

Lateral root primordia are generally initiated through the commencement of cell divisions in the pericycle opposite the xylem poles of the root vasculature. Auxin stimulates root pericycle cells arrested in the G_2 phase of the cell cycle to enter into mitosis thereby promoting the formation of lateral root primordia (Blakely and Evans, 1979). It has been reported that auxin increases both Cdc2 protein kinase and cyclin mRNA levels in roots with induction of cell divisions (Ferreira et al., 1994).

Cell Cycle Control

The classic experiments of Skoog and Miller (1957) on the growth of tobacco pith explants in culture demonstrated a role for both auxins and cytokinins in the control of cell division. These experiments encouraged research into the effects of auxins and cytokinins on cell proliferation *in vitro*. Molecular work on the link between hormone action and cell division has led to the cloning of several genes that are responsive to auxin.

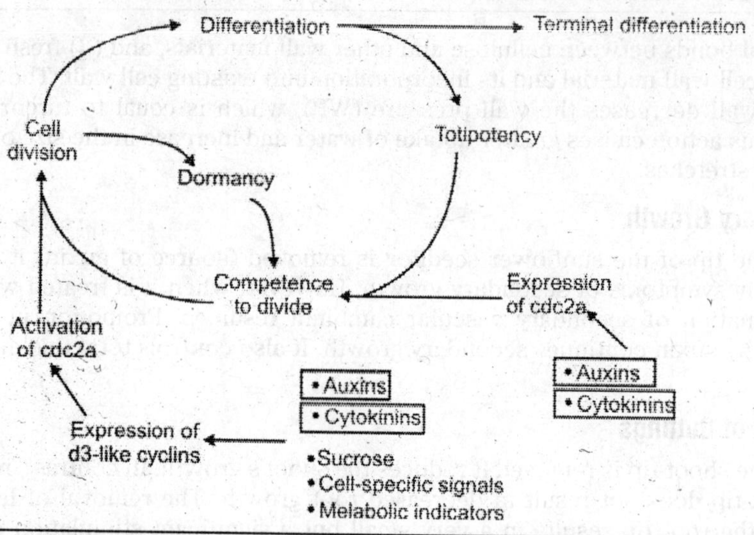

FIG. 20.6: Speculative model for control of the cell cycle through auxins and cytokinins. Both cytokinin and auxin have been reported to regulate the expression of cdc2 kinases and the cyclins that are required for their activation. The interaction between auxin and cytokinin in regulating the cell cycle is synergistic in undifferentiated cells, such as callus or protoplasts, with both auxin and cytokinin stimulating expression of the cdc2 kinase and cytokinin treatment increasing expression of a cyclin. In lateral root primordial the interaction is antagonistic: auxins stimulate and cytokinins reduce levels of the cdc2 kinase and the expression of at least one cyclin is increased by auxin.

Like other eukaryotes, the cell cycle mechanism in plants depends on serine–threonine kinases whose activity is modulated by their association with regulatory cyclin subunits. Auxin increases the expression of a protein related to the Cdc2 class of cyclin–dependent kinases in tobacco pith explants. In tobacco leaf protoplasts, expression of the gene encoding β–glucuronidase (GUS) under the control of an *Arabidopsis* Cdc2 promoter reveals that the corresponding gene is inducible by auxin. The activation of cyclin–dependent kinases is based on the synthesis of cyclin. Cloning of a number of D–type cyclins from *Arabidopsis* shows that the expression of one member of this group, cyclin δ3, is highly dependent of cytokinins. Coenen and Lomax (1997) concluded that once cells have become competent to divide because of auxin–induced expression of a Cdc2 kinase, cytokinin–induced expression of cyclin δ3 may permit entry into the mitotic cycle.

Cell Expansion

Auxin and GA promote cell elongation. It was thought that auxin–stimulated growth could only be observed in excised, auxin–depleted tissue whereas GA could induce the growth of intact plants. However, exception to this generalization has recently been reported. It has been shown that the continuous supply of auxin via a cotton wick stimulates growth of two dwarf mutants of pea, **le** and **ikb**. The first is a GA–deficient mutant whose growth can be fully restored by treatment with GA_1 (Ingram et al., 1983). The latter is a dwarf mutant with a two to threefold reduction in IAA level whose growth is not promoted by applied GA. On the basis of these data, it was concluded that auxin and GA control separate processes in cell elongation. GA may act preferentially in younger cells compared with auxin and induce cell division and cell elongation whereas auxin may act by promoting cell extension.

Cell expansion is driven by water uptake, which in turn results from stress relaxation of the cell wall. Cells respond to auxin by making their cell walls more plastic (more extensible). The auxin effect was partly explained in the early 1970s by the discovery of "acid growth". Plant cells grow faster and their walls become more extensible at acidic pH. Auxin was hypothesized to stimulate growth, in part, by inducing plant cells to acidify their extracellular space. However, the acid-growth theory of auxin action is not universally accepted. Critics point out that the wall pH in auxin–treated tissues is not low enough to elicit the growth. The discovery of the expansins, a family of proteins that exhibit wall–loosing activity at pH 4.5 may open new approaches to the problem of auxin–induced stress relaxation of the cell wall.

It has recently been shown that phospholipases are involved in auxin–stimulated cell elongation. In hypocotyl segments of *Zucchini* and sunflower auxin induces rapid activation of phospholipase A_2 activity and inhibitor of phospholipase A_2 block auxin–stimulated elongation. The auxin–specific activation of phospholipase A_2 in microsomes from suspension–cultured soybean cells is receptor–mediated and influenced by guanine nucleotides. The release of fluorescently–labelled free fatty acids by auxin–activated phospholipase A_2 is quantifiable within 5 min in cultured parsley and soybean cells with low concentrations of auxin. The fatty acids and/or lysophospholipids possess growth-promoting activity and support the emerging role for phospholipase A_2 in auxin–mediated cell elongation.

Tropisms

Tropisms are defined as directional responses of a plant organ to a directed stimulus in the environment. Tropic movements allow the plant to adjust itself or orient its organs in response to its surrounding environment for optimal uptake of energy, nutrients or water. Auxin has been implied as the major signal mediating tropic stimuli. In 1920s the Cholodny–Went hypotheses was formulated to explain the gravitropic response of plant roots and shoots. This hypothesis proposes that the differential growth rate during the gravitropic response is based on auxin redistribution and altered auxin transport within plant tissues.

PIN2 gene controls root gravitropisum in *Arabidopsis* and has recently been cloned. The agravitropic phenotype shows the absence of **PIN2** protein in cortical and epidermal cells. As predicted by the Cholodny–Went hypothesis auxin supplied from the shoot is transported acropetally through the stele in the meristematic region of the root tip where is redistributed in the cells of the root cortex and the epidermis and basipetally transported towards the elongation and differentiation zone. The preferential localization of **PIN2** to the anticlinal basal side of the epidermal cell files oriented towards the elongation zone strongly supports the concepts formulated in the Cholodny–Went hypothesis.

Apical Dominance

Classic studies on the mechanism of suppression of lateral bud growth by the plant apex have shown that auxin produced at the apex represses the outgrowth of lateral buds. The recent work has demonstrated that decapitation (removal of auxin source) leads to a large increase (up to 40–fold) in the cytokinin content of xylem exudates. This increase can be eliminated by application of the synthetic auxin ($\propto$ – NAA) to the apex of decapitated plants. The effect of auxin on cytokinin concentrations in the xylem suggests that auxin can influence apical dominance via inhibition of cytokinin synthesis or export from the roots. However, bud outgrowth can also be inhibited by apically applied auxin in isolated stem segments, which indicates that this is not only mechanism available.

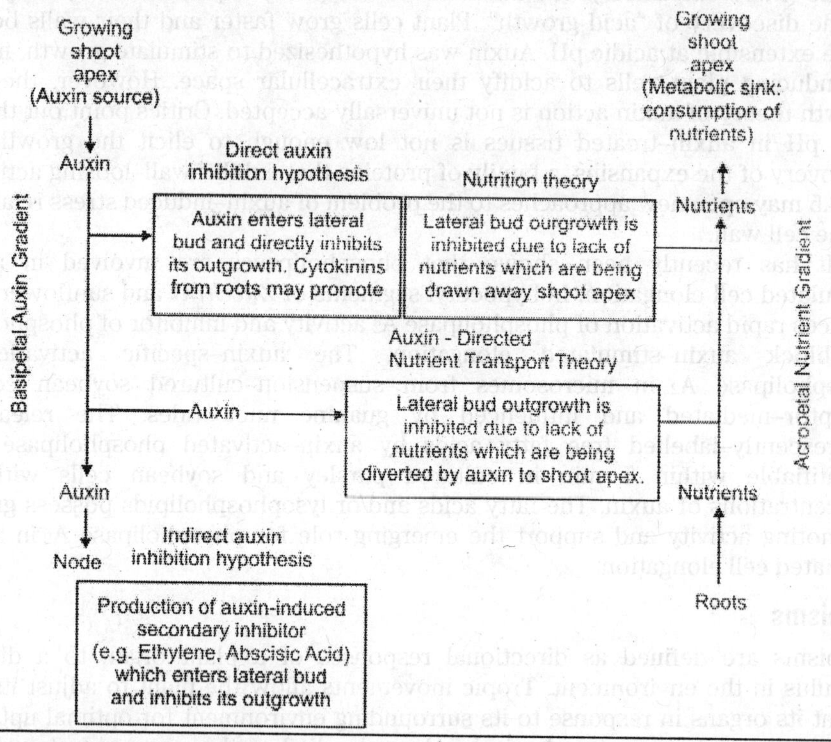

FIG. 20.7: Some hypotheses to explain the mechanism of apical dominance.

Parthenocarpy

Fruit setting starts after pollination and subsequent fertilization of the ovule of a flower. Growth of the ovary wall and, in some cases, the tissue associated with receptacle is enhanced. The acceleration in growth•is due to the cell enlargement induced by endogenous auxin. The pollination and fertilization events are associated with the proper development of fruit; probably these events release some stimulus. Fruit development does not take place without *pollination-action.* Such fruit will be than without seed. The development of fruit in this manner is called *parthenocarpic development* and the fruit is known as parthenocarpic fruit.

Fertilization is the natural phenomenon of all seed plants. Fruit development does not occur without fertilization. How does fertilization of the ovule trigger off responses leading to fruit set? Various possible explanation have been presented from time to time, they are as follows:

1. Swelling of ovary wall of orchids could be stimulated by dead pollen grains (Massart, 1902).
2. Water extracts of pollen are capable of inhibiting floral abscission and stimulating ovary wall swelling in orchid (Fitting, 1909).
3. Parthenocarpic fruits developed after application of pollen extracts of cucumber flowers, which contained auxin (Yashuda, 1934).
4. Application of IAA in linolin paste to stigma of flower could able to induce parthenocarpic fruits (Gustafuson, 1939).
5. A sharp rise in auxin content of tobacco ovaries has been observed just after pollination; however, in absence of pollination such increase was not observed

(Muir, 1942). He further demonstrated that pollen tube increases the amount of extractable auxins in the style of tobacco plants. It shows that pollen tube contains certain enzymes which may catalyses the production of auxin.

6. Pollen tube secretes an enzyme, which can convert tryptophan to auxin (Lund, 1956).

The above experiments reveal that auxin is an important factor in the fruit development. Pollination, growth of pollen tube and fertilization all contribute to the cause that auxins are responsible for fruit development.

Fruit Size

Exogenous application of auxin can alter fruit size and the period of maturity in plums, pineapples and fig. Auxin can increase the size of these fruits and rate of maturity. Auxin-palm application stimulated growth of strawberry fruits also.

Abscission

Abscission is the shading, detachment, separation or dehiscence (a specialized type of abscission) of plant parts, that is, organs, other structures, or tissues.

It has been known for nearly 55 years now that auxin retards abscission. The level of auxin in leaf blades (*Gossypium*) decreases sharply with age and immediately prior to their shading. Jacobs (1962) showed that it is possible to substitute the abscission-retarding effect of the lamina by an appropriate amount of exogenously applied IAA. Applying IAA to explant systems usually retards abscission and a number of workers have either demonstrated or confirmed that when applied distal to the abscission zone, there is an increasing retardation of abscission with increasing levels of IAA. What terms a monophasic response that sustains the inhibition elicited by endogenous auxin? However, in certain systems and under certain conditions, IAA when applied distally results in a diphasic response: acceleration of abscission at low, and retardation at high levels. Such effect has been shown with the synthetic auxin- NAA. Different physiologists regarding have proposed various theories the involvement of auxin in abscission process are as follows:

Auxin Gradient Hypothesis (Addicott et al., 1955)

The proximal-distal gradient of auxin across the abscission zone is the principal factor in the regulation of the onset and rate of abscission. In the young leaf of a healthy, vigourously growing plant the level of auxin in the tissues distal to the abscission zone is characteristically higher than that in positions proximal to it and abscission does not occur. When through age, shading, deblading, treatment with accelerants, etc. there is a decrease in the ratio of distal to proximal auxin, abscission occurs. Lower to this ratio, the greater will be acceleration of the process. The ability of IAA when applied proximally to the abscission zone to elicit this monophasic acceleration of abscission has been shown often.

Auxin Concentration Hypothesis (Gaur and Leopold, 1956)

In contrast to the concepts involving gradient and balance, low concentrations of auxin are thought to promote and high concentrations to retard abscission.

Auxin-Senescence Factor Hypothesis (Osborne, 1955)

Leaf abscission is controlled not only by endogenous auxins but also by some substances derived from senescing petioles (leaves of *Phaseolus*). This substance, called senescence factor, which accelerates abscission, has been detected in *Gossypium* and *Coleus*. It was thought by Osborne (1968) not to be ABA, although this was the working hypothesis of Jacob's laboratory.

The understanding of the role of auxin in abscission on the one hand is complicated by its diverse effects such as those of stimulating respiration and ethylene evolution, retarding the development of the separation layer but than, once formed, enhancing cell division in it and maintaining cell membrane integrity. There are two stages of abscission. During **Stage I** applied auxin prevent abscission, an indication that one of the many physiological changes taking place in explants presumably is a decline or inactivation of endogenous IAA. These changes appear to lie in tissue distal to the abscission zone *per se* and characterize a stage spanning that from full maturity of the organ (leaf, petal, fruit, etc.) to just before the onset of final senescence. During **Stage II** the tissue distal to the abscission zone undergo the final stage of senescence while in the zone itself there is intense biochemical activity as witnessed by the synthesis of RNA, protein and various cell wall degrading enzymes. Application of auxins and other hormones during this stage, accelerate the completion of abscission.

There is as yet no clear explanation of auxin's dual role, but Addicott (1970) suggested that applied auxin maintains the ongoing biochemical pattern in plant. Auxin is also an inhibitor of the synthesis activity of the cell wall degrading enzyme-cellulase, it the principal retardant of abscission.

Respiration and Auxin

Auxin increases respiratory rate of the plant cell. Its activity only takes place in the presence of oxidative metabolism. A correlation between increased growth and auxin treatment has been established. How auxin induces the rate of respiration is a matter of further research?

Organogenesis and Auxin

Auxin causes swelling in the callus tissue (rapidly dividing parenchyma cell) in vitro. Auxin is added in the culture medium for procurement of continued callus growth. Skoog and Miller (1957) established that in interaction between kinetin and the auxin- IAA, IAA alone, or high ratio of IAA to kinetin lead to adventitious root formation. The application of NAA to tobacco explant and on many other plants promotes root formation. Exogenous auxin can have two major effects, therefore, in the initiation and/or enhancement of root organogenesis and as an inhibitor of cytokinin-induced shoot initiation. Another application of auxin is in the rooting and the shoot obtained through micropropagation. Removal of auxin may induce organogenesis.

Crown Galls

Callus like tissue, known as crown galls, is formed in sunflower, marigold, tomatoes and beans, etc. These are produced by an external stimulus, namely by parasitic bacteria. Mature crown gall tissues produce large amount of auxins to continue the growth.

AGRICULTURAL USES OF AUXINS

Auxins as Herbicides, Ripeners, and Rooting Inducers: The most widespread commercial application of auxins as herbicide is well known. Auxins like (2,4-dichlorophenoxy) acetic acid (2,4-D) and (2,4,5-trichlorophenoxy) acetic acid (2,4,5-T), which act as herbicides when used in large doses.

Dicotyledonous weeds are particularly sensitive to very high auxin level. Some of the most valuable and widely used selective herbicides in the weed control are auxins, particularly the phenoxyacetic acid analogous (e.g., 2,4-D, 2,4,5-T, and

McPA). 2,4-D is one of the most widely used herbicides. It is highly selectively, non-corrosive, effective at low concentrations, safe to handle, relatively easy to formulate, and economical to use. Several benzoic acid analogs (e.g., dicamba, chloramben, and substituted picolinic acid, picloram [Tandon]) are also important herbicides.

R. J. Weaver (1972) proposed another use of auxin in agriculture by using the principle of inhibition of abscission layer formation, auxin (e.g., NAA, 2,4-D) are effective to control fruit drop in apple and pear. Auxin including 2,4-D, induced ethylene formation and fruit set in pineapple.

Biennial bearing (light and heavy crops in alternate year) is common to many tree crops. This is a great problem for agriculturists. By corrected and timely thinning in the heavy years and timely application of NAA or other auxins can solve the problem.

2,4-D enhances fruit set and fruit ripening particularly in pineapple cultivation. Induction, or forcing of flowering is the major problem in pineapple. This was achieved initially in the field by 1-naphthyl-acetic acid (NAA), which is also used as a rooting agent. Ethephon is also preferred for this purpose.

Commercial use of rooting auxin compounds is very common. They promote callus and root formation, which can improve establishment from cuttings. Dipping the cut surface into rooting compounds enhances species and cultivars difficult to root. Auxins are also used in the prevention of sprouting of stored potatoes by dipping them in NAA solution or by sprinkling with talc or fuller's earth containing auxin, or stored with strip of paper impregnated with auxin solutions. Indolyl butyric acid (IBA) is also used in inducing the rooting of cuttings. In vitro rooting of shoots of Antonovka 313 apples was achieved by using a medium containing 0.25 mM indolyl butyric acid and 1.5% sucrose. Promotion of rooting by auxins is sometimes enhanced by the presence of some phenolic compounds, which have a synergistic effect. More roots were produced in cuttings of *Phaseolus vulgaris* L. in the presence of both indole butyric acid and 4-chlororesorcinol, an effect explained by the inhibition of polyphenol oxidase (PPO) activity.

Glyphosine [N, N-bis(phosphonomethyl) glycine], an auxin-like substance is also used as sugarcane ripeners. Its action, primarily, is as herbicides, to prevent late growth of the cane so as to cause a major part of the assimilate to be stored as sugar.

$$HOOC\text{-}CH_2\text{-}N[CH_2\text{-}PO(OH)_2]_2$$

The compound has enabled a marked increase in the yield of sugar. Other important sugarcane ripeners are ethephon, 2-chloroethyl-phosphonic acid, and glyphosate N-(phosphonomethyl) glycine.

$$HOOC\text{-}CH_2\text{-}NH\text{-}CH_2\text{-}PO_3H_2$$

The use of ethephon has enabled an increase in the yield of rubber from *Hevea brasiliensis* by 200% to 300%.

Synthetic auxins or auxin-synergists like 2,4-D, 4-CPA (4-chlorophenoxyacetic acid), and N-metatolylphathalamic acid are used widely to increase fruit set in vegetables, for example, with tomatoes grown in glasshouses or under plastic tunnels, or with aubergines. Highly effective synthetic auxins, which are derivatives of benzotriazoles, were recently synthesised, using a design approach based on QSARs.

Menadione Sodium Bisulphite (MSB)

Since the presence of auxin is necessary for plant growth, an inhibition of the degradation of this hormone may enhance growth. Decreased activities of the enzymes involved in IAA oxidation were observed after the application of Menadione Sodium Bisulphite (MSB). Following the addition of MSB to tomato, cucumber, corn, and capsicum plants, the concentrations of IAA in the plant cells were increased by up to four fold relative to the untreated plants. Enzymatic studies in tomato plants sprayed with 10^{-5} M MSB indicated significant decreases in the activities of indole acetic acid oxidase and of peroxidase, but had almost no effect on ascorbic acid oxidase and of peroxidase, but had almost no effect on ascorbic acid oxidase and polyphenol oxidase.

The application of MSB also resulted in remarkable increases in the growth of tomato plants, alfalfa callus, and the rooting of mung bean cuttings. Spraying tomato plants twice during their growth with a 10^{-5} M solution of this compound caused an increase of 40-80% in fruit yield per plot, relative to the untreated plants. The water-soluble part of MSB is the sodium sulphite adduct of menadione (vitamin K_3, 2-methyl-1,4-naphthaquinone). Soil application of a powder containing a mixture of menadione bisulphite. Tween 20, and diatomaceous earth enhanced the growth of roots and aerial parts of rice by 32% and 22%, relative to the control.

CHAPTER 21

Plant Hormone: Gibberellins

Discovery of gibberellins (GAs) is mainly based on experimental work of Japanese plant physiologists. In 1926 Kurosawa discovered a substance in cultures of the fungus *Gibberella fujikuroi* that caused elongation of rice seedlings. The discovery resulted from an investigation of "foolish seedling disease" in which rice plants infected with the fungus elongated abnormally. In 1938 Yabuta and Sumike reported the isolation from cultures of *G. fujikuroi* of crystals of a substance they called gibberellin. The determination of the structure of gibberellin had numerous false starts that were probably caused by the fact that there are many gibberellins with closely related structures–120 at the last count (Silverstone and Sun, 2000).

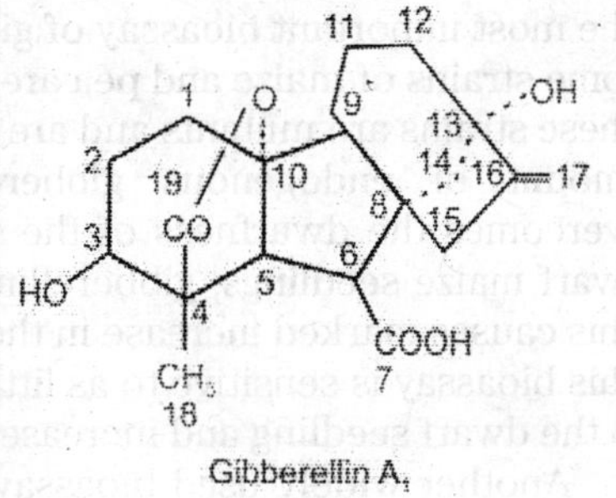

Gibberellin A_1

Because of the origin of gibberellin, the substances were studied for several decades in the context of causing abnormal growth of plants. It was not until Bernie Phinney and Percy Brian who discovered that application of the compounds to dwarf corn and pea seedlings restored stature that it was generally accepted that the gibberellins are natural growth substances. MacMillan, Radley, Sumike and others who purified gibberellins from plants substantiated this idea. In addition to the effect of gibberellin on elongation, it could induce flowering (Anton Lang, 1957) and in 1959, L. Paleg and H. Yomo independently discovered that it activated the expression of amylase during germination of barley seeds.

CHEMISTRY OF GIBBERELLINS

At present about 74 gibberellins are known which have been isolated from plant sources, e.g., GA_{17}, GA_{38}, GA_{44}, GA_9, GA_{29}, and GA_{51}, have been isolated from *Pisum sativum*.

All gibberellins have derivation from the *ent*-gibberellane skeleton. They are acidic and are named GA with a different subscript to distinguish them. Almost all GAs are chemically related to a large group of naturally occurring compounds-*terpenoids* (sterol and carotenoids are occur in plant tissues). The terpenoids are built from 5-carbon *isoprene units*, two unit from *monoterpene* (C-10), three units from a *sesquiterpene*, (C-15) and four units from *diterpene* (C-20). *Kaurene* is the precursor of gibberellin. Gibberellins are made of basic skeleton of *ent*-gibberellane (20-carbons) or *ent*-20 norgibberellane (C-19). They are differentiated from one another by the presence or absence of the lactone configuration (*internal ester*) in the ring A and the substituents, mainly -OH (hydroxyl) groups, about the whole ring structure.

1. Gibberellins are synthesised from mevalonic acid in young tissues of the shoot.

Its synthesis in the roots is still uncertain.

2. GAs are transported in the phloem and xylem tissues.
3. Gibberellins have several effects in plants and some of these include: hyper-elongation of stems by stimulating cell division and cell elongation; bolting in long day plants; induction of seed germination especially in seeds which require stratification or light; stimulate production of several enzymes notably hydrolases in germinating cereal seeds; fruit setting and growth; and induction of maleness in dioecious flowers.
4. Gibberellins differ from indoleacetic acid in structure and properties.

BIOASSAY

The most important bioassay of gibberellins is increased growth of dwarf seedlings. Some strains of maize and pea are shorter (dwarf) in length than the normal strains. These strains are mutants and are believed to be dwarf because they have negligible amount of endogenous gibberellins. Application of exogenous gibberellins overcomes the dwarfness of the seedling by increasing the length of the stem. In dwarf maize seedlings, gibberellin solution is applied to the ligule of the first leaf. This causes marked increase in the length of next node or leaf sheath, in a few days. This bioassay is sensitive to as little as 1.0 nanogram of GA. The hormone is applied to the dwarf seedling and increase in epicotyl length is measured after 4-5 days.

Another widely used bioassay of gibberellins is the barley endosperm test. By this method, as little as 0.2 nanogram of GA can be detected within 24 h. Barley aleurone (de-embryonated half seeds) layers are incubated in a solution of GA for 24 h. The gibberellins stimulate amylase activity which causes hydrolysis of starch to glucose. The glucose is secreted out in the incubation medium. This accelerated secretion of glucose (reducing sugar) solution.

Bioassays are primarily designed to detect the presence of hormone in plant tissue extract. However, now a days more sophisticated biochemical techniques are available to detect and quantitate the plant hormones. For example, gibberellins can be determined by using gas chromatography, high performance liquid chromatography (HPLC) and other such analytical techniques.

BIOSYNTHESIS OF GAS

Efforts to elucidate the biosynthetic pathway have been greatly facilitated by the availability of GA deficient (dwarf) mutants. Metabolic studies have been conducted with systems that are rich source of GAs such as the fungus *Gibberella fujikuroi* and immature seeds of pumpkin, pea and bean. However, maize is the only higher plant in which the entire biosynthetic pathway has been demonstrated in vegetative tissues by feeding various intermediates. These and other studies have shown that the GA biosynthetic pathway can be divided into three stages. These stages are:

Stage 1: From Geranylgeranyl Diphosphate to Ent–Kaurene

The first committed step in GA biosynthesis is the cyclization of geranylgeranyl diphosphate to ent–copalyl diphosphate which in turn is converted to ent–kaurene. The enzymes that catalyse these reactions have been called ent–kaurene synthase A and B respectively but MacMillan (1997) has proposed the appropriate terms ent-copalyl diphosphate synthase and ent–kaurene synthase. The genes encoding these enzymes have been cloned from *Arabidopsis* and pumpkin endosperm respectively. Biochemical evidences indicate that both enzymes are located in proplastids of meristematic shoot tissue but not in mature chloroplasts (Aach et al., 1997). In

Arabidopsis the expression of GA1 is highly regulated during growth and development. Promoter studies with β–glucuronidase as a reporter gene indicate that GA1 expression is highest in shoot apices, root tips and the vascular tissues of leaves.

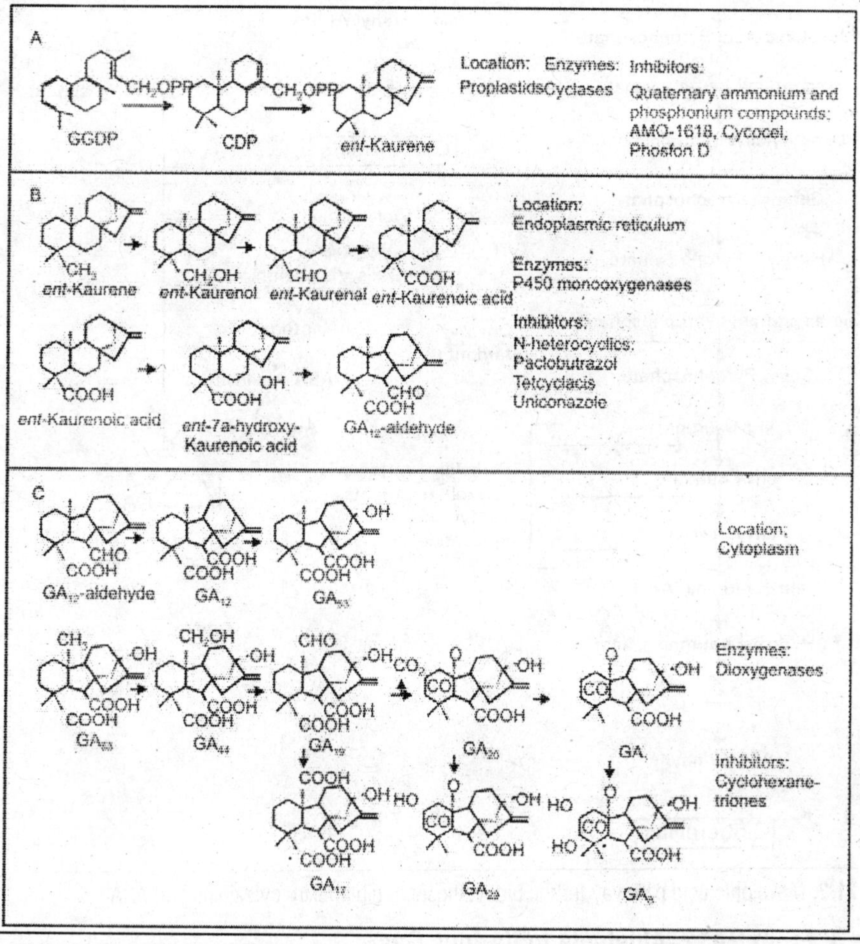

FIG. 21.1: GA Biosynthetic Pathway. (A): Stage 1: conversion of geranylgeranyl diphosphate (GGDP) to *ent*–kaurene, CDP, *ent*–copalyl diphosphate. (B) State 2: from *ent*–kaurene to GA12–aldehyde. (C) Stage 3: represented here by the early–13-hydroxylation pathway from GA12–aldehyde to GA1 and GA8. For each stage, the nature of the enzymes, their location, and specific inhibitors are indicated. In (B) and (C), the modifications taking place at each step are highlighted in boldface.

Stage 2: From Ent – Kaurene to GA12–Aldehyde

The enzyme in the second stage of the pathway are membrane–bound P450 monooxygenases which are thought to be located in the endoplasmic reticulum. The sequential oxidation of C–19 of ent–kaurene via ent–kaurenol and ent–kaurenal to ent–kaurenoic acid is probably catalysed by a single enzyme and the activity of which is impaired in a ga–3 mutant of *Arabidopsis*.

Ent–kaurenoic acid is further oxidated to ent–7∝–kaurenoic acid. Contraction of the β–ring with extrusion of C–7 gives GA12–aldehyde. The Dwarf 3 (D3) gene of maize encodes a P450 monooxygenase but it is not known which step of the pathway the D3 protein catalyses.

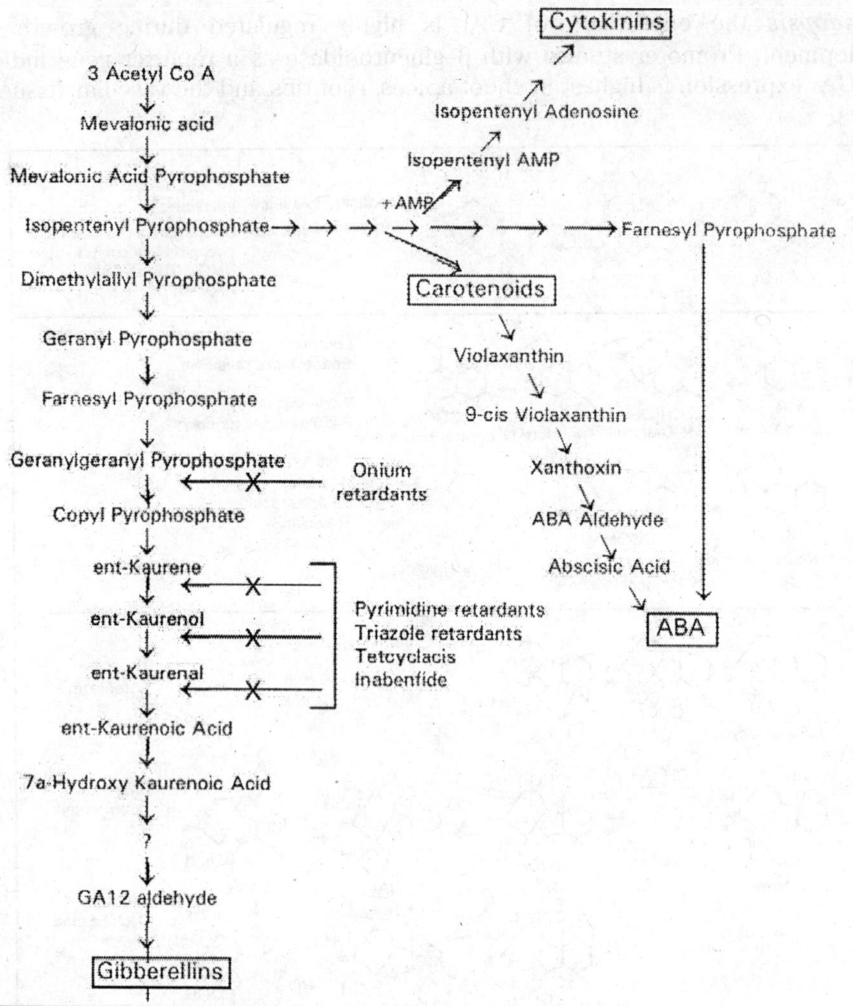

FIG. 21.2: Mevalonic acid pathway for the biosynthesis of gibberellin, cytokinins and ABA.

Stage 3: From GA12–Aldehyde to Various Gas

The first step in stage 3 of the pathway involves oxidation of GA_{12}–aldehyde to GA_{12}. Further metabolism of GA_{12} varies among species or organs of the same species with respect to the position and sequence of oxidative steps. The early 13-hydroxylation pathway, which involves hydroxylation at C–13 to give GA_{53} and leading to the hydroxylated C19 – GAs, GA_1, GA_3 and GA_5. There is indirect evidence for the presence of a second pathway from GA_{12}, the non–early 3,13–hydroxylation pathway. The pathway originates from GA_{12} and leads via GA_9 to the 3β–hydroxylated C_{19}–GAs, GA_4 and GA_7.

Not all GAs have high biological activity. Many of the GAs within a plant are precursors or deactivation products of the active GA. A 3β–hydroxyl group (as in GA_1 and GA_4) is required for activity. It was originally demonstrated with the dwarf **le** mutant of pea and **dwarf 1** (d1) of maize. These mutants are impaired in 3β–hydroxylase activity. At the end of the pathway, bioactive GAs are generally deactivated by 2β–hydroxylation. For example GA_1 is converted to GA_8 and GA_4 is converted GA_{34}. Conjugation to glucosyl esters or glucosides takes place predominantly in maturing seeds. There is only one example of a deactivation mutant, the **Slends** (Sln) mutant in pea, in which 2β–hydroxylation is blocked and GA_{20} accumulates in maturing seeds.

Gibberellic acid (GA_3), which is end–product of GA metabolism in *G. Fujikori* has been commercially available for many years. It has high biological activity and its application to dwarf or rosette plants, dormant buds or dormant and germinating seeds can result in dramatic and diverse effects on growth.

SITE OF BIOSYNTHESIS

All growing, differentiated tissues are potential sites of GA biosynthesis. Work with legumes indicated that GA biosynthesis occurs mainly in actively growing tissues. There is evidence that developing fruits and seeds are sites of GA biosynthesis. They contain enzymes, which can convert mevalonic acid to C19–GAs. In immature seeds there are two main phases of GA biosynthesis. The first phase which results in the production of GA_1 and GA_3 are associated with seed development and pod growth. Both endosperm and testa are potential sites of synthesis. The second phase occurs in the developing embryo and has no physiological function. As seeds approach maturing there is often increase in 2–oxidation activities in embryo and testa. The physiological significance of this deactivation mechanism is demonstrated by the **slender** (sln) mutant of pea which accumulated high concentrations of GA_{20} in the mature seeds due to lack of 2–oxidase activities. On germination the GA_{20} is 3β–hydroxylated to GA_1 resulting in overgrowth of the first 10–12 internodes. Gibberellin production in the pericarp of developing fruit may be important for fruit growth particularly in the absence of seeds. Work with pea indicates that the presence of seeds stimulates GA biosynthesis in the pericarp. Both seeds and 4–chloroindole–3–acetic acid stimulate the conversion of GA_{19} to GA_{20} in pea pods. GAs have been identified in root extracts but evidence for GA biosynthesis in roots is still unknown.

GAs are probably transported in the phloem and xylem.

BIOLOGICAL EFFECTS

Since the 1930, 120 GAs have been identified in plants and fungi (http:/www.plant–hormones.bbsrc.ac.uk/gainplants/gibberellin–nomenclature.htm.) and the complex biosynthetic pathway of this diterpenoid family has been well studied.

GAs control a variety of plant growth and development processes including seed germination, stem elongation, flower initiation and flower and fruit development. Commercial application of GAs includes increased malting of barley during beer production and the treatment of seedless grapes to increase fruit size. GA inhibitors are applied to retard the growth of nursery plants.

Biologists now recognize that gibberellins are concerned in the biogenetic differences between tall and dwarf peas. The phenomenon of bolting in rosette plants is caused naturally by endogenous GAs while dwarfism due to a deficiency in the natural GAs may be reversed by the application of exogenous GAs. The vigorous shoot growth obtained with maize hybrids has been shown to be due to the production of higher than normal levels of GAs. Flowering is also induced by GAs although in some fruit trees flowering may be reduced in the year following application. GAs may modify sex expression of flowers; induce the parthenocarpic development of fruit and delay senescence. They obviate the need for exposure to red–light in the germination of seeds and spores and the need for vernalization in the growth of bulbs and tubers. They are associated with the breaking of winter dormancy and stimulate the formation of hydrolytic enzymes in germinating cereal grains.

Several GA response mutants have been isolated in a range of species including maize, rice, wheat, barley, pea, tomato and *Arabidopsis*. These mutants have been extremely useful to understanding the roles of GA in plant growth and development. Mutants deficient in GA synthesis display a number of phenotypes including poor germination, dwarfed growth habit due to reduced cell expansion, underdeveloped petals and stamen development, delayed flowering and delayed senescence.

Cell Growth in Vegetative Tissues

Cell growth responses to GAs occur in a range of vegetative tissues. GA–regulated growth responses in vegetative tissues is the elongation growth of shoots, particularly stem internodes as well as hypocotyls, coleoptiles, mesocotyls and epicotyls. Other cell growth responses to GA occur only during specific stages of growth of the plant. For example, GA affects the transition from juvenile to adult leaf shape in a wide range of plant species. Cell growth polarity in root tips and shoot apical meristems is also markedly influenced by GAs. During fruit set, GA may stimulate pollen germination and tube growth. GAs promotes elongation growth through effect on both cell division and expansion. But the primary action of GA on these processes is not known (Katsumi and Ishida, 1991). GA has been shown to induce transverse orientation of microtubules in a range of different cell types including some in which the hormone does not affect the rate of elongation growth. GA regulation of the orientation of microtubules is an auxin–dependent process, although the nature of this cooperative response is not clear. Biochemical study indicates that early changes in microtubule dynamics in response to GA are associated with modification of a specific tubulin isotype.

Flower Induction

The transition from vegetative growth to flowering in long–day or cold–requiring plants and in conifers can be promoted by application of GAs. It is thought that vernalisation and long–day photoperiod promote stem elongation and flowering by stimulating GA biosynthesis. A definitive role for GA during flower induction has been established in *Arabidopsis* through studies with the extreme GA–deficient mutant **ga1–3**. In short days this mutant does not flower unless GA is applied. Because other less extreme GA–deficient mutants show only a delay in, rather than prevention of, flowering under short days, yet still have a dwarf phenotype, it seems possible that flower induction may have a lower GA response threshold than stem elongation. Whether or not GAs are required for flowering of *Arabidopsis* in long days is not clear. A delay in flowering in the extreme GA–deficient **ga1–3** grown under long days may be interpreted in favour of this, although the phenotype of other GA–deficient mutants and the GA–insensitive **gai** mutant do not. In short days GAs are important regulators of *Arabidopsis* flower initiation and may influence the expression of floral homeotic genes.

Flower Development

GAs affect the development of anthers which are a source of GAs that appear to influence corolla development and pigmentation (Martin and Gerats, 1993). Inhibition of GA biosynthesis can delay anthesis while exogenous GAs have the opposite effect. The GA–deficient **gib–1** and **gib–2** mutants of tomato show normal floral initiation but are arrested in anther development at a premeiotic G_1 phase.

The extreme GA–deficient mutants of *Arabidopsis*, **ga1–3**, is male sterile and has poorly developed stamens and petals. GA–deficient mutants of maize exhibit abnormal flower development. In the terminal flowers, stamen development is not suppressed resulting in bisexual flowers with pistils and anthers. Flowers of lateral shoots develop anthers but not pistils. The pollen is formed in these mutants but is not released.

Fruit Development

GAs are involved at a number of stages in fruit development. Exogenous GAs can promote fruit set and influence development in a range of horticultural species. Developing seeds contain the highest cellular concentrations of GAs. Whether or not these endogenous GAs are essential for fruit growth and development is not entirely clear. There is evidence for promotive effects on cell division and expansion and on the development of ovaries.

Seed Development and Germination

The clearest evidence that endogenous GAs are important for seed development has come from the **ih** mutant of pea which is GA–deficient in the shoot and developing seeds. Seeds of the **ih** mutant abort unless their endogenous GA level is restored to near that of the wild–type. GA has a strong positive effect on germination. Mutations that decrease levels of active GA or sensitivity to GA also influence germination potential. Studies have indicated that in *Arabidopsis* both GA biosynthesis and perception are required for germination. Therefore, antagonism between remnant ABA levels in the dry seed, and GA levels and perception on imbibition might contribute to the developmental decision between dormancy or germination (Hilhorst and Karssen, 1992).

Mobilization of Storage Reserves

It has become widely accepted that during the early hours of germination in cereals, GAs produced by the embryo and/or its scutellum are released into the endosperm where they diffuse to the aleurone layer. Here they induce the production and secretion of a number of hydrolysing enzymes capable of breaking down the reserves in the endosperm. The most abundant of these enzymes is α–amylase. Removal of the embryo dramatically slows down endosperm hydrolysis, while incubating isolated embryos and embryectomised seed together restores α–amylase levels. A low–molecular weight diffusible factor has been isolated and appears to be a GA. It is now clear that additional signals other than GAs are also involved in controlling aleurone cells.

MECHANISM OF ACTION OF GA

GA and Growth

Role in Cell Division: It has been argued that GA can promote growth of plants by affecting either cell expansion or cell division or both. It must be emphasized that cell division alone cannot result in growth of organisms. Cell division contributes to growth by producing more cells, which can undergo expansion. The effect of GA_3 on cell division can be readily accounted by an effect on cell cycle. Jacqmard (1966) has proposed that one of the effects of GA_3 is to promote the onset of DNA synthesis in cells which are arrested a G_1 phase of cell cycle. It has been established that capacity of cell to respond to GA_3 by undergoing division is the function of their state of differentiation. According to Cosgrov (1961) the cell expansion after

GA_3 treatment is based on biophysical phenomenon. The knowledge is further expanded on the basis osmotic potential. The mechanism for GA_3 activity in growth involving the conversion of starch to sugar was inferred by analogue with the known effects of GA_3 by increasing in length and allowing auxin treatment, there was increase in cell wall extensibility but GA-treatment did not result in a change in cell wall yielding properties. Since GA_3 induced elongation was not correlated with cell wall elongating properties.

Extensibility Changes in GA_3 Treated Tissues

Although the early experiments of Cleaveland (1968) suggested that GA_3 did not affect extensibility property of cucumber hypocotyl; evidence from several laboratories using many plant species indicate that GA_3 have marked influence on extensibility, using the osmotic method to determine the extensibility of tissues examined that effect of light, GA_3, KCl on wall yielding property of hypocotyl sections. Light reduced the extensibility of hypocotyle tissues which darkness and GA_3 overcome this inhibition. Although cell expansion can be explained in biophysical terms, the events leading to cell wall extensibility are dependent on metabolism. It is well-established example that cell expansion rapidly inhibited by respiratory poisons. This shows a direct relation between cell wall extensibility and readily available metabolites of plant system.

The plant cell wall contains proteins and inorganic ions in addition to complete polysaccharides. There have been several proposals that both proteins and ions specially Ca^{2+}; play a role in wall extensibility. However, Ca^{2+} has long been known to cause decrease in wall extensibility in lettuce hypocotyl.

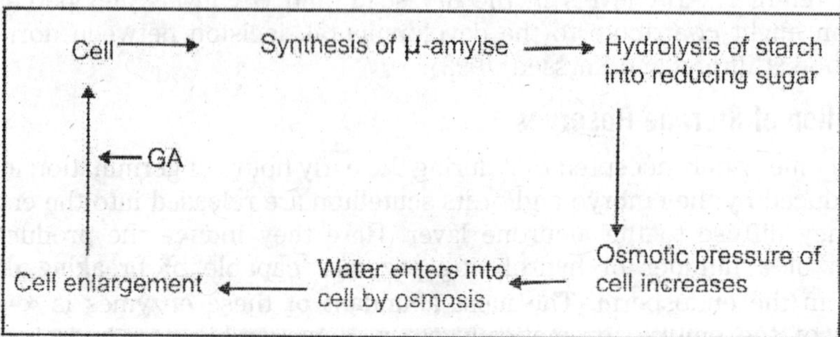

FIG. 21.3: Mechanism of action of GA on osmotic potential of a cell.

Moll and John (1981) indicated that GA_3 promotes uptake of Ca^{2+} ion upto cytoplasm. It was proposed that in this tissue Ca^{2+}activity in cell wall governs tissue extensibility and that extensibility could be regulated by uptake of ion into cytoplasm. Fary (1980) proposed that cell wall extensibility could be influenced by formation of cross bridge between adjacent ferulic acid (phenolic acid) moieties in the wall. The formation of diferuloyl bridges could be catalysed by peroxidase resulting in gelatinization of cell wall. Fery further demonstrated that GA_3 promotes cell expansion and inhibits release of peroxidase of cell. However this hypothesis may not explain how extensibility is controlled in dicot tissues since the primary cell walls of dicot contain negligible amounts of phenols. Atleast two cell wall loosening factors have been identified and have shown to involve in hormone-induced cell wall expansion in plants. The role of proton (H^+) is described by acid growth hypothesis, which is based on the observation that auxin-treatment results in rapid proton pumping from elongating tissue and that H^+ alone can promote growth of auxin treated tissues. Hydrolytic enzymes have also shown to be involved in modification of higher plants cell walls and their activity is thought to be modified by hormones.

GA_3 and Synthesis of RNA Proteins

It is well established that there is an increase in level of RNA and protein in GA-treated plant tissue, which causes an increase in total metabolites in plant cell.

The Cereal Aleurone Layer

Alurone layer of cereal seeds is responsible for degradation of starchy endosperm. Paleg in Australia and Yamo in Japan showed simultaneously that endosperm breakdown in deembryonated Barley caryopsis could be stimulated by addition of GA_3 to the embryo less endosperm. The function of GA_3 is to control production of hydrolyting enzymes by aleurone layer. In barley, the aleurone layer is 2-4 cells layer thick white oats, rice and wheat; it is one cell layer thick. The aleurone layer responds to GA_3 by producing a wide spectrum of hydrolytic enzymes when isolated aleurone layers are incubated in optimal GA_3 concentration (10-5 -10-4m) containing 10-20 minimol Ca^{2+} numerous hydrolytic enzymes are produced. Among newly synthesised enzymes is a-amylase which makes up to 60-70% of total protein produced by these cells. After 24 hours of incubation in GA_3 there occurs ultra-structural changes in aleurone cells, which accompany hydrolases synthesis and secretion. There is no evidence for an increase in total amount of lipid in GA_3 treated barley aleurone.

The ER is site of α-amylase synthesis and GA_3 treated tissues in ER contain 4 iso-enzymes of α-amylase while in untreated tissues it contains only 2. The ER is also involved in transport of α-amylase from barley aleurone cell. There is compelling evidence that GA_3 effects the synthesis of hydrolytic enzymes in cereal aleurone and that it does by controlling the transcription and possibly translation of new messenger RNA. The synthesis of RNA is a pre required for the production of GA-induced enzymes. The α-amylase of barley produced in vivo has a molecular weight of 44000 while that produced in vitro has a mol weight of 46000.

Dicotyledonous Fruits and Seeds

The effect of GA_3 on germination of dormant and non-dormant dicotyledonous seeds and on growth of seedlings is well known. The studies are based on fat storing seeds, in particular to those of almond, castor bean. In the endosperm of *Ricinus communis* seedlings; GA_3 stimulates the activity of about 16 enzymes involved in gluconeogenesis but not the activity of a other enzymes of intermediately metabolites that are not involved in gluconeogenesis.

PHYSIOLOGICAL EFFECTS AND PRACTICAL APPLICATIONS OF GIBBERELLINS

Since the discovery of GAs in the 1920s a considerable body of information has emerged regarding the roles these hormones play in plant growth and development. As these compounds and analogs became available, either by fermentation or chemical synthesis, a number of potential practical applications were identified. Gibberellins have been primarily used for manipulating production practices and insuring the quality of high value specialty crops such as grapes, citrus, cherries, and apples. Increasing market demands for a greater variety of quality fruit may provide new opportunities for the expanded use of Gas (i.e. tropical fruits, thinning and sizing wine and new table grape varieties, and as components of integrated pest management programs for postharvest quality of fruits). Despite considerable research, GAs have not found utility in major agronomic crops. However, increased understanding of the interrelationship between GAs and other PGRs, concurrent with the development of new GAs or GA-mimics, may offer the unique activities necessary to penetrate agronomic markets.

However, despite extensive research with GAs the number of commercial uses has fallen short of their predicted impact and growth in agriculture. Unique opportunities may still result from new research and development with GAs to meet the future demands for increased food quality and production.

Although over 97 different GAs have been identified in plants, only three (GA_3 and a mixture of GA_{4+7}) are available via fermentation by the fungus *Fusarium monilifome* (= *Gibberella fujikuroi*) in reasonable quantity for extensive applied research. The complexity of GA chemistry has precluded the entirely synthetic routes of production. With the diversity of physiological processes in which GAs are involved, it is not surprising that their practical application spans the entire range of plant growth and development. The majority of these uses however, represent specially niche markets in high cash value horticultural commodities. Despite the extensive research with GA_3 over the past 30 years, no defined GA product has been registered for use on a major agronomic crop. This may be due in part to a lack of specificity of response where multiple effects, often detrimental (i.e. lodging), are manifested from exogenous application of GA.

Floriculture

GA_3 is also used to thin approximately 60% of the flowers because almost every flower tends to set, resulting in a tightly packed cluster. The physiological basis for grape flower thinning by GA is unknown, but could be due to pollenicidal activity. The result of adequate thinning almost appears excessive, but with two subsequent applications of GA, berry size is increased by 60%. The final product is a loose cluster that does not require hand thinning and is less susceptible to *Botrytis* bunch rot.

GA_3 also inhibits flowering on several stone fruits. In young sweet and tart cherries, GA_3 is applied to prevent excessive flowering to minimize the competitive effect of early fruiting on vegetation growth. Since cherry has true flower buds, the node following flowering will not support vegetative growth. If flowering is excessive, fewer spurs develop, thereby resulting in lower yield potential during the production life of the tree. The manipulation of flowering by GA_3 has also been integrated into a management program to suppress a pollen transmitted ring spot virus infection as well as the maintenance of the desired ratio of vegetative to flower bud formation in mature tart cherries. The consequence of continuous GA_3 applications over a five years period has been a 22% increase in yield. A mixture of GA_{4+7} is commercially available. It affects on varying structure in the flowering of *Lolium temulentum*, a LD-requiring plant and *Pharbitis nil*, a SD-requiring plant, as well as for flowering of Cupressaceae and Pinaceae conifers. While it is too early to predict commercial implications of such specificities in flowering response for GAs.

Fruit

An additional application of GA_3 to sweet cherry when fruit is light green to straw coloured will delay harvest, increase fruit size and firmness. While this application is not new, the recent development of an export market to Asia for cherries has significantly increased this use.

Fruit Reset

The other commercially used GAs produced by *Fusarium moniliforme* cultures are a mixture of GA_4 and GA_7. An application of GA_{4+7} of increasing importance is the prevention of fruit russet, a superficial disorder in which the exocarp is interrupted by raised corky outgrowths over the surface of green and yellow apple cultivars. Although somewhat obscure, a number of factors have been implicated for russet, such as high humidity, presence of free water, light frosts, pesticides and powdery mildew. The multiple applications on 'Golden Delicious' effectively increase the fruit finish quality and this corresponds with higher packout gardens.

Grape Industry

GA_3 has been used in the seedless table grape industry since 1960 to manipulate three physiological events; rachis cell elongation, flower thinning, and berry enlargement. The first is a pre-bloom spray to promote elongatión of the rachis to provide a larger and looser framework for the grape cluster. In wine grapes, there is a need for cluster loosening by either rachis elongation and/or flower thinning to prevent *Botrytis* bunch rot. To date, there is no chemical thinning agent for wine grapes. Results with GA_3 have been variable with some varieties exhibiting a dramatic reduction of return bloom the year following application; others are relatively insensitive to increasing rates of GA_3. The response of wine grapes to other GAs certainly needs to be examined.

Citrus

The effect of delayed fruit senescence by GA_3 has also been exploited in the citrus industry. GA_3 delays maturity of citrus rind without altering the maturation process of the internal fruit, thus preventing several rind disorders associated with fruit ripening. In lemons, this facility scheduling of harvesting to synchronize with market demands. As harvest date is delayed in grapefruit, there is added benefit in combining 2,4-D to retain the fruit on the tree. Recent findings suggest that the delay in citrus rind maturity is associated with the quantity and quality of epicuticular wax deposition and its influence on respiratory gas exchange.

Apple

More recent results suggest that GA_4 may promote apple flowering. The differential effect of GA_4 and GA_7 in apple serves to illustrate the specificity associated with a specific GA. However, separation of these compounds on a large scale is presently cost prohibitive.

Seed Production

There are also minor practical uses of GAs in seed production of commercially important conifers. GA_3 can be used to promote flowering, and thereby significantly increase female cones and seed per tree in seed production orchards of Cupressaceae and Taxodiaceae species.

Protection against Plant Pathogens

The use of GA to modify growth for protection against plant pathogens was described previously in cherries and grapes and is becoming an increasingly recognized potential benefit of PGRs in a number of crops.

Insect Control

Recent laboratory studies have shown that GA_3 application significantly reduces the number of probes and egg fecundity of the fruit fly due to increased exposure to toxic peel oils contained in the juvenile rind. If this effect is manifested in the field, a high level of resistance may preclude the need for post harvest treatment with an insecticide/larvicide.

Applications on Agronomic Crops

The successful introduction of a GA product in an agronomic crop is dependent upon the specificity of a physiological response over a wide range of environmental/cultural conditions, and is complicated by the inherent genetic variability amongst cultivars. For example, the genetic reduction of plant height has led to dramatic increases in cereal grain yield, due principally to increased reduction of lodging. Consequently, these varieties have poor emergence rates and poor seedling vigor. In rice, GA application promotes seedling establishment in the principle semi-dwarf commercial varieties utilized.

CHAPTER 22

Plant Hormone: Cytokinins

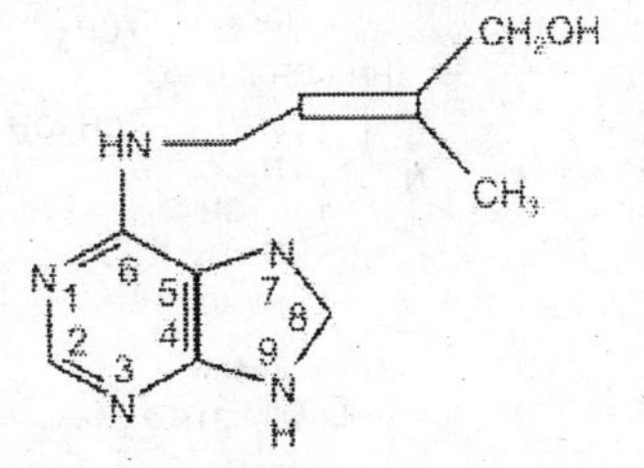

The discovery of cytokinins in the 1950s by F. Skoog, C.O. Miller and collaborators originated in studies of the growth requirements and differentiation of cell cultures (Skoog and Miller, 1957). The work was stimulated by the observation that application of excised vascular tissues to tobacco pith explants in auxin–containing media caused a shift from cell expansion to cell division (Chen, 1998). This stimulated a search for compounds from more readily available sources that caused similar effects. Because of a positive effect of phosphate on growth of cells the effect of organic phosphates was tested. It was observed that in the presence of auxin, autoclaved yeast extract and DNA stimulated cell division of cultures. The active component was found to be 6–furfuryl–aminopurine (kinetin), a substance that was formed during degradation of DNA but which has not been found to occur in plants. Variants of this structure were subsequently discovered and in 1954 a natural cytokinin, Zeatin, was isolated from corn by D.S. Letham and colleagues in New Zealand.

Naturally occurring cytokinins are N^6–substituted adenine derivatives. In addition to higher plants, several bacteria including *Agrobacterium* produce cytokinins. The key biosynthesis step in *Agrobacterium* is the addition of an isopentenyl group from isopentenyl diphosphate to N^6 of AMP, which is catalysed by AMP–isopentenyl transferase **(IPT)**. A similar enzyme activity has also been observed in extracts from plant sources but because of its instability the enzyme has only been partially purified. The **IPT** gene from *Agrobacterium* has been cloned and expressed in transgenic plants. However, there are no reports of plant DNA sequences with similarity to the bacterial **IPT** genes.

CHEMICAL NOMENCLATURE

All naturally occurring cytokinins (CK) like zeatin L (Z) and isopentyl adenine (iP) and many synthetic CKs, like kinetin (K) and 6-benzyl-amino purine (BAP), are 6 substituted amino purines. However, there are reports that BAP and its derivatives exist in plants. A number of synthetic non-purines including substituted ureas, amides and pyrimidines are known to exhibit CK-like activity, e.g. N-phenyl-N-1, 2, 3-thio-dizol-5-yl urea (thydiazurone) shows CK activity comparable to that of Z. Even though dihydro-coniferyl alcohol (Lee *et al.*, 1981) and dehydro-diconiferyl glucosides are naturally occurring non-purines, which promote cell division, they are not classified as CKs.

All the known naturally occurring CKs are virtually substituted purines. The system used for increasing the purine nucleus in also used in CK nomenclature to define the position of substituents. Many systems of abbreviations for CKs, including those based on established system for nucleic acid derivatives IUPACIUB commission on Biochemical Nomenclature (1970) have been used. The nomenclature proposed by Letham (1978) is now widely adopted. Zeatin (Z), isopentenyl adenine (P) kinetin (K) and 6-benzyl aminopurine (BAP) are regarded as parent compounds, and substitutions on the purine nucleus are shown in square brackets, e.g. [9R] Z for zeatin riboside, and substitutions on the side chain are denoted in round brackets e.f. (diH)Z for dihydrozeatin and (OG) [9R] Z for O-Glucosyl derivative of zeatin riboside.

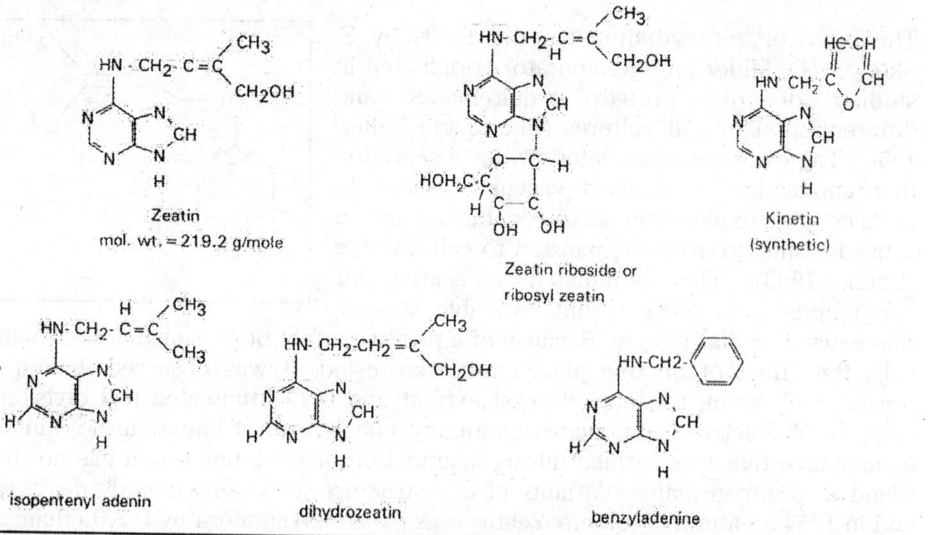

FIG. 22.1: Structures of common natural and synthetic (kinetin) cytokinins. All these are adenine derivatives in which the purine ring is numbered as shown for zeatin (upper left). Zeatin and zeatin riboside can exist with groups arranged about the side-chain double bond either in the *trans* (as shown) or *cis* (with CH_3 and CH_2OH groups interchanged) configuration. The *cis* form predominates in tRNA-bond cytokinins, but the *trans* form exits in free zeatin and zeatin riboside.

DISTRIBUTION IN PLANTS

Cytokinin levels in leaves and xylem sap: Analysis of CK levels in xylem exudate and leaves of plants during development revealed that leaf senescence might be under the regulation of CKs in sites. The induction of leaf senescence is possibly due to the environment or soil bacteria that reduce CK production in total leaves in xylem. The endogenous levels of CKs in the xylem sap seem to be an important factor in regulating leaf senescence.

CYTOKININS BIOSYNTHESIS

Site of CK Biosynthesis

CK activity has been detected in extracts of almost all plant hormones and in many organisms. Although root apices have been recognized as a major site of CK biosynthesis, there are circumstances to indicate that developing seeds, growing buds and shoot apices and leaves are additional sites of CK biosynthesis for the more cambium and embryonic axis have also been suggested as possible sites of biosynthesis. Developing leaves have been shown to contain high CK activity.

Cytokinins biosynthesis is through the biochemical modification of adenine. It occurs in root tip and developing seeds. Isopentenyladenosine–5–monophosphate is the precursor of all forms of cytokinins. Through hydroxylation of the isopentenyl side chain and reduction of the double bond, the ribotides of zeatin and dihydrozeatin are formed. It is generally thought that the freebases such as isopentenyl adenine, zeatin and dihydrozeatin are the active forms of cytokinins. Cytokinins with a hydroxylated side chain can be glucosylated to form the O–glucoside or O–xyloside. These reactions are reversible because O–glucosylated cytokinins have biological activity. Zeatin O–xylosyl transferase has been isolated from bean embryos and antibodies have been prepared. The enzyme is predominantly localized in the endosperm.

Biosynthesis of Free Cytokinins

Two mechanisms have been proposed for free CK biosynthesis, one resulting from the turnover of CK containing t-RNA molecules and the other-de novo biosynthetic pathway.

Biosynthesis via tRNA turnover: Transfer RNA is a potential source of free CKs. It has been suggested that t-RNA hydrolysis may result in release of kinins in dying, autolyzing cells of differentiating vascular tissue. The free CK of bacteria may arise from t-RNA turn over. CK can also be produced from the culture of *Agrobacterium tumefaciens* and from *Lactobacillus acidophilus* and primary roots of *Zea mays*.

De novo Biosynthesis: Free CK biosynthesis may also occur independently of tRNA breakdown obtained a cell free preparation from saline mold *Dictyostium discoideium* and showed adenosine 5'aminophosphate, to be the acceptor molecule of isopentyl group resulting in formation of [R-5'P] is the first step of CK biosynthesis. Enzymes 3-isopentyl pyrophosphate-3-isopentyl transferase community referred as CK synthase or isopentyl transferase, catalysing the above reaction, is considered to be key reaction of biosynthetic pathway to free CK.

CYTOKININS METABOLISM

Cytokinins levels in plant cells are dependent on cytokinins biosynthesis and/or uptake from extracellular sources, metabolic inters–conversions, inactivation and degradation. Several processes have been identified that play a role in cytokinin metabolism.

These include the formation of cytokinin nucleotides and nucleosides from free bases, N–glucosylation, N–alanyl conjugation, O–glucosylation, O–acetylation, and side–chain reduction and side–chain oxidation. It is agreed that the oxidative breakdown of cytokinins is an important process by which plant tissues might control the levels and distribution of cytokinins. Cytokinin oxidase catalyses the oxidation of cytokinin substrates bearing unsaturated isoprenoid side chains. This widely distributed enzyme constitutes a crucial point of control in the metabolism of cytokinins both endogenously or externally applied to plants or plant cultures. The current state of knowledge regarding its biochemical and molecular properties is still poor. This is largely because of the difficulties encountered purifying it to homogeneity and in cloning its gene.

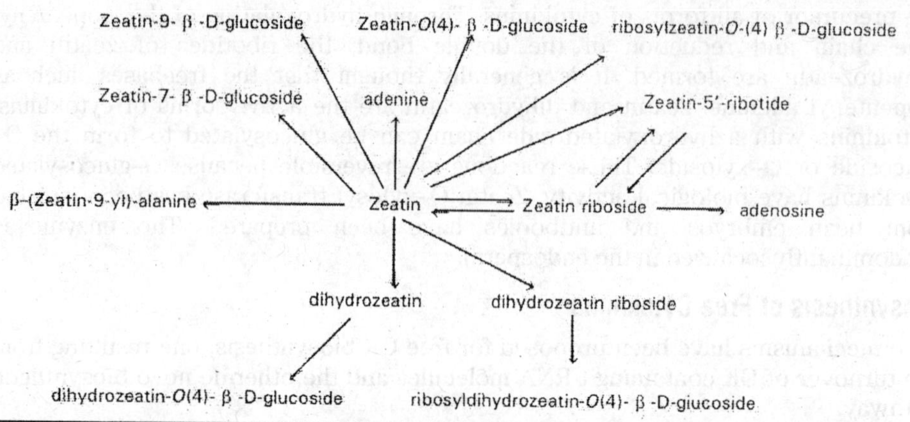

FIG. 22.2: Some of the known routs by which plants can metabolize zeatin. The pathways illustrates do not all occur in any single plant tissue but represent the results of metabolic experiments with several species of plant.

BIOLOGICAL EFFECTS

The cytokinin groups of plant hormones play an essential role in the regulation of plant growth and differentiation including the release of lateral buds from apical dominance and the delay of senescence. Moreover, increasing evidence supports the hypothesis that cytokinins interact with auxins at multiple levels in a complex manner to control various physiological and developmental responses. Cytokinins also appear to antagonize many physiological processes thought to be mediated, all or in part, by ABA. For example, ABA–induced stomata closure, leaf senescence and leaf and fruit abscission are reversed by cytokinins while cytokinin–mediated release of seed dormancy contrasts with ABA inhibition of germination (Salisbury, 1994). This implied antagonism might be the result of metabolic interaction. Cytokinins share at least in part a common biosynthetic origin with ABA.

Cell Cycle Control

Cytokinins were discovered during the 1950s because of their ability to induce cell division in concert with auxin in non–dividing plant cells. Until recently the molecular targets of this hormone in the cell cycle regulatory machinery have remained elusive. Cloning genes encoding for plant cyclin–dependent kinases (Cdc2), cyclins (cycA, B and D families) and cyclin–dependent protein kinase inhibitors (ICKs) has made the cell cycle accessible to molecular studies. The cytokinin–dependent cell division activity in *Arabidopsis* is mediated by the transcriptional regulation of the cyclin D3 (Cyc D3) gene at the G1/S transition. In *Arabidopsis* cell culture it has been found that D3 is induced in the G1 phase shortly before the S phase by cytokinin and sucrose (Riou–Khamlichi et al., 1999). *In situ* hybridizations show Cyc D3 expression in proliferating tissues of the shoot meristem, young leaf primordia, auxillary buds, procambium and vascular tissues of the maturing leaves. Cytokinin treatment enhances the expression in these responsive tissues. The cytokinin–overproducing **amp1** mutant has steady–state Cyc D3 mRNA levels that are enhanced two to three fold (Riou–Khamlichi et al., 1999). Cytokinins have also been reported to have a regulating function at the mitotic control point of G2/M transition (Zhang et al., 1995). In excised stem pith parenchyma of *Nicotiana tabacum* cytokinin activates a Cdc2–like kinase that accumulates in its inactive form after auxin treatment. Cytokinin stimulates a specific dephosphorylation of the kinase, a reaction that is carried out by the Cdc 25 phosphatase. This indicates that cytokinins might control G2/M transition by Cdc25 mediated dephosphorylation of cyclin–dependent kinase. Recently, cyclin–dependent protein kinase inhibitor, ICK1, has been identified in *Arabidospsis* (Wang et al., 1998), which interacts in vitro with Cdc 2a, and Cyc D3. ICK1 expression is induced by ABA and leads to a decrease in Cdc2a–like kinase activity. These results are compatible with the antagonistic functions of ABA and cytokinins and suggest a molecular mechanism in which plant cell division is inhibited by ABA.

Cell Division and Development

Auxin and cytokinin regulate cell division synergistically in callus cells and protoplast and antagonistically during the formation of lateral root primordia. Molecular analyses of cyclins and cyclin–dependent kinases indicate that both types of regulation occur at the level of gene expression. The formation of lateral roots is reduced in auxin–resistant mutants and dramatically increased in the auxin–overproducing **sur1** (super root) mutant but no analyses have been presented on the interaction between auxins and cytokinins in lateral root initiation in these plants. Molecular studies addressing the expression of Cdc2–like proteins and cyclins in these mutants will provide the answer.

Cyclin (Cyc D3) provides a link between cytokinin and plant development. Plants over-expressing Cyc D3 show extensive leaf curling, disorganized meristems, increased leaf number, late flowering and delayed senescence. This indicates a role for Cdc D3 in plant differentiation and contrasts with the results obtained with plants such as Cdc 2a and Cyc B1. The ectopic expression of these genes leads to an altered cell number but fails to perturb plant growth or development. This suggests that morphogenic control in plants is independent of the rate of cell division. Whether or not Cdc D3 does play a privileged role at the interface of cell division and development needs further analysis. In *Drosophila* early development Cdc 25 plays a particular role. Cdc25 is the limiting factor for entry into mitosis. Analysis of its promoter has shown that patterning information directly controls it. Thus, Cdc 25 is a single factor between patterning and cell division. A detailed analysis of plant Cdc 25 has not yet been carried out.

It is not known how merely varying the ratio of auxin and cytokinin can switch the developmental programme in tissue culture from root differentiation to shoot differentiation or continued division of undifferentiation cells. Homeodomain proteins such as **KNAT 1** and **STM**, which prevent differentiation and activate cell division, might play a role in determining cell fate through interaction with the cell cycle machinery. It has been reported that the **KNAT1** and **STM** mRNA levels are increased in both the cytokinin over–producing transgenic *Arabidopsis* and ***amp1*** mutant (Rupp et al., 1999). This indicates that links between cytokinin, cell cycle gene and developmental genes exist. The challenge is to decipher the nature of these links.

Anti–Ageing Effects

Cytokinin is known for its anti–ageing effects in plants. Recently, its strong anti–ageing effects on human skin cells and fruit flies have also been reported. It has been shown that kinetin delays the onset of several cellular and biochemical characteristics associated with cellular ageing in human skin fibroblast cultures. These ageing characteristics include alterations in morphology, increase in cellular protein content and accumulation of protein–lipid–peroxide conjugates in the cytoplasm. The inhibition of growth of human fibroblasts, epithelium and mammary carcinoma has also been reported. In fruitfly *Zaprionus paravittiger* kinetin slows down development and ageing. An increase in the specific activity of catalase during developmental stage and in adult insects has been observed. These results and from dermatological tests against photo–ageing, several skin care products and potential cosmeceutical formulations containing kinetin are being developed.

Cytokinins and Increased Levels of Auxin

The exogenous application of cytokinin or by transformation with the bacterial cytokinin biosynthetic gene **ipt** (isopentenyl transferase) results in more auxin. Increase in free IAA is observed both in cytokinin overproducing lines of *Nicotiana glutinosa* transformed with the **ipt** gene or after exogenous application of cytokinins to maize or pea roots. Although the basis of these cytokinin–induced changes in auxin levels is not fully understood. A putative mechanism is cytokinin inhibition of the enzyme that conjugates free IAA. Such a phenomenon has been observed in mung bean hypocotyls where conversion of exogenously applied IAA into the inactive conjugate is inhibited following the application of cytokinin.

Gall/Tumor/Nodule Formation

Together with IAA, cytokinin is responsible for stimulating growth and development of insect–incited gall and bacteria–induced tumours and nodules. Lipo–chitooligosaccharides (nod factor) secreted by the bacteria, is responsible for the induction of cell diversion in the root cortex leading to the formation of root nodules. The induction of pseudonodules in alfalfa roots by auxin–transport inhibitors and by *E. coli* carrying a trans–zeatin secretion gene from *A. tumefaciens* promoted the notion that nod factors may use hormonal signal transduction mechanisms to induce the formation of nodules. It has been shown that nod factors can substitute for auxin and cytokinin to induce cell divisions in protoplasts derived from tobacco leaves. The cell division is also accompanied by increased expression of the axil gene, which is auxin inducible and may have a function in the control of cell division. The pseudonodules induced by auxins and auxin–transport inhibitors derive mainly from divisions of endodermis and contain many knots of vascular cells however, nodules induced by cytokinins are similar in many aspects to those induced by nod factors. Both cytokinins and nod factor–induced pseudonodules have similar pattern of cortical cell division, amyloplast accumulation and gene expression.

Nodulins genes such as MsENOD12A from alfalfa and SrENOD2 from *Sesbania rostrata*, are induced by both nod factors and cytokinin. Both genes encode a ten amino acid oligopeptide. Overexpression of this peptide in tobacco protoplasts eliminates the inhibitory effects of high auxin concentrations on cell division. Expression of ENOD40 in alfalfa roots has recently been shown to increase in response to exogenously applied cytokinin.

Four classes of phytopathogenic bacteria, *A. tumefaciens*, *Erwinia herbicola*, *Pseudomonas savastanoi* and *Rhodococcus fascians* produce neoplastic or hyperplastic diseases in plants. The resulting galls display either completely unorganized growth or incomplete organogenesis with production of abnormal shoots and roots. The galls incited by all four classes of bacteria are similar but there is a fundamental difference between those incited by *A. tumefaciens* and those incited by the other pathogens.

Anti–Stress Effects

Cytokinin functions as an anti–stress agent. It has been shown that kinetin treatment lowers the lipoxygenase activity and acts by preventing formation of reactive oxygen species or as a direct radical scavenger. A direct effect of kinetin on superoxide dismutase activity has been observed in the plant. There are two superoxide dismutases (SODs) and are encoded by the nuclear genes SodB and SodCp. The sodCp mRNA level decreased by 50% and SodB mRNA level was threefold higher in response to kinetin. The involved mechanism is unknown but kinetin could prevent the hydroxyl radical–mediated DNA damage either by acting as a radical scavenger or by binding iron in such a way that it is no longer a Fenton reductant or in a way that prevents iron from associating with DNA.

MODE OF ACTION OF CYTOKININ

It comes as no surprise to find that cytokinin have been evaluated extensively for their role in nucleic acid metabolism beginning with the study of Guttman (1957) who found that the kinetin treatment of onion roots caused a quick increase in the amount of RNA in the nuclei. Kinetin caused a doubling of the RNA levels within 30 min in onion root tip cells although it reduced DNA.

Cytokinin-Dependent Plastid Development

Cytokinins are shown to stimulate plastid development. The general inference is that cytokinins do not control plastogenesis in cotyledons though hormones control cotyledon development superimposed by cytokinin promoted plastid differentiation through activated nuclear genes that code for organelle polypeptide. Thus CK regulate etiolated tissues whereas in light gene expression is regulated by phytochrome.

The general inference is that cytokinins and ABA do not regulate plastogenesis in cotyledons via selective transcription of plastid genes. However, this may take place in the CK-dependent cell cultures.

By the contrast, kinetin alone induces a rapid increase in the DNA of tobacco pith cells although the effect can be duplicated by auxin.

In leaves of tobacco, kinetin causes marked increase not only in RNA but DNA and protein synthesis as well, which implies that in the deferral of senescence, kinetin acts on the over all synthetic activities of the leaf. That cytokinin plays a role in the protein synthesis. Kinetin stimulates increase in protein synthesis in the leaf discs and felt that a primary action of kinetin might be to regulate DNA synthesis with a resulting regulation of one of the steps leading to protein synthesis. Inhibitors of RNA and protein synthesis prevent the cytokinin action effectively.

PHYSIOLOGICAL EFFECTS OF APPLIED CYTOKININS

Externally applied CKs are known to exert spectacular effects on the growth and development of plants and excised plant tissue. The regulation of all organization in cultured tissue is one of the most dramatic effect of CKs and an example of delicate interaction of auxins and CK in mosses beed formation is induced markedly by CKs. Exogenous CKs have also a pronounced effect on stimulation of cell enlargement in cotyledons of diverse species and on leaf disc. They have been found to substitute for light in some cases e.g. CKs can induce germination of light sensitive seeds. Pigment [Beta-cyanin] synthesis in cotyledons of *Amaranthus* seedling, chlorophyll synthesis in detached yellow leaves and etiolated cotyledons, chloroplast development and opening of stomata, reduction of leaf senescence and release of lateral buds by apical dominance by CK application are some instances which found commercial use. CK can also induce nutrient mobilization and recently have been found to influence unloading of phloem derived photosynthate into cytoplasm of bean seed coats (Palni, 1990).

Role of Cytokinins in Plant Development

Externally applied CKs markedly influence a diversity of physiological processes and thus affect plant growth and development. Selected areas of plant development, which has been studied in some, detail and appear to be influenced by CKs are as follows:

Seed Germination and Early Seedling Growth

Phytohormones play a regulatory role in the imposition of seed dormancy, in the release of seeds from dormancy, in the reserve mobilization during seed germination and in the subsequent development of the seedling. CKs are important in seed germination. The following evidences indicate this:

1. In cultured embryos of *Zea mays*, the radical growth was retarded by the removal of the endosperm. This could however be partially restored by the addition of a plant extract reported to contain CK-O- fkucoside. (Smith and Van Staden, 1979).
2. Pinfield and Stobart (1972) observed that the elongation of the radicle of sycamore seeds could be induced by the external application of CK but not GA.
3. The endogenous CK levels increased markedly in the embryo sporophyte prior to onset of radicle elongation in germinating *Podocarpus henkelii* seeds.

Thus the endogenous CKs would appear to be one of the key factors in the initiation of radicle growth. Endogenous CK appears to be important in reserve mobilization in germinating seeds, particularly in dicots. Seeds store large amount of food reserves, which are localized in the endosperms or in the cotyledons. These reserves are degraded by a series of hydrolytic enzymes following imbibition. The embryonic axis is known to play an important role in the formation of these enzymes for the mobilization of reserve. The axial effect can be replaced, at least in part, by CK's - an ability not usually shown by other phytohormones with possible exception of GA in few cases.

Cotyledon Expansion

The most dramatic and well-studied response to CKs in immature seedlings is cotyledon expansion in dicots. Some cotyledons respond to both GA and CK (Xanthium) and some markedly to CK but only weakly to GA (e.g., Radish) and some only to CK (e.g., Lupin) and Trigonella and cucumber in darkness.

Chlorophyll Synthesis

Another important effect of CKs is their ability to stimulate chlorophyll synthesis and accelerate chloroplast differentiation in the detached cotyledons in light. The stimulus required for chlorophyll formation and cotyledon expansion in the germinating seeds would appear to be a CK emanating from the axis.

Cell Division and Organ Differentiation

CKs are associated with the induction of cell division in cultured tissues and in combination with auxin, and exhibit a striking quantitative relationship in the regulation of morphogenesis. The following evidences can provide a good correlation between hormone level and morphogenesis:

1. A large number of buds are produced in several moss species *Physcomitrella pateus*, when they are exposed to high CK levels. The enhanced bud formation seen in a mutant of this moss is indeed associated with an increased level of endogenous CK.
2. The levels of endogenous auxins and CKs appear to control the altered

morphology of shoot/root bearing crown gall tumour in agreement with the concept of Skoog and Miller, 1957).

3. In Begonia plants and cuttings it has been shown that a reduction in temperature or day length results in elevated CK and lowered auxin levels and also induced bud formation. Under non - inductive conditions (long days and high temperatures) externally applied CKs can induce bud formation.

Control of Leaf Senescence

All the phytohormones are probably involved at some stage of leaf senescence and may interact in its control, but much attention has been given to the role of CKs because of their dramatic effect when applied exogenously. Chibnall (1954) was the first to postulate that some hormone like factors produced in roots may be involved in control of leaf senescence. Today **Chibnall's factor** is generally equated with CK. The overall evidence for the involvement of CKs in the control of leaf senescence is outlined as follows:

1. **Application of CK to excised leaves and leaf discs:** CKs are very effective in retarding senescence of excised leaves of various genera. They induce regreening of detached yellow tobacco leaves. Bioassays for CKs have been based on their senescence retarding effect with excised oat, and barley leaves which were kept in darkness. This effect of CKs is known as **Richmond and Large effect.**
2. **Application of CK to intact and detopped plants:** Frequently CKs have no effect on attached leaves that have a normal connection to a root system. In the classical study of Kulaeva (1962), kinetin when applied directly to the blades of excised tobacco leaves, retarded senescence. However, it had no effect on the leaves of intact plants or detopped plants. Such lack of response is attributable to supply of CK by the root system. CKs retard senescence of attached leaves of some plants for e. g., soybean, French beans, rice, wheat and oat. These results show that in the intact plant, leaves possess sub- optimal levels of CK before the onset of leaf senescence.
3. Studies involving root development or root excision: Root produced CK's appear to play a significant role in retardation of leaf senescences.
4. Cytokinin levels in leaves and xylem sap: The endogenous levels of CKs in xylem sap seem to be an important factor in regulating leaf senescence. The CK activity in the xylem exudate of sunflower plants has been reported to decline at the commencement of flowering, which precedes induction of leaf senescence. The environment or soil factors reduce CK production by roots levels in xylem sap usually causes induction of leaf senescence.

When considered collectively, the above evidence of diverse types provides strong support for the control of leaf senescence in many species by endogenous CK, particularly root derived CK moving in the xylem

Root-Shoot Interactions

In plants roots affect many aspects of shoot development markedly and these may be mediated by CKs produced in the roots and present in significant amounts in the xylem sap. CKs often substitute for the effect of roots on the shoot development and physiology, convincingly in some cases GAs appears to act in this way. Since root derived CKs play an important role in the control of leaf senescence, it is suggested that root produced CK may also regulate other aspects of shoot development including release of lateral buds from apical dominance, development of inflorescence, DNA synthesis in shoot apex, expression in dioecious plants and photosynthetic activity (Goodwin, 1978).

PRACTICAL USES OF CYTOKININS

Mixture of BA and GA_{4+7}

The synergistic combination of BA with other GAs (especially GA_5) has also been demonstrated since 1977. A mixture of BA and GA_{4+7} is used for commercial purpose of a cytokinin in agriculture. Shortly thereafter, the tetrahydropyranyl derivative of BA (PBA) was registered on ornamentals for induction of lateral branching. The mixture of GA_{4+7} and BA is used to induce the extension of apple calyx lobes for the increased size and length/diameter ratio associated with high calyx lobes for the increased size and length/diameter ratio associated with a high quality 'Red Delicious' apple fruit. Additional research with this mixture demonstrated that applications to non-bearing apple trees increased lateral bud break and improved branch angles to provide a better framework and bearing surface for early cropping. Bud break and branch angles were primarily due to the BA component, but the combination with GA_{4+7} promoted shoot growth to a much greater extent than either of the two compounds alone. The mixture is used in the promotion of flowering of *Chrysanthemum* under non-inductive conditions.

Fascicular Bud Break in Conifers

BA is registered for use in promoting fascicular bud break in conifers to reduce the required mechanical shearing and altering the rotation period required to grow marketable Christmas trees. While this use has not been commercially significant, the compound's availability has encouraged research in a number of areas.

Extension of Vegetable Shelf Life

Potential applications such as the extension of vegetable shelf life, prevention of crop senescence, and increased soybean pod set have not been pursued due to cultivar differences or lack of field efficacy.

Commercial Uses of Cytokinin-Like Activity Compounds

In contrast to GAs, a number of compounds have been made synthetically which exhibit cytokinin-like activity including N^6-substituted adenine analogues, like, pyridylureas and thiadiazolylureas and pyrimidines. Of these the highly active thiadiazolyl urea, thidiazuron (TDZ) was found to have cotton defoliant properties and was subsequently the first cytokinin-like product to be successfully developed for widespread major crop use. TDZ is unique as a harvest aid in cotton because it promotes green leaf abscission and prevents regrowth, thus allowing growers more flexible harvesting schedules. Its mode of action in the cotton leaf abscission process appears to be associated with a sustained increase in ethylene production coupled with the disruption of auxin transport.

Thinning of Apples

Recently, promising results were reported for thinning of apples with BA alone or in combination with carbaryl. Another cytokinin-like compound currently under development, which also thins apples, is 2-chloropyridylphenylurea (CPPU). A number of potential applications have been examined with CPPU which overlaps with several GA effects including grape berry enlargement, and increased fruit size and weight in apple. The future of this compound as a commercial product is yet to be determined and may require an identified use in a major agronomic crop to justify further development.

Agronomic Crops

The successful introduction of a CK product in an agronomic crop is dependent upon the specificity of a physiological response over a wide range of environmental/cultural conditions, and is complicated by the inherent genetic variability amongst cultivars. For example, the genetic reduction of plant height has led to dramatic increases in cereal grain yield, due principally to increased reduction of lodging. The current characterization of endogenous CKs in correlation with developmental events in major crop plants will certainly aid in the identification of application timings for candidate compounds.

A key factor in the successful development of thidiazuron (TDZ) in cotton was the availability of a large number of synthetic cytokinin-active compounds. With different classes of cytokinin chemistry, a systematic approach toward structure activity relationships could be employed.

CHAPTER 23

Plant Hormone: Ethylene

HISTORICAL BACKGROUND

One centaury ago, Anton Nicolyovitch NelJubov determined that the gaseous olefin ethylene is capable of altering the development of higher plants in variety of ways. He used the triple response of etiolated pea epicotyls (diageotropism, inhibition of elongation, stimulation of lateral expansion) to assay the various components of coal gas for biological activity. The constituent with the greatest activity was found to be ethylene and Gane positively identified the gas as a natural plant product in 1934. Since then numerous ethylene effects have been described in light–grown plants.

$$H_2C = CH_2$$
Ethylene

Ethylene is a plant hormone, however, about 15 years ago, it was unclear as to whether it was a plant hormone or a metabolite having unusual effects on growth. In 1968, plant physiologists recognized it as a true growth regulator or plant hormone. Pratt and Goeschl (1969) reviewed ethylene biology in 1968 and applied the term plant hormone to ethylene. The rediscovery of auxin-induced ethylene formation was important for this recognition.

BIOSYNTHESIS OF ETHYLENE

Plants in vegetative growth phase, flowers and immature fruits produce barely detectable levels of ethylene until they are subjected to stress or undergo maturation events after which ethylene production accelerates in a spatially and temporally specific pattern. All vascular plants analysed to date synthesise ethylene via the Yang cycle wherein S–adenosyl methionine (AdoMet) is diverted from alternative fates to make the ethylene precursor 1–aminocyclopropane–1–carboxylic acid (ACC). Expression of the enzyme catalysing this reaction, ACC synthase, is induced by stimuli that lead to increased ethylene production. It suggests that ACC synthase activity is rate–limiting for ethylene synthesis. ACC can be readily converted to ethylene, CO_2 and HCN by ACC oxidase or it can be conjugated to a malonyl or glutamyl group to limit its availability for ethylene production.

The gene encoding ACC synthase was first isolated in *Zucchini* and cloning of homologs from other plants quickly followed. A multigene family whose members are differentially expressed in response to developmental, environmental and hormonal factors encodes ACC synthase. For example, by using gene–specific probes the differential expression of tomato ACC synthase family members has been investigated. Transcript of one isoform increased during fruit ripening, those of another increased in response to wounding and those of a third form increased in response to treatment with auxin. ACC synthase genes expressed in response to a particular stimulus are more similar to genes controlled by the same stimulus in other species than they are to other ACC genes in the same species.

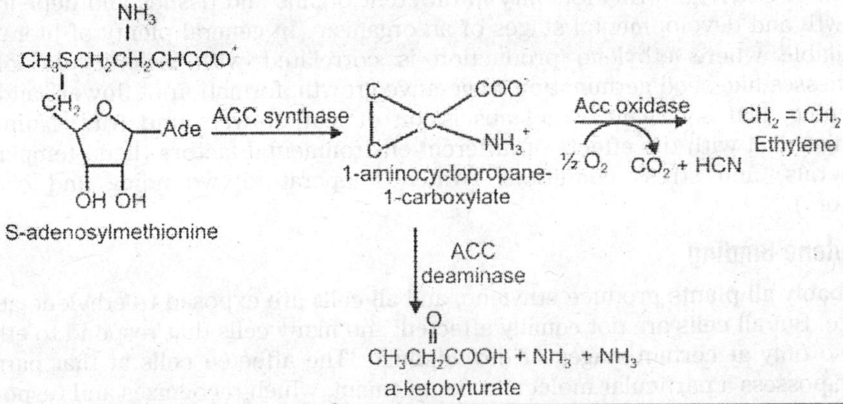

NH_3
$CH_3SCH_2CH_2CHCOO^-$
CH_2 O Ade
OH OH
S-adenosylmethionine
ACC synthase
COO^-
NH_3^+
1-aminocyclopropane-1-carboxylate
Acc oxidase
½ O_2 CO_2 + HCN
$CH_2 = CH_2$
Ethylene
ACC deaminase
O
$CH_3CH_2COOH + NH_3 + NH_3$
a-ketobyturate

FIG. 23.1: Enzymes involved in ethylene biosynthesis. Ethylene is synthesised from S–adenosylmethionine via the intermediate 1–aminocyclopropane–1–carboxylate (ACC). Genes encoding the ACC synthase and ACC oxidase enzymes have been cloned from several plant species. Synthesis of ethylene can be inhibited by expression of the gene encoding ACC deaminase in transgenic plants. This enzyme shunts ACC to α–ketobutyric acid and ammonia, thus preventing its conversion to ethylene.

The final step in ethylene biosynthesis, the conversion of ACC to ethylene, is catalysed by ACC oxidase. ACC oxidase was first identified by expressing the tomato cDNA **pToM 13** in an antisense orientation, which resulted in greatly reduced ethylene production in tomato. ACC oxidase requires CO_2 and ferrous ion for activity but the mechanism by which CO_2 potentiates ACC oxidase enzyme activity is currently in dispute. CO_2 has been postulated to covalently modify an internal lysine side chain as a carbamylate moiety in the ACC oxidase protein. There are seven conserved lysine residues in ACC oxidases and independent substitution of each of them suggests that some of the lysine residues are required for full enzyme activity. It has not yet been demonstrated whether the substitutions directly interfere with carbamylation.

Small multigene families encode ACC oxidase. The differential expression patterns of ACC oxidase gene suggest that oxidases contribute to regulation of ethylene production as well. Periods of ACC oxidase induction correlate with ethylene–regulated events in several instances including senescence, fruit ripening and wounding. In addition, ACC oxidase and ACC synthase together contribute to the positive feedback loop wherein ethylene treatment leads to increased ethylene production.

Ethylene Formation in Model Systems

The Cu^{2+}-ascorbate rapidly converts methonine into ethylene and methionol seems to be intermediate, and ethylene is produced from carbons-3 and 4 of methionine. In this reaction Cu^{2+} is reduced to Cu^{2+} by ascorbic acid and reoxidation of Cu^+ to Cu^{2+} by O_2. Linolenic acid also forms ethylene with Cu-ascorbate system in some situations. The conversion of linolenic acid yields not only ethylene but also ethane. A non-enzymatic conversion of methionine to ethylene is demonstrated by photochemical reaction based on FMN. However, under natural conditions, enzymes play the role of catalysts.

Regulation of Ethylene Biosynthesis

Rates of ethylene formation vary in different organs and tissues and depend upon growth and developmental stages of an organism. In general plenty of literature is available where ethylene production is correlated with several physiological processes like seed germination, vegetative growth, formation of flowers and fruits ripening, leaf senescence and abscission of leaf, flowers and fruit. Number of studies deal with the effects of different environmental factors (light, temperature, minerals) and stress conditions (water, temperature, wounding and chemical factors).

Ethylene Binding

Probably all plants produce ethylene, and all cells are exposed to ethylene at some stage. But all cells are not equally affected; and many cells that respond to ethylene do so only at certain stages of development. The affected cells at that particular time possess a particular molecular complement, which recognizes and responds to ethylene. Ethylene is thought to bind to this receptor, enabling a primary reaction that then initiates other reactions, including modification of gene expression.

Ethylene Receptors

Although direct proof is lacking, some indirect evidence suggests that ethylene binds reversibly to a metal containing receptor site.

ETHYLENE METABOLISM

Besides being converted to ethylene, ACC can also be irreversibly conjugated to form N–malonyl–ACC. Malonylation of ACC regulates the level of ACC and thus production of ethylene. Plant tissues to ethylene oxide and ethylene glycol can metabolize ethylene but the physiological significance of this metabolism remains to be established. As a gas, ethylene can readily diffuse from plant tissues so metabolism is not essential for its removal.

BIOCHEMICAL ACTION

Ethylene has various biochemical actions on plant. They are RNA synthesis, protein synthesis, protein loss, enzymes (cellulase, glucanase, pectinase, polygalacturonase, α-amylase, catalase, chitinase, phenylalanine ammonia lyase, malic enzyme, peroxidase, leakage of peroxidase, release of alfa-amylase, stabilization of enzyme activity), modification of ribosomes, mitochondria, ATPase stimulation, swelling, volume change, inducement of alternative path of respiration, chloroplast (chromoplast), chlorophyll synthesis, chlorophyll degradation, degreening, carotenoid synthesis, pigments and formation of anthocyanin.

The pigment metabolisms result from the ethylene-enhancement of related enzymes. The modification of endoplasmic reticulum indicates a change in protein synthesis or transportation of proteins. Eventually ethylene action results from DNA, RNA and protein synthesis of efficient utilization. Knowledge of molecular biology has facilitated the determination of the action mechanisms of plant hormones such as ethylene, auxin and gibberellins. However, the rapid response to the plant hormones as in the case of IAA, if it actually occurs, cannot be explained by the molecular biological concept involving enzyme biosynthesis.

BIOLOGICAL EFFECTS

Ethylene is a powerful elicitor of morphological changes during all stages of the plant life cycle. At germination, ethylene causes the hypocotyl to swell and broaden, girding it as it penetrates through soil. As the plant matures, ethylene influences sex determination and promote fruit ripening. Ethylene also protects plants from the onslaughts of nature, strengthening trunks of wind-baltered trees, assisting flooded plants to survive submergence and fending off pathogenic attack. Finally, ethylene plays a role in senescing flowers and leaves.

Ethylene and Defense Responses

Plants have a diverse arsenal of weapons with which to fend off pathogenic organisms. The initial response to bacterial or fungal pathogens called the hypersensitive response (HR) is the formation of necrotic lesions at the site of infection. A secondary, long lasting response, called systemic acquired resistance (SAR) develops throughout the plant in infected and uninfected tissues. Plant tissues displaying signs of pathogenic infection or wounding also produce elevated levels of ethylene. Recent experimental evidence using the ethylene-insensitive mutants etr 1 and ein 2 and the ethylene over producer eto 1 has indicated that ethylene is not essential. None of the mutants affect the resistance response in *Arabidopsis* and both the HR response and inhibition of bacterial growth occur when the mutants are infected with avirulent bacteria. Interestingly, all mutants behave as wild-type plants when infected with virulent bacteria except for the ein 2 mutant. $EIN2^-$ shows far fewer necrotic lesions than its wild-type counterpart and no restriction in bacterial proliferation. The ethylene over-producer eto1 neither confer greater disease resistance with avirulent bacteria nor greater disease susceptibility with virulent bacteria. These results indicate that EIN2 plays a essential role in causing necrotic lesions during pathogen attack. During SAR response, numerous PR proteins are induced, one of which is known to confer increased tolerance to pathogen infection. Using the ethylene-insensitive mutants etr1 and ein2, it was shown that the development of SAR response does not require ethylene.

Senescence, Fruit Ripening and Abscission

As a plant develops over time, organs that once were critical for early stages of growth become obsolete. In a prime example, newer, large leaves that shade them, rendering the old leaves ineffective for photosynthesis, supersede early leaves. To avoid squandering nutrients on non-functional structures, plants reclaim nutrients from the dying structure and jettison it by the ethylene-regulated processes of senescence and abscission. A progressive yellowing of leaves, followed by necrosis and death, marks its decline. Several senescence-associated genes have been identified whose expressions increase as leaves become yellow. Several catabolic enzymes or proteins have been implicated in catabolic events including cysteine and aspartic proteases, components of the ubiquitin-proteolytic pathway and ribonucleases. Senescence is also characterized by chloroplast breakdown and remobilization of lipids and nutrients from the dying organ to the rest of the plant. Ethylene induces senescence in dicot. The effect of ethylene is modulatory rather than a absolute requirement. Application of ethylene causes premature expression of senescence markers only in mature leaves. Ethylene mutants such as **etr 1** and **ein 2** show a significant delay in senescence of older leaves but eventually they do yellow and die. These data suggest that there are mandatory developmental events that must transpire before ethylene can potentiate the senescence process.

Leaf senescence is closely linked to fruit maturation so that energy can be shunted from the drying leaves to produce more robust fruits. Fruit ripening in the climacteric tomato fruit has been one of the most intensely studied ethylene-mediated developmental processes. The tomatoes produced by the "Never–ripe" mutant fail to mature into red fruit and the mutation has been shown to map to a homolog of the ERS, ethylene receptor gene.

A third phenomenon tied to senescence and fruit ripening is the process of abscission. It is complex series of molecular and morphological changes that result in the release of leaves, flower petals or fruits. The hydrolytic enzymes that act on cell wall components are expressed in the cells of abscission zone. The cells of abscission zone form a fracture plane and then develop into scar tissue once the organ has been released. Like senescence, ethylene promotes abscission.

Chlorophyll Degradation

Degreening does not occur during exposure to ethylene in the peel of satsuma mandarin fruits, but does by exposing the ethylene-treated fruits to air. Chlorophyllase enzyme is responsible for degradation of chlorophyll in plants. Its activity is low during exposure to ethylene and during the first 6 h of the subsequent exposure to air. These observations indicate that a lag phase is a prerequisite for the activation of chlorophyllase and also that oxygen is required for the activation of chlorophyllase and also that oxygen is required for ethylene-enhanced chlorophyll degradation and chlorophyllase activity. The constructing effect of these two inhibitors of protein synthesis indicates that de novo synthesis of cytoplasmic chlorophyllase is essential for ethylene-enhanced chlorophyll degradation. Ethylene degreening of *Citrus* fruits is a common commercial practice in USA, Japan, Israel, and Australia to insure product quality in the market. Practical degreening methods are generally complete, but the role of ethylene is still not certain.

Ethylene and Chloroplast Structure

The number of chloroplast decreases with ethylene treatment and tree ripening. Reduction in chloroplast size was a characteristic feature of ethylene-treated fruits lost grana and lamellar systems.

Stress and Ethylene

An upsurge in ethylene production can be observed in response to a wide variety of environmental stresses such as mechanical stress, toxic chemicals, radiation, temperature extremes, flooding, drought, wounding and pathogen attack. Mechanical stress can induce morphological changes in both roots and shoots by inhibition of elongation and promotion of radial growth. Exposing non–stressed plants to ethylene can mimic these changes in morphology. This led to hypothesis that ethylene action might be casually related to changes in patterns of plant growth under impeded conditions. The stimulated ethylene production of both roots and shoots of etiolated maize seedlings upon pressure–stimulated impedance is regulated at both the ACC synthase and ACC oxidase level.

Various phytotoxic chemicals induce the production of stress ethylene. These chemicals include metals, inorganic compounds, pollutants and herbicides/pesticides. The rapid elongation of many semi aquatic plants upon submergence is mediated by ethylene, which accumulates in the submerged tissue. It has been shown that ethylene acts by increasing the tissues responsiveness to GA and that GA is the immediate growth–promoting hormone. Further support for this hypothesis comes from experiments with deep–water rice in which ethylene treatment led to a rapid decline of endogenous ABA levels. It has been suggested that the increased responsiveness to GA is based on reduced ABA content. It has shown that elongation of rice coleoptiles which growth is also promoted by ethylene is stimulated by fluridone, an inhibitor of carotenoid and ABA biosynthesis.

Physical Stress

Since the discovery of mechanically stimulated changes in growth by Darwin, many investigators have observed this phenomenon. Jaffe (1973) observed a significant retardation in internode elongation by rubbing (a gentle mechanical stimulus), and termed this morphogenetic response as **thigmomorphogenesis** and the participation of ethylene in thigmomorphogenesis.

Gene Expression and Disease Resistance

While resistance to disease is likely the result of a number of different processes, ethylene does appear to play a part in a number of instances. One of the earliest events in the plant pathogen interaction is a rapid increase in ACC synthetase. It has been suggested that the resulting increase in ethylene production may signal the plant to mobilize its defense systems.

Ethylene Formation in Soils

Ethylene is produced in anaerobic soils by microbial activity, and can affect the growth of crop plants in waterlogged soils. They were of the opinion that ethylene acts as a soil fungistasis, and developed the concept of the O_2-C_2H_4 cycle (a self-regulating microbial cycle in soil). This cycle helps in regulation of organic matter turn over in soil, better availability of essential plant nutrients, and development of effective biological control of soil borne plant pathogens.

Stimulation of Germination

Germination of a number of plant species is enhanced by ethylene.

Induction of Female Flowers

Cucumber and melon normally produce male flowers earlier than female flowers. Ethylene stimulates the early production of female flowers. Often man flower formation is completely inhibited following ethylene treatment. Treatment of plants with Ethephon has the same effect. Female flower induction in Cucurbitacae is of commercial interest to achieve an earlier harvest and a yield increase.

Male Sterility

In plant breeding of cereals the production of hybrid seed is of great interest to improve quality and to increase yields. Ethylene applied at a certain time of pollen development causes male sterility.

Prevention of Lodging in Cereals

High level of cereal production is intensified by use of herbicides, fungicides and high amounts of fertilizer. Under such conditions cereals tend to lodge when rainfalls and storms occur later in the growing season. Mechanical harvest of lodging cereals is very time consuming. Furthermore, the quality of the crop is lower because of prolonged ripening. Under wet weather conditions the drying of the grains is severely affected, especially in rye germination of grains on the culm may occur. ACC inhibits the growth of cereals only at very high concentrations. Amazingly N-Formyl-ACC and similar derivatives of ACC are much more effective and prevent lodging at relatively low concentrations.

Growth Inhibition

The growth inhibiting properties of ethylene in many plant species are moderate compared with those of plant growth inhibitors interfering with gibberellic acid biosynthesis. In water-plants even a stimulation of growth observed by ethylene. The use of ethylene growth regulators in cereal production to prevent lodging is based on growth inhibition. Some investigations show the usefulness of Ethephon in other crops for growth inhibition. In grape vigorous vegetative growth interferes with reproductive growth, berry size, coloration and ripening of the fruits. Therefore, special pruning systems have been developed for grape production. Ethephon may be useful in inhibiting vegetative growth of grape vines resulting in higher yields and better quality.

ETHYLENE RELATED PLANT GROWTH REGULATORS

Ethylene as a gas can only be applied in more or less closed systems like glass houses or incubation chambers, limiting the usefulness of the hormone.

TABLE 23.1: Common and chemical names of some important ethylene-releasing chemicals

Common name	Chemical Name
Ethephon	(2-chloroethyl) phosphonic acid
Silaid	(2-chloroethyl) methylbis (Phenylmethoxy) silane
Alsol	(2-chloroethyl) tris (2-methoxyethoxy) silane
ACC	1-aminocyclopropane-1-carboxylic acid

Ethephon: 2-chloroethylphosphonic acid (CEPA, Ethephon) is the best-known compound of this group. It releases ethylene above pH 4.0 in aqueous solutions but is relatively stable below pH 4.

Etacelasil: 2-chloroethyl-tris- (2-methoxyethoxy)-silan (Etacelasil) belongs to the group of 2-chlorosilanes, also releasing ethylene in aqueous solutions. The chemistry of ethylene evolution is comparable to that of ethephon.

Haloethylsulphinic acids: A third group of ethylene releasing compounds is the 2-haloethylsulphinic acids and derivatives. These compounds also have a similar mechanism of ethylene release as compared with ethephon.

2-Hydroxyethylhydrazine: The growth regulator 2-hydroxyethylhydrazine causes typical ethylene reactions in plants.

Others Ethylene Releasing Compounds: Three chemically unrelated growth regulators should be mentioned; cycloheximide (trade name Act-Aid), glyoxaldioxim (trade name Pik-off), and 5-chloro-3-methyl-4-nitro-1H-pyrazol (trade name Release). These compounds are useful only for a very limited field of application, namely citrus fruit abscission.

PRACTICAL USE OF ETHYLENE-RELEASING COMPOUNDS

The diversity of present commercial uses of ethephon and other ethylene-releasing compounds indicates the substantial contribution that stimulation of ethylene production is making to agricultural productivity. For many years ethylene itself has been used commercially to elicit some of these effects, for example, fruit ripening. Compounds, which initiate ethylene responses by inducing the plant to produce its own ethylene, (a wound response) have been used in the field.

GROWTH AND DEVELOPMENT

Prevention of Lodging

Neljubow observed prevention of lodging. It is the basis of one of the most important uses of ethephon. The regulator treatment not only reduces length of the stems, but also stiffens the straw.

Carbohydrate Content of the Straw

Ethephon-treated plants contained more water soluble carbohydrates, however, anticipated increase in structural carbohydrates was not observed.

Reduction in Scape Length

The floral axis in bulb-type flowers is often too long for their use as potted flowers. Application of ethephon to hyacinth and narcissus plants when the leaves were 10 cm long reduced the length of the floral scape by 20-30%. The reduction in scape length was associated not only with a reduction in cell size, but, surprisingly, also with an increase in cell number, indicating that ethephon application had actually stimulated cell division.

Reduced Yield and Mean Tuber Size

Hollow heart and brown centre of potatoes are disorders that accompany rapid plant and tuber growth. Although application of ethephon at tuber initiation reduced yield and mean tuber size, it also substantially reduced the incidence of these disorders, leading to increased yield of marketable tubers, relative to control fields which had high disorder incidence.

Flowering Control

Stimulation of Flowering in Pineapples

For many years it has been known that ethylene stimulates flowering of Bromeliads, and treatment of pineapple plants with calcium carbide (which releases acetylene on hydration) was widely used commercially to stimulate flowering of pineapples. This response is now elicited in pineapples and flowering bromeliads by application of ethephon. A comparison of the effectiveness of ACC and ethephon in eliciting this response suggested that ACC was the preferred ethylene-releasing compound, but because of differences in application method and concentrations, this interpretation needs confirmation.

Other Tropical Fruit Crops

Flowering is also stimulated in many other fruit crops after ethylene treatment. Smoky fires are, for example, commonly lit in mango orchards to stimulate flowering. Application of ethephon has also been shown to stimulate flowering in *Litchi*, although the results vary widely depending on application time and cultivar. Ethylene stimulates or accelerate flowering of many geophytes. In *Iris*, for example, a short treatment of harvested bulbs with ethylene will stimulate flowering in small-sized bulbs that would not normally flower. e.g., *Iris* and *Freesia*. Ethephon application has been used to extend the flowering season of plantations of *Leucospermum*, a woody proteaceous ornamental.

Leaf and Branch Abscission

The effects of ethylene in stimulating abscission are the basis of important commercial uses of ethylene-releasing compounds. Enhancement of fruit abscission in sour cherry, nut fall in macadamia, the removal of leaves from potato plants, cotton, and nursery stock, and the dehiscence of the shucks of walnuts and pecans are but a few examples of the potential uses of these materials in agriculture. Defoliation of rutabaga required very high concentrations of ethephon; addition of 1% ammonium peroxydisulphate greatly increased the effectiveness of the ethephon treatment, perhaps due to accelerated oxidation of IAA in the abscising organ.

Fruit Ripening

Ethylene gas is widely used commercially for ripening a variety of climacteric fruits and decolouring non-climacteric citrus fruits. However, the use of ethylene-releasing compounds to affect this response is confined to relatively new crops, and normally when they are used as a preharvest application. This is probably due to the ease with which ethylene gas can be applied to harvested fruit, and the dangers of applying aqueous solutions to these disease-susceptible organs.

Fruit Abscission

Promotion of abscission as an aid in thinning or harvesting fruit crops is an important horticultural use of ethylene-releasing compounds. A recent study by Takeda and Peterson on mechanical harvesting of blackberries provides a good example of the benefits ethylene-releasing compounds provide in stimulating this response.

Combination with other Chemicals

The rapid production of ethylene from the neutralized ethephon solution stimulated fruit abscission, but elevated ethylene concentrations were not presents for long enough to cause leaf abscission. A mixture of 1% glycerin + ethephon is the most successful ethephon formulation for abscission uses. Combination with other PGRs has also proved to be useful in modifying plant responses to ethephon. Application to pecans of a spray combining 6 mM NAA with 3 mM ethephon resulted in accelerated shuck dehiscence without major leaflet abscission. Similarly, bloom-time sprays of ethephon combined with an auxin have achieved successful thinning of apples without undesirable leaf drop.

CHAPTER 24

Plant Hormone: Abscisic Acid

The systematic search for an endogenous plant growth inhibitor was initiated by the work of Bennet–Clark and his colleagues in the early 1950s. They discovered that the acid fraction of an alcoholic extract from a variety of plant tissues displayed growth inhibitory properties. This fraction was called **inhibitor B** and promoted a number of research groups to focus their attention on its identification. Little progress was made until the early 1960s when a series of publications appeared reporting the isolation of compounds, which had the ability to inhibit growth of coleoptile segments. These compounds were given the trivial names **dormin** and **abscisin II** since they had been obtained from dormant sycamore buds and dehiscing cotton balls (Addicott et al., 1965) respectively. The molecular structure of these two compounds was found to be identical and this "putative" growth inhibitor was christened abscisic acid **(ABA)**.

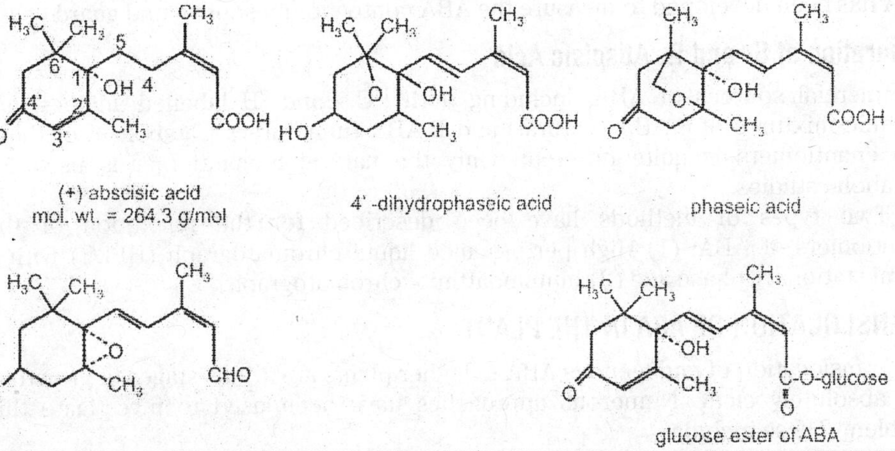

(+) abscisic acid
mol. wt. = 264.3 g/mol

4'-dihydrophaseic acid

phaseic acid

glucose ester of ABA

FIG. 24.1: Structure of ABA and related compounds.

ABA is a sesquiterpeniod composed of three isoprene residues. In addition, the carbon skeleton of the molecule bears a close similarity to the terminal rings of carotenoids such as violaxanthin. It can exist in either the *cis* or the *trans* configuration and the cis has a (+) and (–) enantiomer. Both the *cis* and the *trans* isomers of ABA may be extracted from plant tissues, although only the *cis* isomer exhibits biological activity. Light isomerizes ABA to a mixture of the *cis* and *trans* structures and this photolytic conversion is thought to account for the presence of the inactive isomer in plant extracts.

DISTRIBUTION OF ABA IN NATURE

Abscisic acids (ABA) is ubiquitous in higher plants, and is produced by algae and fungi. However, the source of ABA in animals may be dietary in origin. ABA has been identified in number of *angiosperms* both monocotyledonous and dicotyledonous plants. It is also identified in many gymnosperms (e.g. Yew, Silver fir, Pine, Sitka spruce) and a few *pteridophytes* (e.g., *Anemia phyllitides*, *Equisetum arvense*). Among *bryophytes* it has been found in a moss, but not in any liver worts. It has not been found in any *algae*. The place of ABA as a growth inhibitor in liverworts and algae appears to have been taken by lunularic acid.

ABA has been found in many higher plants tissues like roots, stems, buds, leaves, fruits and seeds. It has also been found in phloem and xylem sap and in nectar. Its concentration varies from tissue to tissue and also within a given tissue during the course of its development. Most tissues contain 20-100 mg/g-wet weights but concentration of 10 mg and 20 mg per g fresh weight have been found in avocado fruit pulp and dormant cocklebur (*Xanthium*) buds respectively.

QUANTIFICATION OF ABA

For purification of ABA in crude extracts, HPLC has largely been replaced TLC. Actual quantification is usually accomplished by Gas Chromatography with an electron capture detector, as well as by radio-immuno assays and enzyme-immunoassays. Antibodies of high specificity and affinity against ABA have been produced by number of laboratories and a monoclonal antibody against S-ABA is available commercially. Several laboratories have developed an indirect enzyme-linked immunoassay in which much less antibody is used than required in the Idetek protocol. An enzyme-amplified immunoassay with a sensitivity of 0.05-2.5 pg ABA has been developed to measure the ABA content of mesophyll and guard cells.

Separation of S- and R- Abscisic Acid

Commercial sources of ABA, including both ^{14}C - and ^{3}H labelled analogs, are racemic mixtures of S-ABA (natural) and R-ABA (unnatural). Catabolism of these two enantiomers is quite different. Only the natural S-enantiomer is used for metabolic studies.

Two types of methods have been described for the resolution of the enantiomers of ABA: (1) High performance liquid chromatograph (HPLC) with a chiral stationary phase and (2) Immunoaffinity chromatograph.

TRANSLOCATION OF ABA IN THE PLANT

The translocation of endogenous ABA in higher plants at various stages of growth is not absolutely clear. Numerous approaches have been used to investigate this problem. These include:

1. Experiment involving the local application of (^{14}C) ABA to the exterior surface of one part of the plant followed by determination of the distribution of radioactivity some time later.
2. Experiments involving the application of (^{14}C) ABA to the cut end of stem, petiole and root segments followed by the determination of how much radioactivity reached the other end.
3. The examination of phloem and xylem sap from various vascular organs for the presence of ABA.
4. From these experiments it appears that externally applied ABA is able to get into tissues rapidly and is distributed freely in all directions within the plant. Cell to cell transport of ABA is slow and non-polar. ABA is present in the phloem and xylem sap and is probably transported throughout the plant in this way.
5. ABA synthesised in the root cap, moves basipetally in the tissues of the steles.

6. There is also some evidence for lateral ABA movement in roots.

BIOASSAY OF ABA

ABA has been estimated by many bioassays and physical methods. Bioassays based on following phenomenon have been developed.

1. The acceleration of abscission of cotyledonary petioles in explants of 14-day old cotton seedlings.
2. The inhibition of IAA-induced straight growth of Oat (*Avena*) coleoptiles.
3. The inhibition of germination of isolated wheat (*Triticum*) embryos.
4. The inhibition of the growth of a duckweed species (*Lemna minor*)
5. The inhibition of the GA-induced synthesis of α-amylase by barley (*Hordeum*) aleurone tissue.
6. The stimulation of stomatal closure.

Physical methods include ultraviolet absorption, optical rotatory dispersion and circular dichroism, gas-liquid chromatography of methyl ester or less frequently the trimethyl silyl ester of ABA on various silicon liquid phases and combined gas liquid chromatography- mass spectrometry.

BIOSYNTHESIS OF ABA

The biosynthetic pathway of ABA is difficult to study using conventional biochemical methods because of its low levels in cells and tissues. For many years it was difficult to establish even the basic framework of the ABA synthetic pathway in higher plants. Since 1987 many *in vivo* and *in vitro* biochemical studies of ABA metabolism, including studies of ABA–deficient mutants, have established an outline of the probable biosynthetic route. A cluster of recent studies based mainly on the cloning of genes responsible for ABA deficiencies have clarified and verified many details of ABA biosynthesis in higher plants.

Initial research focused on two possible routes to ABA. A direct pathway from farnesyl pyrophosphate and second an indirect pathway via cleavage of carotenoid precursor. In each case the ultimate precursor was thought to be mevalonic acid. Several persuasive lines of evidence favoured the indirect pathway to ABA via a C_{40}–carotenoid intermediate. Some reports show that corn viviparous mutants (Vp) defective in carotenoid biosynthesis are also ABA–deficient. In addition, chemical inhibitors of carotenoid biosynthesis such fluridone and norflurazon also inhibit ABA accumulation. *In vivo*–labelling studies using $^{18}O_2$ are consistent with ABA formation from a large precursor molecule probably an oxygenated carotenoid. Existence of the indirect pathway to ABA via carotenoids strongly suggested that the early steps of ABA biosynthesis take place in plastids.

This idea was challenged when the 1–deoxy–D–xylulose–5–phosphate (DOXP) pathway for isoprenoid formation in bacteria was discovered. In the DOXP pathway the C_5 isoprene unit, isopentenyl diphosphate (IPP) is formed via DOXP from pyruvate and glyceraldehyde–3–phosphate rather than mevalonic acid (Lichtenthaler, 1999). Subsequently, this pathway has been shown to provide the precursors for carotenoid biosynthesis in unicellular algae and in higher plants.

Experiments in which both mevalonic acid and pyruvate are incubated with intact spinach chloroplasts confirm that pyruvate is incorporated more efficiently in ABA than mevalonic acid. Pyruvate and glyceral dehyde–3–phosphate are now firmly established as the ultimate precursors of ABA.

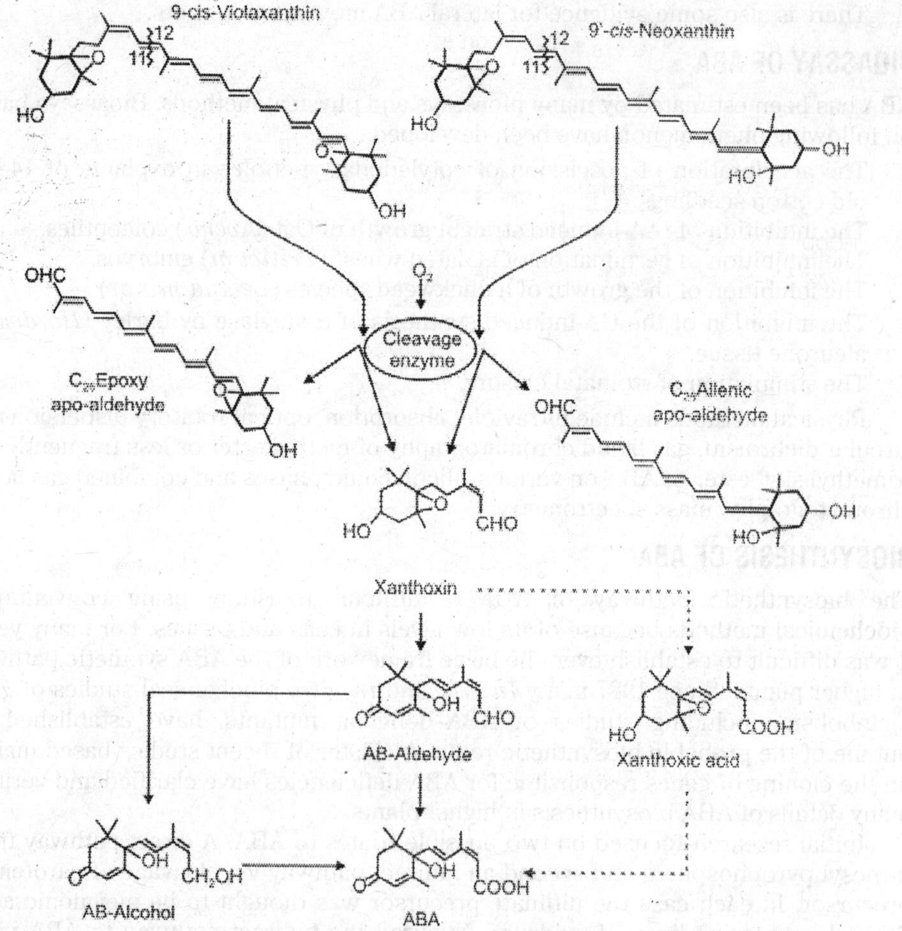

FIG. 24.2: Biosynthesis of (+)–ABA from carotenoid precursors in higher plants. Carotenoids are involved in photosynthesis and therefore epoxycarotenoid (C_{40})–cleavage (catalysed by 9–*cis*–epoxycarotenoid dioxygenase) takes place in the plastid and yields xanthoxin (C_{15}) and a C_{25} derivative. Two–step conversion of xanthoxin to ABA occurs in the cytoplasm. Broken arrows indicate minor or speculative steps. The major route is via abscisic aldehyde (AB–aldehyde), involving the enzymes xanthoxin oxidase and AB–aldehyde oxidase, but there is an alternative route via abscisic alcohol (AB–alcohol) that is especially significant when aldehyde oxidation is blocked. If AB–aldehyde oxidase can use xanthoxin as an alternative substrate, the xanthoxic acid might be an intermediate also.

The ABA–deficient aba1 mutant of *Arabidopsis* is blocked in the epoxidation of zeaxanthin to antheraxanthin and violaxanthin indicating that the epoxycarotenoids violaxanthin and neoxanthin are essential for ABA production. The aba2 mutants of *Nicotiana plumbaginifolia* are orthologous with aba1 of *Arabidopsis* and have been cloned. The corresponding fusion protein has zeaxanthin epoxidase activity. Transgenic experiments provide further evidence for zeaxanthin epoxidase involvement in regulating ABA concentrations and ABA responses in seeds. Overexpression of ABA 2 delays seed germination and antisense expression reduces seed dormancy. The maize vp 14 mutant suggests that VP 14 is responsible for xanthophyll cleavage. The deduced amino acid sequence of the cloned VP 14 gene contains an N–terminal chloroplast transit–peptide sequence and is expressed constitutively in embryos and roots. Wild–type seedling leaves contain low levels of transcript but mRNA accumulates strongly when detached leaves are water-stressed. The cloned VP 14 protein cleaves the 9–cis isomers of both violaxanthin and neoxanthin to cis–xanthoxin and a C_{25} epoxy apo–carotenoid as well as the 9–cis isomer of zeaxanthin at the 11–12 position to form corresponding products. A similar 9–cis–epoxycarotenoid dioxygenase has been characterized in tomato.

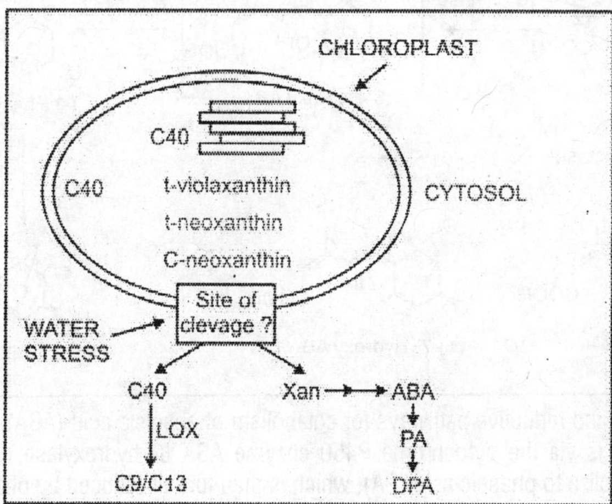

FIG. 24.3: A model showing the possible sub–cellular location of ABA metabolism. The key regulatory step, at least in water–stress–induced ABA biosynthesis, is the cleavage of 9'–*cis*–neoxanthin. LOX, lipoxygenase.

The cleavage product, xanthoxin, leaves the plastid by unknown mechanism and is subsequently converted to ABA in the cytoplasm. The two–step conversion of xanthoxin to ABA occurs in the cytoplasm. The major route is via abscisic aldehyde involving the enzymes xanthoxin oxidase and abscisic aldehyde oxidase but there is an alternative route via abscisic alcohol and is especially significant when aldehyde oxidation is blocked. If abscisic aldehyde can use xanthoxin as an alternative substrate then xanthoxic acid might be an intermediate also.

Carotenoid Biosynthetic Inhibitor

If ABA is derived from a carotenoid, then blocking the carotenoid biosynthetic pathway and reducing the carotenoid pool size should prevent ABA accumulation. Inhibitors of carotenoid biosynthesis such as fluridone and norflurazon, which block the conversion of phytoene to phytofluence, also prevent accumulation of ABA. Fluridone did not inhibit the accumulation of ABA in water-stressed green leaves, but the specific activity was greatly reduced.

ABA METABOLISM

There are several metabolic pathways by which ABA can be removed or degraded in plant tissues. ABA is exported by passive or carrier mediated efflux from the cells. In the majority of the plant tissues catabolic inactivation of ABA proceeds via hydroxylation at the 8 position to form the unstable intermediate, 8–hydroxy ABA which subsequently cyclizes spontaneously and/or enzymatically to form phaseic acid. The phaseic acid is sometimes further reduced at the 4' position to form dihydrophaseic acid. Oxidative catabolism of ABA to phaseic acid has also been demonstrated in *Riccia fluitans* (liverwort) and *Dunaliella* (green alga).

Several conjugates of ABA and its metabolites have been reported of which

ABA glucose ester appears to be the most widespread. These glucose conjugates have little or no biological activity and are not considered to be a reserve or storage form of ABA. In some tissues the formation of ABA glucose ester or of other conjugates appears to be a major pathway for the inactivation of ABA.

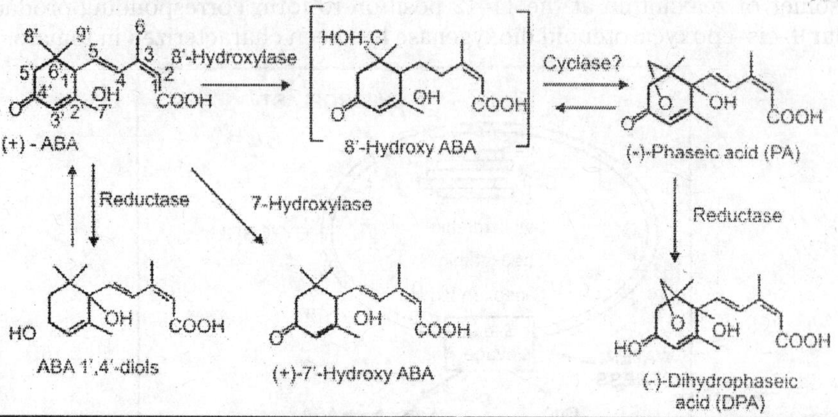

FIG. 24.4: Oxidative and reductive pathways for catabolism of abscisic acid (ABA). The major pathway for ABA breakdown is via the cytochrome P450 enzyme ABA 8'–hydroxylase, which produces 8'–hydroxy ABA that cyclize to phaseic acid (PA), which is then further reduced by phaseic acid reductase to dihydrophaseic acid. Minor routes involving 7'–hydroxylation and reduction of the ketone at 4' are also shown. Curiously, the unnatural isomer (–)–ABA is predominantly metabolized via 7'–hydroxylation (not shown). Formation of glucose esters and other conjugates with ABA and the metabolites shown, is widespread but the nature and range of products made, the site of formation and the transport and relationship to the oxidative and reductive pathways needs a lot more study. Conjugate structures are not shown here. There is no evidence yet that conjugates act as a storage reserve of free ABA. The catabolite 8'–hydroxy ABA, PA and 7'–hydroxy ABA all have some hormonal activity, and so might play a role in mediating ABA–like effects.

MODE OF ACTION OF ABA

The molecular mechanism of ABA is unknown. However, there are two major biochemical effects of the hormone, which are well known. One is an alteration of plasma membrane which is manifested in a change in biometrical potential across the membranes and a leakness and efflux of H^+ which is involved in some action of ABA for e.g., regulation of stomatal opening and closing. This closure of stomata cannot be accounted for by the RNA synthesis and RNA directed protein synthesis as the kinetics show that the response is quite rapid and appears within a minute. ABA may act on plant system as follows:

1. ABA does not exert a general inhibition of DNA transcription but may inhibit the synthesis of a limited number of specific mRNA species, e.g., the a-amylase mRNA of the cells of the aleurone layer in barley seeds, and
2. ABA exerts its main effect by blocking protein synthesis at a post transcriptional level such as mRNA processing or translation rather than by preventing DNA transcription, e.g., ABA has been shown to inhibit the synthesis of protease and isocitrate lyase in the embryos of cotton seeds even though appropriate mRNA's are present.

The other major effect is inhibition of RNA and protein synthesis. This in turn inhibits the synthesis of specific enzymes for e.g. inhibition of a number of enzymes like a-amylase, protease a-glucanase and ribonuclease has been reported to be suppressed by ABA. In most of the studies, ABA has been shown to inhibit the accumulation of labelled RNA precursors into all fractions of RNA as also in specific RNA fractions. This could be due to an increased activity of nucleases, decreased availability of template or inhibition of polymerase activity. In addition, inhibition of polysome formation, possibly through the reduction of mRNA supply, has also been attributed to ABA.

ABA might alter levels of other endogenous GA, since GA and ABA have been found antagonistic in several biological terms as stem growth, synthesis of amylase in barley aleurone cells, dormancy of seedling. The antagonism may be because of the following:

1. The competitive interaction of both GA and ABA for the same site.
2. Inhibition of GA biosynthesis by ABA.
3. Promotion of GA inactivation by binding.

ABA Receptors

The evidence supporting the existence of ABA receptors in guard cell is strong. A high-affinity binding site of ABA on plasmalemma of *Vicia faba* guard cell was found. There were 2×10^6 binding sites per protoplast. Three proteins that were covalently linked to ABA were separated by gel-electrophoresis. One protein had highest affinity for dissociated ABA, while the other 2 bound preferentially to protonated forms. One of these proteins could be modulator of H^+-ATPase, which is inhibited.

Calcium Ions

Strong indications of mode of action of ABA in guard cells have been provided by studies of its interactions with calcium ions, and with known regulators of calcium activity in cells. The essential mechanism behind the modulation of responses by Ca^{2+} is through induced changes in its intracellular concentration (McAinsh et al., 1990). In the cells, Ca^{2+} is thought to bind to the small protein **calmodulin** (CaM) and then the Ca^{2+}-CaM complexes clicit responses of various kinds. Although it is known that the action of ABA is dependent on calcium which Ca^{2+} inhibit stomatal opening in the absence of ABA and the response of stomata to other agents may also be dependent on calcium. Ca^{2+} has the central role in the regulation of the turgor changes of guard cells that are involved in stomatal opening and closing. ABA is just one external agent that is capable of triggering changes in intracellular Ca^{2+} but the importance of ABA in plant water relations is that it provides a means of cell to cell, or long distance signalling in the plant to indicate the need for change in stomatal aperture to regulate transpiration.

BIOLOGICAL EFFECTS

ABA is involved in many physiological and developmental processes throughout the life cycle of plants. ABA was discovered as a result of its ability to promote abscission in explants of cotton seedlings, induced bud dormancy, seed dormancy and germination, tuberization, leaf senescence and fruit ripening, the best authenticated examples of its hormones activity are:

1. Its role as growth inhibitor in the downward growth response of roots to gravity,
2. Its role in causing stomatal closure in response to water stress.

In the early phase of a plant life, ABA regulates seed maturation and the maintenance of embryo dormancy. Later at the onset of ontogenesis, it mediates several adaptational responses towards environmental cues such as desiccation, cold as well as salt stress and acts as a negative growth regulator.

Cell Cycle Control

Mitotic growth arrest is well known in animals and is induced under adverse conditions by preventing the transition of G1 into the S–phase of the cell cycle. The tumor suppressor $\mathbf{P^{53}}$ and the retinoblastoma a protein **pRb** have been identified as control links in the pathway. Extensive phosphorylation of **pRb** at the check–point of the cell cycle (the restriction point) within the mid G1–phase inactivates its growth inhibitory action. The multistep phosphorylation is initiated and is controlled by cell cycle machinery. Central components of the mammalian cell cycle are cyclins and their specific kinases, the proliferating cell nuclear antigen and the cyclin–dependent kinase inhibitors such as $\mathbf{p^{16}}$, $\mathbf{p^{21}}$ and $\mathbf{p^{27}}$. Among the key regulators of S–phase entry are the cyclin D–cyclin–dependent kinase 4/6 as well as cyclin E–cyclin–dependent kinase 2 complexes that promote cell cycle progression by specific protein phosphorylation.

Several different cycline including type A, B, C and D have been isolated from plants. Several cyclins of D–type reveals pRb interaction motifs. The **pRb** proteins of maize have been identified and were capable of interacting with this cyclin type pointing to a similar role for these cyclins in cell cycle control as in animal system. Recently, cyclin–dependent protein kinase inhibitors, which interact in vitro with cyclin–dependent kinase 2a and cyclin D3, have been identified in *Arabidopsis*. Cyclin–dependent protein kinase inhibitor is induced by ABA, which leads to a decrease in cyclin–dependent kinase activity. It suggests a molecular mechanism in which plant cell division is inhibited by ABA.

Growth Control

Approximately 33 years ago, it was observed that ABA inhibits DNA replication in root tips and embryo of *Faxinus excelsior*. Histochemical studies of root tissues also supported these results. Since these investigations it has been well established that ABA generally modifies plant growth by inhibiting shoot and root growth. The role of ABA in modulating the growth rate of roots has been intensely studied and was found to be concentration–dependent whereas a negative correlation between ABA content in the elongation zone of maize roots and growths rate was observed, low concentrations of ABA can even stimulate root growth. It explains the biphasic response of roots upon application of inhibitory ABA concentrations. Exogenons ABA application inhibited root growth after an initial phase of increased growth lasting several hours. Inhibition of ABA biosynthesis by fluoridone or ABA deficiency due to a mutational lesion however, did not affect the root growth of maize (Moore and Smith, 1984). It indicates that endogenous ABA is not a central growth regulator unless ABA levels are enhanced by cold or dessication.

The root response towards inhibitory ABA concentrations is characterized by an inhibition of cell elongation and subsequent osmotic adjustment of the root apex. The reduction in root elongation is reciprocally correlated with an increase in both osmotic potential of the medium and the endogenous ABA level (Ribaut and Pilet, 1991). The same adaptational response is observed under conditions of drought stress. It increases root hair formation and decreases the size of the meristematic zone of the root tip (Robertson et al., 1990). The application of ABA mimics this response of roots.

Seed Development and Dormancy

ABA has been implicated in the control of many events during embryogenesis and seed formation including embryo morphogenesis, desiccation tolerance and the onset and maintenance of dormancy. Typically ABA accumulation in developing seeds is low during the early stages, is greatest during mid–development when storage reserves are being synthesised and declines as the seed undergoes maturing drying. Prevention of germination during development may be due to the endogenous ABA content of the seed, the osmotic environment surrounding the seed or both. Maturation drying and shedding from the parent plant are sufficient to relieve these contraints in nondormant seeds. In these instances, there may be an associated decline in the ABA content of the seed and the sensitivity of the embryo to ABA is much reduced.

Genetic studies indicate that the absence of or insensitive to ABA during seed development results in the production of viviparous or precociously germinating seeds. These include maize viviparous **(VP)**, tomato sitiens **(sit)** and *Arabidopsis* ABA–deficient **(aba)** and ABA–inseasitive **(abi)** mutants. There are also reduced dormancy **(rdo)** mutants of *Arabidopsis* in which neither ABA synthesis nor sensitivity is altered. In **rdo** seeds there is block to some dormancy–inducing process that is normally controlled by ABA but is downstream from ABA synthesis and perception.

Reciprocal crosses between **aba** mutants and wild–type plants show that dormancy is initiated when the embryo itself produces ABA. ABA–deficient seeds **(aba/aba)** on ABA–producing parent plants **(Aba/aba)** are not dormant. Maternal ABA does not play a role in preventing germination in developing maize kernels and embryos of **vp5** and **vp8** mutants are even viviparous in the presence of normal endosperm. *In situ* ABA synthesis in the embryo is necessary for the imposition of dormancy has been demonstrated in sunflower as well.

It has long been suggested that dormancy is associated with gene repression (**dormancy gene** coined by Bewley, 1997) followed by derepression as dormancy is broken. Recently, the genes differentially expressed in dormant and nondromant seeds have been isolated and characterized. In terms of the protein complements, axes of dormant wheat seeds show prolonged synthesis of many heat–stable proteins (e.g. **dehydrin**). In contrast, axes of nondormant seeds initially synthesise a few of the heat–stable proteins but their synthesis declines after a few hours. Many genes expressed preferentially in dormant wheat seeds have been identified using cDNA cloning and differential screening techniques. All of these genes were found to be responsive to ABA. Analyses of sequence homology indicated that most of the clones corresponded to the late–embryogenesis abundant proteins **(LEA)**. These proteins function to maintain embryo viability during repeated cycles of desiccation of dormant seeds and it is unlikely that **LEA** proteins are involved in the regulation of seed dormancy. Recently, a cDNA clone **(pKABA1)** with sequence homology to a serine/threonine protein kinase has been isolated from a cDNA library from ABA–treated wheat embryo. Trarcripts for this gene are hardly detectable in growing seedlings but are induced dramatically by dehydration stress and ABA. Embryonic **pKABA1** transcript levels remain high in hydrated dormant seeds during developmental arrest. However, the transcript level declines upon imbibition in nondormant seeds. The transcript level of **pKABA1** remains high in nondormant seeds well after visible germination and thus it is difficult to envision a regulatory role for this gene in seed dormancy. It is involved in dehydration and ABA–mediated processes, which are related to seed longevity but not to dormancy.

Although several genes that are differentially expressed in imbibed dormant and nondormant embryos have been isolated and expression of many of the corresponding genes are also enhanced by ABA. There is as yet no direct candidate for involvement in the maintenance or termination of seed dormancy.

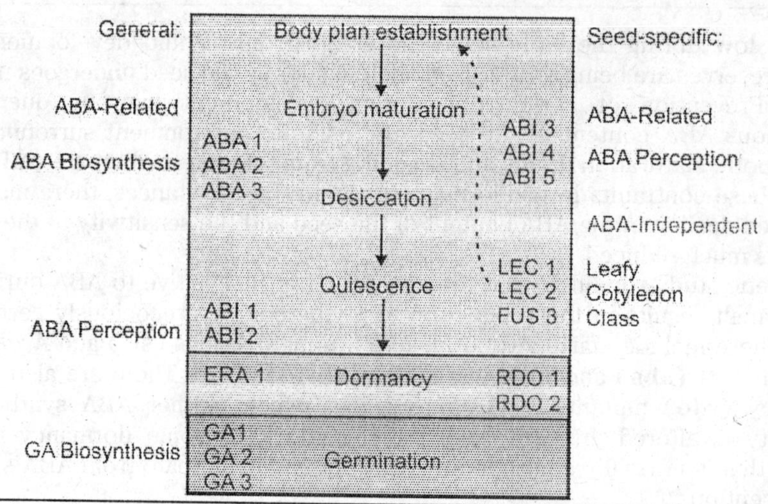

FIG. 24.5: Genetic loci involved in the control of embryo maturation and germination in *Arabidopsis*. Functional positions of genetic loci as revealed by mutant phenotypes are indicated. Abbreviations: ABA, abscisic acid; GA, gibberellic acid; broken arrow indicates the possible involvement of these loci in this process.

Abscisic Acid, Low Temperatures, Salinity and Drought

Plant water deficit occurs when the rate of transpiration exceeds water uptake and is a component of several different stresses including drought, salinity and low temperature. Such stress referred to as water stress has many detrimental effects on plant growth and development and is a major determinant of crop yield. Cellular water deficit can result in a concentration of solutes, changes in cell volume and membrane shape, disruption of water potential gradients, loss of turgor, disruption of membrane integrity and denaturation of protein. Complete loss of free water will result in desiccation or dehydration. The ability of the whole–plant to respond and survive cellular water deficit depends on whole plant mechanisms that can integrate the cellular responses. Responses to water deficit may occur within a few seconds such as a change in the phosphorylation status of a protein or within minutes and hours such as a change in gene expression.

Resistance to water deficit occurs when a plant withstands the imposed stress and may arise from either tolerance or a mechanism that permits avoidance of the situation. Whole–plant mechanisms can contribute to the avoidance of water deficit during the plant's life cycle and avoidance can also occur at the cellular level. The response depends on the species and genotype, the length and severity of water loss, the age and stage of development, the organ and cell type and the subcellular compartment. An example of avoidance at the cellular level is the process of osmotic adjustment where the osmotic potential of the cell is lowered in order for the water potential gradient to favour water uptake and maintenance of turgor. One response is the synthesis of osmolytes. These are compatible solutes that can accumulate to high levels without disrupting protein function. Osmolytes may include amino acids (e.g. proline), sugar alcohols (e.g. pinitol), other sugars (e.g. fructans) and quaternary ammonium compounds (e.g. glycine betaine). The enzymes involved in the synthesis of these compatible solutes permit an osmotic adjustment or the net accumulation of solutes. This results in decreased osmotic potential, which may maintain the favourable water potential gradient and protect the cellular turgor. Proline is synthesised from L–glutamic acid. The reaction is catalysed by two enzymes, pyrroline–5–carboxylate synthetase (P5CS) and P5C reductase (P5CR). In *Arabidopsis* the synthesis and metabolism of proline is controlled by water deficit. The P5CS enzyme is induced by water stress, salinity and ABA but P5CR is not. The breakdown of proline to pyrroline–5–carboxylate (P5C), catalysed by proline dehydrogenase, is repressed by water deficit and induced by proline and rehydration. The transport of proline may also be controlled by water status. A gene encoding a specific proline transporter, ProT2, is induced by water deficit. This indicates that proline distribution throughout the whole plant may be an important aspect of osmolyte function.

Transport proteins, ion channels and carriers also play an important role in water deficit avoidance or osmoregulation. Potassium channels control K^+ uptake and may also regulate Na^+ uptake, which can be an important determinant of salinity tolerance. Aquaporins, a family of membrane proteins that transport water, may be involved in controlling cellular water status in response to water deficit. The apoplastic water potential influences the phosphorylation status of two aquaporins of spinach. The phosphorylation of aquaporins is thought to increase the ability of the channels to transport water and it has therefore been proposed that decreased phosphorylation status during water deficit should slow the loss of water from individual cells.

Many hydrophilic globular proteins accumulate in seeds during the maturation phase when seeds are developing desiccation tolerance. These late embryogenesis abundant (LEA) proteins are also expressed in vegetative organs during periods of water deficit. Two of these classes of proteins have a functional role in stress tolerance. These are HVA1 (a group 3 LEA protein) from barley and LE 25 (a group 4 LEA protein) from tomato. Over-expression of HVA1 improves drought and salinity resistance in transgenic rice plants. The LE 25 protein was expressed in yeast and found to confer improved resistance to high salinity and freezing. One LEA protein, called dehydrin, RAB or D–11, has been identified as an induced gene/protein in response to many different stress conditions as well as during seed maturation.

The integrated response of the whole plant to water deficit must also comprise sensing and signaling mechanisms. ABA is the best-known signal at both the whole plant and the cellular levels. It can move throughout the plant in the vascular system and acts as a signal for changes in stomatal conductance and gene expression in response to soil drying. ABA–induced gene expression has been extensively studied and two classes of DNA elements have been identified. They are ABA–response element (ABRE) and the dehydration–responsive element (DRE). The consensus for the ABRE, RYACGTGGYR, where R refers to a nucleotide with a purine base and Y refers to a nucleotide with the pyrimidine base, is defined by the core element ACGT. This element was first defined in the Em gene from wheat and is bound by the bZIP protein EmBP–1. It has also been shown to be a functional element for ABA–regulated gene expression in maize, barley, rice, tomato and *Arabidopsis*. Additional sequence–specific DNA–binding protein(s) are required for transcriptional activation of ABA–inducible genes via the ABRE. There are also other cis–elements distinct from the ABRE that mediate expression of stress–inducible genes. Dehydration and ABA induce the expression of *Arabidopsis* genes encoding homologues of MYB and MYC proteins. These transcriptional factors then subsequently activate the expression of the dehydration – and ABA–inducible rd22 gene.

The DNA sequence of the DRE from the rd294 gene from *Arabidopsis*, TACCGACAT, has been shown to be involved in the regulation of this gene by an

ABA–independent pathway induced by water deficit, low temperature and salinity. The regulation of this gene is independent of ABA in the first few hours after dehydration but becomes dependent on ABA in the later stages of expression. A gene encoding the transcription factor CBP1, which binds to the DRE has been isolated from *Arabidopsis*.

The enhanced ABA synthesis is probably the key response to water stress. Elevated ABA concentrations lead to rapid stomatal closure, thereby reducing transpirational water loss and result in the synthesis of several proteins that probably have adaptive significance. ABA is biosynthesised via a complex pathway that involves oxidative cleavage of the xanthophyll 9' cis–neoxanthin. The recent cloning of the 9'–cis neoxanthin cleavage enzyme will drive the use of transgenic technology to assess better the role of ABA in water deficit responses.

Abscission

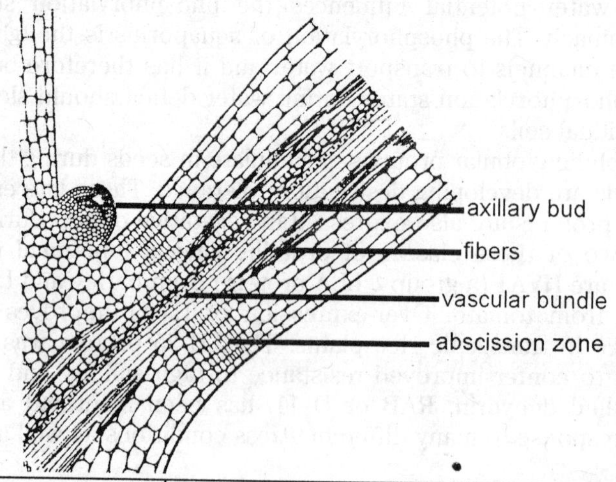

FIG. 24.6: The abscission layer.

Abscission term is synonymous with shading. Two functions are achieved by the abscission. One is, it facilitates vegetative propagation in the case of *Riella* and *Kalanchoe*. Second is the shading of mature plant parts. ABA was first isolated and characterized by abscission-accelerating bioassay using debladed petiole stumps of the cotyledons of cotton seedlings as a convenient explant system. The ABA containing solutions under test were applied dropwise to the cut surfaces of the stumps, and the time taken for separation of petiole from stem was recorded. In all these tests on explants systems, ABA reduced the time to zone cell separation compared with controls.

Some as one of the principal accelerators of abscission and regards ABA by others as having an unclear hormonal role in abscission and bud dormancy. However, the suggestion that ABA's role in abscission is questionable According to Dorffling (1978) *there is no evidence available that ABA plays a role in leaf abscission*. Evidence appears to be conclusive that in the promotion of abscission in young fruit of Lupinus, *Gossypium* and in mature fruits of *Citrus*, ABA plays a hormonal role. Addicott and Wiatr (1977) point out that the relatively few cases where applications of ABA to plants have failed to accelerate abscission were possibly due to the low dosages applied. They draw an analogy to the early work on auxin, where IAA, when applied to intact plants, often failed to promote growth. Dorffling's view (1978) was probably based on the fact that he found no correlation between ABA contents in *Coleus* petioles and the latter's ability to abscise. It may be that tylosis in the xylem vessels induces water stress which in turn causes an increase in the level of ABA. However, the opposite sequence of events is possible since it has not been resolved with certainty whether tylose formation is casual to or a consequence of abscission.

Other Related Effects

Stomatal Closure and Water Relations

The drop in water potential is detected by the chloroplasts of the mesophyll cells of leaf, whose photosynthetic activities are thought to be the most sensitive of all the plant's metabolic processes to water stress. As the water potential decreases there is marked increase in permeability of chloroplast membranes to ABA. The ABA synthesised by and stored in chloroplast then diffuses into mesophyll cytoplasm and moves from cell to cell through plasmodesmata to the guard cells where it induces stomatal closure. The drop in the level of stored ABA in the mesophyll chloroplasts induce fresh biosynthesis of more ABA, which continues to be translocated to the guard cells as long as the water potential remains low. The release of ABA from the mesophyll chloroplast stops when the water potential is restored to normal and its rate of synthesis is markedly reduced. The evidences for this are as follows:

1. The exogenous application of low concentrations (e.g. 1mM) of ABA to leaves causes stomatal closure within 3-9 minutes.
2. The concentration of ABA in water stressed leaves rises markedly.
3. Mesophyll chloroplasts can synthesise ABA but guard cell chloroplasts cannot.
4. The ABA of leaves of well-watered spinach was found to be contained in the chloroplasts but after 4 h of water stress the total ABA level was found to have risen 11 fold whilst that of the chloroplast had risen only 2 fold; this suggests that the newly synthesised ABA is rapidly exported from the chloroplasts.
5. Guard Cells.

Guard cells showed equal closure when exposed to ABA solutions over the pH range 5.0-8.0, while no significant uptake of the hormone took place at pH 8.0. This indicates that the sites of ABA action are on the outer surface of the plasmalemma. ABA prevents stomatal opening by rapidly blocking H^+ extrusion and K^- influx and it initiates closure by the rapid release of osmotica, in particular K^+, Cl^- and malate. The latter phenomenon can also be observed in the shrinking of guard cell protoplasts.

Chemical Signals for Stomatal Closure

Saab and Sharp (1990) have shown that the roots send signals to the shoot and causes stomatal closure and inhibition of leaf growth in Maize when there is even slight reduction in soil water potential with no detectable change in leaf water relations. They observed no effect of soil drying on stomata. In bean plants with oxygen-deficient roots, stomatal closure is influenced by changes in leaf water relations and by increased fluxes of ABA.

ABA and Potassium Uptake

The primary action of ABA on guard cells seems to be its inhibition of potassium uptake by the guard cells. Certain substances, such as a fungal toxin fusicoccin, are known to overcome the effect of ABA by stimulating potassium uptake and proton release. The flow of other solutes also seems to be influenced by ABA during the stimulated closing of stomata.

ABA and other hormones can alter the response of stomata to CO_2. The increase in stomatal sensitivity to CO_2, occurring as an early reaction to the arrival of a chemical signal from the roots, could be important in curtailing transpiration in wind.

Photosynthesis

There are several reports that ABA applied via the transpiration stream to leaves affected photosynthesis both via stomatal closure and via a direct effect on carbon fixation. As a possible mechanism, inhibition of the carboxylation of ribulose -1,5-bisphosphate has been suggested, but extractable ribulose-1,5-bisphosphate carboxylose activity from ABA-treated leaves was as high as that of control leaves.

Root Growth

Reports on the effects of ABA on root growth are contradictory. Since exogenous ABA can inhibit root growth as well as promote growth of this organ.

Elongation of Maize Coleoptile

ABA decreased the elongation rate in maize coleoptile and also inhibited growth-induced by auxin, fusicoccin or acid. In all cases, this effect was due to an inhibition of cell wall loosening.

Root Gravitropism

The root cap responds to light and gravity by synthesizing or accumulating growth inhibitors. It is believed that the inhibitors are produced in the lower portion of root cap in response to gravity and are translocated into the zone of cell enlargement. They may inhibit the cell enlargement of cells on the lower side of the root. Differential growth results in **geotropism**.

Bud Dormancy

Innate or *true dormancy* (i.e. temporarily suspended growth due to internal conditions) is exhibited by the buds of temperate-zone trees during winter, stem tubers (e.g. potato), root tubers (e.g., Dahlia), Corms, bulbs and rhizomes. Hemberg (1949) found that the dormancy of buds of ash (*Fraxinus*) and potato tubers was positively correlated with the concentration in them of growth inhibitors, identified as ABA. The external application of ABA to buds of growing seedlings caused them to become dormant. According to these findings ABA was synthesised in the plant, in the response of some **environmental trigger** and this ABA travelled to the buds and induced them to go dormant.

Seed Dormancy

ABA builds up in embryos of developing seeds. It is believed that increased ABA level comes primarily from *de novo* synthesis. However, there may be translocation of ABA or its precursors from the leaves since the radioactivity from exogenously applied ^{14}C-ABA to leaves can be recovered in developing seeds. ABA in developing ovules inhibits the formation of germination enzymes in the embryo (Dure, 1975). Thus, ABA plays an important role in developing seeds by inhibiting vivipary. In the case of immature seeds, several evidences indicate that ABA prevents precocious germination (vivipary) of the developing embryo.

Heterophylly

Aquatic plants often produce two distinct types of leaves on the same plant. Submerged leaves are highly dissected or linear and they have undifferentiated mesophyll and an epidermis with few stomata, if present. In contrast, aerial leaves are entire and broad and they have differentiated mesophyll and a stomatous epidermis. It is interesting that the ABA treatment can induce the formation of aerial leaves on submerged shoots. Such changes in leaf morphology due to ABA have been reported for a fern, a monocot, and several dicots.

Response Mutants

Several mutants have been isolated that exhibit a decreased response to exogenous ABA. A viviparous mutant of maize, vp 1, has green seedlings and normal ABA and carotenoid levels. The embryos of this mutant are insensitive to ABA. Therefore, it has been classified as a response mutant.

Tuberization

ABA appears to be involved in the process of tuberization. Wareing has shown that ABA formed by the leaf of a one-node cutting taken from a short day-induced plant of *Solanum andigena* (a wild species of potato which requires about 20 short day cycles to fully induce tuberization) is exported to the axillary bud and causes it to develop rapidly into a small tuber. Externally applied ABA has also been shown to promote the formation of these tubers in Dahlia and Jerusalem artichoke.

Senescence

ABA may also be involved in senescence and ripening processes. Externally applied ABA accelerates the senescence of detached leaves; however, its effect on the senescence of leaves still attached to a healthy plant is minimal. Externally applied ABA accelerates the ripening of young fruit; moreover, there is a rise in the level of endogenous ABA during the ripening of strawberries and grapes.

Resistance to Frost Damage

ABA may also be involved in increasing the resistance of temperate-zone plants to frost damage; the external application of ABA has been shown to increase the frost hardiness of seedlings of box elder (*Acer nigundo*), apple and alfalfa (*Medicago sativa*) and to ameliorate chilling injury in cucumber seedlings.

Barley Aleurone Cells

Among the most studied effects of hormones on gene expression is the GA induced production of hydrolytic enzymes and its inhibition by ABA in barley aleurone layers. When ABA is applied 25-fold excess of GA the synthesis of a-amylase, protease b-glucanase and ribonuclease is suppressed.

Developing Embryo

ABA not only prevents premature germination during embryogenesis, but it has also been implicated in the regulation of storage protein synthesis. In embryos of rapeseed, soybean and wheat cultured in vitro, exogenous ABA promoted the synthesis of embryo-specific proteins and mRNAs.

Protein Synthesis

ABA has been called as stress hormone since it enhances adaptation to various stresses. Freezing tolerance could be increased by ABA both in plants and in cell cultures. Increased cold tolerance was not induced in the presence of cycloheximide, which indicates that synthesis of new proteins is involved with stress adaptation.

CHAPTER 25

The Vitamins

The vitamins are cohesive group of organic compounds and have little in common either chemically or in their metabolic functions. They are needed in the diet in small amounts (micrograms or milligrams per day) for the maintenance of normal health and metabolic integrity. Vitamins are essential nutrients found in foods. The requirements are small but they perform specific and vital functions essential for maintaining health.

In 1747, Scottish naval surgeon James Lind discovered that an nutrient (now known to be vitamin C) in citrus foods prevented scurvy.

William Fletcher (1905) was the first scientist to determine that if special factors (vitamins) were removed from food disease occurred.

In 1906, English biochemist Sir Frederick Gowland Hopkins also discovered that certain food factors were important to health. In 1912, Polish scientist Cashmir Funk named the special nutritional parts of food as a "vitamine" after *vita* meaning life and *amine* from compounds found in the thiamine he isolated from rice husks. Vitamine was later shortened to vitamin. Together, Hopkins and Funk formulated the vitamin hypothesis of deficiency disease - that a lack of vitamins could make you sick.

Sir Frederick Gowland Hopkins (1906) conducted experiments by feeding rats on a diet of known nutrients like fats, proteins, carbohydrates, and mineral salts. The animals were unable to grow. Thjereafter, a small amount of milk was added to the diet and observed that animals maintained their normal growth. Hopkins suggested that milk contains one or more *accessory growth factors* and these are the essential nutrients present in small amounts. The accessory food factors were isolated and identified chemically an **amine**. Funk (1912) coined the term **vitamine** (Latin *vita* = life and *amine* = the prominent chemical reactive group).

Year	Scientist	Discovery	Note
1912–1914	V. McCollum and M. Davis Elmer	Vitamin A	butter contained a fat-soluble nutrient soon known as vitamin A. Vitamin A was first synthesised in 1947
1913	Osborne and Lafayette Mendel	Vitamin B	
1912	Casimir Funk	B_1	
1926	D. T. Smith, E. G. Hendrick	B_2	Max Tishler invented methods for synthesizing the essential vitamin B_2.
1937	Conrad Elvehjem	Niacin	
1933	Lucy Wills	Folic acid	
1934	Paul Gyorgy	B_6	
1912	A. Hoist and T. Froelich	Vitamin C	Vitamin C was the first vitamin to be artificially synthesised in 1935. A process invented by Dr. Tadeusz Reichstein, of the Swiss Institute of Technology in Zurich.
1922	Edward Mellanby	Vitamin D	while researching a disease called rickets.
1922	Herbert Evans and	Vitamin E	discovered vitamin E in green leafy vegetables

Year	Scientist	Discovery	Note
1957	Katherine Bishop Frederick Crane	Coenzyme Q10	Japanese manufacturers oriduced cost-effective CoQ10 in the mid-1960s. To this day, fermentation remains the dominant production method around the globe. In 1958, Dr. D.E. Wolf under Dr. Karl Folkers (1958) first described the chemical structure of coenzyme Q10.

The decision as to whether the word should correctly be pronounced *vitamin* or *veitamin* depends in large part on which system of Latin pronunciation one learned – the Oxford English Dictionary permits both.

VITAMINS FUNCTIONS AND SOURCES

As micro nutrients, vitamins are required in small quantities. Generally your body cannot manufacture or synthesise vitamins from scratch. So you need to get them from the food you eat or from supplement pills. Though, vitamins can be synthesised by some bacteria, yeast, mould, algae and some plant species.

Some vitamins like vitamin B_5 have been found to be synthesised by bacteria in the human large intestines. However it is not clear if the vitamins are available for absorption and use by the body.

Vitamins are required by the body for a variety of biological processes. These include growth e.g Vitamin B_6; mental alertness e.g Choline, Niacin; resistance to infection e.g. Vitamin C and Vitamin E.

They also act as catalysts in the body chemistry as well as precursors to vital body factors. This enables the body to use carbohydrates, fats and proteins.

Vitamins in themselves do not have calories and therefore do not give energy to the body. Natural vitamins are found in living things that is plants and animals and are organic food substances. There are many artificially synthesised vitamins.

Vitamins exist in varying quantities in an array of food sources from yeasts, wheat bran, cooked egg that provides Biotin, citrus fruits and milk that provide Vitamin C, green leafy vegetables and legumes that supply Folic acid.

Vitamin Deficiency

The lack of these food factors i.e vitamins results to deficiency diseases. Vitamin defiency diseases are a manifestation of malfunctioning bio-chemical processes due to lack of the vital vitamins. Since the same vitamin may be used in a number of processes, deficiency in some vitamins can be quite serious and even fatal. For example over two hundred enzymes require the niacin vitamin coenzyme. So a lack of niacin makes this two hundred enzymes malfunction.

Vitamin deficiencies have been associated with long term latent deficiency diseases. These diseases occur after a long time of deficiency of some nutritional factors like vitamins. Long term latent deficiency diseases can be as severe as cancer, diabetes and heart disease. In our society today, these diseases are becoming the highest health concern.

The situation is grave because they are not only preventable but also cheaply preventable. A proper nutrition is all it takes to get all required nutrients including sufficient quantities of vitamins. General health benefits of vitamins include protection from a variety of diseases and conditions.

In case of a nutritional gap between what your body and what your diet provides, supplementing is necessary.

Whole foods are and will always be the best source of vitamins. This is because whole foods provide a combination of nutrients to the body including minerals and phytonutrients.

But many people do not receive all nutrients they need from their food choices. Either because they cannot or will not eat enough or they cannot or will not eat right foods. This may be because of medical or physical conditions, your lifestyles, i.e. job, time, availability or even due to poverty. In such cases, supplementation is necessary.

Vitamin Supplementing

Today's lifestyles make it more necessary to supplement our diets. There are groups of people to whom supplementing may really not be a choice. This depends on following:

1. **Sixty five years and older:** At this age some vitamins are not easily absorbed by your body system. Multivitamins may improve your immunity and lower risk of some infections.
2. **Post menopausal woman:** Supplement with calcium and vitamin D to help against osteoporosis.
3. If do not eat the recommended five servings of fruit and vegetables a day.
4. **If one smokes:** The tobacco interferes and diminishes vitamin uptake of vitamins like vitamin B_6, vitamin C, folic acid and even niacin. Ps: This will not make up for the health risk of smoking.
5. If one dirnks excessively that is more than a bottle a day. Long term excessive drinking will reduce absorption of vitamins. Vitamins affected are thiamin (Vitamin B), Folic acid, Vitamin A & D, as well as Vitamin B_{12}.

Selection of Vitamins

Bioavailability of the vitamin: Choose vitamin supplements whose source of ingredients is from the human food chain. Many vitamins are extracted from "natural sources", like algae, which we do not normally eat. This may not be easily available to your body.

Wide spectrum of vitamins: Prefer to use multivitamins that give a number of positively interacting vitamins. Remember we require all nutrients in varying quantities. It is better to have a multivitamin providing daily values instead of megadoses of a single or few vitamins.

Expiry date of the vitamin: Vitamins do expire. Buy to use only for the month. When you need more you can always purchase again.

Storage: Vitamins should be stored safely. Store them in a cool dry place away from sunlight and children.

Health concern: If you have a health condition e.g diabetes or having medication, please first consult your physician.

CLASSIFICATION OF VITAMINS

The two types of vitamins are classified by the materials in which they will dissolve.

1. **Water-soluble vitamins:** They dissolve in water and are not stored; they are eliminated in urine. We need a continuous supply of them in our diets.
2. **Fat-soluble vitamins:** Vitamins A, D, E and K dissolve in fat before they are absorbed in the blood stream to carry out their functions. Excesses of these vitamins are stored in the liver. Because they are stored, they are not needed every day in the diet.

Water Soluble Vitamins

B-complex vitamins and vitamin C are water-soluble vitamins that are not stored in the body and must be replaced each day. These vitamins are easily destroyed or washed out during food storage and preparation. The B-complex group is found in a variety of foods: cereal grains, meat, poultry, eggs, fish, milk, legumes and fresh vegetables. Citrus fruits are good sources of vitamin C. Use of megadoses of vitamins is not recommended.

Vitamin C also called citric acid, Vitamin B_1 (thiamine), Vitamin B_2 (Riboflavin), Vitamin B_3 (niacin), Vitamin B_5 (pantothenic acid), Vitamin B_9 (folic acid), Vitamin B_6 (pyridoxine), Vitamin B_{12} (cobalamin), Cholin, Biotin. Water soluble vitamins are not stored in the body in any appreciable amounts. The surplus is washed out mainly through urine. They requires consistent replenishing using the diet we consume. This makes them safe because they do not collect in the body toxic levels, making large doses of vitamin supplementary safe. But caution should still be practiced because mega doses have side effects and even can be fatal.

For example, there is a low risk of vitamin toxicity from nicotinic acid with mega doses. Nicotinic acid a derivative of vitamin niacin, one may experience flushing, itching, nausea and vomiting, liver cell damage. So supplement with daily value doses. Do not use megadoses unless under ongoing expert advice.

The water-soluble vitamins are the B-complex group and vitamin C. Water-soluble vitamins are easily destroyed or washed out during food storage or preparation. Proper storage and preparation of food can minimize vitamin loss. To reduce vitamin loss, refrigerate fresh produce, keep milk and grains away from strong light, and use the cooking water from vegetables to prepare soups.

Vitamin B-Complex

Eight of the water-soluble vitamins are known as the B-complex group: thiamin (vitamin B_1), riboflavin (vitamin B_2), niacin, vitamin B_6, folate, vitamin B_{12}, biotin and pantothenic acid. These vitamins are widely distributed in foods. Their influence is felt in many parts of the body. They function as coenzymes that help the body obtain energy from food. They also are important for normal appetite, good vision, healthy skin, healthy nervous system and red blood cell formation.

Vitamin B_1: It is water soluble. Vitamin B_1 are thiamine, antiberiberi factor, aneurine and antineuritic factor. Thiamine pyrophosphate is the biochemically active ester of thiamine, involved in many important metabolic processes including: (1) The decarboxylation of alpha-oxoglutaric acid in the citric acid cycle and (2) The conversion of alanine to pyruvic acid then to acetyl coenzyme A.

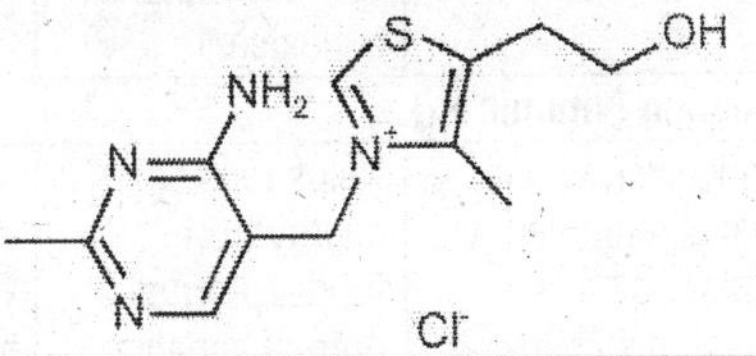

Vitamin B_1 (Thiamine hydrochloride).

Although thiamine occurs widely in foods, it is generally in small amounts. The principal source of thiamine is in dried brewers yeast and there are other good sources including: meats (pork, lamb, poultry, whole grain cereals (the thiamine is present in the germ of the grain), eggs, nuts and legumes.

Thiamine is essential in the body for the releasing of energy from carbohydrate and fat, and is also involved in conduction of nerve impulses. It is not stored well in the body (highest concentrations being in the brain, heart, liver and kidney) and so a continuous supply is required. Deficiency results in two diseases:

1. **Beriberi** is the disorders of the nervous and cardiovascular system. It may be Dry Beriberi (polyneuropathy with severe muscle wasting), Wet Beriberi (characterised by anorexia, muscle weakness, edema, mental confusion and heart failure) and Infantile Beriberi (sudden appearance of vomiting, convulsions, abdominal distention and anorexia).

2. **Wernicke-Korsakoff syndrome** results from a combination of factors including: Inadequate intake, Decreased absorption and Increased requirements.

When grains and grain products are refined, essential nutrients lost during processing are put back into these foods through a process called enrichment. Among the nutrients added during the enrichment process are thiamin, niacin, riboflavin, folate and iron. Some examples of enriched grain products are white rice, many breakfast cereals, white flour, breads, and pasta.

For some populations, rice is the main dietary staple. When "polishing" rice (removing its outer layers) became popular, thiamin deficiency, or beriberi, increased significantly.

TABLE 25.1: Water-soluble vitamins and their characteristics

Common food sources	Major functions	Deficiency symptoms	Overconsumption symptoms	Stability in foods
Vitamin C (abscorbic acid)				
Citrus fruits, broccoli, strawberries, melon, green pepper, tomatoes, dark green vegetables, potatoes.	Formation of collagen (a component of tissues), helps hold them together; wound healing; maintaining blood vessels, bones, teeth; absorption of iron, calcium, folacin; production of brain hormones, immune factors; antioxidant.	Bleeding gums; wounds don't heal; bruise easily; dry, rough skin; scurvy; sore joints and bones; increased infections.	Nontoxic under normal conditions; rebound scurvy when high doses discontinued; diarrhea, bloating, cramps; increased incidence of kidney stones.	Most unstable under heat, drying, storage; very soluble in water, leaches out of some vegetables during cooking; alkalinity (baking soda) destroys vitamin C.
Thiamin (vitamin B$_1$)				
Pork, liver, whole grains, enriched grain products, peas, meat, legumes.	Helps release energy from foods; promotes normal appetite; important in function of nervous system.	Mental confusion; muscle weakness, wasting; edema; impaired growth; beriberi.	None known.	Losses depend on cooking method, length, alkalinity of cooking medium; destroyed by sulphite used to treat dried fruits such as apricots; dissolves in cooking water.
Riboflavin (vitamin B$_2$)				
Liver, milk, dark green vegetables, whole and enriched grain products, eggs.	Helps in releasing energy from foods; promotes good vision, healthy skin.	Cracks at corners of mouth; dermatitis around nose and lips; eyes sensitive to light.	None known.	Sensitive to light; unstable in alkaline solutions.
Niacin (nicotinamide, nicotinic acid)				
Liver, fish, poultry,	Energy production	Skin disorders;	Abnormal liver	

Common food sources	Major functions	Deficiency symptoms	Overconsumption symptoms	Stability in foods
meat, peanuts, whole and enriched grain products.	from foods; aids digestion, promotes normal appetite; promotes healthy skin, nerves.	diarrhea; weakness; mental confusion; irritability.	function; cramps; nausea; irritability.	
Vitamin B_6 (pyridoxine, pyridoxal, pyridoxamine)				
Pork, meats, whole grains and cereals, legumes, green, leafy vegetables.	Aids in protein metabolism, absorption; aids in red blood cell formation; helps body use fats.	Skin disorders, dermatitis, cracks at corners of mouth; irritability; anemia; kidney stones; nausea; smooth tongue.	None known.	Considerable losses during cooking.
Folacin (folic acid)				
Liver, kidney, dark green leafy vegetables, meats, fish, whole grains, fortified grains and cereals, legumes, citrus fruits.	Aids in protein metabolism; promotes red blood cell formation; prevents birth defects of spine, brain; lowers homocystein levels and thus coronary heart disease risk.	Anemia; smooth tongue; diarrhea.	May mask vitamin B_{12} deficiency (pernicious anemia).	Easily destroyed by storing, cooking and other processing.
Vitamin B_{12}				
Found only in animal foods: meats, liver, kidney, fish, eggs, milk and milk products, oysters, shellfish.	Aids in building of genetic material; aids in development of normal red blood cells; maintenance of nervous system.	Pernicious anemia, anemia; neurological disorders; degeneration of peripheral nerves that may cause numbness, tingling in fingers and toes.	None known.	
Pantothenic acid				
Liver, kidney, meats, egg yolk, whole grains, legumes; also made by intestinal bacteria.	Involved in energy production; aids in formation of hormones.	Uncommon due to availability in most foods; fatigue; nausea, abdominal cramps; difficulty sleeping.	None known.	About half of pantothenic acid is lost in the milling of grains and heavily refined foods.
Biotin				
Liver, kidney, egg yolk, milk, most fresh vegetables,	Helps release energy from carbohydrates;	Uncommon under normal circumstances;	None known.	

Common food sources	Major functions	Deficiency symptoms	Overconsumption symptoms	Stability in foods
also made by intestinal bacteria.	aids in fat synthesis.	fatigue; loss of appetite, nausea, vomiting; depression; muscle pains; anemia.		

TABLE 25.2: The major food sources of water-soluble vitamins

Vitamin	Grains	Fruits	Vegetables	Meats, Eggs	Legumes, Nuts, Seeds	Milk, Dairy
Thiamin	X			X	X	
Riboflavin	X					X
Niacin	X			X	X	
Biotin			X	X	X	
Pyridoxine	X			X		X
Pantothenic acid	X	X	X	X	X	X
Vitamin B_{12}				X		X
Folate	X		X			
Vitamin C		X	X			

Vitamin B₂

The earlier names for this water soluble vitamin were lactoflavin, ovoflavin, hepatoflavin and verdoflavin, indicating the sources (milk, eggs, liver and plants) from which the vitamin was first isolated. Riboflavin is often used as a colourant because of its bright yellow colour and is sensitive to light but is heat stable.

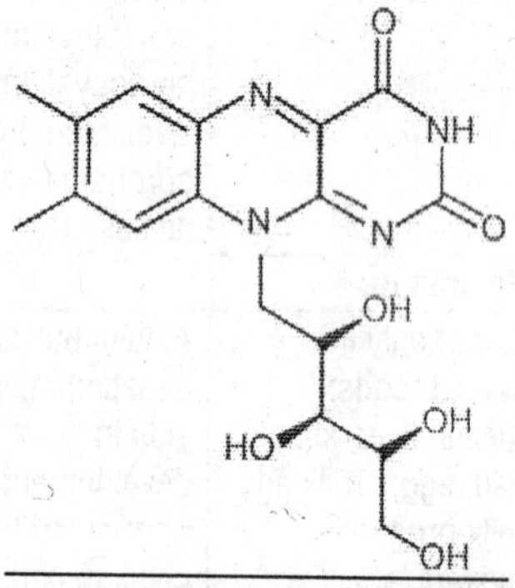

Chemical Structure of Riboflavin (Vitamin B_2).

Biosynthesis: Vitamin B_2 or riboflavin is an intermediary the transfer of electrons in the cellular oxidation-reduction reactions which generate energy from protein, carbohydrate and fat. The riboflavin coenzymes are also important for the transformation of vitamin B_6 and folic acid into their respective active forms, and for the conversion of tryptophan into niacin. There are two methods of manufacture of riboflavin: (1) Fermentation and (2) Synthesis from oxylene, D-ribose and alloxan. (usually the processes used are refinements.

Riboflavin is the precursor of flavoproteins. They may be: **Flavin mononucleotide** which is produced in the mucosal cells of the intestine, FMN is a coenzyme and **Flavin adenine dinucleotide** which is a coenzyme and is synthesised in the liver.

Sources: Riboflavin is very widely occurring in nature, present in all animal and plant cells. However, there are few sources which are rich in riboflavin. The highest concentrations are in yeast and liver. The good sources are milk and milk products, egg whites, meats, leafy green vegetables and egg yolks.

Deficiency symptoms: Riboflavin causes deficiencies symoptoms like glossitis (magenta tongue), angular Stomatitis (fissures at the corner of the mouth), iItching and skin rash.

Vitamin B_3

Niacin (nicotinic acid and the amide derivative nicotinamide) is one of the water soluble B-vitamins. An earlier name for niacin was PP factor (pellagra-preventative factor). Nicotinic acid and nicotinamide are both stable to light, heat, air and alkali.

Biosynthesis: Niacin can be synthesised by the oxidation of 5-ethyl-2-methylpyridine. Niacin is synthesised by producing nicotinamide via 3-methylpyridine uses the starting products acrolein and ammonia. In the blood, brain, kidney and liver it is converted to the coenzymes nicotinamide adenine dinucleotide (NAD) and nicotinamide adenine dinucleotide phosphate (NADP), both of which are involved in the generation of energy in cells. Tryptophan is an amino acid which is a provitamin of niacin.

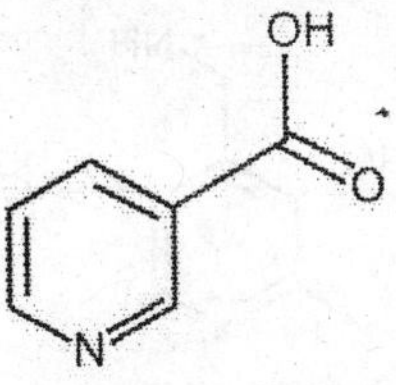

OH NH2 NH_3^+ CO_2^- N N N H

Nicotinic Acid 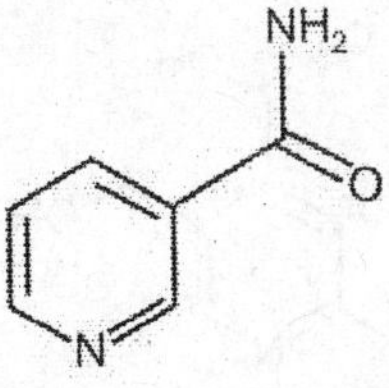 Nicotinamide 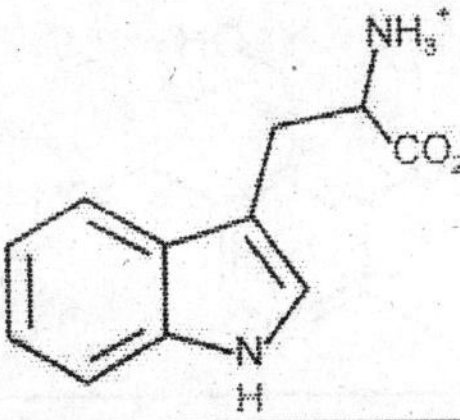 Tryptophan

Sources: Niacin is widely found in nature. Nicotinic acid is the predominant form in plants and nicotinamide in animals. It is commonly found in yeast extract, liver, meats, oily fish, nuts, legumes, green leafy vegetables, milk and milk products. The important dietary sources of tryptophan are meat, milk and eggs.

Deficiency symptoms: Deficiency of niacin results in a disease called Pellagra, the symptoms of which include: dermatosis, dementia, diarrhoea and nervous disorders which can lead to paralysis of the extremities.

Vitamin B_5

Pantothenic acid, one of the water soluble B-vitamins, is essential in the body as a constituent of coenzyme A. Pure pantothenic acid is a viscous hygroscopic oil that is chemically not very stable. In supplements it is usually included as calcium pantothenate (a solid, highly water soluble salt) or as pantothenol. Pantothenol is added to hair products as a moisturiser and conditioner. Pantothenic acid is easily destroyed by heat in acidic or alkaline conditions, but is stable in a neutral solution.

Biosynthesis: Coenzyme A (coenzyme of acetylation) is required for many acyl group transfer reaction pathways. Together with mercaptoethylamine, pantothenic acid makes up pantetheine. Pantetheine and adenosine-3',5'-diphosphate combine to give coenzyme A, which is important for synthesis of fatty acids, polyketides, terpenoids and steroids.

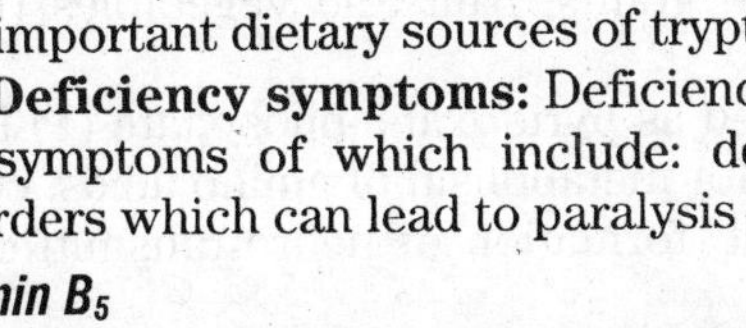

OH H N OH HO O O OH H N OH HO O O

Pantothenic Acid 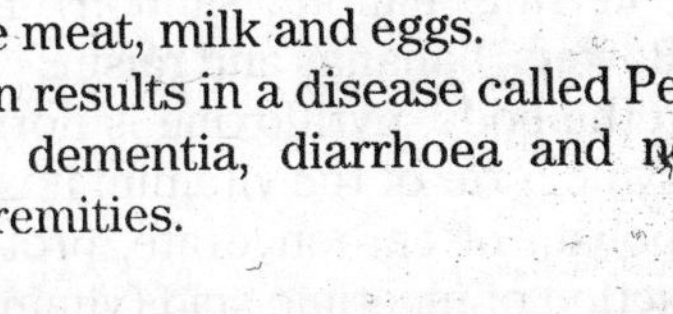 Pantothenol

It is synthesised by condensation of D-pantolactone with P-alanine produces pantothenic acid and the addition of a calcium salt forms colourless calcium pantothenate crystals. As part of coenzyme A, panthothenic acid is essential for Cellular metabolism and production of Sterols such as cholesterol, hormones, neurotransmitters (e.g. acetylcholine), phospholipids, porphyrin (constituent of haemoglobin) and antibodies.

Sources: Pantothenic acid is present in highest concentrations in organ meats such as heart, brain, liver, kidney, and yeast extract and in common dietary sources (dairy products, pulses, legumes, eggs, vegetables and whole grain cereals).

Vitamin B_6 (Pyridoxine)

Vitamin B_6, or pyridoxine is one of the water soluble B-vitamins. Vitamin B_6 is quite stable to heat but is sensitive to air, UV light and alkali. The full chemical name of pyridoxine is 2-methyl-3-hydroxy-4,5-bis(hydroxy-methyl) pyridine. By replacing the $-CH_2OH$ group on position 4 of the pyridoxine molecule with $-CH_2NH_2$ and -CHO respectively, two related compounds, pyridoxamine and pyridoxal can be formed which are also vitamin active.

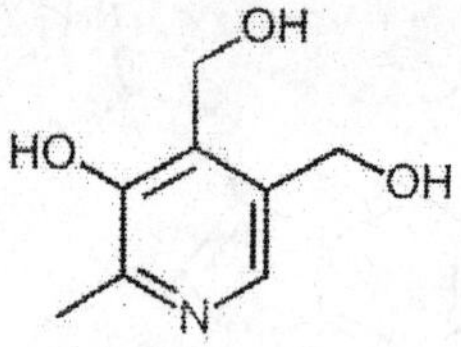

Pyridoxol (Pyridoxine) 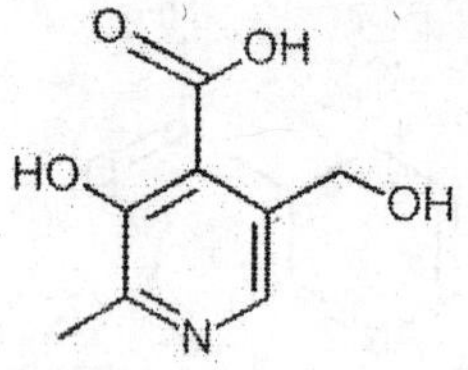 | Pyridoxal 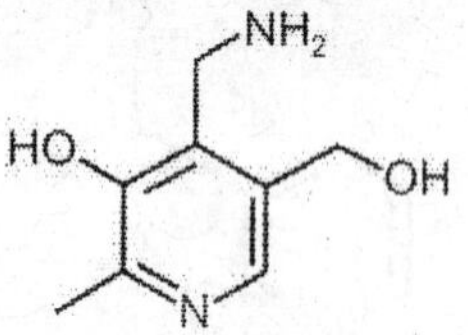 | Pyridoxamine

Sources: Vitamin B_6 is usually bound to protein, pyridoxol being the prominent form in plants, and pyridoxal and pyridoxamine in animal products. It is present in major dietary sources of pyridoxine like chicken, liver, yeast extract, fishes (tuna, trout, herring, halibut, salmon), nuts, whole grains, fruit and vegetables (beans, cauliflowers, bananas and raisins).

In the body, pyridoxine is normally stored as pyridoxal-5-phosphate (PLP), the coenzyme form of the vitamin. It is required for metabolism of amino acids, cellular metabolism of carbohydrate, protein and fat, formation of neurotransmitters and production of nicotinic acid (vitamin B_3).

Vitamin B_9

Folic Acid, or pteroyl-L-glutamic acid, is one of the water soluble B-vitamins. The other names for folic acid are folacin, vitamin Bc, vitamin B_9 and *Lactobacillus casei* factor. Folic acid in food is very unstable and considerable losses occur during short storage and cooking. Folic acid is essential in the body for the formation of new cells. It is involved in the metabolism of DNA and RNA, deoxyribonucleic acid and ribonucleic acid respectively, and is required for normal growth, development and fuctioning of the foetus, nervous system and bone marrow.

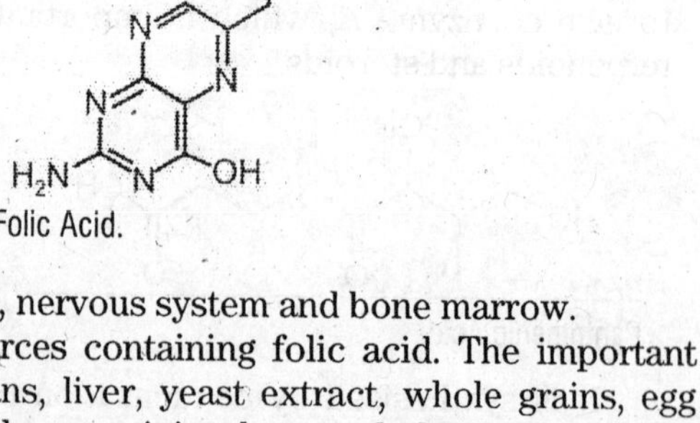

Folic Acid.

Sources: There are many food sources containing folic acid. The important sources are: green leafy vegetables, beans, liver, yeast extract, whole grains, egg yolk, milk and milk products, oranges and orange juice, beets, wholemeal bread and beer.

Deficiency Symptoms: Deficiency of folic acid can cause problems with absorption and metabolism. The important deficiency symptoms are loss of appetite, abdominal pain, sickness, diarrhoea, ulcers in the mouth, in pregnancy - premature birth and/or malformation and in children - growth retardation.

Vitamin B_{12}

The term vitamin B_{12} refers to a group of molecules called cobalamins which all contain cobalt. Vitamin B_{12} is stable to heat but is sensitive to light, oxygen, acid and alkali. Bacteria in the gut synthesise vitamin B_{12} but in different area to where absorption takes place, thus not providing a source of the vitamin. These include hydroxocobalamin, adenosylcobalamin (active coenzyme form), methylcobalamin (active coenzyme form) and cyanocobalamin (synthetic form of vitamin B_{12} that can be transformed by the body into the active coenzymes). The industrial production of vitamin B_{12} involves bacterial fermentation.

Like the other members of the B-complex, vitamin B_{12} is involved in the cellular metabolism of carbohydrate, protein and fat and in the production of blood cells in the bone marrow, nerve sheaths and proteins.

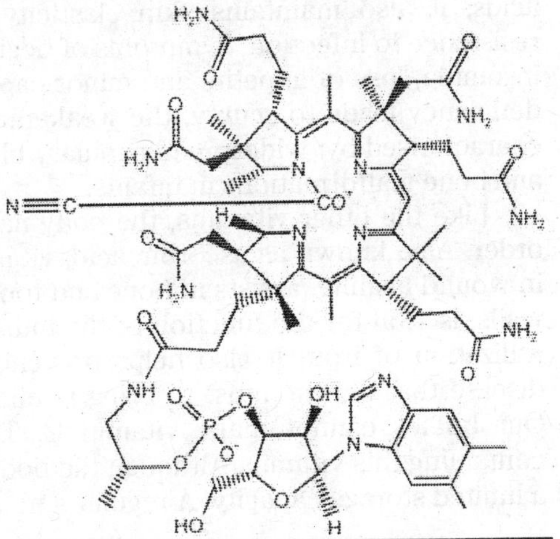

Sources: It is commonly found in animal products like Organ meats, Fish, Eggs and Dairy products.

Deficiency Symptoms: Vitamin B_{12} deficiency symptoms are Megaloblastic anaemia, neuropathy, defective DNA synthesis in cells and damage to the nervous system.

Vitamin B_{12}.

Vitamin C

Vitamin C, or ascorbic acid is one of the water soluble vitamins. Most animals can synthesise their own ascorbic acid within their liver, with the exception of fish, primates (including humans) and Guinea pigs. Vitamin C is commonly added to foods as an antioxidant to protect colour and aroma. Vitamin C in an alkaline solution can be used as a photographic developing agent. As a reducing agent, ascorbic acid is used industrially in metallurgy.

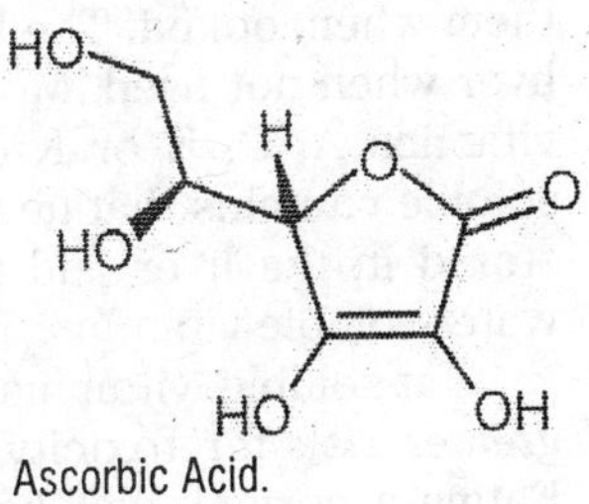

Biosynthesis: Vitamin C can be extracted from plant sources, such as rose hips or blackcurrants, but is more commonly synthesised from glucose. The biosynthetic pathway of L-ascorbic acid in animals is well understood, the plant pathway has remained unknown-one of the few primary plant metabolic pathways for which this is the case. L-ascorbate is abundant in plants (found at concentrations of 1-5 mM in leaves and 25 mM in chloroplasts) and may have roles in photosynthesis and transmembrane electron transport. D-mannose and L-galactose are efficient precursors for ascorbate synthesis and are interconverted by GDP-D-mannose-3,5-epimerase. In pea and *Arabidopsis thaliana*, L-galactose dehydrogenase, that catalyses oxidation of L-galactose to L-galactono-1,4-lactone. The ascorbate biosynthesis pathway involves GDP-D-mannose, GDP-L-galactose, L-galactose and L-galactono-1,4-lactone, and synthesise ascorbate from GDP-D-mannose by way of these intermediates in vitro.

Ascorbic Acid.

Sources: The common foods contain vitamin C, the best food sources are citrus fruits. The body needs Vitamin C due to environmental stress, use of certain drugs (such as oral contraceptives), tissue healing of wounds, growth (children and pregnant women), fever and infection, and smoking. Vitamin C supplements will not necessarily provide extra energy, clear up skin problems, or prevent and cure the common cold, heart disease, and cancer. Unlike pellagra, beriberi, or scurvy, these problems are not the result of a vitamin deficiency. What is known is that the only disease a vitamin will cure is the one caused by a deficiency of that vitamin. Vitamin C also serves as an antioxidant. It works with vitamin E as a free-radical scavenger.

Deficiency Symptoms: Vitamin C is vital for the production of Collagen. Collagen is the intercellular substance that gives bones, teeth, cartilage, blood vessels and muscles their structure. Ascorbic acid is needed for synthesis of bile acids; it also maintains skin elasticity, aids in iron absorption, and improves resistance to infection. Symptoms of beginning vitamin C deficiency include fatigue, insomnia, loss of appetite and minor capillary bleeding. A long period of vitamin C deficiency leads to scurvy, the weakening of body structures containing collagen, characterised by: widespread capillary bleeding, bleeding gums and loosening teeth and bone malformations in infants.

Like the other vitamins, the body needs vitamin C to keep it in good working order. Also known as ascorbic acid, vitamin C helps hold body cells together, aids in wound healing, assists in bone and tooth formation, strengthens the blood vessel walls, is vital for the function of the immune system, and improves absorption and utilization of iron. It also helps prevent nutritional ailments such as scurvy, the disease that did the most to bring public attention to vitamin deficiency diseases. Our bodies cannot make vitamin C. Therefore, every day we must eat foods containing this vitamin. Although the body has a constant need for vitamin C, it has a limited storage capacity. A regular and adequate intake is essential.

Fat Soluble Vitamins

Vitamin A, Vitamin D, Vitamin E and Vitamin K. Being fat soluble these vitamins are stored in the fat in our bodies. They can easily accumulate to toxic levels so if you choose to supplement be cautious. Small amounts of vitamins A, D, E and K are needed to maintain good health. Foods that contain these vitamins will not lose them whencooked. The body does not need these every day and stores them in the liver when not used. Most people do not need vitamin supplements. Megadoses of vitamins A, D, E or K can be toxic and lead to health problems. Unlike water-soluble vitamins that need regular replacement in the body, fat-soluble vitamins are stored in the liver and fatty tissues, and are eliminated much more slowly than water-soluble vitamins.

Fat-soluble vitamins can be stored for long periods. They generally pose a greater risk for toxicity than water-soluble vitamins when consumed in excess. Eating a normal, well-balanced diet will not lead to toxicity in otherwise healthy individuals. However, taking vitamin supplements that contain mega doses of vitamins A, D, E and K may lead to toxicity. Remember, the body only needs small amounts of any vitamin.

While diseases caused by a lack of fat-soluble vitamins are rare in the United States, symptoms of mild deficiency can develop without adequate amounts of vitamins in the diet. Additionally, some health problems may decrease the absorption of fat, and in turn, decrease the absorption of vitamins A, D, E and K. Consult your doctor about this.

Vitamin A

It is also known as **retinol**, has many functions in the body. In addition to helping the eyes adjust to light changes, vitamin A plays an important role in bone growth, tooth development, reproduction, cell division and gene expression. Also, the skin, eyes and mucous membranes of the mouth, nose, throat and lungs depend on vitamin A to remain moist.

Vitamin A can be produced by extraction from fish-liver oil although the more common process is synthesis from beta-ionone.

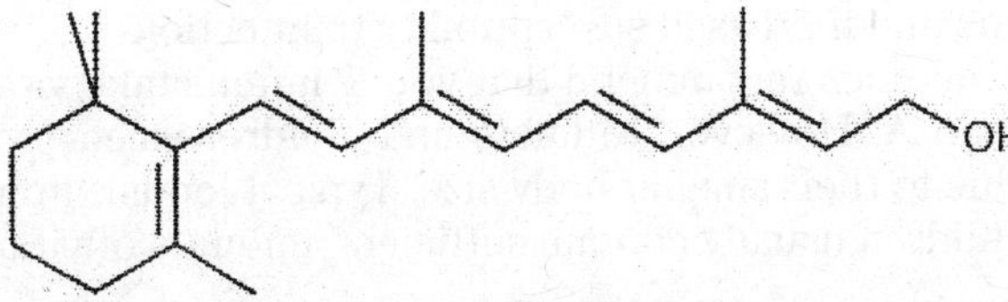

OH

Vitamin A: Chemical Structure of Retinol.

Retinol is sensitive to oxidation by the air. It is present in animal fats and oxidation of these can result in a loss of the fat soluble vitamin. Beta carotene is one of the more stable vitamins in foods. The compound is sensitive to heat but significant losses only occur after long periods of boiling.

Beta Carotene: Carotenoids in plants act as light-absorbing pigments during photosynthesis and also help to protect cells against photosensitization. Beta carotene, a hydrocarbon, is one of a family of dark pigments called provitamin A carotenoids, some of which can be converted to retinol by the body. In the liver, one molecule of beta carotene can be cleaved by an intestinal enzyme into molecules of retinol, so good food sources of beta carotene can also be good sources of retinol. The conversion of carotenoids to retinol will decrease when body stores are full thus preventing a retinol overdose, which can lead to toxic symptoms.

Principal sources of vitamin A, or retinol are liver, egg yolk, oily fish (herring, pilchards, tuna, sardines), whole milk, butter, cheese. The good sources of beta carotene are carrots, yellow and dark green leafy vegetables (e.g. spinach, broccoli, watercress) and herbs (basil, coriander, parsley, apricots, pumpkin, melon, sweet potatoes, mango, papaya, red pepper and guava).

The recommendation for vitamin A intake is expressed as micrograms (mcg) of retinol activity equivalents (RAE). Retinol activity equivalents account for the fact that the body converts only a portion of beta-carotene to retinol. One RAE equals 1 mcg of retinol or 12 mcg of beta-carotene.

Vitamin A in the Body: Vitamin A is essential in the body for maintaining the immune system and vision and also for growth and tissue differentiation. It is also thought to aid prevention of cancer.

- **Vision:** The retina of the eye is made up of light receptor cells, called rods, which allow us to 'see' light and dark. Within these cells is a pigment called rhodopsin which is a complex of opsin, a protein, and retinol. On exposure to light this complex disintigrates and releases electrical charges to the brain via the optic nerve. New rhodopsin is then formed from opsin and retinol.
- **Immune System:** Vitamin A aids health by maintaining the surface linings of the eyes and the intestinal, respiratory and urinary tracts, thus preventing bacteria from entering. It has also been suggested that vitamin A helps lymphocytes (a type of white blood cell) to function effectively, and that carotenoids may act as antioxidants in the body.

- **Tissue Differentiation:** In the body there are many different cells with many different functions. The process by which a cell becomes a specific type of cell is called differentiation, and vitamin A is essential for differentiation of epithelial cells (i.e those of the skin, the mucous membranes, the cornea and the blood vessel walls).

Deficiency of vitamin A can result in night blindness (impaired vision in dim light), Xerophthalmia (partial or total blindness due to damage of the cornea from a severe deficiency), Skin lesions, inhibited growth and bone development and loss of weight in the young and increased susceptibility to infection.

Physicians sometimes recommend that young infants take vitamin supplements that contain vitamin A. However, toddlers and children need protection from too much vitamin A due to their smaller body size. Typical foods eaten in large amounts by toddlers and children usually contain sufficient amounts of vitamin A.

TABLE 25.3: Fat soluble with sources

Vitamin	Source
Vitamin A (retinol) (provitamin A, such as beta carotene)	Vitamin A: liver, vitamin A fortified milk and dairy products, butter, whole milk, cheese, egg yolk.
Provitamin A.	Carrots, leafy green vegetables, sweet potatoes, pumpkins, winter squash, apricots, cantaloupe
Vitamin D	Vitamin D-fortified dairy products, fortified margarine, fish oils, egg yolk. Synthesised by sunlight action on skin.
Vitamin E	Vegetable oil, margarine, butter, shortening, green and leafy vegetables, wheat germ, whole grain products, nuts, egg yolk, liver.
Vitamin K	Dark green leafy vegetables, liver; also made by bacteria in the intestine.

Vitamin D

Vitamin D, or calciferol, is the general name for a collection of steroid-like substances including vitamin D_2, ergocalciferol and vitamin D_3, cholecalciferol as one of the fat soluble vitamins, vitamin D is stable in food storage, cooking and processing. Found only in animal sources, vitamin D is fat soluble and can be manufactured by the body on exposure to UV radiation.

Biosynthesis: In the body, the vitamin D is converted to its active hormone form by the liver and the kidney. The calciferol is converted to 25-hydroxycholecalciferol, calcidiol, and then metabolized by the kidney to several active forms including 1,25-dihydroxycholecalciferol, calcitriol. The main function of vitamin D in the body is to maintain blood levels of calcium and phosphorus. Vitamin D is essential for the absorption of these substances from the small intestine, and also for the mineralization process of the bones. Cholecalciferol is manufactured by irradiating 7-dehydrocholesterol with ultraviolet light. The production of ergocalciferol is similar, using a starting product of ergosterol, a yeast extract.

The result of vitamin D deficiency is rachitis (rickets) in children and osteomalacia in adults.

Vitamin D plays a critical role in the body's use of calcium and phosphorous. It increases the amount of calcium absorbed from the small intestine and helps form and maintain bones. Children especially need adequate amounts of vitamin D to develop strong bones and healthy teeth.

Sources: Vitamin D is only found naturally in animals and animal products; fruit and nuts contain no vitamin D at all. Richest sources of Vitamin D are Fish liver oils (halibut, cod, saltwater fish (herring, mackerel, salmon, sardines, eel), milk, butter, cheese, margarine, cream, yoghurt and egg.

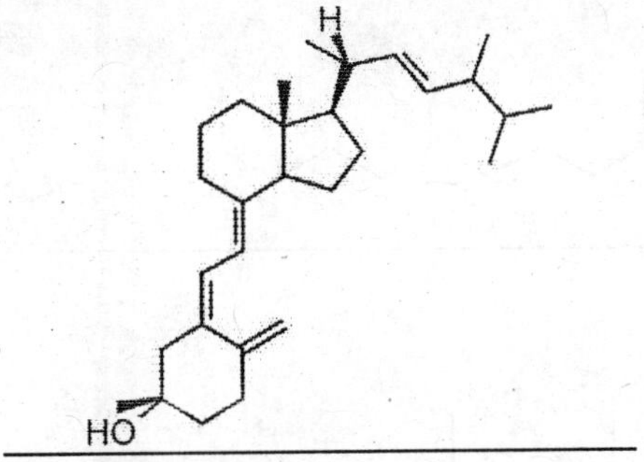

Vitamin D_2 (Ergocalciferol)

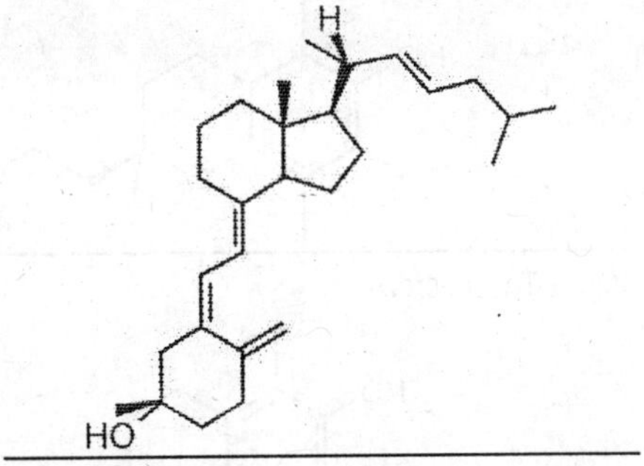

Vitamin D_3

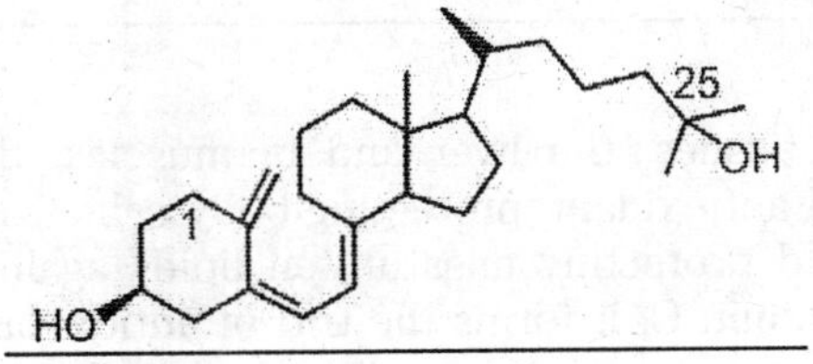

Calcidiol

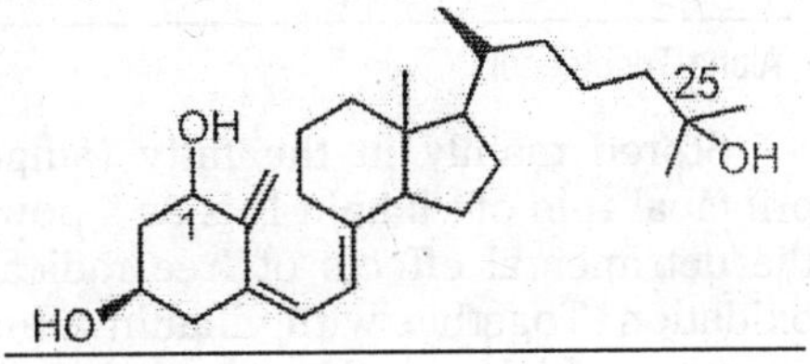

Calcitriol

Deficiency Symptoms: The symptoms of vitamin D deficiency in growing children include rickets (long, soft bowed legs) and flattening of the back of the skull. Vitamin D deficiency in adults is called osteomalacia, which results in muscular weakness and weak bones. High doses of vitamin D supplements coupled with large amounts of fortified foods may cause accumulations in the liver and produce signs of poisoning. Signs of vitamin D toxicity include excess calcium in the blood, slowed mental and physical growth, decreased appetite, nausea and vomiting.

It is important that infants and young children do not consume excess amounts of vitamin D regularly. Children exposed to the sun for 5 to 10 minutes daily will produce enough vitamin D. An **Adequate Intake** (AI) for has been established for vitamin D. The AIs for vitamin D appear as micrograms (mcg) of cholecalciferol. Ten mcg of cholecalciferol equals 400 International Units (IU).

Vitamin E

The name Vitamin E covers a collection of eight fat soluble compounds, tocopherols (methyl derivatives of tocol) and tocotrienols. These are: alpha-Tocopherol (the most common and biologically active (5,7,8,-trimethyltocol), beta-Tocopherol (5,8,-trimethyltocol), gamma-Tocopherol (7,8,-trimethyltocol), delta-Tocopherol (8,-trimethyltocol), alpha-Tocotrienol, beta-Tocotrienol, gamma-Tocotrienol, and delta-Tocotrienol.

The word tocopherol comes from the Greek words *tocos*= childbirth and *pherein* = to bring forth, after the essential role vitamin E plays in animal reproduction. The ending -ol is the standard chemical name ending for an alcohol. The antioxidant property of vitamin E is exploited when it is used as a stabliser in oil and fat containing foods. Vitamin E is used in cosmetics and skin products to prevent cell damage by UV light. Tocopherol is used in many pharmaceutical products as a stabilizer. Some plastics, technical oils and greases contain alpha-Tocopherol as an antioxidant. Vitamin E is sensitive to heat, light and oxygen and significant losses have been found after relatively short times of food storage.

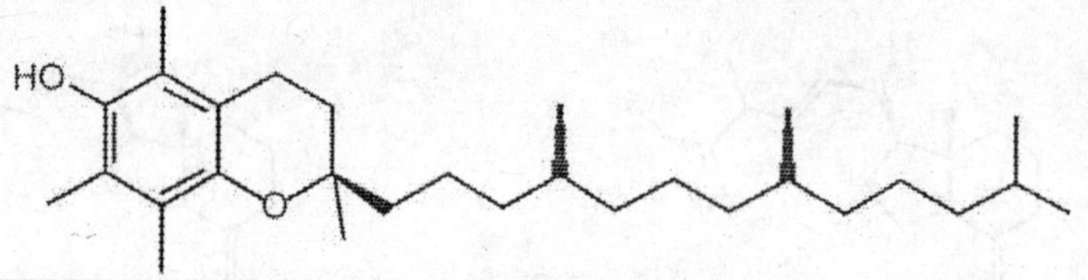

Alpha-Tocopherol.

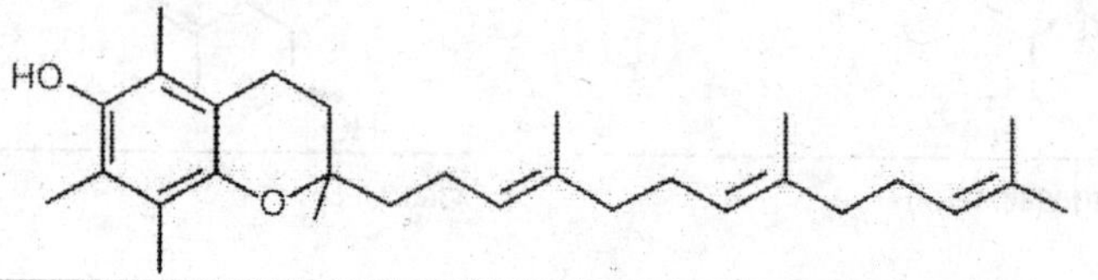

Alpha-Tocotrienol.

Stored mainly in the fatty (adipose) tissues, the liver and in muscles, the principal role of vitamin E is as a powerful antioxidant, protecting body cells from the detrimental effects of free radicals and protecting unsaturated lipids against oxidation. Together with vitamin A and vitamin C, it forms the trio of antioxidant vitamins which are thought to help prevent cancer and cardiovascular disease. Vitamin E is also important in prolonging the life of red blood cells, helping the body make the best use of vitamin A, protecting body membranes and reproduction.

Biosynthesis: Vitamin E derived from natural sources is obtained by molecular distillation and, in most cases, subsequent methylation and esterification of edible vegetable oil products. Synthetic vitamin E is produced from fossil plant material by condensation of trimethylhydroquinone with isophytol.

Sources: The most important sources of vitamin E are Vegetable Oils, including (soya, palm, corn, safflower, sunflower, wheat germ, nut oils), nuts, seeds, whole grains, leafy green vegetables, chick peas, avocados, sweet potatoes, sweetcornm, red peppers, carrots, parsnips, milk, eggs and cheese.

Vitamin E acts as an antioxidant, protecting vitamins A and C, red blood cells and essential fatty acids from destruction. Use of antioxidant supplements, vitamin E in particular, may help to prevent heart disease and cancer. However, newer findings indicate that people who take antioxidant supplements are not better protected against heart disease and cancer than non-supplement users. On the other hand, there are many studies that show a link between regularly eating antioxidant-rich fruits and vegetables and a lower risk for heart disease, cancer and several other diseases. Large doses of vitamin E pose a hazard to people who take blood-thinning medications. People taking statin drugs are also not advised to take supplemental vitamin E because it may interfere with how the medication works.

Deficiency Symptoms: Depletion of body stores of vitamin E takes a long time so deficiency is rare.

Vitamin K

There are three forms of the fat soluble vitamin K. The first two, vitamin K_1 (phylloquinone) and vitamin K_2 (a group of compounds called menaquinones) are found naturally and the third, vitamin K_3 (menadione) is a synthetic compound which is converted to vitamin K_2 in the intestine. Vitamin K_1 is 2-methyl-3-phytyl-1,4-naphthoquinone and is found in plants. The Menaquinones differ in the number of isoprene units in the side chain and in their degree of unsaturation, and are synthesised by bacteria in the human and animal intestines.

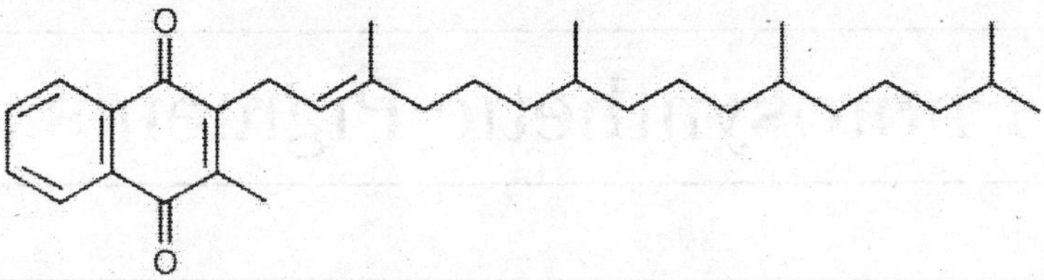

Vitamin K_1 (phylloquinone).

Sources: Principal dietary sources of vitamin K are Green leafy vegetables (spinach, broccoli, cabbage, soya beans, liver, green tea) egg yolks, oats, whole wheat, potatoes, tomatoes, asparagus, butter, cheese, some meats, milk, fruits and vegetables.

Deficiency Symptoms: Vitamin K is essential for blood clotting, as it is involved in production of the protein prothrombin, which converts soluble fibrinogen in the blood to insoluble fibrin which makes up the greater part of blood clots. Deficiency of vitamin K is rare due to the bacterial synthesis in the gut. Naturally produced by the bacteria in the intestines, vitamin K plays an essential role in normal blood clotting and helps promote bone health. Without sufficient amounts of vitamin K, hemorrhaging can occur. Deficiencies may appear in infants, or in people who take anticoagulants or antibiotic drugs. Newborn babies lack the intestinal bacteria to produce vitamin K and need a supplement for the first week. People on anticoagulant drugs (blood thinners) may become deficient in vitamin K. People taking antibiotics may lack vitamin K temporarily because intestinal bacteria are sometimes killed as a result of long-term use of antibiotics. People with chronic diarrhea may have problems absorbing sufficient amounts of vitamin K through the intestine and should consult their physician to determine if supplementation is necessary.

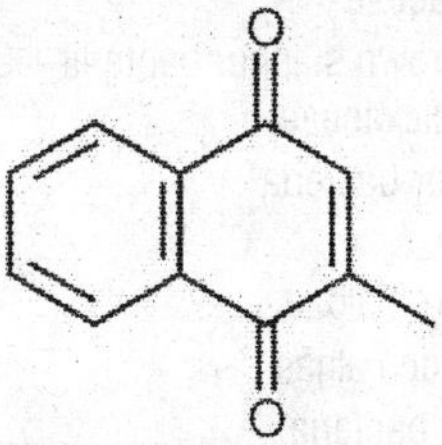

Menadione - the base structure of Vitamin K_3.

Although a tolerable upper intake level has not been established for vitamin K, excessive amounts can cause the breakdown of red blood cells and liver damage. Large doses are not advised.

CHAPTER 26

Photosynthetic Pigments

At global level, it has been estimated that each year the total chlorophyll production across the planet exceed 10^9 tonnes with 75 per cent from terrestrial plants, the reminder from aquatic organisms, largely marine phytoplankton. Production of carotenoids, the second most abundant plant pigment, is probably of the order of about 2×10^8 tonnes. The distribution of the various chlorophylls and bacteriochlorophylls is shown in following table:

TABLE 26.1: The distribution of chlorophylls and bacteriochlorophylls

Pigment	Organism	Light absorbed
Chlorophyll a	All plants, algae and the photosynthetic bacteria Cyanophyta and Prochlorophyta	Red and Blue-violet
Chlorophyll b	All plants, the algal Chlorophyta and Euglenophyta and bacterial Prochlorophyta	Red and Blue violet
Chlorophyll c	Algal Phaeophyta (brown algae), Pyrrophyta (dinoflagellates), Bacillariophyta (diatoms), Chrysophyta, Prasinophyta, Cryptophyta	Red and Blue violet
Chlorophyll d	Some Rhodophyta (red algae) and Chrysophyta	Red and Blue-violet
Chlorophyll e	Algal Xanthophyta	Near-red and Blue-violet
Bacteriochlorophyll a and b	Purple bacteria including Chromatiaceae and Rhodospirillaceae	Near-red and Blue-violet
Bacteriochlorophyll c, d and e	Green and brown sulphur bacteria – Chlorobiaceae and Chloropflexaceae	
Bacterioviridin	Green sulphur bacteria	Near-red and Blue-violet
Phycocyanin	Blue-green, red algae	Orange red
Phycoerythrin	Red, blue-green algae	Green
Carotenoids	Most plants, bacteria	Blue, Blue-green

TABLE 26.2: Chemical composition of intact spinach chloroplasts

Component	Percentage Chloroplast (Dry Weight)
Total Protein	60
Water insoluble protein	27
Water soluble protein	33
Total lipids	20
Chlorophylls	4.3
Carotenoids	0.9
Colourless lipids	14.8
RNA	1.7-3.5
DNA	?
Free Amino acids	2.5

Component	Percentage Chloroplast (Dry Weight)
Orthophosphatase	0.7
Ions (K^+, Na^+, Mg^{2+}, Ca^{2+}, (Cl^-)*	3.6
Polysaccharides**	0.6
Hexosamine	0.2
Starch	variable

*Data of *Beta vulgaris* chloroplast, ** Data of *Allium porrum* chloroplast.

TABLE 26.3: Lipid composition of spinach chloroplasts (based on 100 mol chlorophyll)

Lipid	Mol.
Chlorophylls	100
Chlorophyll *a*	70
Chlorophyll *b*	30
Carotenoids	39
ß-carotene	9
Lutin	14
Viloxanthin	10
Neoxanthin	4
Antheraxanthin	2
Quniones	11
Plastoquinone	4
Lipid	Mol.
α-tocopherol	0.5
Vitamin K_1	0.5
Tocoquinone	0.5
Glycolipids	447
Monoglactosyldiglyceride	270
Digalactosydiglyceride	154
Sulpholipid	23
Phospholipids	43
Phosphatidylglycerine	29
Phosphatididylethanolamine	2
Phosphatidylinosite	2

Chlorophyll a is found in all photosynthetic eukaryotes. The restricted distribution of the other chlorophylls (b, c, d, or e) may have evolutionary significance.

Most biological pigments are relatively large molecules with, typically, molecular masses ranging from about 200 (quinines), 300 (flavonoids), 400 (betalains) and 500 (carotenoid) to 800 (chlorophylls). Most pigments are composed, at least in part of unsaturated hydrocarbons. Many are ring (cyclic) structures and of the 17 or so elements found in pigments of biological materials, just four: H, C, N and O predominate.

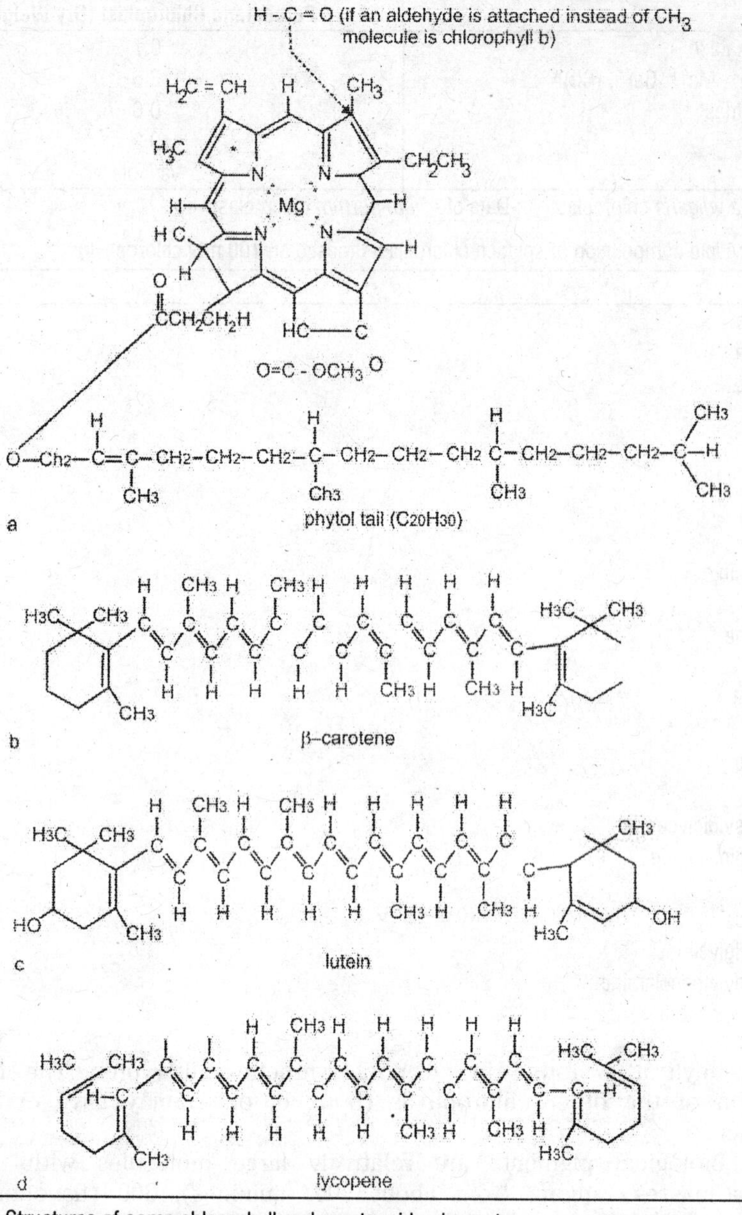

H - C = O (if an aldehyde is attached instead of CH_3 molecule is chlorophyll b)

a — phytol tail ($C_{20}H_{30}$)

b — β-carotene

c — lutein

d — lycopene

FIG. 26.1: Structures of some chlorophyll and carotenoids pigments.

PROPERTIES OF CHLOROPHYLL MOLECULE

The empirical formula of the chlorophyll molecule is $C_{55}H_{72}O_5N_4Mg$. The molecule is made up of five small rings, together forming a larger ring. In the centre of the large ring there is a magnesium atom. Together they form a flat square measuring 1.5 × 12.5 nm. The square is equipped with a long *tail* of carbon atoms which can be thought to act as a handle or fixture which keeps the molecule properly positioned in the plant.

There are many double bonds between the atoms in the ring system. It is possible to circle the large ring in such a way that every second bond encountered is a double bond. Such a system of bonds is termed a *closed conjugated system.* It is this that provides the molecule with two important properties:

- Great light-absorbing power, and
- Great power of stability.

The tail, the magnesium atom and some peripheral atoms may break or loose if the molecule is denatured with heat, acids, etc., but the ring system is intact even in old fossils such as petroleum. A similar ring system as in chlorophyll is present in our red blood pigment, haemoglobin and in the cytochromes.

In addition to the pyrrole rings, *a-b* chlorophyll has a cyclopentanone ring with a methyl ester. A phytol ester is linked by a propionic acid with pyrrole ring D. Chlorophyll *a* is having formyl group at C atom 7 instead of a methyl group. Other chlorophylls, such as chlorophyll C_1 and C_2 are found in brown algae and chlorophyll *d* in red algae.

Photoautotrophic bacteria possess bacteriochlorophyll *a*, which differs from chlorophyll in two features: it is hydrated in ring B and carries a vinyl group instead of the acetyl group on ring A. The green photobacterium, *Chlorobium* has been found to contain the bacteriochlorophylls *c*, *d* and *e* which have formerly been called *Chlorobium* chlorophylls.

Chlorophyll Synthesis

Much progress has been made in our understanding of the biosynthesis of chlorophyll with much of the recent inputs coming from enzymology and molecular biology. Nevertheless, although the main pathway leading to Chl a is fairly well understood, our knowledge of the biosynthetic interrelationship of the different Chl and their intermediates, formation of bChl, regulation and degradation is still limited.

Most of the recent work on the formation of 5-aminolevulinic acid (ALA) has been concerned with regulatory aspects and enzymologic studies of the C_5 pathway utilized in plants, algae and some bacteria for ALA synthesis destined for Chl. The sequence of events leading to the formation of ALA is separated by the subsequent action of the three enzymes involved: glutamyl-t-RNA synthetase, glutamyl-t-RNA reductase and glutamate-1-semialdehyde (GSA) aminotransferase.

The first biosynthetic step involves charging of glutamate with a glutamate-specific t-RNA by the aminoacyl-tRNA synthetase. The enzyme has been isolated and purified from *Chlamydomonas reinhardtii*. It has a native molecular weight of 60 KDa and migrates under denaturating conditions as a single band of 32 KDa, indicating an $\propto_2$ homodimer structure. The same t-RNA is utilized for protein biosynthesis and ALA formation.

The charged Glu (glumate) is next reduced to GSA by an NADPH-dependent glutamate-tRNA reductase. The enzyme was isolated from C. reinhardtii and shown to be a monomer of 130 KDa. The final step in ALA formation is the conversion of GSA by aminotransferase-type enzyme that has been isolated from barley and *Synechococcus* as 43 KDa protein.

The barley enzyme has been cloned and sequenced and was found to be a dimer of identical subunits with a molecular weight of 46 KDa. The enzymes from barley and *Synechococcus* had 72% homology. The enzyme from *C. reinhardtii* was found to be a monomer of 43 KDa, which in the presence of pyridoxal-5'-phosphate converted GSA to ALA. The involvement of DOVA (4,5-dioxovaleric acid) in the biosynthesis of ALA has also been proposed for *Scenedesmus*, where it is a better precursor of ALA than GSA.

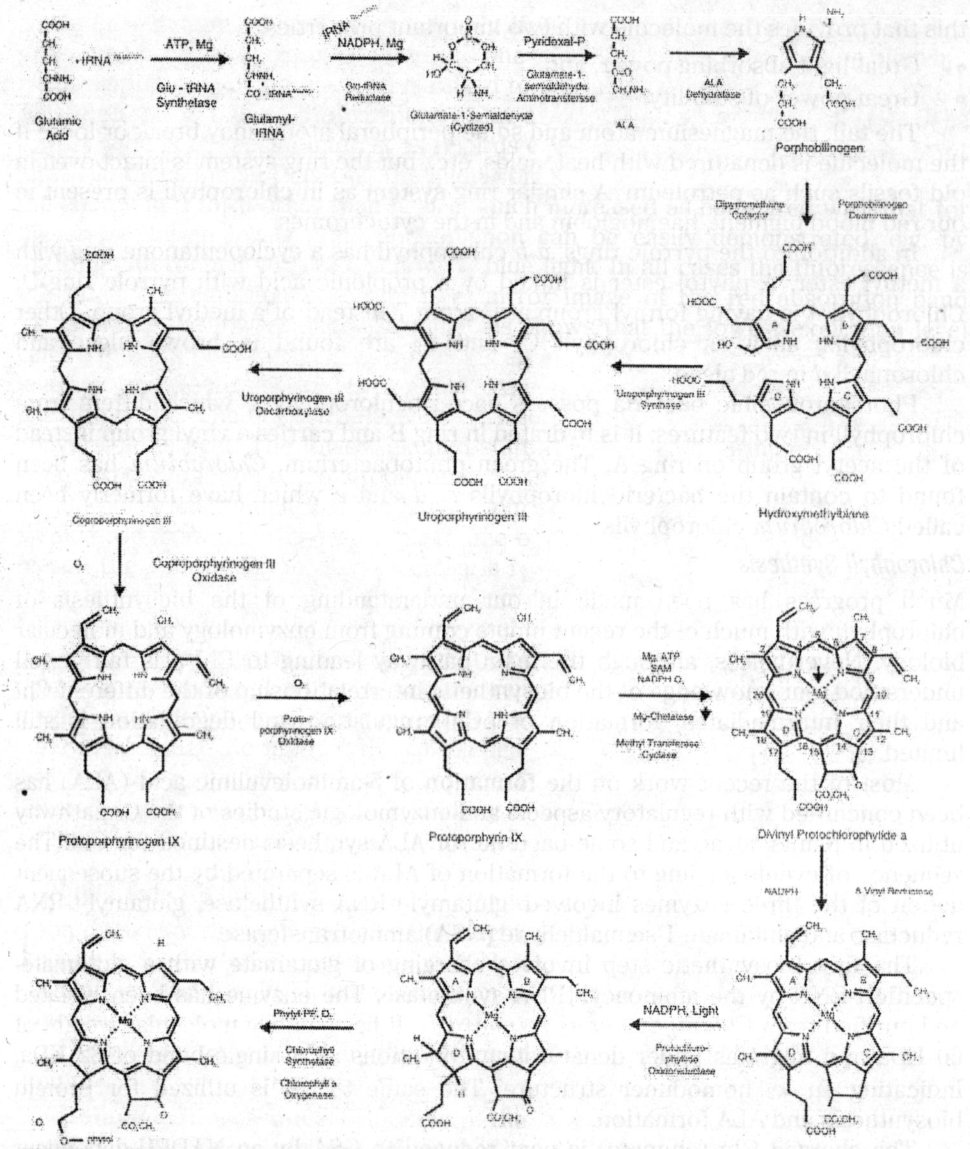

FIG. 26.2: The biosynthetic pathway of chlorophyll.

As a next step, two molecules of ALA are condensed to yield porphobilinogen. This reaction is catalysed by ALA dehydratase. Then, four molecules of porphobilinogen are converted into hydroxymethylbilane by porphobilinogen deaminase. Subsequent reactions include ring closure and simultaneous isomerization of the acetyl and propinoyl groups at the pyrrole ring D, leading to uroporphyrinogen III. This compound is then converted into coproporphyrinogen, followed by two successive oxidations which eventually lead to protoporphyrin IX. In the first reaction, the acetic acid side chains of uroporphyrinogen III are shortened to methyl groups by decarboxylation. Then, two propionic acid side chains are trimmed into vinyl groups, followed by oxidation to establish the system of conjugated double bonds in protoporphyrin IX.

At protoporphyrin IX there is a branch point in the C_5 pathway. One route is directed to the synthesis of heme and phytochromobilin whereas the other branch gives rise to the formation for Chl. Ferrochelatase and Mg-chelatase respectively catalyse the first unique reactions in the Fe and Mg branches of the C_5 pathway. In the Mg branch subsequent esterification of Mg-protoporphyrin IX leads to Mg-protoporphyrin IX monomethyl ester followed by the formation of the isocyclic ring of the macrocycle. 3,8-Divinyl P chlide a as the final product of these reactions is then reduced at the vinyl group at C_8 to yield 3-monovinyl P chlidea. Subsequently, the double bond between C_{17} and C_{18} in ring D of the macrocycle is reduced to yield chlide a. The final steps of chl biosynthesis include the esterification of chlide a with phytol or earlier alcohal precursors which must then be reduced to the stage of the phytyl derivative, and oxygenation of the Chla to Chlb.

Biosynthesis of Bacterial Chlorophyll

Not much progress has been made in studies on the biosynthesis of the bchl, the main reason being the difficulties in finding conditions for satisfactory incorporation of labeled precursors and the development of cell-free enzyme systems.

The first bacterial biosynthetic gene to be sequenced was the one coding for 3-desacetyl-3-hydroxyethyl bchlide dehydrogenase from *Rhodobacter capsulatus*. It was shown to be a 33 KDa protein, which is not an integral membrane protein. From the same organism the genes coding for the activity that converts proto to Mg-proto monomethyl ester, pchlide to chlide and Mg-proto monomethyl ester to pchlide were identified by site-directed mutagenesis. The ALA-synthase enzyme was characterized subsequently by cloning and sequencing of the hemA gene from *R. capsulatus*. A pps gene has been isolated from *R. sphaeroides* that suppresses both bchl and carotenoid synthesis in *R. sphaeroids* and *R. capsulatus*, suggesting a coupled regulation of bchl and carotenoid biosynthesis.

Physical Properties of Chlorophylls

The chlorophyll is a dark green waxy solid; but chlorophyll *a* is bluish black while chlorophyll *b* is dark-green solid. Both chlorophylls are soluble in organic solvents giving green solution; chlorophyll *a* melts between 117-129° C and chlorophyll *b* between 120-130° C.

Chlorophyll *b* contains one oxygen atom more and two hydrogen atoms less than the chlorophyll *a*. The difference is due to the presence of -CHO group on position-3 in place of methyl group.

Ordinary pigments preferentially absorb visible light of a certain spectral region and, upon de-excitation release the absorbed energy as heat and, possibly, fluorescence. In contrast, a photopigment can be defined as a chemical compound, capable of light absorption and subsequent conversion of excitation quanta into a form of energy to be used in some physical or chemical process. In the following description only photopigments of biological interest will be considered.

Apart from heat production and fluorescence emission, two phenomena may happen upon excitation: (1) converted energy is stored and used later on for driving energy requiring processes at a given moment, or (2) this energy is instantaneously used for chemical or physical processes. In case (2) there are three possibilities:

1. Energy conversion is used for triggering a reaction,
2. This conversion results in the transformation of the photopigment into a catalyst and
3. A product suitable for *use* in chemical reactions is generated by the energy conversion.

Light quantum can be absorbed only if its energy content equals the energy difference between the ground state and one of the excitation levels of the pigment. If absorption occurs, an electron is raised to the level corresponding to its energy. This electron is one of the partners of an electron pair in the ground state. The lowest level is of special interest as regards excited states. Its lifetime is usually at least 103 times longer than that the higher ones. Thus the probability for release of the energy in a form other than heat is much increased as compared with that for higher excitation levels. This phenomenon can be easily demonstrated, e.g. by exciting chlorophyll with red, yellow or blue light. In all cases the fluorescence is the same: the emission spectrum is a mirror image of the red absorption band representing the lowest excited state. This shows that the lowest excitation level can be filled in two ways; either by raising an electron from the ground state to this level, or by electrons *tumbling down* from higher energy states to the lowest excitation one. The energy difference between these levels is released by internal conversion as heat. The lowest excited state is therefore of primary interest with regard to the function of a photopigment.

So far, it is assumed that the law of spin conservation holds for the transfer of an electron from a low-energy orbit to one of higher energy. If, however, for some reason the spin of the excited electron is reversed, the triplet state, a metastable state, is formed. This has a slightly lower energy than the above-mentioned lowest excited singlet state. Owing to spin reversal the direct transition from the triplet to the ground state is forbidden, i.e. the probability of its occurrence is small. Thermal quanta may reverse the spin again and raise the electron to the lowest excited singlet state, S(1), hence it can return to the ground state. Population of the triplet state means that the lifetime of the excited molecule is lengthened considerably, and thus the probability of converting the absorbed quanta to chemical energy is enhanced. Moreover, this triplet state may function as a new ground state and can be excited to higher-energy triplet levels.

Photopigments: Functions

Upon de-excitation, conversion of the electronic energy can occur in various ways. Leaving aside internal conversion, which occurs concomitantly with the other possibilities and results in the production of heat, the following results of energy conversion can be envisaged:

Fluorescence, emitted by the return of an electron from the S_1 level to the ground state.

1. **Delayed fluorescence:** This kind of the fluorescence is emitted if an electron is raised, with concomitant reversal of its spin momentum, from the triplet into the S_1 state by thermal energy, and returns from the latter level to the ground state. In this case the emission spectrum is identical with that of fluorescence.
2. **Phosphorescence:** This light emission occurs in the case of the forbidden transition from the triplet state to the ground state with reversal of the spin momentum. As the triplet state is of lower energy than the S_1 level, the emission consists of quanta of energy lower than those emitted when the S_1 state is emptied. Therefore, the phosphorescence spectrum is shifted to longer wavelengths as compared with that of fluorescence.

Type of Biological Photoprocess

The major biological processes in which photopigments operate are:

- **Photosynthesis.** Light energy is converted into chemical energy to be used for metabolism.

- **Vision.** Light is used for providing information for the organism.
- **Photomotive processes.** Light causes displacement in the following ways:
- **Phototaxis.** Light induces motion or change of motion of a free organism or a cellular constituent. Two kinds of phototaxis occur:
- **Topophototaxis.** The direction of motion is determined by the direction of the light beam. If the motion proceeds either towards or away from the light source the taxis is called positive and negative respectively. For animals, Kuhn distinguished three kinds of topophototaxis which will not be further considered.
- **Phobophototaxis.** The direction of motion is not determined by the direction of the light beam, but at the boundary of two light fields of different intensities-a shock reaction of the object occurs. This prevents it from crossing the boundary. If it occurs at a transition from strong to dim light, the taxis are called positive, whereas it is called negative at a transition in the reverse sense.
- **Photokinesis.** The motility of an organism is changed by light. Photokinesis is indicated as positive or negative, if the motility is increased or decreased in the light respectively.
- **Photodinesis.** The effect of light on protoplasmic streaming. Light may affect either the rate of streaming, or the amount of protoplasma taking part in it. The effect may be temporary or tonic. In case of acceleration or retardation, photodinesis is called positive or negative respectively.
- **Phototropism.** The light-induced movement of a part of the organism resulting in bending towards the light source (positive), away from it (negative), or occupation of a certain angle to the light beam (diaphototropism), or in a bending not related to the light gradient, but to the structure of the organelle (photonasty). In most cases, the latter process is due to light-induced changes in turgidity, whereas the former ones are based on photic growth reactions.
- **Photomorphogenesis.** Light also control certain formative proceses. It may do so either by itself, or in combination with endogenous rhythms, as occur in photoperiodicity. The reaction may be restricted to activation of secretary activities affecting the behavioural pattern of animals, or it may affect the entire organism as in seed germination.

Since it may not be one of the major photoprocesses, photorespiration, though it differs from dark respiration, is omitted from consideration. The same is true of photodynamic effects. Secondary photo-effects, such as light-induced changes of membrane properties are not considered either.

Chlorophyll and Human Health

Science has known for years the benefits of chlorophyll; without chlorophyll, the green blood of plant / photosynthesis, life as we know it could not exist. Until recently, most research involving plants has been geared towards better plants for better meat'. For years it was thought that the protein we needed could only come from meat.

Chlorophyllin and Cancer: Chlorophylls and related chemicals can inhibit the ability of certain DNA-damaging chemicals to cause mutations in bacteria. How might this kind of anti-mutagenic activity be important in cancer prevention?.

People in certain parts of Africa, China, and other developing countries with similarly warm, damp climates have the highest rates of liver cancer in the world. The two major risk factors are chronic hepatitis B viral infection, and exposure to aflatoxin B1 (AFB1) in the food supply. Peanuts, corn, rice, and other grains and nuts stored under warm damp conditions can be infected with the mold *Aspergillus flavus*, which produces AFB1 as a secondary metabolite. AFB1 is one of the most potent cancer-causing chemicals, or carcinogens, ever discovered. Interestingly, AFB1 came to be recognized as a potential human liver carcinogen.

Chlorophyll chlorophyllin can inhibit cancers other than liver or caused by carcinogens other than AFB1. Vibeke, Michael Schimerlik, and Tetsu Hayashi showed that chlorophyll chlorophyllin associated tightly with AFB1, even in the acidic environment of the stomach and at the temperature of the human body. Chlorophyll chlorophyllin have ability to reduce bioavailability or uptake of AFB1. The dietary chlorophyll chlorophyllin is effective in preventing AFB1-DNA damage in the liver of rats. Similar safe protective mechanism is almost sure to apply to humans. Therefore, chlorophyll chlorophyllin reduces Aflatoxin indicators among people at risk for liver cancer.

Green-food concentrates are chlorophyll-rich supplements. Most are derived from microalgae (such as *Spirulina*, blue-green algae, or *Chlorella*) or from the young grasses of cereal plants (such as barley, wheat, and alfalfa). Chlorophyll acts as an antioxidant and may have medicinal benefits against cancer and immune problems.

Spirulina prevents and treat cancers while boosting immune system function. Similarly, *Chlorella* is one of the powerful weapons against cancer. *Chlorella* attacks against cancer with two folds. Cancer cells develop in all of our bodies, even in those of us who have never had cancer. When cancer cells develop, our immune systems naturally and efficiently destroy them before cancer symptoms appear. *Chlorella* is required as 1 to 2 teaspoons two or more times daily. It has dense and nutrient rich in vitamins, minerals and easily assimilable protein and it supports the immune system of patients undergoing chemo and radiation therapy, increases phagocytes, T-cells and cancer killing B cells.

Japanese scientists widely studied *Chlorella*'s benefits, since the atomic bombings of Hiroshima and Nagasaki. *Chlorella pyrenoidosa* does not directly kill cancer cells, the scientists concluded that its effects were caused by boosting the immune response. Strengthening the immune system's response to cancer cells, *Chlorella* acts as a preventative measure against cancer by raising blood levels of the protein albumin.

CAROTRNOID PIGMENTS

Carotenes are carotenoids containing no oxygen. Carotenoids containing some oxygen are known as xanthophylls. Carotenoids are nonpolar compounds, which are divided into two subclasses, i.e., more polar compounds called xanthophylls, or oxycarotenoids, and the nonpolar hydrocarbon carotenes. Both classes have at least nine conjugated double bonds, which absorb specific wavelengths of visible light and thus provide carotenoids their characteristic colors. At the end of the polyene chain, at least one unsubstituted B-ionone ring must be present to have provitamin A activity. Xanthophylls have ring structures at the end of the conjugated double-bond chain with polar functions, such as hydroxyl or keto groups.

Structure and properties of carotenoids in relation to function depends on the conjugated **polyene chromophore** determines not only the light absorption properties, and hence color, but also the photochemical properties of the molecule and consequent light-harvesting and photoprotective action.

The polyene chain is also the feature mainly responsible for the chemical reactivity of carotenoids toward oxidizing agents and free radicals, and hence for any antioxidant role. *In vivo*, carotenoids are found in precise locations and orientations in subcellular structures, and their chemical and physical properties are strongly influenced by other molecules in their vicinity, especially proteins and membrane lipids. In turn, the carotenoids influence the properties of these subcellular structures. Structural features such as size, shape, and polarity are essential determinants of the ability of a carotenoid to fit correctly into its molecular environment to allow it to function. A role for carotenoids in modifying structure, properties, and stability of cell membranes, and thus affecting molecular processes associated with these membranes, may be an important aspect of their possible beneficial effects on human health (Britton, 1995).

Examples of xanthophylls include lutein, zeaxanthin, capsanthin, canthaxanthin, astaxanthin, echionine and ß-cryptoxanthin. Small quantities of mono- and diester forms of lutein and zeaxanthin have been identified in foods, and some dietary lutein supplements are sold in the diester form. Papaya, peaches, prunes and squash all contain lutein diesters, whereas squash also contains lutein monoesters and peaches also contain zeaxanthin diesters. When fed at physiologic levels, mono- or diester forms must be deesterified before lutein or zeaxanthin absorption. Hydrocarbon carotenes include ß-carotene, α-carotene, γ -carotene and lycopene.

Carotenes and xanthophylls are well known to act as electron donors in redox processes. This ability is thought to be associated with the inhibition of oxidative reactions in reaction centers and light-harvesting pigment-protein complexes of photosystem II (PSII).

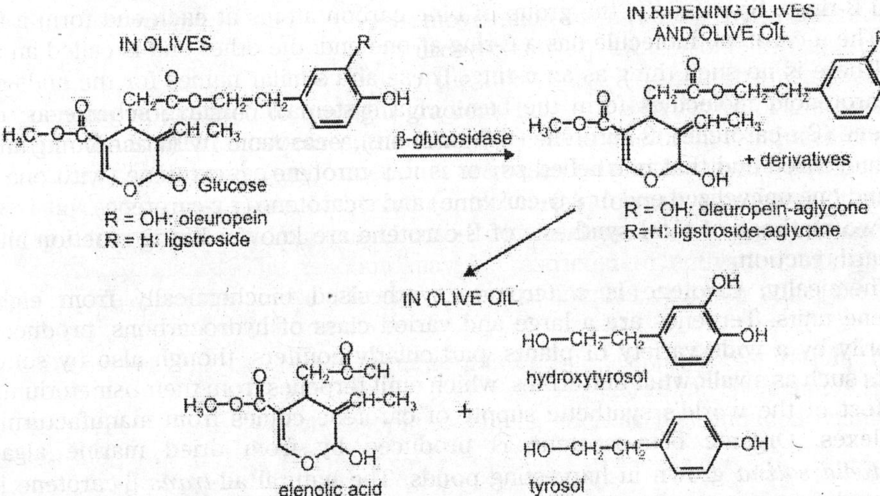

IN OLIVES

R = OH: oleuropein
R = H: ligstroside

ß-glucosidase

IN RIPENING OLIVES AND OLIVE OIL

+ derivatives

R = OH: oleuropein-aglycone
R=H: ligstroside-aglycone

IN OLIVE OIL

elenolic acid

hydroxytyrosol

tyrosol

FIG. 26.3: Chemical structures of the selected carotenoids.

Carotenoids and Human Health: The macula of the human retina contains high amounts of the xanthophyll carotenoids lutein and zeaxanthin [a mixture of (3R,3'A)-zeaxanthin and (3K,3'S-meso)-zeaxanthin]. The potential health benefits of carotenoids, such as their ability to act as antioxidants, immunoenhancers and inhibitors of premalignant lesions, have stimulated investigators' interest. One of the most biologically plausible roles for carotenoids is the potential effect of dietary lutein and zeaxanthin for protection of the macula from degeneration. These possible health benefits seem to relate to the unique geometry of carotenoids.

Carotene

The term **carotene** is used for several related substances having the formula $C_{40}H_{56}$. Carotene is an orange photosynthetic pigment important for photosynthesis. It is responsible for the orange colour of the carrot and many other fruits and vegetables. It contributes to photosynthesis by transmitting the light energy it absorbs to chlorophyll.

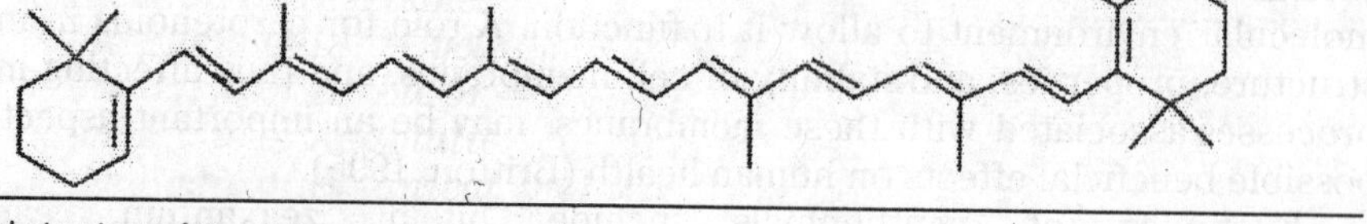

Chemical structure of β-catotene.

It comes in two primary forms designated by characters from the Greek alphabet: alpha-carotene (α-carotene) and beta-carotene (β-carotene). Gamma, delta and epsilon (γ, δ and ε-carotene) also exist. As hydrocarbons, carotenes are fat-soluble and insoluble in water. Beta-carotene is composed of two retinyl groups, and is broken down in the mucosa of the small intestine by beta-carotene dioxygenase to retinal, a form of vitamin A. Carotene can be stored in the liver and converted to vitamin A as needed, thus making it a provitamin.

Biosynthesis: Carotene arc commonly found in two forms α-carotene, ß-carotene. The two primary isomers of carotene, α-carotene and ß-carotene, differ in the position of double bonds in the cyclic group at the end.

The two ends of the ß-carotene molecule are structurally identical, and are called ß-rings. Specifically, the group of nine carbon atoms at each end form a ß-ring. The α-carotene molecule has a ß-ring at one end; the other end is called an ε-ring. There is no such thing as an α-ring. These and similar names for the ends of the carotenoid molecules form the basis of a systematic naming scheme as: α-carotene (ß,ε-carotene), ß-carotene (ß,ß-carotene), γ-carotene (with one ß ring and one uncyclized end that is labelled *psi* or is ß,ψ-carotene), δ-carotene (with one ε ring and one uncyclized end or ε,ψ-carotene) and ε-carotene (ε,ε-carotene).

Two methods of total synthesis of ß-carotene are known: Wittig reaction and Grignard reaction.

Chemically, carotene is a terpene, synthesised biochemically from eight isoprene units. Terpenes are a large and varied class of hydrocarbons, produced primarily by a wide variety of plants, particularly conifers, though also by some insects such as swallowtail butterflies, which emit terpenes from their osmeterium.

Most of the world's synthetic supply of carotene comes from manufacturing complexes. Organic beta-carotene is produced by from dried marine algae *Dunaliella salina* grown in harvesting ponds. The natural all-*trans* β-carotene is produced from a non genetically modified bacteria of the *Sphingomonas* genus isolated from soil.

Sources: ß-Carotene is the more common form and can be found in yellow, orange, and green leafy fruits and vegetables. As a rule of thumb, the greater the intensity of the orange colour of the fruit or vegetable, the more ß-carotene it contains. Carotenes are rich in sweet potatoes, carrots, goji berries, cantaloupe melon, mango, apricots, spinach, kale, chard, turnip greens, dandelion greens, beet greens, mustard greens, collard greens, watercress, cilantro, fresh thyme, broccoli, parsley, romaine lettuce, ivy gourd, rose hips, winter squash, pumpkin, abd cassava. Carotene is also found in palm oil, corn, and in the milk of Guernsey dairy cows, causing their milk to turn yellow. It is also found in some species of termites.

Carotenes and Human body: 6 μg of dietary β-carotene supplies the equivalent of 1 μg of retinol, or 1 RE (Retinol Equivalent). This is equivalent to 3 1/3

IU of vitamin A.

Absorption from these foods is enhanced, if eaten with fats, as carotenes are fat soluble, and if the food is cooked for a few minutes until the plant cell wall splits and the colour is released into any liquid. Carotene protects plant cells against the destructive effects of ultraviolet light. β-Carotene is an anti-oxidant.

Beta-carotene and cancer: It has been shown in trials that the ingestion of beta carotene at about 30 mg/day increases the rate of lung cancer and prostate cancer, and increases mortality in smokers and people with a history of asbestos exposure. There is a relationship between daily supplementation of beta carotene and vitamin E (alpha-tocopherol) and the incidence of lung cancer. A correlation between carotenoid-rich fruits and vegetables and lower lung cancer rates has been found. However, no reduction in lung cancer was found in the participants using these supplements (beta-carotene), and furthermore, these supplements may, in fact, have harmful effects.

Beta Carotene and Cognition: 50 mg of beta carotene every other day prevented cognitive.

Carotenemia: Carotenemia or hypercarotenemia is excess carotene, but unlike excess vitamin A, carotene is non-toxic. Although hypercarotenemia is not particularly dangerous, it can lead to a yellowing of the skin (carotenodermia). It is most commonly associated with consumption of an abundance of carrots.

Xanthophylls

Xanthophylls (originally phylloxanthins) are yellow pigments from the carotenoid group. Xanthophyll has a chemical formula of $C_{40}H_{56}O_2$. Their molecular structure is based on carotenes; contrary to the carotenes, some hydrogen atoms are substituted by hydroxyl groups and/or some pairs of hydrogen atoms are substituted by oxygen atoms. They are found in the leaves of most plants and are synthesised within the plastids. They are involved in photosynthesis along with green chlorophyll, which typically covers up the yellow except in autumn, when the chlorophyll is denatured by the cold.

In plants, xanthophylls are accessory pigments, along with anthocyanins, carotenes, and sometimes phycobiliproteins. Xanthophylls, along with carotenic pigments are seen when leaves turn orange in the autumn season. Animals cannot produce xanthophylls, and thus xanthophylls found in animals (e.g. in the eye) come from their food intake.

Biosynthesis: The group of xanthophylls includes lutein, zeaxanthin, neoxanthin, violaxanthin, and α- and β-cryptoxanthin. Xanthophylls are oxidized derivatives of carotenes. They contain hydroxyl groups and are more polar than carotenes; therefore, carotenes travel further than xanthophylls in paper chromatography.

The xanthophyll cycle involves conversions of pigments from a non-energy-quenching form to energy-quenching forms. This is a way to reduce the absorption cross-section of the light harvesting antenna, and thus to reduce the amount of energy that reaches the photosynthetic reaction centers. Reducing the light harvesting antenna is one of the main ways of protecting against photoinhibition and changes in the xanthophyll cycling takes place on a time scale of minutes to hours. In higher plants there are three carotenoid pigments that are active in the xanthophyll cycle: violaxanthin, antheraxanthin and zeaxanthin. During light stress violoxanthin is converted to antheraxanthin and zeaxanthin, which functions as photoprotective pigments. This conversion is done by the enzyme violaxanthin de-epoxidase.

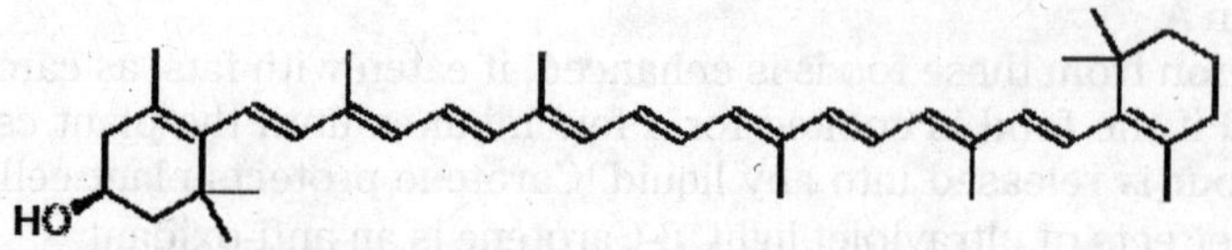

Chemical structure of cryptoxanthin.

Preparation of Xanthophyll Crystals: Xanthophyll crystals can be prepared which contain at least 85% total xanthophylls with at least 90% trans-lutein and/or zeaxanthin, trace amounts of cis-lutein and other carotenoids which comprises admixing and heating xanthophyll ester concentrate using excess alcoholic alkali solution, maintaining the resulting mixture at a temperature in the range of 65°C. to about 80°C, for a period sufficient to saponify the xanthophyll esters; removing the aliphatic alcohol from the mixture under reduced pressure to get saponified crude concentrate; admixing the saponified crude concentrate obtained with water at room temperature to form a diluted oily mixture; extracting the diluted oily mixture obtained with ethyl acetate to get xanthophyll extract which is processed by conventional methods to get xanthophyll crystals containing high amounts of trans-lutein and/or zeaxanthin along with trace amounts of cis-lutein and other carotenoids.

In diatoms and dinoflagellates, the xanthophyll cycle consists of the pigment diadinoxanthin, transformed into diatoxanthin (diatoms) or dinoxanthin (dinoflagellates), at high light.

Source: The yellow color of chicken egg yolks also comes from ingested xanthophylls.

Human Health: The xanthophylls lutein (L) and zeaxanthin (Z) form the macular pigment with the highest density in the macula lutea and is relevant to health and disease of the retina. Concentrations of carotenoids, retinoids, and tocopherols are measured by highperformance liquid chromatography (HPLC).

Most consumers associate colour of food with age and health status of animals and food quality in consequence. Xanthophylls are widely used as feed additives to generate products meeting consumers' demands. Although a lot of xanthophylls with interesting biological properties are found in nature, only a few are actually of industrial importance as feed additives. Xanthophylls are also used in poultry farming and in aquaculture (trout and salmon), referring to natural (*e.g.* lutein from *Tagetes erecta*) as well as synthetic (*e.g.* canthaxanthin) xanthophylls.

Zeaxanthin and meso-zeaxanthin as a dietary supplement: Zeaxanthin is a yellow carotenoid pigment found widely distributed in plants. Like lutein, it contains two alcohol groups in its molecule, and is considered to be a xanthophyll (an oxygen-containing carotenoid). It is not a precursor of vitamin A. In foods zeaxanthin is often found with lutein; it is especially high in corn, persimmons, and spinach. Meso-zeaxanthin, an isomer of zeaxanthin, is found in a variety of animals and in the human eye but does not appear to be a dietary carotenoid. Rather, it is biosynthesised in the retina from either lutein or zeaxanthin. The major commercial sources of zeaxanthin are extracts of marigold petals and chemical synthesis, with pure zeaxanthin available only from the latter source.

Zeaxanthin, meso-zeaxanthin and lutein are the major components of the macular pigment, the colored material in the center of the retina in primates, an area known as the macula lutea. The macula is packed with color-sensitive photoreceptor cells (cones) and is responsible for sharp vision. Proteins that specifically bind xanthophylls, but not other carotenoids, have been isolated from human retinas.

Lutein and zeaxanthin have been shown to assume different spatial arrangements in experimental lipid membranes, and also have slightly different light absorption spectra (6A). In addition to this possible difference in their arrangement in retinal cell membranes, the three xanthophyll pigments are not distributed uniformly in different areas of the retina. They differ quantitatively also; the overall ratio of retinal zeaxanthins to lutein is about 2.4 : 1.

Based on their structural similarity, the absorption of zeaxanthin is probably comparable to that of lutein. Heat processing and addition of some fat probably improves absorption from foods (7A). Recent studies have shown higher levels of absorption in humans fed eggs enriched in lutein and zeaxanthin (from chickens fed marigold petals) than from equal amounts in spinach.

Age-related macular degeneration: The high levels of the xanthophylls in the macula may indicate a protective role in reducing damage from oxidative photochemical reactions (particularly those caused by the blue part of the visible spectrum) in the photoreceptor cells. Epidemiological studies have shown that the density of these pigments is inversely related to the incidence of age-related macular degeneration (ARMD), a major cause of blindness in the elderly. It has been shown experimentally that the density of the pigments in humans can be increased by feeding foods rich in the xanthophylls or by supplementation with lutein and/or zeaxanthin. Human consumption of the xanthophylls as high as 27 mg/day are known, while an eye study reported the upper-fifth of its subjects consumed >5.6 mg/day and had a lower risk of ARMD.

PHOTOPIGMENTS: PROPERTIES

Chlorophylls

Chlorophylls are esters of dibasic acids, the chlorophyllins.

These photopigments are active in (a) photosynthesis either by collecting light quanta or by the generation of free electrons leading to the production of reducing compounds and ATP, (b) phototaxis, (c) photokinesis.

Chlorophyll a

It occurs in all green plant's its formula is $C_{55}H_{72}O_5N_4Mg$. Chlorophyll a, like other chlorophylls, contains four pyrrole nuclei (I-IV), arranged in prophyrin ring structure. In addition to these nuclei an isocyclic, or cyclopentanone, ring (V) occurs adjoining to nucleus III. A Magnesium atoms is located in the centre of the porphyrin ring. In most cases, including chlorophyll a, the esterifying alcohols are methanol and phytol.

It is emphasized that the indication of double and single bonds in the Figure does not depict the actual situation. In reality, like other conjugated bond systems, chlorophyll is a resonance hybrid. According to there are major contributions of three resonating structures in this particular case.

A key position in determining various properties of this photopigment is occupied by the central Magnesium atom. Its presence is required for the oxidative process of allomerization. This reaction occurs upon storage of alcoholic solutions of chlorophyll under air. Allomerization consumes molecular oxygen in the molar ratio 1/1 with respect to chlorophyll. The presence of the central magnesium atom is indispensible for this process. The oxidation occurs in the cyclopentanone ring (V) at C-10 a follows: $H\text{-}C^{10} + O_2\text{-}HO\text{-}O\text{-}C^{10}\text{-}$, given that allomerization yields at least three products. This conclusion is based on the absence of the Molisch phase test with allomerized chlorophyll. It may be mentioned that this process is believed to be based upon a reversible shift of the hydrogen atom at C-10 to C-9. The sum total of these observation, namely (a) the absence of Mg, and of allomerization, (2) the fact that allomerization means fixation of the H atom at C-10, (3) that it also prevents the phase test, and (4) that this test implies a reversible shift of the H atom in question, suggests that the central magnesium atom is responsible for the mobility of the H atom at C-10.

Still, hydrogen exchange proceeds only very slowly. However, with the proton magnetic resonance technique, showed that hydrogen exchange occurs at C-8 exclusively under their experimental conditions.

Bacteriochlorophyll a

This is the main photosynthetic pigment in both sulphur and non-sulphur purple bacteria. Its formula is $C_{55}H_{74}O_6N_4$ Mg. The structure differs from that of chlorophyll **a** in that C-3 and C-4 are hydrogenated, whereas the side group of C-2 consists of H_3C-C=O. Obtained evidence is that the primary intermediate formed upon excitation is the triplet state of bacteriochlorophyll complexed with the reaction centre.

The following pigments will be mentioned only briefly.

Bacteriochlorophyll b

This pigment was described by it occurs in a *Rhodopseudomonas* species as the main pigment.

Chlorophyll-770

Chlorophyll-770 is found in green sulphur bacteria. It turned out to be identical with bacteriochlorophyll. It is suggested that this chlorophyll rather than the chlorobium chlorophylls functions as a main pigment.

Chlorobium Chlorophyll-660

Chlorobium chlorophyll-660 was previously called bacterioviridin. It is found in most of the green sulphur bacteria. It contains neither a methoxyl group at C-10, not phytol. Determined the magnesium content to be 2.92% on a dry weight basis.

Chlorobium Chlorophyll-650

Chlorobium chlorophyll-650 occurs less commonly than the previous pigment. The methoxyl group at C-10 and phytol are absent. Its magnesium content are 2.59% on a dry weight basis.

In green plants, including algae, chlorophyll a is usually accompanied by auxiliary chlorophylls. Such chlorophylls are:

Chlorophyll b

Chlorophyll b, $C_{55}H_{70}O_6N_4Mg$, its structure is like that of chlorophyll **a**, except for a replacement of the methyl group at C-3 by a formyl group.

Chlorophyll c

It is synonyms of chlorophyll g and chlorofucine and occurs in diatoms and other brown algae. It does not contain a phytol group, two kinds, namely, chlorophyll c_1 and c_2 have been observed.

Chlorophyll d

It is found in most of the red algae in rather small amount. It was suggested that this pigment is 2-desvinyl-2-formyl-chlorophyll a.

Chlorophyll e

It was observed in yellow-green algae in very small amounts.

Precursors of chlorophylls might also be considered photopigments. Such may be the case with protochlorophyll, as this precursor needs light for conversion to chlorophyll by hydrogenation at C-7 and C-8. In vivo chlorophylls are complexed with lipoprotein carriers. Absorption spectra of chlorophylls are characterized by four independent electronic transitions responsible for bands between 350 and 900 nm.

Carotenoids

These compounds act as photopigments; in photosynthesis as auxiliary pigments; the absorbed energy is transferred to chlorophyll. It has been suggested that part of the carotenoids, the xanthophylls, act as oxygen transporters in photosynthesis. However, such function seems highly improbable. Their possible protective function in preventing photodestruction of chlorophyll will not be considered. Carotenoids function possibly as photoreceptors in phototaxis, photokinesis, and photomorphogenesis. The structure of xanthophyll lutein carotenoid provides an example of carotenoid structure. There are two classes of carotenoid: polyene hydrocarbons, called carotenes, and their oxygen derivatives, the xanthophylls. All are lipochromes; they are soluble in lipids and fats. The pigments are composed of isoprene units. At one or both ends of the isoprene chain carotenoids carry an ionone ring which may be closed b or open d Carotenoids usually occur in the all-trans configuration, whereas cis isomers are encountered less frequently. Carotenoids participating in photosynthesis are as follows:

Carotenes

α-carotene, $C_{40}H_{56}$, in green plants and algae, β-carotene, $C_{40}H_{56}$, in green plants and algae, γ-carotene, $C_{40}H_{56}$, in green plants and algae, and lycopene, $C_{40}H_{56}$, in purple bacteria are present.

Xanthophylls

Lutein, $C_{40}H_{56}O_2$, in green plants and algae, violaxanthin, $C_{40}H_{56}O_4$, in green plants and algae, neoxanthin, $C_{40}H_{56}O_4$, in green plants and algae, cryptoxanthin, $C_{40}H_{56}O$, in green plants and algae, zeaxanthin, $C_{40}H_{56}O_2$, in green plants and algae, fucoxanthin, in diatoms and other brown algae, (the exact structure is not yet established), spirilloxanthin, $C_{40}H_{54}(OCH_3)_2$, in purple bacterial and rhodopin, $C_{40}H_{58}O$ or $C_{40}H_{56}O$, occurring in purple bacteria.

The absorption spectra of these pigments generally show two or three bands located at wavelengths in between 350 and 550 nm. Carotenoids occur in complexes with lipoproteins in the vivo state.

Phycobiliproteins

These pigments act as photoreceptors in connection with (a) photosynthesis as auxiliary pigments by transfer of their excitation energy to chlorophyll, (b) phototaxis, (c) photokinesis (to a minor degree), and (d) photomorphogenesis as regards the special biliprotein phytochrome, capable of phototransformation. As their names indicate, these pigments are proteins with a chromophoric group. This consists of four pyrrole nuclei linked together by C atoms in a linear or open-cyclic arrangement. There are two major kinds of phycobiliproteins: the blue phycocyanins and the red phycoerythrins. Their chromophoric groups are phycocyanobilins and phycoerythrobilins, respectively. Both pigments coexist in blue-green algae, the Cyanophyceae, and red ones, the Rhodophyceae, but in considerably different proportions. Since phycocyanin as well as phycoerythrin from either blue-green or red algae show different absorption properties, the prefixes C or R are used to indicate their origin. Phycobilins are also observed in Cryptomonads.

Phycobiliproteins are complexes which leach readily out of damaged cells. A complex unit is known to be composed of two or more subunits. This phycocyanin from *Oscillatoria agardhii* consists of two polypeptide chains, each of which carries at least one chromophoric group. The chains, of which the amino acid compositions are given below, are connected by a disulphide bridge. Assuming that each chain carries a single chromophoric group, the molecular weights for the chains were found to be 12,100 and 14,100 respectively.

The major phycobiliproteins are: C-phycocyanin, C-phycoerythrin, R-phycocyanin and R-phycoerythrin.

B-phycoerythrin, occur in the red alga *Porphyra naiadum* allophycocyanin, and is also present in many red and blue-green algae, whereas phytochrome occurs in plants, including algae. It is active in photomorphogenesis and phototaxis.

Flavoproteins

Probably flavoproteins are the photopigments of phototropism. Flavoproteins may also be responsible for the photosensitivity of fungi. It is less certain that these pigments function as photoreceptors for photomotion in microorganisms. The difficulty in assigning the role of biological photopigment to flavoproteins unambiguously is due to the fact that their absorption spectra showing bands around 450 nm, overlap those of the carotenoids. Action spectra, therefore, are hard to interpret. It thus remains a challenge to try to obtain conclusive evidence about the proposed functions of flavoproteins in photobiology.

CHAPTER 27

Hemoglobin

Blood is an amazing and vitally important part of the body, because it contains many finely-tuned chemical systems that allow it to maintain the chemical environment needed for the body's metabolism. Each person has about 5 liters of blood which is 5000 mLs of blood. There are around 5 million RBCs in one cubic millimeter. One of the most important functions of blood is delivering O_2 to all parts of the body by the hemoglobin protein. O_2 is carried in the hemoglobin protein by the heme group.

The oxygen-carrying protein hemoglobin was discovered by Otto Funke in 1851. In 1959, Max Perutz determined molecular structure of hemoglobin and awarded Nobel prize in chemistry in 1962. Hemoglobin (abbr. Hb or Hgb) is the iron-containing oxygen-transport conjugated metalloprotein in the red blood cells of vertebrates, and the tissues of some invertebrates. It has heme as a prosthetic group and the protein, the globin. It is the red coloring matter of blood.

Hemoglobin ⟶ Heme (prosthetic group) + globin (Protein)

or

Hemoglobin = 4 globins + 4 heme groups

Heme is present as a prosthetic group in Hb, as well as in myoglobin, cytochromes, peroxidase, and tryptophan, etc. Our blood stream contains about 150 g/L of the protein known as **hemoglobin** (Hb). It is an oxygen-carrier. An iron atom in the ferrous state (Fe^{2+}) sits in the middle of the heme. Ferric hemoglobin (Fe^{3+}) cannot bind oxygen. The concentration of O_2 in the blood stream reaches 0.01 M - the same concentration as air. Once the Hb-O_2 complex reaches the tissue that consumes oxygen, the O_2 molecules are transferred to another small protein - **myoglobin** (Mb) - which transports oxygen through the muscle tissue. Myoglobin is a sensitive marker for muscle injury and is used as a marker of myocardial infarction (MI) in patients with chest pain.

The **hematocrit** (Ht or HCT) or packed cell volume (PCV) or erythrocyte volume fraction (EVF) is the proportion of blood volume that is occupied by red blood cells. It is normally about 46% for men and 38% for women. It is considered an integral part of a person's complete blood count results, along with hemoglobin concentration, white blood cell count, and platelet count.

FUNCTIONS

1. It can transport large quantities of oxygen. Hb has the dual role of obtaining oxygen from air in the lungs and delivering it to body tissues.
2. High hemoglobin count indicates an above-average concentration of oxygen-carrying proteins in your blood.
3. Hb not only furnishes oxygen to tissue but also transports the waste products of metabolism (H^+ and CO_2) from the tissues.
4. 2, 3, bisphosphoglycerate (2, 3 BPG) is found in red blood cells at nearly the same concentration as Hb. It is responsible for lowering the oxygen affinity of

Hb and allowing Hb to more efficiently release oxygen at the typical PO_2 of tissues. The higher affinity of HbF arises because the γ subunits do not bind 2,2, BPG well. It means that the oxygen affinity binding curve of HbF is shifted to the left as compare to HbA. In the absence of 2, 2 BGP HbF has a slightly lower oxygen affinity than HbA. HbF has a higher affinity for O_2 than does HbA and this difference is physiologically important because it helps to ensure the transfer of oxygen from HbA (maternal circulation) to HbF (fetal circulation).

5. The oxygen binding affinity of Hb is decreased by an increase in hydrogen ion concentration and also increase in PCO_2.
6. Hemoglobin is a powerful buffer.
7. It has a great solubility.
8. Hemoglobin plays an important role in regulating the acid base balance of the blood.
9. Oxygen enters the cells by diffusion and the availability of oxygen is limited by the distance. Over which diffusion must occur. Oxygen binding molecule in circulating system carry more oxygen to the cell and sub cellular as the size of cell and organism increased.

CHEMISTRY

At the core of the molecule is a heterocyclic ring, known as a porphyrin which holds an iron atom; this iron atom is the site of oxygen binding. An iron containing porphyrin is termed a heme. The name hemoglobin is the concatenation of heme and globin, a globin being a generic term for a globular protein. Since a single subunit of hemoglobin is, in fact, made of a heme imbedded in a globular protein, the name makes sense. There is a number of heme containing proteins. Hemoglobin is by far the best known.

1. Hemoglobin was the first protein whose molecular weight was determined.
2. Hemoglobin was the first to be assigned a specific function - dioxygen transport.
3. Hemoglobin has a prosthetic group (non-amino acid) heme group (protoporphyrin IX with a ferrous ion) covalently attached to the protein. The heme group binds dioxygen.
4. Hemoglobin was the first protein in which a point mutation (single base pair change) causes a single amino acid change in the protein, marking the start of molecular medicine.
5. Hemoglobin was the first protein with high resolution x-ray structure.
6. Theory for dioxygen binding explains control of enzyme activity.
7. The binding of dioxygen is regulated by binding of H^+, CO_2, and bisphosphoglcyerate which bind to sites (allosteric) distant from oxygen binding site.
8. Crystals of deoxy-Hb shatter on binding dioxygen, indicating significant conformational changes on binding.

Heme is present as a prosthetic group in Hb, as well as in myoglobin, cytochromes, peroxidase, catalase, tryptophan puyrrolase etc. It is coloured compound of Hb and is a ferrous complex of portoporphyrin. It is same in all the species Hb. Porphyries are cyclic compounds formed by fusion of 4 pyrzole rings linked by methenyl bridges. Since an atom of ferrous is present, heme is a ferroprotoporphyrin.

Globin: Four different globin chains are involved in HB synthesis- alpha (α), bete (β), gamma(ϒ) and delta (δ). In one mole of Hb two kinds of chains are present. All normal Hb have a pair of alpha chains, the other pair may be beta, gamma or delta. Since, there is a difference in the globin part, therefore, there occurs differences in physical and chemical properties of various species of Hb. Each of the globins is folded into a secondary and tertiary structure. Then, all four are put together into the hemoglobin molecule's quaternary structure.

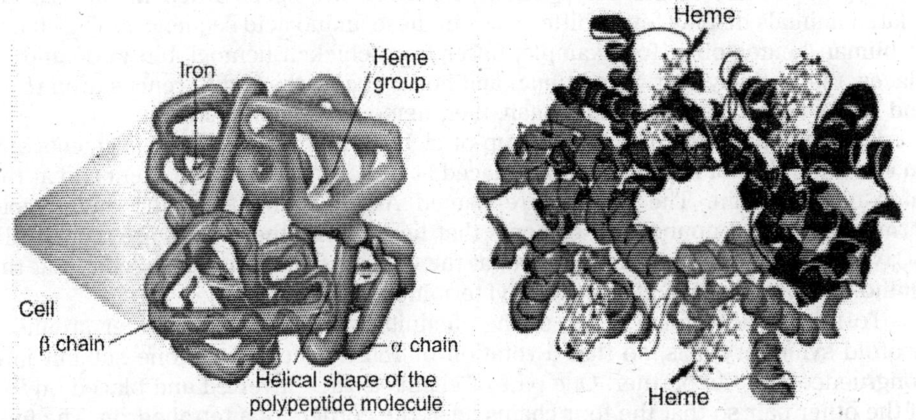

FIG. 27.1: Hemoglobin molecule showing α and β chains and heme.

The hemoglobin molecule is made up of four polypeptide chains: two alpha chains of 141 amino acid residues each and two beta chains of 146 amino acid residues each. Each subunit has a molecular weight of about 16,000 Daltons, for a total molecular weight in the tetramer of about 64,000 Daltons. Each subunit of hemoglobin contains a single heme. The alpha and beta chains have different sequences of amino acids, but fold up to form similar three-dimensional structures. The four chains are held together by noncovalent interactions.

There are four binding sites for oxygen on the hemoglobin molecule, because each chain contains one heme group. In the alpha chain, the 87th residue is histidine and in the beta chain the 92nd residue is histidine. A heme group is attached to each of the four histidines. The heme consists of an organic part and an iron atom. The iron atom in heme binds to the four nitrogens in the center of the protoporphyrin ring. The hemoglobin molecule is nearly spherical, with a diameter of 55 Å. The four chains are packed together to form a tetramer. The heme groups are located in crevices near the exterior of the molecule, one in each subunit. Each alpha chain is in contact with both beta chains. However, there are few interactions between the two alpha chains or between the two beta chains.

Polypeptide Chains

The polypeptide chains of hemoglobin are of two kinds; in the hemoglobin of an adult human. It may be with two polypeptides of one kind with 141 amino acid residues and two of another with 146 residues. The amino acid sequences (the primary structure) of the polypeptides, however, are species dependent. Closely related animals display fewer differences in these amino acid sequences. The chains of human hemoglobin, for example, differ from chicken hemoglobin in 35 and 45 places, respectively. Between human and horse chains these differences drop to 18 and 25. As for chimpanzee and human, their hemoglobins are identical.

Each polypeptide chain is made up of eight or nine alpha-helical segments and an equal number of nonhelical ones placed at the corners between them and at the ends of the chain. The helices are named A-H, starting from the amino acid terminus, and the nonhelical segments that lie between the helices are named AB, BC, CD, etc. The nonhelical segments at the ends of the chain are called NA at the amino terminus and HC at the carboxyl terminus.

To form the tetramer, each of the subunits is joined to its partner around a twofold symmetry axis, so that a rotation of 180 degrees brings one subunit into congruence with its partner. One pair of chains is then inverted and placed on top of the other pair so that the four chains lie at the corners of a tetrahedron. The four subunits are held together mainly by nonpolar interactions and hydrogen bonds. There are no covalent bonds between subunits. The twofold symmetry axis that relates the pairs of alpha and beta chains runs through a water-filled cavity at the center of the molecule. This cavity widens upon transition form the R structure to the T structure_to form a receptor site for the allosteric effector DPG (2,3 diphosphoglycerate) between the two beta chains. The heme group is wedged into a pocket of the globin with its hydrocarbon side chains interior and its polar propionate side chains exterior.

This **cooperative interaction** between different binding sites makes hemoglobin an unusually good oxygen-transport protein because it enables the molecule to pick up as much oxygen as possible once the partial pressure of this gas reaches a particular threshold level, and then give off as much oxygen as possible when the partial pressure of O_2 drops significantly below this threshold level. The hemes are much too far apart to interact directly. But, changes that occur in the structure of the globin that surrounds a heme when it picks up an O_2 molecule are mechanically transmitted to the other globins in this protein. These changes carry the signal that facilitates the gain or loss of an O_2 molecule by the other hemes.

Like all proteins, the "blueprint" for hemoglobin exists in DNA (the material that makes up genes). Normally, an individual has four genes that code for the alpha protein, or alpha chain. Two other genes code for the beta chain. (Two additional genes code for the gamma chain in the fetus). The alpha chain and the beta chain are made in precisely equal amounts, despite the differing number of genes. The protein chains join in developing red blood cells, and remain together for the life of the red cell.

HEME SYNTHESIS

Heme is synthesised in a complex series of steps involving enzymes in the mitochondrion and in the cytosol of the cell.

1. The first step in heme synthesis takes place in the mitochondrion, with the condensation of succinyl CoA and glycine by ALA synthase to form 5-aminolevulic acid (ALA).

2. This molecule is transported to the cytosol where a series of reactions produce a ring structure called coproporphyrinogen III. This molecule returns to the mitochondrion where an addition reaction produces protoporhyrin IX.
3. The enzyme ferrochelatase inserts iron into the ring structure of protoporphyrin IX to produce heme.
4. Iron deficiency, the world's most common cause of anemia, impairs heme synthesis thereby producing anemia.

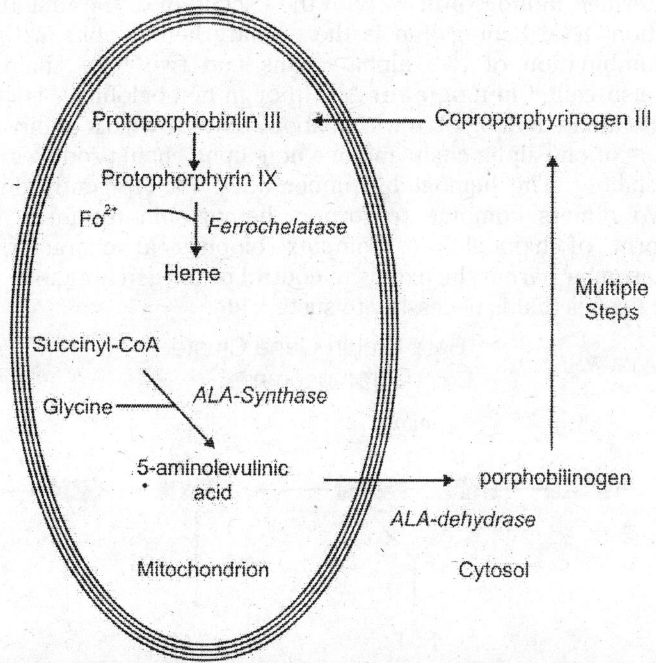

FIG. 27.2: Heme biosynthesis: The synthesis of heme is a complex process that involves multiple enzymatic steps. The process begins in the mitochondrion with the condensation of succinyl-CoA and glycine to form 5-aminolevulinic acid. A series of steps in the cytoplasm produce coproporphrynogen III, which re-enters the mitochondrion. The final enzymatic steps produce heme.

5. A number of drugs and toxins directly inhibit heme production by interfering with enzymes involved in heme biosynthesis.
6. Lead commonly produces substantial anemia by inhibiting heme synthesis, particularly in children.
7. The biosynthetic pathway contains primarily eight enzymes. They are hemA (Delta-aminolevulinate synthase), hemB (Delta-aminolevulinic acid dehydratase), hemC (porphobilinogen deaminase), and hemD (uroporphyrinogen III synthase) in our system. These genes overproduce *precursors* to heme in our cells because over-accumulation of heme itself would result in toxicity. After successful subcloning experiments, the bacterial cell pellets would become reddish-brown due to the accumulation of porphyrins and heme.

Globin Synthesis

Two distinct globin chains (each with its individual heme molecule) combine to form hemoglobin. One of the chains is designated **alpha**. The second chain is called **non-alpha**. With the exception of the very first weeks of embryogenesis, one of the globin chains is always alpha. A number of variables influence the nature of the non-alpha chain in the hemoglobin molecule. The fetus has a distinct non-alpha chain called gamma. After birth, a different non-alpha globin chain, called beta, pairs with the alpha chain. The combination of two alpha chains and two non-alpha chains produces a complete hemoglobin molecule (a total of four chains per molecule).

The combination of two alpha chains and two gamma chains form fetal hemoglobin, termed **hemoglobin F**. With the exception of the first 10 to 12 weeks after conception, fetal hemoglobin is the primary hemoglobin in the developing fetus. The combination of two alpha chains and two beta chains form adult hemoglobin, also called **hemoglobin A**. Although hemoglobin A is called adult, it becomes predominate hemoglobin within about 18 to 24 weeks of birth.

The pairing of one alpha chain and one non-alpha chain produces a hemoglobin dimer (two chains). The hemoglobin dimer does not efficiently deliver oxygen, however. Two dimers combine to form a hemoglobin tetramer, which is the functional form of hemoglobin. Complex biophysical characteristics of the hemoglobin tetramer permit the exquisite control of oxygen uptake in the lungs and release in the tissues that is necessary to sustain life.

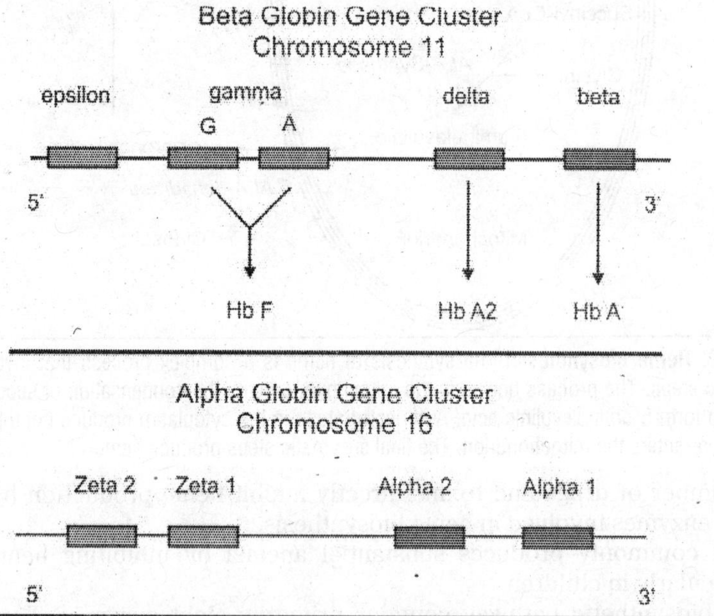

FIG. 27.3: Schematic representation of the globin gene loci. The lower panel shows the alpha globin locus that resides on chromosome 16. Each of the four alpha globin genes contributes to the synthesis of the alpha globin protein. The upper panel shows the beta globin locus. The two gamma globin genes are active during fetal growth and produce hemoglobin F. The "adult" gene, beta, takes over after birth.

Genes for Alpha Globin Chains

The genes that encode the alpha globin chains are on chromosome 16. Those that encode the non-alpha globin chains are on chromosome 11. Multiple individual genes are expressed at each site. Pseudogenes are also present at each location. The alpha complex is called the **alpha globin locus**, while the non-alpha complex is called the **beta globin locus**. The expression of the alpha and non-alpha genes is closely balanced by an unknown mechanism. Balanced gene expression is required for normal red cell function. Disruption of the balance produces a disorder called thalassemia.

Alpha Globin Locus: Each chromosome 16 has two alpha globin genes that are aligned one after the other on the chromosome. For practical purposes, the two alpha globin genes (termed alpha-1 and alpha-2) are identical. Since each cell has two chromosomes 16, a total of four alpha globin genes exist in each cell. Each of the four genes produces about one-quarter of the alpha globin chains needed for hemoglobin synthesis. The mechanism of this coordination is unknown. Promoter elements exist 5' to each alpha globin gene.

Beta Globin Locus: The genes in the beta globin locus are arranged sequentially from 5' to 3' beginning with the gene expressed in embryonic development (the first 12 weeks after conception; called episolon). The beta globin locus ends with the adult beta globin gene. The sequence of the genes is: epsilon, gamma, delta, and beta.

Difference between Hb and Mb

Mb (a monomer containing 8 α–helices, A-H) and Hb (a heterotetramer with two α- and two β– subunits, each which also contains 8 α–helices) are both oxygen binding proteins. Dioxygen is transported from lungs, gills, or skin of an animal to capillaries, where it can be delivered to respiring tissue. It has a low solubility in blood (0.1 mM). Whole blood, which contains 150 g Hb/L, can carry up to 10 mM dioxygen. Invertebrate can have alternative proteins for oxygen binding, including hemocyanin, which contains Cu and hemerythrin, a non-heme protein. On binding dioxygen, solutions of Hb change color to bright red. Solutions of hemocyanin and hemerythrin change to blue and burgundy colored, respectively, on binding dioxygen. Some Antarctic fish don't require Hb since dioxygen is more soluble at low temperature. Mb is found in the muscle, and serves as a storage protein for oxygen transported by Hb.

Hb is a tetramer of two α and two β subunits held together by IMF's (an example of quarternary protein structure), and 4 bound hemes, each of which can bind a dioxygen. In a fetus, two other subunits make up Hb (two zeta - ζ and two epsilion- ε subunits -analogous to the two α and two β subunits, respectively). This changes in fetuses to two α and two γ subunits. Fetal Hb has higher affinity for dioxygen than adult Hb.

Mb is a single polypeptide chains which has higher affinity for dioxygen than Hb. The α and two β chains are similar to Mb, which is unexpected since only 24 of 141 residues in the α and β chains of Hb are identical to amino acids in Mb. This suggests that different sequences can fold to similar structures. The globin fold of Mb and each chain of Hb is common to vertebrates and must be nature's design for dioxygen carriers. A comparison of the sequence of Hb from 60 species shows much variability of amino acids, with only 9 identical amino founds. These must be important for structure/function. All internal changes are conservative (e.g. changing a ŋonpolar for a nonpolar amino acid). The two active sites His are conserved, as is Gly B6 (required for a reverse turn).

Development of Abnormal Hemoglobin

The composition of hemoglobin is the same in all people. The genes that code for hemoglobin are identical throughout the world. Occasionally, however, one of the genes is altered by any of a variety of "accidents" that can occur in nature. These alterations in the genes (called mutations) are very rare. Since genes are inherited, and they contain the information needed to make a protein, if a mutation produces an abnormal hemoglobin gene in a person, the gene will be passed on to his or her children. The children will produce a modified hemoglobin identical to that of the parent. Most mutations in hemoglobin produce no problem. Occasionally, however, the alteration in the protein changes aspects of its behavior. The types of disorders that can result include sickle cell disease and thalassemia.

Genes control the synthesis of all the chemicals needed by organisms. Genes can suffer damage to an extent that they no longer produce normal amounts of hemoglobin. Usually, only one of the sets of hemoglobin genes is affected, that is the alpha gene set or the beta gene set. For example, one of the two beta globin genes may fail to produce a normal quantity of beta chain protein. The alpha globin gene set will continue to produce a normal quantity of alpha chain protein. An imbalance develops in the amount of alpha chain and beta chain protein in the cell. There is too much alpha chain for the amount of beta chain that is present. This imbalance is called "thalassemia ". In this example, it would be beta thalassemia, because it is the beta chain gene that has failed. An analogy would be cars coming out of the factory. Engines and bodies must be made in equal numbers to have functional automobiles. If the engine plant goes on strike (thalassemia), the bodies produced by the body plant are useless.

DEGRADATION OF HEMOGLOBIN

When red cells reach the end of their life, they are broken down, and the hemoglobin molecule broken up and the iron recycled. When the porphyrin ring is broken up, the fragments are normally secreted in the bile by the liver. The major final product of heme degradation is bilirubin. Increased levels of this chemical are detected in the blood if red cells are being destroyed more rapidly than usual. Improperly degraded hemoglobin protein or hemoglobin that has been released from the blood cells can clog small blood vessels especially the delicate blood filtering vessels of the kidneys, causing kidney damage.

Death of Red Blood Cell and Fait of Hemoglobin

When red blood cells die hemoglobin leaks out (typically into the blood) and hemoglobin disassembles outside of the RBC into 4 globins plus 4 heme groups. the globins can be broken down and re-used for its amino acids and the heme groups unfold and the iron atom leaks out: this produces one biliverdin molecule plus one iron atom for each heme group. The iron quickly binds to a plasma protein called **transferrin**, which carries the iron through the blood either to bone marrow for more hemoglobin production or to the liver for storage (it is called transferrin not because it transfers iron, but because ferrin stands for the Latin name for iron, that's why it iron is represented by "Fe").

HEMOGLOBINS AND DIFFERENT BLOOD TYPES

Blood cells are made up of two components. The hemoglobin is in solution inside the cell. The cell is surrounded by a membrane that holds in the hemoglobin. A rough analogy would be a rubber water balloon. The rubber would be the membrane, and the water would be the hemoglobin. The blood types that most of us know as A, B, O, and Rh, are properties of the membrane. The hemoglobin inside the red cells of a person with type O blood and that inside the red cells of a person with type A blood are identical. The analogy would be of water balloons made from blue and red balloons. The color of the ballon would differ, but the material inside (water) would be the same.

ABNORMAL HEMOGLOBINS

Although the changes that produce abnormal hemoglobins are rare, several hundred abnormal (or more precisely, "variant") hemoglobins exist. These have accumulated over the millions of years of human existence. Most variant hemoglobins function normally, and are only found through specialized research techniques. Some hemoglobins, however, do not function normally and can produce clinical disorders, such as sickle cell disease.

HEMOGLOBIN/OXYGEN BINDING

The primary function of human erythrocytes is to transport oxygen to the body's tissues and remove CO_2. This is accomplished by high concentrations of the protein hemoglobin. However, functional expression of hemoglobin requires the co-expression of the small molecule (heme) that specifically binds oxygen; proteins that promote the expression, folding, and addition of heme to hemoglobin, and proteins that maintain the oxidation state of hemoglobin and prevent the accumulation of toxic oxidizing species.

Hemoglobin (Hb) oxygen transport protein is to transport oxygen because oxygen is not very soluble in water (the major constituent of blood), therefore, oxygen transport protein is used which allows oxygen to be soluble. Hb is the used in the blood of vertebrates. It is composed of 4 polypeptide chains (represented in this diagram by different colours. Each chain contains one heme group (coloured orange), each of which contains one iron ion (not shown). The iron is the site of oxygen binding; each iron can bind one O_2 molecule thus each hemoglobin molecule is capable of binding a total to four (4) O_2 molecules.

In humans, the average hemoglobin concentration is 16 g/100 ml that is there are approximately 150,500,000,000,000,000,000 hemoglobin molecules in 100 ml of whole blood.

Oxygen Binding to Hemoglobin

Hb picks up oxygen at the lungs and to deliver it to the tissues. Therefore, Hb binds and releases oxygen. The primary factor that determines how much oxygen is actually bound to hemoglobin is the partial pressure of oxygen (pO_2) in the hemoglobin solution.

Stepwise Reaction:

$$Hb + O_2 \Leftrightarrow HbO_2$$
$$HbO_2 + O_2 \Leftrightarrow Hb(O_2)_2$$
$$Hb(O_2)_2 + O_2 \Leftrightarrow Hb(O_2)_3$$
$$Hb(O_2)_3 + O_2 \Leftrightarrow Hb(O_2)_4$$

Summary Reaction:

$$Hb + 4O_2 \rightarrow Hb(O_2)_4$$

There is small amount that is dissolved in the plasma of the blood, however, this amount is physiologically insignificant. Amount of Hb in blood will determine the amount of oxygen binding. All the Hb molecules are occupied by oxygen and the blood is said to be 100% saturated and the blood cannot carry any more oxygen. When half of the sites are filled with oxygen, the blood is said to be 50% saturated. The following graph demonstrates the effect that pO_2 has on the per cent saturation of Hb. This curve was created by a scientist who exposed Hb to different pO_2 then determining the % saturation of Hb at each pO_2.

Oxygen Transport

Humans are obligate aerobes and oxygen is needed for survival. Oxygen is transported from the lungs to peripheral tissues by the hemoglobin that is densely packed in our red blood cells (erythrocytes). In the lung, oxygen diffuses across the alveolar membrane, and then the red cell membrane in lung capillaries. When it encounters a molecule of hemoglobin, it wedges itself between the iron atom and nitrogen attached to the globin chain, which helps to hold the heme group in place in the protein. One molecule of hemoglobin with its four heme groups is capable of binding four molecules of diatomic oxygen, O_2. The loaded pigment is called **oxyhemoglobin**. The pressure from dissolved oxygen in plasma and in the surroundings in the red cell helps to keep the oxygen on its binding site.

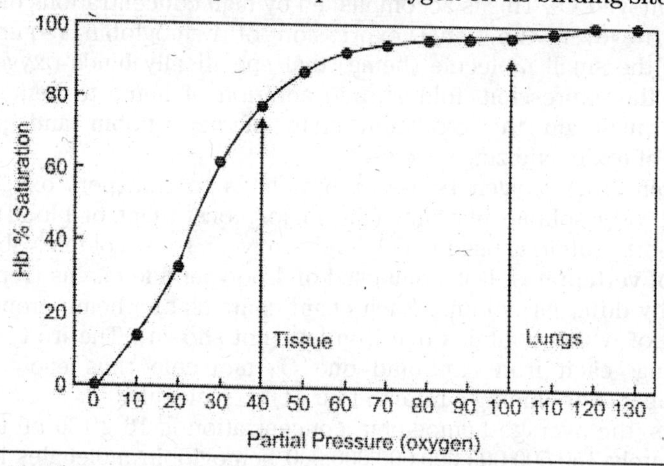

FIG. 27.4: Hb saturation curve. More active will use more oxygen.

As the blood circulates to the periphery, the small amount of dissolved oxygen is consumed first by cells in organs and tissues. This release in pressure makes available the much larger reservoir of heme-bound oxygen, which begins a sequential unloading of its four oxygen molecules. At the most, under normal circumstances only 3 molecules of oxygen are unloaded. Partially or fully unloaded hemoglobin is called **deoxyhemoglobin**. It is a dark blue to purplish colour as in venous blood.

During oxygen unloading, the hemoglobin tetramer undergoes subtle intramolecular conformational changes called **cooperativity**. Oxygen lack is known as hypoxia; the complete absence of oxygen is called anoxia.

Carbon monoxide binds preferentially to hemoglobin over oxygen, smoking tobacco can greatly impact oxygen transport, blocking up to 20% of the oxygen active sites on hemoglobin. Hemoglobin' s affinity for oxygen is decreased in the presence of carbon monoxide because both gases compete for the same binding sites on hemoglobin, carbon monoxide binding preferentially to oxygen. Carbon dioxide occupies a different binding site on the hemoglobin. Carbon dioxide reacts with water to give bicarbonate, carbonic acid freed protons via the reaction, which is catalysed by carbonic anhydrase:

$$CO_2 + H_2O \qquad HCO_3^- + H^+$$

CHAPTER 28

Plasma Proteins and Immunoglobulins

PLASMA

TABLE 28.1: Constituents of plasma as percentages of the total volume

Constituent	(%)
Water	90.0
Protein	8.0
Inorganic ions	0.9
Organic substances	1.1

Plasma is the clear, yellowish fluid portion of blood, lymph, or intramuscular fluid in which cells are suspended. Its total volume in a normal adult is around 2.5 to 3 litres. It differs from serum in that it contains fibrin and other soluble clotting elements. The blood plasma is the clear, liquid part of the blood, composed mainly of water and proteins, in which the blood cells (erythrocytes, leukocytes, and thrombocytes) are suspended. The blood plasma of mammals also contains platelets. Any of the proteins in present in blood plasma is called plasma protein.

Plasma contains clotting agents and on exposure to air it will form a clot. The clear fluid exuded from clotted whole blood and clotted plasma is called serum. The quantities of the constituents of serum are the same as those for plasma minus the small amount of clotting agents, such as fibrinogen, destroyed during the clotting process.

Blood plasma comprises around 20 % of the boy's extracellular fluid and is very similar in composition to interstitial fluid. The main difference is in the protein content. Interstial fluid contains much less protein at around 2% by volume.

This is because most of the plasma protein molecules are too large to pass through the capillary walls into the interstitial area. The small amount of protein which does leak through is eventually taken up, by the lymph and then ultimately returned to the blood.

PLASMA PROTEINS

Plasma proteins form three major groups and these have various functions. Plasma is almost identical to interstitual fluid except that it contains about 7% protein while interstitual fluid contains about 2% protein. The plasma proteins are divided into three major classes:

Protein type	% in serum	% in blood	g/l blood	Primary function/s
Albumin	53%	61	35-50	capillary colloid osmotic pressure, reduces fluid leakage out of capillaries transport
Globulin				

Protein type	% in serum	% in blood	g/l blood	Primary function/s
Alpha globulin	14%	34	23-35	Transport, substrates for formation of other substances
Beta globulin	12%			Transport, substrates for formation of other substances
Gamma globulin				immunity [antibodies]
Fibrinogen	NA	4.1	1.9-3.6	Blood clotting
Total			**60-84**	

All the plasma proteins contribute to developing a colloid osmotic pressure. Fibrinogen is a soluble protein that polymerises to form the insoluble protein *fibrin* during blood clotting. All (except gamma globulins) are synthesised in the liver.

Synthesis and secretion of plasma proteins: Broad spectrum of plasma proteins are synthesised in chick embryo hepatocytes in primary monolayer culture secreted into the culture medium. The secreted proteins are stable and consist almost exclusively of plasma proteins.

Functions of the plasma proteins are varied. It has intravascular osmotic effect. This is important in maintaining fluid and electrolyte balance and is discussed more fully in those sessions. It also maintains the viscosity of the plasma and transport of insoluble substances around the body by allowing them to bind to protein molecules. Plasma serves as reserve for the body and helps in clotting of wounds and wound healing. It has inflammatory response again and protect the body from infection. Gamma globulins function as antibodies. Plasma protein maintains acid-base balance. The relative proportions of plasma proteins can vary in certain diseases and electrophoretic tracings showing such changes can be a useful diagnostic aid.

Source: Most of the plasma proteins are produced by the liver. The gamma globulins are produced by cells of the body's immune system.

Pharmaceutical compositions containing plasma protein: Certain water-soluble products and pharmaceutical formulations in solid or liquid forms or therapeutically active substance having low aqueous solubility and a substantial binding affinity to plasma proteins are present. A plasma protein fraction in controlled aggregation state whereby the said active substance and the said protein fraction are bound to each other by way of non-covalent bonds. It also covers processes for the preparation of the products and pharmaceutical formulation. It is done by dissolving the water-insoluble active substances in a water-miscible, pharmaceutically acceptable solvent, combining said solution with the aqueous solution of a plasma protein fraction in controlled aggregation state whereby a true solution is obtained containing the said active substance and the said protein fraction bound together by way of non-covalent bonds.

Optionally a further pharmaceutically acceptable auxiliary additive–such as a protein aggregation controller and/or a stabilizer–may be present. The organic solvent is eliminated by dialysing, ultrafiltrating, diafiltrating and/or lyophilising. The solid products consisting of the active substance and the protein are also protected. On optional dissolution in clear water, liquid compositions are obtained which are suitable for direct parenteral or other administration.

Types of Plasma Proteins

Fresh frozen plasma: Fresh frozen plasma (FFP) is prepared from a single unit of blood or by apheresis, drawn from a single person. It is frozen after collection and can be stored for one year from date of collection. FFP contains all of the coagulation factors and proteins present in the original unit of blood. It is used for treatment of coagulopathies from warfarin overdose, liver disease, or dilutional coagulopathy. FFP which has been stored more than the standard length of time is sometimes re-classified as simply frozen plasma, which is identical except that the coagulation factors are no longer considered completely viable. Plasma is used as a source of Cryoprecipitate (Plasma, Cryoprecipitate Reduced) is more limited, but still acceptable for many uses. The term "FFP" used informally to mean any frozen transfusable plasma product, including products, which do not meet the standards for FFP.

Dried plasma: Dried plasma was developed and first used in World war- II, using liquid plasma and whole blood. Dried plasma package was developed for for the armed forces to reduce breakage and easy for transportation, packaging, and storage much simpler.

TABLE 28.2: Properties of human albumin

Molecular mass (calc'd)	66438.0 Da
Molecular dimensions	30 X 30 X 80 Å
Isoelectric point, 0.15 M NaCl	5.16
Isoelectric point, defatted	5.8
Absorbance 279 nm, 1 mg/mL	0.531

Albumin

Albumin is the smallest of the plasma proteins and can pass through the capillary walls. In normal circumstances the small amount is leaked into the interstitial fluid. In svere kidney disease large amounts of albumin are able to leak out through the damaged kidney tubules and can be detected in the urine. Because the liver can quickly and easily replace lost albumin the body may lose large amounts of the protein without showing signs of disease.

Chemistry

Inorganic ions: The blood plasma contains inorganic ions which are important in regulating cell function and maintaining homeostasis. For example depletion of potassium may occur following severe diarrhoea and vomiting. Potassium affects cell excitability and severe loss will cause muscle weakness and abnormalities of the cardiac impulse. The same problems may result in severe sodium depletion. Lack of sodium in the plasma will result in a reduction in the overall volume of extracellular fluid which in turn leads to a drop in blood pressure causing weakness, dizziness, mental confusion and fainting.

TABLE 28.3: The normal range of concentration of inorganic ions in the blood plasma

Ion	Symbol	Concentration (mmol/l)
Sodium	Na^+	135-146
Potassium	K^+	3.5-5.2
Calcium	Ca^{++}	2.1-2.7
Chloride	Cl^-	98-108
Hydrogen Carbonate	HCO_3-	23-31
Phosphate	PO_4^-	0.7-1.4

Biomolecules: Human serum albumin is a single peptide chain of 585 amino acids, held in three homologous domains by 17 disulphide bonds. Within each domain are two long loops plus one shorter loop. The S-S bonds provide stability while the intervening peptide strands allow for flexibility. The configuration includes 67% alpha helix and 10% beta turn.

Other substances: A wide range of substances in transit to various tissues throughout the body. These include dissolved gases (mainly carbon dioxide). Oxygen is not very soluble in water hence the need for the specialised oxygen transporting red blood cells.

1. Nutrients are carried in the blood plasma. The most abundant is glucose which is the primary source of energy for cell metabolism. Other nutrients in transit in the plasma include amino acids, fatty acids, triglycerides, cholesterol and vitamins.
2. Waste products of metabolism are also transported by the plasma including urea, uric acid and cratinine from the kidneys and bilirubin from the gall bladder.
3. Hormones, such as cortisol and thyroxine are also transported around the body in plasma attached to plasma proteins.
4. Other substances can be transported in the plasma the most obvious examples being drugs and alcohol.

Structure of Albumin

The structure of albumin from the physical chemical standpoint was discussed by C. Tanford Cohn and Edsall (2003) and reviewed protein model of albumin. The hydrodynamic data reveal that albumin has classic linear, cigar-shaped model. According to Ferrer et al. (2001) that in solution, albumin has heart-shaped structure. There is a rotational correlation time of bovine albumin complexed to the dye and erythrosine. The albumin in neutral solution is rigid and very similar to the 80 × 80 × 80 × 30 heart-shaped structure derived from crystallography of human albumin. Albumin structure is affected even by concentrations of urea found in vivo.

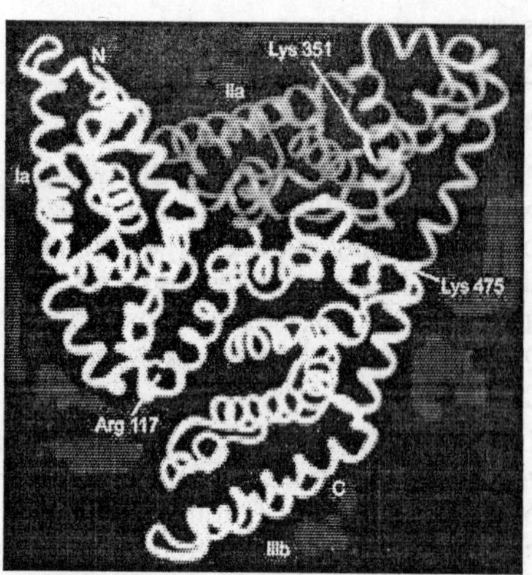

FIG. 28.1: Structure of Human albumin.

Bovine Serum Albumin (BSA)

Bovine serum albumin (BSA) is a large globular protein (66,000 Dal) with a good essential amino acid profile. It has been well characterized and the physical properties of this protein are well known. BSA binds free fatty acids, other lipids and flavour compounds, which can alter the heat denaturation of the protein. Isolated BSA has been reported to be a very functional protein. It is reported to partially unfold between 40° and 50° C, exposing non-polar residues on the surface and facilitating reversible protein-protein interactions. Phospholipid-protein-calcium complexes are formed at pH levels below the isoelectric point of the BSA. Whether these interactions have any biological function is not known.

Biological Function: Bovine serum albumin has been given little attention in respect to its role in the functional properties of whey protein concentrates, and makes up only about 5% of the protein in whey protein concentrates. Its primary biological function has been associated with its lipid binding properties, but the mechanism of this role has not been clearly elucidated. It may play a role in mediating lipid oxidation, since BSA has been shown in vitro to protect lipids against phenolic induced oxidation.

BSA may reduce the probability of a person acquiring certain diseases, such as insulin dependent diabetes or auto-immune disease.

Bosselears *et al.* (1994) compared the anti-mutagenic effect of BSA, soy protein, total whey protein, β-lactoglobulin and pepsin-hydrolysed casein. Of these proteins, only the enzyme-hydrolysed casein and BSA were effective against genotoxic compounds. Bovine serum albumin has been used as a component of cell media to regenerate plants from cultured guard cells and to provide for enhancement of production of plasminogen activator.

BSA can be used as a diluent or a blocking agent in numerous applications including ELISA's (Enzyme-Linked Immunosorbent Assay), blots and immunohistochemistry. It is also used as a nutrient in cell and microbial culture. In restriction digests, bovine serum albumin is used to stabilize some enzymes during digestion of DNA and to prevent adhesion of the enzyme to reaction tubes and other vessels. Bovine serum albumin does not affect other enzymest that do not need it for stabilization.

Human Serum Albumin (HSA)

Human albumin is the most abundant protein in human blood plasma. Human Serum Albumin is produced in liver. Basically albumin is the main protein of plasma and binds water, cations fatty acids, hormones, bilirubin and drugs.

It shows a typical blood concentration of 5 g/100 ml. HSA has a high affinity to a very wide range of materials, including metals such as Cu^{2+} and Zn^{2+}, fatty acids, amino acids, metabolites such as bilirubin and for many drug compounds. The most important physiological role of the protein is therefore thought to be to bring such solutes in the bloodstream to their target organs, as well as to maintain the pH and osmotic pressure of plasma. In addition to its ordinary clinical applications, such as hypovolemic shock treatment, many investigators have attempted to utilize HSA as a carrier to deliver various drugs to their specific targets.

The primary sequence of HSA shows that the protein is a single polypeptide with 585 residues containing 17 pairs of disulphide bridges and one free cysteine. Human serum albumin, as well as serum albumin from other species, has been found to consist of three homologous domains probably derived through gene multiplication. By using hydrodynamics, small-angle X-ray diffraction, electron microscopy and structural prediction, its three-dimensional molecular structure has remained largely unknown.

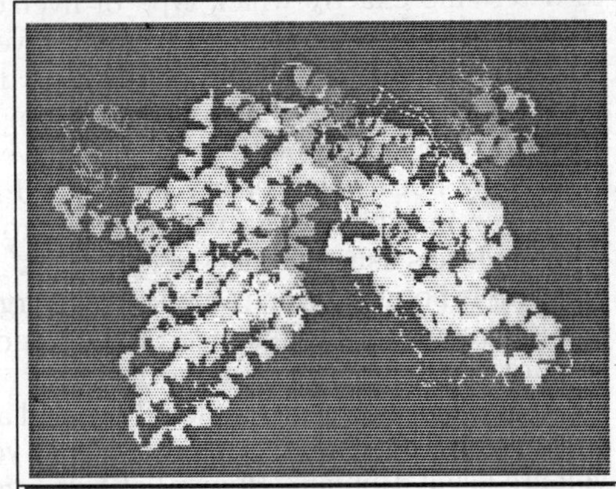

FIG. 28.2: The HSA molecule.

Although all three domains of the HSA molecule have similar three-dimensional structures, their assembly is highly asymmetric). Domains I and II are almost perpendicular to each other to form a T-shaped assembly in which the tail of subdomain IIa is attached to the interface region between subdomains Ia and Ib by hydrophobic interactions and hydrogen bonds. In contrast, domain III protrudes from subdomain IIb at an angel of about 45° to form a Y-shaped assembly for domains II and III. Domain III interacts only with subdomain IIb. These features make the HSA molecule heart-shaped.

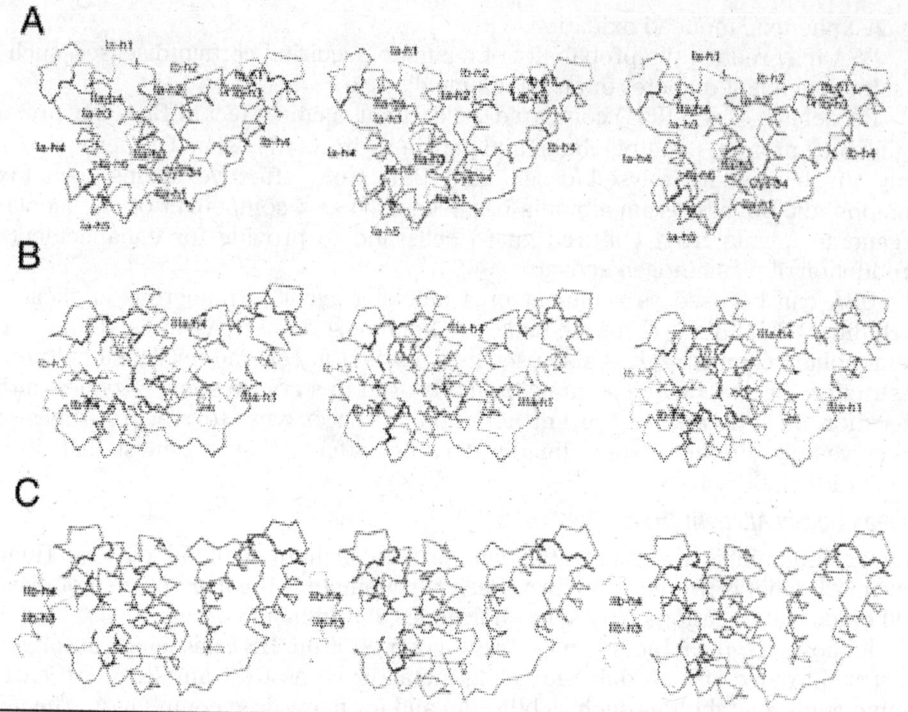

FIG. 28.3: Stereographic illustration of Cα traces of three domains (**A**, domain I; **B**, domain II; **C**, domain III).

Albumin and Human Health

Low albumin levels can suggest liver disease. Other liver enzyme tests are ordered to determine exactly which type of liver disease. Low albumin levels can reflect diseases in which the kidneys cannot prevent albumin from leaking from the blood into the urine and being lost. In this case, the amount of albumin (or protein) in the urine also may be measured. Low albumin levels can also be seen in inflammation, shock, and malnutrition. Low albumin levels may also suggest conditions in which your body does not properly absorb and digest protein (like Crohn's disease or sprue) or in which large volumes of protein are lost from the intestines. High albumin levels usually reflect dehydration.

HAS is not used when a person is allergic to any ingredient in albumin (Human) and a person has ever had heart failure, kidney failure or stable chronic anemia, or you are at risk for fluid overload.

Use Albumin (Human) should be used as per doctor's advise. Albumin (Human) is usually administered as an injection at your doctor's office, hospital, or clinic. If you are using Albumin (Human) at home, carefully follow the injection procedures taught to you by your health care provider. If Albumin (Human) contains particles or is discolored, or if the vial is cracked or damaged in any way, do not use it. Do not infuse Albumin (Human) faster than the rate specified by your doctor.

Globulin

Any of a class of proteins that are found extensively in blood plasma, milk, muscle, and plant seeds and that are insoluble in pure water and half saturated ammonium sulphate, soluble in dilute salt solution, and coagulable by heat.

Globulins can also be measured quantitively and qualitatively with electrophoresis. Radial immunodiffusion is used for accurate quantification of immunoglobulins and has also replaced immunoelectophoresis for determining the immunoglobulin comprising a monoclonal gammopathy.

Five legume species including white bean, mottled bean (Taylor bean), chickpea, lentil, lupin, and a modified mottled bean seeds contain globulins composed by G1 fraction only. Globulins (G1+G2) are present in the seeds and can be analysed for their *in vitro* iron and zinc dialysability.

Alpha globulin: It is a type of globulin in blood plasma that exhibits great colloidal mobility in electrically charged neutral or alkaline solutions. It consists of alpha-1 and alpha-2 globulins. Alpha-1 globulins include alpha-1 antitrypsin, alpha-1 antichymotrypsin, orosomucoid (acid glycoprotein), serum amyloid A, and alpha-1 lipoprotein (HDL). Alpha-2 globulins include alpha-2 macroglobulin (protease inhibitor), haptoglobin (binds free hemoglobin), protein C (inhibitor of activated coagulation factors FVIII and FV), ceruloplasmin (carrier of copper) and alpha-2 lipoprotein (VLDL).

Alpha-1-microglobulin is present in human urea which can be identified by gel chromatography and immunoaffinity chromatography with additional purification by activated charcoal.

Alpha 2u-Globulin (alpha 2u) has been shown to accumulate in the kidneys of male rats treated with 2,2,4-trimethylpentane (TMP). 2,4,4-Trimethyl-2-pentanol (TMP-2-OH), a metabolite of TMP, is found reversibly bound to alpha 2u isolated from the kidneys of these treated rats.

Beat globulins: It consists of beta-1 and beta-2 globulins. Beta-1 globulins include transferrin (binds iron) and hemopexin. Beta-2 globulins include complement factors 3 and 4, C-reactive protein, plasminogen, beta-2 lipoprotein (LDL), hemopexin, beta-2 microglobulin and some proportion of IgA (especially) and IgM. Fibrinogen also migrates in this region.

Gamma globulin: Gamma globulin a group of globulin proteins in human blood plasma, including most antibodies. These antibody substances are produced as a protective reaction of the body's immune system to the invasion of disease-producing organisms (see immunity). Injections of gamma globulin are used to create a rapid but temporary immunity in patients who have been exposed to certain diseases. Children who have been exposed to, but are not immunized against, measles and patients with hepatitis receive some protection from gamma globulin when it is administered during the incubation period of the infection. The gamma globulin used for such purposes is extracted from blood plasma from a large, diverse adult population; the resulting mixture is thus likely to contain antibodies from individuals who had been exposed to the appropriate infections.

Fibrinogen

Presence of fibrinogen in bosy is genetically determined. The soluble protein fibrinogen circulates in the blood and provides the material from which the insoluble fibrin clot is formed during blood coagulation. High plasma fibrinogen concentration predicts future coronary heart disease in men and women. The adult fibrinogen level appears to be determined by environmental factors operating throughout the life course, and levels are inversely related to socioeconomic status. Fibrinogen, as an acute phase reactant, responds to infection and other short-term inflammatory stressors such as tobacco smoke.

The three unique polypeptide chains of human fibrinogen differ significantly in molecular weight. Cross-linkage of fibrin by fibrin-stabilizing factor results in the rapid formation of cross-links between γ-chains and a slower formation of cross-links between α-chains. β-Chains are not involved directly in the cross-linking of fibrin. Reduced, cross-linked fibrin contains uncross-linked β-chains, dimers of γ-chain, and higher polymers of α-chain. Although it is uncertain whether the γ-γ dimers are formed by chains in different molecules of fibrin, the polymers α-chain in fibrin can only be accounted for by cross-linkage of α-chains in different molecules. The nature of cross-linkage among the subunits in fibrin can account well for the three-dimensional, covalent structure of cross-linked, insoluble fibrin.

FIG. 28.4: Small unappreciated crystals of fragment D from human fibrinogen.

Fibrinogen level is higher among smokers than non-smokers and lower among moderate consumers of alcohol compared with non-drinkers. The nature of the alcohol-fibrinogen relationship is congruent with the apparent protective influence of alcohol on risk of myocardial infarction.

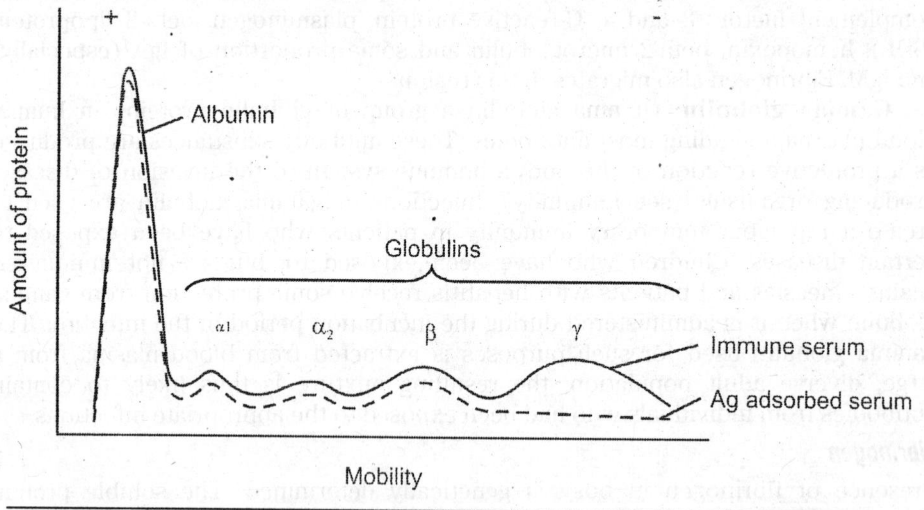

FIG. 28.5: Electrophoretic separation of serum proteins.

The Whitehall II study shows that low control at work is associated with a high fibrinogen level, thus providing evidence that chronic psychosocial stress is indexed by plasma fibrinogen. A single measure of plasma fibrinogen is a candidate allostatic load marker capturing information in particular on the adaptation of the haemostatic system to cumulative lifetime behavioural, biological and psychosocial exposures.

High plasma fibrinogen concentration in adulthood is associated with elevated risk of coronary heart disease and stroke.

IMMUNOGLOBULINS

Immunoglobulin is a protein produced by plasma cells and lymphocytes and characteristic of these types of cells. Immunoglobulins play an essential role in the body's immune system. They attach to foreign substances, such as bacteria, and assist in destroying them. Immunoglobulin is abbreviated Ig. The classes of immunoglobulins are termed immunoglobulin A (IgA), immunoglobulin G (IgG), immunoglobulin M (IgM), immunoglobulin D (IgD) and immunoglobulin E (IgE).

Immunoglobulins (Ig) are Glycoprotein molecules that are produced by plasma cells in response to an immunogen and which function as antibodies. The immunoglobulins derive their name from the finding that they migrate with globular proteins when antibody-containing serum is placed in an electrical field. Immunoglobulins generally assume one of two roles:

1. Immunoglobulins may act as plasma membrane bound antigen receptors on the surface of a B-cell (B Cell or B lymphocytes is a major class of lymphocytes that produce immunoglobulins and are primarily involved in humoral immunity, that is the production of antibodies), or

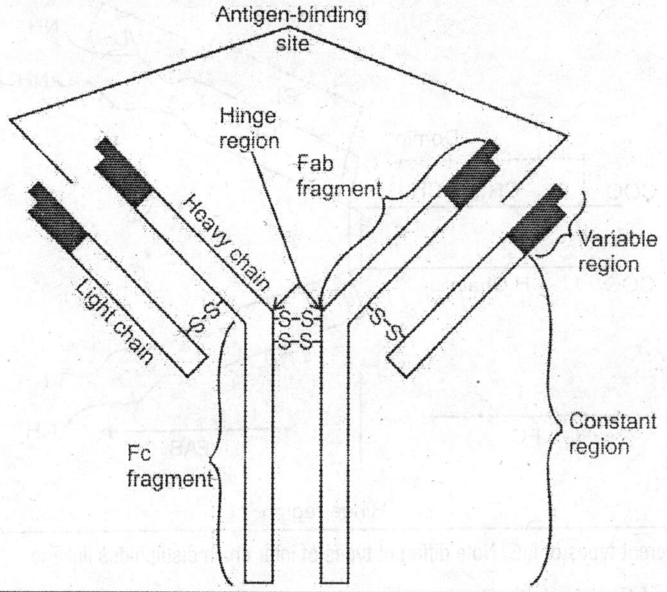

FIG. 28.6: Basic structure of immunoglobulin.

2. Act as antibodies free in cellular fluids functioning to intercept and eliminate antigenic determinants. In either role, antibody function is intimately related to its structure and this page will introduce immunoglobulins (antibodies) and relate their structure to their function in host defense. Antibody is a immunoglobulin whose formation is induced by the introduction of an antigen in an animal body. It reacts with the corresponding antigen specifically in some observable way, that each antibody has binding sites for an identified antigen.

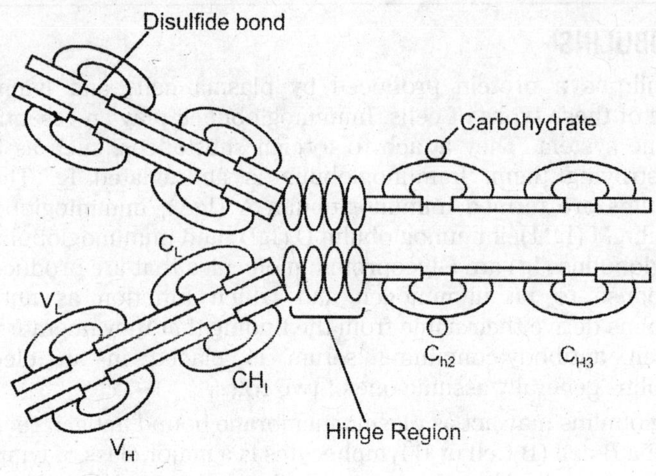

FIG. 28.7: The basic structure of immunoglobulins.

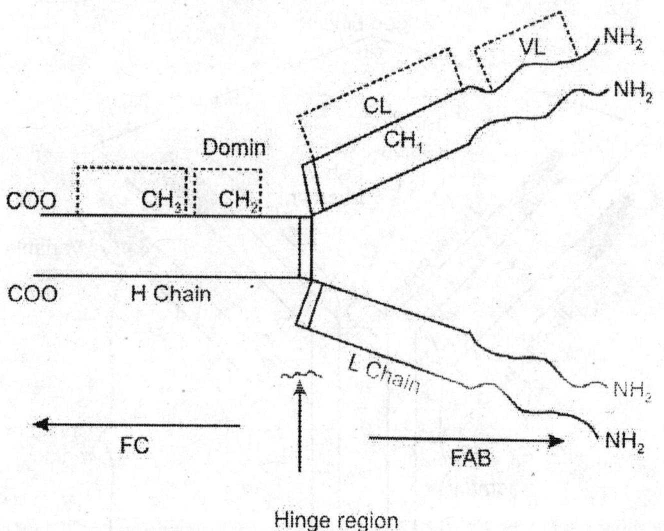

FIG. 28.8: Different types of IgG. Note different types of inter chain disulphides linkage.

Variable (V) and Constant (C) Regions

All immunoglobulins have a four chain structure as their basic unit. They are composed of two identical light chains (23kD) and two identical heavy chains (50-70kD). Each chain has two following regions:

1. **Constant regions:** Constant regions, comprising amino acids 111-220 (or 440-550), are rather uniform, in comparison, from one antibody to another, within the same isotype.

   Hypervariable regions or Complementarity Determining Regions (CDRs) are found within the variable regions of both the heavy and light chains. These regions serve to recognize and bind specifically to antigen. The four polypeptide chains are held together by covalent disulphide (-S-S-) bonds.

2. **Variable regions:** They have amino (NH_2) terminal end of the polypeptide chain (amino acids 1-110). When comparing one antibody to another, these amino acid sequences are quite distinct.

a) **Two light chains** (lambda or kappa): Light chains are composed of 220 amino acid (V_L - 110 amino acids and C_L - 110 amino acids) residues while heavy chains are composed of 440-550 amino acids.
b) **Two heavy chains** (alpha, delta, gamma, epsilon or mu): It determines the immunoglobulin isotype (IgA, IgD, IgG, IgE, IgM, respectively). It contains V_H (110 amino acids) and C_H (330-440 amino acids).

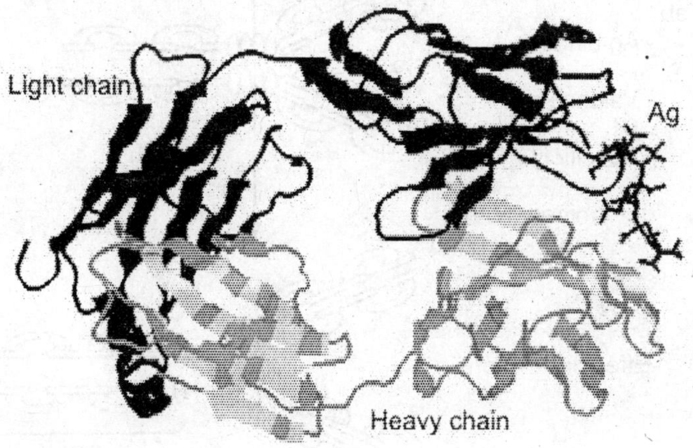

FIG. 28.9: Antigen binding within Fab fragment.

Disulphide bonds

Inter-chain disulphide bonds: The heavy and light chains and the two heavy chains are held together by inter-chain disulphide bonds and by non-covalent interactions. The number of inter-chain disulphide bonds varies among different immunoglobulin molecules.

Intra-chain disulphide binds: Within each of the polypeptide chains there are also intra-chain disulphide bonds.

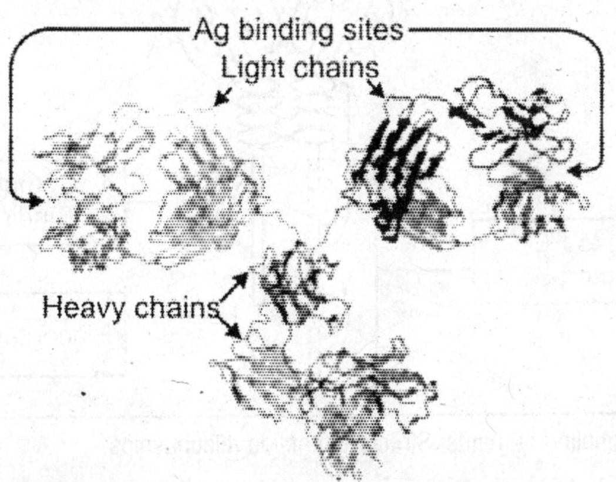

FIG. 28.10: Molecular structure of IgG.

Domains

Three dimensional images of the immunoglobulin molecule show that it is not straight rather it is folded into globular regions each of which contains an intra-chain disulphide bond. These regions are called domains, they are: light chain domains (V_L and C_L) and heavy chain domains (V_H, C_{H1} - C_{H3} or C_{H4}).

Oligosaccharides

Carbohydrates are attached to the C_{H2} domain in most immunoglobulins. However, in some cases carbohydrates may also be attached at other locations.

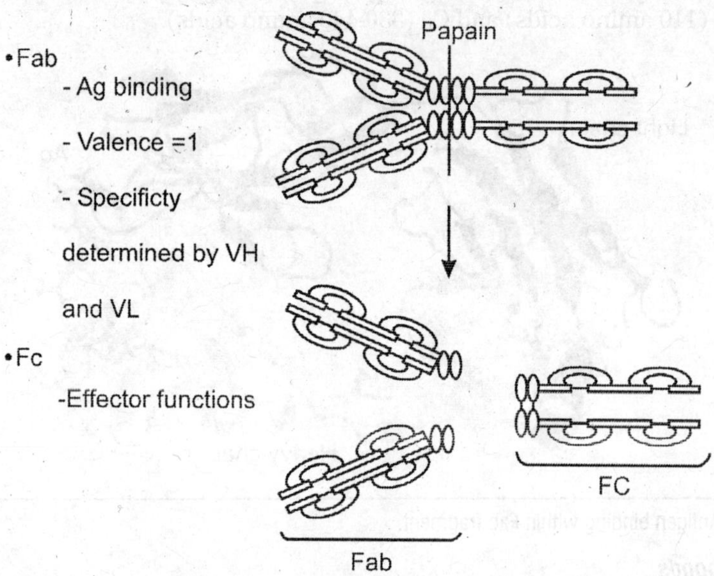

FIG. 28.11: Immunoglobulin fragments: Structure/function relationships.

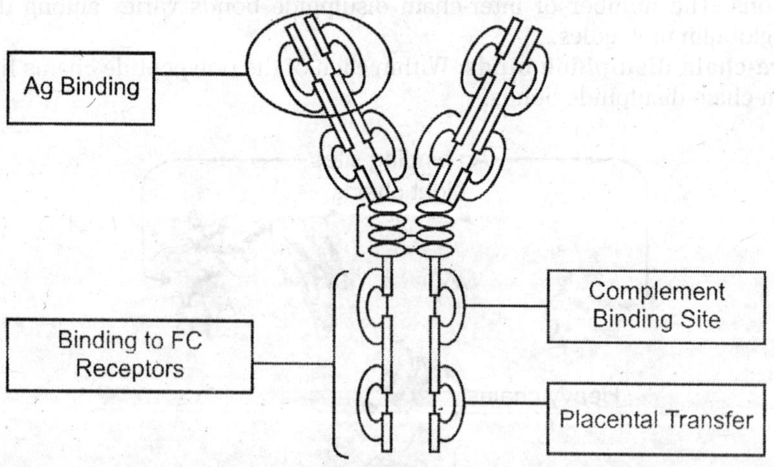

FIG. 28.12: Immunoglobulin fragments: Structure function relationships.

Classification

IgG .... Gamma heavy chains
IgM .... Mu heavy chains
IgA .... Alpha heavy chains
IgD .... Delta heavy chains
IgE .... Epsilon heavy chains

Immunoglobulins considered as a population of molecules are normally very heterogeneous because they are composed of different classes and subclasses each of which has different types and subtypes of light chains. In addition, different immunoglobulin molecules can have different antigen binding properties because of different V_H and V_L regions. Following are the major classes of immunoglobulins.

Classes Immunoglobulin

The immunoglobulins can be divided into five different classes, based on differences in the amino acid sequences in the constant region of the heavy chains.

IgG Subclasses
IgG1 .... Gamma 1 heavy chains
IgG2 .... Gamma 2 heavy chains
IgG3 .... Gamma 3 heavy chains
IgG4 .... Gamma 4 heavy chains
IgA Subclasses
IgA1 .... Alpha 1 heavy chains
IgA2 .... Alpha 2 heavy chains

Subclasses of Immunoglobulin

The classes of immunoglobulins can de divided into subclasses based on small differences in the amino acid sequences in the constant region of the heavy chains. All immunoglobulins within a subclass will have very similar heavy chain constant region amino acid sequences. The differences are detected by serological means.

Immunoglobulin Types

Immunoglobulins can also be classified by the type of light chain that they have. They are: (1) Kappa light chains (2) Lambda light chains. Light chain types are based on differences in the amino acid sequence in the constant region of the light chain. These differences are detected by serological means.

Immunoglobulin Subtypes

The light chains can also be divided into subtypes based on differences in the amino acid sequences in the constant region of the light chain. The Lambda subtypes are: (i) Lambda 1, (ii) Lambda 2, (iii) Lambda 3, and (iv) Lambda 4.

Basis of structural differences

Immunoglobulins are classified on the basis of structural differences.

1. The type of heavy chain determines the immunoglobulin isotype which is based on the primary sequence of amino acids in the constant region of the heavy chain. It also determines the three-dimensional structure of the molecule.
2. Immunoglobulins are proteins and act as an antigen and show immune response which generates anti-immunoglobulin antibodies. However, the structural (three-dimensional) features that define isotypes are not immunogenic in an animal of the same species, since they are not seen as foreign. In human five isotypes (IgA, IgD, IgG, IgE and IgM) has been identified and after injection of human IgG into another human will not generate antibodies directed against the structural features (determinants) that define the IgG isotype. However, injection of human IgG into a rabbit *would* generate antibodies directed against those same structural features.

Basis of Allotype

Immunoglobulins are also defined by the term allotype. Like isotypes, allotypes are determined by the amino acid sequence and corresponding three-dimensional structure of the constant region of the immunoglobulin molecule. Allotypes show genetic differences between members of the same species. This means that not all members of the species will possess any particular allotype. Therefore, injection of any specific human allotype into another human could possibly generate antibodies directed against the structural features that define that particular allotypic variation.

Basis of Idiotype

Immunoglobulins are also classified on the basis of idiotype. Idiotypes are also determined by the amino acid sequence and corresponding three-dimensional structure of the variable region of the immunoglobulin molecule. In this regard, idiotypes reflect the **antigen binding specificity** of any particular antibody molecule. Idiotypes are so unique that an individual person is probably capable of generating antibodies directed against their own idiotypic determinants. This probability forms the basis of the *Idiotypic Network Hypothesis.*

Basis of Antigenic Character

Based on the antigenic character of heavy chains the immunoglobulins are classified into various types, the important are: Immunoglobulin G (IgG), Immunoglobulin M (IgM), Immunoglobulin D (IgD), Immunoglobulin A (IgA), Immunoglobulin E (IgE).

IgG: It is the major immunoglobulin in human serum, accounting for 70 to 75% of the immunoglobulin pool. IgG is presents in blood plasma and tissue fluids. The IgG class acts against bacteria and viruses by opsonizing the invaders and neutralizing toxins. It is also one of the two immunoglobulin classes that activate complement by the classical pathway. IgG is the only immunoglobulin molecule able to cross the placenta and provide naturally acquired immunity for the newborn.

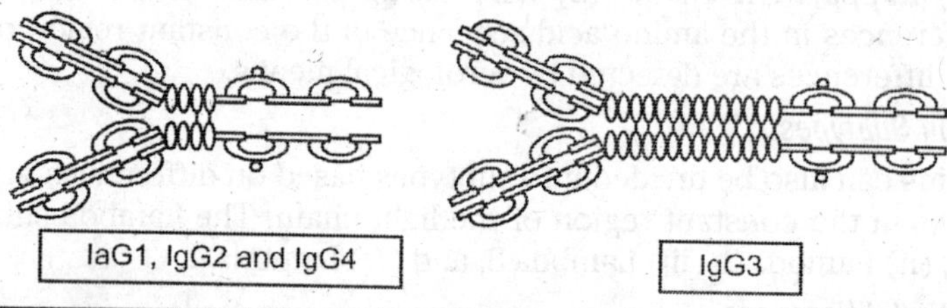

FIG. 28.13: IgG structure.

IgG subclasses: There are four subclasses of IgG: IgG1, IgG2, IgG3, and IgG4. These subcleases vary in their chemical composition of their chain and the number and arrangement of interchain disulphide bonds. About 65% of the total serum IgG is IgG1, and 23% is IgG2. Differences in biological function have been noted in these subclasses. For example, IgG2 antibodies are opsonic and develop in response to antitoxins. Anti-Rh antibodies are of the IgG1 or IgG3 subclass. IgG1 and IgG3 also bind best to monocytes and macrophages and activate complement most effectively. The IgG4 antibodies function as skinp sensitizing immunoglobulins.

IgM: It accounts for about 10% of the immunoglobulin pool. It is usually a polyer (pentamer) of five monomeric units, each composed of two heavy chains and two light chains. The monomers are arranged in a pinwheel array with the Fc ends in the center, held together by a special **J** (joining) **chain**. IgM is the first immunoglobulin made during B-cell maturation and the first secreted into serum during primary antibody response. Since IgM is so large, it does not leave the bloodstream or cross the placenta. IgM agglutinates bacteria, activates complement by the classical pathway, and enhances the ingestion of pathogens by phagocytic cells. This class also contains special antibodies such as red blood cell agglutinins and heterophile antibodies.

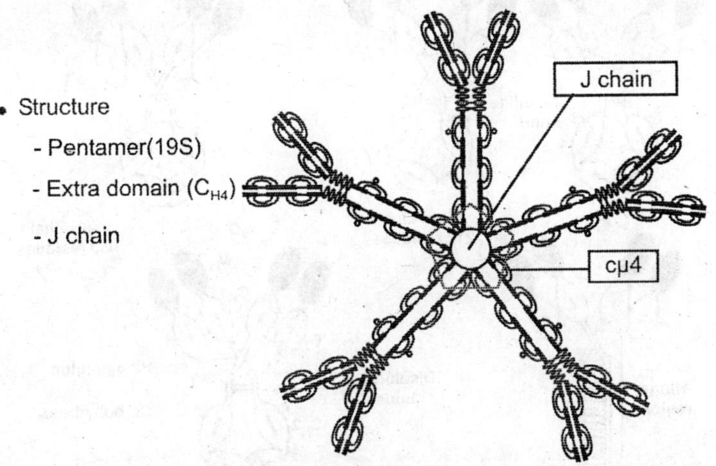

FIG. 28.14: Pentameric serum IgM structure.

Although most IgM appears to be pentameric, around 5% or less of human serum IgM exists in a hexameric form. This molecule contains six monomeric units but seems to lack a J chain. Hexameric IgM activates complement up to twenty fold more effectively than does the normal pentameric form. It has been suggested that bacterial cell wall antigens such as gram-negative lipopolysaccharides may directly stimulate B cells to form hexameric IgM without a J chain. If this is the case, the immunoglobulins formed during primary immune responses are less homogenous than previously thought.

IgA: It accounts for about 15% of the immunoglobulin pool. Some IgA is present in the serum as a monomer of two heavy and two light chains. Most IgA, however, occurs in the serum as a polymerized dimer held together by a J chain. IgA has special features that are associated with secretory mucosal surfaces. IgA, when transported from the mucosa-associated lymphoid tissue to mucosal surfaces, acquires a protein termed the secretory component.

Secretory **IgA (sLgA)** as the modified molecule is now called, is the primary immunoglobulin of the secretory immune system. This system is found in the gastrointestinal tract, upper and lower respiratory tracts, and genitourinary system. Secretory IgA is also found in saliva, tears, and breast milk. In these fluids and related body areas, sLgA plays a major role in protecting surface tissues against infectious microorganisms by the formation of an immune barrier. For example, in breast milk sLgA helps protect nursing newborns. In the intestine, sLgA attaches to viruses, bacteria, and protozoan parasites such as *Entamoeba histolytica*. This prevents pathogen adherence to mucosal surfaces and invasion of host tissues, a phenomenon known as immune exclusion. In addition, sLgA binds to antigens within the muscosal lamina propria, and the antigen-sLgA complexes are excreted through the adjacent epithelium into the gut lumen. This rids the body of locally formed immune complexes and decreases their access to the circulatory system. Secretory IgA also may neutralize viruses and other intracellular pathogens that reside within epithelial cells. Secretory IgA also plays a role in the alternate complement pathway.

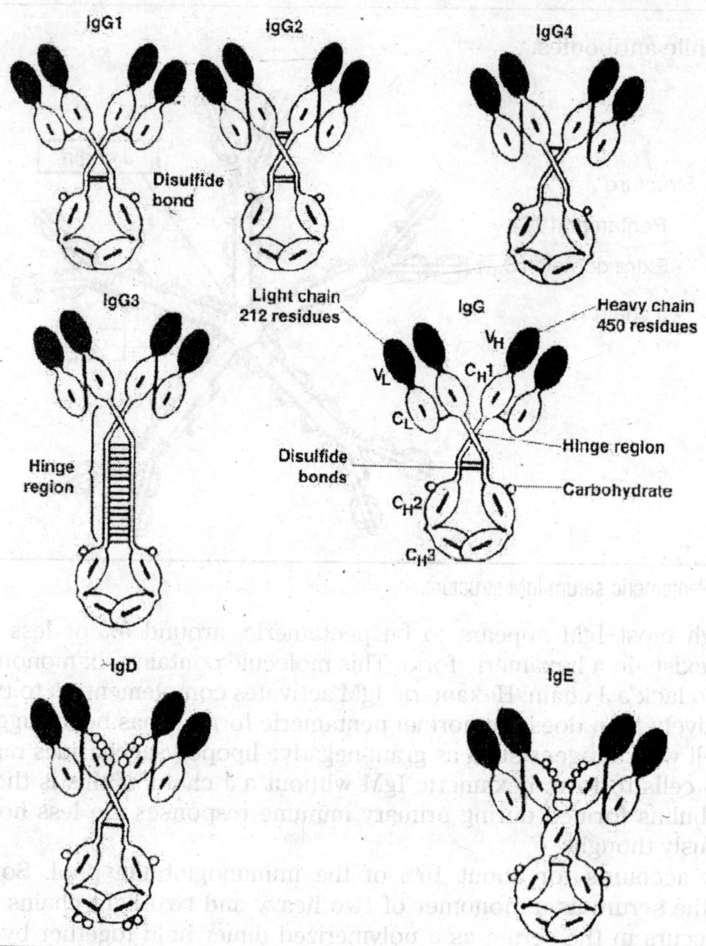

FIG. 28.15: Different types of immunoglobulins.

IgD: It is an immunoglobulin found in trace amounts in the blood serum. It has a monomer structure similar to IgG. IgD antibodies do not fix complement and cannot cross the placenta, but they are abundant on the surface of B cells and bind antigens, thus signaling the B cell to start antibody production.

IgE: It makes up only 0.00005% of the total immunobulin pool. The classic skin-sensitizing and anaphylactic antibodies belong to this class. IgE molecules have four constant region domains ($C_\varepsilon 1$, $C_\varepsilon 2$, $C_\varepsilon 3$ and $C_\varepsilon 4$) and two light chains.

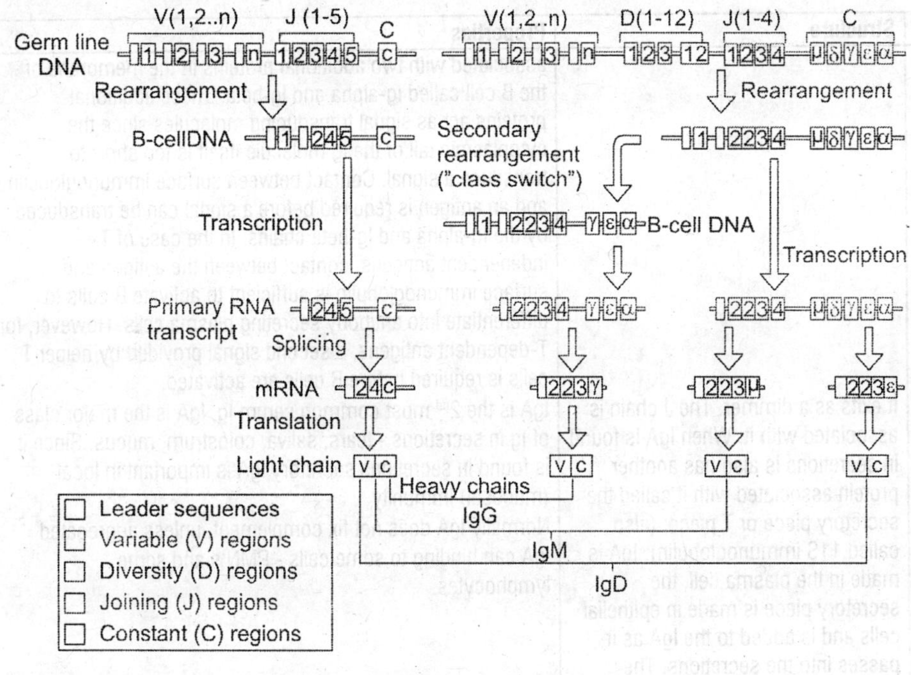

FIG. 28.16: Immunoglobulin chain differentiation and production.

TABLE 28.4: Properties of Ig classes and subclasses

IgG	Structure	Properties
	All IgG's are monomers (7S immunoglobulin). The subclasses differ in the number of disulphide bonds and length of the hinge region.	It is versatile immunoglobulin because it is capable of carrying out all of the functions of immunoglobulin molecules. IgG is the major Ig in serum - 75% of serum Ig is IgG IgG is the major Ig in extra vascular spaces Placental transfer - IgG is the only class of Ig that crosses the placenta. Transfer is mediated by receptor on placental cells for the Fc region of IgG. Not all subclasses cross equally Fixes complement - Not all subclasses fix equally well; IgG4 does not fix complement **Binding to cells** - Macrophages, monocytes, PMN's and some lymphocytes have Fc receptors for the Fc region of IgG.
IgM	IgM normally exists as a pentamer (19S immunoglobulin) but it can also exist as a monomer. In the pentameric form all heavy chains are identical and all light chains are identical. Thus, the valence is theoretically 10. IgM has an extra domain on the mu chain (C_{H4}) and it has another protein covalently bound via a S-S bond called the J chain. This chain functions in polymerization of the molecule into a pentamer.	IgM is the third most common serum Ig IgM is the first Ig to be made by the fetus and the first Ig to be made by a virgin B cells when it is stimulated by antigen It has pentameric structure, IgM is a good complement fixing Ig. Thus, IgM antibodies are very efficient in leading to the lysis of microorganisms IgM is also a good agglutinating Ig. IgM antibodies are very good in clumping microorganisms for eventual elimination from the body IgM binds to some cells via Fc receptors **B cell surface Ig**: Surface IgM exists as a monomer and lacks J chain but it has an extra 20 amino acids at the C-terminus to anchor it into the membrane. Cell surface IgM functions as a receptor for antigen on B cells. Surface IgM is noncovalently

IgG	Structure	Properties
		associated with two additional proteins in the membrane of the B cell called Ig-alpha and Ig-beta. These additional proteins act as signal transducing molecules since the cytoplasmic tail of the Ig molecule itself is too short to transduce a signal. Contact between surface immunoglobulin and an antigen is required before a signal can be transduced by the Ig-alpha and Ig-beta chains. In the case of T-independent antigens, contact between the antigen and surface immunoglobulin is sufficient to activate B cells to differentiate into antibody secreting plasma cells. However, for T-dependent antigens, a second signal provided by helper T cells is required before B cells are activated.
IgA	It exits as a dimmer. The J chain is associated with it. When IgA is found in secretions is also has another protein associated with it called the secretory piece or T piece; (also called 11S immunoglobulin). IgA is made in the plasma cell, the secretory piece is made in epithelial cells and is added to the IgA as it passes into the secretions. The secretory piece helps in transporation of IgA across mucosa and protects it from degradation in the secretions.	IgA is the 2nd most common serum Ig, IgA is the major class of Ig in secretions - tears, saliva, colostrum, mucus. Since it is found in secretions secretory IgA is important in local (mucosal) immunity Normally IgA does not fix complement, unless aggregated IgA can binding to some cells - PMN's and some lymphocytes.
IgD	It exists only as a monomer.	IgD is found in low levels in serum; its role in serum uncertain IgD is primarily found on B cell surfaces where it functions as a receptor for antigen. IgD on the surface of B cells has extra amino acids at C-terminal end for anchoring to the membrane. It also associates with the Ig-alpha and Ig-beta chains IgD does not bind complement.
IgE	IgE exists as a monomer and has an extra domain in the constant region.	IgE is the least common serum Ig since it binds very tightly to Fc receptors on basophils and mast cells even before interacting with antigen Involved in allergic reactions - As a consequence of its binding to basophils an mast cells, IgE is involved in allergic reactions. Binding of the allergen to the IgE on the cells results in the release of various pharmacological mediators that result in allergic symptoms IgE also plays a role in parasitic helminth diseases. Since serum IgE levels rise in parasitic diseases, measuring IgE levels is helpful in diagnosing parasitic infections. Eosinophils have Fc receptors for IgE and binding of eosinophils to IgE-coated helminths results in killing of the parasite IgE does not fix complement.

TABLE 28.5: Various immunoglobulins

IgM	IgD	IgG	IgE	IgA
	B-cell		mast cell	

TABLE 28.6: Clinical Implications of human immunoglobulin classes

Human immunoglobulin classes	In increased level	In decreased level
IgG	Chronic granulomatous infections Disease associated with hypersensitivity granulomas, dermatologic disorders, and IgG myeloma Dysproteinemia Hyperimmunization Infections of all types Liver disease Malnutrition (severe) Rheumatoid arthritis	Agammaglobulinemia Bence Jones proteinemia Chronic lymphoblastic leukemia IgA myeloma Lymphoid aplasia Selective IgG, IgA deficiency
IgM	Actinomycosis Carrión's disease (bartonellosis) Dysgammaglobulinemia (certain cases) Infectious mononucleosis Lupus erythematosus Malaria Rheumatoid arthritis Trypanosomiasis Waldenström's macroglobulinemia. In the newborn, a level of IgM above 20 ng./dl is an indication of in utero stimulation of the immune system and stimulation by the rubella virus, the cytomegalovirus, syphilis, or toxoplasmosis.	Agammaglobulinemia Lymphoproliferative disorders (certain cases) Chronic lymphoblastic leukemia Dysgammaglobulinemia IgG and IgA myeloma Lymphoid aplasia.
IgA	Certain stages of collagen and other autoimmune disorders such as rheumatoid arthritis and lupus erythematosus Chronic infections not based on immunologic deficiencies Cirrhosis of the liver (most cases) IgA myeloma Wiskott-Aldrich syndrome.	Acute lymphoblastic leukemia Chronic lymphoblastic leukemia Hereditary ataxia telangiectasia IgG myeloma Immunologic deficiency states (*e.g.*, dysgammaglobulinemia, congenital and acquired agammaglobulinemia, and hypogammaglobulinemia) Lymphoid aplasia Malabsorption syndromes.
IgD	Chronic infections IgD myelomas	
IgE	Anaphylactic shock Asthma	Congenital agammaglobulinemia Hypogammaglobulinemia due to faulty

Human immunoglobulin classes	In increased level	In decreased level
	Atopic skin diseases such as eczema Hay fever IgE-myeloma	metabolism or synthesis of immunoglobulins

The Fc portion of the Cε4 chain can bind to special Fc_ receptors on mast cells and basophils. When two IgE molecules on the surface of these cells are corss-linked by binding to the same antigen, the cells degranulate. This degranulation releases histamine and other pharmacological mediators of anaphylaxis. It also stimulates eosinophilia and gut hypermotility (increased rate of movement of the intestinal contents) that aid in the elimination of helminthic parasites. Thus, through IgE is present is small amounts, this class of antibodies has very potent biological capabilities.

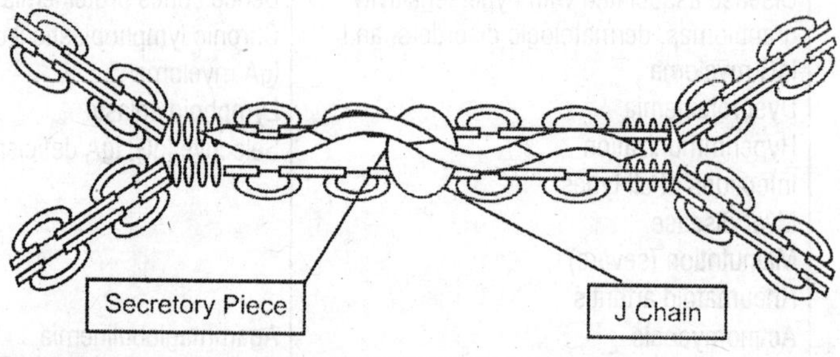

FIG. 28.17: IgA structure.

FUNCTIONS OF IMMUNOGLOBULIN

All immunoglobulin molecules are bifunctional. The Fab region is concerned with binding to antigen, whereas the Fc region mediates binding to host tissue, various cells of the immune system, some phagocytic cells, or the first component of the complement system. The binding of an antibody with an antigen usually does not cause destruction of the antigen or of the microorganism, cell or agent to which it is attached. Rather the antibody serves to mark and identify the target for immunologic attack and to activate non-specific immune responses that can destroy the target for phagocytosis by neutrophils and macrophages. This ability of an antibody to stimulate phagocytosis is termed **opsonization.** Immune destruction also is promoted by antibody-induced activation of the complement system.

Sources of Antibodies

An antibody is a immunoglobulin whose formation is induced by the introduction of an antigen in an animal body. It reacts with the corresponding antigen specifically in some observable way, that each antibody has binding sites for an identified antigen. The need for pure homogeneous antibodies has increased dramatically in recent years. Currently antibodies are produced either naturally by immunization or artificially through hybridoma formation.

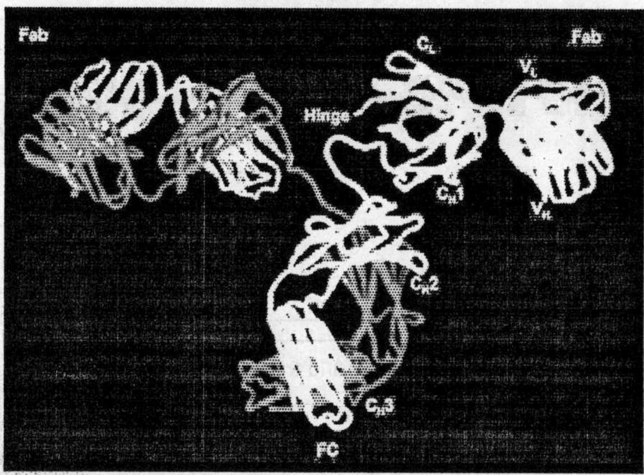

FIG. 28.18: Structure of antibody.

Biosynthesis of Antibodies

Evidences are still lacking to explain the production of antibodies by cells following antigenic stimulation. Two theories have been introduced:

1. **Directive or Template Theory:** This theory explains that antigen in the antibody forming cells acts as a templet for biosynthesis of antibody having complementary configuration. The antibody formed than dissociates from the antigen molecule which further acts as templet for next molecule of antibody. It has been suggested that antigen brings about genetic changes in the cell so that it and its daughter cells continue to produce specific antibodies.
2. **Selective Theory:** The theory was first proposed by Paul Ehrlich in 1880 and is also known as **Ehrlich side chain theory**. It explains that antibody forming cells have antibody molecules as side chains on their surface and the antigen selects its corresponding side chain, attaches to it and finally knocks it down. This acts as a trigger and then the cell starts synthesizing similar side chains repeatedly which are dislodged to free circulation and unite with the antigen.

Diversity of Antibodies

One unique property of antibodies is their remarkable diversity. According to current estimates each human or mouse can synthesise more than 10 million different kinds of antibodies. How is this diversity generated? The answer is threefold:

1. rearrangement of antibody gene segments,
2. somatic mutations, and
3. generation of different codons during antibody gene splicing.

Generation of Antibody Diversity

The immune system has the capacity to recognize and respond to about 10^7 different antigens. This extreme diversity can be generated in at least three possible ways:

1. Multiple genes in the germ line DNA.
2. Variable recombination during the differentiation of germ line cells into B-cells.
3. Mutation during the differentiation of germ line cells into B-cells.

It is known that all three of these possibilities take place to produce antibody diversity. The following figures illustrate these possibilities:

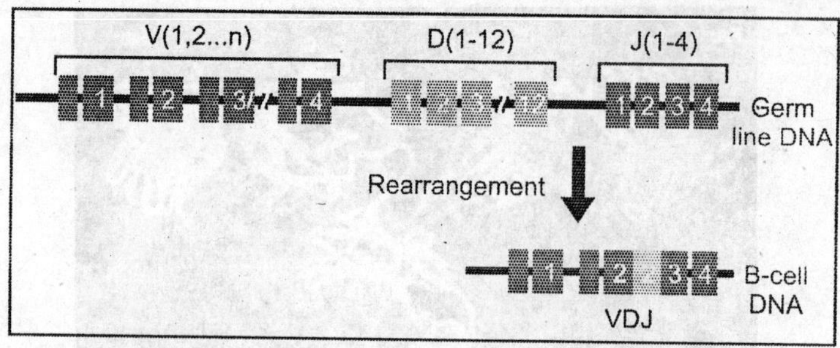

FIG. 28.19: Genetic makeup of a germ line cell and a mature B-cell at the loci controlling heavy chain production.

The figure shows the genetic makeup of a germ line cell and a mature B-cell at the loci controlling heavy chain production. Germ line DNA has many (up to 200) different variable (V) region genes, in addition to 12 diversity (D) region genes and four joining (J) region genes. During differentiation of this cell into the B-cell, rearrangement of the DNA occurs. This rearrangement aligns one of the many V genes with one of the D genes and one of the J genes, producing a functional VDJ recombinant gene. Since any of the genes may recombine with any others, this rearrangement has the potential to generate 200 × 12 × 4 = 9600 different possible combinations. The same type of event occurs in the genes encoding the immunoglobulin light chains where about 200 different V regions may recombine with about 5 different J regions giving rise to 200 × 5 = 1000 possible light chains. Since in any particular B-cell, any light chain combination can occur along with any heavy chain combination, the total possible immunoglobulin combinations approaches 10^7 (9600 × 1000).

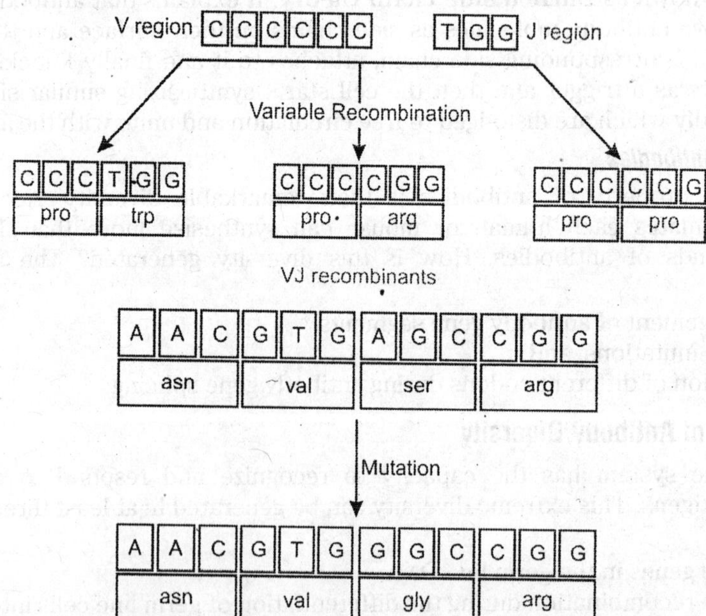

A **second way that diversity** can result is through a process of variable or "inaccurate" recombination. The figure illustrates three possible recombination events between the variable (V) and joining (J) regions of an immunoglobulin light chain. In the first event, a proline-tryptophan dipeptide sequence is produced in the resulting protein. However, in the second and third events, differential recombination places proline-arginine or proline-proline sequences into the resulting immunoglobulin. These types of events may also occur between the V and D regions and the D and J regions of the heavy chain DNA sequence.

A third way that diversity can result is through a process of mutation. This process simply involves changes in DNA sequence that occur during differentiation of the B-cell. The figure illustrates how an A:T to G:C transition mutation could change a serine residue into a glycine residue in the resulting immunoglobulin. This process may, in part, explain the diversity observed in hypervariable (CDR) regions.

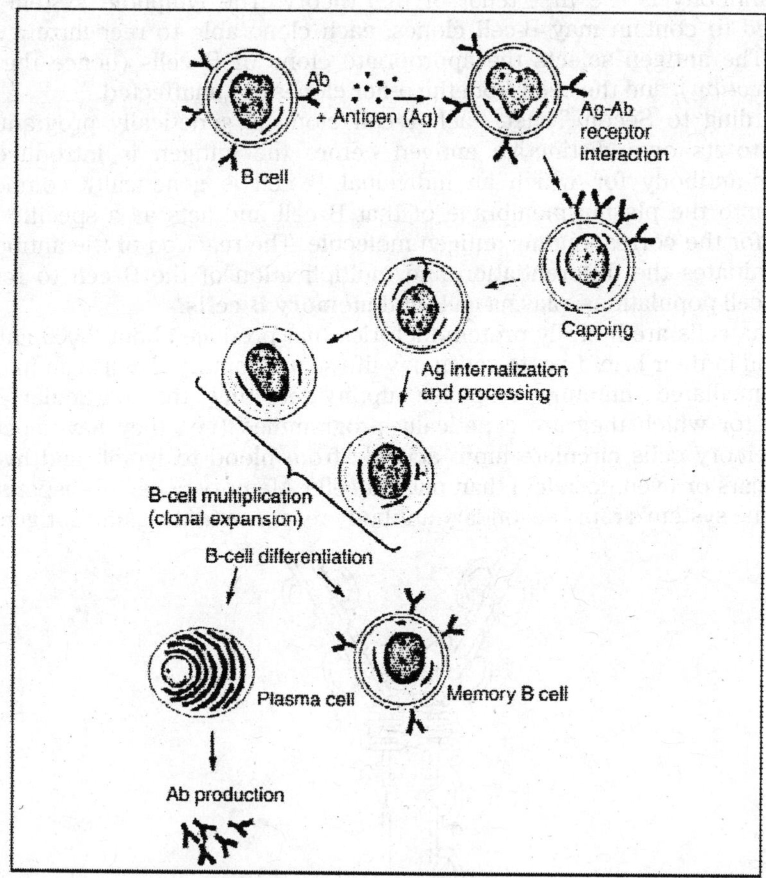

FIG. 28.20: Clonal Selection. It is through clonal selection that the immune system can respond specifically to myriad of possible antigens, whether they are individual molecules or are attached to pathogens and abnormal cells such as cancer cells. B cells or B lymphocytes constantly roam the body, particularly the blood and lymphoid tissues. Each B cell synthesises only one of the millions of possible antibodies and displays this antibody of the proper specificity (top left), it complexes with the antibody and capping occurs. (Capping is the regional aggregation of antibodies on the surface of the following Ag-Ab interaction.) The antigen is then internalized; the B cell swells and begins to divide rapidly, producing a B-cell clone. The activated B-cell clone differentiates into plasma cells and memory cells. Plasma cells form the specific antibody that immediately attack the antigen that provoked its formation. Memory B cells persist in the body and boost the immune system's readiness to eliminate the same antigen if it present itself in the future.

Specificity of Antibodies

As noted previously, combinatorial joining, somatic mutations, and variations in the splicing process generate the great variety of antibodies produced by B cells. From a large, diverse B-cell pool, specific cells are stimulated by antigens to reproduce and form a B-cell clone that contains the same genetic information. This is known as the **clonal selection theory**, a hypothesis to explain immunologic specificity and memory.

The existence of a small B-cell **clone** (a population of cells derived asexually from a single parent) that can respond to one or a few antigens by producing the correct antibody is the first tenet of this theory. The lymphoid system is thus considered to contain may B-cell clones, each clone able to recognize a specific antigen. The antigen selects the appropriate clone of B cells (hence the pharse *clonal selection*), and the cells from the other clones are unaffected.

According to Secon Tenet, each B-cell clone is genetically programmed to respond to its own distinctive antigen before the antigen is introduced. The particular antibody for which an individual B cell is genetically competent is ingrated into the plasma membrane of that B cell and acts as a specific surface receptor for the corresponding antigen molecule. The reaction of the antibody and antigen initiates the differentiation and multiplication of the B cell to form two different cell populations: plasma cells and **memory B cells.**

Plasma cells are literally protein factories that produce about 2,000 antibodies per second in their brief five- to seven-day life span. Memory B cells can initiate the antibody-mediated immune response upon detecting the particular antigen molecule for which they are genetically programmed (i.e., they have specificty). These memory cells circulate more actively from blood to lymph and live much longer (years or even decades) than plasma cells. Memory cells are responsible for the immune system's rapid secondary antibody response to the same antigen.

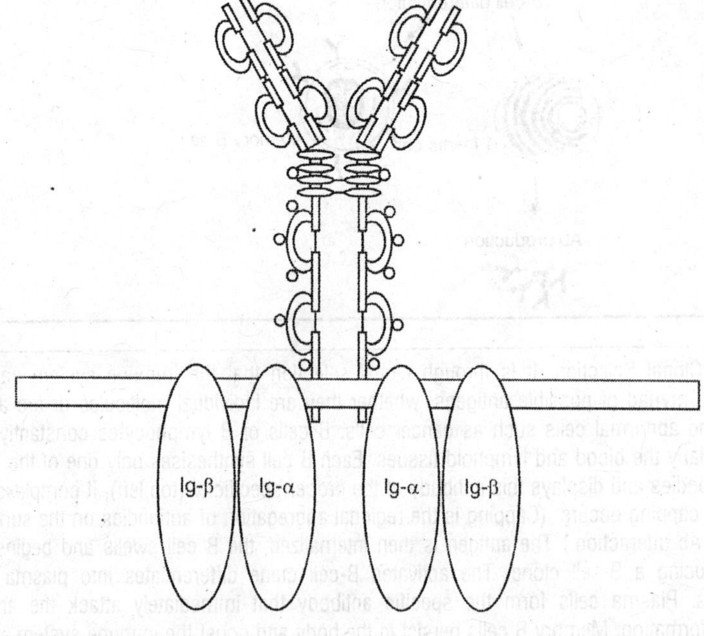

FIG. 28.21: B cell antigen receptor (BcR).

Immunization

Specific antibodies can be produced naturally by the immunization of domestic animals or human volunteers. Whatever the source, purified antigen is injected into the host. The host's immune system recognizes and responds to the antigen, and its B cells proliferate and differentiate to produce specific antibodies. To promote the efficiency of antigen stimulation of antibody production, the antigen is mixed with an **adjuvant** (Latin *adjuvans*, aiding), which enhances the rate and quantity of antibody produced.

Following repeated antigen injections at regular intervals, blood is withdrawn from the post and allowed to clot. The fluid that remains after the blood clots is the **serum**. This serum has been obtained from an immunized host and contains the desired antibodies. It is called **antiserum**.

Antiserum is a major and convenient source of antibodies, however, its usefulness is limited in following ways:

1. Antibodies obtained by this method are polyclonal; they are produced by several B-cell clones and have different specificities. The decreases their sensitivity to particular antigens and results in some degree of cross-reaction with closed related antigen molecules.
2. Second or repeated injections of antiserum from one species to another can cause serious allergic or hypersensitivity reactions.
3. Antiserum contains a mixture of antibodies, not all of which are of interest in a give immunization.

The Primary Antibody Responses

During immunization procedures (and also in naturally acquired immunity) there is an initial lag phase of several days following a primary challenge with an antigen. During the *lag phase no antibody can be detected.* The antibody **titer**, which is the reciprocal of the highest dilution of an antiserum that gives a positive reaction in the test being used, rises logarithmically to a plateau during the second or log phase. In the plateau phase the antibody titer stablizes. This is followed by a decline phase, during which antibodies are naturally metabolized or bound to the antigen and cleared from the circulation. During the primary antibody response, IgM appears first, then IgG. The affinity of the antibodies for the antigen's determiants is low to moderate during this primary antibody response.

The Secondary Antibody Response

The primary antibody response primes the immune system so that it possesses specific immunologic memory through its clones of memory B cells. Upon secondary antigen challenge, the B cells mount a heightened or **anamnestic** [Greek *anamnesis*, remembrance] **response** to the same antigen. Compared to the primary antibody response, the secondary antibody response has a shorter lag phase, a more rapid log phase, persists for a longer plateau period, attains a higher IgG titer, and produces antibodies with a higher affinity for the antigen (affinity maturation).

Catalytic Antibodies

In the past several years immunologists have applied the principles of enzymology to create a new class of antibody molecules- **catalytic antibodies**. Catalytic antibodies accelerate specific chemical reactions by lowering the free energy of transition states. They accomplish this by binding reactants in a cleft or crevice on their surfaces and inducing structural changes in the substrate molecules. Because antibodies directed against a huge array of biopolymers, natural products, and synthetic molecules can be formed, catalytic antibodies offer a unique approach for generating tailor-made, enzyme-like catalysts.

Catalytic antibodies are made by coupling a transitionstate analogue to a carrier protein and injecting the combination into an experimental animal. Antibody-secreting spleen cells are taken from the animal and fused with myeloma cells. The hybrid antibody-secreting cells divide indefinitely and generate clone of cells, each hybrid clone secreting a monoclonal catalytic antibody with a unique antigen-binding pocket. A clone that makes catalytic antibody specific for the analogue is then selected.

Currently available catalytic antibodies transform relatively simple compounds. Much of the potential of catalytic antibodies for biotechnology and molecular biology depends on the development of catalytic antibodies able to act on proteins or nucleic acids. If this can be accomplished, catalytic antibodies could extend the immune system's innate capacity to defend the body. For example, one might stimulate the immune system of a patient with heart disease to produce antibodies that would break up the proteins in blood clots, forestalling heart attacks.

ANTIGENS

The name antigens (Gk. *anti* = against, *genos* = genus) is given to organic substances of a colloid structure (proteins and different proteins complexes in combination with lipids of polysaccharides). It upon injection into the body (subcutaneously, intracutaneously, cutaneously, into the mucous membranes, intramuscularly, intravenously and orally) are capable of causing the production of antibodies and reacting specifically with them.

Proportion of Antigens

An antigen may be soluble substance such as horse serum proteins or a bacterial toxin, or it may be present on particulate matter like red blood cells, a bacterial cell, or a virus. An antigen is always a foreign substance for the host. An antigen must be capable of inducing an antibody response. The molecular weight of any antigen should be more than 6000 Daltons. The portion of antigen that specifically combines with antibody is called its determinant group. Antigenic properties are pertinent to toxins of a plant origin (ricin, robin, abrin, cortin, etc.), toxins of an animal origin (toxins of snakes, spiders, scorpions etc.), enzymes, native foreign proteins, various cellular component, bacteria and their toxins and viruses etc.

Antigens are of two types: (a) Complete antigens and (b) Partial antigens:

1. **Complete antigens:** They cause the production of antibodies in the body, and react with them in vivo as well as in vitro.
2. **Partial antigens:** They are also known as haptens and do not cause the production of antibodies, but can react with them. Haptens includes lipids, complex carbohydrate and other substances.

The immunological specificity of antigens is linked with a determinant group found on the surface of the antigen as one or more active areas.

Isoantigens: These substances have antigenic properties and are contained in some individuals of a given species. Isoantigens are found in erythrocytes of animals and man.

Antigen-Antibody Binding

Every antigen monomer has two similar antigen combining sites and each site is formed by `V' region of `H' and `L' chains. This helps the antibody molecule to link 2 similar antigens together. When many antibody and antigen unit join together, it results in the formation of a *lattice*. A large antigen and the cell surface of the microorganisms have several antigenetically active sites. A large number of antibody molecules are usually linked with antigenic sites and form an aggregates.

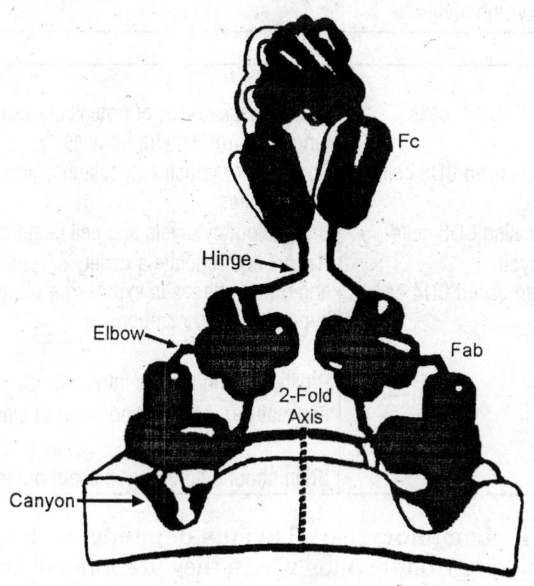

FIG. 28.22: Antigen-antibody binding: an example of antigen binding represented in the model.

THE IMMUNE RESPONSE

In immune mechanism there is an involvement of special types of cells called **lymphocytes**. The immunological response in an individual is brought about by the introduction of antigens or immunogens. The immune response mechanism is of two types: **humoral immunity** - involving production of antibodies, and **cell-mediated immunity** in which the lymphocytes react directly with foreign material. An immunologic response to an immunogen consists of one or more of the following components:

1. **Antibody production:** It involves `B' cell division in immune system. Development and specific cell mediated immunity which involves the 'T' cell division of the immune system.
2. **Immunological memory:** Which involves one or both components of the immune system, is responsible for an accelerated immune response when the same antigen enters or is administered in the body after it has induced the first response (primary response).
3. **Immunological tolerance:** It is the specific immunological unresponsiveness brought about by certain antigens in certain circumstances.

LYMPHOCYTES

Origin of Lymphocytes: Lymphocytes are small, nonphagocytic, mononuclear leukocytes that are immunologically competent, or are precursors of such cells. They lack stainable cytoplasmic granules and are formed in lymphatic tissues. Lymphocytes are the pivotal cells of the specific immunologic response. Undifferentiated lymphocytes are derived from bone marrow stem cells. They are produced in the bone marrow at a very high rate (10^9 cells per day). Some lymphocytes migrate through the circulatory or lymphatic systems to the secondary lymphoid tissue (thymus, spllen, aggregated lymph nodules in the intestines, and lymph nodes) where they prôduce lymphocyte colonies.

TABLE 28.7: Classes of Lymphocytes

Lymphocyte	Role
T Cells	
T_H (helper) cells; also called CD4[1] cells	Provide *assistance*, or potenitiate expression of immune function by other lymphocytes
T_S (suppressor) cells; also called CD8 cells	Suppress or impair expression of immune function by other lymphocytes
T_C (cytotoxic) cells; also called CD8 cells	Bring about cytolysis and cell death of "targets"
T_D cells or T_{DTH} (delayed type hypersensitivity) cells; also called CD4 cells	Recruit and regulate a variety of non-specific blood cells and macrophages in expression of delayed (Type IV) hypersensitivity reactions
B Cells	
B lymphocytes	Proliferate and mature into antibody-producing cells
Plasma cells	Are mature, active antibody-producing cells
Null Cells	
Natural killer cells	Bring about cytolysis and death of target cells

T cells or T Lymphocytes: Thymus-dependent lymphocytes or T lymphocytes and migrate from thymus where they are influenced by the hormone- **thymosin** and become immunological competent. T lymphocytes are mainly involved in cellular type of immunological response that is with **cellular immunity**, such as rejection of foreign tissue.

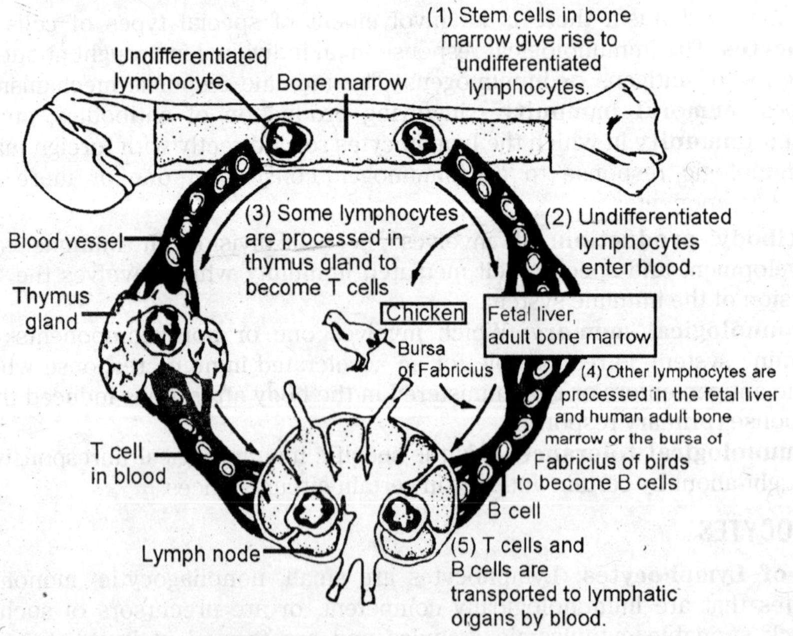

FIG. 28.23: Schematic presentation of lymphocyte development. Bone marrow releases undifferentiated lymphocytes which after processing become T and B cells.

Some T cells are transported away from the thymus and enter the bloodstream where they comprise 70 to 80% of the circulating lymphocytes. Other T cells tend to reside in various organs of the lymphatic system, such as the lymph nodes and spleen. This thymus-depended differentiation of T cells (or **theymoyctes**) occurs during early childhood, and by adolescence the secondary lymphoid organs of the body generally contain a full complement of T cells.

B Cells or B Lymphocytes

B-lymphocytes: The origin of these lymphocytes is from the *bursa of Fabricus* of birds (a mass of lymphoid tissue near the cloaca). The letter **B** was originally derived from the *busa of Fabricius*, a specialized appendage of the cloaca of chickens where these lymphocytes differentiate. B cells are distributed by the bloom and make up 20 various lymphoid organs along with the T cells. These lymphocytes are mainly involved in the production of antibodies- **humoral immunity**. B lymphocytes differentiate the fetal liver and adult bone marrow.

1. **Plasma Blasts:** These cells are produced by B lymphocytes following antigenic stimulation. In fact, the lymphocytes change themselves into plasma blasts which multiply and differentiate into plasma cells. These are the large cells and contain relatively more basophilic cytoplasm, less developed endoplasmic reticulum, large nucleus with nucleoli and can multiply and differentiate themselves in plasma cells.
2. **Plasma Cells:** The plasma cells are uninucleated with amphophillic cytoplasm, rich in ribosomes, rough endoplasmic reticulum, with prominent Golgi bodies, acentric nucleus with cart wheel type chromatin.
3. **Null Cell:** There is a population of lymphoid cells that do not have characteristics of either T or B cells. These are called null cells because they lack the specific surface markers of B or T cells, and can be distinguished from them by the presence of cytoplasmic granules. It is currently believed that this population of cells contains most **natural killer (NK) cells** and antibody-dependent cytotoxic T cells. These cells are probably of bone marrow origin, however, their exact lineage in uncertain.

Function of Lymphocytes

Plasma cells are fully differentiated antibody-synthesizing cells that are derived from B lymphocytes. They respond to antigens by secreting antibodies into the blood and lymph. Antibodies are glycoproteins produced by plasma cells after the B cells in their lineage have been exposed to antigens. Antibodies are specifically directed against the antigen that caused their formation. Because *antibodies are soluble in blood and lymph fluids*, they provide **humoral** [Latin *humor*, a liquid] **immunity** or **antibody-mediated immunity**.

The humoral immune response defends mostly bacteria, bacterial toxins, and viruses that enter the body's various fluid systems. T cells do not secrete antibodies. Instead they attack (1) host cells that have been parasitized by viruses or microorganisms, (2) tissue cells that have been transplanted from one host to another, and (3) cancer cells. They also produce **cytokines**, chemical mediators that play specific augmenting and regulatory roles in the immune system. Since T cells must physically contact foreign cells or infected cells in order to destroy them, they are said to provide **cell-mediated immunity**.

Null cells (and particularly natural killer cells) destroy tumor cells and virus- and other parasite-infected cells. They also help regulate the immune response. Null cells often exhibit antibody-dependent cellular cytotoxicity.

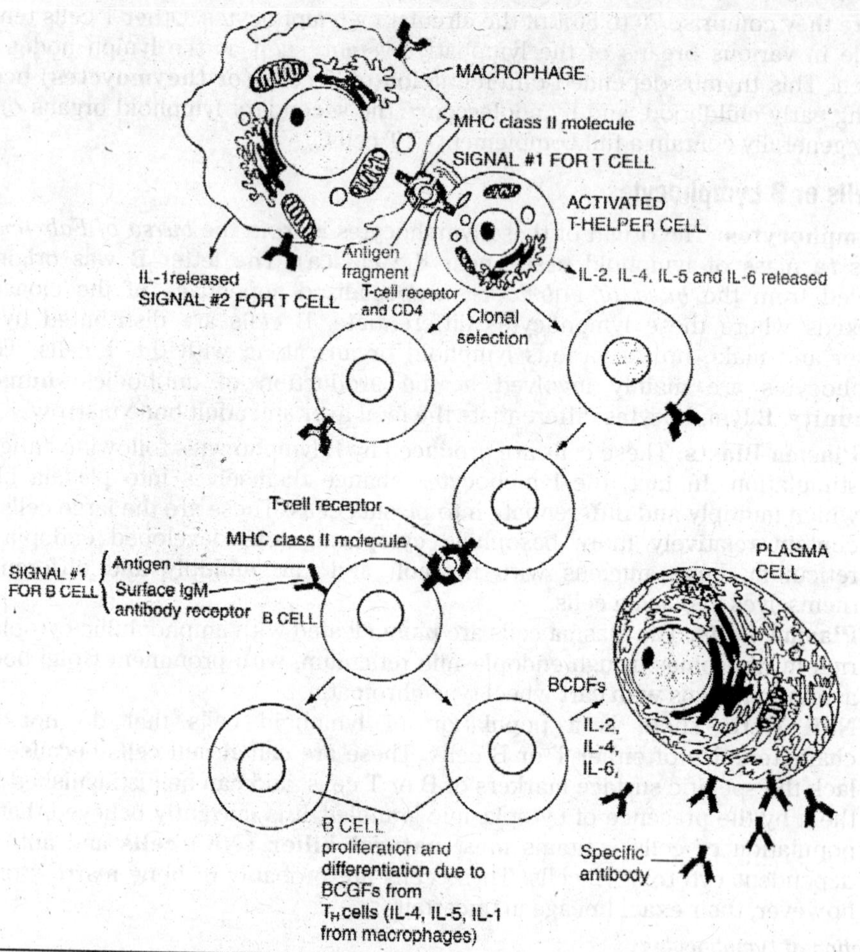

FIG. 28.24: Schematic diagram of the events occurring in the interactions of macrophages, T-helper cells, and B cells that produce cell-mediated immunity.

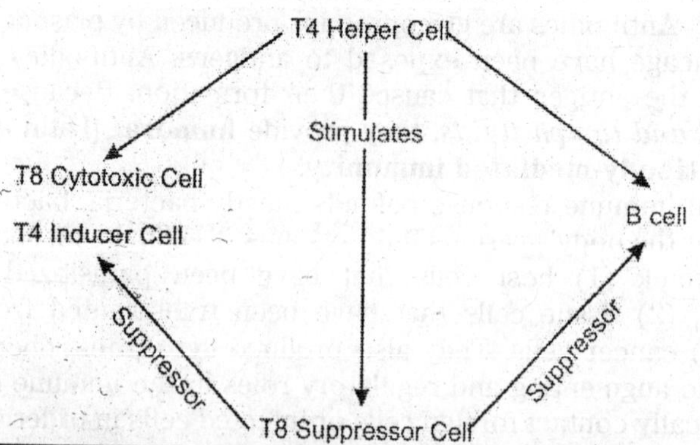

FIG. 28.25: Interaction between T- and B-lymphocytes.

Interaction of T and B Lymphocytes: Both the lymphocytes are quite distinct and independent in their function, even they help each other in antibody production. The production of antibody increases in presence of T lymphocytes.

How T lymphocyte increases the production of antibody in presence of B lymphocytes is poorly understood. However, T lymphocytes in this reference are known as helper cells. It has been suggested that T lymphocytes act by focussing antigen on `B' lymphocytes. Sometimes 'T' cells also act as suppressor cells and may be involved in the maintenance of a state of immunological unresponsiveness.

Accessory Cells (A Cells): Macrophages: Macrophages are large uninuclear phagocytic cells and contain antigen. Those macrophages which do not take part directly in immune response are called **accessory cells** or A-cells of the immune system.

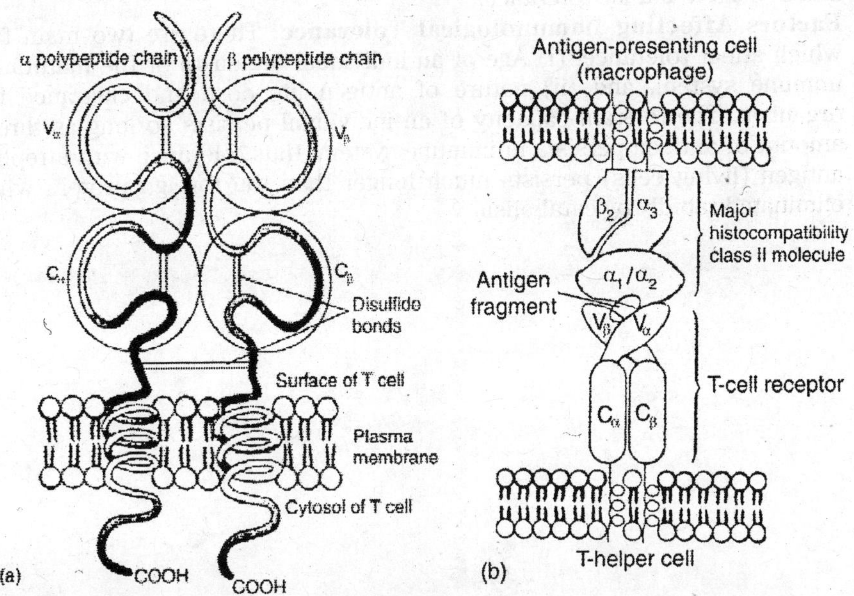

FIG. 28.26: T-cell CD4 receptor and its activation. (a) A schematic illustration of the proposed overall structure of the antigen receptor site on a T-cell plasma membrane. (b) An antigen-presenting cell begins the activation process by displaying an antigen fragment (e.g., peptide) on its surface as part of a complex with histocompatibility molecules. A T-helper cell is triggered when the variable region of its CD4 receptor (designated V_α and V_β) reacts with the antigen fragment on the presenting cell surface.

IMMUNE SYSTEM

Immunological Tolerance

The important characteristics of immunological tolerance are: (i) it is induced by an antigen, (ii) it is specific, (iii) it is associated with immunological memory, and (iv) it is usually requires persistence of the antigen. Either both T and B cells or individual cells are involved in immunological tolerance.

Mechanisms of Immunological Tolerance Development

These are as follows:

1. **Clonal Abortion:** When an antigen comes in contact with immature immunologically competent cells (T or B cells), maturation of these cells is aborted, which results in lack of corresponding active T and B cells to react with antigen in the later life.
2. **Clonal Exhaustion:** When an individual is challenged with repeated doses of T

cells independent antigen, all the corresponding B cells are stimulated to differentiate into short lived antibody forming cells (plasma cells). Thereafter no competent cell is left to react with the antigen.

3. **Functional Deletion:** This has been explained on the basis of two aspects (i) by T cell dependent antigens, and (ii) by T cell independent antigen.
4. **Antibody forming Cell Blockade:** T cell independent antigen can also produce tolerance by blocking the antibody forming cells. It, however, requires lesser amount of antigen.
5. **Generation of `T' Suspensor Cells:** Sometimes antigen may stimulate generation of `T' suspensor cells which directly suppresses corresponding `T' and B' cells to induce tolerance.
6. **Factors Affecting Immunological Tolerance:** There are two main factors which affect tolerance: (i) Age of an individual organism or the maturation of immune system, and (ii) nature of antigen, its dose and clearance in the organism. The tolerant-capacity of an individual persists so long as threshold amount of antigen persists in immune system, thus tolerance with a replicated antigen (living cells) persists much longer than the foreign protein which is eliminated rapidly by catabolism.

CHAPTER 29

Glycoproteins

Glycoproteins are proteins that contain covalently attached sugar residues. The **hydrophilic** and polar characteristics of sugars may dramatically change the chemical characteristics of the protein to which they are attached. The addition of sugars is often required for a glycoprotein to function properly and reach its ultimate destination in the cell or organism.

Mucins are glycoproteins found in the body. They are secreted in the mucus of the respiratory and digestive tracts. The sugars attached to mucins give them considerable water-holding capacity and make them resistant to proteolysis by digestive enzymes. Soluble glycoproteins often show a high viscosity, for example, in egg white and blood plasma.

Glycoproteins are frequently present at the surface of cells where they function as membrane proteins or as part of the extracellular matrix. These cell surface glycoproteins play a critical role in cell–cell interactions and the mechanisms of infection by bacteria and viruses.

ANALYSIS

A variety of methods used in detection, purification, and structural analysis of glycoproteins are:

TABLE 29.1: Methods used to study glycoproteins

Method	Use
Periodic acid-Schiff stain	Detects glycoproteins as pink bands after electrophoretic separation.
Incubation of cultured cells with glycoproteins as radioactive decay bands	Leads to detection of a radioactive sugar after electrophoretic separation.
Treatment with appropriate endo- or exoglycosidase or phospholipases	Resultant shifts in electrophoretic migration help distinguish among proteins with N-glycan, O-glycan, or GPI linkages and also between high mannose and complex N-glycans.
Agarose-lectin column chromatography	To purify glycoproteins or glycopeptides that bind the particular lectin used.
Compositional analysis following acid hydrolysis	Identifies sugars that the glycoprotein contains and their stoichiometry.
Mass spectrometry	Provides information on molecular mass, composition, sequence, and sometimes branching of a glycan chain.
NMR spectroscopy	To identify specific sugars, their sequence, linkages, and the anomeric nature of glycosidic chain.
Methylation (linkage) analysis	To determine linkage between sugars.
Amino acid or cDNA sequencing	Determination of am

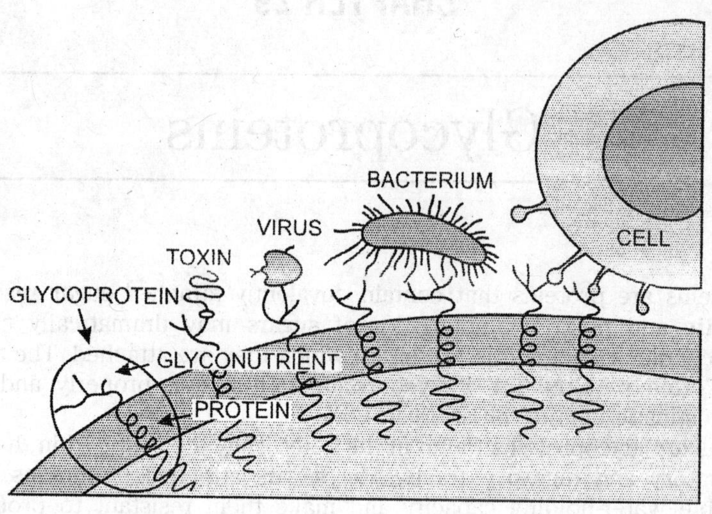

FIG. 29.1: Glycoprotein receptors. The surface carbohydrate on cells serve as point attachment for other cells, infectious bacteria, viruses, toxins, hormones and many other molecules.

TYPES

There are three types of glycoproteins based on their structure and the mechanism of synthesis:

1. N-linked glycoproteins.
2. O-linked glycoproteins.
3. Nonenzymatic glycosylated glycoproteins.

N-Linked Glycoproteins

N-linked glycoproteins are synthesised and modified within two membrane-bound organelles in the cell, the rough **endoplasmic reticulum** and the Golgi apparatus. The protein component of the glycoprotein is assembled on the surface of the rough endoplasmic reticulum by the sequential addition of amino acids, creating a linear polymer of amino acids called a **polypeptide**. Twenty different amino acids can be used for the synthesis of polypeptides. The specific order of the amino acids in the polypeptide is critical to its function and is referred to as the amino acid sequence. One of the twenty amino acids used for the synthesis of polypeptides, asparagine ($C_4H_8N_2O_3$), is essential for the synthesis of N-linked glycoproteins.

N-linked glycoproteins have carbohydrates attached to the R side chain of asparagine residues within a polypeptide. The carbohydrate is always located in amino acid sequences, where the asparagine is followed by some other amino acid and then a serine or threonine residue (-Asn-Xaa-Ser/Thr). Carbohydrate is not attached to the polypeptide one sugar at a time. Rather, a large preformed carbohydrate containing fourteen or more sugar residues is attached to the asparagine as the protein is being translated in the rough endoplasmic reticulum. The carbohydrate on the glycoprotein is then modified by enzymes that remove some sugars and attach others as the newly formed glycoprotein moves from the rough endoplasmic reticulum to the Golgi apparatus and other locations in the cell. Many N-linked glycoproteins eventually become part of the cell membrane or are secreted by the cell.

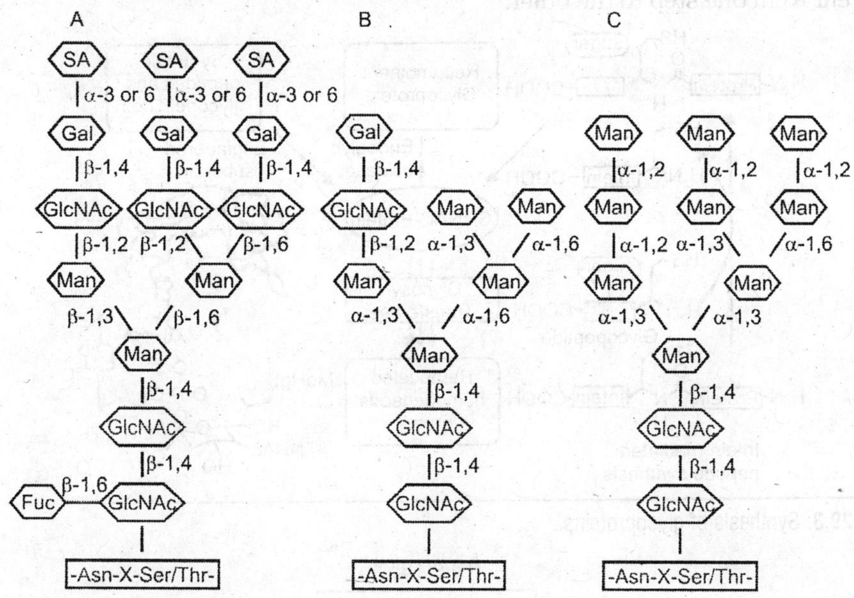

FIG. 29.2: N-linked glycoproteins.

O-Linked Glycoproteins

O-linked glycoproteins are usually synthesised by the addition of sugar residues to the hydroxyl side chain of serine or threonine residues in polypeptides in the Golgi apparatus. Unlike N-linked glycoproteins, O-linked glycoproteins are synthesised by the addition of a single sugar residue at a time. Many O-linked glycoproteins are secreted by the cell to become a part of the extracellular matrix that surrounds it.

Nonenzymatic Glycosylation

Nonenzymatic glycosylation or glycation creates glycoproteins by the chemical addition of sugars to polypeptides. Since this type of glycosylation is nonenzymatic, the factors that control glycosylation are simply time and the concentration of sugar. Older proteins are more glycosylated, and people with higher circulating levels of glucose experience higher levels of nonenzymatic glycosylation. This is the basis of the glycosylated hemoglobin Alc diagnostic test used for the monitoring and long-term maintenance of blood sugar levels in diabetics.

GLYCOSYLATION

Glycosylation is the process or result of addition of saccharides to proteins and lipids.

In general most of the proteins that pass through ER get glycosylated. Glycosylation is site specific, protein specific and tissue specific; for the same protein, glycosylation is different in different tissues.

Many proteins are at first glycosylated in ER and then they are transported into Golgi membranes, there they may be further modified in different ways in different tissues.

There are specific enzymes for glycosylation in both ER and Golgi complex.

In general glycosylation builds up at cytosolic surface from monosaccharides to disaccharides and then to oligo saccharides. The kind of monosaccharides added is different from one step to the other.

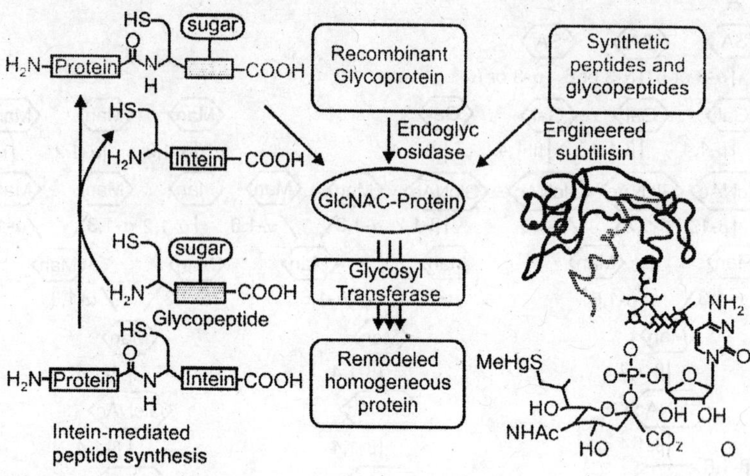

FIG. 29.3: Synthesis of glycoproteins.

Asparagine

N-linked

CH_2OH — O — H — N — C(=O) — CH_2 — CH — NH — C—O

Polypoplide backbone

OH, H, $NHCOCH_3$

Serine (X-H)
Threonine (X=CH_3)

O-linked

CH_2OH — O — O — CH(X) — CH(NH) — C=O

OH, H, $NHCOCH_3$

N-Acetylgalactosamine

FIG. 29.4: Glycoprotein.

Once such blocks of oligo saccharides are built, they are transferred to specific proteins to specific sites, which are determined by its sequence.

The most common amino acid to which the oligos are added is Aspargine; such glycosylation is called asn-linked or. N-linked glycosylation. This N-linked glycosylation starts in ER. In Golgi, it may undergo further modifications.

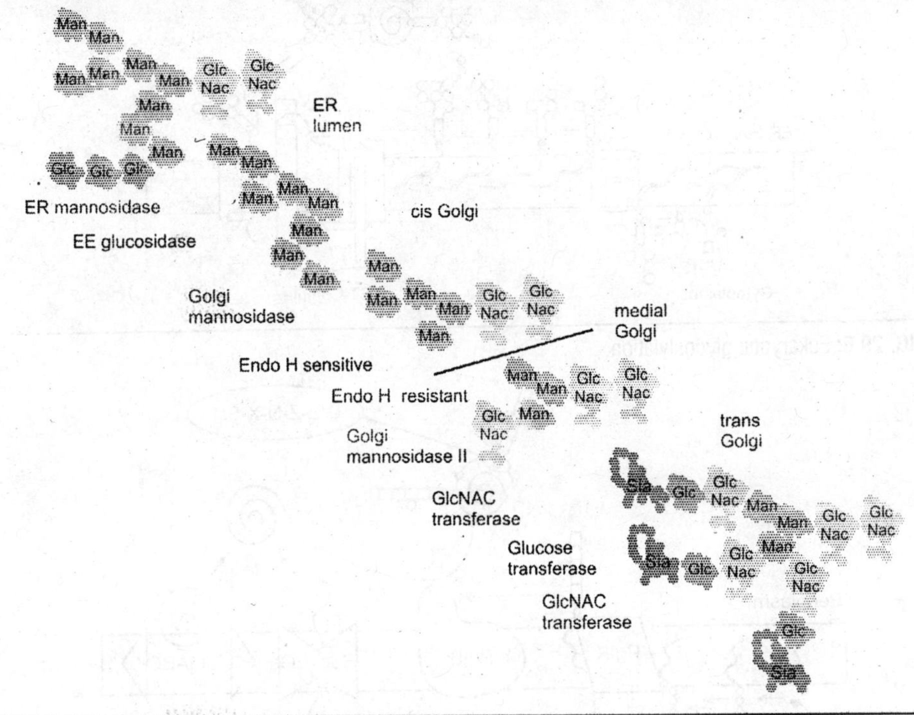

FIG. 29.5: Mechanism of action of glycoproteins.

In Golgi complex, another type of glycosylation takes place, which is called O-linked glycosylation, where sugars are added to the OH group of serine or threonine. In some cases the C-terminal regions of proteins are covalently bonded to the amino group ethanolamine of GPI, which is the resident of membranes.

The process is one of four principal co-translational and post-translational modification steps in the synthesis of membrane and secreted proteins and the majority of proteins synthesised in the rough ER undergo glycosylation. It is an enzyme-directed site-specific process, as opposed to the non-enzymatic chemical reaction of glycation. Two types of glycosylation exist:

1. *N*-linked glycosylation to the amide nitrogen of asparagine side chains.
2. *O*-linked glycosylation to the hydroxy oxygen of serine and threonine side chains.

Mechanisms of Action

There are various mechanisms for glycosylation, although all share several common features:

1. Glycosylation is an enzymatic process;
2. The donor molecule is an activated nucleotide sugar;
3. The process is site-specific.

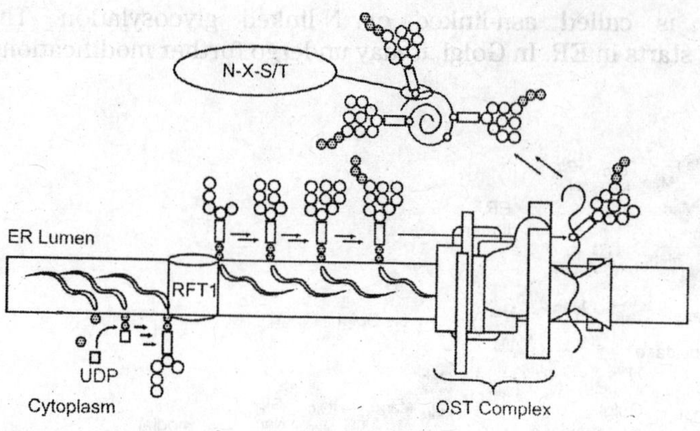

FIG. 29.6: Eukaryotic glycoslylation.

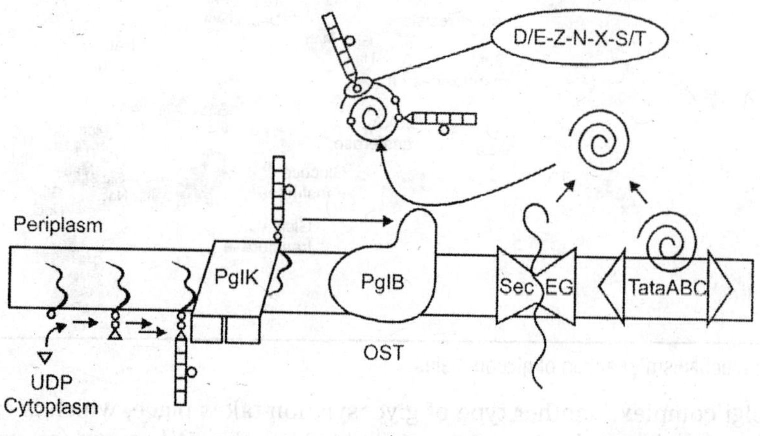

FIG. 28.7: Prokaryotic glycosylation.

N-Linked Glycosylation

N-linked glycosylation is important for the folding of some eukaryotic proteins. The *N*-linked glycosylation process occurs in eukaryotes and widely in archaea, but very rarely in bacteria.

For *N*-linked oligosaccharides, a 14-sugar precursor is first added to the asparagine in the polypeptide chain of the target protein. The structure of this precursor is common to most eukaryotes, and contains 3 glucose, 9 mannose, and 2 *N*-acetylglucosamine molecules. A complex set of reactions attaches this branched chain to a carrier molecule called dolichol, and then it is transferred to the appropriate point on the polypeptide chain as it is translocated into the ER lumen. Therefore, **Dolichol** refers to any of a group of long-chain mostly unsaturated organic compounds which are made up of varying numbers of isoprene units terminating in an ⟨-saturated isoprenoid group, containing an alcohol functional group.

Addition of Glyco components to Dolichol-P~P (pyrophosphate) starts at cytosolic side of ER. Once the basic carbohydrate complex or tree is built it is flipped towards the luminal side of the ER; where more carbohydrates are added and modified. Such carbohydrate complexes are specific-to-specific proteins and they are linked at specific sites of the protein in sequence specific and motif dependent manner. All the required enzymes and components are found in cytoplasm and in ER lumen. Once the components are built at cytosol they are then flipped towards lumen side.

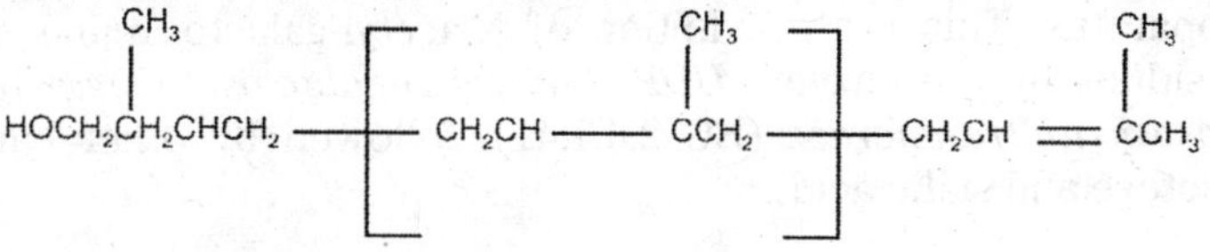

$$HOCH_2CH_2\overset{CH_3}{\overset{|}{C}}HCH_2 - \left[CH_2CH = \overset{CH_3}{\overset{|}{C}}CH_2 \right] - CH_2CH = \overset{CH_3}{\overset{|}{C}}CH_3$$

FIG. 29.8: Dolichol is a multimer of isoprene (pentameric carbon skeleton) units, the number of units vary from 20 to 52.

There are two major types of *N*-linked saccharides:

High-mannose oligosaccharides: High-mannose is, in essence, just two *N*-acetylglucosamines with many mannose residues, often almost as many as are seen in the precursor oligosaccharides before it is attached to the protein.

Complex oligosaccharides: Complex oligosaccharides are so named because they can contain almost any number of the other types of saccharides, including more than the original two *N*-acetylglucosamines.

Proteins can be glycosylated by both types of oligos on different portions of the protein. Whether an oligosaccharide is high-mannose or complex is thought to depend on its accessibility to saccharide-modifying proteins in the Golgi. If the saccharide is relatively inaccessible, it will most likely stay in its original high-mannose form. If it is accessible, then it is likely that many of the mannose residues cleaved off and the saccharide will be further modified by the addition of other types of group as discussed above.

Sequon

A sequon is a sequence of three consecutive amino acids in a protein that can serve as the attachment site to a polysaccharide (sugar) called an N-linked-Glycan. This is a polysaccharide linked to the protein via the nitrogen atom in the side chain of asparagine (Asn). A sequon is either Asn-X-Ser or Asn-X-Thr, where X is any amino acid except proline. The term was probably first used by Derek Marshall. Occasionally, such as in the leukocyte surface protein (CD69), the amino acid sequence Asn-Xaa- Cys is an acceptable sequon for the addition of N-linked glycans.

The oligosaccharide chain is attached by oligosaccharyltransferase to asparagine occurring in the tripeptide sequence Asn-X-Ser, Asn-X-Thr or Asn-X-Cys, where X could be any amino acid except Pro. This sequence is known as a glycosylation **sequon**.

After attachment, once the protein is correctly folded, the three glucose residues are removed from the chain and the protein is available for export from the ER. The glycoprotein thus formed is then transported to the Golgi where removal of further mannose residues may take place. However, glycosylation itself does not seem to be as necessary for correct transport targeting of the protein, as one might think. Studies involving drugs that block certain steps in glycosylation, or mutant cells deficient in a glycosylation enzyme, still produce otherwise-structurally-normal proteins that are correctly targeted, and this interference does not seem to interfere severely with the viability of the cells. Mature glycoproteins may contain a variety of oligomannose *N*-linked oligosaccharides containing between 5 and 9 mannose residues. Further removal of mannose residues leads to a 'core' structure containing 3 mannose, and 2 *N*-acetylglucosamine residues, which may then be elongated with a variety of different monosaccharides including galactose, N-acetylglucosamine, N-acetylgalactosamine, fucose and sialic acid.

O-Linked Glycosylation

O-linked glycosylation occurs at a later stage during protein processing, probably in the Golgi apparatus. This is the addition of N-acetyl-galactosamine to serine or threonine residues by the enzyme *UDP-N-acetyl-D-galactosamine:polypeptide N-acetylgalactosaminyltransferase* (EC 2.4.1.41), followed by other carbohydrates (such as galactose and sialic acid).

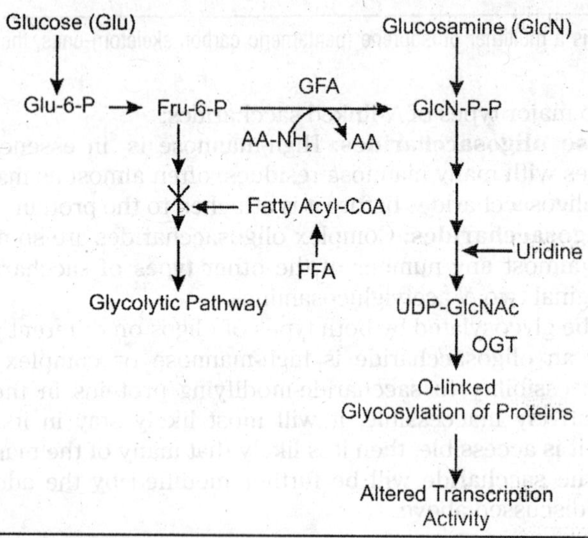

FIG. 29.9: O-linked glycosylasion of protein.

This process is important for certain types of proteins such as proteoglycans, which involves the addition of glycosaminoglycan chains to an initially unglycosylated proteoglycan core protein. These additions are usually serine *O*-linked glycoproteins, which seem to have one of two main functions. One function involves secretion to form components of the extracellular matrix, adhering one cell to another by interactions between the large sugar complexes of proteoglycans.

The other main function is to act as a component of mucosal secretions. It the high concentration of carbohydrates that tends to give mucus its "slimy" feel. Proteins that circulate in the blood are not normally *O*-glycosylated, with the exception of IgAl and IgD (two types of antibody) and C1-inhibitor.

O-Fucose

O-fucose is added between the second and third conserved cysteines of EGF-like repeats in the Notch protein, and possibly other substrates by GDP-fucose protein O-fucosyltransferase 1, and to Thrombospondin repeats by GDP-fucose protein O-fucosyltransferase 2. In the case of EGF-like repeats, the O-fucose may be further elongated to a tetrasaccharide by sequential addition of N-acetylglucosamine (GlcNAc), galactose, and sialic acid, and for Thrombospondin repeats, may be elongated to a disaccharide by the addition of glucose. Both of these fucosyltransferases have been localized to the endoplasmic reticulum, which is unusual for glycosyltransferases, most of which function in the Golgi apparatus.

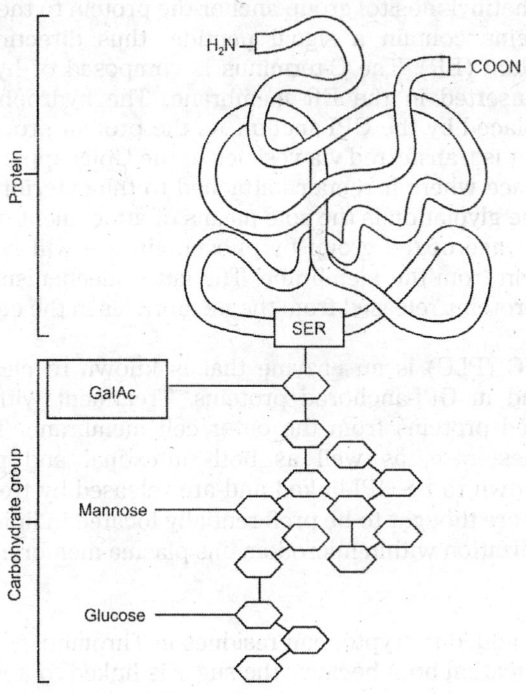

FIG. 29.10: Showing O-linked glycosylation.

O-Glucose

O-glucose is added between the first and second conserved cysteines of EGF-like repeats in the Notch protein, and possibly other substrates by an unidentified *O*-glucosyltransferase.

O-N-Acetylglucosamine (O-Glcnac)

O-GlcNAc is added to serines or threonines by *O*-GlcNAc transferase. *O*-GlcNAc appears to occur on serines and threonines. that would otherwise be phosphorylated by serine/threonine kinases. Thus, if phosphorylation occurs, *O*-GlcNAc does not, and *vice versa*. This is an incredibly important finding because phosphorylation/dephosphorylation has become a scientific paradigm for the regulation of signaling within cells. A massive amount of cancer research is focused on phosphorylation. Ignoring the involvement of this form of glycosylation, which clearly appears to act in concert with phosphorylation, means that a lot of current research is missing at least half of the picture. *O*-GlcNAc addition and removal also appear to be key regulators of the pathways that are deregulated in diabetes mellitus. The gene encoding the *O*-GlcNAc removal enzyme has been linked to non-insulin dependent diabetes mellitus. It is the terminal step in a nutrient-sensing hexosamine signaling pathway.

Glycosylphosphatidylinositol (GPI Anchor)

It is a glycolipid, which is attached to the C-terminus of a protein during posttranslational modification. It is composed of a hydrophobic phosphatidylinositol group linked through a carbohydrate containing linker (glucosamine and mannose glycosidically bound to the inositol residue) to the C-terminal amino acid of a mature protein. The two fatty acids within the hydrophobic phosphatidyl-inositol group anchor the protein to the cell membrane.

Glypiated proteins contain a signal peptide, thus directing them into the endoplasmic reticulum (ER). The C-terminus is composed of hydrophobic amino acids which stay inserted in the ER membrane. The hydrophobic end is then cleaved off and replaced by the GPI-anchor. As the protein processes through the secretory pathway, it is transferred via vesicles to the Golgi apparatus and finally to the extracellular space where it remains attached to the exterior leaflet of the cell membrane. Since the glypiation is the sole means of attachment of such proteins to the membrane, cleavage of the group by phospholipases will result in controlled release of the protein from the membrane. The latter mechanism is used *in vitro*, i.e. the membrane proteins released from the membranes in the enzymatic assay are glypiated protein.

Phospholipase C (PLC) is an enzyme that is known to cleave the phosphoglycerol bond found in GPI-anchored proteins. Treatment with PLC will cause release of GPI-linked proteins from the outer cell membrane. The T-cell marker Thy-1, acetylcholinesterase, as well as both intestinal and placental alkaline phosphatase are known to be GPI-linked and are released by treatment with PLC. GPI-linked proteins are thought to be preferentially located in lipid rafts, suggesting a high level of organization within microdomains plasma membrane.

C-Mannosylation

A mannose sugar is added to tryptophan residues in Thrombospondin repeats. This is an unusual modification both because the sugar is linked to a carbon rather than a reactive atom like a nitrogen or oxygen and because the sugar is linked to a tryptophan residue rather than an asparagine or serine/threonine.

FUNCTIONS

The polysaccharide chains attached to the target proteins serve various functions. For instance, some proteins do not fold correctly unless they are glycosylated first. Also, polysaccharides linked at the amide nitrogen of asparagine in the protein confer stability on some secreted glycoproteins. Experiments have shown that glycosylation in this case is not a strict requirement for proper folding, but the unglycosylated protein degrades quickly. Glycosylation may play a role in cell-cell adhesion (a mechanism employed by cells of the immune system), as well.

Glycoproteins are important for white blood cell recognition, especially in mammals. Examples of glycoproteins in the immune system are:

1. Molecules such as antibodies (immunoglobulins), which interact directly with antigens.
2. Molecules of the major histocompatibility complex (or MHC), which are expressed on the surface of cells and interact with T cells as part of the adaptive immune response.

Glycoproteins are the components of the zona pellucida, which surrounds the oocyte, and is important for sperm-egg interaction. Structural glycoproteins occur in connective tissue. These help bind together the fibers, cells, and ground substance of connective tissue. They may also help components of the tissue bind to inorganic substances, such as calcium in bone.

TABLE 29.2: Functions of glycoproteins

Function	Type of Glycoproteins
Structural molecule	Collagens
Lubricant and protective agent	Mucins

Function	Type of Glycoproteins
Transport molecule	Transferrin, ceruloplasmin
Immunologic molecule	Immunoglobins, histocompatibility antigens
Hormone	Chorionoic gonadotropin, thyroid-stimulating hormone (TSH)
Enzyme	Various, eg, alkaline phosphatase
Cell attachment-recognition site	Various proteins involved in cell-cell (eg, sperm-oocyte), virus-cell, bacterium-cell, and hormone cell interactions
Antifreeze	Certain plasma proteins of coldwater fish
Interact with specific carbohydrates	Lectins, selectins (cell adhesion lectins), antibodies
Receptor	Various proteins involved in hormone and drug action
Affect folding of certain proteins	Calnexin, calreticulin
Regulation of development	Notch and its analogs, key proteins in development
Hemostasis (and thrombosis)	Specific glycoproteins on the surface membranes of platelets

Hormones that contain glycoproteins include:

Follicle-stimulating hormone: FSH is a glycoprotein. Each monomeric unit is a protein molecule with a sugar attached to it; two of these make the full, functional protein. Its structure is similar to those of LH, TSH, and hCG. The protein dimer contains 2 polypeptide units, labeled alpha and beta subunits. The alpha subunits of LH, FSH, TSH, and hCG are identical, and contain 92 amino acids. The beta subunits vary. FSH has a beta subunit of 118 amino acids (FSHB), which confers its specific biologic action and is responsible for interaction with the FSH-receptor. The sugar part of the hormone is composed of fucose, galactose, mannose, galactosamine, glucosamine, and sialic acid, the latter being critical for its biologic half-life. The half-life of FSH is 3-4 hours.

Luteinizing_hormone: LH is a glycoprotein. Each monomeric unit is a sugar-like protein molecule; two of these make the full, functional protein. Its structure is similar to the other glycoproteins, follicle-stimulating hormone (FSH), thyroid-stimulating hormone (TSH), and human chorionic gonadotropin (hCG). The protein dimer contains 2 polypeptide units, labeled alpha and beta subunits that are connected by two disulphide bridges:

The different composition of these oligosaccharides affects bioactivity and speed of degradation. The biologic half-life of LH is 20 minutes, shorter than that of FSH (3-4 hours) or hCG (24 hours).

Thyroid-stimulating hormone: Thyroid-stimulating hormone (also known as TSH or thyrotropin) is a hormone synthesised and secreted by thyrotrope cells in the anterior pituitary gland which regulates the endocrine function of the thyroid gland.

Human chorionic gonadotropin: Human chorionic gonadotropin (hCG) is a peptide hormone produced in pregnancy that is made by the embryo soon after conception and later by the syncytiotrophoblast (part of the placenta). Its role is to prevent the disintegration of the corpus luteum of the ovary and thereby maintain progesterone production that is critical for a pregnancy in humans. hCG may have additional functions; for instance, it is thought that hCG affects the immune tolerance of the pregnancy. Early pregnancy testing, in general, is based on the detection or measurement of hCG. Because hCG is produced also by some kinds of tumor, hCG is an important tumor marker, but it is not known whether this production is a contributing cause or an effect of tumorigenesis.

Alpha-fetoprotein: Alpha-fetoprotein (AFP) is a molecule produced in the developing embryo and fetus. In humans, AFP levels decrease gradually after birth, reaching adult levels by 8 to 12 months. Normal adult AFP levels are low, but detectable; however, AFP has no known function in normal adults. In normal fetuses, AFP binds the hormone estradiol. AFP is measured in pregnant women, using maternal blood or amniotic fluid, as a screening test for subset developmental abnormalities, principally open neural tube defects. It is also measured in pregnant women, other adults, and children, serving as a biomarker to detect a subset of tumours, principally hepatocellular carcinoma and endodermal sinus tumours.

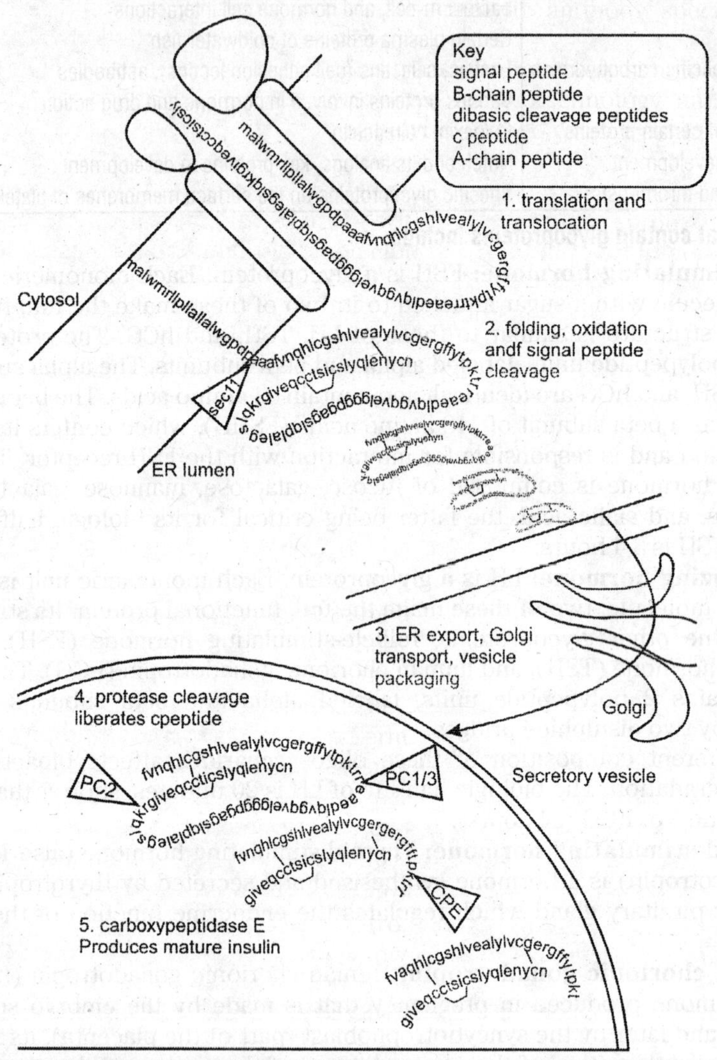

FIG. 29.11: Post-translational modification (PTM).

Erythropoietin (EPO): It is a glycoprotein hormone that is a cytokine for erythrocyte (red blood cell) precursors in the bone marrow. Also called hematopoietin or hemopoietin, it is produced by the kidney, and is the hormone that regulates red blood cell production. As a class of drug EPO is a Erythropoietin Stimulating Agent (ESA).

They may be synthesised and administered exogenously. However, such molecules can at a latter stage be detected, since they differ slightly from the endogenous ones in e.g. features of posttranslational modification.

Posttranslational modification (PTM) is the chemical modification of a protein after its translation. It is one of the later steps in protein biosynthesis for many proteins.

Also, enzymes may remove amino acids from the amino end of the protein, or cut the peptide chain in the middle. For instance, the peptide hormone insulin is cut twice after disulphide bonds are formed, and a propeptide is removed from the middle of the chain; the resulting protein consists of two polypeptide chains connected by disulphide bonds. Other modifications, like phosphorylation, are part of common mechanisms for controlling the behavior of a protein, for instance activating or inactivating an enzyme.

Glycoprotein IIb/IIIa Inhibitors

Glycoprotein IIb/IIIa inhibitors are medicines used in treating patients who have unstable angina, certain types of heart attacks, and in combination with angioplasty with or without stent placement. They are given in combination with heparin or aspirin (blood-thinning agents) to prevent clotting before and during invasive heart procedures.

Glycoprotein IIb/IIIa inhibitors are classified as potent platelet inhibitors. These agents are used to prevent platelets from binding together, which can occur in patients with heart attacks and after angioplasty with or without stent placement.

Glycoprotein IIb/IIIa inhibitors are administered only under the supervision of a doctor, through an intravenous injection or IV infusion during hospitalization.

Patients with an acute coronary syndrome, such as unstable angina or certain kinds of heart attacks, are eligible to receive one of these agents. Other eligible patients include those undergoing coronary angioplasty with or without stent placement.

Patients who do not have low platelet counts and who are not at high risk for severe bleeding might be good candidates for the glycoprotein inhibitors. Patients with active internal bleeding should not take glycoprotein IIb/IIIa inhibitors.

As with any medicine, an allergic reaction might occur, but the most common side effect of glycoprotein IIb/IIIa inhibitors is bleeding. Before, during, and after this medicine is taken, patients should have blood tests that indicate platelet count and clotting time of the blood.

CHAPTER 30

Hemostatis and Thrombosis

The term hemostasis comes from the Greek roots *heme*, blood + *stasis*, halt = halt of the blood. It is the stoppage of bleeding or hemorrhage or stoppage of blood flow through a injured blood vessel or organ of the body. Hemostasis is the arrest of bleeding, whether it be by normal vasoconstriction (the vessel walls closing temporarily), by an abnormal obstruction (such as a plaque) or by coagulation or surgical means (such as ligation).

Hemostasis needs the combined activity of vascular, platelet, and plasma factors. Regulatory mechanisms counterbalance the tendency of clots to form. Hemostatic abnormalities can lead to excessive bleeding or **thrombosis**.

Hemostasis is a complex process. It changes blood from a fluid to a solid state. Intact blood vessels are central to moderating blood's tendency to clot. The endothelial cells of intact vessels prevent thrombus formation by secreting tissue plasminogen activator (t-PA) and by inactivating thrombin and adenosine diphosphate (ADP). Injury to vessels overwhelms these protective mechanisms and hemostasis ensues. Hemostasis proceeds in two phases: primary and secondary hemostasis.

TYPE OF HEMOSTASIS

Normal Hemostasis

Hemostatis is a balance of the physiological processes. It prevents excessive bleeding after vessel injury and maintains a viable circulation by keeping the blood in an uncoagulated state. These processes are part of a complex system with many feedback loops and parallel circuits.

Primary Hemostasis

It is characterized by vascular contraction, platelet adhesion and formation of a soft aggregate plug. It begins immediately after endothelial disruption. Injury causes temporary local contraction of vascular smooth muscle. Vasoconstriction slows blood flow, enhancing platelet adhesion and activation. Primary hemostasis is short lived. The immediate post injury vascular constriction abates quickly. If flow is allowed to increase, the soft plug could be sheared from the injured surface, possibly creating emboli.

Adhesion occurs when circulating von Willebrand factor (**vWf**) attaches to the subendothelium. Next, glycoproteins on the platelet surface adhere to the sticky von Willebrand factor (**vWf**).

Von Willebrand factor is a blood glycoprotein involved in hemostasis. It is deficient or defective in von Willebrand disease and is involved in a large number of other diseases, including:

1. Thrombotic thrombocytopenic purpura (TTP or *Moschcowitz disease*): It is a rare disorder of the blood-coagulation system, causing multiple blood clots to

form in blood vessels around the body. Most cases of TTP arise from deficiency or inhibition of the enzyme ADAMTS13, which is responsible for cleaving large multimers of von Willebrand factor. This leads to hemolysis (the breaking open of red blood cells and the release of hemoglobin into the surrounding fluid that is plasma in body) and end-organ damage, and may require plasmapheresis therapy (the removal, treatment, and return of (components of) blood plasma from blood circulation. Plasmapheresis is thus an extracorporeal therapy (a medical procedure which is carried outside the body). The method can also be used to collect plasma for further manufacturing into a variety of medications.

2. Heyde's syndrome: It is a syndrome of aortic valve stenosis associated with gastrointestinal bleeding from colonic angiodysplasia. It is named after Edward C. Heyde, who first described the association in 1958. It is due to the induction of von Willebrand disease type IIA by the valvular stenosis.
3. Hemolytic-uremic syndrome: It is also called hemolytic-uremic syndrome (or haemolytic-uraemic syndrome, abbreviated HUS). It is a disease characterized by microangiopathic hemolytic anemia, acute renal failure and a low platelet count (thrombocytopenia).

Platelets collect across the injured surface. These platelets are then activated by contact with collagen.

Collagen is the most abundant protein in mammals. About one quarter of all of the protein in human body is collagen. Collagen is the main protein of connective tissue. Collagen has an unusual amino acid composition. It contains large amounts of glycine and proline, as well as two amino acids that are not inserted directly by ribosomes – hydroxyproline and hydroxylysine It has great tensile strength, and is the main component of ligaments and tendons. It is responsible for skin elasticity, and its degradation leads to wrinkles that accompany aging. Collagen also fills out the cornea where it is present in crystalline form. It is also used in cosmetic surgery, for example lip enhancement.

Collagen-activated platelets form pseudopods which stretch out to cover the injured surface and bridge exposed fibers. The collagen-activated platelet membranes expose receptors which bind circulating fibrinogen to their surfaces. Fibrinogen has many platelet binding sites. An aggregation of platelets and fibrinogen build up to form a soft plug. Platelet aggregation occurs about 20 seconds after injury.

Secondary Hemostas

It is responsible for stabilizing the soft clot and maintaining vasoconstriction. Vasoconstriction is maintained by platelet secretion of serotonin, prostaglandin and thromboxane. The soft plug is solidified through a complex interaction between platelet membrane, enzymes, and coagulation factors.

The coagulation system is tuned to arrest bleeding at the site of vascular injury and quickly remove clots that obstruct blood flow. In the fetus, components of the coagulation system show unique developmentally regulated patterns and times for maturation to normal adult protein quantities and functions. In addition, several coagulation proteins contribute to cellular proliferation and differentiation uniquely during fetal life. In spite of this, results of most screening tests of hemostasis vary modestly from adult normal values in the healthy term infant, and both hemorrhage and thrombosis are rare in the well infant.

Coagulation Factors

Coagulation factors are also known as blood clotting factors, clotting factors, or by the individual factor number (Factor I, Factor II, etc.) or name (Fibrinogen,

Prothrombin, etc.). Coagulation factors are produced by the liver and circulate in an inactive form until the coagulation cascade is initiated. The cascade occurs in steps. The completion of each step activates another coagulation factor in a chain reaction which leads to the conversion of fibrinogen to fibrin.

BLOOD COAGULATION

Blood coagulation is part of an important host defense mechanism termed hemostasis (the cessation of blood loss from a damaged vessel). Upon vessel injury, platelets adhere to macromolecules in the subendothelial tissues and then aggregate to form the primary hemostatic plug. The platelets stimulate local activation of plasma coagulation factors, leading to generation of a fibrin clot that reinforces the platelet aggregate. Later, as wound healing occurs, the platelet aggregate and fibrin clot are broken down. Mechanisms that restrict formation of platelet aggregates and fibrin clots to sites of injury are necessary to maintain the fluidity of the blood.

Hemorrhage results from trauma, vascular defects (e.g., esophageal varices, peptic ulcer disease), platelet abnormalities, or deficiencies of one or more of the plasma coagulation factors.

Thrombosis

Thrombosis is a pathologic process in which a platelet aggregate and/or a fibrin clot forms in the lumen of an intact blood vessel or in a chamber of the heart. It may occur at various sites of the body as described follows:

In an artery: If thrombosis occurs in artery, the tissue supplied by the artery may undergo ischemic necrosis (e.g., myocardial infarction due to thrombosis of a coronary artery).

In a vein: If thrombosis occurs in a vein, the tissues drained by the vein may become edematous and inflamed. When thrombosis occurs in a deep vein (in the lower extremity) it may be complicated by pulmonary embolism, in which all or a portion of the thrombus breaks loose, is carried in the bloodstream through the vena cava and the right side of the heart, and becomes lodged in a pulmonary artery. Massive pulmonary embolism can cause hypoxemia, shock, and death.

Extrinsic and Intrinsic Pathways

Extrinsic Pathways

The terms Extrinsic and Intrinsic can be more than a little bit confusing. Extrinsic means from outside the blood. In terms of the **extrinsic clotting pathway**, this means that the clotting mechanism is initiated (prothrombin activation) by factors from the blood vessel rather than from the blood, itself.

In the extrinsic clotting pathway, Factor III released by the traumatized tissue (the wetable surface) activates a simple series of biochemical reactions involving factors produced by platelets. Note that activation of the extrinsic pathway leads to the conversion of prothrombin to thrombin.

Subsequent steps in the pathway will lead to the formation of a blood clot within 15-20 seconds in severe trauma and within 1-4 minutes with minor trauma. For this reason, the extrinsic clotting pathway is often called the fast pathway. The extrinsic pathway is faster than the intrinsic pathway simply because the extrinsic pathway is shorter meaning, the extrinsic pathway has fewer chemical steps than the intrinsic pathway.

Intrinsic Pathways

Initiation of coagulation via the intrinsic clotting pathway is the slow pathway (as clotting will not begin for 3-4 minutes), even with moderate to severe trauma in which prothrombin activation begins within the blood; that is, in this pathway, prothrombin activation is driven by platelets, which are formed elements of the blood. The pathway includes a series of biochemical reactions that lead to clotting, where the process is driven by factors within the blood that is, the process is driven by platelets. Factor XII from platelets keeps the pathway going on. Though factor XII is released from platelets but they need the activation of factor XII which they get after contact with traumatized tissue. Once the pathway is initiated, activated platelets continue to drive the reactions in the pathway until bleeding stops.

The platelets in the intrinsic pathway work even the wetable surface is covered. This means that the intrinsic pathway will continue to run until the hole in the vessel wall is plugged up, well sealed and no longer leaking.

Coagulation *In Vitro*

When blood is removed from the body and placed in a glass test tube, it clots fairly quickly. Calcium ions are required for this process. Addition of EDTA or citrate prevents clotting by binding calcium.

Recalcified plasma (minus platelets and other blood cells) will clot in 2-4 min. The clotting time after recalcification can be shortened by adding an emulsion of negatively-charged phospholipids (PL). The reaction initiated by kaolin, PL, and calcium is termed the activated partial thromboplastin time (aPTT) test. The reaction initiated by thromboplastin and calcium is termed the **prothrombin time (PT)** test.

TABLE 30.1: Coagulation *in vitro*

Type	Clotting time
Whole blood	4-8 min
Whole blood + EDTA or citrate	infinite
Citrated platelet-poor plasma + Ca^{++}	2-4 min
Citrated platelet-poor plasma + PL + Ca^{++}	60-85 sec
Citrated platelet-poor plasma + kaolin + PL + Ca^{++}	21-32 sec (aPTT)
Citrated platelet-poor plasma + thromboplastin + Ca^{++}	11-12 sec (PT)

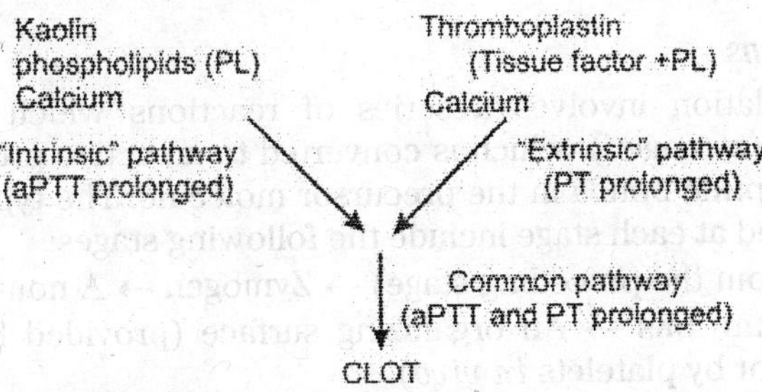

FIG. 30.1: Intrinsic and extrinsic coagulation pathways of coagulation *in vitro*.

In vitro by knowing the distinct coagulation factors and their deficiencies (e.g., the hypothetical orange and blue factors, coagulation can be controlled. There are 11 plasma coagulation factors as follows:

1. Six factors in the intrinsic pathway (factors VIII, IX, XI, XII, prekallikrein, and high-molecular weight kininogen).
2. One factor in the extrinsic pathway (factor VII).
3. Four factors in the common pathway (factors II, V, X, and fibrinogen).

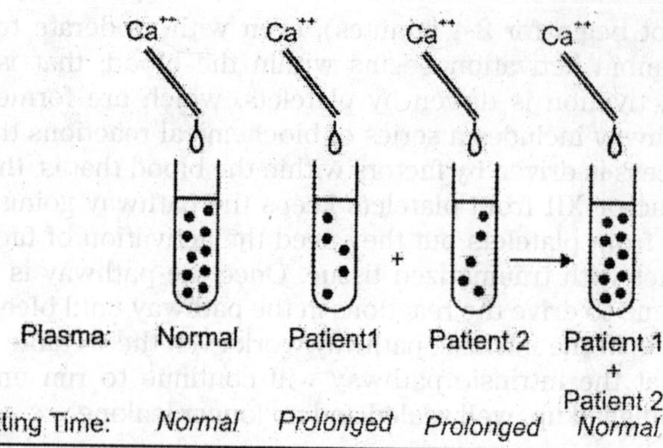

FIG. 30.2: Differentiation of congenital bleeding disorders.

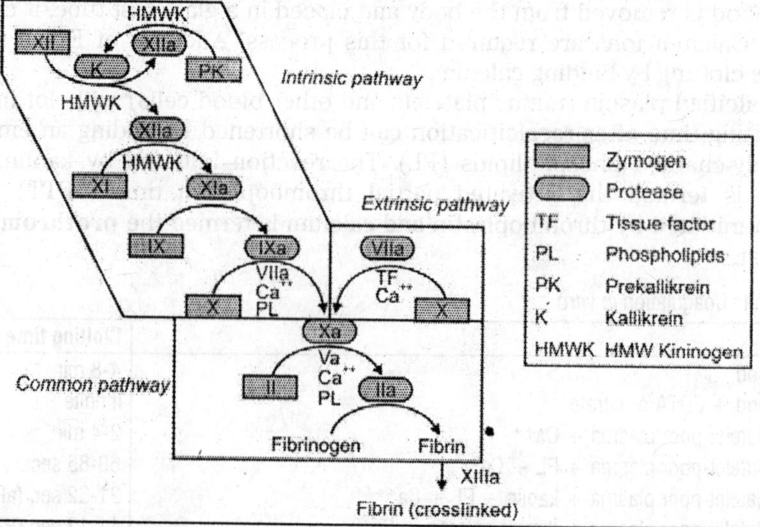

FIG. 30.3: Coagulation reactions.

Coagulation Reactions

The blood coagulation involves a series of reactions which are activated by precursor protein (zymogen), which is converted to an active protease by cleavage of one or more peptide bonds in the precursor molecule. The types of components that can be involved at each stage include the following stages:

A protease (from the preceding stage) → Zymogen → A non-enzymatic protein cofactor → Calcium ions → An organizing surface (provided by a phospholipid emulsion *in vitro* or by platelets *in vivo*)

The final protease generated is thrombin (factor IIa). Thrombin converts the soluble protein fibrinogen into an insoluble fibrin gel, which is strengthened further by covalent cross-linking catalysed by factor XIIIa.

Coagulation *In Vivo*

Initially the concept of coagulation was based on intrinsic and extrinsic pathways for normal hemostasis. The recent evidence has shown that exposure of plasma to tissue factor initiates coagulation. The tissue factor is a **non-enzymatic lipoprotein**. Exposure of plasma to these cells initiates coagulation outside a broken blood vessel. Endothelial cells also express tissue factor when stimulated by endotoxin, tumor necrosis factor, or interleukin-1, and may be involved in thrombus formation under pathologic conditions.

TABLE 30.2: Various plasma coagulation factors

Factor	Molecular Weight	Plasma Concentration (μg/ml)	Required for Hemostasis (% of normal concentration)
Fibrinogen	330,000	3000	30
Prothrombin	72,000	100	40
Factor V	300,000	10	10-15
Factor VII	50,000	0.5	5-10
Factor VIII	300,000	0.1	10-40
Factor IX	56,000	5	10-40
Factor X	56,000	10	10-15
Factor XI	160,000	5	20-30
Factor XIII	320,000	30	1-5
Factor XII	76,000	30	0
Prekallikrein	82,000	40	0
HMWK	108,000	100	0

Tissue Factor Pathway

Tissue factor binds factor VIIa and accelerates factor X activation about 30,000-fold. Although factor VII is activated by its product, factor Xa, a trace amount of factor VIIa appears to be available in plasma at all times to interact with tissue factor.

Factor VIIa also activates factor IX in the presence of tissue factor, providing a connection between the extrinsic and intrinsic pathways. Factors IXa and Xa assemble with their non-enzymatic protein cofactors (VIIIa and Va, respectively) on the surface of aggregated platelets. This leads to local generation of large amounts of Xa and thrombin (IIa), followed by conversion of fibrinogen to fibrin.

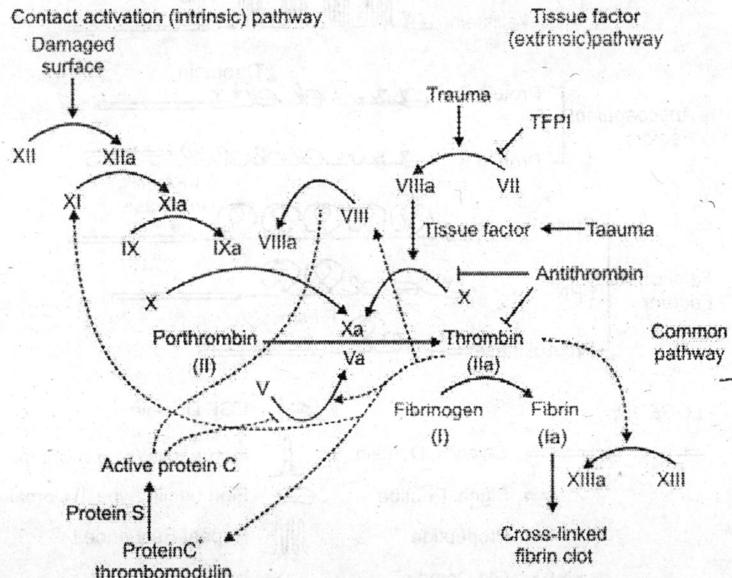

FIG. 30.4: Relationship between contact activation (intrinsic) and tissue factor (extrinsic) pathways.

Endothelial cells
Platelets
X
VIIa
II
X
IX
IXa
Xa
Va
Fibrinogen
IIa
VIIa
TF
Fibrin
Fibroblasts/ Macrophages
Zymogen
Protease
Intravascular
TF Tissue Factor Extravascular

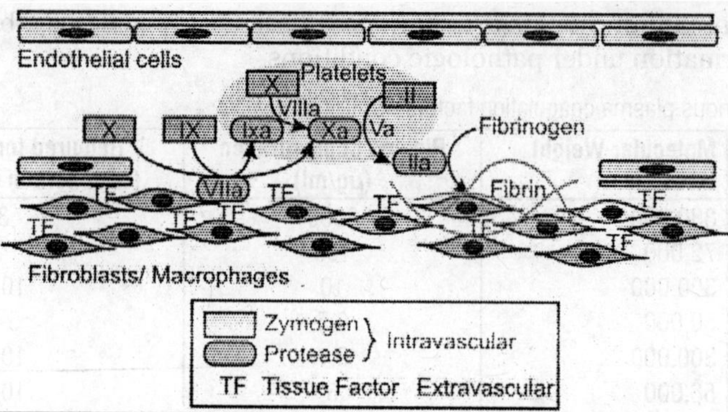

FIG. 30.5: Initiation of coagulation *In Vivo.*

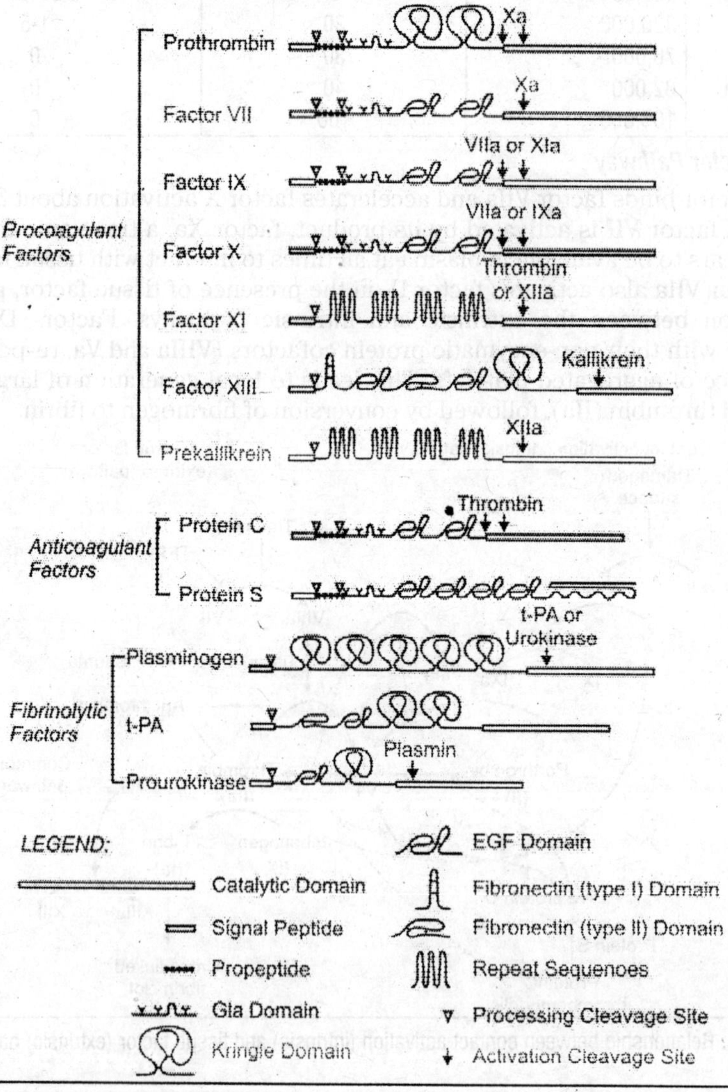

Prothrombin
Xa
Factor VII
Xa
Factor IX
VIIa or XIa
Factor X
VIIa or IXa
Thrombin or XIIa
Factor XI
Factor XIII
Kallikrein
Prekallikrein
XIIa
Procoagulant Factors
Protein C
Thrombin
Protein S
Anticoagulant Factors
Plasminogen
t-PA or Urokinase
t-PA
Prourokinase
Plasmin
Fibrinolytic Factors
LEGEND:
Catalytic Domain
Signal Peptide
Propeptide
Gla Domain
Kringle Domain
EGF Domain
Fibronectin (type I) Domain
Fibronectin (type II) Domain
Repeat Sequenoes
Processing Cleavage Site
Activation Cleavage Site

FIG. 30.6: Structures of coagulation factors.

Tissue factor pathway inhibitor (TFPI) is a 34-kDa protein associated with plasma lipoproteins and with the vascular endothelium. It binds to and inhibits factor Xa. The Xa-TFPI complex then interacts with VIIa/tissue factor and inhibits activation of factors X and IX. TFPI may prevent coagulation unless the VIIa/tissue factor initially present generates a sufficient amount of factor IXa to sustain factor X activation via the "intrinsic" pathway. Thus, VIIa/tissue factor may provide the initial stimulus to clot (in the form of relatively small amounts of IXa and Xa) and then be rapidly turned off, while IXa and VIIIa may be responsible for generating the larger amounts of Xa and thrombin required for clot formation. Activation of factor XII is not required for hemostasis.

Biochemistry of Coagulation

Structure of Coagulation Protease Zymogens

The structures of the protease zymogens involved in coagulation (i.e., factors II (prothrombin), VII, IX, X, XI, XII, and kallikrein). Hepatocytes secrete signal peptide protein into the bloodstream. It is removed during transit into the endoplasmic reticulum. About 200 amino acid residues at the C-terminal end of each zymogen are homologous to trypsin. They have active site Ser, Asp, and His residues of the protease (catalytic domain). The proteins also contain a variety of domains that are homologous to portions of other proteins such as epidermal growth factor (EGF) and fibronectin. These domains appear to be involved in specific interactions between the proteases and their substrates, cofactors, and/or inhibitors.

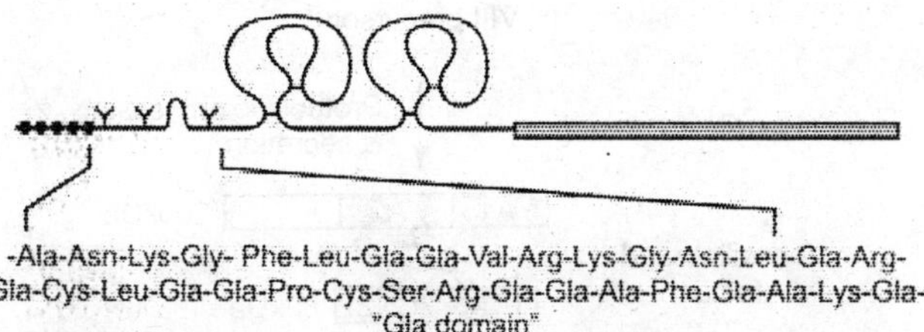

FIG. 30.7: γ-Carboxylation of Prothrombin (Factor II).

Glutamate
γ-Carboxylutamate
Ca^{++}
COO^-
CH_2
CH_2
^-COC COO^-
CH
CO_2, O_2
Vitamin K (reduced)
Vitamin K (oxidized)
NAD^+
Warfarin (inhibits)
NADH

FIG. 30.8: Role of Vitamin K in Biosynthesis of Factors II, VII, IX, and X.

Factors II, VII, IX, and X are homologous to each other at their N-terminal ends. After removal of the signal peptide, a carboxylase residing in the endoplasmic reticulum or Golgi binds to the propeptide region of each of these proteins and converts ~10-12 glutamate (Glu) residues to γ-carboxyglutamate (Gla) in the adjacent "Gla domain". The propeptide is removed from the carboxylated polypeptide prior to secretion.

Non-Enzymatic Protein Cofactors

These factors are V and VIII, tissue factor, and high-molecular weight kininogen (HMWK).

Functions of Factors V and VIII: These factors are large plasma proteins and made of repeated sequences homologous to the copper-binding protein ceruloplasmin (A1, A2 & A3 domains. **Thrombin** cleaves V and VIII to yield activated factors (Va and VIIIa) that have at least 50 times the coagulant activity of the precursor forms. Va and VIIIa have no known enzymatic activity. Instead, they serve as cofactors that increase the proteolytic efficiency of Xa and IXa, respectively. Factor VIII circulates in plasma bound to von Willebrand factor; thus, a patient with von Willebrand factor deficiency will generally have a concomitant decrease in the plasma concentration of factor VIII.

TABLE 30.3: Non-enzymatic Protein Cofactors

Cofactor	Location	Mol Wt	Activated by	Cofactor for
Factor V	Plasma	300,000	Thrombin	Factor Xa
Factor VIII	Plasma (bound to vWF)	300,000	Thrombin	Factor IXa
Tissue factor	Cell membranes	40,000		Factor VIIa
HMWK	Plasma	110,000		Factor XIIa

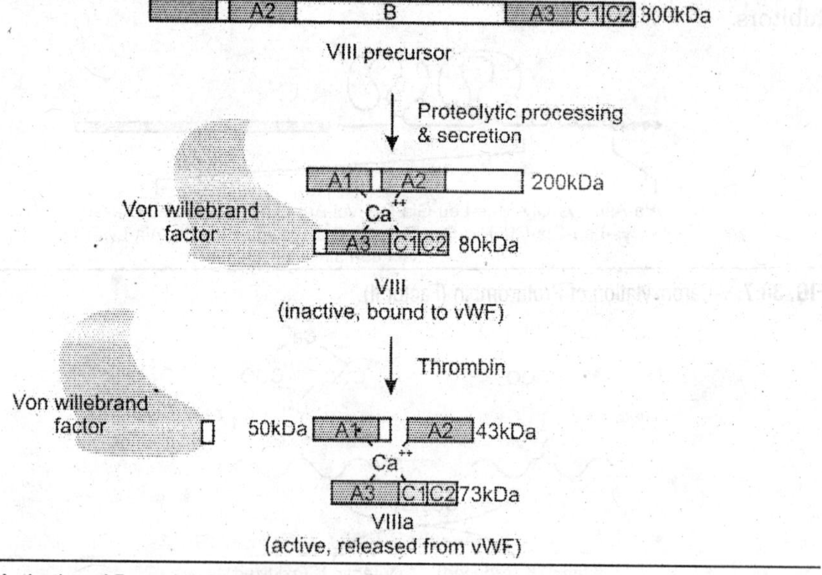

FIG. 30.9: Activation of Factor VIII by Thrombin

Prothrombin Activation

Factor Xa converts prothrombin (factor II) to thrombin (factor IIa) by cleaving two peptide bonds in the zymogen. Activation of prothrombin by Xa is accelerated by Va, platelets (or phospholipids), and calcium ions. The complete system activates prothrombin at a rate about 300,000 times greater than that of Xa and calcium alone.

Fibrinogen

It consists of three pairs of polypeptide chains, which are covalently linked by disulphide bonds and has a molecular weight of approximately 330,000. Thrombin converts fibrinogen to **fibrin monomers** by cleaving fibrinopeptides A (16 amino acid residues) and B (14 amino acid residues) from the N-terminal ends of the Aα and Bβ chains, respectively. Removal of the fibrinopeptides allows the fibrin monomers to form a gel consisting of long polymers. Formation of the fibrin gel constitutes the end point of the aPTT and PT tests. At this stage, the fibrin monomers are bound to each other non-covalently.

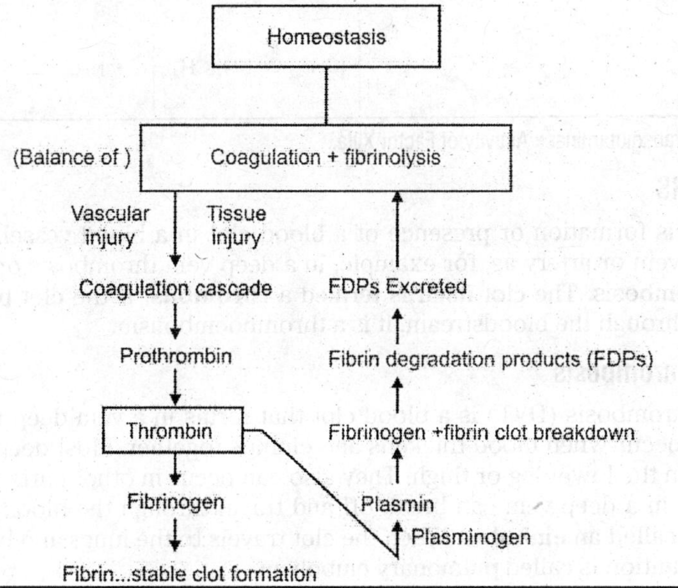

FIG. 30.10: Pathway of coagulation and thrombin.

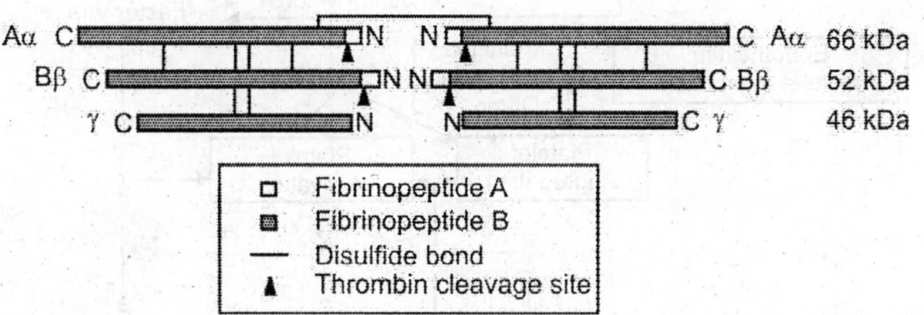

FIG. 30.11: Structure of Fibrinogen.

Factor XIII

Covalent cross-linking of fibrin polymers by activated factor XIII (XIIIa) is required for adequate clot strength and normal wound healing *in vivo*.

Amplification and Localization of Coagulation Reactions

The catalytic nature of coagulation reactions allows tremendous amplification of the initial stimulus. Each VIIa/tissue factor complex can produce many Xa molecules, which subsequently produce an even greater amount of thrombin.

$$-(CH_2)_2-\overset{O}{\overset{\|}{C}}-NH_2 \quad :NH_2(CH_2)_4-$$

Glutamine *Lysine*

Factor XIII $\xrightarrow{\text{Thrombin}}$ Factor XIIIa

$$-(CH_2)_2-\overset{O}{\overset{\|}{C}}-NH_2(CH_2)_4- \quad + NH_3$$

FIG. 30.12: Transglutaminase Activity of Factor XIIIa.

THROMBOSIS

Thrombosis is formation or presence of a blood clot in a blood vessel. The vessel may be any vein or artery as, for example, in a deep vein thrombosis or a coronary (artery) thrombosis. The clot itself is termed a **thrombus**. If the clot breaks loose and travels through the bloodstream, it is a thromboembolism.

Deep Vein Thrombosis

Deep vein thrombosis (DVT) is a blood clot that forms in a vein deep in the body. Blood clots occur when blood thickens and clumps together. Most deep vein blood clots occur in the lower leg or thigh. They also can occur in other parts of the body. A blood clot in a deep vein can break off and travel through the bloodstream. The loose clot is called an **embolus**. When the clot travels to the lungs and blocks blood flow, the condition is called pulmonary embolism.

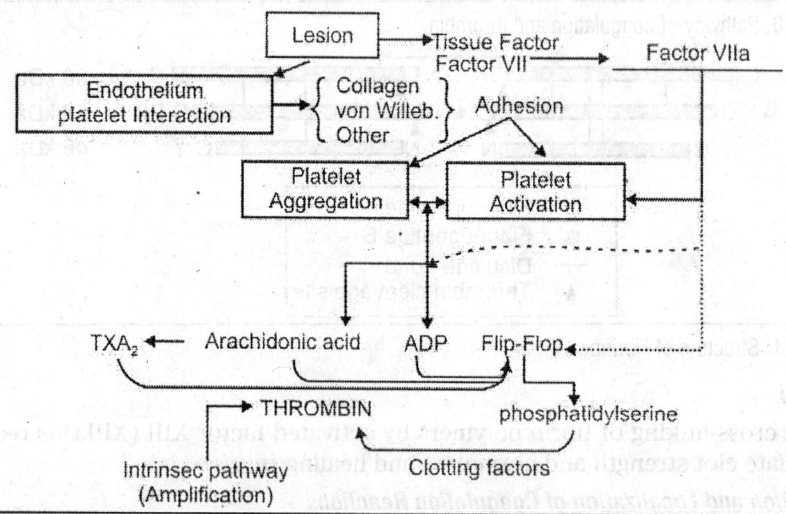

FIG. 30.13: Relation between various factors of coagulation and thrombin pathways.

A pulmonary embolism (PE) is a sudden blockage in a lung artery, usually due to a blood clot that traveled to the lung from a vein in the leg. A clot that forms in one part of the body and travels in the bloodstream to another part of the body is called an **embolus**. Under serious condition PE can cause: (1) Permanent damage to part of your lung from lack of blood flow to lung tissue, (2) Low oxygen levels in your blood and (3) Damage to other organs in your body from not getting enough oxygen. If the blood clot is large, or if there are many clots, PE can cause death.

Mechanism of Hemostasis

Damage to small blood vessels and capillaries frequently occurs. When these vessels are damaged, there are three basic mechanisms that promote hemostasis or the stoppage of bleeding.

Following damage, there is an immediate reflex that promotes vasoconstriction, thus diminishing blood loss. Exposed collagen from the damaged site will promote the platelets to adhere.

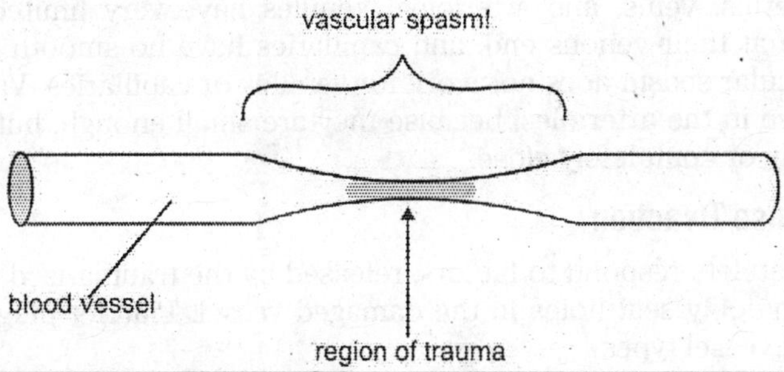

FIG. 30.14: Vasoconstrictive mechanism.

Vascular Spasm

It is also referred to as **vascular response**. Vascular spasm refers to a local vasoconstrictive mechanism. That is, trauma to the vessel wall causes vasoconstriction in both directions along the traumatized vessel. The region of vasoconstriction of the traumatized vessel extends several centimeters in each direction from the area of trauma.

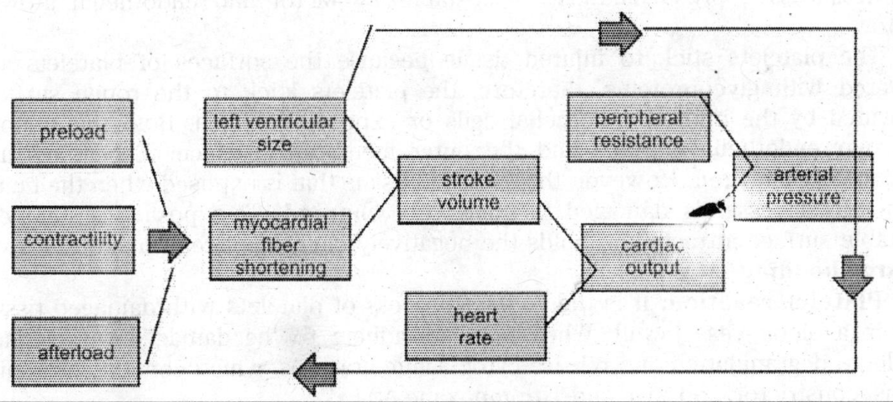

FIG. 30.15: Vasoconstrictive mechanism is dependent on various factors.

Vascular spasm reduces blood flow through the traumatized vessel and, thus, reduces blood loss. Also, keep in mind that a number of factors that lead to coagulation (clotting) are released from the traumatized blood vessel wall. Reducing blood flow through the traumatized vessel wall reduces the dilution of the clotting factors in the traumatized area. If the factors were diluted, coagulation time would be increased (it would take longer to stop bleeding).

A stimulus (trauma) from within the wall of the traumatized vessel causes the smooth muscle of the wall to constrict, reducing blood flow (loss) from the ruptured vessel. Local vasoconstriction results partly from nervous reflexes (nociceptive or associated with the sensation of pain) but mostly from local myogenic spasm that is originates from within the muscle. The direct damage to the vessel wall causes action potential transmission and contraction through the vascular smooth muscle for several centimeters in each direction. The sharper the cut, the greater the blood loss as less tissue is traumatized in a clean cut than in a rough or serrated cut or in a crushing wound. The local vascular myogenic response (spasm) can last for hours; this restricts blood flow, reducing blood loss but also reducing dilution of the clotting factors during clot formation.

Vascular spasm occurs only in those blood vessels with smooth muscle, which means the arteries, veins, and arterioles. Venules have very limited amounts of smooth muscle at their venous end, and capillaries have no smooth muscle at all. Therefore, vascular spasm does not work for venules or capillaries. Vascular spasm is most effective in the arterioles, because they are small enough, but with enough muscle to nearly or completely close.

Platelet Response/Reaction

At this stage platelets respond to factors, released by the traumatized vessel wall in an attempt to quickly seal holes in the damaged vessel. Platelet plug formation is important in all vessel types.

A nick in a blood vessel results in exposure of the **wetable surface**. The wetable surface consists of injured endothelial cells and exposed collagen where platelets stick.

The platelet reaction consists of platelets sticking to the wetable surface, swelling and attracting other platelets. When punctures in the blood vessels are small enough, platelet plugs will stop the blood flow and no blood clot is required. Platelets contain contractile proteins (including actin and myosin), residual endoplasmic reticulum (Ca^{2+}) and Golgi bodies; this means that platelets can synthesise ATP, prostaglandins, fibrin-stabilizing factor and endothelial growth factor.

The platelets stick to injured tissue because the surfaces of platelets are covered with glycoproteins, therefore, the platelets stick to the rough surface provided by the injured endothelial cells or exposed collagen. Both the smooth vascular endothelial surface and the outer membrane surface of platelets are **negatively charged**. However, the damaged tissue that is exposed when the lining of a blood vessel is damaged is **positively charged**. This positively charged, wetable surface attracts and holds the negatively charged platelet membranes via **charge bonding.**

Platelet reaction: It is the contact process of platelets with damaged tissue inside a blood vessel wall. When platelets adhere to the damaged vessel, they undergo degranulation and release cytoplasmic granules, which contain serotonin, a vasoconstrictor, and ADP and Thromboxane A2.

The platelets swell up and become increasingly sticky due to the release of glycoproteins. The platelets begin to release a bunch of arachidonic acid metabolites (arachidonic is a fatty acid). Among the arachadonic acid metabolites released by platelets are thromboxanes (TX's) and prostaglandins (PG's). Activated platelets also release serotonin (5-HT) and thrombopoietin.

The ADP attracts more platelets to the area, and the thromboxane A2 promotes platelet aggregation, degranulation, and vasoconstriction. Thus ADP and thromboxane A2 promote more platelet adhesion and therefore more ADP and thromboxane. The positive feedback promotes the formation of a platelet plug.

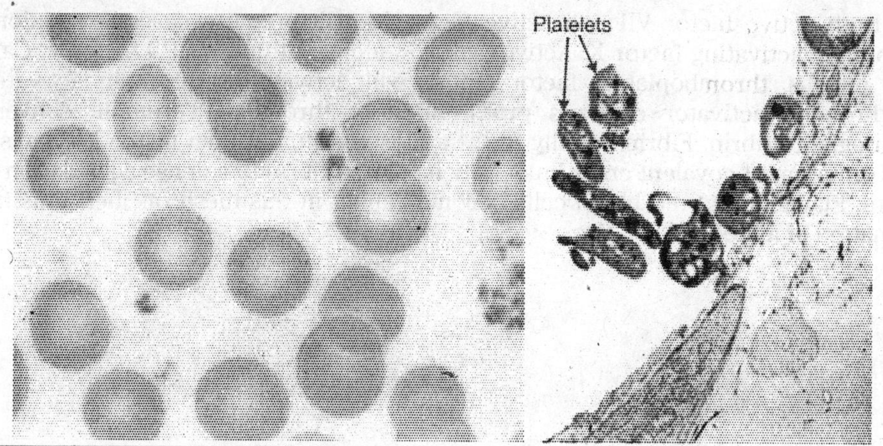

FIG. 30.16: Showing active platelets.

Factors released by platelets: Following are the factors responsible for release of platelets:

1. Attract more platelets into the region of the damaged tissue.
2. Causes vasoconstriction.
3. Increases the production rate of new platelets.
   a) Thromboxanes (TX's) and progstaglandins (PG's) attract (positive chemotaxis) and activate surrounding platelets, causing platelets to clump together and form a loose platelet plug.
   b) Serotonin (5-HT) induces vascular spasm.
   c) Thrombopoietin stimulates the production of more platelets.

Blood Coagulation (Clotting)

Damaged tissue releases factor III, which with the aid of Ca^{++} will activate factor VII, thus initiating the extrinsic mechanism. Factor XII from active platelets will activate factor XI, thus initiating the intrinsic mechanism.

Clotting is initiated by the platelet reaction and a network of fibrin molecules. Both make a stronger patchwork over a traumatized area. In fact, platelets appear to have specific binding sites for attachment to fibrin.

Fibrin threads are strong protein filaments. Clotting is the patchwork which form a bunch of blood cells get trapped in the middle of the net in which a number of RBC's are trapped within the fibrin network.

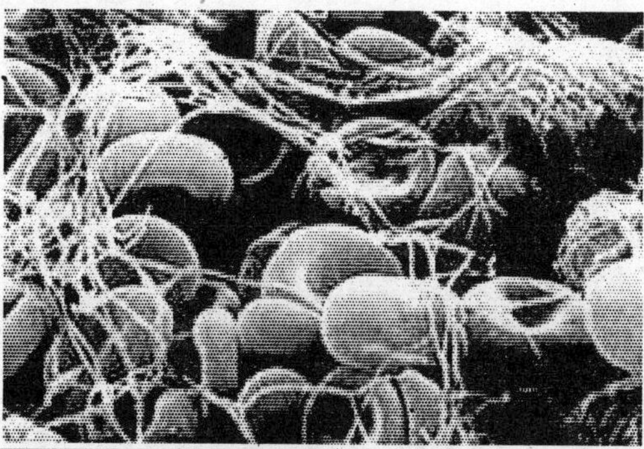

FIG. 30.17: Presentation of platelets, fibrin and clotting.

Both active factor VII and active factor XI will promote cascade reactions, eventually activating factor X. Active factor X, along with factor III, factor V, Ca^{++}, and platelet thromboplastic factor (PF3), will activate prothrombin activator. Prothrombin activator converts prothrombin to thrombin. Thrombin converts fibrinogen to fibrin. Fibrin initially forms a loose mesh, but then factor XIII causes the formation of covalent cross links, which convert fibrin to a dense aggregation of fibers. Platelets and red blood cells become caught in this mesh of fiber, thus the formation of a blood clot.

CHAPTER 31

Metabolism of Xenobiotics

Xenobiotics (*xenos* = foreing; *bios* = life) are organic compounds which have no counterpart in the natural world, i.e. they are not naturally occurring compounds which endanger the flora and fauna of the ecosystem. An examination of the fate of foreign compounds (xenobiotics) in biological systems is a natural outgrowth of man's curiosity about his environment, and his susceptibility to it. While most modern studies concern the fate of drugs in man and animals, there are also investigations into the fate of organic compounds in plants, fruit flies, nematodes and easts. The term "xenobiotic" was coined to cover all compounds that are foreign to the organism under study. In some situations, this is loosely defined to include naturally present compounds administered by alternate routes or at unnatural concentrations.

Xenobiotics are foreign anti biotic (*anti* = against) (*biotic* = life) or foreign chemical substances. These anti-biotic/foreign chemical substances, once in the cell, can block certain enzyme systems and thereby interfere with the normal energy producing and functional mechanisms of those cells. Our special line of xenobiotic combination formulas are specially created homeopathic remedies designed to assist in the removal of specific toxic and anti-biotic elements and substance in the system. These formulas combine well-known homeopathic drainage remedies with multiple potencies of the specific xenobiotic isodes.

An isode is a homepathic remedy that uses the actual poison or substance, microorganism or parasite that is causing or associated with the disease condition, in a minute dose to cause the cure. (*isos* = a combining form, meaning same, equal or alike) A Nosode is any disease product or substance used as a remedy in homeopathic medicines. (Greek, *nosos* = disease)

Xenobiotics are chemicals found in organisms, but not expected to be produced or present in them; or they are chemicals found in much higher concentrations than usual. Organs transplanted cross-species are also called xenobiotic. An example of a xenobiotic is an antibiotic; the human body does not produce them, and they are not expected to be present in the normal human diet. It is more commonly used to describe dioxins and polychlorinated biphenyls, which affect biological organisms in a variety of ways both fatal and nonfatal.

Sustainable chemistry aims at an improved efficiency of using natural resources which are used to meet human needs for chemical products. Chemists in science and industry, have become aware of the importance to design environmentally benign chemicals. One aspect is the biological persistence is the degradation of xenobiotics in the environment. Different structural reasons for chemical and biological persistence are described in this chapter and strategies to use single bacterial isolates or microbial communities for the elimination of xenobiotic pollutants in the environment. Perspectives and limitations to evolve and use this catabolic potential are critical and the complexity of mixtures of xenobiotics often found in nature. An interdisciplinary approach for the prospective design of environmentally benign substances is required.

Xenobiotics are eliminated from the body through xenobiotic metabolism, in which the xenobiotic is deactivated and secreted away from the body. The liver is typically the primary focus of xenobiotic metabolism, and secretion can occur through urine, feces, breath, and sweat. Hepatic enzymes metabolize xenobiotics. An understanding of xenobiotic metabolites is critical for the pharmaceutical industry because they are responsible for the breakdown of drugs.

Sewage is increasingly contaminated with xenobiotics of all sorts, from human waste to antibiotics to dioxins. Sewage treatment systems have more trouble sorting out some of these xenobiotics than they do with other contaminants. This is because many newer xenobiotics are derived from plants, and closely mimic these natural substances. It can be nearly impossible to remove these new pollutants.

Xenobiotics are found when the body absorbs chemicals that are not nutrients, or when normally occurring substances (nutrients or endogenously produced biochemicals) become denatured (modified from original molecular structure) or are in excess. Some authors suggest that we are burdened by as many as 100,000 man-made xenobiotics - and that does not take into account the natural ones.

A considerable number of clinical studies are demonstrating that xenobiotic chemicals, which were once thought to have a very precise and direct toxic effect, may actually affect different parts of the body in a variety of ways. One study has suggested a deficiency in detoxification pathways is involved in symptoms of Parkinson's disease. Other studies have associated the development of respiratory tract and food allergies with insufficient detoxification. Given the increasing number of individuals suffering from various recognized allergies, as well as the direct or indirect implication of allergies in a variety of disorders (they may cause the initial symptoms, or as is generally understood, exacerbate them), this has extraordinary ramifications. According to Dr. William Holub, Treatment of allergy, then, requires far more than the identification and isolation from the allergen, but requires, in addition, an applicable program of dietary regulation and food supplementation to insure adequate detoxification. Xenobiotics and food allergies are associated with many physical and psychological disorders.

HISTORY OF XENOBIOTICS

The earliest history of drug metabolism is interwoven with the birth of Organic Chemistry. In the early 1800's the belief prevailed that the composition of the human body and indeed, of all living things, was a result of a *vital force* or *internal flame*" and that mere mortals were incapable of understanding these workings and would be particularly unable to synthesise constituent compounds of the human body. Faced with this daunting concept, scientists were discouraged from thinking about the processes that dominated everyday digestion and nutrition. Yet, it was this problem which came to dominate the mind of a young scientist, Friedrich Woehler.

FIG. 31.1: Friedrich Woehler.

Woehler was born in the early nineteenth century near Frankfort. He was interested in minerals in his youth and developed a keen interest in chemistry. Many of these early investigations were performed upon his dog and upon himself. It is a continuing theme of early metabolism studies that the major subjects studied were also the authors of the studies. In 1828, Woehler was able to synthesise urea, one of the compounds he had examined in his studies on urinary waste products and sent a note to Berzelius "*I must tell you that I can prepare urea without requiring a kidney of an animal, either man or dog*".

In modern day terminology this was a true *paradigm shift* and led to revised thinking on what chemists could accomplish with regards to *organic* compounds. Woehler became interested in potential chemical transformations occurring in the body and began experiments in dogs toward this end. He had speculated on the conversion of benzoic acid to hippuric acid but his initial experiments in dogs led to the conclusion that only benzoic acid was excreted in the urine.

The first human metabolism study was performed in 1841 by Alexander Ure (born in Scotland), who observed the conversion of benzoic acid to hippuric acid and proposed the use of benzoic acid for the treatment of gout.

$$C_6H_5COOH \text{ (Benzoic Acid)} + H_2NCH_2COOH \text{ (Glycine)} \longrightarrow C_6H_5CONHCH_2COOH \text{ (Hippuric Acid)}$$

W. Keller, in Woehler's laboratory provided the confirmation of Ure's experiments. Bernhard Naunyn (1839-1925, Berlin) worked in Frerichs' lab developed his keen interest in metabolic pathology. He recognized the contrafermentation property of benzene. He became curious about the effects of benzene on the patient and soon came to realize that humans excrete phenol after having received benzene.

Eugen Baumann (1846-1896, born in Wurtemberg in 1846) studied the formation of sulphates in the body led to the finding that urinary ethereal sulphates related to putrefactive decomposition occurring in the intestine. Baumann and Preusse also described the formation of bromophenylmercapturic acid. Five years later Baumann reported the structure of mercapturic acids as consisting of acetyl derivatives of cysteine conjugates.

Oswald Schmiedeberg (1838-1921) born in 1838 in Kurland, a Baltic province of Russia (later Latvia). He made initial pharmacological studies on some of the most important drugs and poisons of that era including muscarine, nicotine, the digitalis glycosides, and various toxic heavy metals. He then was able to show that glucuronic acid was a component of cartilage and occurred as a disaccharide of chondroitin sulphate. In his many years of studying the chemistry of mucopolysaccharides and mucoproteins he elaborated the components of hyaluronic acid and explored it's relation to chondroitin sulphate, collagen and amyloid.

XENOBIOTIC COMPOUNDS

Several xenobiotic compounds have been synthesised for agricultural and industrial purposes. These include fungicides, herbicides, pesticides, and insecticides, chloro-organic chemicals, biphenyl compounds, environmental pollutants etc. Fertilizers, pesticides, and herbicides enter the environment through various paths through direct application of polycyclic aromatic hydrocarbons, dibenzo-p-dioxins, and dibenzofurans, released by combustion processes. Xenobiotics are present in the waste effluents produced by synthetic products industries. Their range in environment is form parts per thousand (ppt) to parts per billion (ppb) and their concentration depends on the amount of release, extent of dilution, the mobility of the compound (for example, in soil) and its rate of degradation in a particular environment.

Biomagnifications or Biological Magnification

Xenobiotics ate progressively concentrating in each link of food chain; the process in called biomagnifications or biological magnification. The increase in concentration of a substance, such as the pesticide DDT, that occurs in a food chain as a consequence of:

1. Food chain energetics.
2. Low (or nonexistent) rate of excretion/degradation of the substance.

It is an important concept in ecology, environmental science, and ecotoxicology: it says that *the solution to certain* types of pollution is not dilution, because food chains will concentrate the pollutant. Although sometimes used noninterchangeably with bioaccumulation, an important distinction is drawn between the two. Bioaccumulation occurs within a trophic level, and is the increase in concentration of a substance in an individual's tissues due to uptake from food and sediments in an aquatic milieu.

Bioconcentration

Bioconcentration is a process that results in an organism having a higher concentration of a substance than is in its surrounding environmental media, such as stream water. Bioconcentration of a substance is correlated to the octanol-water partition coefficient (Kow) of the substance. For more information see the definition of the related term bioaccumulation. Bioconcentration differs from bioaccumulation because it refers to the uptake of substances into the organism from water alone. Bioaccumulation is the more general term because it includes all means of uptake into the organism. Thus bioconcentration and bioaccumulation occur within an organism, and biomagnification occurs across trophic (food chain) levels.

Lipid soluble (lipophilic) substances cannot be excreted in urine, a water-based medium, and so accumulate in fatty tissues of an organism if the organism lacks enzymes to degrade them. When eaten by another organism, fats are absorbed in the gut, carrying the substance, which then accumulates in the fats of the predator. Since at each level of the food chain there is a lot of energy loss, a predator must consume many prey, including all of their lipophilic substances.

For example, though mercury is only present in small amounts in seawater, it is absorbed by algae (generally as methylmercury). It is efficiently absorbed, but only very slowly excreted by organisms (Croteau et al., 2005). Bioaccumulation and biomagnification result in buildup in the adipose tissue of successive trophic levels: zooplankton, small nekton, larger fish etc. Anything which eats these fish also consumes the higher level of mercury the fish have accumulated. This process explains why predatory fish such as swordfish and sharks or birds like osprey and eagles have higher concentrations of mercury in their tissue than could be accounted for by direct exposure alone. For example, herring contains mercury at approximately 0.01 ppm and shark contains mercury at greater than 1 ppm.

Removal of these xenobiotics involves physical and chemical process like land filling, excavation, incineration, etc, which are expensive and sometimes difficult to execute. Microorganisms are naturals original recyclers, converting toxic organic compounds to harmless products, often carbon dioxide and water.

Fate and Impact of Xenobiotics in the Environment

Information about the biodegradability, metabolic pathways, conjugation and de-conjugation, sorption, transport in the environment, and persistence of xenobiotics are needed to predict their fate and impacts and to help in the creation of policy to protect humans and ecosystems from their effects.

There is a need for the development and implementation of experimental methods and analytical protocols to examine the degradation of selected compounds, identification and characterization of degraded products, elucidation of degradation mechanisms, screening for health and ecological impacts, and determination of the extent to which degradation processes and the presence of problematic compounds in the environment. These thematic areas will provide the necessary information to satisfy.

There is a need for development of new scientific knowledge of the pollution potential of xenobiotics and their degradation products. It will also help to standardized protocols for assessing the health and environmental impacts of xenobiotics.

BIOTRANSFORMATION OF XENOBIOTICS

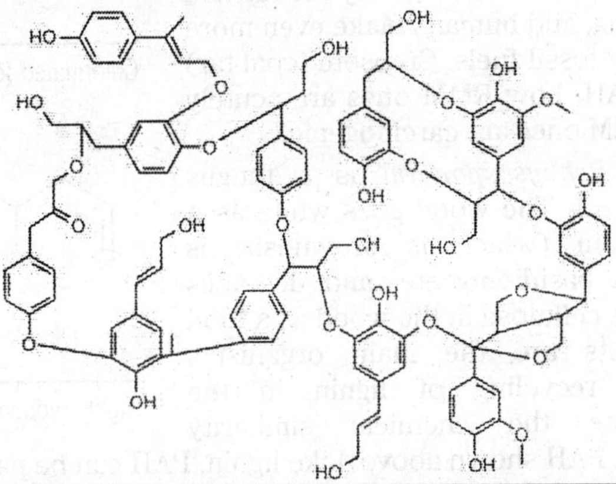

Lignin.

Biotransformation is the chemical alteration of compounds by living things; in particular, the microbial alteration of anthropogenic xenobiotics. Xenobiotics are chemicals made by humans that are not formed by 'natural' processes. It may also include 'natural' compounds produced *primarily* by human beings. The word means 'foreign to life'. Xenobiotics include such infamous things as:

1. Polychlorinated aromatics: DDT, dioxins.
2. Organophosphates and carbamates.
3. Polycyclic aromatics: pyrene (creosote).

Xenobiotics are often polycyclic, halogenated, highly reduced and lipophilic. As a result of the latter property, some can bioaccumulate up food chains. Because the liver (in mammals) cannot break many xenobiotics down fully, they accumulate in fat. Fats ingested by carnivores contain xenobiotics, which they then accumulate themselves. And so on up the foodchain. Top predators can receive very damaging doses of xenobiotics from their food, despite the environmental concentrations being quite low. Many xenobiotics are highly toxic. All are 'unusual' so far as ordinary metabolic processes are concerned, because they are generally not things that Life has been exposed to in its 3500 million year history. However, they are not *that* unusual: often, xenobiotics are degraded by enzymes

Benzo[a]pyrene.

that act on something else that occurs naturally. This is usually something chemically similar (for some definition of similar). *e.g.* polyaromatic hydrocarbons can be degraded by *Phanerochaete*, a white rot fungus, which usually degrades lignin, which is chemically somewhat similar.

Xenobiotic compounds often sequester in (adsorb to) soil. Although they disappear from the aquatic phase, they have *not* been degraded, they have merely concentrated into a phase (such as sediment) in which they are not very available for analysis or degradation. PCP, DDT, etc., have very long half lives because they are not bioavailable.

Xenobiotics are frequently degraded by oxidases and dehalogenases. This probably reflects the nature of the compounds humans tend to synthesise, rather than anything more profound.

Chlorinated (dehalogenase required).

Polyaromatic hydrocarbons (PAH) are a well understood group of xenobiotics. They are formed naturally by burning, and humans make even more of them by burning fossil fuels. Creosote (coal-tar) is 90% or more PAH. Low RMM ones are acutely toxic, and high RMM ones are carcinogenic.

Phanerochaete chrysosporium is a fungus that causes white rot. The wood goes white as it rots because lignin (which is brown-ish) is destroyed. It is a basidiomycete, and degrades lignin to access the cellulose in the wood as a food source. White rots are the main organisms involved in the recycling of lignin in the environment. Note the chemical similarity between lignin and PAH shown above. Like lignin, PAH can be partially degraded by ligninase. Ligninase contains a cytochrome-P450 that has extremely high oxidising power. It removes electrons from veratryl alcohol (VA) and dumps them onto peroxide. VA^+ radical ions then pull electrons out of aromatic rings.

Highly reduced (oxidase required).

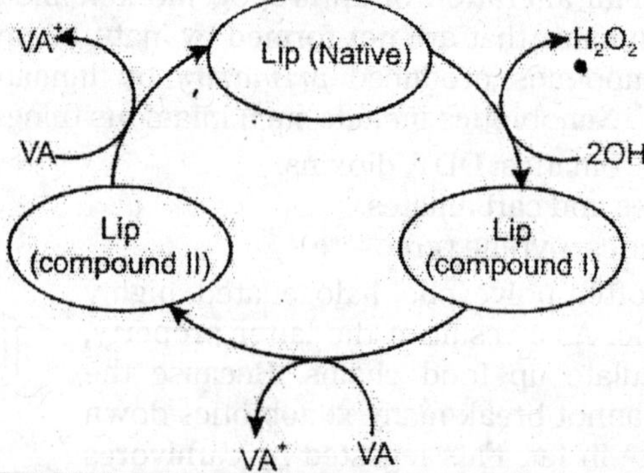

FIG. 31.2: VA pulls electrons out of PAH rings, generating quinones by nucleophilic attack.

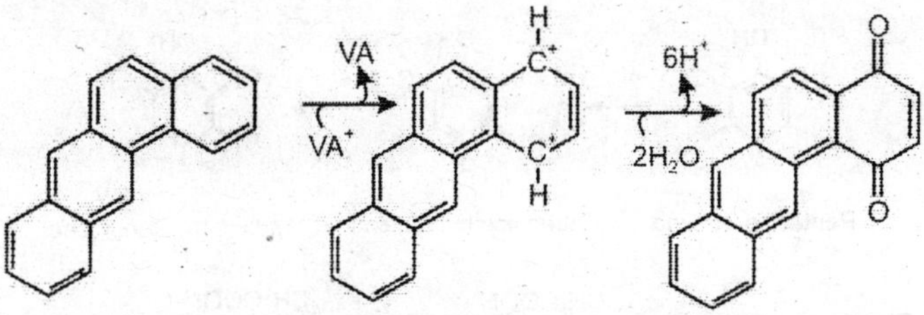

Bacteria can also actively metabolise PAH < 4 rings, by dioxygenation followed by ring cleavage. Aliphatic bits can be metabolised by Krebs cycle.

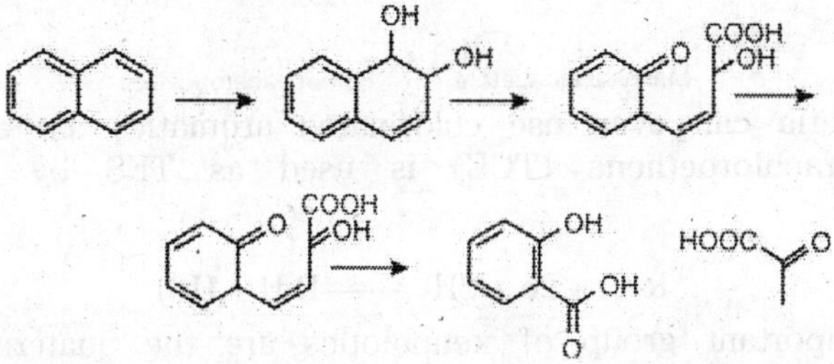

*Mono*oxygenation of BAP by mammalian liver enzymes produce an extremely carcinogenic epoxide. This epoxide can form an adduct with DNA, causing mutation.

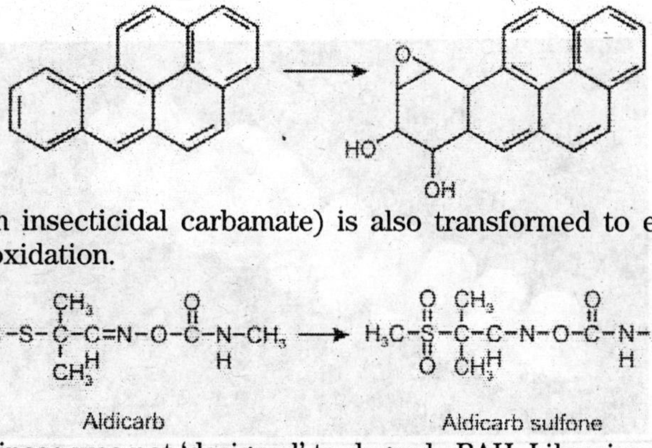

Aldicarb (an insecticidal carbamate) is also transformed to even more toxic compounds by oxidation.

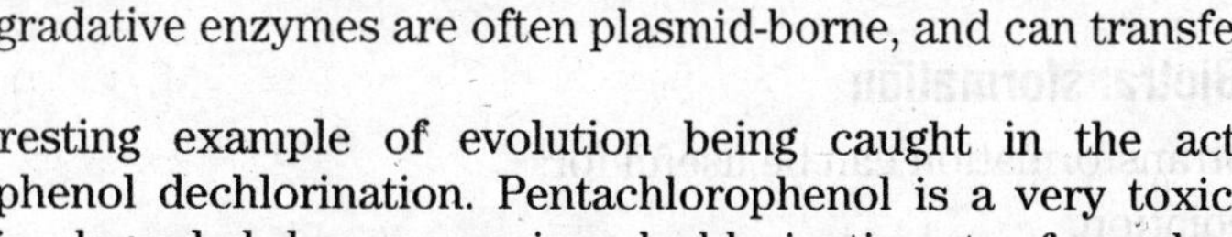

Aldicarb → Aldicarb sulfone

Clearly, ligninase was not 'designed' to degrade PAH. Likewise, the degradation of other xenobiotics is a happy accident that may provide a selective advantage. Bacterial degradative enzymes are often plasmid-borne, and can transfect into other bacteria.

An interesting example of evolution being caught in the act is that of pentachlorophenol dechlorination. Pentachlorophenol is a very toxic chlorinated phenol. It is degraded by successive dechlorination to form hydroquinones. Seemingly unrelatedly, the synthesis of tyrosine requires a glutathione (GSH)-mediated *cis/trans* isomerase. This enzyme also catalyses GSH- reductive dechlorination of tetrachlorohydroquinone.

Pentachlorophenol → Tetrachlorohydroquinone → Trichlorohydroquinone

Maleylacetoacetate Fumarylacetoacetate

Some bacteria can even use chlorinated aromatics as terminal electron acceptors. Tetrachloroethene (TCE) is used as TES by *Dehalococcoides ethenongenes*.

$$\text{R-Cl} + 2e^- + 2H^+ \longrightarrow \text{R-H} + \text{HCl}$$

Another important group of xenobiotics are the quaternary ammonium compounds (QACs), used in detergents. Molecularly, these are like aberrant membrane lipids, with a long fatty acid chain and a polar head group. When degraded, the head group is cleaved, and the tail is degraded by β -oxidation, just like fats.

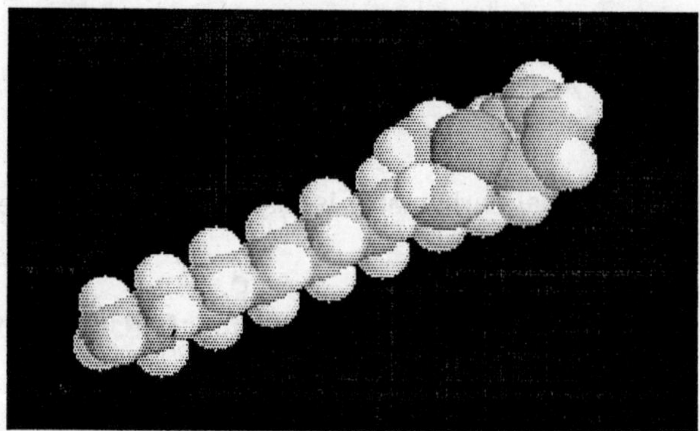

FIG. 31.3: *Benzalkonium chloride* is a quaternary ammonium compound found in many antiseptic creams.

Bacteria in Biotransformation

Bacterial biotransformation can be useful for:

1. Bioremediation.
2. Industrial biotech applications.
3. Understanding the fate of biocides in the environment.

Bacteria that degrade xenobiotics can be isolated using enrichment culture. This provides the xenobiotic as the sole source of an element (usually carbon). This selects for bacteria able to metabolise the chemical, over those that just modify or resist it.

Bacterial degradation can (somewhat arbitrarily) be divided into metabolic *vs.* cometabolic activity. Metabolic degradation actually uses the xenobiotic as a source of carbon, nitrogen, energy, or some combination of the above. Cometabolic degradation is 'fortuitous' transformation that does not lead to energetic/nutritional gain. Cometabolic degradation can still be adaptive, as it may reduce local toxicity.

$CH_3-(CH_2)_{11}-C_6H_4-SO_2^-Na^+$
Dodecylbenzenesulfonate Sodium salt (an alkyl benzene sulfonate)

COO^-
Banzoate

Biphenyl

Camphor

Cl–
4-Chlorobiphenyl

OCH_2COOH, Cl, Cl
2,4-Dichlorophenoxyacetic acid (herbicide)

Dibenzothiophene

$CH_3-CH_2-S-C(=O)-N(CH_2CH_2CH_3)_2$
S-Ethyl-N,N-Dipropylthiocarbamate (herbicide)

Naphthalene

$CH_3-(CH_2)_6-CH_3$
Octane

$O_2N-C_6H_4-O-P(=S)(OCH_2CH_3)_2$
Parathion(o,o-diethyl-o-nitrophenyl-phosphorothioate;insecticide)

$CH=CH_2$
Styrene

CH_3
Toluene

FIG. 31.4: Some of the examples of herbicides and insecticides degraded by encoded on catabolic plasmids.

IPBC

IPBC (wood wood preservative) is an example of cometabolic degradation is the degradation of the wood preservative IPBC.

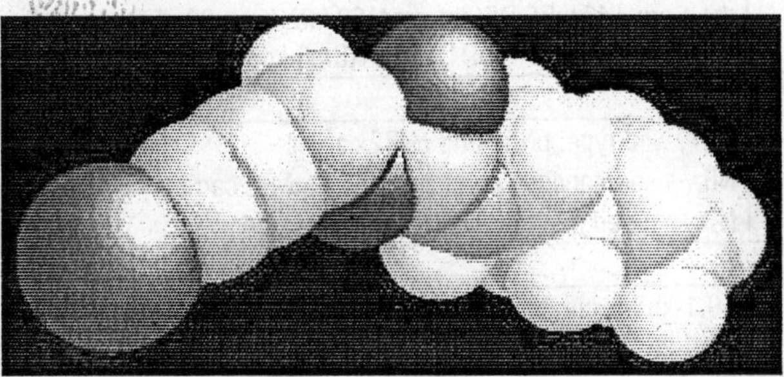

FIG. 31.5: IPBC (wood wood preservative).

IPBC appears to be degraded cometabolically by bacteria: the bacteria cannot survive on IPBC as a sole source of carbon, and degradation appears to be limited to reductive dehalogenation.

Bacteria like *Pseudomonas*, *Alcaligenes*, *Mycobacterium* and Actinomycetes (*Rhodococcus*). Fecund gene transfer, enzymatic competence and thick cell walls are probably important in some or all of these bacteria. They are also (sometimes facultatively) aerobic, and easy to culture. Sometimes, several bacteria act together to degrade a compound. This is a form of cross-feeding.

BIODEGRADATION: GENETIC APPROACHS

Release of massive amounts of nitroaromatic compounds into the environment has enabled a number of microorganisms to evolve the ability to mineralize these xenobiotic compounds. Nitrobenzene (NB), one of the recalcitrant nitroaromatic compounds, has been reported to be mineralized by microorganisms through oxidative and partial reductive pathways. In addition, construction of a hybrid pathway for the degradation of NB was attempted with *Pseudomonas putida* TB103.

Catabolic Plasmids

Catabolic plasmids have the potential to proliferate in unrestricted environment and can use pesticides and other potentially toxic compounds. They also have potential to eliminate toxic compounds and degrade the pesticides spreads among the different bacterial strains in the natural population by interspecies transfer of a catabolic plasmid that carries the genes for the degradative enzymes. Catabolic plasmids encode enzymes for degrading naturally occurring compounds such as camphor ($C_{10}H_{16}O$), octane (C_8H_{18}), naphthalene ($C_{10}H_8$) and toluene ($C_8H_5CH_3$), parathion, ($C_{10}H_{14}O_5NPS$). S-ethyl-N, N-dipropylthiocarbamate, 2,4 dichlorophenoxyacetic acid ($C_6H_3Cl_2OCH_2COOH$) etc.

The degradative ability of catabolic plasmids is due to cooerative interaction between the genes carried on the plasmid and those on chromosomes of the host cell. Many plasmids have genes that encode enzymes for part of the catabolic pathway for a given compound. Plasmid mediated products of transportation can then be utilized by chromosomal encoded enzymes functioning in the central energy producing metabolic pathway of the host cell.

TABLE 31.1: Bacterial strains and plasmids

Strain or plasmid	Description
Escherichia coli	
JM109	recA1 supE44 hsdR17 gyrA96 relA1 thi (lac-proAB) (F' traD36 proAB$^+$ lacI^q lac M15)
NFR502	thr-1 ara-14 leuB6 DE(gpt-proA)62 lacY1 tsx-33 nfsB-2 qsr'-0 glnV44 galK2 LAM-nfsA1 Rac-0 hisG4 rfbD1 rpsL31(Sm^r) kdgK51 xylA5 mtl-1 argE3 thi-1
Pseudomonas putida	
HS12	NB$^+$, wild-type, pNB1 and pNB2 carrier
HS121	NB$^-$, a spontaneous mutant of HS12, pNB2 carrier
HS124	NB$^-$, a cured mutant of HS12, pNB1 carrier
Plasmids	
pNB1	59.1-kb NB catabolic plasmid of P. putida HS12
pNB2	43.8-kb NB catabolic plasmid of P. putida HS12
pACYC184	Cm^r Tc^r
pBR328	Ap^r Cm^r Tc^r
pBluescript SK(+)	Ap^r
pBluescript KS(+)	Ap^r
pHS289	pBluescript KS carrying a 1.1-kb SmaI-SnaBI fragment from pNB1
pHS291	pBR328 carrying a 6.6-kb SnaBI-SmaI fragment from pNB1
pHS2910	pBR328 carrying a 2.3-kb SnaBI-AsnI fragment from pNB1
pHS2911	pBluescript SK carrying a 4.8-kb SnaBI-PstI fragment from pNB1

Strain or plasmid	Description
pHS2912	pBluescript KS carrying a 2.4-kb EcoRI-PstI fragment from pNB1
pHS2916	pBluescript SK carrying a 1.8-kb PstI fragment from pNB1
pHS35	pACYC184 carrying a 9.0-kb EcoRI fragment with nbzB from pNB2
pHS3951	pBR328 carrying a 1.1-kb SalI-SphI fragment from pNB2
pHS3952	pBR328 carrying a 1.0-kb SspI-SphI fragment from pNB2
pHS941	pBR328 carrying a 2.5-kb SspI-EcoRV fragment with nbzB from pNB2

Abbreviations: NB$^+$, growth on NB; NB$^-$ no growth on NB; Ap^r, ampicillin resistance; Cm^r, chloramphenicol resistance; Tc^r, tetracycline resistance.

Pseudomonas pseudoalcaligenes

Pseudomonas pseudoalcaligenes is an aerobic, Gram negative soil bacterium that was first isolated from swimming pool water. It is able to use cyanide as a nitrogen source, and as a result it may be used for bioremediation. Based on 16S rRNA analysis, *P. pseudoalcaligenes* has been placed in the *P. aeruginosa* group. *Pseudomonas pseudoalcaligenes* JS45 was reported to have a partial reductive pathway, and several enzymes involved in NB catabolism were purified and characterized.

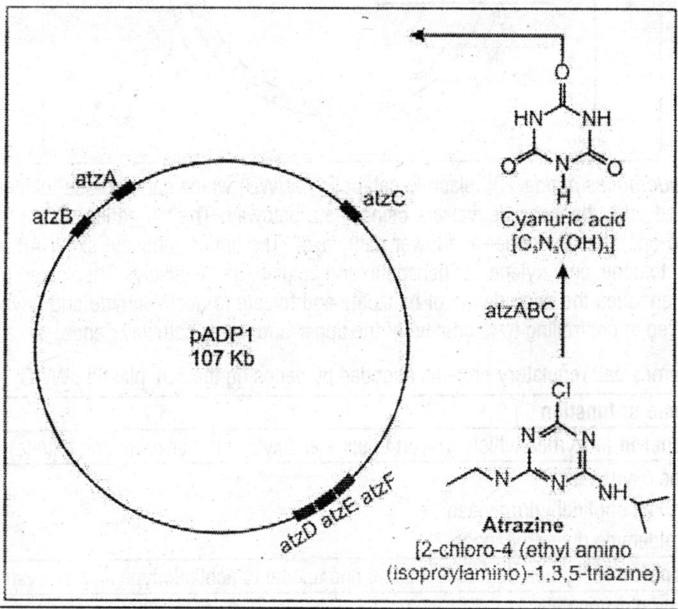

FIG. 31.6: Catabolic pathway for atrazine by *Pseusomonas* species ADP and plasmid pADP1. The plasmid contains respective genes for encoding required enzyme.

In the partial reductive pathway, NB is first reduced to hydroxylaminobenzene (HAB) by the NB nitroreductase and then HAB is rearranged to 2-aminophenol (2-AP) by HAB mutase. 2-AP undergoes a *meta* ring cleavage to 2-aminomuconic 6-semialdehyde (2-AMS) by 2-AP 1,6-dioxygenase. 2-AMS dehydrogenase converts the resulting 2-AMS to 2-aminomuconate (2-AM), which is, by the action of 2-AM deaminase, finally deaminated to yield 4-oxalocrotonate, a well-known intermediate in the catechol *meta* cleavage pathway. Some of the NB catabolic enzymes, such as NB nitroreductase, 2-AP 1, 6-dioxigenase, 2-AMS dehydrogenase, and 2-AM deaminase in *P. pseudoalcaligenes* JS45, were purified and characterized, and DNA sequences of the genes coding for HAB mutase, 2-AP 1, 6-dioxygenase, and 2-AMS dehydrogenase in *Pseudomons pseudoalcaligenes* JS45 were also released.

Pseudomonas putida

Pseudomonas putida TOL plasmid has the capacity to degrade not only toluene but also m-and p-xylene and other benzene derivatives degradation. The chromosomal genes encode the ortho pathway and the TOL (pWWO) plasmid encodes the meta pathway. In this pathway the oxidative attack on benzene is hydroxylation followed by incorporation of two atoms of oxygen from and oxygen molecule in the ring and the product is the enzymatically converted to catechol. The plasmid genes encod catabolic enzymes and named as xyl genes. The xyl genes of TOL (pWWO) are organized into two operons referred to as the upper and lower (meta) pathways.

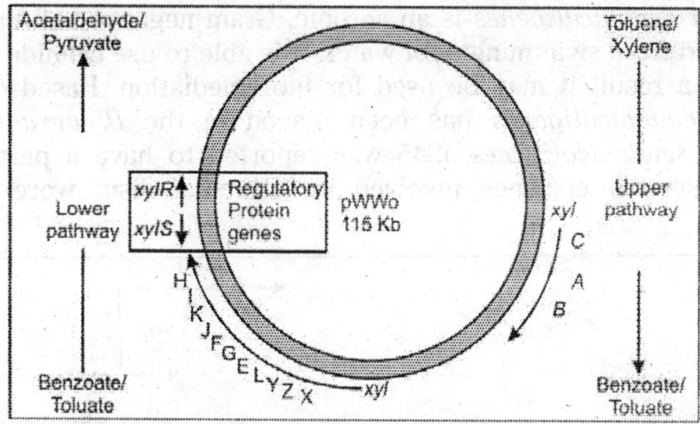

FIG. 31.7: *Pseudomonas putida* TOL plasmid catabolic (pWWO) which can degrade toluene, m- xylene and p-xylene and other benzene derivatives using meta pathway. The xyl genes of TOL (pWWO) are located into two operons (the upper and lower pathways). The upper pathway, sxy ICAB, encodes the degradation of toluene and xylene to benzoate and toluate, respectively. The lower pathway, xyl IXYZLEGFJKIH, encodes the degradation of benzoate and toluate to acetaldehyde and private. The xylS genes are involved in controlling transcription of the upper-and lower-pathway genes.

TABLE 30.2: Enzymes and regulatory proteins encoded by genes on the TOL plastid pWWO

Gene	Enzyme or funstion
Upper-pathway operon (enzymes which convert toluene and xylene to benzoate and toluate)	
XylA	Xylene oxygenase
XylB	Benzyl alcohol dehydrogenase
XylC	Benzaldehyde dehydrogenase
Lower-pathway operon (Enzymes degradebenzoate and toluate to acetaldehyde and pyruvate)	
XylX, Y, Z	Toluate dioxygenase
XylE	Catechol 2, 3-dioxygenase
XylF	2-Hydroxymuconic semialdehyde hydrolase
XylG	2-Hydroxymuconic semialdehyde dehydrogenase
XylH	4-Oxalocrotonate tautomerase
XylI	4-Oxalocrotonate decarboxylase
XylJ	2-Oxopent-4-enoate hydratase
XylK	2-Oxo-4-hydroxypentenoate aldolase
XylL	Dihydroxycyclohexadiene carboxyl ate dehydrogenase
	Proteins involved in controlling the transcription of the upper-and lower pathway genes
XylR	Regulatory protein
xylS	Regulatory protein

Many catabolic pathways involved in the degradation of aromatic compounds have been found to be encoded on catabolic plasmids, and biochemical and genetic studies of these catabolic plasmids have been extensively investigated. In the well-studied catabolic TOL plasmid, the genes encoding the enzymes for the oxidation of toluene via the *meta* cleavage of catechol are organized into two operons. The upper pathway operon is responsible for the conversion of toluene to benzoate, and the lower pathway operon is responsible for further conversion of benzoate to pyruvate and acetaldehyde. NAH7, another well-known catabolic plasmid, also carries two operons encoding catabolic enzymes for the oxidation of naphthalene via salicylate.

Although the NB catabolic pathway and the corresponding catabolic enzymes have been studied extensively, little is known about the molecular basis of NB catabolism, such as regulation and genetic organization of the catabolic genes.

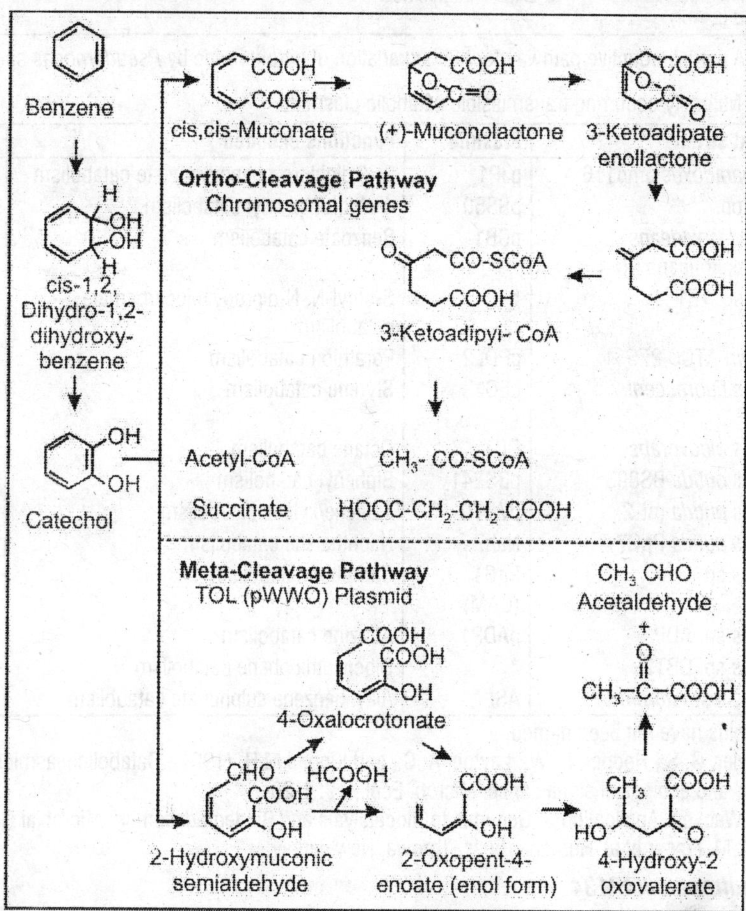

FIG. 31.8: Benzene degradation pathways by *Pseudomonades* species.

Pseudomonas species involved for its ability to catabolism the herbicide atrazine ($C_8H_{14}N_5Cl$) may contain as many as five distinct catabolic plasmids that together comprise approximately 107 KB of DNA. Catabolic plasmids have also been found in organisms belonging to many other genera isolated form a wide variety of environments. The majority of catabolic plasmids are self-transmissible. And many have abroad host range. These transmissible catabolic plasmids represent a pool of metabolic potential, available to many strains in a microbial community through interspecies transfer of genetic information.

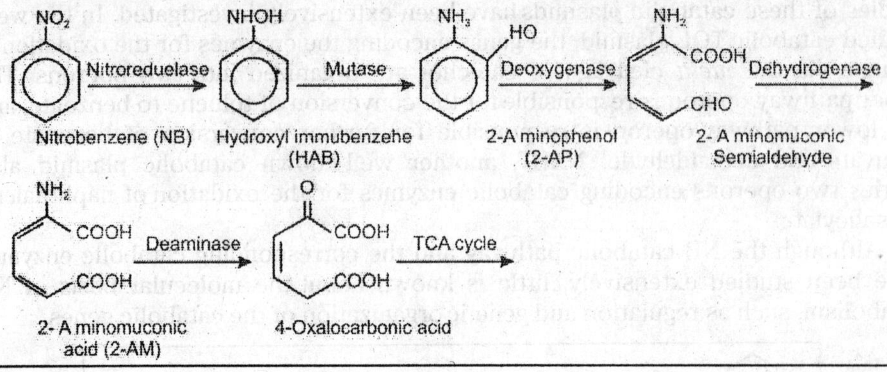

Nitrobenzene (NB) → Nitroreduelase → Hydroxyl immubenzehe (HAB) → Mutase → 2-A minophenol (2-AP) → Deoxygenase → 2-A minomuconic Semialdehyde → Dehydrogenase → 2-A minomuconic acid (2-AM) → Deaminase → 4-Oxalocarbonic acid → TCA cycle

FIG. 31.9: A partial reductive pathway for the degradation of nitrobenzene by *Pseudomonas* species.

TABLE 31.3: Naturally occurring transmissible catabolic plasmids

Bacterial host strain	Plasmid	Functions encoded	Size (kb)
Alcaligenes paradoxus Jmp116	pJP1	2, 4-dichloro-phenoxyacetate catabolism	87
Alcaligenes spp.	pSS50	4-chlorobiphenyl catabolism	53.2
Alcaligenes xylosoxideans subspecies denitrificans PN-1	pCB1	Benzoate catabolism	17.4
Arthrobacter sp. TE1	*	S-ethyl-N, N-dipropylthiocarbamate catabolism	75.7
Flavobacterium ATCC 27551	pPDL2	Parathion catabolism	43
Pseudomonas fluorescents PAW340	pEG	Styrene catabolism	37
Pseudomonas oleovorans	OCT	Octane catabolism	500
Pseudomonas putida BS893	pBS241	Biphenyl catabolism	195
Pseudomonas putida mt-2	pWWO	Toluene/xylene catabolism	117
Pseudomonas putida PpG7	Nah7	Naphthalene catabolism	83
Pseudomonas sp.	PpG1 (CAM)	Camphor catabolism	500
Pseudomonas sp. ADP	pADP1	Atrazine catabolism	107
Pseudomonas sp. DBT2	*	Dibenzothiophene catabolism	82.5
Pseudomonas *testosteroni*	ASL	Alkyl benzene sulphonate catabolism	91.5

*These plasmids have not been named.

Sources: Sayler, G. S., Hooper, S. W., Layton, A. C., and King, J.M.H. (1990). Catabolic plasmids of environmental and ecological significance. Microb. Ecol., 19; 1-20.

Lawrence P. Wackett. Application of Genomic to Biocatalysis and Biodegradation. In: Microbial Genomics edn. 2004. C. M. Fraser et al. Humana Press. Totowa, New Jersey.

Ralstonia eutropha JMP134

Two chlorocatechol catabolic gene modules on plasmid pJP4 have been studied by Michael Schlömann of TU Bergakademie Freiberg, 09599 Freiberg, Germany (J Bacteriol. 2002 August; 184(15): 4049–4053). The environmentally relevant strains, *Ralstonia eutropha* JMP134 has different sets of enzymes which are involved in the degradation of 3-chlorobenzoate and 2,4-dichlorophenoxyacetate (2,4-D). (JMP134, which was originally assigned to the species *Alcaligenes eutrophus* (now *Ralstonia eutropha*). Chlorophenoxyacetates may be degraded via chlorophenols to chlorocatechols, which are then cleaved between the hydroxy groups to give chloro-*cis*, *cis*-muconates.

The characterization of the enzymes with respect to their substrate specificities show that the enzymes induced during growth with 3-chlorobenzoate convert chlorosubstituted substrates, or the dienelactones derived from them, better than the enzymes induced during growth with benzoate. It thus became clear that a special set of enzymes is responsible for the degradation of chlorocatechol intermediates, and these enzymes were considered to have relaxed or broad substrate specificity.

Sphingomonas Aromaticivorans F199

Sphingomonas strains have 17 xenobiotic-degrading plasmids. *Sphingomonas aromaticivorans* F199 possess the novel ability to degrade a variety of aromatic compounds including toluene, all isomers of xylene, *p*-cresol, naphthalene, biphenyl, dibenzothiophene, fluorene, salicylate, and benzoate. *Sphingomonas* genus is well adapted for the degradation of high-molecular-weight polycyclic aromatic hydrocarbons and other aromatic contaminants. The inability to detect *Sphingomonas* biodegradative genes via hybridization with catabolic genes from phylogenetically distinct bacteria suggested that biodegradative genes from *Sphingomonas* sp. evolved independently from phylogenetically distinct bacteria such as those within the genus *Pseudomonas*. Some *Sphingomonas* strains are further distinguished in that the genes necessary for degradation of one type of aromatic compound are distributed into multiple operons that also possess genes for the degradation of other aromatic compounds. This unusual gene arrangement suggests that a highly complex regulatory network is responsible for the expression of aromatic degradative pathways in some *Sphingomonas* spp.

The catabolic plasmid pNL 1, form *Sphingomonas aromaticivorans* strain F199, has been completely sequenced contains genes encoding enzymes for the metabolism of biphenyl $(C_6H_5)_2$, naphthalene $(C_{10}H_8)$, m-xylene (C_8H_{10}), and p-cresol $(CH_3C_6H_4OH)$ (Romine et al. 1999). The complete 184,457-bp sequence of the aromatic catabolic plasmid, pNL1, from *Sphingomonas aromaticivorans* F199 has a total of 186 open reading frames (ORFs) are predicted to encode proteins, of which 79 are likely directly associated with catabolism or transport of aromatic compounds. Genes that encode enzymes associated with the degradation of biphenyl, naphthalene, *m*-xylene, and *p*-cresol are predicted to be distributed among 15 gene clusters. The unusual coclustering of genes associated with different pathways appears to have evolved in response to similarities in biochemical mechanisms required for the degradation of intermediates in different pathways. *S. aromaticivorans* F199 possess two plasmids, which are designated pNL1 (~180 kbp) and pNL2 (~480 kbp).

BIOREMEDIATION OF XENOBIOTICS

Bioremediation is the use of microorganisms to remove chemical pollutants from an area, or (equivalently), the human manipulation of environmental biotransformation. It is easier to remediate point sources of pollutants, such as contaminated soil near a chemical plant or oil-spills and other accidental releases. It is more difficult to bioremediate distributed sources, which must be collected for remediation, such as pallet boards and sewage waste. There are many targets for bioremediation. These include soil contaminated with pesticides, biocides, radioisotopes, heavy metals, or hydrocarbon (oil) pollution; water contaminated with any of the above; and sewage. There are essentially to approaches to bioremediation.

In situ techniques	Bioaugmentation: encourage specific microflora using nutrients, oxygenation, etc. Inoculation: add microorganisms of our choosing to soil. Phytoremediation: use plants to sequester toxins (especially heavy metals).
Ex situ techniques	Compost heap Activated sludge Trickle over biofilters Fermenter

Ex situ **bioremediation** is said to occur when we excavate contaminated soil (or pump out contaminated water), then clean it up in a bioreactor.

In situ **bioremediation** is said to occur when we encourage microbial biotransformation in soil or water without moving it elsewhere. The intrinsic rate of degradation is that achieved without human help. The augmented rate of degradation is that achieved with human intervention to encourage the bioremediation. This generally involves a bioreactor, which is a vessel in which biotransformation occurs. This does not necessarily need to be very high tech.

These two approaches are not exclusive: *in situ* may be used to treat contaminated soil, whilst *ex situ* is used to treat associated groundwater contamination.

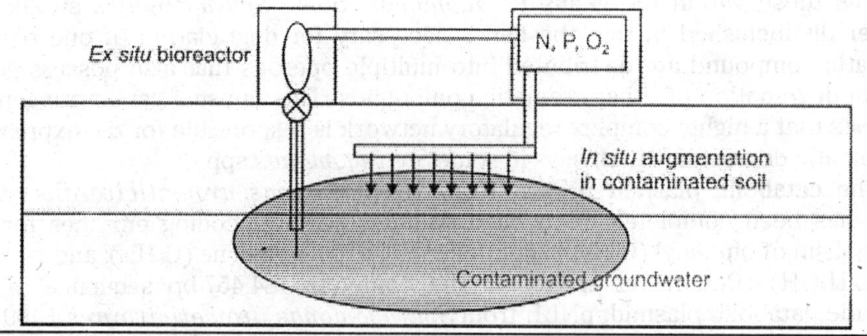

FIG. 31.10: *In situ* bioreactor in soil.

An important consideration of how we encourage bioremediation is whether the degradation of the xenobiotics in question proceeds most quickly under aerobic or anaerobic conditions. PCP is degraded by reductive dehalogenation and this proceeds better under anaerobic conditions. Anaerobic conditions may be encouraged by the addition of electron acceptors. Degradation of DCP (dichlorophenol) proceeds better under aerobic condition, which may be encouraged by sparging with air.

Intrinsic - Remediation of Heavy Metals

Bacteria can be used to remove metal contaminants from the acidic effluent streams of metal mines *in situ*. Sulphate-reducing bacteria reduce sulphate to sulphide, and metal ions are precipitated as sulphide salts. The pH increases as sulphuric acid is removed.

Bioaugmentation - Oil Slick Surfactants

Surfactants can be used to increase the solubility of oil *in situ*, but this has had mixed results. The increased surface area may lead to increased degradation rate, but the decreased attachment to particles any decrease the degradation rate.

Inoculation - Rhodococcus and PCP

In vitro studies indicated *Rhodococcus chlorophenolicus* could degrade PCP very rapidly. This bacterium can be inoculated into PCP-contaminated soil, and mineralisation of PCP (conversion to inorganic CO_2, H_2O and Cl^-) can be demonstrated to occur *in situ*.

Inoculation - Pseudomonas and DCP

However, other studies have been abject failures. *Pseudomonas* can degrade dichlorophenol readily in culture, but it cannot do this in soil (*in situ*). It appears that there are many factors besides enzymatic competence that determine whether inoculation will succeed.

Composting - Explosives Waste

This is an example of high temperature bioremediation: soil is excavated and treated *ex situ* in a big compost heap. Chlorophenols and nitroarene explosives degrade well in composting systems with *Pseudomonas*. The system requires aeration: turning or pumping.

Composting - Phanerochaete and PCP

This fungus degrades PCP in *ex situ* composting situations, but not very quickly. Is it possible to use it *in situ*? It can be encouraged by adding wood substrate to the soil, and seems promising with PAH, but this is still in early stages.

Activated Sludge - Sewerage

Contaminated water is collected *ex situ*. Sludge is strongly aerated to encourage growth of aerobic microbes, which degrade faecal waste. This may be adapted for PAH remediation by the inoculation of the sludge with PAH-degradative bacteria.

FIG. 31.11: Activated sludge – sewerage treatment.

Biofilters - Sewerage

Trickle bed filters in sewerage are a form of these. They have a solid substrate with a biofilm of degradative organisms growing on it. Contaminated water (*ex situ*) is trickled through. Organic pollutants such as acetone and toluene are degraded. This soil bed can be used to degrade pollutants in groundwater.

Biofilters - Contaminated Air and Water

Contaminated water may be treated by sparging if it contains volatile toxins. Gas can then be collected and subjected to *ex situ* treatment by forcing them through a biofilter to remove the pollutants.

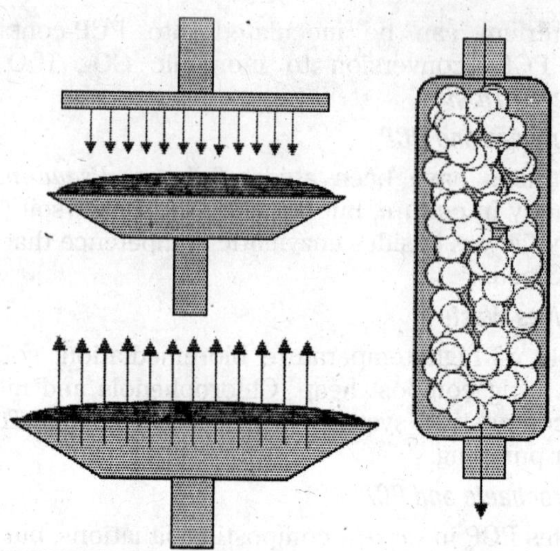

FIG. 31.12: Biofilters.

Biofilters - Immobilised Bacteria

Bacteria may also be immobilised in microparticles and packed into columns. Polluted water can then be trickled through the bioreactor to effect *ex situ* degradation.

Fermenters - Methanotrophs and TCE

Methanotrophic bacteria can be used to remediate TCE (tetrachloroethene, an industrial solvent) in an *ex situ* bioreactor. TCE is oxidised cometabolically by methane monooxygenase. Cells are grown in a fermenter then drip fed into a reactor supplied with TCE-contaminated water and methane. You can also use them immobilised. *In situ* remediation may be achieved by bioaugmentation with methane.

HUMAN HEALTH AND XENOBIOTICS

A detoxification regimen has been found to be safe for use by individuals exposed to recreational (abused) and medical drugs, patent medicines, occupational and environmental chemicals.

Over four million distinct chemical compounds have been reported in the literature since 1965, with 6,000 new compounds added to the list each week. Human exposure to these chemicals is both direct an indirect. More than 3,000 chemicals are deliberately added to food and over 700 have been identified in drinking water. Along with pharmaceuticals and recreational street drugs, the direct exposure to humans is considerable. In addition, the biomagnification of chemicals discharged into the environment has resulted in human accumulation, generally due to the partitioning of these xenobiotics from water into lipids. Additional partitioning from one form of lipid to another leads to accumulation of these chemicals in lipid deposits throughout the body, but especially in adipose tissue.

Over 400 chemicals have been identified in human tissues, with some 48 found in adipose, 40 in milk, 73 in the liver, and over 250 in blood plasma. The characters of chemicals found in adipose tissue are diverse, but tend to reflect biologically persistent or often used materials such as DDT, PCB, dioxin, nalkanes, PCP and THC. Chemicals stored in adipose and other tissues pose a continuing physiological and psychological threat to human health. Dioxin has been associated with ischemic vascular disease and with other physiological as well as psychological effects as long as ten years after initial exposure. Oncological studies have shown a significant association between PCB and DDE levels in fat and increased cancer incidence. In addition, PCB exposures have resulted in increased plasma triglycerides, even in the absence of overt symptoms of PCB intoxification. PCB's in monkeys not only resulted in increased blood lipids, but negatively affected the ability to maintain pregnancy. Further, they have been related to personality and cognitive functioning of persons unexpectedly exposed. Phencyclidine (PCP, Angel Dust, etc.), also shown to be stored in adipose, has been demonstrated to have long-lasting behavioral effects as well.

Concern over the potential health effects associated with lipid and other tissue stored xenobiotics has resulted in public upset of remarkable proportions. With exposure to chemicals significant, long-term health effects likely, and public concern continuing, the need for detoxification of xenobiotics takes on increasing importance.

Approaches to detoxification generally exploit pathways which lead to excretion of chemicals and their metabolites in urine and feces, or extrarenal excretion in sweat or sebum.

Lipid mobilization serves as the basis for promotion of xenobiotic metabolism through the action of detoxifying chemicals such as ascorbic acid, niacin, and phenobarbital. Typically, lipid mobilization is not enhanced prior to use of bioactive chemicals, however techniques have been studied which accomplish increased mobilization, especially through starvation or exercise. A fecal associated route exists which is not dependent upon bioactive chemicals and partitions xenobiotics through the intestine wall into non-absorbable materials such as paraffin.

The second major class of pathways is extrarenal excretion via sweat or sebum. One of these pathways (it is unclear which) has been identified as a route for loss of n-alkanes, paraffinic hydrocarbons, methadone, amphetamines and antiepileptics, among others.

REDUCTION OF TOXIC LOAD IN HUMAN BODY

All traditional health care systems have believed that toxin accumulation is a contributor to health care problems. Therapies that support cleansing and detoxification are a part of most traditional healing traditions. Over the last 100 years there has been growing scientific evidence to support this concept.

When talking about Toxic Load we are referring to the *accumulation of chemicals or molecules that are foreign to our biological systems*. These products are referred to as Xenobiotics. Xenobiotics may originate externally from toxic chemicals in the environment. They may originate internally from foods we eat or medications or supplements we take. Man may absorb xenobiotics from the living organisms in our gut when there is Dysbiosis and Leaky Gut Syndrome.

The issue of Toxic Load contributes to many health problems. A scientific article from the Journal of the National Cancer Institute suggested that more than 75% of cancer is caused by diet and environmental factors. In addition toxic exposure is a contributing factor to cardiovascular diseases, (heart attack), cerebrovascular diseases, (stroke), and neurodegenerative diseases, (Parkinson's disease and Alzheimer's disease). Environmental pollutants stockpile in the body contributing to the chronic diseases.

Common Intoxicants Include

1. Heavy metals such as lead, mercury and cadmium,
2. Arsenic, which is found in pesticides, herbicides, fungicides, wood preservatives, paint, tobacco, and manufacturing residues,
3. Polyaromatic Hydrocarbons, which come from diesel exhaust and jet fuel,
4. Plastics are ubiquitous in our environment. Many scientists believe that plastic molecules compete with the hormone binding sites of cells. This may be a contributing factor to infertility, precocious puberty and cancer.

Signs and Symptoms of Toxic Load

The signs and symptoms of toxic load are all inclusive. The patient and the clinician must have a high index of suspicion. When normal people subject themselves to a biopsy of fat tissue, toxic chemicals are always present. Simply stated, we are all intoxicated, some of us are more symptomatic than others. Some of us handle the intoxication better than others. Ultimately, to optimize health we must attend to reduction of the toxic load.

Testing for Toxic Load

Neurotoxins

Low molecular weight molecules that are fat soluble have been referred to as neurotoxins. These lipid soluble, low molecular weight molecules can come from biological sources such as unhealthy bacteria or yeast in our gut, toxins produced from food contaminants, (ciguatera poisoning), toxins produced from organisms in the environment (pfeisteria, molds), or toxins produced from infectious diseases (lyme disease). There is a simple and inexpensive test for this type of intoxication called the **Visual Contrast Sensitivity Test**.

Heavy Metals

Heavy metals are ubiquitous in the environment. They include mercury, cadmium, lead and arsenic. Heavy metal burden has been implicated in arteriosclerosis, chronic renal insufficiency/failure, cataracts, cancer, dementia/alzheimer's disease, fatigue syndromes, and hypertension. Toxic substances can be pulled out from the body and can be collected withib 6 hour as urine specimen. The routine blood tests do not measure the toxic burden in body tissue other than blood.

Treatment of Toxic Load

The first strategy is to reduce exposure to additional toxins in the environment. Simple measures include a good water filtration system, the consumption of organic foods when possible, and food preparation practices that do not create carcinogens; (barbecuing is an example of a process that does create pathogens).

Detoxification and Cleansing

Our bodies invest an enormous amount of energy in removing the chemical byproducts of normal metabolic processes. One estimate is that 80% of our basal energy expenditure goes to detoxification and cleansing. This process is what allows for order and efficiency in all of our bodily systems. Disease and a loss of well-being will result when the detoxification process is impaired. The most dramatic examples of impaired detoxification processes are end stage renal disease and end stage liver disease, (cirrhosis). As in all biological phenomena these processes happen along a continuum that begins with optimal function and ends with total malfunction. Conventional medicine addresses these problems when they manifest as overt diseases. Preventive Medicine aims to optimize the detoxification processes in order to prevent illness and maximize health throughout the life span.

In addition to the production of normal metabolic byproducts man is exposed to external chemicals that are not produced by the body (xenobiotics). External toxins include heavy metals, solvents, pesticides, herbicides, agricultural and industrial chemicals, and terpenes. Foods may be injurious because of individual sensitivity or because of unbalanced nutritional content. Microorganisms such as bacteria, viruses and parasites can contaminate the food we eat. Environmental extremes may be toxic. Finally, the emotional experiences of a lifetime may produce changes in our internal chemical environment that enhance toxin productions and are disease producing.

Strategies to Support Detoxification and Cleansing

Avoidance Strategies: There are strategies for the home, work place and automobile. They involve air purification, dust and mold mitigation. It should be obvious that it is desirable to minimize toxins in the environment from cleaning materials, pesticides and herbicides. Individuals may have food intolerances, food sensitivities or true food allergy. Blood testing for antibodies and elimination diet programs are excellent ways to determine whether a food is creating a problem.

Hydration: Dilution is the solution to pollution. Adequate hydration with high quality water is essential to good health. People who drink 50-60 ounces of water per day have a much lower incidence of bladder cancer and stroke.

Diet and Nutrition: Simplification of diet supports detoxification processes in the body. It is commonly recommend a supplement that is hypoallergenic and has nutritional constituents that support the detoxification enzyme systems in the liver. When the supplement is used with an elimination diet protocol many people experience a marked improvement in a variety of health problems.

Intravenous therapies with Glutathione, Phosphatidyl Choline and Vitamin C may be very useful.

Physiological Tests and Observations: Laboratory analysis of blood cholesterol and triglycerides should be conducted.

Herbal Therapy: Herbal therapy program should take into account the unique constitution of the individual, background health problems, the potential for interaction with prescription medications and other nutritional supplements. There are many possible therapeutic agents. Some commonly used herbs are: burdock root, chamomile, dandelion root, ginseng, milk thistle, nettles, red clover, sage, uva ursi, and yellow dock.

Exercise Therapy: Regular exercise will improve detoxification processes. It does not have to be extreme. Walking will do. Yoga practice can support detoxification as well.

Baths and Saunas: The idea behind the benefit of baths and saunas is that the heat will increase blood flow to the skin, create a sweat and facilitate the elimination of toxins through the skin. The use of baths and saunas for detoxification can be very effective if basic safety measures are followed. A person with cardiovascular disease or diabetes should consult a health care practitioner before using these techniques. The techniques are:

1. Fill the bath with water at a temperature of 106-108° F. Gradually increase the time of the soak from 5 minutes to 30 minutes.
2. The bath experience may be improved by using two cups of Epsom Salts and or essential oils such as lavender, (10-20) drops. Other oils that have a desirable effect include rosemary and tangerine oils.
3. Drinking a diaphoretic, (sweat inducing), tea prior to the bath will improve the

experience. Teas such as elder, yarrow, peppermint, boneset and ginger can be used. Drinking water during the bath is recommended.
4. Taking 1-5 grams of Vitamin C can improve the detoxification process. After the bath wash off with a shower. A bath can be done 1-3 times per week.
5. Sauna is a very effective detoxification adjunct. Individuals with cardiovascular problems should proceed with their physicians' advice. A sauna taken 2-3 times per week is a very effective support therapy.

Chelation Therapy: Chelation therapy may be useful if there is a documented heavy metal burden. It should be done under the supervision of a qualified health professional.

Antimicrobial Xenobiotics

Antimicrobials may have a cidal (killing) effect or a static (inhibitory) effect (useful to induce active immunization against recurrent infections, but dangerous in immunocompromised patients) on a range of microbes (antibiosis): sometimes a static compound may become a cidal compound just by increasing its dose. The range of Bacteria or other microorganisms that is affected by a certain antibiotic is expressed as its activity spectrum / range of activity.

- **Broad spectrum**: kill or inhibit a wide range of Gram-positive and Gram-negative *Bacteria*.
- **Narrow spectrum**: effective mainly against Gram-positive or Gram-negative *Bacteria*.
- **Limited spectrum**: effective against a single organism or disease.

Parameters that influence antimicrobial choice are as follows:

1. Effectiveness as determined by antibiogram.
2. Pharmacokinetics as determined by experimental *in vivo* studies.

Side effects [organ toxicity must be evaluated if individual already has organ failure(s)]. Antimicrobial therapy usually lasts >5 days (>5 months in prostatis) to avoid cronicization of infection. Antimicrobials can be classified according to source as follows:

Antibiotics

These are low-molecular weight substances that are produced as secondary metabolites by certain groups of microorganisms, especially *Streptomyces* spp., *Bacillus* spp., and a few molds (*Penicillium* and *Cephalosporium*) that are inhabitants of soils. They are produced at the same time that the cells begin sporulation. Antibiotics may require as many as 30 separate enzymatic steps to synthesise. The maintenance of a substantial component of the bacterial genome leads to the conclusion that the process (or molecule) is important, if not essential, to the survival of these organisms in their natural habitat. Most of the microorganisms that produce antibiotics are resistant to the action of their own antibiotic, although the organisms are affected by other antibiotics, and their antibiotic may be effective against closely-related strains (e.g., penicillin G).

Semisynthetic Antibiotic

These antimicrobial are produced by the microbe subsequently modified by the chemist to achieve desired properties.

Chemotherapeutic Agents

These are completely chemically-synthesised antimicrobials (e.g., sulphonamides).

Penicillin G is a synthetic antimicrobial compounds, originally discovered as products of microorganisms, that can be synthesised entirely by chemical means.

CHAPTER 32

Muscle

In animals muscles are used for the conversion of stored chemical energy (ATP) into the mechanical work. In vertebrate animals following are the types of muscles:

1. **Heart or cardiac muscle:** These muscles make up the wall of the heart. Throughout life, it contracts some 70 times per minute pumping about 5 liters of blood each minute.
2. **Smooth muscles:** They are present in the walls of all the hollow organs of the body (except the heart).
3. **Skeletal or striated muscle:** These muscles are attached to the skeleton. The contraction of skeletal muscle is under voluntary control.

SKELETAL MUSCLE

Skeletal muscle is a rather non-specific term usually applied to the striated muscle of vertebrates that is under voluntary control. The muscle fibres are syncytial and contain myofibrils, tandem arrays of sarcomeres. Smooth muscle is muscle tissue in vertebrates made up from long tapering cells that may be anything from 20-500m long. Smooth muscle is generally involuntary and differs from striated muscle in the much higher actin/myosin ratio, the absence of conspicuous sarcomeres and the ability to contract to a much smaller fraction of its resting length. Smooth muscle cells are found particularly in blood vessel walls, surrounding the intestine (especially the gizzard in birds) and in the uterus. The contractile system and its control resemble those of motile tissue cells (for example fibroblasts, leucocytes) and antibodies against smooth muscle myosin) will cross react with myosin from tissue cells, whereas antibodies against skeletal muscle myosin will not.

A single skeletal muscle, such as the **triceps** muscle, is attached at its **origin** to a large area of bone; in this case, the **humerus.** At its other end, the **insertion**, it tapers into a glistening white **tendon** which, in this case, is attached to the **ulna**, one of the bones of the lower arm.

Skeletal muscles are excitable by neuronal stimulation and as result they contract. Skeletal do not have an intrinsic mechanism for elongation but have elastic properties that restore them to resting length. Additionally skeletal muscles can be extended by the contraction of the antagonistic muscle.

On contraction of triceps, the insertion is pulled toward the origin and the arm is straightened or extended at the elbow. Thus the triceps is an **extensor**. Because skeletal muscle exerts force only when it contracts, a second muscle - a **flexor** - is needed to flex or bend the joint. The **biceps** muscle is the flexor of the lower arm. Together, the biceps and triceps make up an **antagonistic pair** of muscles. Similar pairs, working antagonistically across other joints, provide for almost all the movement of the skeleton.

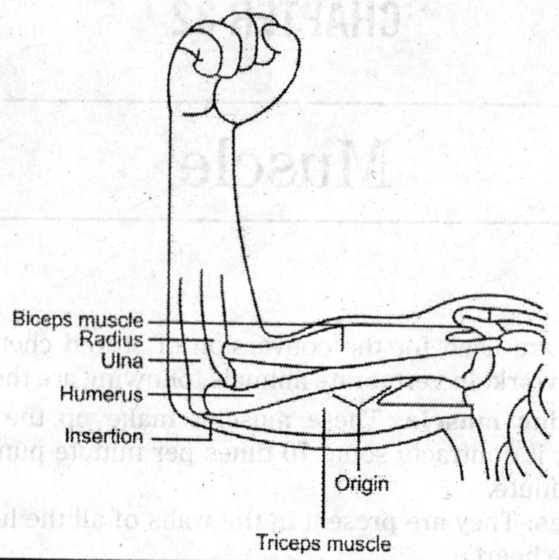

FIG. 32.2: Presentation of muscle of a hand, note biceps and triceps muscles.

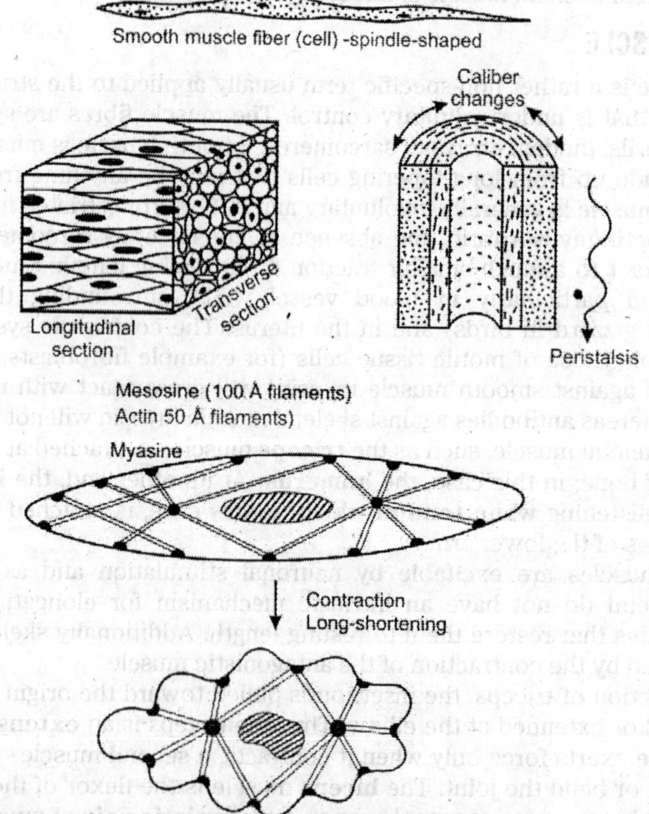

FIG. 32.1: Smooth muscles.

Skeletal muscle cells are multinucleated because they consist of several cells that have fused together. Muscle cells contain many mitochondria_and the characteristic protein filaments (myosin and actin) which are responsible for the muscle contraction. The cells also contain an extensive smooth endoplasmic reticulum (SER) and many nuclei. The multiple nuclei develop because each muscle fiber develops from the fusion of many cells (called **myoblasts**). The number of fibers is probably fixed early in life.

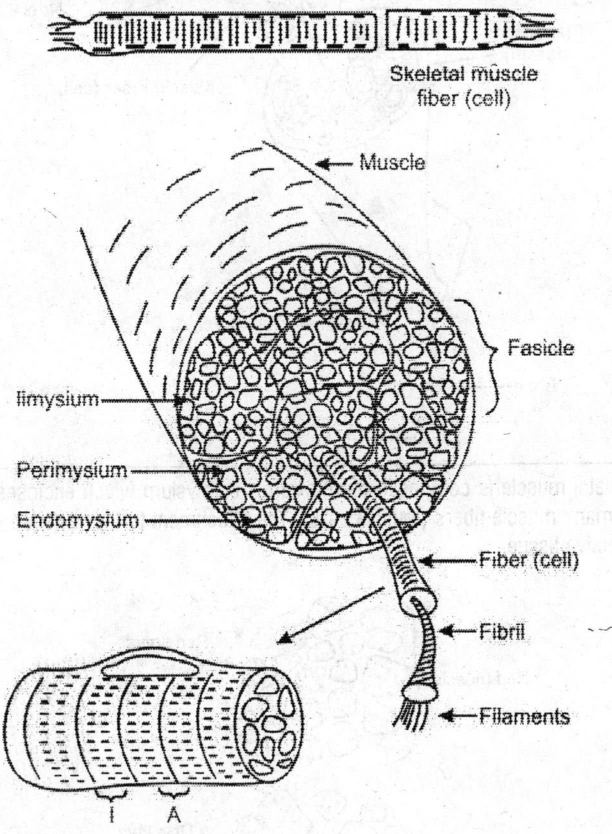

FIG. 32.3: Sleletal muscle.

Muscle Fiber

Skeletal muscle is made up of thousands of cylindrical muscle **fibers** often running all the way from origin to insertion. They are bound together by connective tissue through which run blood vessels and nerves. Each muscle fiber is made of an array of myofibrils that are stacked lengthwise and run the entire length of the fiber.

Myostatin: The fixing of fibers is regulated by myostatin. It is cytokine and is synthesised in muscle cells, which circulates as a hormone later in life. Myostatin is responsible for the suppression of skeletal muscle development. Cattle and mice with inactivating mutations in their myostatin genes develop much larger muscles. Some athletes and other remarkably strong people have been found to carry one mutant myostatin gene. These discoveries have already led to the growth of an illicit market in drugs supposedly able to suppress myostatin.

An adult has increased strength and muscle mass. This is due to an increase in the thickness of the individual fibers and in the amount of connective tissue. In the mouse, at least, fibers increase in size by attracting more myoblasts to fuse with them. The fibers attract more myoblasts by releasing the cytokine interleukin 4 (IL-4). Anything that lowers the level of myostatin also leads to an increase in fiber size.

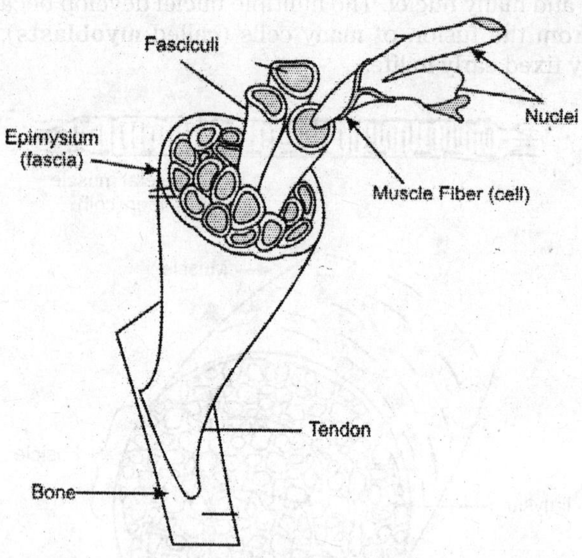

FIG. 32.4: A skeletal muscle is composed of an external epimysium which encloses several fasciculi each containing many muscle fibers (muscle cells). The attachment to the bone is via a tendon which is made of connective tissue.

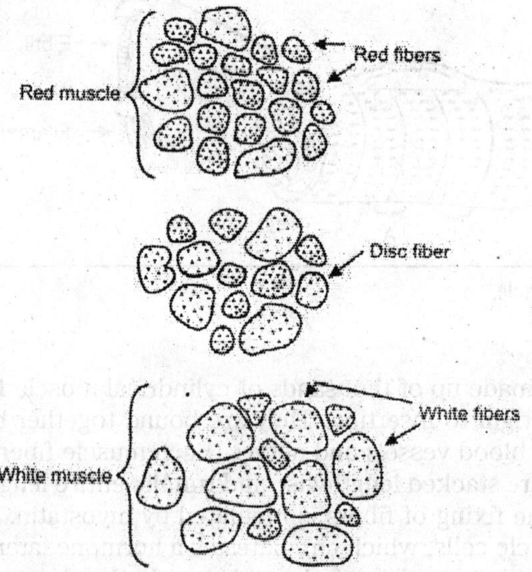

FIG. 32.5: Fiber type.

Since, the muscle fiber is not a single cell, its parts are often given special names such as (1) sarcolemma (for plasma membrane), (2) sarcoplasmic reticulum (for endoplasmic reticulum), (3) sarcosome (for mitochondrion), and (4) sarcoplasm (for cytoplasm).

Although this tends to obscure the essential similarity in structure and function of these structures and those found in other cells. The nuclei and mitochondria are located just beneath the plasma membrane and the endoplasmic reticulum extends between the myofibrils.

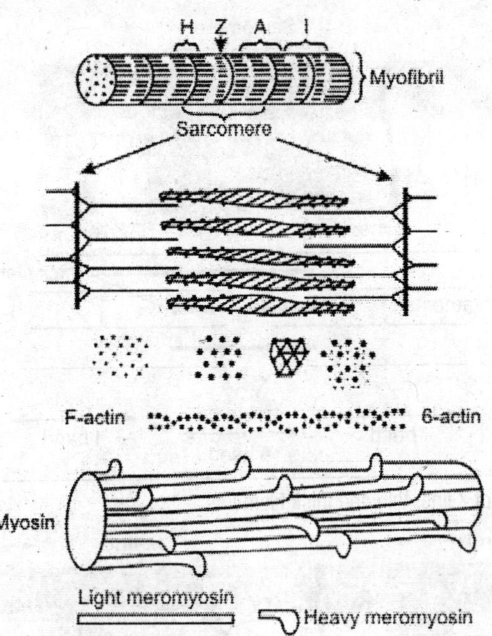

FIG. 32.6: Structural hierarchy of skeletal muscle.

Striated muscle: striated muscle innervated by a single motoneuron and having an electrically excitable membrane that exhibits an all or none response (c.f. Tonic muscle): in mammals almost all skeletal muscles are twitch muscles. Physiologists often divide muscles into fast and slow twitch types, the fast twitch muscles being associated with fast motor units.

Under the microscope, skeletal muscle fibers show a pattern of cross banding, which gives rise to the other name:

The Muscle Fiber

The complex of elongated muscle fibers are arranged in parallel order and held together with connective tissue. Muscle fibers are typically 2 to 3 cm long and 100 µm in diameter. Fibers are built from smaller units called **myofibrils** (1-2 µm in diameter), which primarily contain thick myosin filaments (12-18 nm diameter) and thin actin filaments (5-8 nm diameter). During muscle contraction, metabolism of ATP induces movement of cross bridges between thick and thin filaments, leading to a shortening of the muscle fiber.

The striated appearance of the muscle fiber is created by a pattern of alternating dark **A bands** and light **I bands**. The A bands are bisected by the **H zone.** The I bands are bisected by the **Z line**.

Each myofibril is made up of arrays of parallel filaments. The thick filaments have a diameter of about 15 nm. They are composed of the protein myosin. The thin filaments have a diameter of about 5 nm. They are composed chiefly of the protein actin along with smaller amounts of two other proteins: (1) **troponin** and (2) **tropomyosin.**

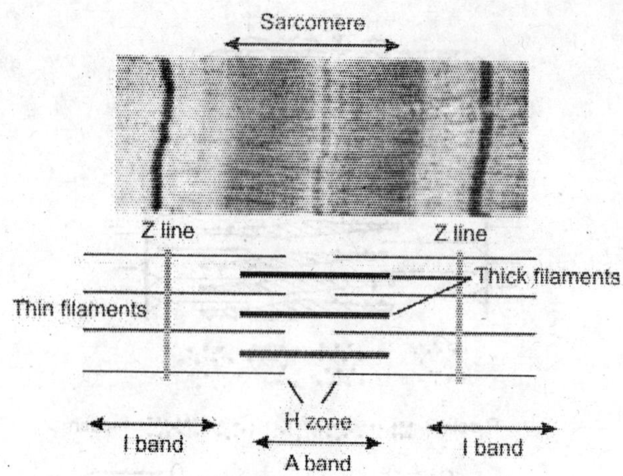

FIG. 32.7: Presentation of Z line, thin and thick filaments.

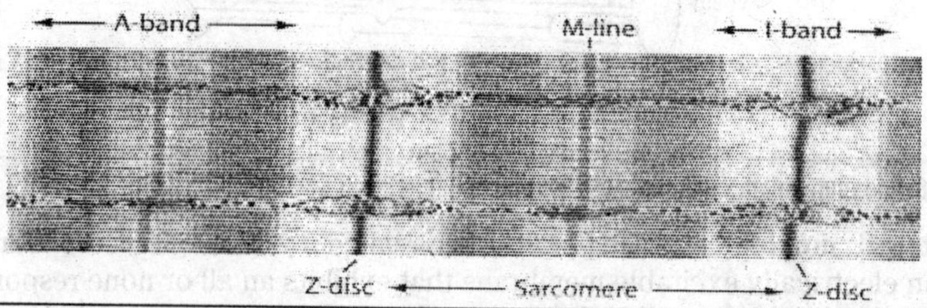

FIG. 32.8: Electron microscope image of striated muscle represents multiple myofibrils aligned in parallel, and the distinguishing features are labeled. The **A-band** (dark band) consists mainly of thick myosin filaments (joined at the **M-line**), whereas the **I-band** (light band) are composed of thin actin filaments. Each **sarcomere** is flanked by a **Z-disc** region.

A band: It is a set of stacked thick filaments centered on the M line, overlapped on both sides by thin filaments, which compose the **I band**. In between each I band, a dense vertical region called the **Z line** divides the myofibrils into a functional unit called the **sarcomere.**

Sarcomer: It is the smallest self-contained contractile unit (resting length 2 to 3 µm) of the striated muscle. It is the basic unit of a muscle's cross-striated myofibril. Sarcomeres are multi-protein complexes composed of three different filament systems. A muscle cell, from a biceps, may contain 100,000 sarcomeres. The myofibrils of smooth muscle cells are not arranged into sarcomeres. The thick filament system is composed of myosin protein which is connected from the M-line to the Z-disc by **Titin.**

Titin: It is also known as **connectin** and made of is a protein that is important in the contraction of striated muscle tissues. It also contains myosin-binding protein C which binds at one end to the thick filament and the other to **actin**. The thin filaments are assembled by actin monomers bound to Nebulin.

Actin: It is a globular, roughly 42-kDa protein found in all eukaryotic cells (except for nematode sperm) where it may be present at concentrations of over 100 µM. It is also one of the most highly-conserved proteins, differing by no more than 20% in species as diverse as algae and humans. It is the monomeric subunit of microfilaments, one of the three major components of the cytoskeleton, and of **thin filaments**, which are part of the contractile apparatus in muscle cells. Thus, actin participates in many important cellular functions, including muscle contraction, cell motility, cell division and cytokinesis, vesicle and organelle movement, cell signaling, and the establishment and maintenance of cell junctions and cell shape.

A monomers bound to Nebulin also involve tropomyosin; a dimer which coils itself around the F-actin core of the thin filament. Nebulin and Titin give stability and structure to the sarcomere.

The Elasticity of Muscle

Elasticity is a mechanical property necessary for the reversible contractile and stretching movement of sarcomere. This elasticity permits myofibrils to be stretched to twice their resting length without damaging their structure. The physiological sarcomere length ranges from 1.9 to 2.5 μm in cardiac muscle and 2.2 to 3.8 μm in soleus, a skeletal calf muscle.

Titin is the largest protein identified in nature, with a molecular weight of nearly ~4 MDa and a contour length of ~1 μm. Anchored at the sarcomeric Z and M lines, a string-like titin molecule spans half of the sarcomere in rigid contact with the thick filament along the A band, but flexible along the I band.

Electron micrographs and gene sequence analysis reveal that the architecture of titin is built upon a linear assembly of 300 immunoglobulin-like (Ig) or fibronectin III-like (FN-III) domains, as well as a few other unique domains such as PEVK regions. These domains control the passive elasticity of sarcomere and regulate a number of tension-mediated activities. In the A-band and Z-line regions, titin intimately associates with proteins located in the thick and thin filaments, respectively.

The only part of titin that appears to move freely is its I-band region, which provides the major contribution to passive elastic properties for the sarcomere. Titin I-band domains of variable composition, called isoforms, have been found in different types of muscle tissue, indicating that the specific distribution of domain types within the titin spring determine its stiffness and extensibility.

The mechanical properties of titin Ig domains have been experimentally studied using AFM and optical tweezers and corroborated with MD simulations. These studies on the elastic properties of titin modules have demonstrated that titin functions as a mechanical force bearing protein crucial for the stability of the muscle fibers.

The Titin Z1Z2/Telethonin Complex

While the structural role of Ig and FN-III domains near the center of titin molecule have been studied extensively, the contribution that N- and C-terminal titin domains lend to the mechanical stability of titin is less well understood. The N-terminal of titin interacts with several sarcomeric proteins at the Z-disc. These interactions provide structural support for the development and function of myofibrils and also regulatory roles such as responding to stretching forces developed in titin. Unique among them is the structural protein telethonin, which is believed to anchor the ends of two separate titin molecules to the Z-disc.

Biochemical studies of telethonin have shown that its presence in conjunction with the two N-terminal titin modules Z1 and Z2 is required for initial muscle growth, and point mutations to telethonin resulting in premature stop codons have been correlated to a form of limb girdle muscular distrophy (type 2A), suggesting that telethonin plays a major role in the initial assembly and stabilization of N-terminal titin at the Z-disc.

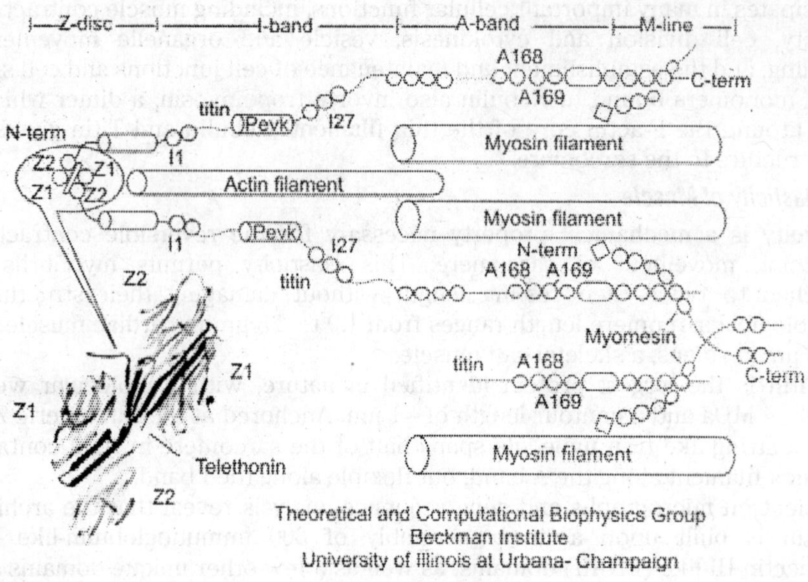

FIG. 32.9: Schematic cross-section of the sarcomere showing major macroscopic components. In particular, the large protein **titin** is illustrated along with its various domains such as I1, I27, and A168-169. The Z-disc protein **telethonin** is also shown interacting with two separate titin molecules at their **Z1** and **Z2** domains.

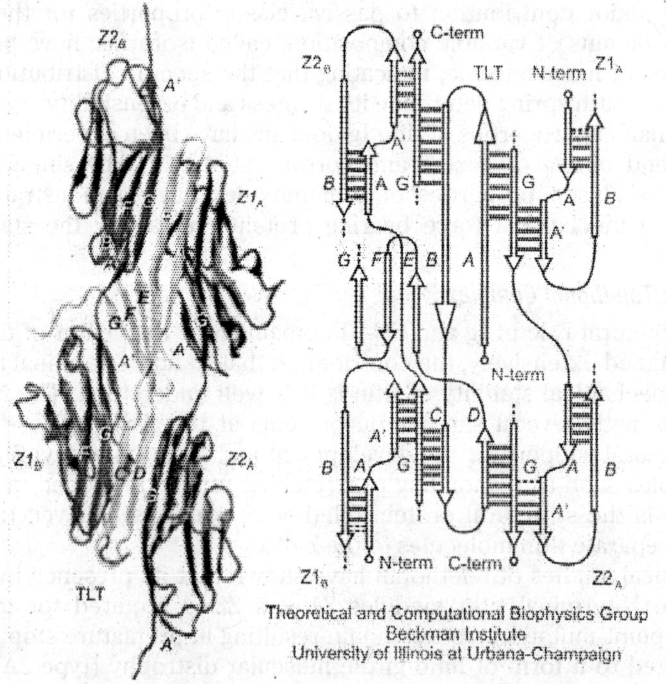

FIG. 32.10: Overview of the Titin Z1Z2-Telethonin complex. Left shows pairing of Z1Z2 dimers on each side of telethonin. Right depicts the overall β-strand alignment and associated interstrand hydrogen bonds.

Destabilization of the Z-disc region of titin has also been associated with dilated cardiomyopathy, indicating that disruption of the telethonin-titin Z1Z2 interface may be a direct cause of this pathology.

Activation of Skeletal Muscle

The contraction of skeletal muscle is controlled by the nervous system. In this respect, skeletal muscle differs from smooth and cardiac muscle. Both cardiac and smooth muscle can contract without being stimulated by the nervous system. Nerves of the **autonomic branch of the nervous system** lead to both smooth and cardiac muscle, but their effect is one of moderating the rate and/or strength of contraction.

Tetanus

The process of contracting takes some 50 msec; relaxation of the fiber takes another 50–100 msec. Because the refractory period is so much shorter than the time needed for contraction and relaxation, the fiber can be maintained in the contracted state so long as it is stimulated frequently enough (e.g., 50 stimuli per second). Such sustained contraction is called **tetanus**. Clonus and tetanus are possible because the refractory period is much briefer than the time needed to complete a cycle of contraction and relaxation. Note that the amount of contraction is greater in clonus and tetanus than in a single twitch.

The Sliding-Filament Model

Each molecule of myosin in the **thick filaments** contains a globular subunit called the **myosin head**. The myosin heads have binding sites for the **actin** molecules in the **thin filaments** and **ATP**.

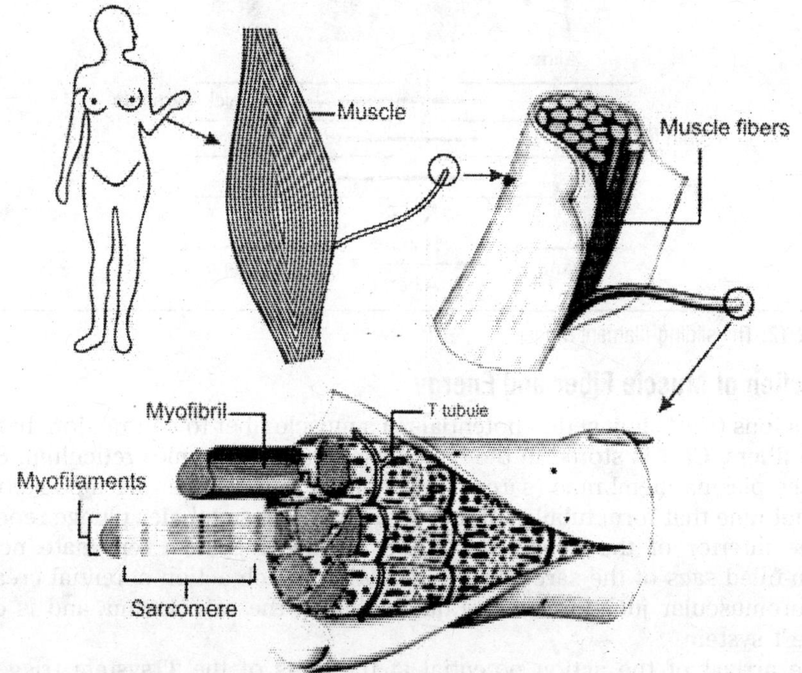

FIG. 32.11: The sliding-filament model of muscular contraction.

Contraction of skeletal muscle is initiated whan an action potential traveling down a motor neuron reaches the neuromuscular junction. The motor neuron releases acetylcholine, which binds with receptors on the muscle fiber and depolarizes the cytoplasmic membrane of the muscle fiber.

Activation of the muscle fiber causes the myosin heads to bind to actin. An allosteric change occurs which draws the thin filament a short distance (~10 nm) past the thick filament. Then the linkages break (for which ATP is needed) and reform farther along the thin filament to repeat the process. As a result, the filaments are pulled past each other in a ratchetlike action. There is no shortening, thickening, or folding of the individual filaments.

As a muscle contracts: (1) the Z lines come closer together, (2) the width of the I bands decreases. (3) the width of the H zones decreases, but there is no change in the width of the A band. Conversely, as a muscle is stretched, the width of the I bands and H zones increases but there is still no change in the width of the A band.

Testing the Sliding-Filament Model

The model postulates that:

1. When skeletal (or cardiac) muscle contracts, the thin and thick filaments in each **sarcomere** slide along each other without their shortening, thickening, or folding.
2. The strength of the relative motion between the thick and thin filaments is determined by the number of cross-bridges that can form between the two.

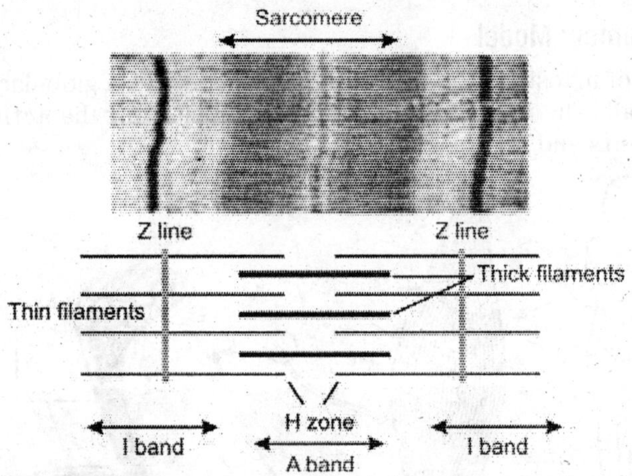

FIG. 32.12: The sliding-filament model.

Contraction of Muscle Fiber and Energy

Calcium ions (Ca^{2+}) link action potentials in a muscle fiber to contraction. In resting muscle fibers, Ca^{2+} is stored in the endoplasmic (sarcoplasmic) reticulum. Spaced along the plasma membrane (sarcolemma) of the muscle fiber are inpocketings of the membrane that form tubules of the "T system". These tubules plunge repeatedly into the interior of the fiber. The tubules of the T system terminate near the calcium-filled sacs of the sarcoplasmic reticulum. Each action potential created at the neuromuscular junction sweeps quickly along the sarcolemma and is carried into the T system.

The arrival of the action potential at the ends of the T system triggers the release of Ca^{2+}. The Ca^{2+} diffuses among the thick and thin filaments where it binds to **troponin** on the thin filaments. This turns on the interaction between actin and myosin and the sarcomere contracts. Because of the speed of the action potential (milliseconds), the action potential arrives virtually simultaneously at the ends of all the tubules of the T system, ensuring that all sarcomeres contract in unison. When the process is over, the calcium is pumped back into the sarcoplasmic reticulum using a **Ca^{2+} ATPase.**

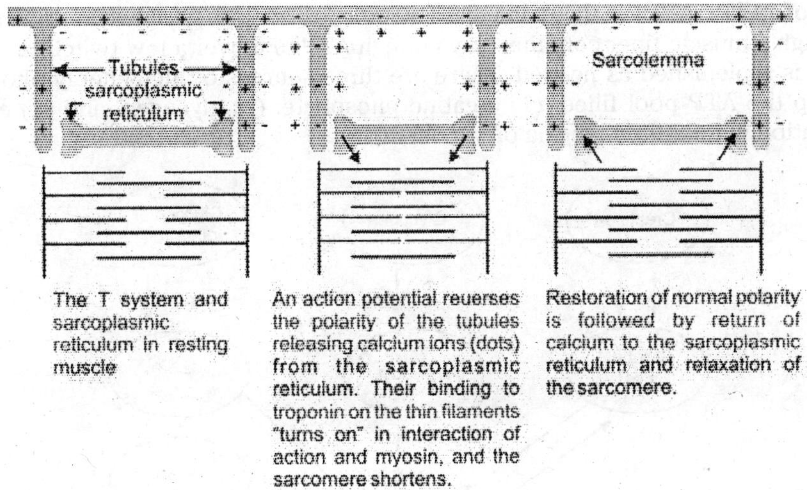

The T system and sarcoplasmic reticulum in resting muscle

An action potential reuerses the polarity of the tubules releasing calcium ions (dots) from the sarcoplasmic reticulum. Their binding to troponin on the thin filaments "turns on" in interaction of action and myosin, and the sarcomere shortens.

Restoration of normal polarity is followed by return of calcium to the sarcoplasmic reticulum and relaxation of the sarcomere.

The Ca^{2+} ATPases

A Ca^{2+} ATPase is located in the plasma membrane of all eukaryotic cells. It uses the energy provided by one molecule of ATP to pump one Ca^{2+} ion out of the cell. The activity of these pumps helps to maintain the ~20,000-fold concentration gradient of Ca^{2+} between the cytosol (~ 100 nM) and the ECF (~ 20 mM).

In resting skeletal muscle, there is a much higher concentration of calcium ions (Ca^{2+}) in the sarcoplasmic reticulum than in the cytosol. Activation of the muscle fiber allows some of this Ca^{2+} to pass by facilitated diffusion into the cytosol where it triggers contraction. After contraction, this Ca^{2+} is pumped back into the sarcoplasmic reticulum. This is done by another Ca^{2+} ATPase that uses the energy from each molecule of ATP to pump 2 Ca^{2+} ions.

Isotonic v/s Isometric Contractions

If a stimulated muscle is held so that it cannot shorten, it simply exerts tension. This is called an isometric (same length) contraction. If the muscle is allowed to shorten, the contraction is called isotonic (same tension).

Motor Units

All motor neurons leading to skeletal muscles have branching axons, each of which terminates in a neuromuscular junction with a single muscle fiber. Nerve impulses passing down a single motor neuron will thus trigger contraction in all the muscle fibers at which the branches of that neuron terminate. This minimum unit of contraction is called the motor unit. The size of the motor unit is small in muscles over which the human body has precise control because a single motor neuron triggers fewer than 10 fibers in the muscles controlling eye movements and the motor units of the muscles controlling the larynx are as small as 2–3 fibers per motor neuron. In contrast, a single motor unit for a muscle like the gastrocnemius (calf) muscle may include 1000–2000 fibers (scattered uniformly through the muscle). Although the response o f a motor unit is all-or-none, the strength of the response of the entire muscle is determined by the **number of motor units** activated.

The skeletal muscles are in a state of partial contraction called **tonus**. Tonus is maintained by the activation of a few motor units at all times even in resting muscle. As one set of motor units relaxes, another set takes over.

Fueling Muscle Contraction

Role of ATP: ATP is the immediate source of energy for muscle contraction. Although a muscle fiber contains only enough ATP to power a few twitches, its ATP "pool" is replenished as needed. There are three sources of high-energy phosphate to keep the ATP pool filled: (1) creatine phosphate, (2) glycogen, and (3) cellular respiration in the mitochondria of the fibers.

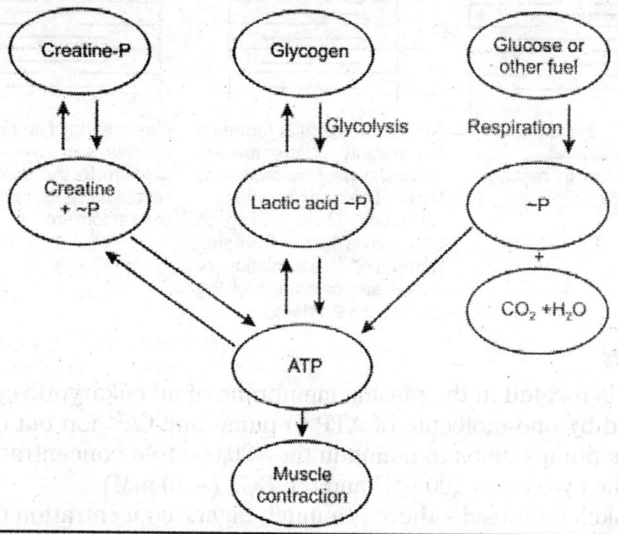

FIG. 32.13: Role of ATP in fueling muscle contraction.

Creatine phosphate: The phosphate group in creatine phosphate is attached by a high-energy bond like that in ATP. Creatine phosphate derives its high-energy phosphate from ATP and can donate it back to ADP to form ATP.

Adenosine Triphosphate
ATP

Creatine phosphate + ADP ↔ creatine + ATP

The pool of creatine phosphate in the fiber is about 10 times larger than that of ATP and thus serves as a modest reservoir of ATP.

Glycogen: Skeletal muscle fibers contain about 1% **glycogen.** The muscle fiber can degrade this glycogen by glycogenolysis producing glucose-1-phosphate. In **glycogenolysis:**

- phosphate groups – not water – break the 1 → 4 linkages.
- the phosphate group must then be removed so that glucose can leave the cell.

The liver and skeletal muscle are major depots of glycogen. There is some evidence that intense exercise and a high-carbohydrate diet (carbo-loading) can increase the reserves of glycogen in the muscles and thus may help marathoners work their muscles somewhat longer and harder than otherwise. But for most of us, carbo loading leads to increased deposits of fat. In glycolytic pathway 2 molecules of ATP ae generated for each pair of lactic acid molecules produced. Not much, but enough to keep the muscle functioning if it fails to receive sufficient oxygen to meet its ATP needs by respiration. However, this source is limited and eventually the muscle must depend on cellular respiration.

Cellular respiration: Cellular respiration is the process of oxidizing food molecules, like glucose, to carbon dioxide and water. The energy released is trapped in the form of ATP for use by all the energy-consuming activities of the cell. The process occurs in two phases: (1) **glycolysis**, the breakdown of glucose to pyruvic acid (2) the complete **oxidation of pyruvic acid** to carbon dioxide and water. Cellular respiration not only is required to meet the ATP needs of a muscle engaged in prolonged activity (thus causing more rapid and deeper breathing), but is also required afterwards to enable the body to resynthesise glycogen from the lactic acid produced earlier (deep breathing continues for a time after exercise is stopped). The body must repay its **oxygen debt**.

Type I *vs.* Type II Fibers: Two different types of muscle fiber can be found in most skeletal muscles. The Type I and Type II fibers differ in their structure and biochemistry. Most skeletal muscles contain some mixture of Type I and Type II fibers, but a single motor unit always contains one type or the other, never both. The ratio of Type I and Type II fibers can be changed by endurance training (producing more Type I fibers).

TABLE 32.1: Differences between Type I and Type II fibers

Type I Fibers	Type II Fibers
They are loaded with mitochondria and depend on cellular respiration for ATP roduction. They are resistant to fatigue They are rich in myoglobin and hence red in colour. They are activated by small-diameter, thus slow-conducting, motor neurons and also known as "slow-twitch" fibers. They are dominant in muscles that depend on tonus, e.g., those responsible for posture	Presence of few mitochondria and rich in glycogen which are dependent on glycolysis for ATP production. They fatigue easily and are low in myoglobin hence whitish in colour. They are activated by large-diameter, thus fast-conducting, motor neurons. They are also known as fast-twitch fibers. They are dominant in muscles used for rapid movement

CARDIAC MUSCLE

Cardiac or heart muscle resembles skeletal muscle in some ways: it is striated and each cell contains sarcomeres with sliding filaments of actin and myosin.

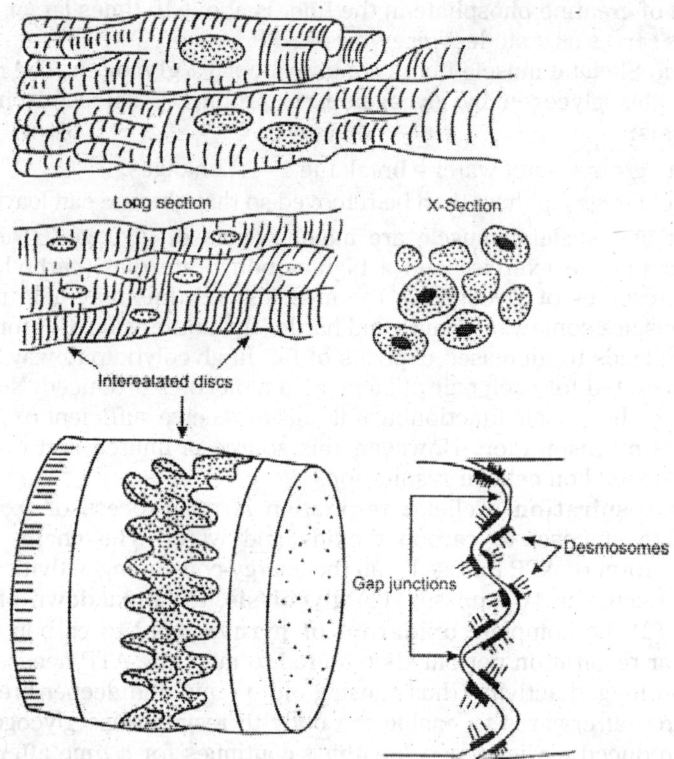

FIG. 32.14: Cardiac muscle.

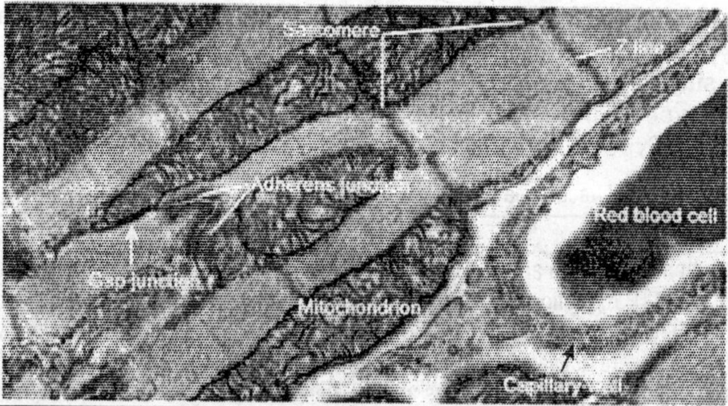

FIG. 32.15: Electron micrograph of cardiac muscle.

However, cardiac muscle has a number of unique features that reflect its function of pumping blood. The myofibrils of each cell (and cardiac muscle is made of single cells – each with a single nucleus) are branched. The branches interlock with those of adjacent fibers by adherens junctions. These strong junctions enable the heart to contract forcefully without ripping the fibers apart.

Adherens Junctions

Adherens junctions provide strong mechanical attachments between adjacent cells. They hold cardiac muscle cells tightly together as the heart expands and contracts.

They hold epithelial cells together and seem to be responsible for contact inhibition. Some adherens junctions are present in narrow bands connecting adjacent cells. Others are present in discrete patches holding the cells together.

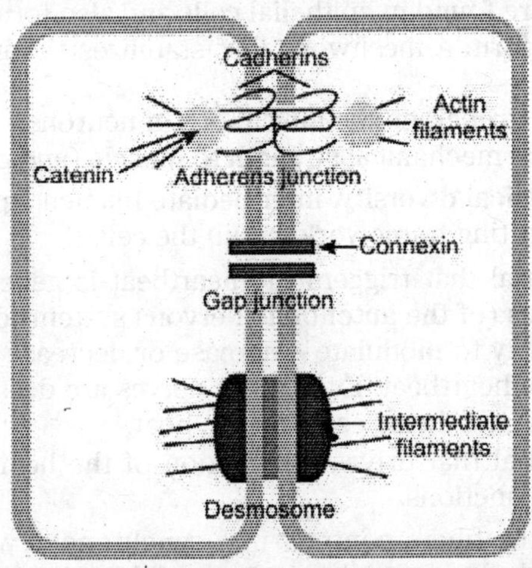

FIG. 32.16: Illustration of cardiac muscle showing adherens junction, gap junction and Desmosomes.

Adherens junctions are constructed from cadherins. These are transmembrane proteins whose extracellular segments bind to each other and whose intracellular segments bind to catenins. Catenins are connected to actin filaments. Inherited mutations in a gene encoding a cadherin can cause stomach cancer. Mutations in a gene (APC), whose protein normally interacts with catenins, are a common cause of colon cancer. Loss of functioning adherens junctions may also lead to tumor metastasis.

Gap Junctions

Gap junctions are intercellular channels some 1.5–2 nm in diameter. These permit the free passage between the cells of ions and small molecules (up to a molecular weight of about 1000 Daltons). They are cylinders constructed from 6 copies of transmembrane proteins called connexins. Because ions can flow through them, gap junctions permit changes in membrane potential to pass from cell to cell. The action potential in heart (cardiac) muscle flows from cell to cell through the heart providing the rhythmic contraction of the heartbeat. At some synapses in the brain, gap junctions permit the arrival of an action potential at the synaptic terminals to be transmitted across to the postsynaptic cell without the delay needed for release of a neurotransmitter. As the time of birth approaches, gap junctions between the smooth muscle cells of the uterus enable coordinated, powerful contractions to begin. Several inherited disorders of humans such as: (1) certain congenital heart defects and (2) certain cases of congenital deafness have been found to be caused by mutant genes encoding connexins.

Desmosomes

Desmosomes are localized patches that hold two cells tightly together. They are common in epithelia (e.g., the skin). Desmosomes are attached to intermediate filaments of keratin in the cytoplasm.

Intermediate Filaments: These cytoplasmic fibers average 10 nm in diameter (and thus are "intermediate" in size between actin filaments (8 nm) and microtubules (25 nm) (as well as of the thick filaments of skeletal muscle fibers). There are several types of intermediate filament, each constructed from one or more proteins characteristic of it.

1. **Keratins:** They are found in epithelial cells and also form hair and nails.
2. **Nuclear lamins** form a meshwork that stabilizes the inner membrane of the nuclear envelope.
3. **Neurofilaments** strengthen the long axons of neurons;
4. **Vimentins** provide mechanical strength to muscle (and other) cells.

Despite their chemical diversity, intermediate filaments play similar roles in the cell: providing a supporting framework within the cell.

- The action potential that triggers the heartbeat is generated within the heart itself. Motor nerves (of the autonomic nervous system) do run to the heart, but their effect is simply to modulate - increase or decrease - the intrinsic rate and the strength of the heartbeat. Even if the nerves are destroyed (as they are in a transplanted heart), the heart continues to beat.
- The action potential that drives contraction of the heart passes from fiber to fiber through gap junctions.

Significance: All the fibers contract in a synchronous wave that sweeps from the atria down through the ventricles and pumps blood out of the heart. Anything that interferes with this synchronous wave (such as damage to part of the heart muscle from a heart attack) may cause the fibers of the heart to beat at random - called **fibrillation**. Fibrillation is the ultimate cause of most deaths.

The **refractory period** in heart muscle is **longer** than the period it takes for the muscle to contract (systole) and relax (diastole). Thus **tetanus** is not possible. Cardiac muscle has a much richer supply of mitochondria than skeletal muscle. This reflects its greater dependence on cellular respiration for ATP. Cardiac muscle has little glycogen and gets little benefit from glycolysis when the supply of oxygen is limited. Thus anything that interrupts the flow of oxygenated blood to the heart leads quickly to damage - even death - of the affected part. This is what happens in heart attacks.

SMOOTH MUSCLE

They are present in the walls of all the hollow organs of the body (except the heart). The contraction of smooth muscle is generally not under voluntary control. Its contraction reduces the size of these structures. These muscles regulate the flow of blood in the arteries and make possible the movement of food along through the gastrointestinal tract of the body. They also help in expelling urine from urinary bladder and help women to bring out baby out into the world from the uterus. They also regulate the flow of air through the lungs.

Smooth muscle is made of single, spindle-shaped cells. They have one centrally placed nucleus per cell and are usually organized in small clusters of cells. Fine collagenous, reticular connective tissue separates each cell and the bundles. It gets its name because no striations are visible in them. Nonetheless, each smooth muscle cell contains **thick** (myosin) and **thin** (actin) filaments that slide against each other to produce contraction of the cell. The thick and thin filaments are anchored near the plasma membrane (with the help of intermediate filaments).

One way to distinguish smooth muscle is that it is devoid of regular pattern of sarcomeres (no A, I bands or Z lines). Smooth muscle cells also have only one nucleus. In that sense they are like cardiac muscle cells.

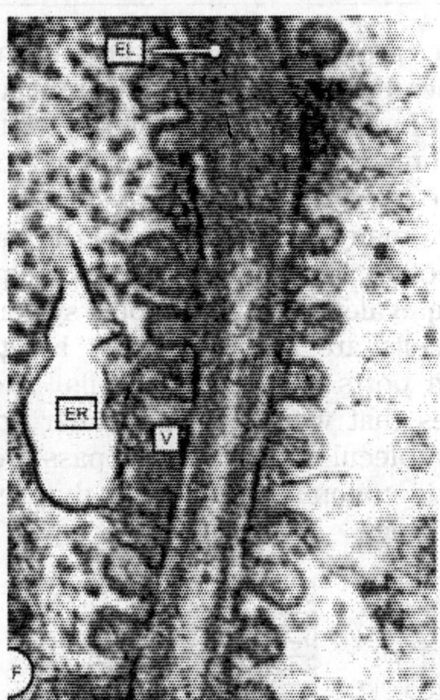

FIG. 32.17: Vesicles (v) called caveolae are numerous at the periphery of smooth muscle cells. These work like the T tubules, increasing the surface area for transfer of calcium into the cytoplasm. The smooth endoplasmic reticulum (SER) is nearby. SER sequesters calcium and also releases it before the muscle contracts.

Smooth muscle cells are different then that of connective tissue by their organized appearance. They are usually in a homogeneous bundle or sheet of cells rather than scattered single cells. Fibroblasts are found in connective tissue in isolated groups.

Distribution in Body

Smooth muscle cells are found as sheets lining the walls of hollow organs. They are found running circularly around blood vessels, especially prominent around muscular arteries and in the wall of the gastrointestinal tract. The cells are found in 2-3 layers. One layer may run circularly around the lumen; another layer may be running longitudinally along the length of the organ. In the stomach, there is a layer that runs more diagonally or obliquely. Each of these works together.

Ultrastructure of Smooth Muscle Cell

Smooth muscle cells have a region around the nucleus that is filled with some smooth endoplasmic reticulum (SER) and lots of mitochondria.

- The mitochondria - provide the ATP needed for contraction.
- The SER - provides a site for calcium storage.

In cytoplasm contractile filaments broken by focal densities, thin filaments (actin with tropomyosin) are present. In **smooth muscle cells protein are absent, however,** calcium binding protein is present. In addition, there are thick filaments that project the heavy meromyosin heads all along their length. Thus, there is a larger surface area for interaction of the myosin with the actin.

The organizational sites include thick and thin filaments (actin and myosin). Desmin or vimentin filaments also bind at these sites to help hold the filaments together. These intermediate filaments help relay the contraction and help to shorten the cell. The dense bodies are also present (alpha actinin, actin binding protein). These dense bodies are similar to the Z lines of the striated muscle.

The densities in the membrane are called adherent junctions. Recall the same types of junctions are seen in epithelial cells. These are focal adhesions or zonula adherents. They help attachment sites to the connective tissue outside the cell and also help the muscle cells to work together.

Plasma membrane have many invaginations on the periphery. These are called vesicular or saccular (caveolae). They are believed to help bring in calcium needed for contraction. They are equivalent to the T- Tubule system in striated muscle.

The smooth muscle cells are interconnected by gap junctions, which are specialized communication ports between the cells. These are regions of many paired connexon molecules that work like molecular pores or channels running between the cells. Small molecules or ions can pass from cell to cell via these connexons and they provide communication links that regulate contraction of the entire bundle of smooth muscle.

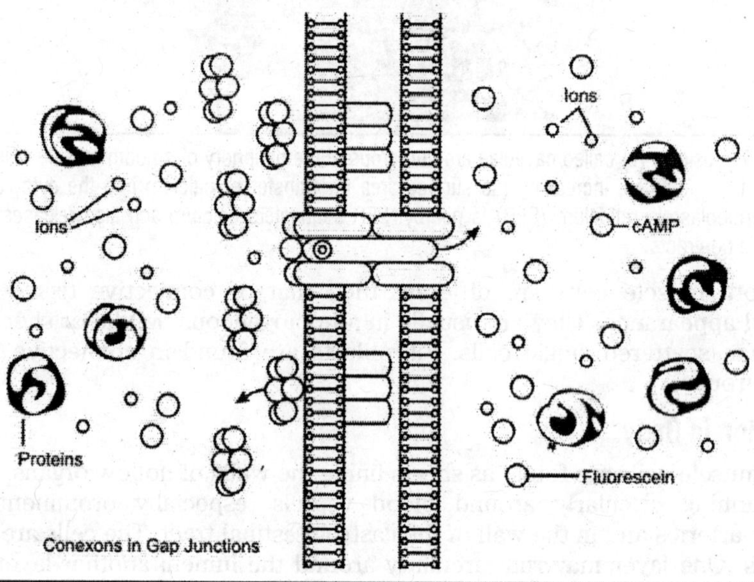

FIG. 32.18: Gap Junction – showing two adjacent cells. These are long, donut like molecules that allow passage of small molecules from one cell to another. Examples of molecules that can pass include calcium ions, cAMP, ATP, etc. Small molecules like fluorescein can also pass, which proves that cells have formed gap junctions.

Regulation of Smooth Muscle

Smooth muscle can either receive innervation via the autonomic nervous system, or hormones through the blood system can regulate them. Some muscle responds to both types of regulation. Innervation involves nerve endings forming synapses with smooth muscle cells. They are usually in the form of swellings of axons which contain synaptic vesicles (norepinephrine or acetylcholine are two neurotransmitters). An example of hormonal control would be found in the uterus, which is under the influence of oxytocin. This would be active particularly during labor and delivery.

Smooth muscle (like cardiac muscle) does not depend on motor neurons to be stimulated. However, motor neurons (of the autonomic system) reach smooth muscle and can stimulate it r relax it depending on the neurotransmitter they release (e.g. noradrenaline or nitric oxide, NO).

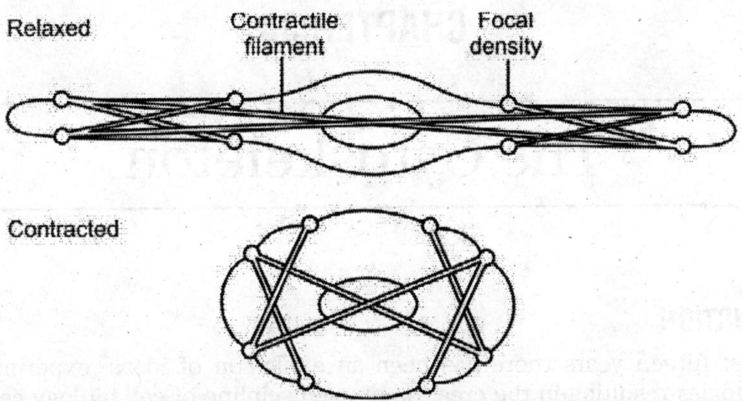

FIG. 32.19: Illustration of relaxed and contracted smooth muscles.

The Autonomic Nervous System: The autonomic nervous system consists of sensory neurons and motor neurons that run between the central nervous system (especially the **hypothalamus** and **medulla oblongata**) and various internal organs such as the: heart, lungs, viscera, glands (both exocrine and endocrine). It is responsible for monitoring conditions in the internal environment and bringing about appropriate changes in them. The contraction of both smooth muscle and cardiac muscle is controlled by motor neurons of the autonomic system.

Smooth muscle can also be made to contract by other substances released in the vicinity (paracrine stimulation). The release of **histamine** causes contraction of the smooth muscle lining our air passages (triggering an attack of **asthma**) by hormones circulating in the blood. Similarly, when oxytocin reaching the uterus, it stimulates it to contract to begin childbirth. The contraction of smooth muscle tends to be slower than that of striated muscle. It also is often sustained for long periods. This, too, is called **tonus** but the mechanism is not like that in skeletal muscle.

The important muscle diseases are: (1) The Muscular Dystrophies (MD), (2) Myasthenia Gravis, (3) The Cardiac Myopathies.

CHAPTER 33

The Cytoskeleton

INTRODUCTION

In the last fifteen years there has been an explosion of ideas, experiments, and methodologies resulting in the creation of subdiscipline of cell biology centered on the cytoskeleton (skeleton of a cell). The various cellular organelles are suspended in the cytosol (cytoplasmic matrix), which fills all the spaces of the cell and constitutes its true internal milieu.

The existence of an organized fibrous array in the structure of the protoplasm was postulated in 1928 by Koltzoff. He concluded that *each cell is a system of liquid components and rigid skeletons which generate the shape, and even though we rarely see the skeletal fibrils in living and fixed cells, that only means that these fibrils are very thin or they are not distinguished by their refractive index from the surrounding colloidal solution'*. The final details of the cytosol were not discernible with light microscope. However, by using electron microscope, existence of some cytoplasmic skeletal network was reported. For example, when tissue culture cells enter mitosis, the usually flat cells become rounded and often detach from the substrate on which they were growing; the dividing cells are shown to contain a highly organized spindle consisting of a framework of microtubules which disappear during interphase. These observations provided graphic evidence that there were structures within the cell that controlled cytoarchitecture and these cytoskeletal elements were capable of rather extensive redistribution.

CYTOSKELETAL ELEMENTS

Cytoskeleton is a network of protein fibers in the cytoplasm that gives shape to a cell, holds and moves organelles, and is typically involved in cell movement.

The term cytosceleton is slightly misleading since no rigid building plan for cytosceletons with a defined place for each bone exists. The cytosceleton is actually a multi-functional structure more accurate described by the terms cytobones and cytomuscles that were suggested by C. de DUVE (1984) but never found their way into scientific literature. One of its main features is the constant restructuring. It does not only determine the shape of cells but is the basis for all intracellular and whole-cell movements, too. It consists of three independent systems, the microtubuli system, the microfilament system and the system of intermediate filaments. Microfilaments and microtubuli have a great deal in common. Both are polymers consisting of globular protein subunits (actin and tubuline, respectively). Polymerisation is reversible and is balanced by depolymerisation. Monomerous actin is called G-actin (globular actin), polymerized actin is termed F-actin (fibrillar actin).

Tubuline is a dimerous protein out of two similar subunits (*alpha* and *beta*). On the protein level could no homologies of actin and tubuline be detected. Both systems are associated with numerous further proteins that guarantee on one hand the movements and participate in the required energy transformation (the splitting of ATP) and that help on the other hand regulating the movements. In this context it becomes the question *how the eliciting signal is recognized* (for example in light induced movements) *and the subsequent transformation into a movement interesting*?. The recognition will be discussed elsewhere, and about the signal transduction to elements of the cytosceleton is so far not much known.

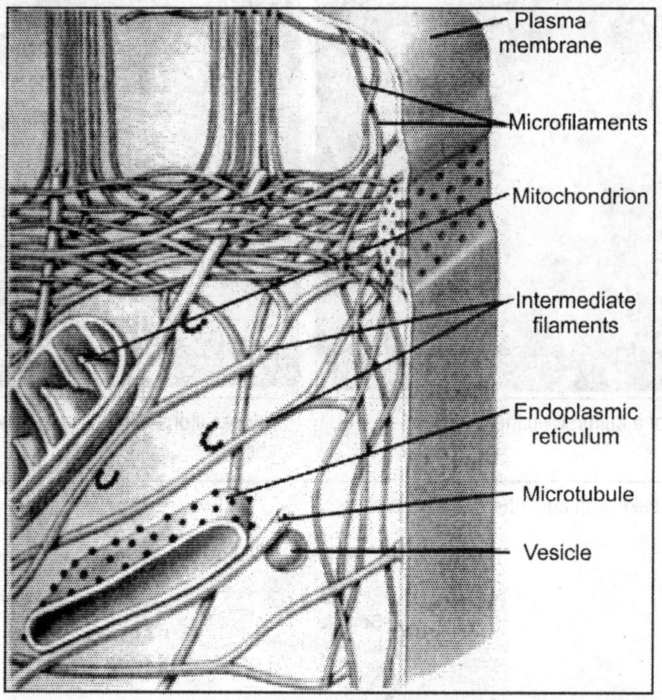

FIG. 33.1: Structure of a cell showing microfilaments.

The cytoskeleton is composed of three well defined fibrous elements:

1. Microtubules.
2. Microfilaments.
3. Intermediate filaments.

These cytoskeletal elements are linear, unbranched polymer of protein subunits that may extend several microns through the cytosol. In addition, there is a less well defined system of fibrous cytoplasmic elements, the microtrabeculae, which appears to be associated with the cytoskeleton and may serve to integrate the entire complex of materials into functional unit. The classification of these elements is mainly based on their diameter; microtubules 25 nm; microfilaments 5-7 nm and intermediate filaments 10-11 nm. The cytoskeletal elements represent a highly dynamic group capable of rapid and dramatic reorganization.

MICROTUBULES

Structure

The microtubules have been observed in all the eukaryotic cells that have been studied with the electron microscope. These are small, hollow, cylindrical structures. Typically a tubule has an outer diameter of approximately 25 nm, a wall diameter of approximately 5 nm leaving the lumen diameter of 15 nm. The wall of the microtubule is a polymer composed of globular subunits. In a cross section the wall nearly always reveals 13 subunits making up the complete circumference of the wall. The subunits are arranged in longitudinal rows, termed protofilaments, that are aligned parallel to the long axis of the tubule. Each microtubule is surrounded by a zone of low electron density from which ribosomes or other particles are absent.

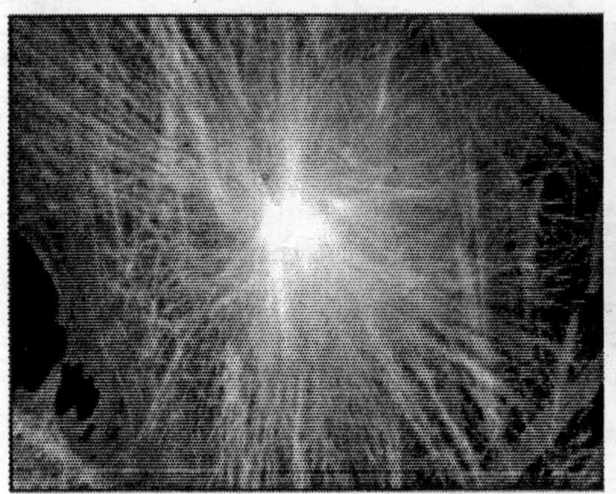

Cytoskeleton of a cultured epithelial cell

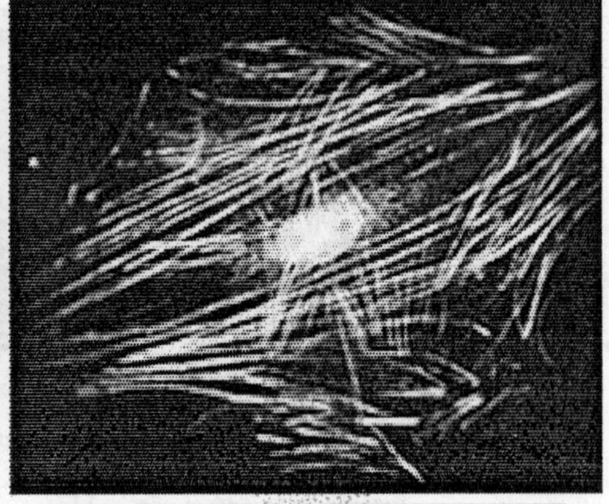

Cytoskeleton: microfilaments with fluorescent label

FIG. 33.2: Cytoskeleton and intermediate filaments.

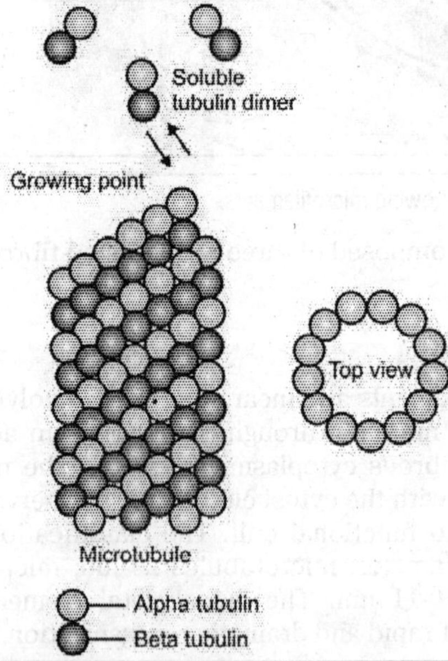

FIG. 33.3: Cross section of microtubule model (side and top view).

Chemically, the microtubules are composed totally of protein, tubulin. Tubulin is a dimer of 110,000 to 120,000 Daltons. Each subunit consists of two molecules (monomer) of protein tubulin; one of alpha-tubulin and one of beta tubulin.

A polymer made of asymmetric subunits, all oriented in the same direction along the axis, is itself a polarized structure with one end distinguishable from the other. In case of microtubules, each protofilament possesses an alpha-tubulin molecule at one end and a beta-tubulin molecule at the other end. This structural polarity enables these organelles to participate in directed mechanical movements. Some tissues may possess multiple subtypes of both the alpha and beta species, thus exhibiting molecular heterogeneity. A number of other proteins, termed microtubule-associated proteins (MAPs) are generally found at the surface of the microtubule and may be present as regularly spaced projections. These constitute about 15-20 per cent of the proteins. The MAPs of mammalian brain cells fall into two groups - a group of high molecular weight species (MAP-1 and MAP-2) and a group of small molecular weight MAPs termed as **tau**. In the cell, the microtubules are in equilibrium with free tubulin. Phosphorylation of the tubulin monomers by a cyclic AMP-dependent kinase favours the polymerization. The assembly and disassembly of tubulin constitute a polarized phenomenon.

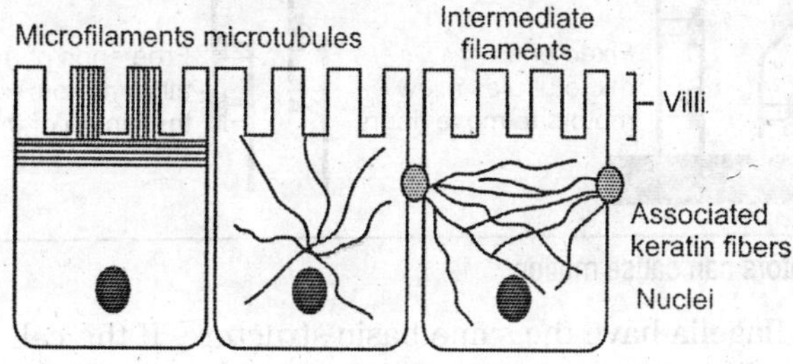

FIG. 33.4: Cytoskelaton components of intestinal epithelial cells.

CENTRIOLES

The centrioles and basal bodies of cilia are microtubule orientation centres from which the polymerization is directed. In a microtubule, the assembly of tubulin dimers takes place at one end while disassembly is prevalent at the other end. The assembly is accompanied by the hydrolysis of guanosine triphosphate (GTP) to guanosine diphosphate (GDP) in the presence of Mg ions.

Functions

When a cell enters the cell cycle, and proceeds from G_1 to S phase, each centriole is duplicated. A "daughter" centriole grows out of the side of each parent centriole. Thus centriole replication - like DNA replication (which is occurring at the same time) - is semiconservative. Once formed, most of the functions of the centrosomes can be accomplished without centrioles. However,

1. Centrioles appear to be needed to organize the centrosome in which they are embedded.
2. Sperm cells contain a pair of centrioles; eggs have none. The sperm's centrioles are absolutely essential for forming a centrosome which will form a spindle enabling the first division of the zygote to take place.
3. Centrioles are also needed to make **cilia** and **flagella**.

Cilia and Flagella

Both cilia and flagella are constructed from microtubules, and both provide either locomotion for the cells (*e.g.*, sperm) or move fluid past the cells (*e.g.*, ciliated epithelial cells that line our air passages and move a film of mucus towards the throat).

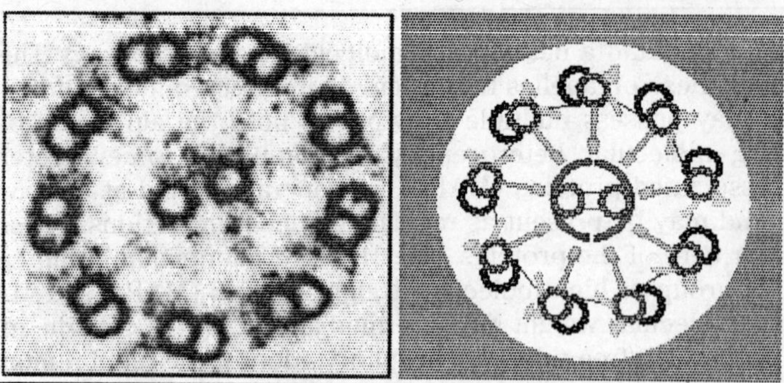

FIG. 33.5: Electron mcrograph of the cross section of a sperm tail.

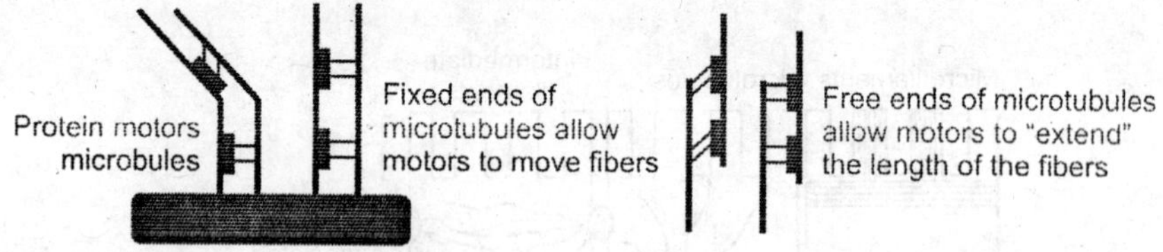

FIG. 33.6: Protein motors can cause motion.

Both cilia and flagella have the same basic structure. If the cell has many short ones, we call them cilia or only one or a few long ones, we call them flagella. Each cilium (or flagellum) is made of:

1. A cylindrical array of 9 evenly-spaced microtubules, each with a partial microtubule attached to it. This gives the structure a "Figure 8" appearance when view in cross section.
2. Two single microtubules run up through the center of the bundle, completing the so-called "9+2" pattern.
3. The entire assembly is sheathed in a membrane that is simply an extension of the plasma membrane.

This electron micrograph (courtesy of Peter Satir) shows the 9+2 pattern of microtubules in a single cilium seen in cross section. Motion of cilia and flagella is created by the microtubules sliding past one another. Motor molecules of **dynein** link adjacent to microtubules together, and the energy of ATP. Each cilium or flagellum grows out from, and remains attached to, a **basal body** embedded in the cytoplasm. Basal bodies are identical to centrioles and are, in fact, produced by them.

Primary Cilia: Motile, "9+2", cilia are found only on certain cells in the vertebrate body, *e.g.*, the epithelia lining the airways. But almost every cell in vertebrates has - or had - a single **primary cilium**. These do not beat because they lack the central pair of microtubules; that is they are "9+0". Where functions have been identified, they all involve sensory reception. The important examples are:

Mechanoreceptors: A primary cilium extends from the apical surface of the epithelial cells lining the kidney tubules and monitors the flow of fluid through the tubules. Inherited defects in the formation of these cilia cause **polycystic kidney disease**.

Chemoreceptors: Odours by receptors on the primary cilium of olfactory neurons.

Photoreceptors: The outer segment of the rods in the vertebrate retina is also derived from a primary cilium.

Shaping of the Cell and Mechanical Support

The orientation and distribution of the microtubules are responsible for shaping some processes arising from the cell. These also provide framework that also determines the shaping of the cell. Microtubules are commonly seen to be lined up parallel with and close to cell surfaces and in general conform to whatever particular shape a cell may exhibit. This arrangement is particularly evident in flattened cells, such as the erythrocytes of lower vertebrates or in cells with highly elongated processes such as neurons or the heliozoan protozoans.

Transport

The microtubules are also associated with the transport of macromolecules from one part of the cell to another. The pseudopodia of Actinosphaerium (a protozoan) contain about 500 microtubules disposed in a helical configuration. The cytoplasmic particles migrate back and forth in these pseudopodia. The pigment granules found in the erythrophores of fish scales move at a speed of 25 to 30 microns per second in between the microtubules.

Morphogenesis

Microtubules play an important role in shaping of the cell during differentiation. For example, the elongation of the nucleus of spermatid during spermiogenesis is accompanied by the production of orderly array of microtubules which become wrapped around the nucleus in double helical manner.

Cellular Motility

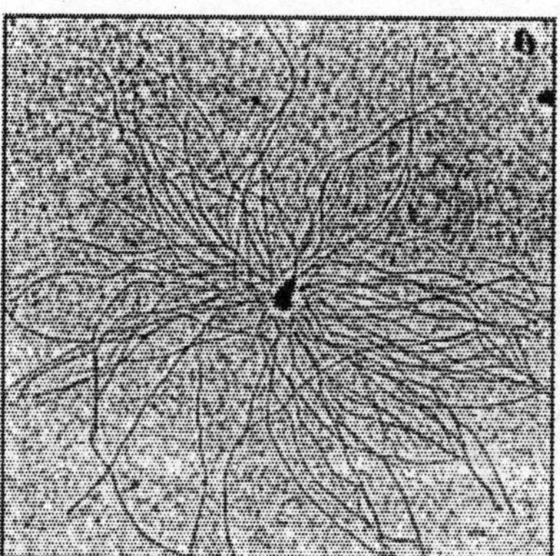

FIG. 33.7: The centrosome showing microtubules growing *in vitro* from an isolated centrosome. These spontaneously assembled into microtubules only in the presence of centrosomes.

Microtubules have also been implicated in the movement of cells themselves. In this capacity, the microtubules are components of more complex structures such as cilia and flagella. Another interesting example is that of chromosomal movement during cell division. The microtubular network of the interphase cell disassembles and then reassembles into mitotic spindles which help in chromosomal movement toward their respective poles.

The Centrosome

The centrosome is located in the cytoplasm attached to the outside of the nucleus. It is duplicated during S phase of the cell cycle. Just before mitosis, the two centrosomes move apart until they are on opposite sides of the nucleus. As mitosis proceeds, microtubules grow out from each centrosome with their plus ends growing toward the metaphase plate. These clusters of microtubules are called

spindle fibers.

Spindle fibers have three destinations:

1. Some attach to one **kinetochore** of a dyad with those growing from the opposite centrosome binding to the other kinetochore of that dyad.
2. Some bind to the arms of the chromosomes.
3. Still others continue growing from the two centrosomes until they extend between each other in a region of overlap.

All three groups of spindle fibers participate in the separation of the chromosomes at anaphase. The sister kinetochores separate and, carrying their attached chromatid, move along the microtubules powered by minus-end motors, **dyneins**, while the microtubules themselves shorten (probably at both ends). The overlapping spindle fibers move past each other (pushing the poles farther apart) powered by plus-end motors, the "bipolar" **kinesins**. In this way the sister chromatids end up at opposite poles.

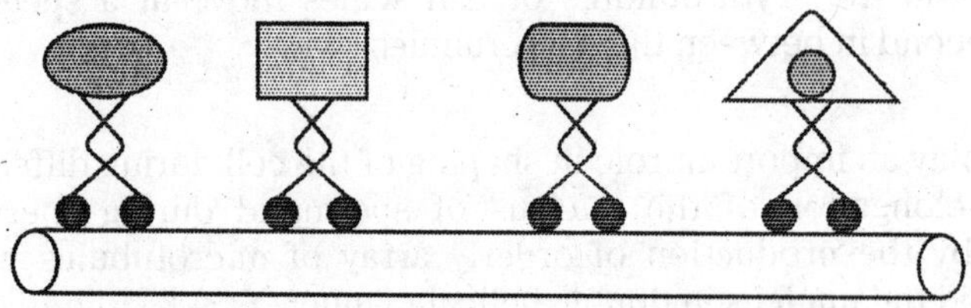

FIG. 33.8: Family of kinesin motors.

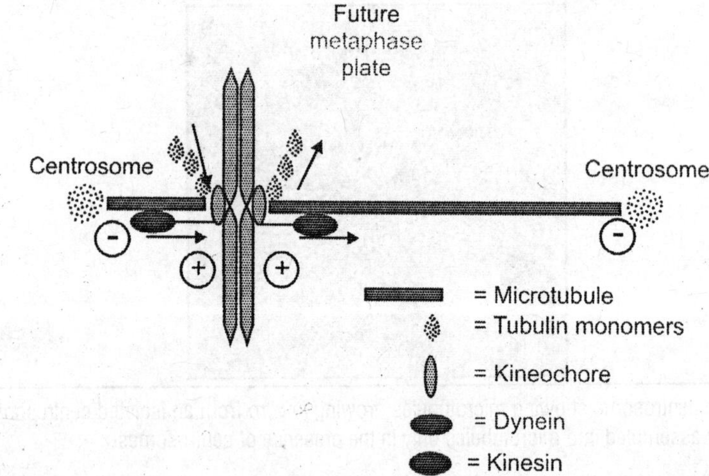

FIG. 33.9: The assembly of the chromosomes at the metaphase plate at metaphase (Proposed mechanism). Microtubules attached to opposite sides of the dyad shrink or grow until they are of equal length. Microtubules motors attached to the kinetochores move them toward the minus end of shrinking microtubules (a dynein); toward the plus end of lengthening microtubules (a kinesin). The chromosome arms use a different kinesin to move to the metaphase plate.

Centrosomes play other important roles in animal cells:

1. Signaling that it is o.k. To proceed to cytokinesis. Destruction of both centrosomes with a laser beam prevents cytokinesis even if mitosis has been completed normally.
2. Signaling that it is o.k. For the daughter cells to begin another round of the cell cycle; specifically to duplicate their chromosomes in the next **s phase**.

Destruction of one centrosome with a laser beam still permits cytokinesis but the daughter cells fail to enter a new s phase.

3. Segregating signaling molecules (*e.g.*, mrnas) so that they pass into only one of the two daughter cells produced by mitosis. In this way, the two daughter cells can enter different pathways of differentiation even though they contain identical genomes.
4. In at least some developing neurons, the position of the centrosome establishes the point at which the axon will grow out.

MICROFILAMENTS

Structure

Microfilaments are twisted double strands consisting of a string of proteins, from 7 nm to several cm long. The protein is actin. Its function helps muscle contraction, cell shape, and movement in cytoplasm.

The term microfilament is generally reserved for a specific type of fibrous element one that has a diameter of approximately 6 nm and is composed of the protein - actin. Actin is a globular protein of 42,000 Daltons. Its monomeric form is termed as G-actin which can be assembled into a double stranded filament termed as F-actin *in vivo* and *in vitro*, this process is termed as **nucleation**. The monomeric and polymeric states exist in a dynamic equilibrium. As with tubulin, actin is encoded by a family of closely related genes whose polypeptide products are highly conserved among all eukaryotes.

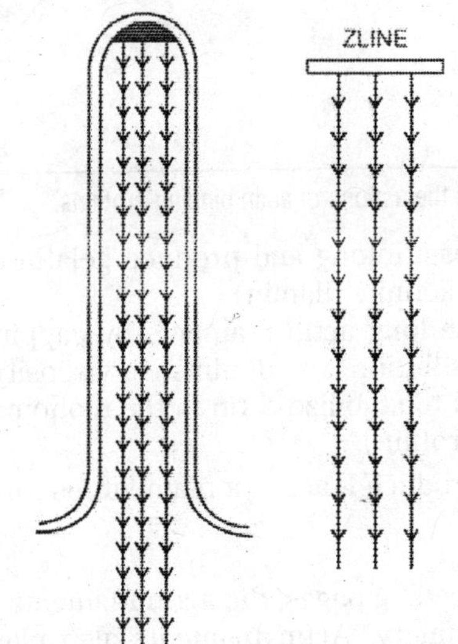

FIG. 33.10: Diagram showing the polarity of the actin filaments in the microvilli and the skeletal muscle.

Actin Filaments

Actin filaments are capable of interacting with another protein - myosin, in a highly specific manner. Heavy meromyosin (HMM), one of the sub fragments of myosin attaches to the actin filaments to form a complex which is identifiable in the electron microscope. Actin filaments can be localized with light microscope by using fluorescently labelled HMM molecules or fluorescently labelled antiactin antibodies. Actin filaments are present in muscle cells along with myosin, troponin, and tropomyosin. These protein molecules constitute the contractile machinery of the muscle cells. In non-muscle cells actin may appear in a variety of structural forms. For example:

1. Actin filaments exhibit well organized and regular polarization in the intestinal microvilli.
2. Bundles of actin filaments constitute stress fibres in spreading moving cells in culture.
3. In pseudopodia, ruffling membranes and in the bulk of the cytoplasm, these are disposed at random and form a cross-linked network.

The cytoplasm also contains some other proteins which are complex in nature in addition to actin, that are designated as actin-binding proteins. These proteins can be classified into three functional groups:

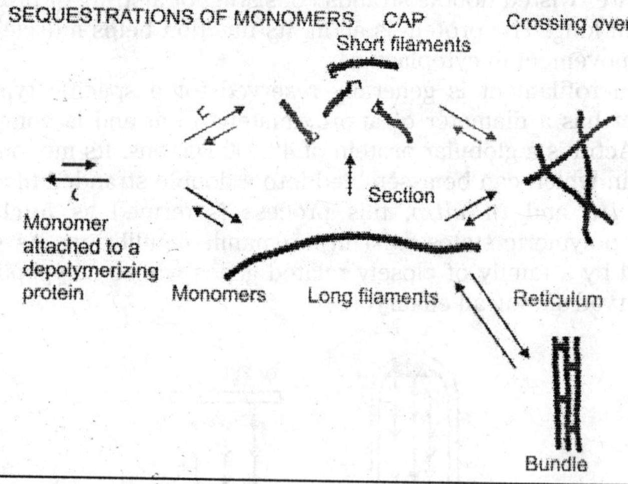

FIG. 33.11: Diagram representing the actions of actin binding proteins.

1. those that promote cross linking and produce gelatin of microfilaments (*e.g.*, fimbrin, spectrin, alpha actinin, filamin),
2. those able to severe the long actin filaments by capping to short filaments, a process that, results in solution, *i.e.*, diminished viscosity, (*e.g.*, gelsolin) and
3. those proteins that tend to stabilize actin in the monomeric form (G-actin) also causing solation (*e.g.*, profilin).

G-actin polymerize to produce F-actin (a filamentous form).

Functions

As also indicated in the preceding pages, the actin filaments form major component of a cells' contractile machinery. Actin filaments also play major roles in other diverse processes such as cytokinesis, cell movement, nerve outgrowth, tubular gland formation, movement of intestinal microvilli, cytoplasmic streaming, gastrulation, etc. In most cases, the evidence of the involvement of actin containing filaments is indirect, but substantial.

INTERMEDIATE FILAMENTS

Structure

The intermediate filaments are long, unbranched filaments having a diameter of approximately 10 nm, which is intermediate between that of microtubules and microfilaments.

These are chemically a heterogeneous group of structures. Inspite of heterogeneity, intermediate filaments exhibits similar structural organization. They contain considerable amounts of alpha-helix and show a periodic structure with a length of 21 nm. They are insoluble in physiological solutions. These filaments form networks that interconnect the nucleus with the cell surface. In fibroblasts intermediate filaments attach to the nuclear envelop and form a cage around the nucleus; at the plasma membrane they sometimes appear associated with microfilaments at the adhesion plaques. The intermediate filaments which radiate through the cytoplasm are in continuation with the filaments forming the cage around the nucleus.

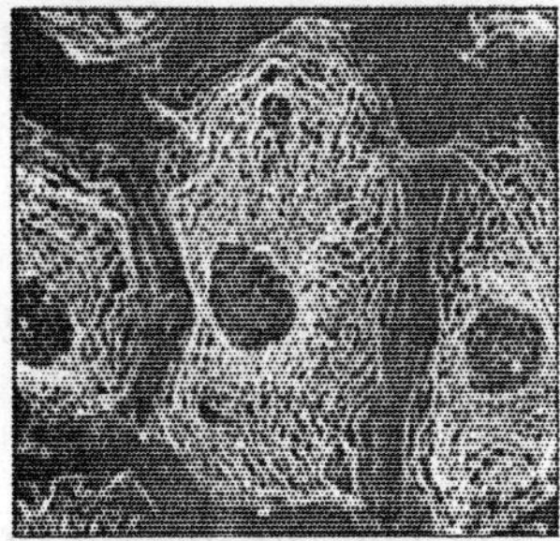

Intermediate filaments

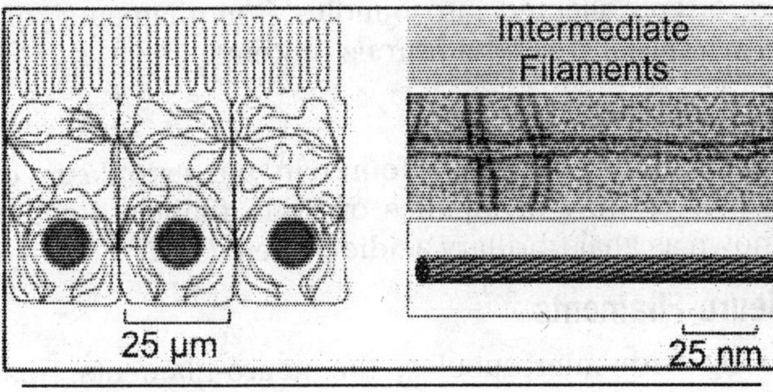

Intermediate filaments

FIG. 33.12: Intermediate filaments. These are rope-like fibers with a diameter of about 10 nm; they are made of intermediate filament protein, which constitute a large and heterogenous family. One type of intermediate filament forms a meshwork called the nuclear lamina just beneath the inner nuclear membrane. Other types extend across the cytoplasm, giving cells mechanical strength and distributing the mechanical stresses in an epithelial tissue by spanning the cytoplasm from one cell = cell junction to another.

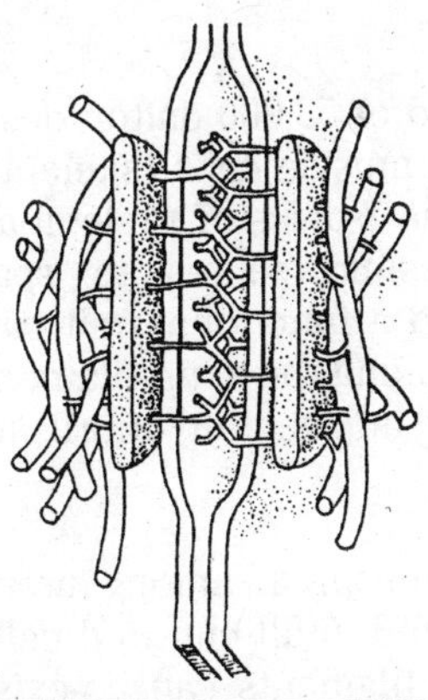

FIG. 33.13: Diagram of a spot desmosome.

Intermediate filaments are formed by protein subunits of a multigene class that are expressed differentially in various cell types.

The intermediate filaments are classified on the basis of their subunit composition. The distribution of these structures shows considerable tissue specificity and unlike the tubulins and actins, the proteins that makeup these filaments reflect the tissue type and state of differentiation of the cells in which they are synthesised. The different types of intermediate filaments may be as follows:

Keratin Filaments

These filaments are the most complex class of intermediate filaments. These are variously known as tonofilaments, prekeratin or cytokeratin filaments. Network of cytoplasmic filaments of 8 nm have been detected in a variety of epithelial cells including epidermal cells, hepatocytes, pancreatic acinar cells and myoepithelial cells. These filaments are composed of a family of proteins (42,000 to 65,000 Daltons) very similar in structure to alpha-keratin. In the epidermis these filaments become increasingly cross-linked as the cells differentiate and eventually contribute to the outer layer of dead cells that comprise the stratum corneum. These filaments are often collected into bundles (tonofilaments) of the desmosomes. Mammalian cytokeratins are alpha-fibrous proteins.

Glial Filaments

The glial filaments are found in the cytoplasm of glial cells. These present a scattered array of bundles of 8 nm filaments composed of 55,000 dalton protein known as glial fibrillary acidic protein (GFA).

Neuro-Filaments

Along with microtubules, the neuro-filaments (diameter 10 nm) are present in axons, dendrites and the neuronal perikarya. These form packed bundles with their long axes oriented parallel to that of the length of the cells. In mammals neuro filaments are composed of three distinct proteins (68,000, 145,000, and 220,000 Daltons) that co-polymerize to form the intact filament. Neuro filaments and the microtubules are connected to one another by side arm appendages and together are believed to form the three dimensional structural framework for the extended axonal and dendritic processes of nerve cells. The three dimensional lattice converts the axoplasm in a highly structured gel.

Desmin Filaments

Desmin filaments are composed of 53,000 dalton desmin protein having a diameter of 10 nm. These are present in muscle cells (skeletal, cardiac, and smooth fibres). These filaments are found at the Z-discs of skeletal and cardiac muscle cells and it has been suggested that desmin serves to spatially integrate the various components of muscle cell into a functional unit capable of shortening. At Z-disc these filaments play a role in linking the actin filaments of adjacent sarcomeres in a lengthwise direction and folding adjacent myofibrils in a lateral direction.

Vimentin Containing Filaments

These filaments are the most widespread among intermediate filaments. Vimentin is a protein of 57,000 Daltons found in all cultured cells, and most cells of the body have been investigated. These filaments can co-exist in the cytoplasm with glial filaments, desmin filaments or keratin filaments. These interconnect the nuclear and plasma membrane and thus play a role in supporting the nucleus and maintaining its position within the cell.

Synemin Containing Filaments

Synemin is a protein of 230,000 daltons. It is found in muscle cells along with desmin and vimentin. Closely associated with other proteins, synemin makes a network of interlinked rings within the plane of the Z-discs in skeletal muscle fibres.

CYTOPLASMIC MICROTRABECULAR SYSTEM (LATTICE)

Keith Porter (1976) reported the presence of a new cytoskeletal network, termed as microtrabecular lattice. Porter suggested that this network contains `microtrabeculae' ranging in diameter from 3 to 10 nm, extending throughout the cell and has associated with it all the organelles, microtubules, microfilaments and intermediate filaments. These filaments branch, fuse, and ramify throughout the cytoplasm. In association with other cytoskeletal elements these constitute a *form-fitting support*, or cytoskeleton, for the outer cell membrane and all the internal cell organelles.

The microtrabeculae remained unnoticed over the years because of its very fine dimensions, its lack of a defined structure, and its relative invisibility in thin plastic sections commonly used in conjunction with the electron microscope. Porter developed a technique in which unembedded and unsectioned cells were observed in the high-voltage electron microscope (HVEM).

Like other cytoskeletal structures, the micro-trabeculae are also subject to disassembly and reassembly undergoing a dynamic change in structural organization during specific physiological processes.

It has also been suggested that the microtrabeculae act as a scaffold to which soluble enzymes of the cytosol and free polyribosomes can attach. This lattice serves to compartmentalize the cytosol much the same way as membranes serve to compartmentalize the cell on a larger scale. The enzymes which catalyse the reactions of glycolysis, gluconeogenesis, amino acid biosynthesis etc. are not simply suspended freely in the fluid matrix of the cell but are organized into some type of three dimensional array around the microtrabeculae much like the components of the electron transport chain organized within the inner membrane of the mitochondrion.

COVALENT MODIFICATION OF CYTOSKELETAL PROTEINS

A large number of cytoskeletal proteins undergo covalent modification after translocation. Phosphorylation is one of the commonest modifications. Proteins undergoing phosphorylation may be heavy and light chains of myosin, filamin, fibronectin, vinculin, ß-tubulin, vimentin, desmin, α-and ß-tropomyosin, and spectrin. The majority of these proteins are phosphorylated on serine, and to a lesser extent on threonine. Vinculin, filamin and vimentin are the only cytoskeletal proteins showing phosphorylation on tyrosine. Some of the functions of this phosphorylation are known:

1. Non-muscle myosin light chains are activated by phosphorylation.
2. Myosin heavy chains disassemble with phosphorylation.
3. Many microtubule-associated proteins are more extensively associated with microtubules, when phosphorylated.

In addition to phosphorylation, a number of other covalent modifications have been detected:

1. Tubulin is post translationally tyrosinylated and it may by glycosylated.
2. Actin shows two types of covalent modifications:
   a) It contains N- methyl histidine, and N-terminal acetyl group; the methyl histidine is shared with myosin.
   b) It also contains trimethyllysine groups, *e.g.* calmodulin, and Acanthamoeba.
3. An acyl group, which may be a 16 carbon side chain is present in vinculin to

varying extents.

CYTOSKELETAL ARCHITECTURE

Common fundamental design of the cytoskeleton finds expression in forms specific to species and tissues. The core structural proteins are in general highly conserved, but specific modifications of the accessory proteins have permitted fine tuning of the cytoskeleton for the functions of each cell. Thus, most cells are characterized by a particular spectrum of cytoskeletal proteins present in a particular stoichiometry and arrangement.

The cytoskeleton confers upon the cell the following features:

1. its shape.
2. its ability to attach to substrates or to other cells.
3. freedom of movement.
4. the capacity to move materials within the cell.
5. the capacity to export material from the cell.

This list of functions would present a formidable task in analysing their structural basis individually.

The cytoskeleton can also undergo regulated changes in architecture. Some of these changes are in response to particular stimuli and may either be autonomous or require either protein synthesis or RNA synthesis. Other such changes occur over the lifetime of the cell, and still other architectural changes are visible during the development of a cell lineage.

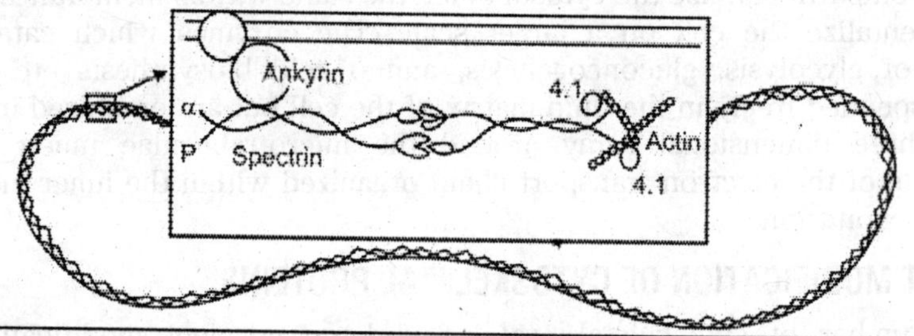

FIG. 33.14: Association between actin, spectrin and band 4.1 protein in RBC.

Red Blood Cells

Among the known cytoskeletal architectures, the mammalian red blood cells exhibit simplest pattern. It consists of a submembranous meshwork of actin and spectrin. The actin is predominantly in the F-form since it contains ADP rather than ATP. The spectrin possesses multiple binding sites. The head to tail tetramers of spectrin can cross link the oligomers of actin by binding to the sides of the actin protofilaments. The spectrin molecules also contain, on the ß spectrin subunit a binding site for ankyrin. Ankyrin binds both to spectrin and to an integral membrane protein, Band 3, to - form a cross bridge between the actin- spectrin meshwork and the membrane. The actin spectrin associations are stabilized by yet another protein, Band 4.1, which binds near the end of the spectrin molecule occupied by the side of the actin filament.

Non-mammalian RBC contain a peripheral band of microtubules called the marginal band. This bundle appears to be cross-linked and is organized around a pair of centrioles. The intermediate filament network consists of vimentin filaments which are decorated with a 230 kD protein, synemin, which is distributed periodically along the vimentin filaments. The intermediate filament network forms a cage around the nucleus within the central region of the RBC. Synemin appears to cross link this network. In non-mammalian RBC submembranous actin- spectrin network also occurs.

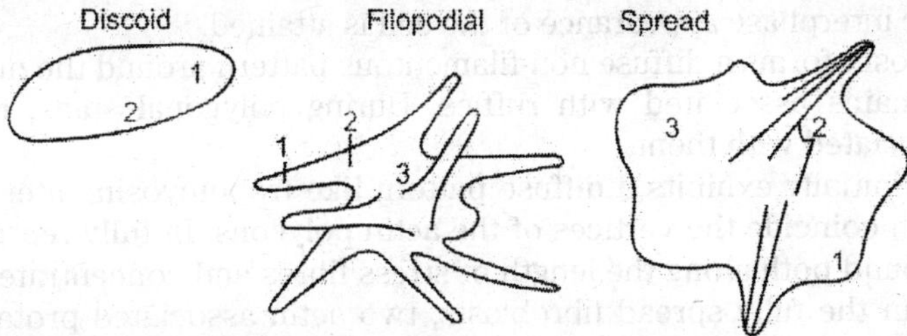

FIG. 33.15: The three forms of platelets.

Platelets

The platelets change shape and attach to the surfaces. The resting platelet is a discoidal, symmetrical cellular fragment. Activation causes numerous filopodia to extend from surface followed by "spread" form which spreads on the surface.

The discoid form contains a marginal band of microtubules cross linked by microtubule associated proteins. In addition, actin and numerous actin associating proteins are also found, which undergo co-ordinated rearrangements during filopodial extension and spreading. Microtubules are found in the marginal band, which undergoes distortion and rearrangements during activation.

The distribution of various proteins during various stages is depicted in following table.

TABLE 33.1: Platelet cytoskeletons

Region	Discoidal		Filopodial			Spread		
	1	2	1	2	3	1	2	3
Tubulin	++	-	-	-	+	±	-	tangles
Actin	granular	granular	++	++	±	+	++	±
Myosin	granular	granular	-	+	granular	±		+
α-Actinin	granular	granular	-	+	±	++		±
Filamin	granular	granular	+	+	±	±		+
Tropomyosin	diffuse	diffuse	++	++	-	±		+

Fibroblasts

Fibroblasts perform various functions:

1. These secrete extracellular matrix and contribute to tissue compaction.
2. Help in wound healing.
3. During embryogenesis, these undergo numerous movements and give rise to mesenchymal derivatives.

Because of these varied functions, the fibroblasts display complex cytoskeletal organization.

When the fibroblast responds to contact with a suitable substrate, numerous filopodia extend. If fibroblasts make contact with a non-adherent particle, frequently the filopodium will attach to the particle and retract with it. As soon as a sufficient number of contacts have been made, the cell begins to ruffle around its periphery; the ruffling and filopodial extensions may inter convert into each other. At this stage, the actin is found extensively associated with the filopodial ruffles and in a number of transnuclear cables. As the cell continues to spread, these transnuclear cables reorganize and begin to be distributed in the interior parts of the cell as a network of intersecting polygons. Over a period of hours these polygonal networks continue to be reorganized into stress fibres, until the characteristic interphase appearance of the cell is attained.

Tropomyosin forms a diffuse non-filamentous pattern around the nucleus when the actin remains associated with ruffles. During polygonal stage, tropomyosin becomes associated with them.

α-actinin initially exhibits a diffuse pattern like tropomyosin, later on it forms patches which coincide the vertices of the actin polygons. In fully reorganized cell, α-actinin is found both along the length of stress fibres and concentrated in ruffling membranes. In the fully spread fibroblasts, two actin associated proteins, fimbrin and vinculin are found.

Epithelial Cells

Epithelial cells cover the body surface, line the gut and form diverse glands within the body. When the epithelial cells form contacts with substratum, focal contacts are produced which contain α-actinin. Tropomyosin and α-actinin are more closely spaced than in fibroblasts, myosin is found periodically among the stress fibres.

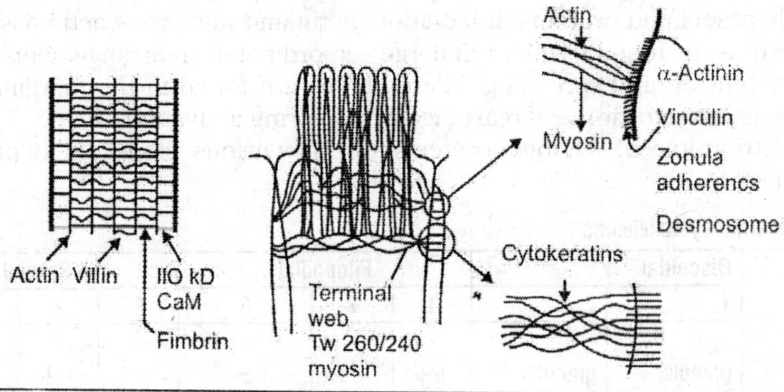

FIG. 33.16: Various types of proteins in epithelial cells.

These cells contain extensive microtubular networks which frequently connect with subcellular organelles. Intermediate filament network is most diverse and highly specialized in epithelia. The intermediate filaments are composed of cytokeratin proteins which are of atleast seven major classes. The cytokeratin fibres bundle and anastamose in extended arrays throughout the cytoplasm and terminate at the desmosome.

Epithelial cells exhibit amplification and specialization of the cortical actin meshwork and its microvilli. The microvilli are found on the luminal surface of these cells. The microvilli are anchored in a terminal web. The core of the microvilli contains actin bundles of approximately 20 filaments cross linked to each other and to the microvillar membrane.

This bundle of actin filaments is capped at the distal end; proximally, it penetrates the terminal web of the cell. The core is filled by actin, villin (95 kD), and fimbrin (68 kD), which cross link the actin filaments and a 105 kD protein which binds calmodulin and appears to cross link the bundle to the membrane. The terminal web consists of actin and tropomyosin, myosin and filaments of spectrin family (TW 240/ 260).

Endothelial Cells

Endothelial cells form sheets like epithelial ones. Their intermediate filaments are made solely of vimentin. Actin forms a diffuse network in the periphery. Stress fibres are seen only below actin network of major branches in the arteries. The microtubules extend from the cell center in radial tracks.

Transformed Cells

In general, transformed cells have a disorganized actin based network. Mutant actin has also been detected in transformed cells. Actin associated proteins may be modified as well; α and ß tropomyosin may be diminished or vinculin may be more highly phosphorylated.

Plants

Plant cells contain both actin filaments and microfilaments. Actin is known to be associated with cytoplasmic streaming, mediated through cables anchored in the plasma membrane. These cables of actin consist of bundles of microfilaments all with the same polarity *i.e.*, with the barbed end anchored in the membrane. Microtubules change location at each stage of the cell cycle:

1. **Interphase**. Microtubules lie in the cortical region parallel to the cellulose microfilaments. Microtubules and plasma membrane are cross linked.
2. **Prophase**. Most of the microtubules are found in a cage around the nuclear material.
3. **Metaphase and Anaphase**. Microtubules become organized and spindles are formed, which are barrel shaped.
4. **Telophase**. Microtubular content of spindles reduces and these begin to appear in the phragmoplast, which participates in cell division by orienting the formation of the cell plate.

After the completion of cell division, microtubules gradually take up their interphase cortical situation.

CHAPTER 34

Water and Human Metabolism

In human body, water remains as a healing agent because a human body contains about 70% of body weight. The quantity is high in organs such as lungs and brain and fluids such as blood, lymph, saliva and secretions by the organs of the digestive system.

Every time you swallow a bite of sandwich or slurp a smoothie, your body works hard to process the nutrients you've eaten. Long after the dishes are cleared and the food is digested, the nutrients you've taken in become the building blocks and fuel needed by your body. Your body gets the energy it needs from food through a process called metabolism.

Metabolism (pronounced: muh-**tah**-buh-lih-zum) is a collection of chemical reactions that takes place in the body's cells to convert the fuel in the food we eat into the energy needed to power everything we do, from moving to thinking to growing. Specific proteins in the body control the chemical reactions of metabolism, and each chemical reaction is coordinated with other body functions. In fact, thousands of metabolic reactions happen at the same time - all regulated by the body - to keep our cells healthy and working. Therefore, it is the sum of the chemical reactions that take place within each cell of a living organism and that provide energy for vital processes and for synthesizing new organic material. Energy obtained from the metabolic nutrients is utilized for various growth and other maintenance processes.

In human body, water remains as a medium for different enzymatic and chemical reactions. It helps in ther movement of various nutrients, hormones, antibodies and oxygen through the blood stream and lymphatic system. All living organisms are unique in that they can extract energy from their environments and use it to carry out activities such as movement, growth and development and reproduction. The proteins and enzymes in our body function more efficiently in solutions of low viscosity. Water also acts as solvent of the body and it regulates different functions, including the activity of everything it dissolves and circulates. The maintenance of toxicity of body fluids, within a very marroe physiologic range is made possible by homeostatic mechanism that controls the intake and excretion of water. Critical to this process are the osmoreceptors in the hypothalamus that control the secretion of antidiuretic hormones (ADH) in response to change in toxicity.

Water as Carrier of Chemical Energy

At any given time, a neutral molecule of water dissociates into a hydrogen ion (H^+) and a hydroxide ion (OH^-). Under normal conditions (neutrality), the concentration of hydrogen ions (acidic ions) is equal to that of the hydroxide ions (basic ions); each are at a concentration of 10-7 moles per litre, which is described as a pH of 7.

In case of shortage of water in our body, histamine a chemical compound present in all cells - initiates a system of water regulation. Histamine directs some neurotransmitters to operate sub-systems to regulate water intake. The ratio of the water content in and outside the cells of the various organs is very important. As age advances, water content in the cells decreases. Since the water content in each cell plays a vital role in maintaining its normal function, inadequate water can lead to loss of some functions resulting in specific symptoms.

Removal of Surplus Salts

The waste products of metabolism and surplus salts get removed from your body through urine. The human kidneys normally produce 0.9-1.5 litres of urine per day, containing some 50-70 g of solids - mostly urea, uric acid, and inorganic salts. Thus water plays a crucial role in our metabolic reactions.

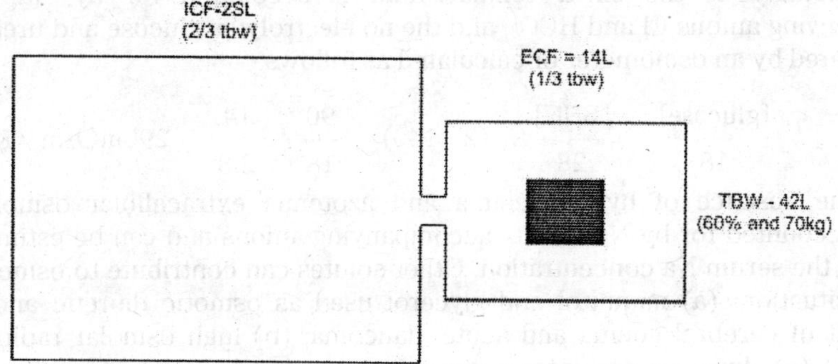

Total Body Water and Water Balance

A lean body mass contains about 60% of water in which about two thirds of total body water is present inside cell as intracellular fluid (ICF), and one third is extra cellular fluid (ECF). The ECF is further divided into an intravascular compartment (25%) and an interstitial compartment (75%).

For example, a man of 70 kg body weight has about 42 liters of water and out of which 28 liters remains intracellular, and 14 liters remains as extra cellular out of which 3.5 liters of the extra cellular fluid is located in the intravascular compartment, while most of the other water - 10.5 liters is located in the interstitial fluid compartment. A small component of extra cellular fluid (about 1-2 liters) is present in the transcellular compartment (in joints, the peritoneal space etc.) but under certain circumstances can account for a substantial component of extra cellular fluid (e.g. peritonitis) where it is termed the thirdspace. Total body water distributions illustrated above.

Under normal physiologic condition, the human body maintains fluid balance and water intake equals water output. Water intake depends on:

1. The thirst and personal habit.
2. Renal fluid losses.
3. Insensible water loss via perspiration and loss via the pulmonary and gastrointestinal routes determine water loss.

Water of metabolism contributes to the water intake column: "Average" intake is tabulated below; obviously the main determinant of balance under steady state conditions will be water intake, which can vary widely.

In		Out	
Oral intake	2,000 ml	Kidney	2000 ml
Metabolism	500 ml	Insensible	500 ml
Total	**2,500, l**	**Total**	**2,500 ml**

If water intake exceeds loss, total body water increases and water balance will be positive. Conversely, if loss exceeds intake, total body water decreases and water balance will be negative. Under both circumstances the two major body fluid compartments share the gain or loss of pure water proportionally so that the gain or loss of pure water does not alter the **relative volumes** of these two compartments.

OSMOLALITY

Osmolality is a count of the number of particles in a fluid sample. The unit for counting is the mole which is equal to 6.02×10^{23} particles (Avogadro's number).

The osmolality of a solution depends on the number of particles per kg water. The osmolality of the extra cellular fluid is accounted for by Na and its accompanying anions Cl and HCO_3, and the no electrolytes glucose and urea. It can be measured by an osmometer or calculated as follows:

$$2\ [Na\ ] \quad \frac{[glucose]}{18} \quad \frac{[BUN]}{28} \quad (2\ \ 140) \quad \frac{90}{18} \quad \frac{14}{2.8} \quad 290 mOsm/KgH_2O$$

In the absence of hyperglycemia and azotemia extracellular osmolality is largely accounted for by Na and its accompanying anions and can be estimated by doubling the serum Na concentration. Other solutes can contribute to osmolality in certain situation: (a) mannitol and glycerol used as osmotic diuretic and in the treatment of cerebral edema and acute glaucoma; (b) high osmolar radiographic dyes; and (c) during acute intoxication or poisoning by ethanol, methanol and ethylene glycoly.

Molarity

Molarity is the number of particles of a particular substance in a volume of fluid (eg mmol/L) and molality is the number of particles dissolved in a mass weight of fluid (mmol/kg). Osmolality is a count of the total number of osmotically active particles in a solution and is equal to the sum of the molalities of all the solutes present in that solution. For most biological systems the molarity and the molality of a solution are nearly exactly equal because the density of water is 1 kg/L. There is a slight difference between molality and molarity in plasma because of the non-aqueous components present such as proteins and lipids which make up about 6% of the total volume. Thus serum is only 94% water and the molality of a substance in serum is about 6% higher than its molarity. Except in unusual circumstances this difference is unimportant and the terms molarity to the molarity are often used interchangeably. Note that the size of the particle is unimportant so that a single ion, eg sodium, contributes as much to the serum osmolality as a single large protein molecule, eg albumin.

Effective Osmolality or Tonicity

As indicated earlier, the major extracellular solute is Na and its associated anions. The major **intracellular** solute is K and its associated anions. These solutes are relatively restricted to their compartments via a number of transport proteins of which Na-K-ATPase is most important. Solutes that are relatively restricted to one particular body fluid compartment are able to exert an osmotic force for water movement from other compartments. Such solutes are called osmotically effective, or simply effective solutes.

Tonicity

Tonicity relates to the osmotic gradient due to solutes that affects a semi-permeable membrane (i.e. a membrane that is permeable to water). Only solutes that do not cross the membrane contribute to this effect. The two characteristics are different and obviously have vastly different consequences on membranes. Weak bases, even though ionized, have some measurable permeability and so their contribution to tonicity must be much less than their contribution to osmolarity. Osmium is hydrophobic and quite permeable through membranes, so contributes nothing to an osmotic gradient that can disrupt the membranes (even assuming it doesn't alter their permeability to tonic agents like sucrose). So osmium contributes nothing to tonicity, but certainly does contribute to osmolarity.

Effective osmolality or tonicity is thus determined by the amount of effective solute in a particular body fluid compartment. It is important to realize that tonicity is a unit less concept that can be expressed only in reference to a physiologic system: a hypertonic solutions one that would shrink cells, while a hypotonic one would cause them to swell. In this context Na and glucose are **effective extracellular solutes** since they can cause movement of water out of cells (hypertonic). Mannitol, glycerol and sorbitol also behave as effective solutes.

Some solutes, notable urea passes freely across cell membranes and do not exerta force for water movement between the two major body fluid compartments. Ethanol, methanol and ethylene glycol are also **non-effective solutes**. Such non-effective solutes contribute to body osmolality but not to tonicity. Importantly in the presence of the aforementioned toxins measured serum osmolality will be increased whereas calculated serum osmolality will be normal. This so called **osmolar gap** serves as an important clue for the presence of an imbibed toxin.

Osmolality	Specific gravity
300 mOsm	1010
650 mOsm	1020
1000 mOsm	1030

Specific gravity: Whereas osmolality of a solution depends on the number of particles per kg water, the specific gravity of a solution is dependent upon **the weight of the particles** relative to the volume it occupies. Urinary specific gravity and osmolality usually change in parallel, but discrepancies occur between the two parameters. For example, for a given molar concentration, the specific gravity of urine tends to rise less with urea than with glucose. Urinary specific gravity can be easily measured at the bedside using a hygrometer and provides rapid and useful information regarding urine concentration and/or dilution. A rough approximation between urine osmolality and specific gravity is as follows:

Osmolality of Physiological Fluids

The osmolality of physiological fluids tends to be dominated by small molecules which are present in high concentrations. For example in serum, sodium, potassium, chloride, bicarbonate, urea and glucose are the only components present in high enough concentrations to individually affect the osmolality. Together these make up over 95% of total osmolality of serum. Large serum components contribute little to the overall osmolality. For example the molar concentration of albumin, the most abundant serum protein, is about 0.6 mmol/L. Only a few exogenous compounds such as ethanol, methanol, ethylene glycol and manitol can be present in the blood at high enough quantities to significantly affect the osmolality.

Osmolality of Plasma

The osmolality of plasma is closely regulated by anti-diuretic hormone (ADH). In response to even small increases in plasma osmolality (usually rises in plasma sodium), ADH release from the pituitary is increased causing water resorption in the distal tubules and collecting ducts of the kidney and correction of the increased osmolality. The opposite happens in response to a low plasma osmolality with decreased ADH secretion and water loss through the kidneys. Note that ADH is also secreted in response to hypovolaemia and this stimulus will over-ride any response to serum osmolality.

Urine Osmolality

Urine osmolality may vary between 50 and 1200 mmol/kg in a healthy individual depending on the state of hydration. The urine osmolality is the best measure of urine concentration with high values indicating maximally concentrated urine and low values very dilute urine. The main factor determining urine concentration is the amount of water which is resorbed in the distal tubules and collecting ducts in response to ADH. In a dehydrated patient with normally functioning pituitary and kidneys, a small volume of highly concentrated urine will be produced. In a patient with fluid overload the opposite will be an appropriate response. Note that there is no reference interval (normal range) for urine osmolality as the interpretation depends on the clinical condition of the patient to determine an appropriate response.

Measurement of Osmolality

The osmolality of a solution can be measured using an osmometer. The most commonly used instrument in modern laboratories is a freezing point depression osmometer. This instrument measures the change in freezing point that occurs in a solution with increasing osmolality. Osmolality can be measured in samples of serum (gold top tube) or heparin plasma (lime top tube).

Plasma osmolality can also be calculated from the measured components. While there are many equations, a simple one is as follows:

Osmolality (calc) = 2 × Na + Glucose + urea (all measurements in mmol/L)

The doubling of sodium accounts for the negative ions associated with sodium and the exclusion of potassium approximately allows for the incomplete dissociation of sodium chloride. The difference between the measured and calculated plasma osmolality is known as the osmolar gap and normally is between 0 and 10 mmol/kg.

Clinical Uses of Osmolality

Serum osmolality: Serum osmolality is used in two main circumstances: investigation of hyponatraemia and identification of an osmolar gap. Urine osmolality is an important test of renal concentrating ability, for identifying disorders of the ADH mechanism, and identifying causes of hyper-or hyponatraemia. Faecal osmolality can be used to assist with diagnosis of the cause of diarrhoea.

Serum osmolality is a useful preliminary investigation for identifying the cause of hyponatraemia. If a patient with significant hyponatraemia (serum sodium < 130 mmol/L) has a normal plasma osmolality, the patient may have pseudohyponatraemia due to excess lipids or proteins, or the sample may have been collected from a drip arm containing dextrose. If the patient has an increased osmolality it is likely the patient has reactive hyponatraemia due to an excess of solute pulling water out of cells. Examples of this include glucose in diabetes mellitus or hyperglycinaemia after trans-urethral resection of the prostate. The finding of a hypo-osmolar hyponatraemia (true hyponatraemia) then leads to further investigation of the cause.

Serum Osmolar Gap: The osmolar gap is determined by subtracting the calculated osmolality from the measured osmolality. While there are many formulae for the calculated osmolality, the most commonly used is:

Calculated osmolality = 2 × serum sodium + serum glucose + serum urea (all in mmol/L)

The normal osmolar gap is up to 10 mmol/L and values in excess of this usually indicate the presence of an exogenous agent. The most common by far is ethanol, but methanol, ethylene glycol, acetone and isopropyl alcohol can occasionally be present in sufficient quantities to produce an increased osmolar gap. The equation to calculate the osmolar gap when ethanol has been measured, as a sceren for the presence of other substances, is as follows:

Calculated osmolality = 2 × serum sodium + serum glucose + serum urea - 270 x ethanol

(all in mmol/L except ethanol in mg/dL or %, ie the units in use at SydPath)

Calculated serum osmolality = 2 x sodium + glucose + urea - 1.25 x ethanol (all in mmol/L including ethanol)

Urine Osmolality: Urine osmolality is an important test for the concentrating ability of the kidney. Interpretation of urine osmolality must always be made in the light of the appropriate physiological response to the state of hydration of the patient. The test is useful in the following areas:

1. For determining the differential diagnosis of hyper- or hyponatraemia.
2. For identifying SIADH (urine osmolality > 200 mmol/kg, urine sodium > 20 mmol/L, low serum sodium, patient not dehydrated and no renal, adrenal, thyroid, cardiac or liver disease or interfering drugs)
3. For differentiating pre-renal from renal kidney failure (high urine osmolality is consistent with pre-renal impairment, in renal damage the urine osmolality is similar to plasma osmolality).
4. For identifying and diagnosing diabetes insipidus (low urine osmolality not responding to water restriction).

Faecal Osmolar Gap: Faecal osmolality is used to determine the faecal osmolar gap. This is defined as the difference between the measured osmolality and an osmolality calculated from 2 × (Na + K). If the gap is greater than 100 mmol/L this is consistent with an osmotic diarrhoea (eg carbohydrate (poor absorption, eg mannitol, sorbitol, lactulose); monosaccharides; short chain fatty acids; magnesium as used in antacids). A normal faecal osmolar gap indicates a secretory diarrhoea indicating damage or irritation of the gastro-intestinal mucosa. This test can only be applied to faecal samples which contain a high fluid content.

Totalbody Water in Human

Cell membranes allows water to move freely across it, however, how the major extracellular and intracellular solutes (Na and K receptively) move across them. Under steady state conditions, intracellular fluid and extracellular fluid osmolalities are equal (about 290 mOsm/kg water), therefore, despite of the nature of the composition of the solute in each compartment, the water concentration on both sides of the cell membrane is identical.

Thus total body water must distribute according to the total amount of solute that is (relatively) restricted to each body fluid compartment. Therefore, under normal conditions, the intracellular fluid contains two-thirds of total body osmotically active solute and the extracellular fluid contains the remaining one-third. Thus the relative volumes of the two major body fluid compartments are determined by their respective effective solute contents.

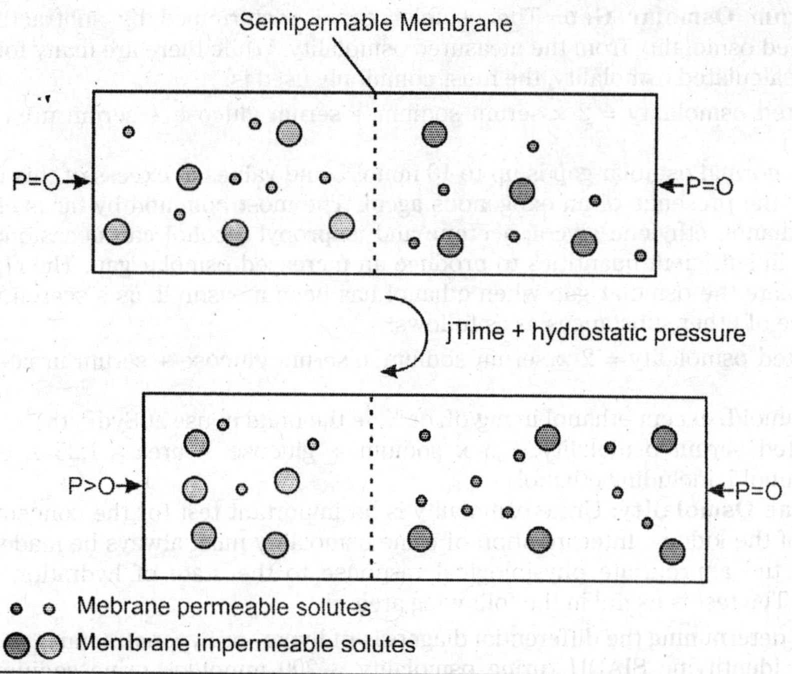

FIG. 34.1: Illustration of membrane permeable to solute and impermeable to solutes.

MECHANISMS OF WATER HANDLING BY THE KIDNEY

The Kidneys

The kidneys belong to the excretory system. They are reddish, bean-shaped organs situated one on each side of the vertebral column, just below the ribs. Often, the left kidney is positioned an inch higher than the right kidney. Each kidney is about 4-5 inches long and about two inches thick, weighing 4-6 grams in the average adult. Because of the presence of many blood vessels, the kidneys are coloured a dark reddish-brown. The mammalian kidney consists of an outer **cortex** and an inner **medulla**. It is composed of units called **nephrons**. Each nephron consists of a **glomerulus**, situated in the cortex of the kidney, and a long U-shaped tubule (the **loop of Henle**) that extends into the medulla of the kidney and connects to collecting tubules which eventually merge to join the ureter. The kidney has an outer region called the cortex and an inner region called the medulla. As one progresses inward from the cortex to the inner medulla, the concentration of solutes increases from about 300 mosm/L to 1200 mosm/L.

Mechanism

The kidneys are responsible for the elimination of most of the waste products from the metabolism of nitrogen-containing compounds such as urea and creatinine from protein catabolism, as well as ketone bodies from fat catabolism. The difficulty of this kidney function is compounded by the fact that the kidney also reabsorbs molecules necessary for normal body function such as glucose, amino acids, and bicarbonate. The kidneys are involved in the maintenance of a constant internal environment (**homeostasis**) of electrolyte (salt) concentrations, fluid balance, and acid-base balance. The fluid volume of the blood is maintained by the reabsorption of 98-99% of the water that leaves the blood in the initial step of urine formation, while the electrolyte and pH balance of the blood is maintained by the selective reabsorption of such ions as Na^+, K^+, Cl^- and HCO_3.

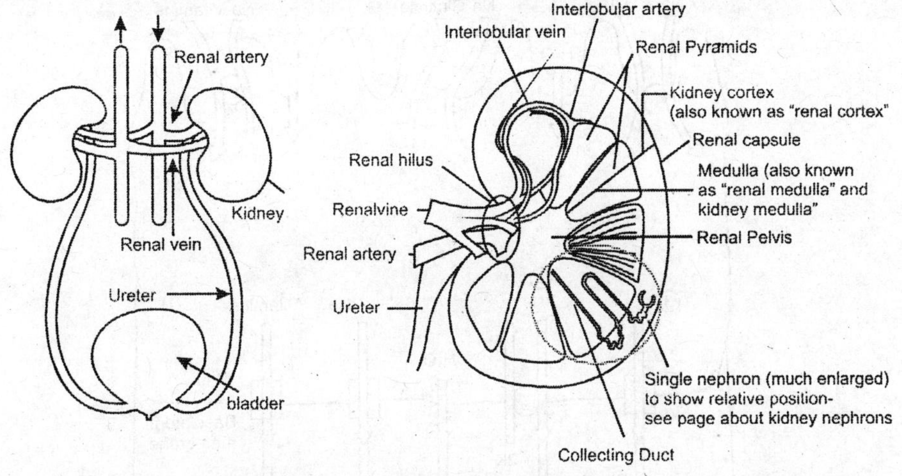

FIG. 34.2: Structure and various parts of human kidney.

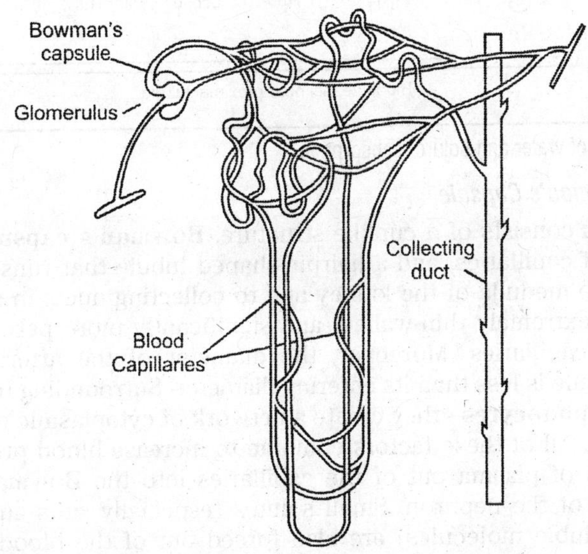

FIG. 34.3: Various parts of kidney showing Bowman's capsule, glomerulus, blood capillaries, conducing ducts.

Blood is brought to the kidney in the renal artery. The kidneys filter the blood and then reabsorb useful materials such as glucose. After it has been purified the blood returns to the circulation through the renal vein. Urine is taken from the kidneys to the bladder by the ureters. The bladder stores the urine until it is convenient to expel it from the body. The main waste product removed in the urine is urea. Urea is a waste product produced in the liver when surplus amino acids are broken down to be used for energy. It is released into the bloodstream until it is filtered out and removed by the kidney. Urine is produced in microscopic structures in the kidney called nephrons. Each kidney has many millions of nephrons.

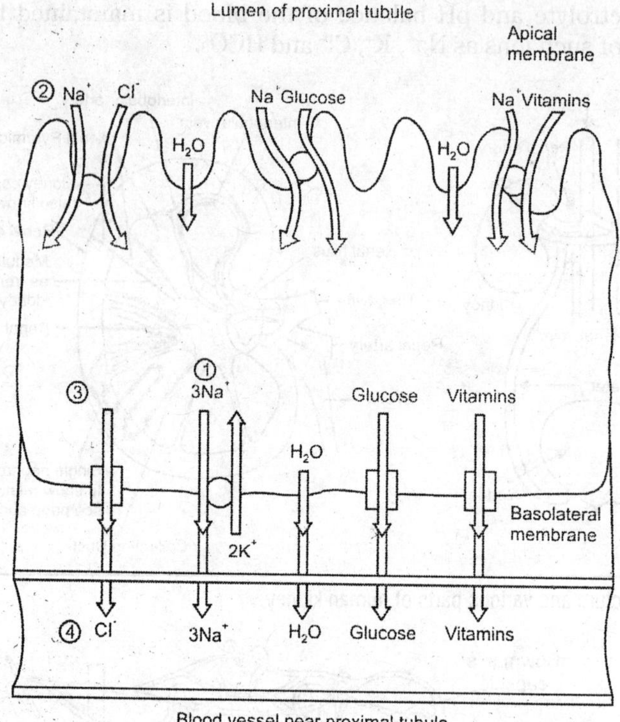

FIG. 34.4: Model of water and solure reabsorption.

Filtration in Bowman's Capsule

The glomerulus consists of a cuplike structure, **Bowman's capsule**, within which lies a cluster of capillaries, and a hairpin-shaped tubule that runs from Bowman's capsule into the medulla of the kidney and to collecting ducts in the medulla. The capillaries are extremely thin-walled, and significantly more permeable to plasma than ordinary capillaries. Moreover, the diameter of the arteriole as it leaves Bowman's capsule is less than its entering diameter. Surrounding the capillaries are cells known as **podocytes** - they create a network of cytoplasmic extensions which aid in filtration. All of these factors combine to increase blood pressure and force large quantities of plasma out of the capillaries into the Bowman's capsule and down the tube of the nephron. Small solutes (especially salts and urea, but also other water soluble molecules) are also forced out of the bloodstream with the plasma. Larger proteins and cells remain in the capillaries, creating a very hypertonic solution.

The general strategy here is this: the blood plasma is full of nutrients, proteins, ions, water, and other dissolved particles, some of which the body needs, some of which the body must remove. To remove the wastes, a large portion of the plasma is filtered from the blood and then the substances the body needs are put back into the blood. The substances that the body doesn't need are left behind and removed during urination.

Fluid that is filtered by the glomerulus and appears in Bowman's space is osmotic with serum (300 mOsm/kg water). The proximal tubule is comprised of a "leaky" epithelium so that fluid is neither concentrated nor diluted in this nephron segment. Quantitatively, 70% of filtered water is reabsorbed in the proximal tubule;

it passively follows solutes that are absorbed in this segment of which Na is the most important. The remaining 30% of fluid entering the loop of Henle is still isosmotic.

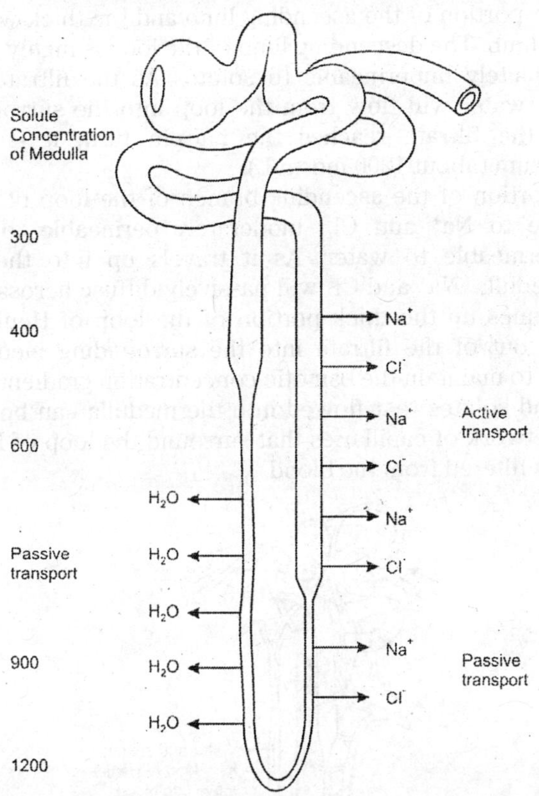

FIG. 34.5: Illustration of passive and active transport in kidney.

Reabsorption in the Proximal Tubule

Blood is filtered in the glomerulus. The filtrate is collected by the bowman's capsule and enters the tubules. Useful substances such as glucose, some salt and water are reabsorbed into the blood. Reabsorption is done by blood capillaries which are closely wrapped round the tubules. The waste, consisting of water, some salt and urea is urine. The urine is collected by the collecting duct, taken to the ureters and then to the bladder.

The lumen wall of the epithelial cells of the proximal tubule is like the lumen wall of the small intestine - both are bordered with millions of microvilli to increase surface area. The role of these epithelial cells is to reabsorb ions, nutrients, and water and transport them to the blood vessels nearby.

1. Na^+/K^+ ATPase is located on the basolateral membrane of the epithelial cell (the side of the cell opposite the lumen) actively pump Na^+ out of the cell into the blood. This sets up a strong concentration gradient in the cell.
2. The gradient created by the Na^+/K^+ ATPase provides a potential to allow Na^+ contransporters in the apical membrane to reabsorb nutrients and electrolytes. Water also flows in via osmosis.
3. Solutes exit the epithelial cells and enter the blood through channels.
4. Water flows from the epithelial cells into the blood via osmosis. Since, osmosis

occurs, the osmolarity of the filtrate remains isotonic. The volume decreases.

Creation of an Osmotic Gradient in the Loop of Henle

The loop of Henle has three distinct regions, the thin-walled descending limb, the thin-walled lower portion of the ascending limb and the thick-walled upper portion of the ascending limb. The descending limb of the loop is highly permeable to water but almost completely impermeable to solutes. As the filtrate travels down the descending loop, water will flow from the loop into the surrounding medium via osmosis. When the filtrate reaches the hairpin turn, it is isotonic with the surrounding medium (about 1200 mosm/L).

The lower portion of the ascending branch of the loop of Henle, however, is highly permeable to Na^+ and Cl^-, moderately permeable to urea, and almost completely impermeable to water. As it travels up into the less-concentrated regions of the medulla, Na^+ and Cl^- will passively diffuse across the membrane. As the filtrate continues up the thick portion of the loop of Henle, Na^+ and Cl^- are actively pumped out of the filtrate into the surrounding medium. This requires energy, but helps to maintain the osmotic concentration gradient in the medulla.

The water and solutes that flowed into the medulla can be reabsorbed by the **vasa recta** - a network of capillaries that surround the loop of Henle and reabsorb water and solutes filtered from the blood.

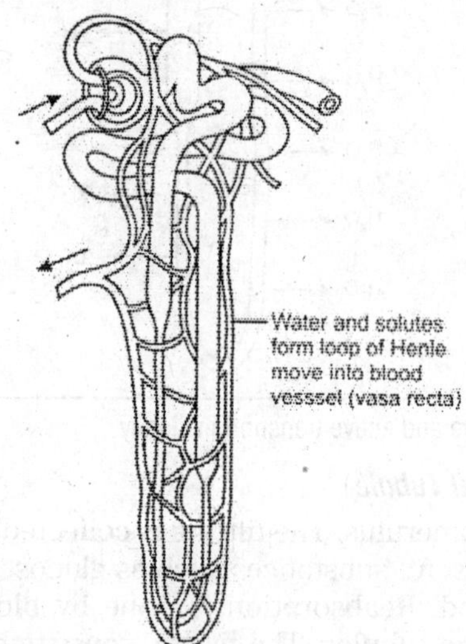

FIG. 34.6: Structure of vasa recta.

No ADH is present and the collecting duct is not permeable to water and large volume of urine is produced.

ADH is present and collecting duct is permeable to water and a small volume of urine is produced.

Regulating Water and Electrolyte Balance in the Distal Tubule and the Collecting Duct

The first three steps in urine formation, filtration, reabsorption, and establishment of an osmotic gradient, result in a fluid that is slightly hypotonic to blood. The major solutes still present in this fluid are urea and other wastes. Electrolytes and water are always absorbed by the distal tubule, the amount of Na^+, Cl^-, and water absorbed is variable. Regulation of these processes are under hormonal control.

If the Na^+ levels in the blood are low, the hormone **aldosterone** is released, which leads to reabsorption of Na^+ and Cl^- in the distal tubule. Water will also flow into the tubule via osmosis. However, if a person is dehydrated, the hormone **ADH (antidiuretic hormone)** is released. This causes aquaporin channels to be inserted in the membrane of the collecting duct so that large quantities of water can be reabsorbed.

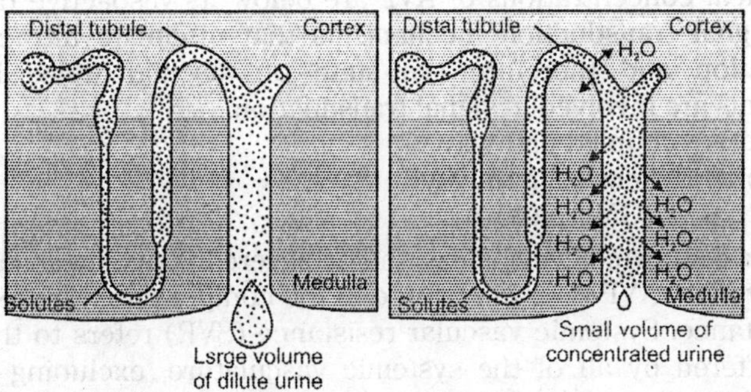

FIG. 34.7: Showing large colume of dilute urine and small colume of concentrated urine.

Regulation of Vasopressin Release

Vasopressin (arginine vasopressin, AVP; antidiuretic hormone, ADH) is a peptide hormone formed in the hypothalamus, then transported via axons to, and released from, the posterior pituitary. On average, normal serum osmolality is about 280 mOsm/kg water at which time plasma vasopressin levels are barely detectable (below 1pg/ml). With small increases in osmolality to 285 or 290 mOsm/kg water vasopressin levels incrementally rise to 3-4 pg/ml. The exquisite sensitivity of this system can therefore be readily appreciated. Interestinglya similar close association.

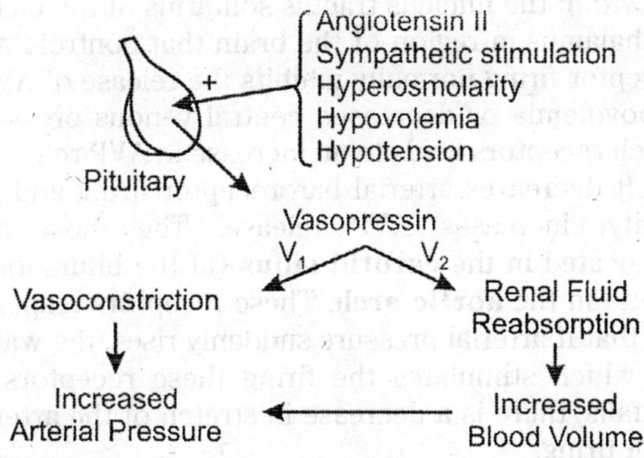

FIG. 34.8: Role of vasopression on blood volume and renal fluid.

The regulation of vasopressin release from the posterior pituitary is dependent primarily on two mechanisms: (a) osmotic and (b) non-osmotic/volume. AVP has two principle sites of action: the kidney and blood vessels.

1. The primary function of AVP in the body is to regulate extracellular fluid volume by affecting renal handling of water, although it is also a vasoconstrictor and pressor agent (hence, the name vasopressin). AVP acts on renal collecting ducts via V_2 receptors to increase water permeability (cAMP-dependent mechanism), which leads to decreased urine formation (hence, the antidiuretic action of antidiuretic hormone). This increases blood volume,

cardiac output and arterial pressure.

2. A secondary function of AVP is vasoconstriction. AVP binds to V_1 receptors on vascular smooth muscle to cause vasoconstriction via the IP3 signal transduction pathway, which increases arterial pressure; however, the normal physiological concentrations of AVP are below its vasoactive range. There are several signal transduction mechanisms that modulate intracellular calcium concentration and therefore the state of vascular tone. Three different mechanisms are involved in signal transduction:
   a) Gs-protein-coupled pathway
   b) Phosphatidylinositol pathway
   c) Nitric oxide-cGMP pathway

Studies have shown, nevertheless, that in severe hypovolemic shock, when AVP release is very high, AVP does contribute to the compensatory increase in systemic vascular resistance. Systemic vascular resistance (SVR) refers to the resistance to blood flow offered by all of the systemic vasculature, excluding the pulmonary vasculature. This is sometimes referred as total peripheral resistance (TPR). SVR is therefore determined by factors that influence vascular resistance in individual vascular beds. Mechanisms that cause vasoconstriction increase SVR, and those mechanisms that cause vasodilation decrease SVR. Although SVR is primarily determined by changes in blood vessel diameters, changes in blood viscosity also affect SVR.

There are several mechanisms regulating the release of AVP, the most important of which are the following:

Hypovolemia: It occurs during hemorrhage and dehydration, results in a decrease in atrial pressure. Specialized stretch receptors within the atrial walls and large veins (**cardiopulmonary baroreceptors**) entering the atria decrease their firing rate when there is a fall in atrial pressure. Afferent nerve fibers from these receptors synapse within the nucleus tractus solitarius of the medulla, which sends fibers to the hypothalamus, a region of the brain that controls AVP release by the pituitary. Atrial receptor firing normally inhibits the release of AVP by the posterior pituitary. With hypovolemia or decreased central venous pressure, the decreased firing of atrial stretch receptors leads to an increase in AVP release.

Hypotension: It decreases arterial baroreceptor firing and leads to enhanced sympathetic activity, increases AVP release. The most important arterial baroreceptors are located in the **carotid sinus** (at the bifurcation of external and internal carotids) and in the **aortic arch**. These receptors respond to stretching of the arterial wall so that if arterial pressure suddenly rises, the walls of these vessels passively expand, which stimulates the firing these receptors. If arterial blood pressure suddenly falls, there is a decrease in stretch of the arterial walls lead to a decrease in receptor firing.

Hypothalamic osmoreceptors sense extracellular osmolarity and stimulate AVP release when osmolarity rises, as occurs with dehydration.

Angiotensin II receptors: These are located in a region of the hypothalamus regulate AVP release – an increase in angiotensin II simulates AVP release. The renin-angiotensin-aldosterone system (RAAS) plays an important role in regulating blood volume and systemic vascular resistance, which together influence cardiac output and arterial pressure. As the name implies, there are three important components to this system: (1) renin, (2) angiotensin, and (3) aldosterone. Renin, which is primarily released by the kidneys, stimulates the formation of angiotensin in blood and tissues, which in turn stimulates the release of aldosterone from the adrenal cortex.

Increased sympathetic activation stimulates AVP release.

Heart failure: It is associated with what might be viewed as a paradoxical increase in AVP. Increased blood volume and atrial pressure associated with heart failure suggest that AVP secretion might be inhibited, but it is not. It may be that sympathetic and renin-angiotensin system activation in heart failure override the volume and low pressure cardiovascular receptors (as well as the hypothalamic control of AVP release) and cause an increase in AVP secretion. Nevertheless, this increase in AVP during heart failure may contribute to the increase in systemic vascular resistance as well as the enhanced renal retention of fluid that accompanies heart failure.

AVP infusion is being used in treating septic shock, a condition that can be caused by a bacterial infection in the blood and the release of bacterial endotoxins such as lipopolysaccharide. Infusion of AVP increases systemic vascular resistance and thereby elevates arterial pressure. Some studies have shown that low-dose infusions AVP (which are used in septic shock) also cause cerebral, pulmonary and renal dilation (mediated by endothelial release of nitric oxide) while constricting other vascular beds.

Between urine osmolality and plasma vasopressin exists as shown in Fig. 34.8. From this correlation it is also evident that when serum vasopressin has risen to 3-4 pg/ml, urine osmolality approaches maximal concentration.

CHAPTER 35

Nutrient Metabolism

NUTRIENT

A nutrient is a substance used in an organism's metabolism which must be taken in from the environment. Non-autotrophic organisms typically acquire nutrients by the ingestion of foods. Methods for nutrient intake vary, with animals and protists having an internal digestive system, while plants digest nutrients externally and then ingested. The effects of nutrients are dose-dependent. Organic nutrients include carbohydrates, fats, proteins (or their building blocks, amino acids), and vitamins. Inorganic chemical compounds such as minerals; water and oxygen may also be considered nutrients. A nutrient is essential to an organism if it cannot be synthesised by the organism in sufficient quantities and must be obtained from an external source. Nutrients needed in relatively large quantities are called macronutrients and those needed in relatively small quantities are called micronutrients. The human diet must provide the following:

Calories	It is required to meet our daily energy needs.
Amino Acids	There are nine, or so, essential amino acids that we need for protein synthesis and that we cannot synthesise from other precursors.
Fatty Acids	There are three "essential" fatty acids that we cannot synthesise from other precursors.
Minerals	Inorganic ions. We probably need 18 different ones: a few like calcium in relatively large amounts; most, like zinc, in "trace" amounts.
Vitamins	A dozen, or so, small organic molecules that we cannot synthesise from other precursors in our diet.

Macronutrients

The chemical elements humans consume in the largest quantities are carbon, hydrogen, nitrogen, oxygen, phosphorus, and sulphur. The classes of chemical compounds humans consume in the largest quantities and which provide bulk energy are carbohydrates, proteins, and fats. Water and atmospheric oxygen also must be consumed in large quantities, but are not always considered food or nutrients. Calcium, salt (sodium and chloride), magnesium, and potassium (along with phosphorus and sulphur) are sometimes added to the list of macronutrients because they are required in relatively large quantities compared to other vitamins and minerals. They are sometimes referred to as the macrominerals.

Micronutrients

The remaining vitamins, minerals, or elements, are called micronutrients because they are required in relatively small quantities. Numerous minerals are required in trace amounts and are usually cofactors for enzymes. Some trace mineral elements (RDA $<$ 200 mg/day) are as follows:

Role of Mineral Nutrients an Body

As with vitamins, mineral requirements vary. Humans as well as other vertebrates require relatively large quantities of Calcium and Phosphorus for the construction and maintenance of bone. Calcium is also necessary for the normal functioning of nerves and muscles, and Phosphorus is also a necessary component of ATP and nucleic acids. Iron is an important component of the cytochromes that function in cellular respiration and of hemoglobin, the oxygen-binding protein of red blood cells. Magnesium, manganese, zinc, and cobalt are cofactors built into the structure of certain enzymes. Iodine is necessary to make thyroxine, a thyroid hormone that regulates metabolic rate. Sodium, potassium, and chlorine are important in nerve function. A healthful diet must supply enough calories for energy needs, carbon chains, organic nitrogen, and ample quantities of the essential nutrients. The roles of minerals in human body are as follows:

Calcium	**Bone and tooth formation; blood clotting; nerve transmission.**
Chlorine	Formation of gastric juice; acid-base balance.
Cobalt	Constituent of vitamin B-12.
Copper	Constituent of enzymes associated with iron metabolism.
Flourine	May be important in maintenance of bone structure
Iodine	Constituent of thyroid hormones.
Iron	Constituent of hemoglobin and enzymes involved in energy metabolism.
Magnesium	Activates enzymes; involved in protein synthesis.
Manganese	Activates several enzymes, including one required for urea production
Phosphorous	Bone and tooth formation; acid-base balance; ATP formation.
Potassium	Acid-base balance; nerve function.
Sodium	Acid Base balance; body water balance; nerve function.
Sulphur	Constituent of tissue compounds, Cartilage, and tendons.
Zinc	Constituent of enzymes involved in digestion.

There are many more roles of Vitamins and Minerals in the body and many more nutrients are beneficial than are listed here. This is just a small sample of the roles of vitamins, minerals, amino acids and essential fatty acids for informational purposes only.

Nutrient Metabolism

The Nutrient Metabolism controls at the organ, cellular, and subcellular levels in normal and diseased states. Therefore, it is essential to understand the physiologic function and mechanism of action/interaction of nutrients within the body; nutrient influence on gene regulation and expression; metabolism and function of nutrient antioxidants; the effects of environment, heredity, stress, drug use, toxicants, and physical activity on problems of nutrient imbalance and nutrient requirements in health and disease; and specific metabolic considerations relating to alternative forms of nutrient delivery and use, such as total parenteral nutrition.

Other trace minerals still lacking solid proof, for example chromium (Chromium III) which is implicated in sugar metabolism in humans, leading to a market for chromium picolinate. Similarly, Vanadium has no specific biochemical function in humans but it is found in other organisms).

Bone Structure and Metabolism

Bone is a dynamic tissue that is remodeled continuously throughout life. Specialized cells called osteoblasts initiate bone formation, osteocytes monitor bone mechanical stresses, and osteoclasts resorb bone. Bone density is determined by the dynamic balance between bone formation and bone resorption. Formation starts with the deposition of an organic matrix by osteoblasts followed by the process of mineralization.

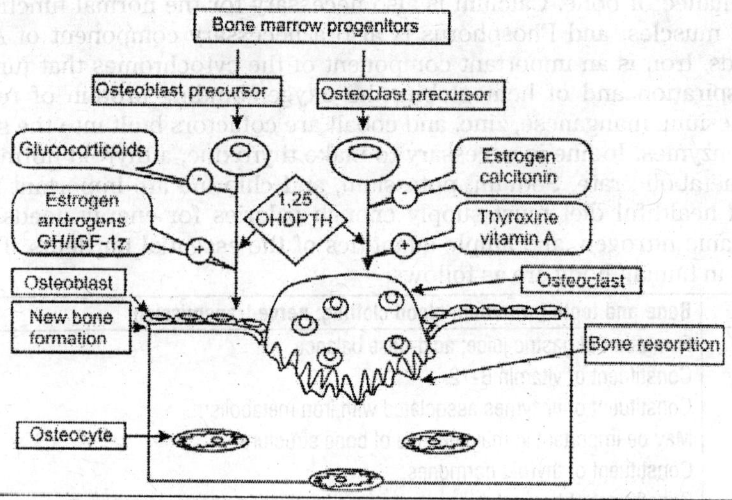

FIG. 35.1: Development of osteoblasts and osteoclasts from bone marrow progenitors. Factors affecting the development and function of these cells, bone resorption by osteoclast and new bone formation by osteoblasts. Abbreviations: GH: growth hormone, IGF: insulin like growth factors, PTH: parathyroid hormone (Valsamis *et al.*, Nutrition & Metabolism 2006 3:36).

The organic matrix is comprised predominantly of type I collagen (90–95%) with contributions from various other proteins including osteocalcin, osteonectin, osteopontin and thrombospondin. The mineral phase of bone, hydroxyapatite, is composed of calcium and phosphorus. The concentration of these ions in plasma and extracellular fluid determines the rate at which hydroxyapatite is formed and deposited. There is a critical limit for the concentrations of calcium and phosphorus ions below which the mineralization of organic matrix does not occur.

Energy Substances

Following are the various substances which provide energy to the body.

1. **Carbohydrates** are compounds made up of sugars. Carbohydrates are classified by their number of sugar units: monosaccharides (such as glucose and fructose), disaccharides (such as sucrose and lactose), oligosaccharides, and polysaccharides (such as starch, glycogen, and cellulose).
2. **Proteins** are organic compounds that consists of the amino acids joined by peptide bonds. The body cannot manufacture some of the amino acids (termed essential amino acids); the diet must supply these. In nutrition, proteins are broken down through digestion by proteases back into free amino acids.
3. **Fats** consist of a glycerin molecule with three fatty acids attached. Fatty acids are unbranched hydrocarbon chains, connected by single bonds alone (saturated fatty acids) or by both double and single bonds (unsaturated fatty acids). Fats are needed to keep cell membranes functioning properly, to insulate body organs against shock, to keep body temperature stable, and to maintain healthy skin and hair. The body does not manufacture certain fatty acids (termed essential fatty acids) and the diet must supply these. Fat has an energy content of 9 kcal/g (~37.7 kJ/g); proteins and carbohydrates 4 kcal/g (~16.7 kJ/g). Ethanol (grain alcohol) has an energy content of 7 kcal/g (20.3 kJ/g).

Substances for Support of Metabolism

1. **Dietary minerals** are generally trace elements, salts, or ions such as copper and iron. Some of these minerals are essential to human metabolism.
2. **Vitamins** are organic compounds essential to the body. They usually act as coenzymes or cofactors for various proteins in the body.
3. **Water** is an essential nutrient and is the solvent in which all the chemical reactions of life take place.

Control of Food Intake

A complex web of signals controls appetite and the intake of food. These include both nerve signals and hormones – both of which signal centers in the brain – chiefly in the hypothalamus. This table lists some of the hormonal signals that have been identified, their effect on appetite (and weight gain), and provides links to pages where they are discussed.

Appetite Stimulants	Appetite Suppressants
Neuropeptide Y	Leptin
Ghrelin	α-MSH and ß-MSH
Agouti-related protein (AgRP)	$PYY_{3\text{-}36}$
	Cholecystokinin (CCK)
	Insulin
	Amylin

CHAPTER 36

Recombinant DNA Technology

Recombinant DNA technology deals with isolation and manipulation of genes. Before nucleic acids can be cut or otherwise manipulated, they must be either isolated and purified or artificially synthesised. The isolated genes are also sometimes sequenced for a better understanding of its structure that needs to be manipulated. In this chapter we will discuss the methods available for isolation, sequencing and synthesis of genes.

ISOLATION OF TOTAL CELLULAR DNA

DNA is very easily damaged by shear forces; even rapid stirring of solution can break even with high DNA molecular weight into much shorter fragments. Consequently, DNA is recovered from cells by gentlest possible method of rupture, in the presence of EDTA to chalate the Mg^{2+} ions needed for DNase activity. Cell walls, if present, are digested enzymatically (*e.g.* lysozyme treatment of bacteria) and the cell membrane is solubilised using detergent. Cell disruption is performed at 4 °C using autoclaved glassware and solutions to destroy DNase activity.

After release of nucleic acids from the cells, RNA can be removed by treatment with ribonuclease (RNase). The other major contaminant, protein, is removed by shaking the solution gently with water-saturated phenol, or with phenol/chloroform mixture, either of which will denature protein but not nucleic acids. Centrifugation of the emulsion formed by this mixture produces a lower, organic phase, separated from the upper aqueous phase by an interface of denatured proteins. The aqueous solution is recovered and deproteinised repeatedly, until no matter is seen at interface. Finally, the deproteinised DNA preparation is mixed with ethanol and allowed to precipitate out in freezer. After centrifugation, the DNA pallet is redissolved in a buffer containing EDTA for protection against DNases and this solution can be stored at 4 °C for at least a month. DNA solutions can be stored frozen, but repeated freezing and thawing tends to damage long molecules by shearing; so preparation in frequent use are normally stored at 4 °C.

This procedure is suitable for total cellular DNA. If the DNA from specific organelle or viral particle is needed, it is best to isolate the organelle or virus before extracting its DNA. Now we will discuss several basic techniques related with gene isolation and identification.

NUCLEIC ACID HYBRIDIZATION

If a double-stranded DNA molecule is exposed to high temperature, or to very alkaline conditions, then the two strands will come apart. The molecule is said to have become denatured. The temperature at which denaturation occurs is termed as melting temperature or T_m. If the denatured DNA is returned to a temperature below its T_m, or to neutral pH when alkali was used to denature it, each strand will, after a time, find its complementary strand. The two strands will 'zipper' back together to re-form a double-stranded DNA molecule. This ability of complementary sequences to anneal, or hybridize, to one another is called nucleic acid hybridization. This technique helps in determining the gene structure and in identifying molecules which contain same sequences of nucleotides. In a complex mixture of nucleic acid molecules, nucleic acid hybridization technique helps in separation of complementary sequence.

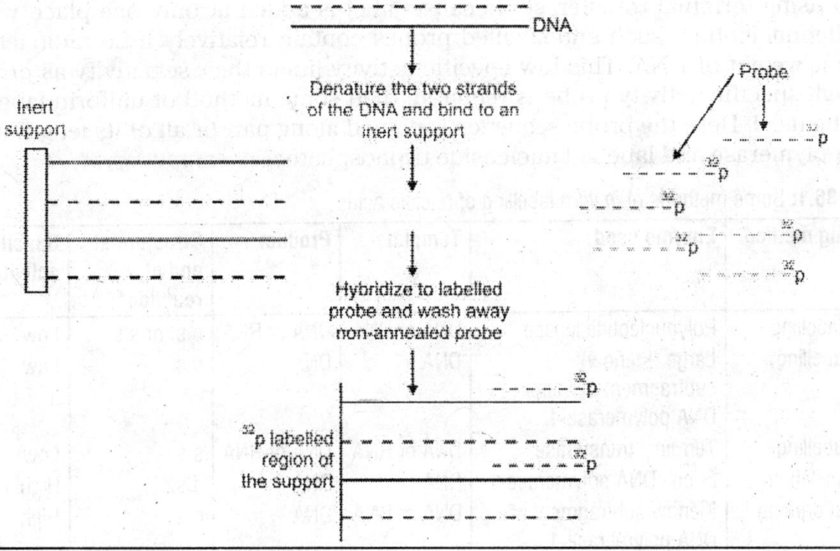

FIG. 36.1: The principle of hybridization using an inert support. The target DNA is most often localized to one part of the support and, in most cases, the aim of the hybridization experiment is to identify this region. The probe is prelabelled by some method, most often by the incorporation of radioactivity. After hybridization, the annealed probe is detected by autoradiography.

When employed analytically, hybridization is normally performed using one labelled sequence, termed the probe, and an unlabelled sequence called the target. **Probe** is a short synthetic oligo deoxyribonucleotide which is complementary to target DNA sequences. Target DNA may be obtained from total cellular DNA preparations. The probe is labelled by incorporation of either radioactively labelled nucleotides or with some chemicals (described later). The probe is the known, pure species in the hybridization and the target is the unknown species to be identified.

The target will most often form part of a mixture of unrelated nucleic acid sequences. The probe is usually added in considerable excess over the target, and its concentration determines the rate of reaction. Hybridization reaction time is extended to ensure complete reaction. The target and probe can be hybridized to each other in solution, or alternatively the target can be immobilized on an inert support such as nitrocellulose. When the target is immobilized the rate of annealing is greatly reduced.

The rate of hybridization depends upon the DNA concentration, the concentration of salt in the solution and the temperature. Hybridization reactions are normally performed at 10-20 °C below the T_m of the duplex. Under any given conditions of ionic strength and temperature, the T_m of a duplex also depends on the base composition of the annealed partners.

Hybridization involving RNA is performed in the presence of formamide as RNA cannot withstand higher temperatures. Formamide disrupts hydrogen bonding and thus reduces the effective T_m of the duplex, allowing annealing to occur at a lower temperature (*e.g.* 37 °C).

METHODS FOR LABELLING NUCLEIC ACIDS

For identification of hybridized nucleic acid duplexes, labelling of probe or target is necessary. As probe is short synthetic oligonucleotide it can be easily labelled. A ^{32}P isotopic label can be added to the 5' hydroxyl group using polynucleotide kinase to a probe (Table 36.1). Alternatively, labelled nucleotides can be incorporated at the 3' end using terminal transferase. Because label is added at only one place within the oligonucleotide, such end-labelled probes contain relatively little radioactivity per unit weight of DNA. This low specific activity limits their sensitivity as probes. If a high specific activity probe is required, then some method of uniform labelling must be used. Here the probe sequence is copied along part or all of its length using DNA polymerase and labelled nucleoside triphosphate.

TABLE 36.1: Some methods of *in vitro* labelling of nucleic acids

Labelling method	Enzyme used	Template	Product	Structure at end of reaction*	Specific activity
5' end labelling	Polynucleotide kinase	DNA or RNA	DNA or RNA	d.s. or s.s.	Low
3'end labelling	Large (Klenow) subfragment of E. coli DNA polymerase-1	DNA	DNA	d.s.	Low
3'end labelling	Terminal transferase	DNA or RNA	DNA or RNA	s.s.	Low
Nick translation	E. coli DNA polymerase-1	DNA	DNA	d.s.	High
Random priming	Klenow subfragment of DNA polymerase-1	DNA or RNA	DNA	d.s.	High
In-vitro transcription	Bacteriophage (e.g. T7) RNA polymerase	DNA	RNA	s.s.	High

*d.s. - double stranded; s.s. - single stranded.

METHODS OF LABELLING NUCLEIC ACID AND PROBES

There are five basic methods for labelling nucleic acids. These are: nick translation, primer extension, methods based on RNA polymerase, end labelling methods and direct labelling methods.

Nick Translation

This is done by making single-strand cuts (**nicks**) in the double-stranded DNA molecule by brief exposure to a dilute solution of an endonuclease (usually deoxyribonuclease I of *E. coli*). DNA polymerase I is then used in the presence of at least one radioactive precursor (*e.g.*, Â-^{32}P-dATP) to "translate" the nick along the molecule in the 5' to 3' direction. This occurs by the concerted action of the 5'→3' polymerase and the 5'→3' exonuclease activities built into DNA polymerase I. The 5'→3' polymerase adds nucleotides to the 3'-OH terminus at each nick, while the 5'-phosphate terminus at the nick. The net result is that a nonradioactive strand of DNA is replaced by a radioactive strand. The DNA is then denatured and used as a radioactive probe in hybridization experiments (Southern blots, northern blots, etc.).

Nick translation can be used with a variety of labels to generate probes suitable for most hybridization applications. It is also appropriate for the generation of biotinylated probes (described later).

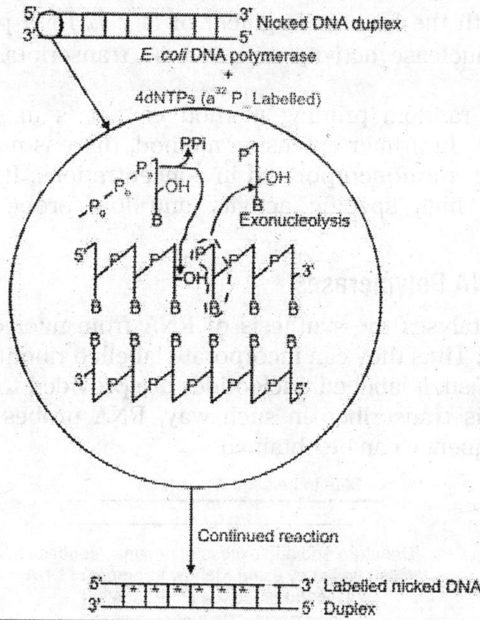

FIG. 36.2: ^{32}P-labelling of duplex DNA by nick translation. Asterisks indicate radiolabelled phosphate groups.

A major advantage of nick translation over other uniform labelling methods is its flexibility with respect to probe size, specific activity and concentration. It is particularly suitable for the production of large quantities of probes and for use in multiple hybridization techniques and/or where a high probe concentration is required.

Primer Extension Method

Primers are synthetic oligodeoxyribonucleotides which are complementary to specific regions of known vector DNA. Some commonly used primers are M13 sequencing primer, pBR322 primer etc. The 3' termini of these primers serve as initiation site for template dependent DNA synthesis by enzymes like DNA polymerase I.

DNA polymerase works by extending a short double-stranded region made by annealing an oligonucleotide primer to the single-stranded template. Thus this method of uniform labelling requires a primer which matches the probe sequence. Radiolabelling of primers can be done with two methods.

If the probe sequence is not known then random oligonucleotide labelling can be used. It is often in the case when natural cellular DNA is used. These primers are made by adding a mixture of all four bases at each step in the chemical synthesis reaction. The DNA is denatured and the two complementary strands are copied in the presence of labelled primers as well as, nucleoside triphosphates. The polymerase used is Klenow fragment derived from DNA polymerase-I of *E. coli*. Chance homology ensures that these primers anneal to the separated DNA strands at many points along their length, thus providing a base for polymerase to initiate DNA synthesis. This is only one of several uniform labelling methods.

The second method uses a unique primer to restrict labelling to a particular sequence of interest. In the primer extension method it is essential to use a polymerase lacking a 5'→3' exonuclease activity otherwise degradation of the primer will occur. Both the Klenow fragment of *E. coli* DNA polymerase I, which lacks the 5'→3' exonuclease activity and reverse transcriptase have been used successfully.

An advantage of random priming method is that is may be used to label relatively impure DNA. In primer extension method, there is not synthesis of DNA and labelled nucleotides are incorporated in concentrations. It is an ideal method for situations where high specific activity and low probe concentration are frequently employed.

Methods Based on RNA Polymerases

RNA polymerases, catalyses the synthesis of RNA from nucleoside triphosphates using a DNA template. Thus they can incorporate labelled ribonucleotides into RNA during transcription if such labelled nucleotides are provided to it. If a specific site of a vector or DNA is transcribed in such way, RNA probes (or transcripts) of defined length and sequence can be obtained.

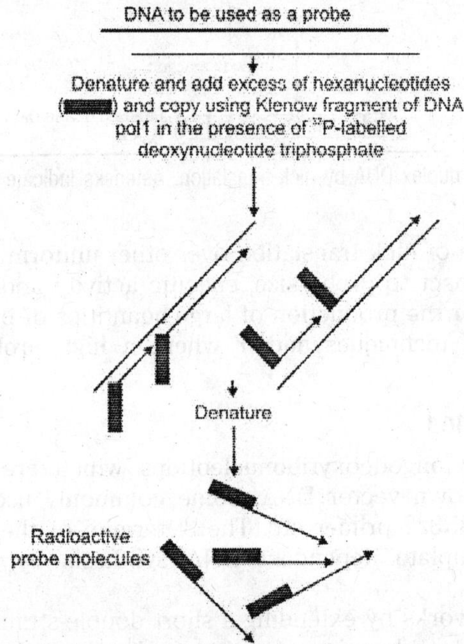

FIG. 36.3: The principle of random primed (oligo-) labelling. The DNA to be used as a probe is denatured by heating and mixed with hexanucleotides of random sequence which then act as primers for DNA synthesis.

Initially RNA probes were obtained from RNA polymerase from *E. coli*. This enzyme when used *in vitro* exhibited very little template and promoter specificity and therefore produces transcripts which have been initiated and terminated at random. Today, RNA polymerases from a number of phages have been used, which possess a high degree of specificity. Thus transcription can be limited to sequences cloned downstream of an appropriate promoter.

RNA probe's specific activity is determined solely by the specific activity of labelled nucleotide and moreover such probes are need not to be denatured before hybridization. These probes are more sensitive than nick translated probes because of avoidance of probe reannealing during hybridization (as RNA-RNA hybrids are more stable than RNA-DNA hybrids). A further advantage of RNA probes is that non-specifically bound probe can be removed with RNase A. RNase A is very specific for single stranded RNA.

The only difficulty with RNA polymerase-based-labelling is that a specific promoter is read by RNA polymerases before transcripting the downstream. This promoter has to be subcloned and inserted into the vector. However, once an insert has been cloned into such a vector, DNA probes may still be prepared using nick translation or random primer labelling.

END-LABELLING OF NUCLEIC ACIDS

A wide variety of techniques is available for introducing label at either the 3' or 5' ends of linear DNA or RNA. Usually only a single label is introduced at the terminus. Nucleic acid can be 5' end labelled using T4 polynucleotide kinase. The enzyme catalyses the transfer of the γ-phosphate of ribonucleoside 5'-triphosphate donor to the 5'hydroxyl group of a poly- or oligo-nucleotide or even to nucleoside-3-triphosphate. Radiolabelled phosphate group is donated by [γ^{32}-P] ATP to DNA or RNA containing a 5'-hydroxyl terminus. This is termed as forward reaction.

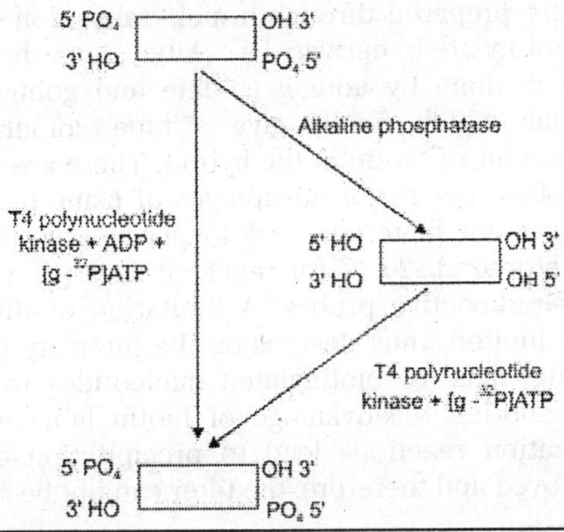

FIG. 36.4: The 5'-end labelling reaction using T4 polynucleotide kinase.

If 5'-phosphate group is present in DNA or RNA, then it is removed with alkaline phosphatase. This reaction is driven by excess ADP which causes the enzyme to transfer the terminal 5'-phosphate from DNA to ADP. This is known as exchange reaction. The DNA is rephosphorylated by transfer of labelled γ-phosphate from [γ^{32}-P] ATP.

The major advantages of 5'-end labelling are:

1. Both DNA and RNA can be labelled.
2. Oligonucleotide can be conveniently labelled *e.g.* for detection of mutants following site-directed mutagenesis.
3. Location of labelled group is known.
4. Very small fragments can be labelled.
5. Restriction digest fragment can be labelled.

DNA fragments also can be labelled at their 3'end by using terminal deoxynucleotidyl transferase, however, this method is not of wide use.

CHOICE OF LABEL

There are two technical parameters, resolution and sensitivity, which determines the success of probe application. High degree of resolution is required to known the relative position of a nucleic acid fragment. High sensitivity is necessary because sequence of interest may be present at low abundance.

Other factors are probe stability, safety and ease of use. Broadly labels can be categorised into radioactive and non radioactive types.

Radioactive Labels

These labels have wider applications as they can be easily detected with autoradiography. Their detection gives two important information, firstly about occurrence of hybridization between probes and target DNA and secondly about their position. Radioactive methods using ^{32}P are easily detectable. They are used often.

Non Radioactive Labels

A number of non radioactive labels for probes are available but biotin is widely used.

Biotin Labelled Probes

Biotinylated probes are prepared through a nick-translation reaction by replacing nucleotides with biotinylated derivatives. After hybridization and washing, detection of hybrids is done by adding avidine and going through a series of cytochemical reactions which finally give a blue colour whose intensity is proportional to the amount of biotin in the hybrid. There are several advantages of using biotinylated probes. For major advantages of using biotinylated probes are: (a) assays employ non- toxic materials, with longer half-life (b) can be prepared in advance in bulk and stored at -20 °C for repeated uses (c) detection of hybrids is much faster than by radioactive probes. A limitation of this technology is that biotinylated sites are limited, thus decreasing the intensity of signal. It has been solved by adding long 'tails' of biotinylated nucleotides to the probes through enzymatic methods. Another disadvantage of biotin labelled probes is that the cytochemical visualization reactions lead to precipitation of insoluble material which can not be removed and therefore, the filter can not be reused.

Digoxigenin Labelled Probes

Another method is to label DNA with digoxigen-labeled deoxyuridive triphosphate (dig-ÂUTP). After hybridization to the target DNA the hybrids are detected by enzyme-linked immunoassay using an antibody conjugate *e.g.* anti digoxigenin alkaline phosphatase. Enzyme reporter molecules with enzymes can also be used as label.

Alternatives to Biotin and Digoxigenin Labelling

Other ligands (*e.g.* hapten determinants, 2, 4-dinitrophenol, arsenate derivatives, etc.) are attached to specific proteins that can be tagged with enzymes or fluorescent molecules. It is possible to monitor many probes simultaneously by using several different ligands since each ligand would yield a different signal. A chemiluminescent (light emitting) probe system has also been developed in which two different probes complementary to a continuous segment of DNA hybridize to adjacent of a gene. The first label is a chemiluminescent complex which emits light at a specific wavelength. This emission excites the label molecule on the second probe to emit light at a different wave length which can be detected using a photomultiplier device. The processes called non-radioactive energy transfer, and can occur only if the two probes hybridize correctly and the two labels are close to each other.

The major disadvantage of biotin, and of other nonradioactive labels, is low sensitivity in comparison to radioactive labels.

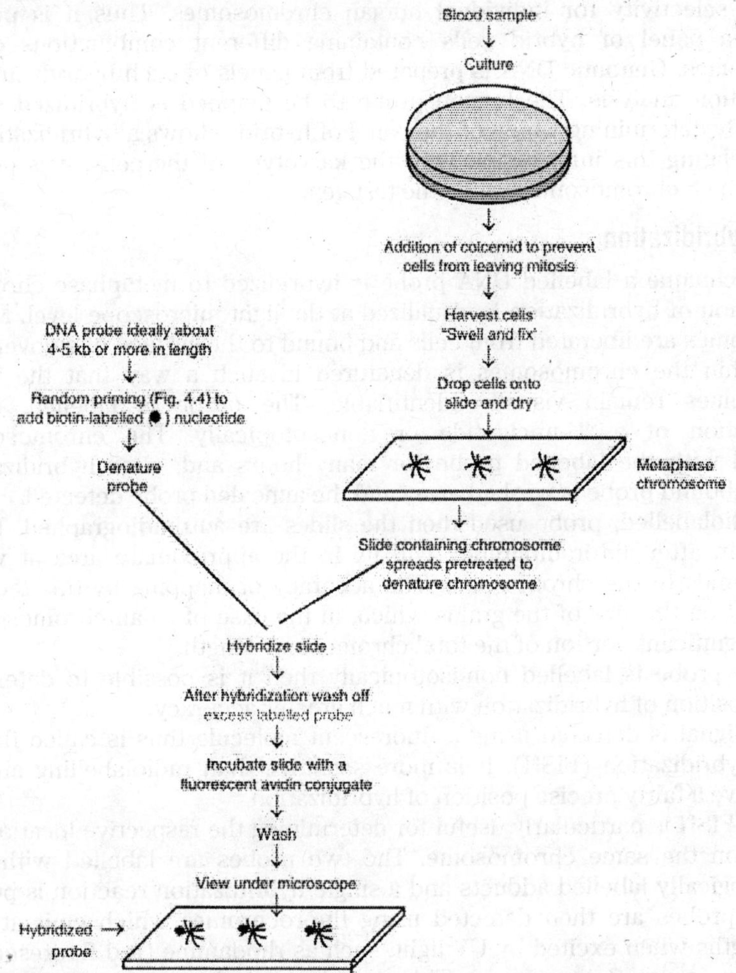

FIG. 36.5: FISH is used where gene mapping is done with in-situ hybridization. It uses a non-radioactive labelling system involving biotin attached. When viewed under special filters the specific sites of DNA probe can be seen on a chromosome.

MAPPING GENES ON SPECIFIC CHROMOSOMES

One of the most important applications of nucleic acid hybridization is construction of genetic maps. This helps in determining the chromosomal location of individual genes. There are three commonly used methods that differ in the degree of resolution they afford viz somatic cell hybrids, *In situ*, hybridization and genetic linkage mapping.

Somatic Cell Hybrids

Somatic cell hybrids are prepared by fusing together human and rodent cells using reagents, such as polyethylene glycol or Sendai virus. The reagents help in membrane fusion. The resulting hybrid cells are then grown in culture medium containing selective agents which allow only the hybrid cells to grow. In hybrids produced using human and rodent cells the rodent chromosomes are retained, while the human chromosomes are lost.

This random chromosome loss occurs during continued culture and shows no apparent selectivity for individual human chromosomes. Thus it is possible to produce a panel of hybrid cells containing different combinations of human chromosomes. Genomic DNA is prepared from panels of such hybrids and used in hybridization analysis. The human probe to be mapped is hybridized with such hybrids. By determining which of the panel of hybrids shows a hybridization signal, and correlating this information with the karyotype of the cells, it is possible to deduce which chromosome carries the test gene.

In Situ Hybridization

In this technique a labelled DNA probe is hybridized to metaphase chromosome. The position of hybridization is visualized at the light microscope level. Metaphase chromosomes are liberated from cells and bound to the surface of a cover slip. The DNA within the chromosomes is denatured in such a way that the individual chromosomes remain visually identifiable. The probe is labelled, either by incorporation of a ^{3}H-nucleotide or non-isotopically. The chromosomes are incubated with the labelled probe for many hours and, after hybridization, the excess unbound probe is washed away and the annealed probe detected.

If radiolabelled, probe used then the slides are autoradiographed. Deposited silver grain after autoradiography will lie in the approximate area at which the probe bounds to the chromosome. The accuracy of mapping by this technique is dependent on the size of the grains which, in the case of small chromosomes, can cover a significant portion of the total chromosome length.

If the probe is labelled non-isotopically then it is possible to determine the precise position of hybridization with much greater accuracy.

The signal is detected using a fluorescent molecule thus is called fluorescent *in-situ* hybridization (FISH). It is more sensitive than radiolabelling and can be used to give a fairly precise position of hybridization.

Thus FISH is particularly useful for determining the respective locations of two markers on the same chromosome. The two probes are labelled with different immunologically labelled adducts and a single hybridization reaction is performed. The two probes are then detected using fluorochromes which emit at different wave lengths when excited by UV light, such as rhodamine (red fluorescence) and fluorescein (green fluorescence). This technique has been used for probes which are separated by few hundred kilobases of DNA on same chromosome. The FISH technique can also be applied to interphase nuclei, which allows accurate ordering of probes along a chromosome when they are separated by no more than 20-30 kilobases.

Transposon Tagging

Isolating a eukaryotic gene by transposon tagging involves inserting into target DNA, fractionating fragments of the latter on southern blots (described later in this chapter) and hybridizing the transposon with a probe. Cleaving out the transposon, together with flanking DNA, reveals the adjoining sequences of a gene. The probes are made for these adjoining sequences to find the fragment having gene of interest.

Presence of transposon is confirmed by altered phenotype. This method was first employed to isolate *white* locus in *Drosophila*. The mutants with *white-apricot* eyes harboured a *copia* retrotransposon as well as accompanying *white* locus.

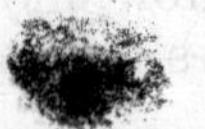

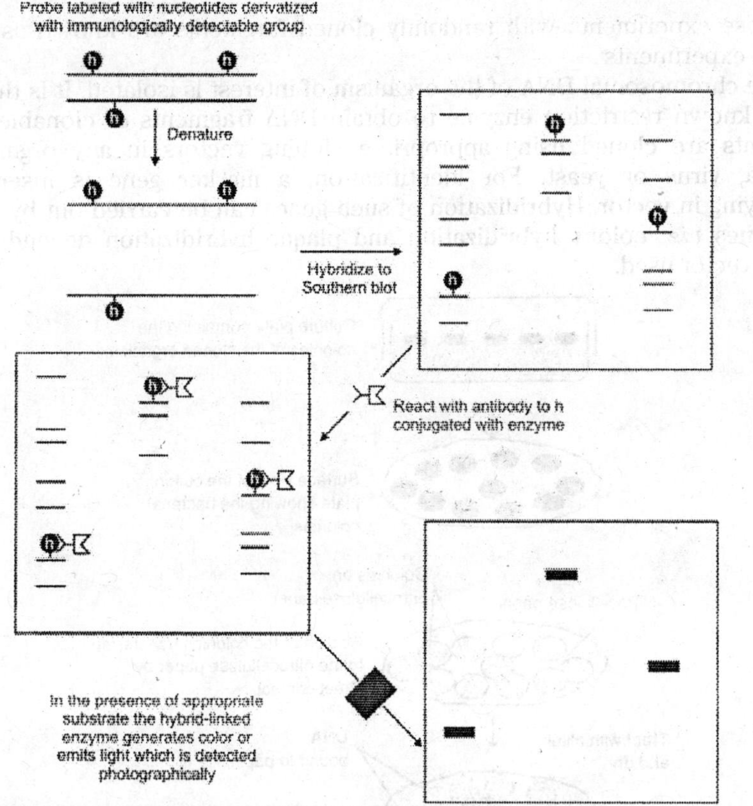

FIG. 36.6: The principle of non-radioactive detection methods for in-situ hybridization.

Genetic Linkage Mapping

This form of mapping is used to construct genetic linkage maps of individual chromosomes and also to localize genes involved in genetic disease. Genetic linkage analysis relies on an ability to estimate the frequency of crossing over (recombination) that occurs between homologous chromosomes during meiosis. The closer two genetic markers are to each other then the lower is the chance that they will be separated during meiosis. The statistical probability that two markers are linked is measured as the logarithm of the odds (Lod). A Lod score of 3 is usually taken as the threshold of significance, while a Lod score below -2 suggests that the two loci are not linked. Lods between -2 and +3 are not significant.

GENOMIC LIBRARY

The complete set of cloned fragments of DNA of an organism is known as genomic library. This is also called gene library. Such a library contains all or most of the genes present in the cell or the organism. It is difficult to identify a particular gene from an organism and then clone it. Hence the organism's entire genome as DNA fragments is cloned and the clone containing the sequence of interest is identified.

Cloning experiments can be divided into two types (1) those where the cloned DNA segment is of unknown composition and function and, (2) those where the cloned DNA segment is of known composition and function. In the first approach the entire cellular DNA has to be fragmented into clonable size by either digestion with restriction endonucleases or, mechanical sheering.

These experiments with randomly cloned fragments are known as 'shot gun cloning experiments'.

The chromosomal DNA of the organism of interest is isolated. It is then treated with a known restriction enzyme to obtain DNA fragments of clonable size. The fragments are cloned using appropriate cloning vectors in any organisms like bacteria, virus or yeast. For identification, a marker gene is inserted while multiplying in vector. Hybridization of such genes can be carried out by two major techniques *viz.* colony hybridization and plaque hybridization depending on the type of vector used.

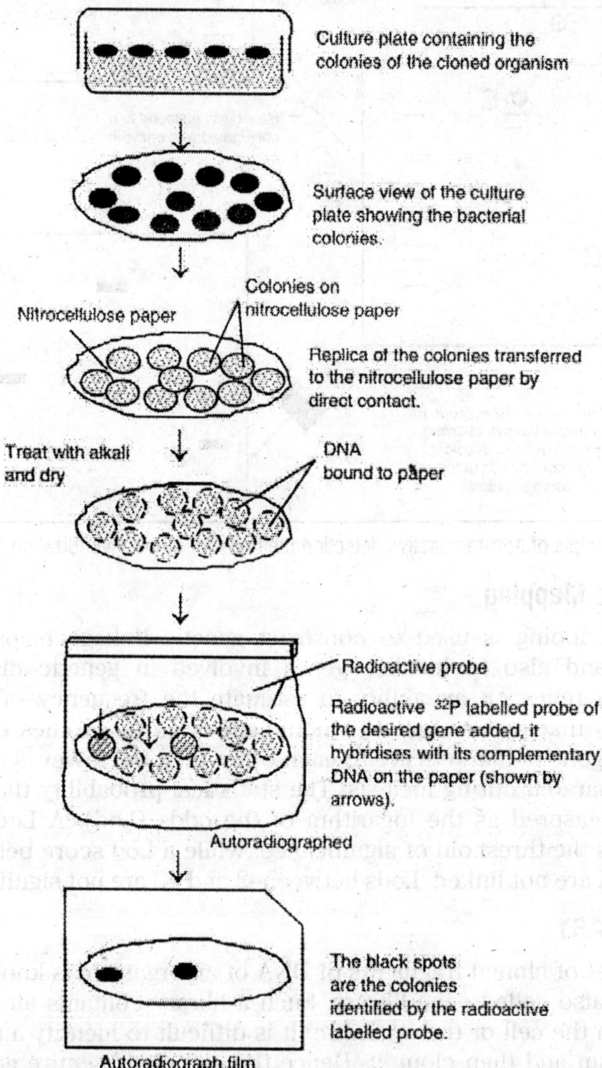

FIG. 36.7: Colony hybridization to identify the clones containing the DNA of interest.

Colony Hybridization

Once a genomic library or cDNA library is available, gene sequence can be isolated by colony hybridization techniques. It involves replica plating the colonies formed by transformed cell into nitrocellulose filter and their *in situ* hybridization with a radioactive probe. This method was first developed by Hogness and Grunestein (1975). Procedure is described in following steps:

1. The cloned yeast colonies, bacterial colonies or phage plaques to be tested are transferred from the culture plate on to a nitrocellulose filter paper by replica plating.
2. The filter with the colony replicas is treated with NaOH to lyses the cells/phages and to denature DNA.
3. The filter is then baked to fix the DNA.
4. It is then treated with a radioactive ^{32}P labelled single stranded DNA probe that is complementary to the ssDNA of interest.
5. The filter is washed to remove the unbound excess probe, dried and then autoradiographed.
6. The DNA fragments hybridized with the radioactive probe can be detected on the autoradiograph film.
7. This indicates that the cells in the corresponding colonies contain the desired gene. The colony is taken out from the master culture plate for mass culture.

Plaque Hybridization

This technique is applicable where a chimeric phage particles are carrying the gene of interest. In such a situation, a bacterial lawn is infected with a mixture of chimeric phage particles (*i.e.* the library) and a large number of plaques develop overnight. These plaques can be treated just like the colonies in colony hybridization to identify and isolate the chimeric phage particle carrying the gene of interest. Two new techniques have shown their potential in gene isolation and characterization. They are chromosome walking and chromosome jumping. Let us have a small account on both of these techniques.

Chromosome Walking

Identification of fragments with overlapping sequences may be a key to the reconstruction or characterization of large chromosome regions. This is achieved by the technique popularly described as chromosome walking.

The technique of chromosome walking involves the following steps:

1. From the genomic library a alone of interest (identified by a probe is selected.
2. One end of the clone is subcloned.
3. The subcloned fragment of the selected clone may be hybridized with other clones in the library and a second clone hybridizing with the subclone of the first clone is identified due to presence of overlapping region.
4. The end of the second is then subcloned and used for hybridization with other clones. This helps in identifying a third clone having overlapping region with the subcloned end of the second clone.
5. Identified third clone is also subcloned and hybridized with clones in the same manner and the procedure may be continued.
6. Restriction map of each selected clone is prepared and compared to know the region of overlapping, so that identification of few overlapping restriction sites resembles the walking along the chromosome or along a long chromosome segment.

Regions of chromosomes approaching 1000 kb have been mapped following the above technique. Restriction maps of entire chromosomes can be prepared in this manner following the technique of chromosome walking.

Chromosome Jumping

In order to bring the molecular marker close to the gene of interest, *chromosome jumping* approach has been utilized. This technique can be described in the following steps:

1. The distance of 'jumps' or 'hopsize' (*e.g.* 100 kb or 200 kb) i s determined.
2. Genomic DNA molecules in the range of selected size are selected through pulsed-field gel electrophoresis.
3. For circularization of DNA segments, ligation between two ends of each long linear DNA molecule is allowed using T4 ligase).
4. These circular DNA circles are digested with *Eco*RI.
5. Phage vector Ch3A lac, is also cut with *Eco*RI and used for cloning small DNA fragments.
6. Such cloned DNA fragments represent the jumping library, which can be plated on a bacterial host and screened through the technique of plaque hybridization.

The technique of chromosome jumping helps in narrowing the gap between the gene and available molecular markers. After several cycles of chromosome jumping followed by cloning the regions broximal to the gene, it will be possible to approach very close to the desired gene capable approached and cloned. The gene for **cystic fibrosis** has recently been isolated using this technique.

CLEAVING DNA WITH RESTRICTION ENZYMES

After locating gene of interest in colony or on chromosome, the DNA is extracted and cleaved with enzymes. Most restriction enzymes cleave DNA molecules in a site specific manner. The genomic DNA from an individual organism is digested with either a single or more restriction enzymes. This results in fragmentation of DNA molecules with variation in length. These fragments are known as restriction fragment length polymorphisms or RFLPs. Since restriction enzymes cut DNA in a sequence specific manner, every homologous DNA molecule will be cut at exactly the same sites. This allows to isolate large quantities of specific DNA fragments for subcloning and isolating the DNA sequence of interest.

The sizes of the restriction fragments can be determined by polyacrylamide or agrose gel electrophoresis. It also helps in separation of various RFLPs. Here few electrophoretic and blotting techniques are described.

Gel Electrophoresis

Nucleic acid molecules can be distinguished by differences in their chain length, and study of gene organization and expression includes nucleic acid hybridization with a chain length measurement.

DNA and RNA have a net negative charge. At neutral pH therefore, these molecules migrate towards the positive terminal when placed in an electric field. This process of electrophoresis is performed in an agarose or polycrylamide gel. Nucleic acids are loaded into slots in the gel and allowed to migrate towards the positive terminal. The pores in the gel act to sieve the molecules. Therefore, the mobility of a particular nucleic acid species depends on its length. All the molecules of a particular size move at approximately the same rate through the gel, forming a band which gradually increases in width during electrophoresis because of diffusion. The size of the molecules can be estimated by running DNA molecules of known size in a parallel track on the gel, and determining their migration position relative to the species of unknown size.

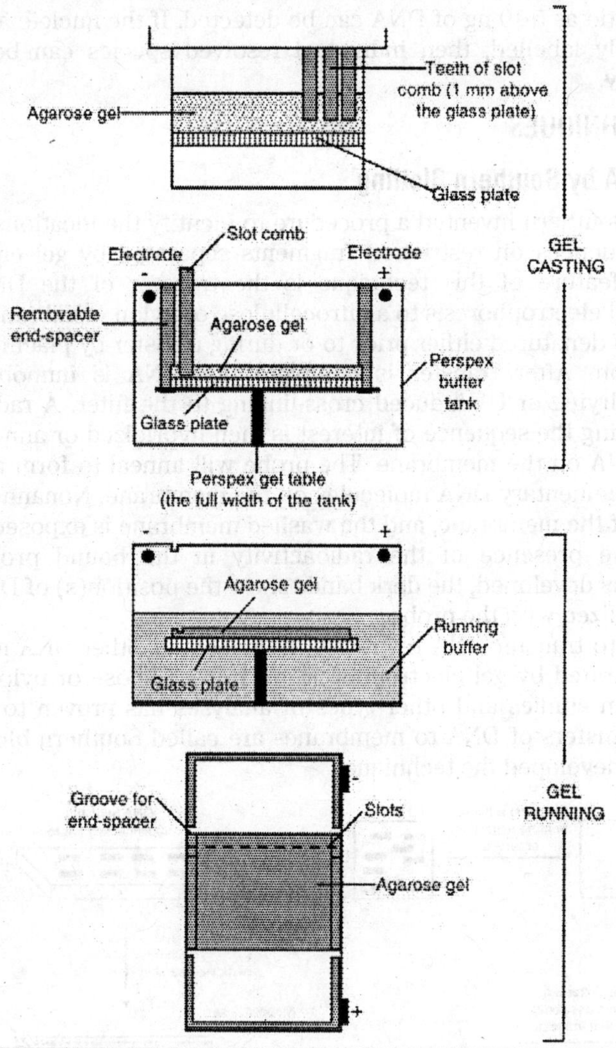

FIG. 36.8: Agarose gel electrophoresis.

Polyacrylamide has a smaller average pore size than agarose and so is effective at separating molecules of 10-1000 nucleotides in length (1000 nucleotides is termed 1 kilobase or 1 kb), with very high resolution. The larger pores in agarose allow it to resolve much bigger molecules, up to 100 kb in length. However, the separated molecules migrate as a wider band so that the resolution (*i.e.* the ability to separate molecules which differ only slightly in size) is lower with agarose than with polyacrylamide. In pulsed-field gel electrophoresis (PFGE) the electrical field across the gel is repeatedly switched back and forth between two or more directions. This allows very long DNA molecules, several million base pairs in length, to be separated, since very big molecules change direction less rapidly as they move through the agarose gel.

After gel electrophoresis, the nucleic acid molecules must be detected. They can be visualized directly by including ethidium bromide in the polyacrylamide or agarose gel. This dye binds to double-stranded nucleic acids by intercalating between adjacent base pairs, and emits a red fluorescence when exposed to UV radiation. As little as 5-10 ng of DNA can be detected. If the nucleic acid molecules are radioactively labelled, then individual resolved species can be detected by autoradiography.

BLOTTING TECHNIQUES

Analysis of DNA by Southern Blotting

In 1975, E. M. Southern invented a procedure to identify the locations of genes and other DNA sequences on restriction fragments separated by gel electrophoresis. The essential feature of this technique is the transfer of the DNA molecules separated by gel electrophoresis to a nitrocellulose or nylon membrane.

The DNA is denatured either prior to or during transfer by placing the gel in an alkaline solution. After transfer is complete, the DNA is immobilized on the membrane by drying or UV-induced cross-linking to the filter. A radioactive DNA (*probe*) containing the sequence of interest is then hybridized or annealed with the immobilized DNA on the membrane. The probe will anneal to form a double helix only with complementary DNA molecules on the membrane. Nonannealed probe is then washed off the membrane, and the washed membrane is exposed to X-ray film that detects the presence of the radioactivity in the bound probe. After the autoradiogram is developed, the dark bands show the position(s) of DNA sequences that have hybridized with the probe.

The ability to transfer DNA restriction fragments or other DNA molecules that have been separated by gel electrophoresis to nitrocellulose or nylon membranes for hybridization studies and other types of analyses has proven to be extremely useful. Such transfers of DNA to membranes are called Southern blots after E. M. Southern, who developed the technique.

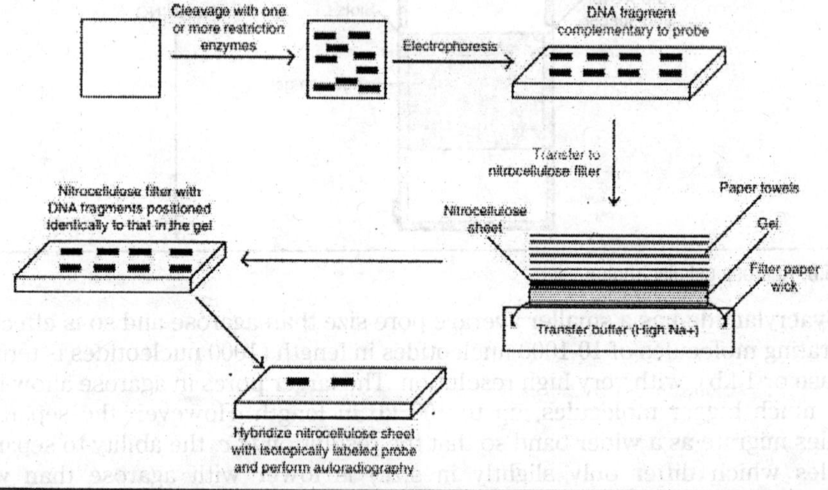

FIG. 36.9: Southern blotting method of analysing DNA segment that share homology with a nucleic acid probe.

Analysis of RNase by Northern Blot Hybridizations

As DNA molecules can be transferred from agarose gels to nitrocellulose or nylon membranes for hybridization studies, RNA molecules can also be separated by agarose gel electrophoresis similarly and transferred for analyses analysed. Such RNA transfers are used routinely in molecular genetics laboratories and are called *northern blots* as it is the mirror image of the Southern blotting technique.

The northern blot procedure is essentially identical to that used for Southern blot transfers however, degradation by RNA molecules are more sensitive to degradation by RNases. Thus, gloves must be worn at all times to prevent contamination of solutions with RNases on one's fingers. In addition, all glassware must be baked overnight to inactivate RNases, and all chemical reagents must be baked or be highly purified to ensure that they are RNase-free. RNase molecules are extremely stable enzymes; they are not inactivated by boiling or other treatments that would totally destroy most enzymes. Most RNA molecules contain considerable secondary structure. Therefore, they must be kept denatured during electrophoresis to separate them on the basis of size. Denaturation is accomplished by adding formaldehyde or some other chemical denaturant to the loading buffer and the electrophoresis running buffer. After transfer to an appropriate membrane, the RNA blot is hybridized to either RNA or DNA probes just as with Southern blots. Now an alternative to nitrocellulose, DBM or diazobenzyloxymethyl paper is used. It may be reused several times for different probes, without appreciable loss of nucleic acid signal.

Northern blot hybridizations are extremely useful in studies of gene expression. They can be used to determine whether a particular gene is transcribed in all tissues of an organism or only certain tissues. They can also be used to study the temporal pattern of expression of individual genes during growth and development. Major disadvantage with northern blot hybridizations is that it only measure the accumulation of RNA transcripts. But explains nothing about why the observed accumulation occurs.

Analysis of Proteins by Western Blot Techniques

Polyacrylamide gel electrophoresis has been used as an important tool for the separation and characterization of proteins.

Since many functional proteins are composed of two or more subunits, individual polypeptides are separated by electrophoresis in the presence of the detergent sodium dodecyl sulphate (SDS), which denatures the proteins.

After electrophoresis, the proteins are commonly detected by treating with Coomassie blue or silver stain. However, the separated polypeptides in the gel can also be transferred to a nitrocellulose membrane, and individual proteins can be detected by using specific antibodies. This transfer of proteins from acrylamide gels to nitrocellulose membranes is called western blotting and is performed by using an electric current-so-called *electroblotting*. After transfer, a specific protein of interest is identified by placing the membrane with the immobilized proteins in a solution containing an antibody to the protein. Non-bound antibodies are then washed off the membrane, and the presence of the initial ("primary") antibody is detected by placing the membrane in a solution containing a "secondary" antibody. This "secondary" antibody reacts with immunoglobulins (the group of protein containing all antibodies) in general. The secondary antibody is conjugated to either a radioactive isotope (permitting autoradiography) or an enzyme that produces a visible product when the proper substrate is added. The western blot procedure is a very powerful tool for identifying and characterizing specific gene-products.

Dot Blots and Slot Blots

Another variety of Southern or Northern blots is dot blots or slot blots. In dot blots, cloned or purified extract of DNAs is tested by spotting them adjacent to each other on a nitrocellulose membrane. DNA blots thus produced are immobilized and denatured so that they are bound on the membrane as single stranded DNA blots. The membrane is then hybridized with readioactively labelled probe (DNA or RNA).

The dot representing sequence related to probe will light up by autoradiography. The intensity of the dot will indicate the relative concentration of a sequence related to the probe in the DNA sample used for a particular dot. No steps involving digestion with enzyme, gel electrophoresis or transfer from gel to membrane are needed in this technique. Dot blot are useful in screening viral genomic RNA, RNA transcripts and especially in samples where restriction sites are not considered.

Sometimes, an apparatus is used for placing spots on the membrane, through slots made in this equipment. The spots made thus are in the form of oblong slots rather than round blots. These slots are used just like dot blots and are described as slot blots.

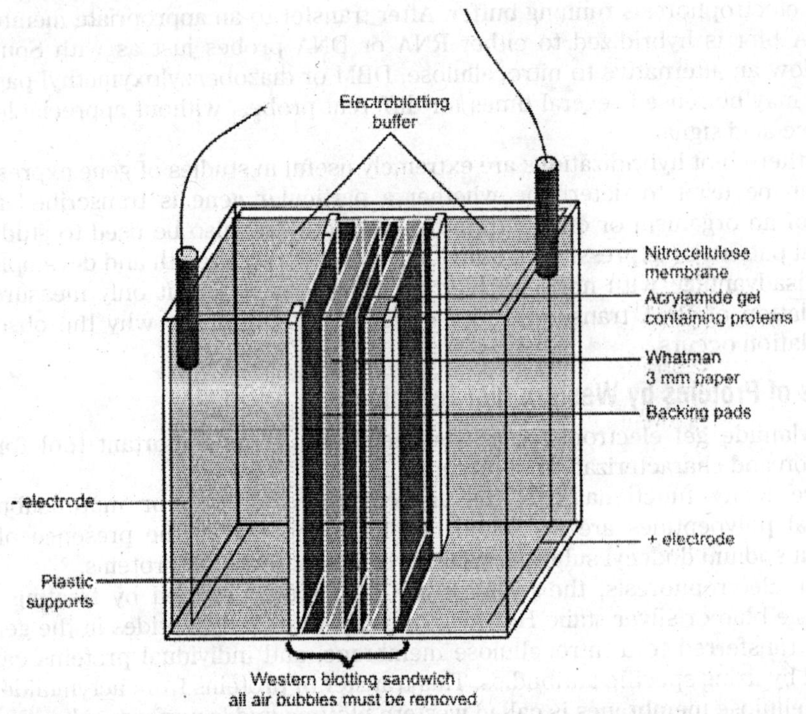

FIG. 36.10: Diagram of a typical Western blotting or electroblotting apparatus. An electric current is used to transfer the proteins from the gel to the nitrocellulose membrane placed next to it in the blotting sandwich. All other components of the sandwich function to provide gentle but firm support; tight contact between the gel and the membrane is essential for good transfer. All bubbles must be carefully removed from the sandwich to assure uniform flow of current and complete transfer of all proteins.

DETECTION OF RFLPS

A RFLP can be demonstrated using the following steps:

1. Extract and purify DNA from several individuals which may differ in some respects among themselves.
2. Digest DNA samples with a restriction endonuclease.
3. Subject each DNA digest to gel electrophoresis separately, but on the same gel slab.
4. The DNA fragments of different lengths resulting from digestion can be separated by gel electrophoresis, but the differences in the distribution patterns of fragments in several DNA samples due to digestion by several endonucleases

can not be detected directly. This is because the number of fragments is large and the range in size is rather continuous, such that they form a continuous smear on the gel. Electrophoresis is, therefore, followed by the following steps.

5. Fragments are transferred from gel to nitrocellulose filters using the technique of **Southern blotting**.
6. The filters with small fragments, fully or partially homologous to the probe, can be detected by autoradiography after hybridization.

With a sufficient number of digests, it is possible to deduce the origin of each band on the gel, that is, which enzyme or combinations of enzymes have generated it. This yields a restriction enzyme cleavage map or restriction map. Obtaining such a map is very often essential first step in analysing a piece of DNA of unknown structure. Another method is by labelling with probe followed by autoradiography. Since DNA clones having repetitive DNA may hybridize with many fragments, they will give a smear on autoradiography. Therefore, unique DNA sequences are generally used as probes for detecting RFLPs. The DNA fragments hybridizing with the probe, can be visualized by autoradiography, and will be called RFLPs at the phenotypic level. RFLPs can be studied in a set of related species, using a random or a specific DNA probe. The similarities and differences be used to infer phylogenetic relationships. This has been done in a number of cases, both in plants and in animals. RFLPs have also been used to prepare chromosome mops in humans mice, fruitfly, maize, tomato, lettuce and rice.

DNA SEQUENCING

Restriction enzyme mapping and Southern blotting help establish the gross organization of a piece of DNA, and allow the approximate position of sequences of interest to be determined. The next step in analysis is normally to determine the nucleotide sequence.

Methods for Analysing DNA Sequence

Our present ability to sequence essentially any DNA molecule is the result of four major developments. (1) The most important breakthrough was the discovery of restriction enzymes and their use in preparing homogeneous samples of specific segments of chromosomes. (2) A second major advance was the improvement of gel electrophoresis procedures to the point where DNA fragments that different in length by a single nucleotide could be resolved. (3) The development of gene cloning techniques greatly facilitated the preparation of large quantities of a particular gene or DNA sequence of interest for use in sequencing studies. (4) Finally, two different approaches to the rapid sequencing of DNA molecules (e.i. by chemical degradation and by chain termination) were developed in the period from 1974 to 1977.

Sequencing by Chemical Degradation

Determination of DNA sequence depends upon the resolution obtained from polycylamide gels which are run in the presence of denaturant *e.g.* urea. If two single-stranded DNA molecules, identical in sequence over most of their length but differing in just one base on one of the two molecules, are subjected to electrophoresis they will be separated. In the chemical degradation procedure a DNA restriction fragment is isotopically labelled with P at either its 5' or its 3' termini and each of the two strands is separated and isolated. The DNA is then partially modified with chemical reagents specific for the different bases, and cleaved at the modified nucleotides. This generates a set of molecules differing in length but with the same isotopically labelled terminus. These fragments are subjected to denaturing gel electrophoresis in parallel slots of high resolution gel, and the sequence of the DNA can be deduced from the resulting autoradiogram.

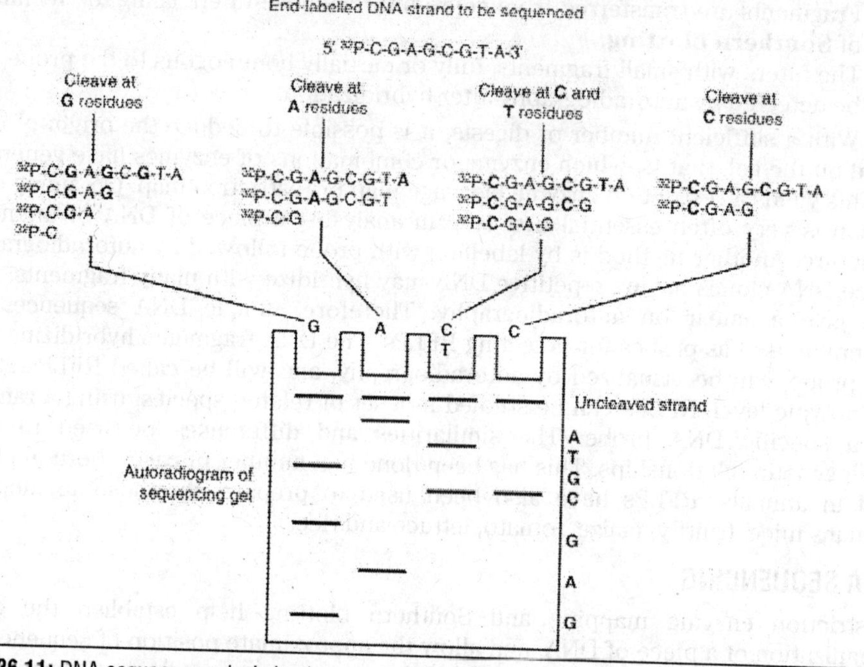

FIG. 36.11: DNA sequence analysis by the chemical degradation (Maxam and Gilbert) technique.

Sequencing by Chain Termination

In practice the chemical degradation procedure has been replaced by chain termination method. In this procedure DNA is synthesised rather than degraded, but the principle is similar to the chemical method. DNA polymerases function by adding a primer, a short synthetic oligonucleotide, to pre-existent DNA strand, to provide a fixed 5' terminus for the growing chain.

The reaction is performed in the presence of chain terminators. These nucleotide derivatives act as blocking group at their 3' hydroxyl terming and once there are incorporated into the growing chains, they prevent further addition of nucleotides. For example, it dideoxy GTP, is mixed with un-modifies GTP and the other three nucleoside triphosphates, terminates at a subset of G residues in the newly synthesised strand. Four parallel reactions are run with dideoxy radio-labelled nucleoside triphosphates of all four kinds are included. The nucleotide sequence of the template strand can be deduced by comparison of the pattern of autoradiographic bands in the four lanes of the gel.

A suitable template for single-strand DNA sequence analysis can be created by inserting the double-stranded template DNA into the genome of a bacterial virus which naturally packages only one of the two strands into viral particles. Another method is denaturing of double-stranded DNA template by treatment with alkali. An excess of primer is added before re-annealing occurs, so that primer will compete with the complementary strand for annealing to the template. This is a preferable method over insertion of DNA into viral vector.

DNA sequence analysis using chain terminators is best performed using a single-stranded template. This allows the primer to be free accessible to the template.

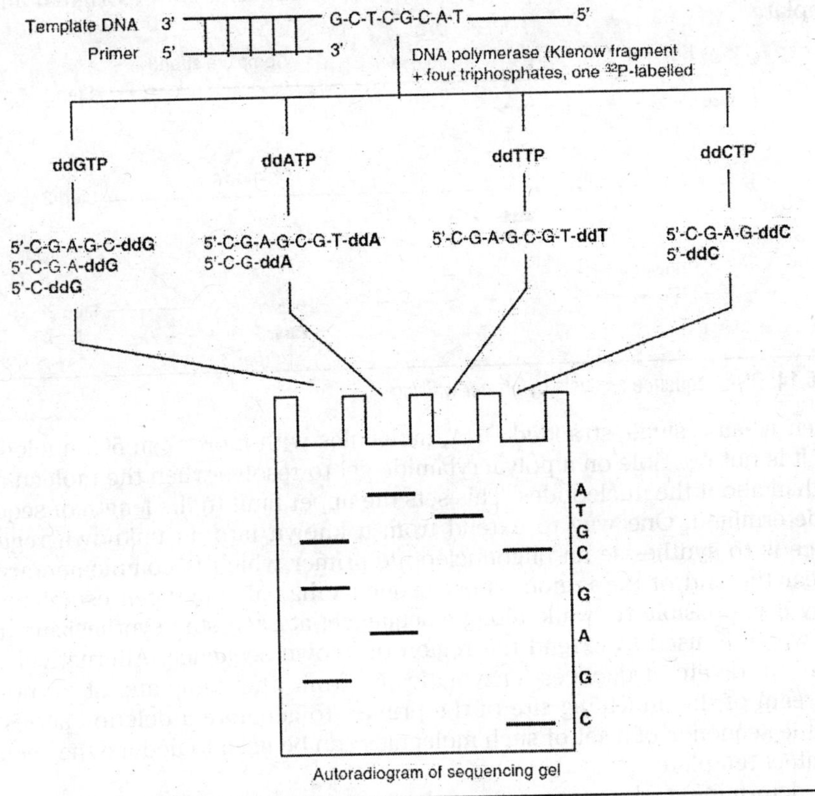

FIG. 36.12: DNA sequence analysis by the chain termination (Sanger) method.

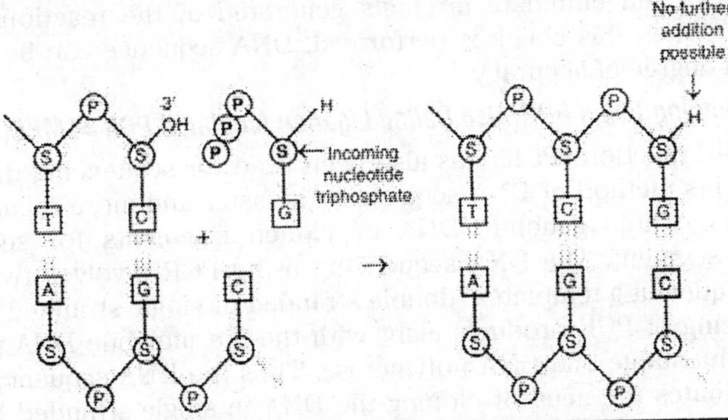

FIG. 36.13: The chemical basis of sequence analysis using chain terminators. This is a representation of a reaction in which a dideoxy G residue incorporated into the growing chain blocks further incorporation.

The major disadvantage of double-stranded DNA sequence analysis is the presence of competing complementary strand. This leads to a higher level of sequencing artifacts, like presence of more than one band. In such situation it may be impossible to decide, which is the correct residue at that position. Artefacts of this kind are not restricted to double-strand DNA sequence analysis; some level of artefacts is a universal problem in sequence analysis.

The DNA sample then re-sequenced using another primer, or by using the other strand (*i.e.* the strand which is complementary to that used in the original analysis) as template.

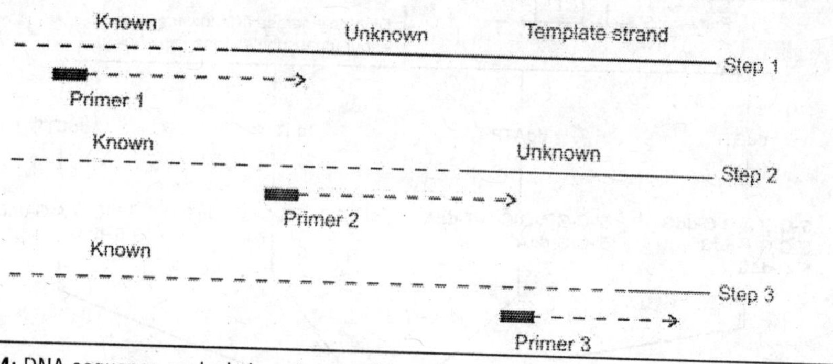

FIG. 36.14: DNA sequence analysis by *primer walking*.

Even when a single-stranded DNA, molecules with more than 500 nucleotides, clearly it is not possible on a polyacrylamide gel to resolve when the molecules are larger than about the nucleotides. This sets the upper limit to the length of sequence to be determined. One way to extend from a known into an unknown region of sequence is to synthesise an oligonucleotide primer which is complementary to a tract near the end of the region whose sequence has already been established. In this way it is possible to 'walk' along a sequence, at each step synthesizing a new primer which is used to extend the region of known sequence. Alternatively, it is possible to delete sequences enzymatically from the template at a position downstream of the annealing site of the primer, to generate a deletion series. The nucleotide sequence of a set of such molecules can be used to deduce the sequence of the intact template.

By determining the sequence of both strands of a DNA molecule, and confirming that the two deduced sequences are perfectly complementary, it is possible to detect and eliminate artefacts generated in the reactions or during electrophoresis. When this check is performed, DNA sequence can be established with a very high degree of accuracy.

Direct DNA Sequencing Using PCR(Also Called Ligation Mediated PCR=LMPCR)

Polymerase chain reaction (PCR) has also been used for sequencing the amplified DNA product. This method of DNA sequencing is faster and more reliable and can utilize either the whole genomic DNA or cloned fragments for sequencing a particular DNA segment. The DNA sequencing using PCR involves two steps: (i) generation of sequencing templates (double stranded or single stranded) using PCR and (ii) sequencing of PCR products eight with the thermolabile DNA polymerase or with the thermostable Taq DNA polymerase. Thus the DNA sequencing method using PCR eliminates the need of cloning the DNA in single stranded DNA phage vector *i.e.* M13. Although double stranded DNA product of PCR can be utilized for sequencing, this may cause difficulty due to reassociation after denaturation thus preventing the sequencing primer from annealing to its complementary sequence to allow extension. To reduce this problem, either a variant of the standard method for sequencing double stranded DNA is employed or single stranded DNA templates are produced following **asymmetric PCR** earlier discussed in Chapter 4.

A number of thermolabile DNA polymerases have been used for sequencing of *in vitro* amplified DNA. Alternatively, **Taq polymerase** (thermostable enzyme) used for PCR can also be used for sequencing reaction. In either case, Sanger's synthetic method involving incorporation of dideozynucleotide (ddNTP) for chain termination is used for sequencing. However, Maxam and Gilbert's chemical degradation method can also be used.

In Sanger's method, as usual, four mixtures are prepared, each using one of the four ddNTP. The sequencing primer is labelled with ^{32}P and the mixtures with amplified DNA, Taq polymerase and appropriate buffer are incubated at 70 °C for 5 min. The reaction is stopped by addition of formamide stop solution in each tube and mixtures are run on polyacrylamide sequencing gel to obtain the ladders, which can be read by computer or manually. This method using PCR helped in automation of DNA sequencing.

Automated DNA Sequence Analysis

Using the chain termination method, DNA sequencing can be mechanized to make the process rapid. In a conventional sequence reaction described above four separate singlestrand extension reactions are performed, with each containing a different chain-terminating dideoxynucleoside triphosphate. The radio-labelled reaction products are separated on four separate tracks of a polyacrylamide gel. In machine-based sequencing a single reaction is run, with each of the four dideoxynucleoside triphosphates containing a different fluorescent label.

When excited by an argon laser, the fluorescent labels emit light at a different wavelength. The same chain-termination reactions are carried out; however, it is no longer necessary to run the products in separate lanes of a gel. The reaction products are mixed and run in a single lane in a gel and as the different length fragments pass the laser each fluorescent label gives a unique signal which is stored in the computer.

Machine-based DNA sequencing is beginning to change the speed and scope of sequencing projects dramatically. The first complete genomic sequences were those of simple viruses such as the SV40 tumor virus, containing around 500 based of DNA. Recently, the DNA sequence of over 106 000 bases of DNA from human chromosome 19 was determined in a few months using DNA sequencing machines. The sequence of a significant fraction of the *E. coli* genome has been established, the entire sequence of yeast chromosome 3 is known, and there are plans eventually to determine the sequence of the entire human genome.

MECHANICAL SHEARING OF DNA

In addition to digestion DNA with restriction endonucleases to produce discrete fragments, there are a variety of treatments which result in non specific breakage. Non specific endonucleases and chemical degradation can be used but the only method that has been much applied to gene manipulation involves mechanical shearing.

The long, thin threads which constitute duplex DNA can not be easily broken by shear forces in solution. Intense sonication with ultrasound can reduce the length to about 300 nucleotide pairs. More controlled shearing can be achieved by high-speed stirring in a blender. Breakage occurs essentially at random with respect to DNA sequence. The termini consist of short, single-stranded regions which may have to be taken into account in subsequent joining procedures.

TRANSCRIPT MAPPING TECHNIQUES

Analysing the nucleotide sequence of a piece of DNA is a necessary preliminary to understanding its precise organization and function, but is not normally sufficient for these purposes. Because eukaryotic gene transcripts are subjected to a complex series of post-transcriptional processing events, the structure of a gene can be understood only by correlating the organization of the genomic DNA with the organization of its mRNA transcript. This has required the establishment of sophisticated cloning and mapping techniques which can, when used in conjunction, give a complete description of the structure of a gene.

It is possible to isolate, by gene cloning, the double-stranded DNA copies of individual mRNA sequences. These cDNA clones will normally contain a DNA copy of the poly(A) sequence at the 3' terminus. Hence, by determining which strand of the cDNA clone contains a poly(A) tract at one end, it is possible to identify the coding strand of the cDNA clone. Provided a sufficient length of sequence has been established to make their presence statistically likely (in practice 300-500 nucleotides), termination codons will occur in two of the three potential coding frames, ruling them out as translational open reading frames (ORFs).

Sometimes this level of analysis, that is, identification of the reading frame and deduction of the encoded protein sequence, will be all that is required. However, it is often desirable to determine the structure of the gene itself. It will have been isolated as part of a genomic clone. This will contain the gene and a variable amount of flanking DNA depending upon the method of cloning used.

By correlating the sequence of the gene with the sequence of the mRNA deduced from the cDNA clone, it is possible to identify the site of poly(A) addition and the positions of introns. However, technical limitations in the cDNA cloning procedure ensure that the cDNA clone will not normally contain sequences derived from the extreme 5' terminus of the mRNA. There are two potential methods to determine the position of the 5' end, primer extension and S1 mapping.

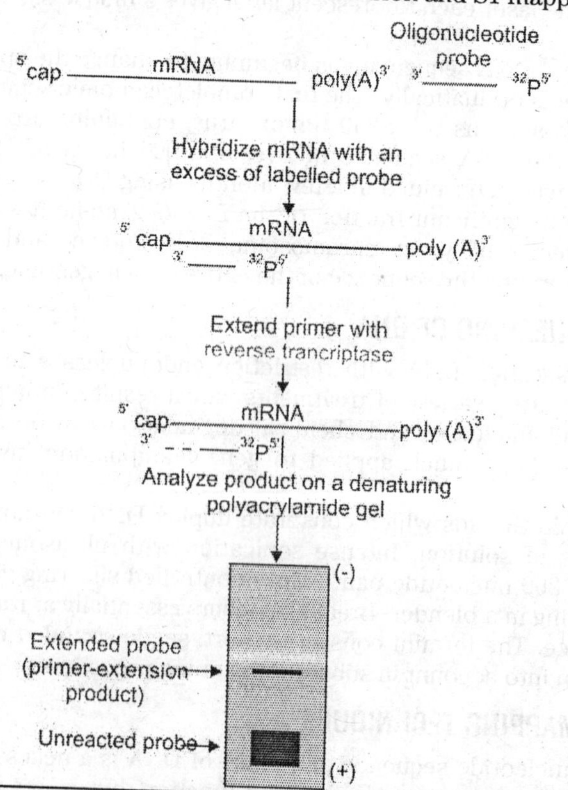

FIG. 36.15: Identification of the 5' end of an mRNA by primer extension.

Primer Extension

In primer extension, a short oligonucleotide complementary to a sequence tract near to the 5' end of the mRNA is annealed to the mRNA. This is then used to prime synthesis with the enzyme reverse transcriptase. This enzyme produces a DNA copy of the mRNA (a cDNA) which extends to the 5' terminus of the mRNA. By correlating the length of the cDNA primer extension product with the sequence of the genomic DNA, it is possible to deduce the start site of transcription.

S1 Mapping

The alternative technique is called S1 mapping. An isotopically labelled probe is generated from the genomic clone. This probe is complementary to the mRNA over a region encompassing part or all of the first exon, and extending upstream of the region where the cap site is expected to be located. The cap site may have already been putatively localized using primer extension. The probe is annealed to the mRNA and digested with an enzyme which destroys single-stranded nucleic acids but leaves double-stranded molecules intact. In early experiments an enzyme called nuclease S1 was generally used, and the name S1 mapping has been retained, even though other nucleases are now sometimes used. During digestion, the 3' proximal region of the probe, which has no complement in the mRNA molecule and which therefore remains single-stranded, is degraded. A molecule is generated which is the complement of the region between the cap site and the position at which the 5' terminus of the probe annealed to the mRNA. Again, by determining the length of this molecule it is possible to infer the position of the cap site.

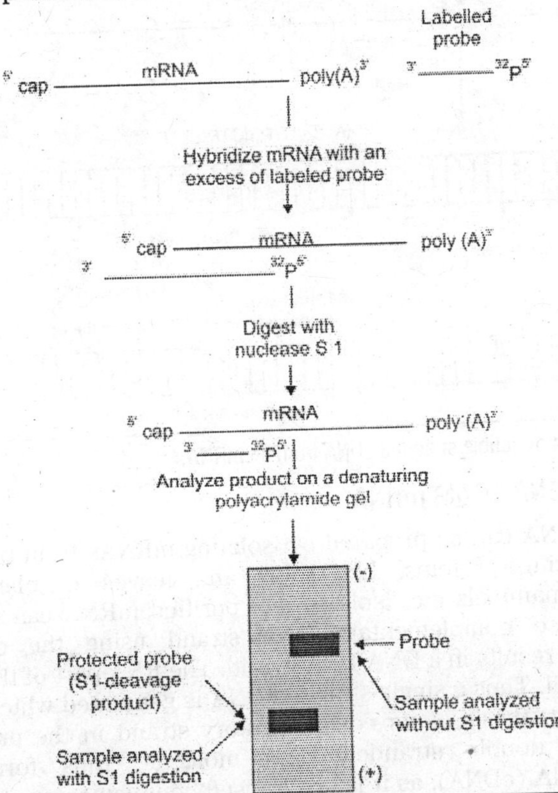

FIG. 36.16: Identification of the 5' end an mRNA by S1 mapping.

These two techniques are often used in parallel to determine the cap site. They are both subject to artefacts, but of different kinds. Therefore if both methods agree, it greatly increases confidence in the assignment of the cap site. If the hybridization is performed with an excess of probe, the amount of primer extension or nuclease-resistant product is proportional to the amount of primer extension or nuclease-resistant product is proportional to the amount of complementary mRNA. Both techniques can therefore be used an alternative to Northern transfer to quantitate mRNA levels.

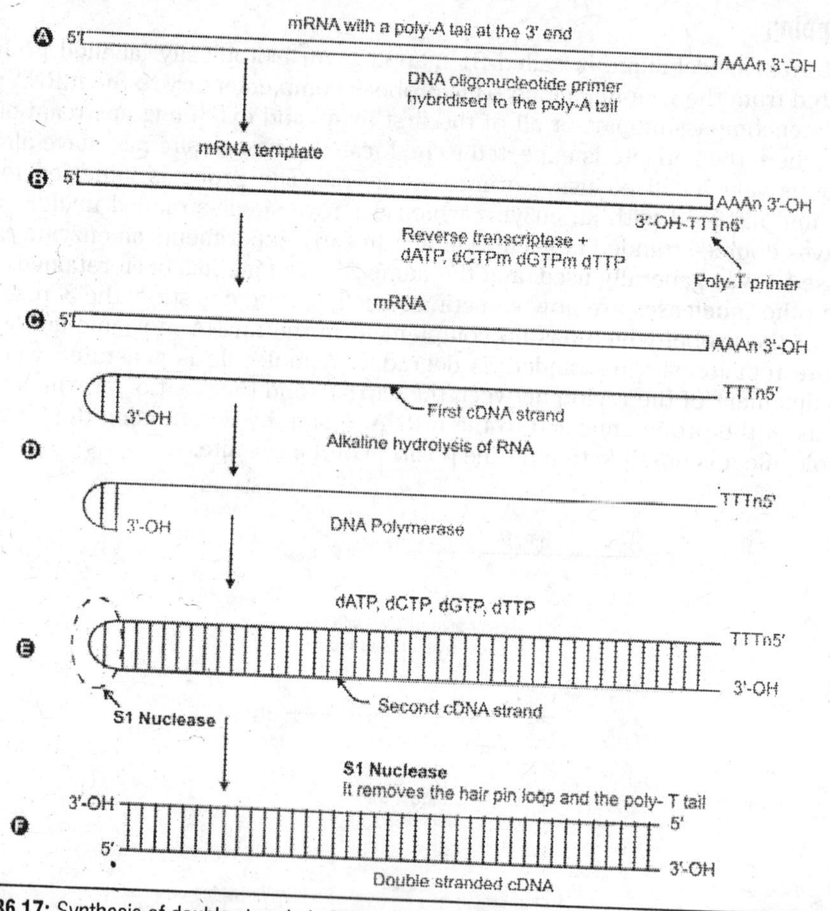

FIG. 36.17: Synthesis of double stranded cDNA from an mRNA.

SYNTHESIS OF CDNA FROM mRNA

Complementary DNA can be prepared by isolating mRNAs from tissues which are actively synthesizing proteins, like roots and leaves in plants, ovaries or recticulocytes in mammals etc. Isolated and purified mRNA can then be directly used to synthesise complementary DNA strand using the enzyme reverse transcriptase. This results in a DNA/RNA hybrid. The RNA part of the hybrid is then removed with alkali. Thus a single-stranded DNA is generated which can act as the template for the synthesis of its complementary strand in the presence of DNA polymerase. The double stranded DNA molecule thus formed is called complementary DNA (cDNA), as it is synthesised complementary to the mRNA. In the prokaryotes the double stranded cDNA will be identical to the gene (DNA) from which the mRNA is formed. But, most of the eukaryotic genes contain intervening sequences (introns) which interrupt the coding sequences. Hence, the cDNA of the eukaryotic cells and their native DNA are not colinear.

As bacterial genes do not have the intron - exon organisation found in the eukaryotic genes they find the expression of eukaryotic genes in them (bacterial) difficult. Therefore one has to depend upon cDNA of eukaryotes, which does not have introns, for cloning.

Preparation of Double-Stranded cDNA

The mRNA that is specific for a particular kind of protein is isolated and purified. All naturally occurring mRNAs of eukaryotes have about 200 adenylate residues at their 3'-termini. This is called polyadenylate tail. The enzyme reverse transcriptase can mediate the synthesis of a new DNA strand only when it has a growing point in the form of a pre-existing primer, which is base paired with the mRNA template and has a free 3' OH group. Such a primer is provided by hybridizing a short complementary oligo (dT) molecule with the poly adenylate (poly-A) sequence.

The 3'OH end of the poly-T primer is made double stranded by reverse transcriptase forming RNA-DNA hybrid. The RNA strand of this RNA - cDNA hybrid is digested by alkaline hydrolysis (DNA is resistant to alkali). At the 3'-OH end of the newly formed single stranded cDNA, a hairpin is formed which is stabilised by base - pairing. This hairpin, which has a 3'-OH free end acts as a self-priming structure. Forming second strand of the cDNA in the presence of the fragment of the *E. coli* DNA polymerase-1 (Kornberg enzyme) and the four deoxynucleotide triphosphates.

The enzyme reverse transcriptase or T4DNA polymerase can also be used for the synthesis of the second strand of cDNA. The synthesised double stranded DNA is treated with the single-strand specific nuclease S1. This enzyme digests the curved region of the hair pin and any polynucleotide tails. Thus, finally a double stranded cDNA molecule is formed. This DNA can now be inserted into a suitable vector and can be further cloned in the desired host.

Complementary DNA of a given mRNA is inserted into a suitable vector can then be cloned in a host. The cDNA is blunt ended. It has no cohesive ends that are complementary to the cohesive ends of any of the vectors. This is overcome by homopolymer tailing. Once inside the host cell, the cDNA in the hybrid will express the gene(s) it carries.

Cdna Library

Complementary DNAs (cDNAs) of the genome of an organism can be synthesised and cloned to generate a cDNA library. Thus, cDNA library is a collection of cDNA clones of an organism. The starting material for the synthesis of cDNA is the mRNA. Hence the resulting cDNAs will represent the mRNAs present in the cells from where these mRNAs were isolated. In the case of multicellular organisms there is specialisation of individual cells. For example, a mammal is made up of a large number of cell types: brain cells, mammary cells, blood cells, liver cells, cells of the pancreas, nerve cells etc. Each type of cell will contain the same complement of genes. However, only a particular set of genes will be expressed in a given type of cell. Other genes will remain silent. Only those genes that are expressed are transcribed into mRNA. Hence the cDNA clones of this type of cell will be only a selection of the total number of genes in a cell. Hence for such organisms, cDNA clones of different cellare prepared.

SEARCHING FOR GENE USING COMPUTERS

Although the series of analyses described above are necessary ultimately to establish the organization and mode of expression of a particular gene, genes can be sought by less laborious means using computer programs. These searches for regions of a DNA sequence which look like coding regions flanked by the signals that direct gene transcription and processing of their RNA products. This procedure is of greatest value when large regions of genomic DNA for which cDNA clones are not available are being analysed. It is thus an essential part of the Human Genome Project.

One of the principles, that of searching for long ORFs (= open reading frames), has already been described. Although the sequence (the Kozak sequence) that precedes the ATG initiation codon is only weakly conserved between different genes it can provide helpful guides to the start site of translation. The start site of transcription often lies downstream of a TATA box, and it is also possible to search for the consensus binding sites of other transcription factors.

RNA processing signals can also be used to help establish the likely gene organization. The sequence that signals polyadenylation is well conserved, and splice junctions have a consensus sequence comprising an almost perfectly conserved core (5' GU—AG 3'), with moderately well conserved flanking sequences.

Identification genes by analysis of DNA sequences is an evolving science. The structures of new genes are constantly adding and the sophistication of the search programs is increasing. The databanks exists which contain many thousands of gene sequences *e.g.* Genbank and the EMBL. Although some scientists initially submit their gene sequence data to one rather than other, the two organizations that compile the databases then exchange data.

CHAPTER 37

Genomics

Every organism has a genome. Genome is a term used to describe all of the genes and genetic material that predict the nature of an organism. Every cell of a specific organism has the same copy of its genome. Genomics is the study of an organism's genome. It includes an examination of number of genes in an organism, function of specific genes, influence of one gene on another and The activation and suppression of genes. Genomics also includes the search for genes that cause diseases. When order of nucleotides in a strand of DNA, it is called **sequencing**.

The **genome** consists of one or more long strands of deoxyribonucleic acid – commonly known as DNA. The DNA, in turn, is composed of a string of molecules called nucleotides. There are four types of nucleotides: adenine (A), guanine (G), cytosine (C), and thymine (T). Like in a series of letters that make up a word (e.g. c-a-t = cat), certain combinations or sequences of nucleotides, when interpreted by the cell, have functional meaning. Such sequences – sequences that have meaning – are called genes. Genes carry out their functional meaning through the production of proteins.

The study of genes first began in the mid-1800s when a monk named Gregor Mendel discovered that he could manipulate the physical characteristics of pea plants from generation to generation. As time passed, the hereditary units described by Mendel were later named genes. Eventually, other discoveries were made including the genome, chromosomes, DNA, and nucleotides. By the 1970s, techniques used to sequence DNA were becoming more advanced and in 1977 a virus became the first organism to have its entire genome sequenced. Genetic laboratory techniques continued to improve and by the early 1980s, discussion of sequencing the human genome had begun. The initial sequencing efforts began in the mid-1980s. Two significant factors during this time accelerated the human genome sequencing. The first was a race between two groups (Celera Genomics and a public multinational research group) to be the first to sequence the entire human genome. The other significant factor was an improvement in the sequencing technology. New automated sequencing devices reached the market in the 1990s, as did DNA microarray technology.

The determination of an unknown DNA sequence, it can either cut or copy the DNA. Cutting and copying can be used in combination or alone, depending on the technique. Scientists' ability to control which nucleotide appears beside a given cut, or the end of a given copy, is the key to sequencing. This means the identity of the nucleotide at the end of the cut or copied DNA piece is known. Each nucleotide is then "lined-up" using lab techniques that separate the DNA pieces. Once the pieces are lined up, the sequence can be read.

Sequence information is recorded as a long string of the same four letters representing the combination of nucleotides in a DNA strand. Certain segments of the nucleotide sequence make up a gene. However, finding a particular gene from a long string of nucleotides can be difficult. Additionally, the function of that gene may be unknown. Two new areas of research, intimately related to genomics address these issues:

Bioinformatics is the coupling of electronic information management and biological information. Among other things, this new field of research enables scientists to effectively search and organize genome information. Bioinformatics is the use of computer software and programs to organize, store and analyse biological information and data to better understand biological systems. It is the field of science where biology, computer science and information technology converge as a single discipline. Researchers use advanced computer and statistical techniques to wade through and analyse large amounts of biological data created through modern biotechnologies referred to as the **omics** technologies – such as genomics and proteomics. These "omics" technologies focus on different aspects of biological systems. For example:

1. **Genomics** is the analysis of the entire genetic make-up of an organism. It includes a study of the functions of genes in cells, organs and organisms.
2. **Proteomics** focuses on the total protein make-up of a cell at any given time.

Determining gene function and expression is another major focus of genomics research. Making this research possible is the DNA microarray. The DNA microarray allows scientists to determine which genes are active in a cell and for what functions they are responsible.

Bioinformatics helps in developing new statistics and data to be studied and used to assess relationships among the different sets of biological data. It also helps in analysing and interpreting the various types of data, which include nucleotide and amino acid sequences and the properties of various proteins and developing and implementing tools to enable efficient access and management of different types of pertinent information.

DNA MICROARRAY

In the past, scientists have only been able to conduct genetic analyses on a few genes at once. The development of DNA microarray technology allows scientists to examine thousands of genes at the same time, an advance that will help them determine the complex relationships between individual genes.

One of the most important benefits of DNA microarrays lies in their ability to measure the expression of a large number of genes with great efficiency. Microarrays permit scientists to assess the relative expression levels of hundreds, even thousands, of genes in a single experiment. DNA microarrays are also called DNA chips, gene chips, and bio chips.

DNA microarrays are stamp-sized glass or silicon microchips that are imbedded with thousands of single-stranded DNA, each placed at a specific location on the chip in grid formation. The identities and locations of the single-stranded DNAs are known. The chip is immersed in a solution prepared from a patient's sample (usually blood), and if a particular DNA strand from the solution is complementary to the ones on the chip, a bond will occur on that specific spot. Since scientists know that each grid represents a certain DNA, a bond indicates that that particular sequence of DNA is present in the sample given by the patient.

DNA microarray technology is possible because scientists have found a way to use the natural characteristics of DNAs to bond in a complementary fashion. DNA is made of smaller parts called nucleotides which are strung together into a strand. A fully functioning DNA has two of these strands that are attached to each other at every nucleotide point. There are four different types of nucleotides: adenine, guanine, cytosine, and thymine, commonly known as A, G, C, and T. these nucleotides can only bond from one strand to the other in one way: A always bonds with T, and C always bonds with G. It is this complementarity that makes them useful in diagnostics, like the DNA microarray. DNA microarrays are used extensively to measure the expression level of a particular gene or genes.

Gene Expression

Expressed genes are those that have been turned into proteins, which are the functional products of genes. A gene may code for something to happen, but in order for that something to happen, a protein must be produced to do the work.

Every single cell in the body has the same set of genes. However, different genes are active (and therefore "expressed") in different kinds of cells and tissues. In the process of having certain genes expressed, the cell uses several proteins. One of these is called a messenger RNA (mRNA for short) and it is only produced when a gene is being turned into a protein. This is how a DNA microarray can detect the level of expression of a certain gene - by testing for the presence and level of mRNAs associated with that particular gene.

Changes in a cell's environment can trigger the production of a protein. Once the cell is ready to make a certain protein, a segment of the double-stranded DNA, which is responsible for that code, unravels so that it temporarily becomes two single strands. Then, mRNAs make a complementary copy of one single strand of DNA, nucleotide by nucleotide. The mRNA, which is then an exact complementary version of the DNA strand, travels to another part of the cell where its genetic information is used to construct a particular protein. The more a gene is translated into protein, the more copies of mRNA are present inside the cell. This level of mRNA can be measured by microarrays which are imbedded with a known set of DNA.

DNA Chip

A specific tissue sample is taken from the patient. The mRNA from the tissue is changed into complementary DNA (cDNA), which is easier to work with. At this point, the mRNA and cDNA represent the same information from the patient, just in a different form. This cDNA is then tagged with radioactive molecules so that they can be detected later. This is called **labelling**.

When the DNA microarray is dipped into the solution of cDNA made from the patient's sample, the cDNAs that are complementary to the DNA on the chip will bond. Because scientists know which DNAs are imbedded on the chip, a bond with a certain DNA indicates the expression of a particular DNA sequence in the patient's cell or tissue. Finally, the microarray is inserted into an instrument that can read the presence and location of the radioactive molecules.

Uses of DNA Microarrays

DNA microarrays have many uses. If scientists know that an over-expression of a certain gene is the cause of a disease, they can test a tissue sample to see if this is the case. As well, they can compare the DNA microarray results of two samples, such as one from a normal tissue, and another from a diseased tissue and see which genes are expressed, over-expressed or under-expressed. It is an effective way to find out which genes are involved in certain diseases. It is used for identification of tissue-specific genes because different cells and tissues express different genes. Therefore, the gene expression of different tissues and cells can be compared. It helps in analysing gene regulation in cancer. Sincde, genetic diseases can be the result of too much or too little expression of a certain gene. In cancer, diseased cells may be studied for expression levels of certain cancer-related genes.

It is also used in Pharmaceutical analysis It helps in study the relationship between a person's response to specific drugs and the genetic make-up of that person. As well, more effective drug targets can be studied by analysing the gene expression after toxin or drug exposure.

Other Uses

Researchers believe that each person has about 35,000 genes. The complete set of genes together is known as the human genome, commonly referred to as "the instruction manual" for how the body works. Each gene carries instructions for making proteins, which direct the body's cells and functions.

Most cells have 46 chromosomes--23 from each parent. Chromosomes contain thousands of genes, which are made up of deoxyribonucleic acid (DNA), the chemical material that is inherited.

Genomics is the study of an individual's gene structure, including how the genes interact with each other and with the environment. Experts say genomics has the potential to revolutionize the practice of medicine. That revolution, called "personalized medicine," includes the use of genomic information to improve the diagnosis of disease, as well as the prevention and treatment of disease.

An example of a preventive approach is when a genetic test predicts which diseases an individual is likely to develop. For instance, people who have certain mutations in the BRCA1 gene have a high risk of developing breast, ovarian, and possibly prostate, and colon cancers, according to the National Cancer Institute (NCI). Alterations in the BRCA2 gene have been associated with breast, pancreatic, gallbladder, and stomach cancers. An example of a treatment approach is when a genetic test determines whether a person is among the 10 per cent of those for whom a particular drug is likely to work.

Personalized Medicine

Personalized medicine is to get the best medical outcomes by choosing treatments that work well with a person's genomic profile, or with certain characteristics in the person's blood proteins or cell surface proteins," Frueh says. Genetic information isn't usually meant to be used alone to make treatment decisions, but rather is used with other factors such as the patient's family history, medical history, clinical exam, and other non-genomic diagnostic tests.

Pharmacogenomics

It is the combination of drugs (pharmacology) with genomics is known as pharmacogenomics, the science that allows researchers to predict the probability of a drug response based on a person's genetic makeup. It's about getting the right dose of the right drug to the right patient at the right time.

The science of pharmacogenomics has advanced significantly in the last five years, but it's still in its infancy. There are three main ways that pharmacogenomics is applied:

1. The first is to help predict the appropriate dose of a drug.
2. The second is to target therapy to a subset of a disease. This means picking the most effective drug for the disease subset.
3. The third is to test viral genomics, such as in selecting treatment for HIV based on resistance.

The usual doses of drugs work well for most people. They are sometimes based on weight, age, and kidney function. But for someone who metabolizes a drug quickly, the typical dose may be ineffective and a higher dose may be needed. By contrast, someone who is a slow metabolizer may need a lower dose; the typical dose could cause toxic levels of the drug to build up in the blood.

When one takes medicine, it moves through our body, gets broken down by drug-metabolizing enzymes, and interacts with countless proteins. Genes regulate drug metabolism. Differences in the sequence of a gene can cause differences in enzyme activity, which is a result of enzymes appearing in various forms in individuals. This is why different people process the same drug differently.

The children with leukemia, getting the dose wrong can mean the difference between life and death. One wants to keep harsher treatments away from children whose bodies can't tolerate them or don't need them.

The genetic test for defects in the enzyme thiopurine methyltransferase (TMPT) is also important. The defect prevents patients from metabolizing the anti-cancer drug 6-mercaptopurine (6MP). One patient may need a full dose; another, who has a mutation in the gene, may need less than 10 per cent of that dose. Before this test, some patients needed blood transfusions or to be hospitalized for removal of infections. Now we can avoid some of the toxicity of the drug by testing all the children who come in and making dose adjustments based on their genetic test and response to therapy.

Targeted therapy, the second major aspect of pharmacogenomics, is also referred to as "tumor genomics. Tumours have different genomic variations, and genomic tests are helping doctors to identify cancers that are likely to respond to a particular treatment.

Pharmacogenomics also helps in testing for drug resistance. For example, the HIV virus genome is always changing, and resistance testing can help doctors choose the drug that will best match the virus and suppress it. Therefore, pharmacogenomics helps the consumers and availability of drugs that have a greater chance of benefit in terms of treating illness, because, consumers want effective treatment and minimal side effects.

CHAPTER 38

Gene Expression

Gene expression is the process by which the information contained in genes is converted into molecules that determine the properties of cells and viruses. The transfer of genetic information from DNA into protein constitutes gene expression. The information transfer is accomplished by a series of events in which the sequence of bases in DNA is first copied into an RNA molecule and then the RNA is used, either directly or after some chemical modification, to determine the amino acid sequence of a protein molecule. The principal steps in gene expression can be summarized as follows:

RNA molecules are synthesised enzymatically by *RNA polymerase*, which uses the base sequence of a segment of a single strand of DNA as a template in a polymerization reaction similar to that used in replicating DNA. The overall process by which the segment corresponding to a particular gene is selected and an RNA molecule is made is called transcription.

In eukaryotes, the RNA usually undergoes chemical modification in the nucleus called processing.

Protein molecules are then synthesised by the use of the base sequence of a processed RNA molecule to direct the sequential joining of amino acids in a particular order, and so the amino acid sequence is a direct consequence of the base sequence. The production of an amino acid sequence from an RNA base sequence is called translation, and the protein made is called the gene product.

PROTEINS AND AMINO ACIDS

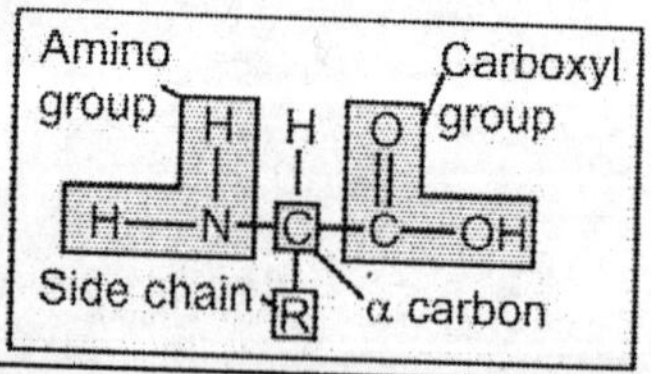

FIG. 38.1: A general structure of an amino acid.

Proteins are the molecules responsible for catalysing most intracellular chemical reactions (enzymes), for regulating gene expression (regulatory proteins), and for determining many features of the structures of cells, tissues, and viruses (structural proteins). The proteins are composed of one or more chains of amino acids. Each of these chains is a series of covalently jointed amino acids that constitute a polypeptide. The 20 different amino acids commonly found in polypeptides can be joined in any number and in any order. Because the number of amino acids in a polypeptide usually ranges from 100 to 1000, an enormous number of different protein molecules can be formed from the 20 common amino acids. Each amino acid contains a carbon atom (the α carbon) to which is attached one carboxyl group (-COOH), one amino group ($-NH_2$), and a side chain commonly called an R group. The R groups are generally chains or rings of carbon atoms bearing various chemical groups. The simplest side chains are those of lycine (-H) and of alanine (-CH_3). For each amino acid, the R group is indicated by a gold rectangle.

Polypeptide chains are formed when the carboxyl group of each amino acid becomes joined with the amino group of the next amino acid in line; the resulting chemical bond is an ordinary covalent bond called a peptide bond. Thus the basic unit of a protein is a polypeptide chain in which α-carbon atoms alternate with peptide groups to form a backbone that has an ordered array of side chains.

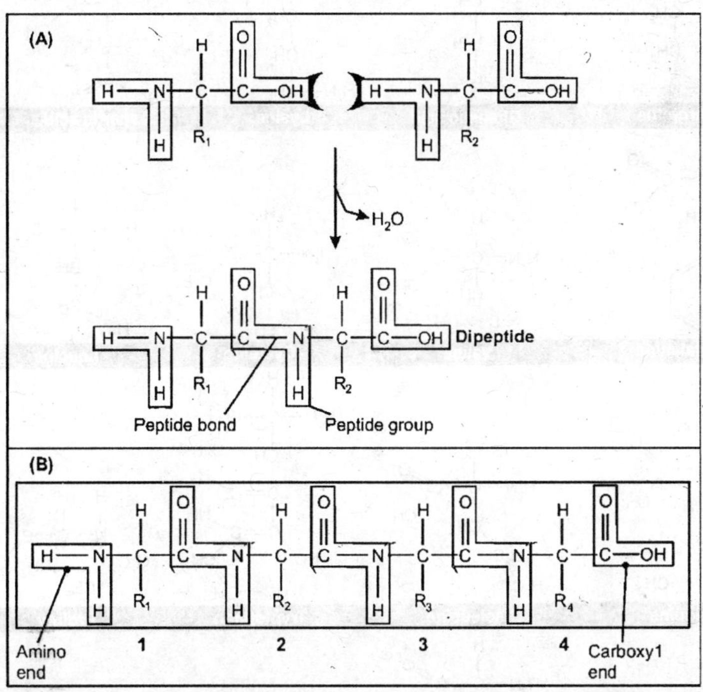

FIG. 38.2: Properties of a polypeptide chain. (A) Formation of a dipeptide by reaction of the carboxyl group of one amino acid (left) with the amino group of a second amino acid (right). A molecule of water (HOH) is eliminated to form a peptide bond (red line). (B) A tetrapeptide showing the alternation of α-carbon atoms (black) and peptide groups (blue). The four amino acids are numbered below.

The two ends of every polypeptide molecule are distinct. One end has a free –NH_2 group and is called the amino terminus; the other end has a free – COOH group and is the carboxyl terminus. Polypeptides are synthesised by adding successive amino acids to the carboxyl end of the growing chain. Conventionally, the amino acids of a polypeptide chain are numbered starting at the amino terminus. Most polypeptide chains are highly folded and a variety of three-dimensional shapes have been observed. The manner of folding is determined primarily by the sequence of amino acids – in particular, by noncovalent interactions between the side chains – so each polypeptide chains tends to fold into a unique three-dimensional shape as it is being synthesised.

In some cases, protein folding is assisted by interactions with other proteins in the cell called chaperones. Generally speaking, the molecules fold so that amino acids with charged side chains tend to be one the surface of the protein (in contact with water) and those with uncharged side chains tend to be internal. Specific folded configurations also result from hydrogen bonding between peptide groups. Two fundamental polypeptide structures are the α helix and the β sheet. The α helix, represented as a coiled ribbon, is formed by interactions between neighbouring amino acids that twist the backbone into a right-handed helix in which the N-H in each peptide group is hydrogen-bonded with the C-O in the peptide group located four amino acids further along the helix.

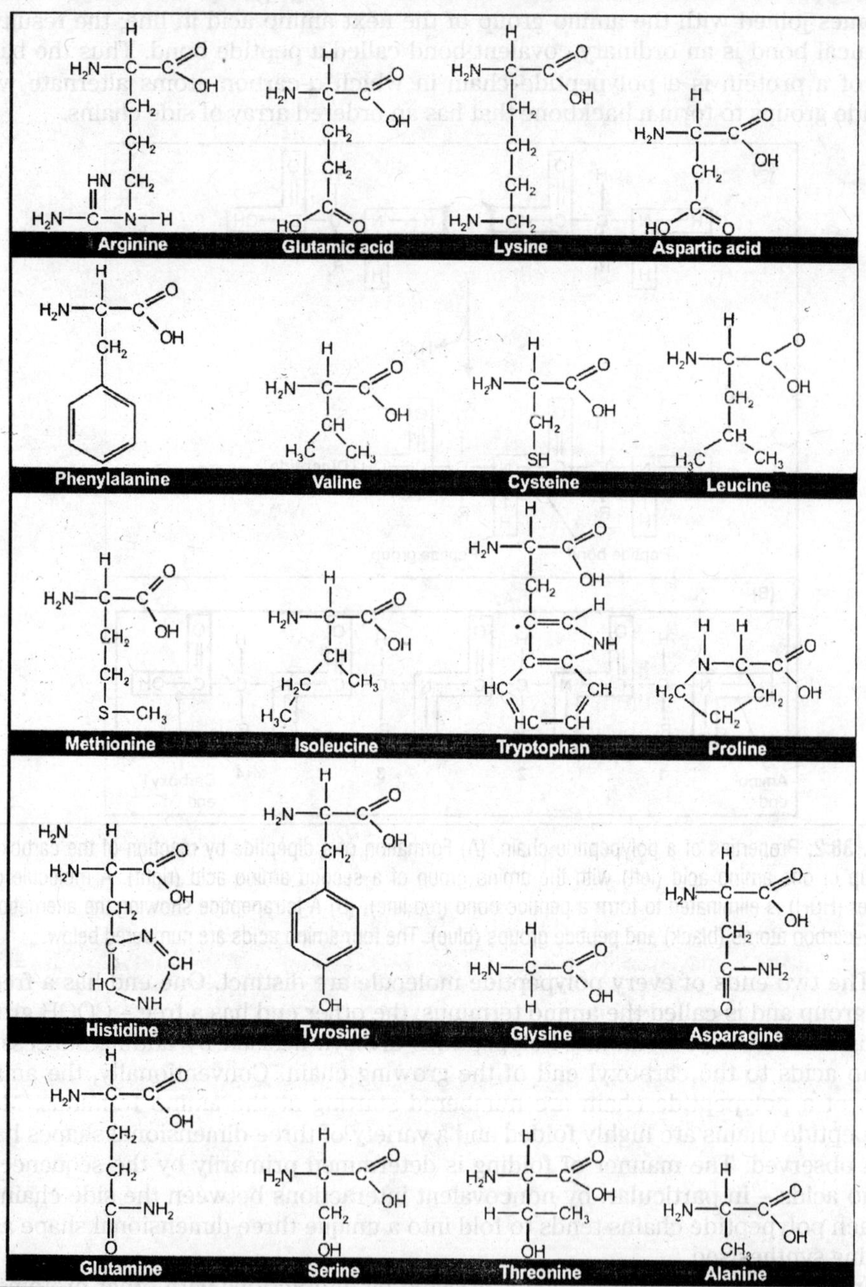

FIG. 38.3: Chemical structures of the amino acids. Note that proline does not have the general structure.

In contrast, the β sheet, represented as parallel "flat" ribbons 4, is formed by interactions between amino acids in distant parts of the polypeptide chain; the backbones of the polypeptide chains are held flat and rigid (forming a "sheet"), because alternate N-H groups in one polypeptide backbone are hydrogen-bonded with alternate C-O groups in the polypeptide backbone of the adjacent chain. In each polypeptide backbone, alternate C-O and N-H groups are free to form hydrogen bonds with their counterparts in a different polypeptide backbone on the opposite side, so a β sheet can consist of multiple aligned segments in the same (or different) polypeptide chains. Other types of interactions also are important in protein folding – for example, covalent bonds may form between the sulphur atoms of pairs of cysteines in different parts of the polypeptide. However, the rules of folding are so complex that, except for the simplest proteins, the final shape of a protein cannot usually be predicted from the amino acid sequence alone.

Many protein molecules consist of more than one polypeptide chain. When this is the case, the protein is said to contain subunits. The subunits may be identical or different. For example, hemoglobin, the oxygen carrier of blood, consists of four subunits: two copies of each of two different polypeptides, which are, designated the α chain and the β chain. (The use of α and β as names of the polypeptide chains has nothing to do with the α helices and β sheets that are found within the polypeptides).

GENES AND POLYPEPTIDES

In the early 1900s, Archibald Garrod suggested that hereditary human diseases, such as phenylketonuria, result from inborn errors of metabolism. Support for this idea came in the 1940s when George Beadle and Edward Tatum demonstrated, using *Neurospora crassa*, that genes govern the ability of the fungus to synthesise amino acids, purines, and vitamins. That genes control metabolism by determining protein structure was demonstrated when it was shown, in the early 1950s, that the allele for sickle-cell anemia brings about a change in the charge of the hemoglobin molecule by causing substitution of an uncharged valine for a negatively charged glutamic acid at residue number 6 in the β-globin chain.

Most genes contain the information for the synthesis of only one polypeptide chain. Furthermore, the *sequence* of nucleotides in a gene determines the *sequence* of amino acids in a polypeptide. This point was first proved by studies of the tryptophan synthase gene *trpA* in *E. coli*, a gene in which many mutations had been obtained an accurately mapped. The effects of numerous mutations on the amino acid sequence of the enzyme were determined by directly analysing the amino acid sequences of the wildtype and mutant enzymes.

Each mutation was found to result in a single amino acid substituting for the wild type amino acid in the enzyme; more important, *the order of the mutations in the genetic map was the same as the order of the affected amino acids in the polypeptide chain*. This attribute of genes and polypeptide is called colinearity, which means that the sequence of base pairs in DNA determines the sequence of amino acids in the polypeptide in a collinear, or point-to-point, manner. Colinearity is found almost universally in prokaryotes. However, we will see later that, in eukaryotes, non-informational DNA sequences interrupt the continuity of most genes; in these genes, the order of mutations along a gene (but not their spacing) correlates with the respective amino acid substitutions.

TRANSCRIPTION

The first step in gene expression is the synthesis of an RNA molecule copied from the segment of DNA that constitutes the gene. The essential chemical characteristics of the enzymatic synthesis of RNA resemble those of DNA synthesis.

1. The precursors in the synthesis of RNA are the four-ribonucleoside 5'-triphosphates-adenosine triphosphate (ATP), guanosine triphosphate (GTP), cytidine triphosphate (CTP), and uridine triphoshpate (UTP). They differ from the DNA precursors only in that the sugar is ribose rather than deoxyribose and the base uracil (U) replaces thymine (T).
2. In the synthesis of RNA, a sugar-phosphate bond is formed between the 3'-hydroxyl group of one nucleotide and the 5'-triphosphate of the next nucleotide in line. This is the same chemical bond as in the synthesis of DNA, but the enzyme is different. The enzyme used in transcription is RNA polymerase rather than DNA polymerase.
3. The linear order of bases in an RNA molecule is determined by the sequence of bases in each DNA template. Each base added to the growing end of the RNA chain is chosen for its ability to base pair with the DNA template strand. Thus the bases C. T. G. and A in a DNA strand cause G, A, c, and U, respectively, to be added to the growing end of an RNA molecule.
4. Nucleotides are added only to the 3'-OH end of the growing chain; as a result, the 5' end of a growing RNA molecule bears a triphosphate group. Note that the 5'-to-3' direction of RNA chain growth is same as that in DNA synthesis.

A significant difference between DNA polymerase and RNA polymerase is that *RNA polymerase is able to initiate chain growth without a primer*. Furthermore, Each RNA molecule produced in transcription derives from a single strand of DNA, because in any particular region of the DNA, only one strand serves as a template for RNA synthesis.

The synthesis of RNA can be described as consisting of four discrete stages.

Promoter Recognition

RNA polymerase binds to DNA within a base sequence from 20 to 200 bases in length called a promoter. Many promoter sequences have been isolated and their base sequences determined. Although there is substantial sequence variation among promoter regions (in part corresponding to different strengths of the promoters in binding with the RNA polymerase) certain sequence patterns or "motifs" are quite frequent. Each pattern is defined by a consensus sequence of bases determined from the actual sequences by majority rule: Each base in the consensus sequence is the base most often observed at that position in actual sequences. Any particular sequence may resemble the consensus sequence very well or very poorly. The consensus sequences in the promoter regions in *E. coli* are TTGACA, centered approximately 35 base pairs upstream from the transcription start site (+1), and TATAAT, centered approximately 10 base pairs upstream from the +1 site. The 10 sequences, which are called the TAT box, are similar to sequences found at corresponding positions in many eukaryotic promoters. The positions of the promoter sequences determine where the RNA polymerase begins synthesis, and an A or G is often the first nucleotide in the transcription.

The strength of the binding of RNA polymerase to different promoters varies greatly, which causes differences in the extent of expression from one gene to another. Most of the differences in promoter strength result from variations in the –35 and –10 promoter elements and in the spacing between them. Promoter strength among *E. coli* genes differs by a factor of 10^4, and most of the variation can be attributed to the promoter sequences themselves. In general, the more closely the promoter elements resemble the consensus sequence, the stronger the promoter. Mutations that change the base sequence in a promoter can alter the strength of the promoter; changes that result in less resemblance to the consensus sequence lower the strength, whereas those with greater resemblance to the consensus increase the strength. Furthermore, there are promoters that differ greatly from the consensus sequence in the –35 region. These promoters typically require accessory proteins to activate transcription by RNA polymerase. In eukaryotes, in addition to the promoter sequences, there are also other DNA sequences called *enhancers* that interact with the promoter to determine the level of transcription.

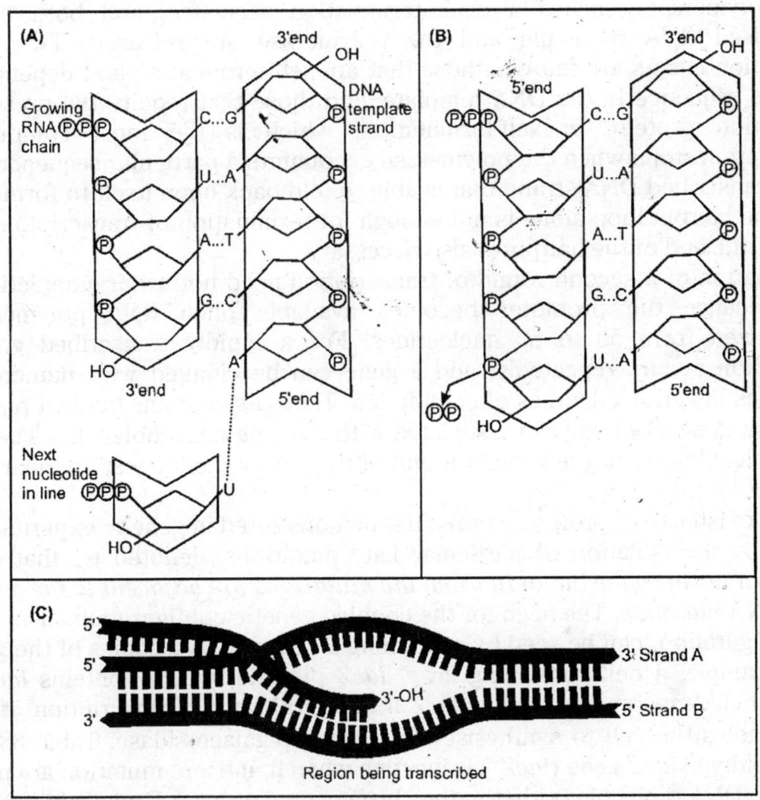

FIG. 38.4: RNA synthesis. (A) The polymerization step in RNA synthesis. The incoming nucleotide forms hydrogen bonds (red dots) with a DNA base. The – OH group pin the growing RNA chain reacts with the orange P in the next nucleotide in line (B). (C) Geometry of RNA synthesis. RNA is copied from only one strand of a segment of a DNA molecule – in this example, strand B – without the need for a primer. In this region of the DNA, RNA is not copied from strand A. However, in a different region (for example, in a different gene) strand A might be copied rather than strand B. Because RNA elongates in the 5'-to-3' direction, its synthesis move along the DNA template in the 3'-to-5' direction; that is, the RNA molecule is antiparallel to the DNA strand being copied.

Chain Initiation

After the initial binding step, the RNA polymerase "melts" (locally denatures) the DNA double helix, causing the strand that is to be transcribed to separate from it s partner strand and become accessible to the polymerase. The RNA polymerase then initiates RNA synthesis at a nearby transcription start site, denoted the +1 site. The first nucleoside triphosphate is placed at this site, and synthesis proceeds in a 5'-to-3' direction.

Chain Elongation

RNA polymerase moves progressively along the transcribed DNA strand, adding nucleotides to the growing RNA chain. Only one DNA strand, the template strand, is transcribed.

Chain Termination

RNA polymerase reaches a chain-termination sequence, and both the newly synthesised RNA molecule and the polymerase are released. Two kinds of termination events are known: those that are self-terminating and depend only on the base sequence in the DNA template, and those that require the presence of a termination protein. In self-termination, which is the most common case, transcription stops when the polymerase encounters a particular sequence of bases in the transcribed DNA strand that is able to fold back upon itself to form a hairpin loop. The hairpin loop alone is not enough for termination of transcription; the run of U's at the end of the hairpin is also necessary.

Initiation of a second round of transcription need not await completion of the first, because the promoter becomes available once RNA polymerase has polymerized from 50 to 60 nucleotides. For a rapidly transcribed gene; such reiinitiation occurs repeatedly, and a gene can be cloaked with numerous RNA molecules in various degrees of completion. *Triturus* contains tandem repeats of a particular gene. Each gene is associated with growing RNA molecules. The shortest RNA molecules are at the promoter end of the genes; the longest are near the gene terminus.

The existence of promoters was first demonstrated in genetic experiments with *E. coli* by the isolation of particular Lac^- mutations, denoted p^-, that eliminate activity of the *lac* gene but *only when the mutations are adjacent to the gene in the same DNA molecule.* The need for the coupled genetic configuration, also called the *cis* configuration, can be seed by examining a cell with two copies of the gene *lacZ* – for example, a cell containing an F' *lacZ* plasmid, which contains *lacZ* in the bacterial chromosome as well as *lacZ* in the F' plasmid. Transcription of the *lacZ* gene enables the cell to synthesise the enzyme β-galactosidase. Table 38.1 shows that a wildtype *lacZ* gene ($lacZ^+$) is inactive when it and a p^- mutation are present in the same DNA molecule (either in the chromosome or in an F' plasmid); this can be seen by comparing entries 4 and 5. Analysis of the RNA shows that, in a cell with the genotype p^- $lacZ^+$, the $lacZ^+$ gene is not transcribed, whereas if the genotype is p^+ *lacZ*, a mutant RNA is produced. The p^- mutations are called promoter mutations.

TABLE 38.1: Effect of promoter mutations on transcription of the lacZ gene

	Genotype	Transcription of $lacZ^+$ gene
1.	p^+lacZ^+	Yes
2	p^-lacZ^+	No
3	p^+lacZ^+ / p^+lacZ^-	Yes
4.	p^-lacZ^+ / p^+lacZ^-	No
5.	p^+lacZ^+ / p^-lacZ^-	Yes

Note: $lacZ^+$ is the wildtype gene; $lacZ^-$ is a mutant that produces a nonfunctional enzyme.

Mutations have also been instrumental in defining the transcription-termination region. Mutations have been isolated that create a new termination sequence upstream from the normal one. When such a mutation is per cent, an RNA molecule is made that is shorter than the wildtype RNA. Other mutations eliminate the terminator, resulting in a longer transcript. The best understood RNA polymerase is that of the bacterium *E. coli*. This enzyme consists of five protein subunits and can be easily seen by electron microscopy. In *E. coli*, this enzyme catalyses all transcription. Eukaryotic cells have three distinct RNA polymerases, denoted I, II, and III, each of which makes a particular class of RNA molecular. RNA polymerase I catalyses synthesis of all ribosomal RNA species except 5S RNA: RNA polymerase III catalyses synthesis of 5S and all of the transfer RNAs. The RNA polymerase II is the enzyme responsible for the synthesis of all RNA transcripts that contain information specifying amino acid sequences. These transcripts are called messenger RNA molecules, which are discussed in the next section. RNA polymerase II also catalyses synthesis of most small nuclear RNAs involved in RNA splicing.

Messenger RNA

Amino acids do not bind directly to DNA. Consequently, intermediate steps are needed for arranging the amino acids in a polypeptide chain the order determined by the DNA base sequence. This process begins with transcription of the base sequence of the template strand of DNA into the base sequence of an RNA molecule. In prokaryote, this RNA molecule, which is called messenger RNA, or mRNA, is used directly in polypeptide synthesis. In eukaryotes, the RNA molecule is generally processed before it becomes mRNA. The amino acid sequence is then determined by the base sequence in mRNA by the protein-synthesizing machinery of the cell.

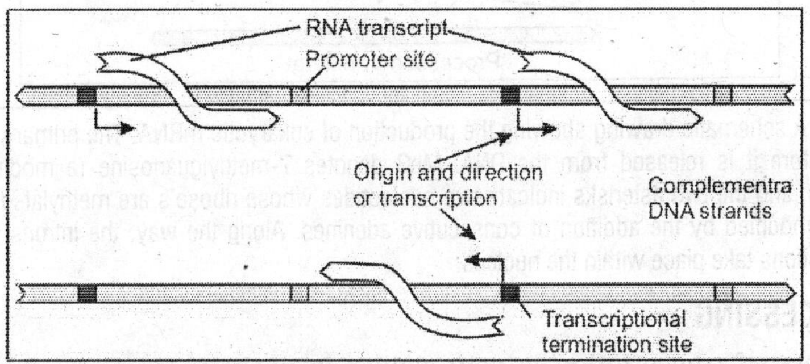

FIG. 38.5: A typical arrangement of promoters (green) and termination sites (red) in a segment of a DNA molecule. Promoters are present in both DNA strands. Termination sites are usually located such that transcribed regions do not overlap.

Not all base sequences in an mRNA molecule are translated into the amino acid sequences of polypeptides. For example, translation of an mRNA molecule rarely starts exactly at one of its ends and proceeds to the other end; instead, initiation of polypeptide synthesis may begin many nucleotides downstream from the 5' end of the mRNA. The untranslated 5' segment of RNA is called a leader and in some cases contains regulatory sequences that affect the rate of protein synthesis. The leader is followed by a coding sequence, also called an open reading frame, or ORF, which specifies the order in which the amino acids are present in the polypeptide chain. A typical coding sequence in an mRNA molecule is between 500 and 3000 bases long (depending on the number of amino acids in the protein). Like the leader sequence, the 3' end of an mRNA molecule following the coding sequence is not translated. The template strand of each gene is only one of the two DNA strands present in the gene, but which DNA strand is the template can differ from gene to gene along a DNA molecule. That is, except in some small viruses, not all mRNA molecules are transcribed from the same DNA strand. Thus in an extended segment of a DNA molecule, mRNA molecules would be seen growing in either of two directions, depending on which DNA strand functions as a template.

In prokaryotes, most mRNA molecules are degraded within a few minutes after synthesis. In eukaryotes, a typical lifetime is several hours; although some last only minutes, and others persist for days. In both kinds of organisms, the degradation enables cell to dispose of molecules that are no longer needed. The short lifetime of prokaryotic mRNA is an important factor in regulating gene activity.

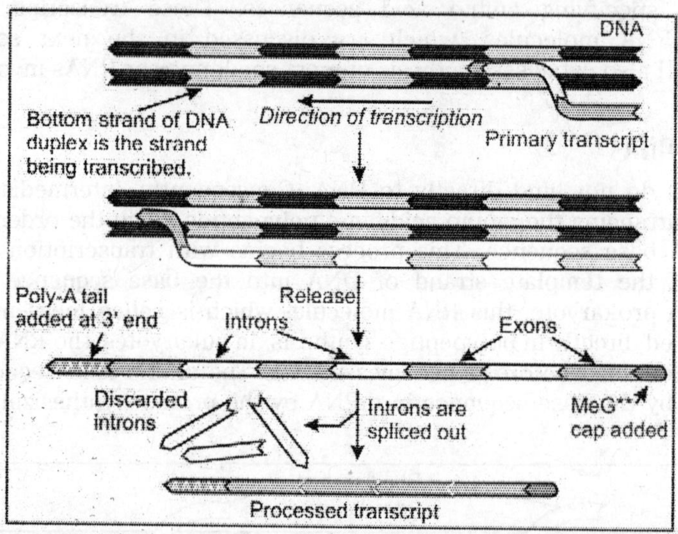

FIG. 38.6: A schematic drawing showing the production of eukaryotic mRNA. The primary transcript is capped before it is released from the DNA. MeG denotes 7-methylguanosine (a modified form of guanosine), and the two asterisks indicate two nucleotides whose ribose's are methylated. The 3' end is usually modified by the addition of consecutive adenines. Along the way, the introns are excised. These reactions take place within the nucleus.

RNA PROCESSING

The process of transcription is very similar in prokaryotes and eukaryotes, but there are major differences in the relation between the transcript and the mRNA used for polypeptide synthesis. In prokaryotes, the immediate product of transcription (the primary transcript) is mRNA; by contrast, *the primary transcript in eukaryotes must be converted into mRNA.* This conversion, which is called RNA processing, usually consists of two types of events: modification of the ends and excision of untranslated sequences embedded within coding sequences.

Each end of a eukaryotic transcript is processed. The 5' end is altered by the addition of a modified guanosine, 7-methyl guanosine, in an uncommon 5'-to-5' (instead of 3'-to-5') linkage; this terminal group is called a cap. The 3' terminus of a eukaryotic mRNA molecule is usually modified by the addition of a polyadenosine sequence (the poly-A tail) of as many as 200 nucleotides. The 5' cap is necessary for the mRNA to bind with the ribosome to begin protein synthesis, and the poly-A tail helps to determine mRNA stability. A second important feature peculiar to the primary transcript in eukaryotes, is the presence of segments of RNA, called introns or intervening sequences, that are excised from the primary transcript. Accompanying the excision of introns is a rejoining of the coding segments (exons) to form the mRNA molecule. The excision of the introns and the joining of the exons is called RNA splicing. The symbols are N, any nucleotide; R, any purine (A or G); Y, any pyrimidine (C or U); and S, either A or C. In the first step of splicing, the 2'-OH of the adenosine (A) at the branch site, which is located a short distance upstream from a run of pyrimidines (Y) near the acceptor site, attacks the phosphodiester bond at the donor splice site junction. The attack result in cleavage at the donor splices site and formation of a branched molecule known as a lariat because it has a loop and a tail. The A – G linkage at the "knot" of the lariat is unusual in being 2'-to-5' (instead of the usual 3'-to-5'). In the final step of splicing, the 3'-OH of the guanosine of the donor exon attacks the phosphodiester bond at the acceptor splice site, freeing the lariat intron and joining the donor and acceptor exons together. The lariat intron is rapidly degraded into individual nucleotides by nucleases.

RNA splicing takes place in nuclear particles known as spliceosomes. These abundant particles are composed of protein and several types of specialized small RNA molecules ranging from 100 to 200 bases in length. The specificity of splicing comes from the small RNAs, some of which contain sequences that are complementary to the splice junctions, but numerous spliceo-some proteins also are required for splicing. One model for the process is illustrated, in which U1, U2 and U5 are designations for three different types of small nuclear RNAs. The ends of the intron are brought together by U1, which forms base pairs with nucleotides in the intron at both the 5' and the 3' ends. The ends of the exons are brought together by U5, which forms base pairs with nucleotides in the exons at both the donor splice site and the acceptor splice site. The black arrow in Figure indicates the initial attack of the branch site A on the donor splice site junction. Note that the branch site is held in place by U2.

Introns are also present in some genes in organelles, but the mechanisms of their excision differ from those of introns in nuclear genes because organelles do not contain spliceosomes. In one class of organelle introns, the intron contains a sequence coding for a protein that participates in removing the intron that code for it. The situation is even more remarkable in the splicing of a ribosomal RNA precursor in the ciliate *Tetrahymena*. In this case, the splicing reaction is intrinsic to the folding of the precursor; that is, the RNA precursor is *self-splicing* because the folded precursor RNA creates its own RNA-splicing activity. The self-splicing *Tetrahymena* RNA was the first example found of an RNA molecule that could function as an enzyme is catalysing a chemical reaction; such enzymatic RNA molecules are called ribozymes. The existence and the positions of introns in a particular primary transcript are readily demonstrated by renaturing the transcribed DNA with the fully processed mRNA molecule. The DNA-RNA hybrid can then be examined by electron microscopy. The DNA copies of the introns appear as single-stranded loops in the hybrid molecule, because no corresponding RNA sequence is available for hybridization.

The number of introns per RNA molecule varies considerably from one gene to the next. For example, 2 introns are present in the primary transcript of human α-globin, and 52 introns occur in collagen RNA. Furthermore, within a particular RNA molecule, the introns are widely distributed and have many different sizes. In human beings and other mammals, most introns range in size from 100 to 10,000 base pairs, and in the processing of a typical primary transcript, the amount of discarded RNA ranges from about 50 per cent to nearly 90 per cent of the primary transcript. Genes in lower eukaryotes, such as yeast, nematodes, and fruit flies, generally have fewer introns than genes in mammals, and the introns tend to be much smaller.

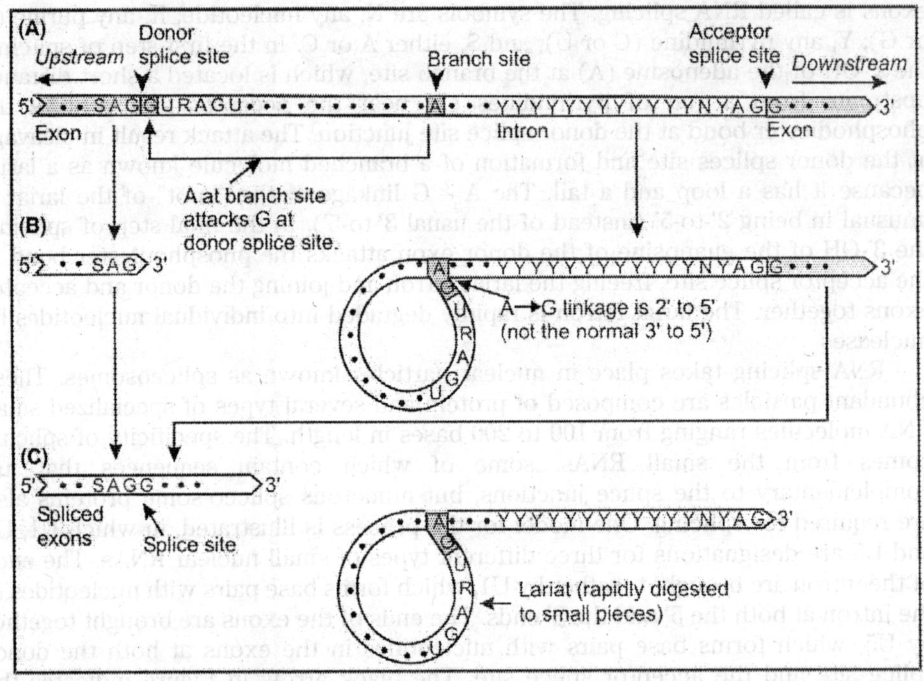

FIG. 38.7: A schematic diagram showing removal of one intron from a primary transcript. The A nucleotide at the branch site attacks the terminus of the 5' exon, cleaving the exon-intron junction and forming a loop connected back to the branch site. The 5' exon is later brought to the site of cleavage of the 3' exon, a second cut is made, and the exon termini are joined and sealed. The loop is released as a lariat-shaped structure that is degraded. Because the loop includes most of the intron, the loop of the lariat is usually very much longer than the tail.

Most introns appear to have no function in themselves. An artificial gene that lacks a particular intron usually functions normally. In those cases in which an intron seems to be required for function, it is usually not because the interruption of the gene is necessary, but because the intron happens to include certain nucleotide sequences that regulate the timing or tissue specificity of transcription. The implication is that many mutations in introns, including small deletions and insertions, should have essentially no effect on gene function, and this is the case. Moreover, the nucleotide sequence of a particular intron is found to undergo changes (including small deletions and insertions) extremely rapidly in the course of evolution, and this lack of sequence conservation is another indication that most of the nucleotide sequences present with in introns are not important.

Mutations that affect any of the critical splicing signals do have important consequences, because they interfere with the splicing reaction. The intron with the mutated splice site fails to be removed, and it remains in the processed mRNA. The result is the production of a mutant protein with a normal sequence of amino acids up to the splice site but an abnormal sequence afterward. Most introns are long enough that, by chance, they contain a stop sequence that terminates protein synthesis, and once a stop is encountered, the protein grows no further. In second case, splicing does occur, but at an alternative splice site. (The example shows the alternative site downstream from the mutation, but alternative sites can also be upstream). The alternative site is called a cryptic splice site because it is not normally used. The cryptic splice site is usually a poorer match with the consensus sequence and is ignored when the normal splice site is available. The result of using the alternative splice site is again an incorrectly processed mRNA and a mutant protein. In some splice-site mutations, both outcomes can occur: Some transcripts leave the intron unspliced, whereas others are spliced at cryptic splice sites.

Although introns are not usually essential in regulating gene expression, they may play a role in gene evolution. In some cases, the exons in a gene code for segments of the completed protein that are relatively independent in their folding characteristics. For example, the central exon of the β-globin gene codes for the segment of the protein that folds around an iron-containing molecule of heme. Relatively autonomous folding units in proteins are known as folding domains, and the correlation between exons and domains found in some genes suggests that the genes were originally assembled from smaller pieces. In some cases, the ancestry of the exons can be traced. For example, the human gene for the low-density lipoprotein receptor that participates in cholesterol regulation shares exons with certain blood-clotting factors and epidermal growth factor. The model of protein evolution through the combination of different exons is called the exon-shuffle model. The mechanism for combining exons from different gene is not known. Although some genes support the model, in other genes the boundaries of the folding domains do not coincide with exons.

On the one hand, introns may be an ancient feature of gene structure. The existence of self-splicing RNAs means that introns could have existed long before the evolution of the spliceosome mechanism, and therefore some introns maybe as old as the genes themselves. Furthermore, the finding that some gene shave introns in the same places in both plants and animals suggests that the introns may have been in place before plants and animals became separate lineages. If introns are ancient, then exon shuffling might have been important early in evolution by creating new genes with novel combinations of exons. It has even been suggested that all forms of early life had introns in their gene sand that today's prokaryotes, which lack introns, lost their introns in their evolution. On the other hand, it has also been argued that introns arose relatively late in evolution and became inserted into already existing genes, particularly in vertebrate genomes.

TRANSLATION

The synthesis of every protein molecule in a cell is directed by an mRNA, which is originally copied from DNA. Protein production includes two kinds of processes: (1) information-transfer processes in which the RNA base sequence determines an amino acid sequence, and (2) chemical processes in which the amino acids are linked together. The complete series of events is called translation. The main ingredients necessary for translation are as follows:

1. ***Messenger RNA***: Messenger RNA is needed to bring the ribosomal subunits together and to provide the coding sequence of bases that determines the amino acid sequence in the resulting polypeptide chain.
2. ***Ribosomes***: These components are particles on which protein synthesis takes place. They move along an mRNA molecule and align successive transfer RNA molecules; the amino acids are attached one by one to the growing polypeptide chain by means of peptide bonds. Ribosomes consist of two subunit particles. In *E. coli*, their sizes are 30S (the small subunit) and 50S (the large subunit). The counterparts in eukaryotes are 40S and 60S. (The S stands for *Svedberg unit*, which measures the rate of sedimentation of a particles in a centrifuge and so is an indicator if size). Together, the small and large particles form a functional ribosome.

3. ***Transfer RNA or tRNA***: The sequence of amino acids in a polypeptide is determined by the base sequence in the mRNA by means of a set of adaptor molecules, the tRNA molecules, each of which is attached to a particular amino acid. Each group of three adjacent bases in the mRNA forms a codon that binds to a particular group of three adjacent bases in the tRNA (an anticodon), bringing the attached amino acid into line for addition to the growing polypeptide chain.
4. ***Aminoacyl tRNA synthetases***: This set of enzymes catalyses the attachment of each amino acid to its corresponding tRNA molecule. A tRNA attached to its amino acid is called an aminoacylated tRNA ora charged tRNA.
5. ***Initiation, elongation, and termination factors***: Polypeptide synthesis can be divided into three stages: (1) initiation, (2) elongation, and (3) termination. Each stage requires specialized proteins.

In prokaryotes, all of the components for translation are present throughout for cell; in eukaryotes, they are located in the cytoplasm, as well as in mitochondria and chloroplasts. In overview, the process of translation is that an mRNA molecule binds to a ribosome. The aminoacylated tRNAs are brought along sequentially, one by one, to the ribosome that is translating the mRNA molecule. Peptide bonds are made between successively aligned amino acids, each time joining the amino group of the incoming amino acid to the carboxyl group of the amino acid at the growing end. Finally, the chemical bond between the last tRNA and its attached amino acid is broken, and the completed polypeptide is removed.

Initiation

The main features of the initiation step in polypeptide synthesis are the binding of mRNA to the small subunit of the ribosome and the binding of a charged tRNA bearing the first amino acid. The ribosome includes three sites for tRNA molecules. They are called the E (exit) site, the P (peptidyl) site, and the A (amino acyl) site. In the initiation of translation, two initiation factors (IF-1 and IF-3) interact with the 30S subunit at the same time that another initiation factor (IF-2) binds with a special initiator tRNA charged with methionine. (In prokaryotes, the initiator tRNA actually carries formylmethionine, yield $tRNA^{fMet}$). These components come together and combine with an mRNA. In prokaryotes, the mRNA binding is facilitated by hydrogen bonding between the 16S RNA present in the 30S subunit and the ribosome-binding site of the mRNA; in eukaryotes, the 5' cap on the mRNA is instrumental. Together, the 30S + $tRNA^{Met}$ + mRNA complex recruits a 50S subunit, in which the $tRNA^{Met}$ is positioned in the P site and aligned with the AUG initiation codon, forming the 70S initiation complex. The tRNA binding is accomplished by hydrogen bonding between the AUG codon in the mRNA and the three-base anticodon in the tRNA. In the assembly of the completed ribosome, the initiation factors dissociate from the complex.

Elongation

The elongation stage of translation consists of three processes: bringing each new aminoacylated tRNA into line, forming the new peptide bond to elongate the polypeptide, and moving the ribosome along the mRNA so that the codons can be translated successively. A key role is played by the elongation factor EF-Tu, although a second protein, EF-Ts, is also required. (The eukaryotic counterpart of EF-Tu is called Ef-1α). The EF-Tu, bound with guanosine triphosphate (EF-Tu-GTP), brings the next aminoacylated tRNA into the A site on the 50S subunit, which in this example is $tRNA^{Phe}$. This processes requires the hydrolysis of GTP to GDP, and once the GDP is formed, the EF-Tu-GDP has low affinity for the ribosome and diffuses away, becoming available for reconversion into EF-Tu-GTP.

Once the A site is filled, a peptidyl transferase activity catalyses a concerted reaction in which the bond connecting the methionine to the tRNA^{Met} is transferred to the amino group of the phenylalanine, forming the first peptide bond. Peptidyl transferase activity is not due to a single molecule but requires several components of the 50S subunit, including several proteins and an RNA component (called 23S) of the 50S subunit. Some evidence indicates that the 23S RNA, which would suggest that 23S is an example of a ribozyme at work, carries out the actual catalysis. The 30S subunit, in light green at the top, binds the mRNA and moves along it in the direction indicated by the arrow. The 50S subunit, in blue at the bottom, contains the tRNA binding sites in the P and A sites. The P site is in red, to the left, and the A site is in dark green, to the right.

Note that the relative positions of the 30S and 50S ribosomal subunits are shifted from one panel to the next. The configuration of the subunits in the top panel is called the pretranslocation state. In the middle panel, the 30S subunit shifts one codon to the right. This event is called translocation. After translocation, the ribosome is said to be in the pos-translocation state. In the next step of polypeptide synthesis, the 50S subunit shifts one step over to the right, which reconfigures the ribosome back into the pretranslocation state.

The term translocation, as applied to protein synthesis, means the movement of the 30S subunit one codon further along the mRNA. With each successive translocation of the 30S subunit, one more amino acid is added to the growing polypeptide chain. The entire cycle of charged tRNA addition, peptide bond formation, and translocation is elongation. Starting with a ribosome in the pretranslocation state, the elongation factor EF-G binds with the ribosome. (The eukaryotic counterpart of EF-G is called EF-2). Like EF-Tu, the EF-G comes on in the form EF-G-GTP and, in fact, binds to the same ribosomal site as EF-Tu-GTP. Hydrolysis of the GTP to GDP yields the energy to shift the tRNAs in the P and A sites to the E and P sites, respectively, as well as to translocate the 30S subunit one codon along the mRNA. The ribosome is thereby converted to the post translocation state, and the EF-G-GDP is released. At this stage, EF-Tu-GTP comes into play again, and four events happen.

- The next aminoacylated tRNA is brought into line (in this case, tRNA^{Val}).
- The uncharged tRNA is ejected from the E site.
- In a concerted reaction, the bond connecting the growing polypeptide chain to the tRNA in the P site is transferred to the amino group of the amino acid in the A site, forming the new peptide bond.
- The ribosome transitions to the pretranslocation state.

Also, the EF-Tu-GDP is released, making room for the EF-G-GTP.

After these steps, the EF-G-GDP is released for regeneration into EF-G-GTP for use in another cycle. The translocation state of the ribosome is not depicted separately because it happens so rapidly. In effect, the ribosome shuttles between the pretranslocation and posttranslocation states. Hence the ribosome is again available for EF-Tu-GTP to start the next round of elongation. Polypeptide elongation may therefore be considered as a cycle of events repeated again and again. The steps B→C→B→C (or, equivalently, C→D→C→D) are carried out repeatedly until a termination codon is encountered.

For example, the configuration of the ribosome is such that the tRNA^{Gly} and tRNA^{Val} are occupying the E and P sites, respectively. The aminoacylated tRNA that corresponds to the codon AGU is brought into line, which is tRNA^{Ser}, and the bond connecting the polypeptide to $tRNA^{Val}$ is transferred to the amino group of Ser, creating a new peptide bond and elongating the polypeptide by one amino acid. At the same time, the ribosome is converted to the pretranslocation state in preparation for translocation. The elongation cycle happens relatively fast. Under optimal conditions, *E. coli* synthesises a polypeptide at the rate of about 20 amino acids per second; in eukaryotes, the rate of elongation is about 15 amino acids per second.

Termination

The elongation steps of protein synthesis are carried out repeatedly until a stop codon for termination is reached. The stop codons are UAA, UAG, and UGA. No tRNA exists that can bind to a stop codon, so the tRNA holding the polypeptide remains in the P site. Specific release factors act to cleave the polypeptide from the tRNA to which it is attached as well as to disassociate the 70S ribosome from the mRNA, after which the individual 30S and 50S subunits are recycled to initiate translation of another mRNA. The release factor RF-1 recognizes the stop codons UAA and UAG, whereas release factor RF-2 recognizes UAA and UGA. A third release factor, RF-3, is also required for translational termination.

Monocistronic and Polycistronic mRNA

The process of selecting the correct AUG initiation codon is of some importance in understanding many features of gene expression. In prokaryotes, mRNA molecules commonly contain information for the amino acid sequences of several different polypeptide chains; such a molecule is called a polycistronic mRNA. (*Cistron* is a term often used to mean a base sequence that encodes a single polypeptide chain). In a polycistronic mRNA, each polypeptide coding region is preceded by its own ribosome-binding site and AUG initiation codon. After the synthesis of one polypeptide is finished, the next along the way is translated. The genes contained in a polycistronic mRNA molecule often encode the different proteins of a metabolic pathway. For example, in *E. coli*, the ten enzymes needed to synthesise histidine are encoded by one polycistronic mRNA molecule. The use of polycistronic mRNA is an economical way for a cell to regulate the synthesis of related proteins in a coordinated manner. For example, in prokaryotes, the usual way to regulate the synthesis of a particular protein is to control the synthesis of the mRNA molecule that codes for it. With a polycistronic mRNA molecule, the synthesis of several related proteins can be regulated by a single signal, so that appropriate quantities of each protein are made at the same time; this is termed coordinate regulation. In eukaryotes, the 5' terminus of an mRNA molecule binds to the ribosome, after which the mRNA molecule slides along the ribosome until the AUG codon nearest the 5' terminus is in contact with the ribosome. Then protein synthesis begins. There is no mechanism for initiating polypeptide synthesis at any AUG other than the first one encountered. Eukaryotic mRNA is always monocistronic.

The definitive feature of translation is that it proceeds in a particular direction along the mRNA and the polypeptide:

The mRNA is translated from an initiation codon to a stop codon in the 5'-to-3' direction. The polypeptide is synthesised from the amino end toward the carboxyl end by the addition of amino acids, one by one, to the carboxyl end. For example, a polypeptide with the sequence NH_2 – Met – Pro - … - Gly – Ser – COOH would start with methionine, and serine would be the last amino acid added to the chain.

In writing nucleotide sequences, by convention, we place the 5' end at the left and, in writing amino acid sequences, we place the amino end at the left.

Polynucleotides are generally written so that both synthesis and translation proceed from left to right and polypeptides are written so that synthesis proceeds from left to right. This convention is used in all of the following sections concerning the genetic code.

THE GENETIC CODE

The four bases in DNA-A, T, G, and C – are sufficient to specify the 20 amino acids in proteins because each codon is three bases in length. Each sequence of three adjacent bases in mRNA is a codon that specifies a particular amino acid (or chain termination). The genetic code is the list of all codons and the amino acid that each one encodes. Before the genetic code was determined experimentally, it was reasoned that if all codons were assumed to have the same number of bases, then each codon would have to contain at least three bases. Codons consisting of pairs of bases would be insufficient, because four bases can form only $4^2 = 16$ pairs; triplets of bases would suffice, because four bases can form $4^3 = 64$ triplets. In fact, the genetic code is a triplet code, and all 64 possible codons carry information of some sort. Most amino acids are encoded by more than one codon. Furthermore, in the translation of mRNA molecules, the codons do not overlap but are used sequentially.

Genetic Evidence for a Triplet Code

Although theoretical considerations suggested that each codon must contain at least three letters, codons having more than three letters could not be ruled out. The first widely accepted proof for a triplet code came from genetic experiments using *rII* mutants of bacteriophage T4 that had been induced by replication in the presence of the chemical *proflavin*. These experiments were carried out in 1961 by Francis Crick and collaborators. Proflavin-induced mutations typically resulted in total loss of function. Because proflavin is a large planar molecule, it was suspected that it caused mutations that inserted or deleted a base pair by interleaving between base pairs in the double helix. Analysis of the properties of these mutations led directly to the deduction that the code is read three nucleotides at a time from a fixed point; in other words, there is a reading frame to each mRNA.

The genetic analysis of the structure of the code began with an *rII* mutation called FC0, which was arbitrarily designated (+), as if it had an inserted base pair. (It could also arbitrarily have been designated (-), as if it had a deleted base pair. Calling it (+) as a lucky guess, however, because when FC0 was sequenced, it did turn out to have a single-base insertion). If FC0 has a (+) insertion, then it should be possible to revert the FC0 allele to "wildtype" by deletion of a nearby base. Selection for r^+ revertants was carried out by isolating plaques formed on a lawn of an *E. coli* strain K12 that was lysogenic for phage λ. The basis of the selection is that *rII* mutants are unable to propagate in K12(λ). Analysis of the revertants revealed that each still carried the original FC0 mutation along with a second (suppressor) mutation that reversed the effects of the FC0 mutation. The suppressor mutations could be separated by recombination from the original mutation by crossing each revertant to wildtype; each suppressor mutation proved to be an *rII* mutation that, by itself, would cause the *r* (rapid lysis) phenotype. If FC0 had an inserted base, then the suppressors should all result in deletion of a base pair; hence each suppressor of FC0 was designated (-). Three such revertants and their consequences for the translational reading frame, illustrated using ordinary three-letter words. The (-) mutations are designated $(-)_1$, $(-)_2$, and $(-)_3$, and those parts of the mRNA translated in the correct reading frame are indicated in green.

Each of the individual (-) suppressor mutations could, in turn, be used to select other "wildtype" revertants, with the expectation that these revertants would carry new suppressor mutations of the (+) variety, because the (-) (+) combination should yield a phage able to form plaques on K12(λ). Various double mutant combinations were made. Usually any (+) (-) combination, or any (-) (+) combination, resulted in a wildtype phenotype, whereas (+) (+) and (-) (-) double mutant combinations always resulted in the mutant combinations always resulted in the mutant combinations always resulted in the mutant phenotype. The truly telling result came when triple mutants were made. Usually, the (+) (+) (+) and (-) (-) (-) triple mutants yielded the wildtype phenotype!

The phenotypes of the various (+) and (-) combinations were interpreted in terms of a reading frame. The initial FC0 mutation, a + 1 insertion, shifts the reading frame, resulting in incorrect amino acid sequence from that point on and thus a non-functional protein. Deletion of a base pair nearby will restore the reading frame, although the amino acid sequence encoded between the two mutations will be different and incorrect. In (+) (+) or (-) (-) double mutants, the reading frame is shifted by two bases; the protein made is still non-functional. However, in the (+) (+) (+) and (-) (-) (-) triple mutants, the reading frame is restored, though all amino acids encoded within the region bracketed by the outside mutations are incorrect; the protein made is one amino acid longer for (+) (+) (+) and one amino acid shorter for (-) (-) (-).

The genetic analysis of the (+) and (-) mutations strongly supported the following conclusions:

- Translation of an mRNA starts from a fixed point.
- There is a single reading frame maintained throughout the process of translation.
- Each codon consists of three nucleotides.

Crick and his colleagues also drew other inferences from these experiments. First, in the genetic code, most codons must function in the specification of an amino acid. Second, each amino acid must be specified by more than one codon. They reasoned that if each amino acid had only one codon, then only 20 of the 64 possible codons could be used for coding amino acids. In this case, most frameshift mutations should have affected one of the remaining 44 "noncoding" codons in the reading frame, and hence a nearby frameshift of the opposite polarity mutation should not have suppressed the original mutation. Consequently, the code was deduced to be degenerate, which means that more than one codon can specify a particular amino acid.

Elucidation of the Base Sequences of the Codons

Polypeptide synthesis can be carried out in *E. coli* cell extracts obtained by breaking cells open. Various components can be isolated, and a functioning protein-synthesizing system can be reconstituted by mixing ribosomes, tRNA molecules, mRNA molecules, and various protein factors. If radioactive amino acids are added to the extract, then radioactive polypeptides are made. Synthesis continues for only a few minutes because mRNA is rapidly degraded by nucleases in the mixture. The elucidation of the genetic code began with the observation that when the degradation of mRNA was allowed to go to completion and the synthetic polynucleotide polyuridylic acid (poly-U) was added to the mixture as an mRNA molecule, a polypeptide consisting only of phenylalanine (Phe-Phe-Phe-...) was synthesised. From this simple result, and knowledge that the code is a triplet code, it was concluded that UUU must be a codon for the amino acid phenylalanine. Variations on this basic experiment identified other codons. For example, when a long sequence of guanines was added at the terminus of the poly-U, the polyphenyulalanine was terminated by a sequence of glycines, indicating that GGG is a glycine codon. A trace of leucine or tryptophan was also present in the glycine-termianted polyphenylalanine. Incorporation of these amino acids was directed by the codons UUG and UGG at the transition point between U and G. When a single guanine was added to the terminus of a poly-U chain, the polyphenylalanine was terminated by leucine. Thus UUG is a leucine codon, and UGG must be a codon for tryptophan. Similar experiments were carried out with poly-A, which yielded polylysine, and with poly-C, which produced poly-proline.

Other experiments led to a complete elucidation of the code. Three codons:

UAA UAG UGA

were found to be stop signals for translation, and one codon, AUG, which encodes methionine, was shown to be the initiation codon. AUG also codes for internal methionines, but it uses a different tRNA to do so.

A Brief Summary of the Code

The *in vitro* translation experiments, which used components isolated from the bacterium *E. coli* have been repeated with components obtained from many species of bacteria, yeast, plants, and animals. The standard genetic code deduced from these experiments is considered to be nearly universal because the same codon assignments can be made for nuclear genes in almost all organisms that have been examined. However, some minor differences in codon assignments are found in certain protozoa and in the genetic codes of organelles.

The standard code is shown in Table 38.2. Note that four codons – the three stop codons and the start codon – are signals. Altogether, 61 codons specify amino acids, and in many cases several codons direct the insertion of the same amino acid into a polypeptide chain. This feature confirms the inference from the *rII* frameshift mutations that the genetic code is redundant (degenerate). In a redundant genetic code, some amino acids are encoded by two or more different codons. In the actual genetic code, all amino acids except tryptophan and methionine are specified by more than one codon. The redundancy is not random.

For example, with the exception of serine, leucine, and arginine, all codons that correspond to the same amino acid are in the same box of Table 38.2; that is, *synonymous codons usually differ only in the third base.* For example, GGU, GGC, GGA, and GGG all code for glycine. Moreover, in all cases, in which two codons code for the same amino acid, the third base is either A or G (both purines) or T or C (both pyrimidines).

The codon assignments are completely consistent with all chemical observations and with the amino acid sequences of wildtype and mutant proteins. In virtually every case in which a mutant protein differs by a single amino acid from the wildtype form, the amino acid substitution can be accounted for by a single base change between the codons corresponding to the two different amino acids. For example, substitution of glutamic acid by valine, which occurs in sickle-cell hemoglobin, results from a change from GAG to GUG in codon six of the β-globin mRNA.

Mutations that change the nucleotide sequence of a gene may differ in their consequences, and a special terminology is used to describe them. A missense mutation results in the replacement of one amino acid by another.

TABLE 38.2: The standard genetic code

First position (5' end)	Second position U		Second position C		Second position A		Second position G		Third position (3' end)
U	UUU Phe	F	UCU Ser	S	UAU Tyr	Y	UGU Cys	C	U
	UUC Phe		UCC Ser		UAC Tyr		UGC Cys		C
	UUA Leu	L	UCA Ser		UAA Stop		UGA Stop		A
	UUG Leu		UCG Ser		UAG Stop		UGG Trp	W	G
C	CUU Leu	L	CCU Pro	P	CAU His	H	CGU Arg	R	U
	CUC Leu		CCC Pro		CAC His		CGC Arg		C
	CUA Leu		CCA Pro		CAA Gin		CGA Arg		A
	CUG Leu		CCG Pro		CAG Gin		CGG Arg		G
A	AUU Ile	I	ACU Thr	T	AAU Asn	N	AGU Ser	S	U
	AUC Ile		ACC Thr		AAC Asn		AGC Ser		C
	AUA Ile		ACA Thr		AAA Lys	K	AGA Arg	R	A
	AUG Met	M	ACG Thr		AAG Lys		AGG Arg		G
G	GUU Val	V	GCU Ala	A	GAU Asp	D	GGU Gly	G	U
	GUC Val		GCC Ala		GAC Asp		GGC Gly		C
	GUA Val		GCA Ala		GAA Glu		GGA Gly		A
	GUG Val		GCG Ala		GAG Glu		GGG Gly		G

For example, the change from GAG to GTG in the DNA of the gene for β-globin results in the replacement of glutamic acid by valine in the sickle-cell hemoglobin molecule; the mutation is therefore a missense mutation. In contrast, a silent mutation is one that does not change the amino acid sequence. Silent mutations often result from changes in the third codon position; for example, a mutation that changes an AAA codon into an AAG codon is silent because both codons specify lysine. A most interesting class of mutations consists of changes that convert a codon that specifies an amino acid into a chain-terminating codon. A mutation of this type is called a nonsense mutation, and it results in premature termination of the polypeptide chain. An example of a nonsense mutation is found in the β-globin gene, in which a mutation from AAG to TAG in the seventeenth codon results in a truncated polypeptide only 16 amino acids in length. This mutation is one of several types associated with the disease β-thalassemia.

Transfer RNA and Aminoacyl-tRNA Synthetase Enzymes

The decoding operation by which the base sequence within an mRNA molecule becomes translated into the amino acid sequence of a protein is accomplished by aminoacylated, or charged, tRNA molecules, each of which is linked to the correct amino acid by an aminoacyl-tRNA synthetase.

The tRNA molecules are small, single-stranded nucleic acids ranging in size from about 70 to 90 nucleotides. Like all RNA molecules, they have a 3'-OH terminus, but the opposite end terminates with a 5'-monophosphate rather than a 5'-triphosphate, because tRNA molecules are cut from a larger primary transcript. Internal complementary base sequences form short double-stranded regions, causing the molecule to fold into a structure in which open loops are connected to one another by double-stranded stems. In two dimensions, a tRNA molecule is drawn as a planar cloverleaf. Its three-dimensional structure is more complex, in which part A shows a skeletal model of a yeast tRNA molecule that carries phenylalanine and part B is an interpretive drawing. Note how the TψC loop and the DHU loop are in close proximity.

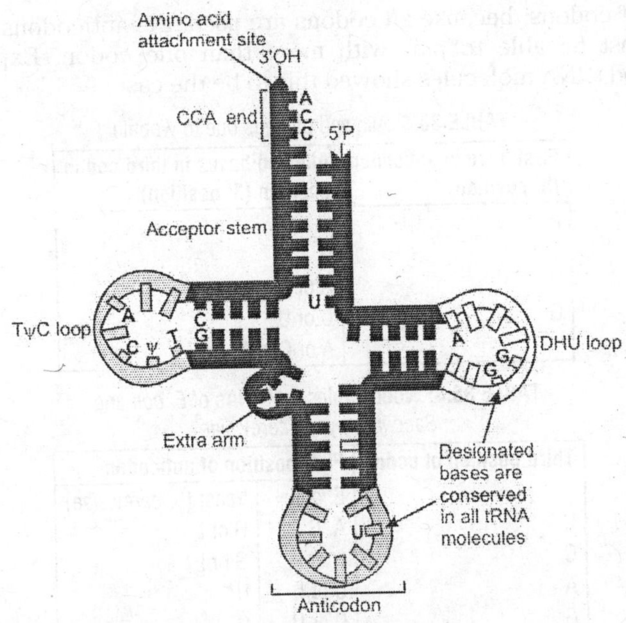

FIG. 38.8: A tRNA cloverleaf configuration. The heavy black letters indicate a few bases that are conserved in the sequence of all tRNA molecules. The labelled loop regions are those found in all tRNA molecules. DHU refers to a base, dihydrouracil, found in one loop; the Greek letter ψ is a symbol for the unusual base pseudouridine.

Particular regions of each tRNA molecule are used in the decoding operation. One region is the anticodon sequence, which consists of three bases that can form base pairs with a codon sequence in the mRNA. No normal tRNA molecule has an anticodon complementary to any of the stop codons UAG, UAA, and UGA, which is why thee codons are stop signals. A second critical site is at the 3' terminus of the tRNA molecule, where the amino acid attaches. A specific aminoacyl-tRNA synthetase matches the amino acid with the anticodon. At least one, and usually only one, aminoacyl-synthetase exists for each amino acid. To make the correct attachment, the synthetase must be able to distinguish one tRNA molecule from another. Recognition regions that encompass many parts of the tRNA molecule provide the necessary distinction.

On binding with the tRNA, a part of the protein makes contact with the variable part of the TψC loop of the tRNA and guides the acceptor stem into the active pocket of the enzyme. These interactions depend primarily on recognition of the shape of the $tRNA^{Ser}$ through contacts with the backbone and only secondarily on interactions that are specific to the anticodon.

Redundancy and Wobble

Several features of the genetic code and of the decoding system suggest that something is missing in the explanation of codon-anticodon binding. First, the code is highly redundant. Second, the identity of the third base of a codon is often unimportant. In some cases any nucleotide will do; examples include proline (Pro), threonine (Thr), and glycine (Gly). In other case, either purine (A or G) or either pyrimidine (U or C) in the third position codes for the same amino acid; examples include histidine (His), glutamine (Gln), and tyrosine (Tyr). Third, the number of distinct tRNA molecules that have been isolated from a single organism is less than the number of codons; because all codons are used, the anticodons of some tRNA molecules must be able to pair with more than one codon. Experiments with several purified tRNA molecules showed this to be the case.

TABLE 38.3: Allowed pairings due to wobble

First base in anticodon (5' position)	Allowed bases in third codon position (3' position)
A	U
C	G
U	A or G
G	C or U
I	A or C or U

TABLE 38.4: Wobble rules for tRNAs of *E. coli* and *Saccharomyces cerevisiae*

Third position of codon	First position of anticodon	
	E. coli	Yeast (*S. cerevisiae*)
U	A, G, or I	G or I
C	G or I	G or I
A	U or I	U*
G	C or U	C

Note: I indicates inosine, which is structurally similar to adenosine except that the – NH_2 is replaced with – OH. U* indicates a modified uridine.

To account for these observations, Francis Crick advanced the wobble concept in 1966. He proposed that the first two bases in a codon form base pairs with the tRNA anticodon according to the usual rules (A-U and G-C) but that the base at the 5' end of the anticodon is less spatially constrained than the first two and can form hydrogen bonds with more than one base at the 3' end of the codon. He suggested the pairing rules.

On the other hand, analysis of tRNAs in the yeast *Saccharomyces cerevisiae* has indicated that wobble is more restricted in yeast than in *E. coli*. The yeast rules may hold for other eukaryotes as well. In yeast, single tRNAs can recognize the pairs of related codons ending in U or C. However, separate tRNAs are needed for codons that end in A or G. Thus at least three tRNAs are required for amino acids such as proline and glycine, which are specified by a set of four codons. A total of 46 tRNAs are needed to decode mRNA molecules in yeast. As noted, there are no tRNAs corresponding to the stop codons.

Nonsense Suppression

Mutations can occur that result in premature chain termination during translation. For whimsical historical reasons, the stop codons are referred to as the amber (UAG), ochre (UAA), and UGA codons. A remarkable observation was that some phage mutants bearing nonsense mutations in the gene encoding the major head protein were able to propagate in some bacterial strains but not in others. How could this happen? On further analysis, it turned out that the strains able to support growth of the phage nonsense mutants carry suppressor mutations that act by changing the way the mRNA is read, not by changing the nucleotide sequence of the phage gene. The suppressor mutations proved to be mutant tRNA genes. These suppressor tRNA genes act by reading a stop codon as though it were a signal for a specific amino acid. The amino acid is inserted at that position and translation continues. So long as the inserted amino acid is compatible with the function of the protein, the effects of the original mutations are suppressed and plaques can be produced. In *E. coli*, "amber suppressors" recognize and suppress only amber mutations because the anticodon of the altered tRNA pairs only with the UAG codon; on the other hand, "ochre suppressors" suppress both ochre and amber mutations because the mutant tRNA can recognize and suppress both UAG and UAA. In yeast there is more specificity because of the more stringent wobble rules: Ochre suppressors suppress only ochre mutations, and amber suppressors suppress only amber mutations.

We can illustrate nonsense suppression by examining a chain-termination codon formed by mutation of the tyrosine codon UAC to the stop codon UAG. A mutant leucine tRNA molecule can suppress such a mutation. In *E. coli*, $tRNA^{Leu}$ has the anticodon 3'-AAC-5', which pairs with the codon 5'-UUG-3'. A suppressor mutation in the $tRNA^{Leu}$ gene produces an altered tRNA with the anticodon AUC; this tRNA molecule is still charged with leucine but responds to the stop codon UAG rather then to the normal leucine codon UUG. Thus, in a cell that contains this suppressor tRNA, the mutant protein is completed and suppression occurs as long as the mutant protein can tolerate a replacement of leucine for tyrosine. Many suppressor tRNA molecules of this type have been observed. Each suppressor is effective against only some nonsense mutations, because the resulting amino acid replacement may not yield a functional protein.

In *E. coli*, there are three classes of tRNA suppressors: those that suppress only UAG (amber suppressors), those that suppress both UAA and UAG (ochre suppressors), and those that suppress only UGA. They share the following properties:

The original mutant gene still contains the mutant base sequence. The suppressor tRNA suppresses all chain-termination mutations with the same stop codon, provided that the amino acid inserted is an acceptable amino acid at the site. A cell can survive the presence of a tRNA suppressor only if the cell contains two or more copies of the same tRNA gene. If only one $tRNA^{Leu}$ gene were present in the genome and if it were mutated, then the normal leucine codon UUG would no longer be read as a sense codon, and all polypeptide chains would terminate wherever a UUG codon occurred. However, multiple copies of most tRNA genes exist, so if one copy is mutated to yield a suppressor tRNA, a normal copy nearly always remains. Any chain-termination codon can be translated by a suppressor tRNA mutation that recognizes that codon. For example, translation of UAG by insertion of an amino acid would prevent termination of all wild type mRNA reading frames terminating in UAG. However, the anticodon of the suppressor tRNA usually binds rather weakly to the stop codon, so the stop codon often results in termination anyway.

Suppressor tRNA mutations are very useful in genetic analysis because they allow nonsense mutations to be identified through their ability to be suppressed. This is important because nonsense alleles usually result in a truncated and completely inactive protein and so are considered true loss-of-function alleles. Suppressor tRNAs have been widely used for genetic analysis in prokaryotes, yeast, and even nematodes, because their other harmful effects are tolerated by the organism. In higher organisms, suppressor tRNAs have such severe harmful effects that they are of limited usefulness.

The Sequence Organization of a Typical Prokaryotic mRNA Molecule

Most prokaryotic mRNA molecules are polycistronic, which means that they contain sequences specifying the synthesis of several proteins. A polycistronic mRNA molecule must therefore possess a series of start and stop codons for use in translation. If an mRNA molecule encodes three proteins, then the minimal coding requirement would be the sequence.

AUG (start)/Protein 1/stop – AUG/protein 2/stop-AUG/protein 3/stop.

The stop codons might be UAA, UAG, or UGA. Actually, such an mRNA molecule is probably never so simple in that (I) the leader sequence preceding the first start signal may be several hundred bases long and (2) spacer sequences containing ribosome-binding sites are usually present between one stop codon and the next start codon.

OVERLAPPING GENES

The idea that two or more reading frames might exist within a single coding segment of DNA was not considered for many years. The reason is that a mutation in a gene that overlaps another gene often produces defects in both gene products, but double mutations are very rare. Furthermore, the existence of overlapping reading frames was thought to place severe constraints on the amino acid sequences of two proteins translated from the same part of an mRNA molecule. However, because the code is highly redundant, the constraints are not so rigid.

If genes contained multiple reading frames, then a single DNA segment would be utilized with maximal efficiency. However, a disadvantage is that evolution might be more difficult because random mutations would rarely improve the function of both proteins. Nevertheless, some cases of overlapping genes have been found. Most examples are in transposable elements or in small viruses in which there is a premium on packing the largest amount of genetic information into a small DNA molecule. Some of the best examples of overlapping genes occur in the *E. coli* phage ϕX174.

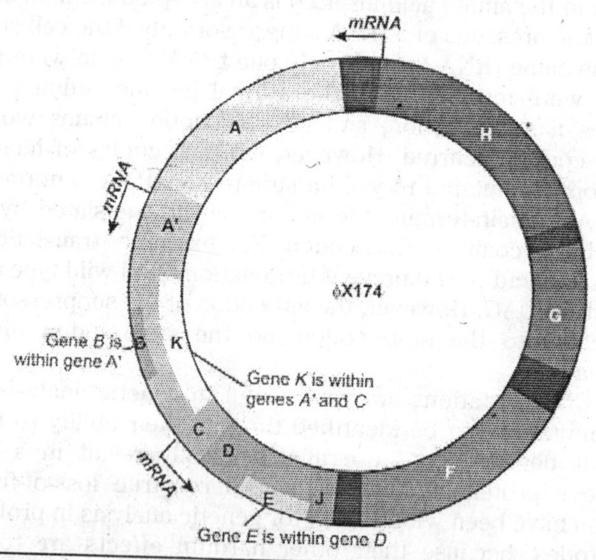

FIG. 38.9: Physical map of E. coli phage ϕX174. The red arrows show the start points for synthesis of the major mRNA transcripts, and the uppercase letters indicate the regions from which different protein products are translated. The solid black regions are untranslated spacers.

Phage ϕX174 contains a single strand of DNA consisting of 5386 nucleotides. If a single reading frame were used, at most 1795 amino acids could be encoded in the sequence, and with an average protein size of about 400 amino acids, only 4 or 5 proteins could be made. However, ϕX174 makes 11 proteins containing a total of more than 2300 amino acids. This paradox was resolved when it was shown that translation occurs in several reading frames from three mRNA molecules.

For example, the sequence for protein B is contained totally in the sequence for protein A' but is translated in a different reading frame; similarly, the sequence for protein E is included within the sequence for protein D. Protein K is initiated near the end of gene *A'*, includes the base sequence of gene *B*, and terminates in gene *C*; synthesis is not in phase with either gene *A'* or gene *C*. Of note is protein A', which is formed by initiating translation within the mRNA for protein A using the same reading frame, so that it terminates at the same stop codon as protein A. Thus, the amino acid sequence of A' is identical with a segment of protein A. In total, six different proteins obtain some or all of their primary structure from shared base sequences in ϕX174.

COMPLEX TRANSLATION UNITS

In Most prokaryotes and eukaryotes, the unit of translation is almost never simply one ribosome traversing an mRNA molecule. Rather, it is a more complex structure, of which there are several forms. Two examples are given in this section. After about 25 amino acids have been joined together in a polypeptide chain, an AUG initiation codon is completely free of the ribosome, and a second initiation complex can form. The overall configuration is that of two ribosomes moving along the mRNA at the same speed. When the second ribosome has moved along a distance similar to that traversed by the first, a third ribosome can attach to the initiation site. The process of movement and reinitiation continues until the mRNA is covered with ribosomes at a density of about one ribosome per 80 nucleotides. This large translation unit is called a polysome, and this is the usual form of the translation unit.

An mRNA molecule being synthesised has a free 5' terminus. Because translation takes place in the 5'-to-3' direction, the mRNA is synthesised in a direction appropriate for immediate translation. That is, the ribosome-binding site (in prokaryotes) and the 5' terminus (in eukaryotes) is transcribed first, followed in order by the initiating AUG codon, the region encoding the amino acid sequence, and finally the stop codon. Thus in prokaryotes, in which no nuclear envelope separates the DNA and the ribosome, the initiation complex can form before the mRNA is released from the DNA. This allows the simultaneous occurrence, or coupling, of transcription and translation. The lengths of the polysomes increase with distance from the transcription initiation site, because the mRNA is farther from that site, because the mRNA is farther from that site and hence of greater length because the process of transcription has been going on for a longer time. *Coupled transcription and translation does not take placein eukaryotes*, because the mRNA is synthesised and processed in the nucleus and later transported through the nuclear envelope to the cytoplasm where the ribosomes are located.

CHAPTER 39

Biotechnology for New Drug Discovery

During the last two decades, new techniques and advancements have added layers of complexity to the research done in biotechnology, drug discovery, and biomedical research. Many of the techniques developed have been computational in nature: in effect, taking their particular disciplines in silico (that is, to be executed within the computer). Often, these techniques are lumped under the term of **biomedical computing**. To visualize these advancements, compare the 'traditional' drug discovery cycle used by many companies until the early-1980's with the current biological target-based drug discovery cycle used by most pharma and biotech concerns today.

In the **traditional cycle**, an initial lead compound was found by isolating a molecule with a certain biological activity, perhaps by ethanobotany or by fortuitous happenstance (as in the discovery of penicillin), modifications of this lead compound were planned using clues provided by a crude analysis of structure-activity relationships (SAR) or by traditional medicinal chemistry techniques. The new, modified compounds were then synthesised and retested. This cycle continued until the particular biological activity of the compound was maximized. This cycle, while successful, was not rapid. It often took 5-6 years to bring a drug to the preclinical phase.

The current biological target-based cycle (also known as structure-based drug design is now recognizing that many drugs are antagonists (inhibitors) of macromolecules (most often an enzyme protein), a **biological target** molecule is chosen before any drug discovery project is begun.

The biological target is a macromolecule which is crucial for the biological activity or process which is to be inhibited. In some cases, biological target selection is simply a matter of common sense and examination of results from basic research.

For example, human immunodeficiency virus-1 (HIV) is expressed as a single polypeptide within an infected host cell. This polypeptide is then processed by a virally-encoded protease, the processed proteins are then packaged and the virus explodes from within the infected cell. It was therefore correctly surmised that HIV protease was critical for virus maturation and was an important biological target for drug discovery and development. This makes possible to develop several highly effective therapeutics for HIV. It is not always that simple, however, especially when the biological activity to be inhibited is not parasitic in nature (as in the case of a viral infection) or when the number of possible targets is enormous.

The computational field of bioinformatics plays a critical role by providing methodology for scanning the current **genomic databases** for an optimal target or targets and for predicting the activities of the as-of-yet uncharacterized genes and proteins. Assuming that one is able to select a target, several technologies come approximate method for the determination of the ground-state wave function and ground-state energy of a quantum many-body system. As the basis set size is increased, the energy and wave function tend towards a limit called the Hartree-Fock limit. Many types of calculations (known as post-Hartree-Fock methods) begin with a Hartree-Fock calculation and subsequently correct for electron-electron repulsion, referred to also as electronic correlation.

In computational chemistry, Post-Hartree-Fock methods are the set of methods developed to improve on the Hartree-Fock (HF), or self-consistent field (SCF) method. They add electron correlation which is a more accurate way of including the repulsions between electrons than in the Hartree-Fock method where repulsions are only averaged.

As these methods are pushed to the limit, they approach the exact solution of the non-relativistic Schrödinger equation. In order to obtain exact agreement with experiment, it is necessary to include relativistic and spin orbit terms, both of which are only really important for heavy atoms. In all of these approaches, in addition to the choice of method, it is necessary to choose a basis set. This is a set of functions, usually centered on the different atoms in the molecule, which are used to expand the molecular orbitals with the LCAO ansatz. Ab initio methods need to define a level of theory (the method) and a basis set.

The Hartree-Fock wave function is a single configuration or determinant. In some cases, particularly for bond breaking processes, this is quite inadequate, and several configurations need to be used. Here, the coefficients of the configurations and the coefficients of the basis functions are optimized together.

The total molecular energy can be evaluated as a function of the molecular geometry; in other words, the potential energy surface. Such a surface can be used for reaction dynamics. The stationary points of the surface lead to predictions of different isomers and the transition structures for conversion between isomers, but these can be determined without a full knowledge of the complete surface.

A particularly important objective, called computational thermochemistry, is to calculate thermochemical quantities such as the enthalpy of formation to chemical accuracy. Chemical accuracy is the accuracy required to make realistic chemical predictions and is generally considered to be 1 kcal/mol or 4 kJ/mol. To reach that accuracy in an economic way it is necessary to use a series of post-Hartree-Fock methods and combine the results. These methods are called quantum chemistry composite methods.

Density Functional Methods

Density functional theory (DFT) methods are often considered to be *ab initio* methods for determining the molecular electronic structure, even though many of the most common functional use parameters derived from empirical data, or from more complex calculations. This means that they could also be called semi-empirical methods. It is best to treat them as a class on their own. In DFT, the total energy is expressed in terms of the total one-electron density rather than the wave function. In this type of calculation, there is an approximate Hamiltonian and an approximate expression for the total electron density. DFT methods can be very accurate for little computational cost. The drawback is that, unlike ab initio methods, there is no systematic way to improve the methods by improving the form of the functional. Some methods combine the density functional exchange functional with the Hartree-Fock exchange term and are known as hybrid functional methods. A hybrid functional is an exchange-correlation functional used in density functional theory (DFT) that incorporates a portion of exact exchange from Hartree-Fock theory with exchange and correlation from other sources (ab initio, such as LDA, or empirical).

Semi-Empirical and Empirical Methods

Semi-empirical quantum chemistry methods are based on the Hartree-Fock formalism, but make many approximations and obtain some parameters from empirical data. They are very important in computational chemistry for treating large molecules where the full Hartree-Fock method without the approximations is too expensive. The use of empirical parameters appears to allow some inclusion of correlation effects into the methods.

Semi-empirical methods follow what are often called empirical methods, where the two-electron part of the Hamiltonian is not explicitly included. For -electron systems, this was the Hückel method proposed by Erich Hückel, and for all valence electron systems, the Extended Hückel method proposed by Roald Hoffmann.

Molecular Mechanics

In many cases, large molecular systems is modeled successfully while avoiding quantum mechanical calculations entirely. Molecular mechanics simulations, for example, use a single classical expression for the energy of a compound, for instance the harmonic oscillator. All constants appearing in the equations must be obtained beforehand from experimental data or ab initio calculations.

The database of compounds used for parameterization, i.e., the resulting set of parameters and functions is called the **force field**, is crucial to the success of molecular mechanics calculations. A force field parameterized against a specific class of molecules, for instance proteins, would be expected to only have any relevance when describing other molecules of the same class.

These methods can be applied to proteins and other large biological molecules, and allow studies of the approach and interaction (docking) of potential drug molecules.

Presently, the above techniques are used for estimating the strength of binding of a small molecule to a biological macromolecule, the ability to do complex electrostatic calculations on macromolecules, and modeling techniques which allow the modeling of novel macromolecules from other macromolecule structures of nominal homology.

The program GROWMOL is capable of generating organic molecules complementary to the binding site of a macromolecule assuming that the three-dimensional structure of the macromolecule is known. It has been successful in generating inhibitors for a variety of macromolecules and promises to add another wonderful tool to the computational chemist's toolbox. Dr. Bohacek is a research scientist at Ariad Pharmaceuticals.

PROTEIN ENGINEERING

There have been significant experimental contributions of protein engineering. Protein engineering is a technique for isolating and studying proteins and creating tailor-made proteins by altering the genes that direct their composition. The technology will often be used in conjunction with genetic modification to improve existing proteins, usually enzymes and helps in the production of proteins with new or artificial amino acid sequences. It also helps in design and construction of new proteins; an enabling technology for nanotechnology. In the early 1980's, researchers began to attempt to rationally engineer protein molecules in such a way as to modify existing structure, stabilize existing proteins against the stresses of heat or salt, add new activities or to modify existing activities, and even to design novel proteins *de novo*. The successes in this area have been quite impressive, especially in the last five years. As one might expect, molecular modeling has always played a pivotal role in these experiments. Often, however, its contribution has been qualitative in nature by allowing researchers to visualize the effect of point mutations within a protein structure.

Recently, researchers have been attacking the problem of *de novo* protein engineering with more quantitative approaches which demand, as one might imagine, a significant amount of computation. *De novo* protein design is essentially an attempt to solve the inverse folding problem: one is attempting to predict the structure of a protein given only its sequence or, vice-versa, given a particular fold or protein structure, to predict the sequences which can adopt that structure.

Over the years an impressive array of successful experiments has been carried out by applying qualitative rules of protein structure and function through inspection using computer graphics. These include changes in substrate specificity, introduction of metal binding sites for affinity purification, stabilization and allosteric contro and even the de novo creation of small proteins that adopt defined secondary structures. Nevertheless, it is clear that as soon as more than a few mutations are required, the design problem rapidly becomes too complex to solve by inspection. For instance, the designed proteins, although having globally correct folds, often resemble molten globules more than the tightly packed unique arrangements of the polypeptide chain found in the interior of natural proteins, unless cofactors such as metals are introduced to decrease the disorder. It is also very difficult to rationally design by inspection thermostable proteins, drastically changed substrate specificity, or create new enzyme activities. All of these problems are manifestations of the same obstacle: it is extremely hard to get all the details in a designed protein correct simultaneously.

Recent advances in molecular evolution have provided a set of experimental tools to generate populations of variants from which proteins with desirable properties can be isolated using various genetic selection and screening techniques.

Currently much of the research in protein engineering is devoted to the development and application of these techniques. Nevertheless, research into structure-based design has persisted, and several significant advances in rational structure-based design, driven by the emergence of automated protein design programs, have occurred recently. One of the latest examples of these new approaches is described in a recent article by Malakauskas and Mayo16 published last month in Nature Structural Biology, which reports the construction of an extremely hyperthermophilic variant of the B1 domain of Streptococcal protein G and clearly illustrates the technological potential of modern computational protein engineering approaches.

Dr. Homme Hellinga (Duke University Dept. of Biochemistry), was the first scientist to successfully introduce a new binding site *de novo* into an existing protein structure, in this case the introduction a copper-binding site into *E. coli* thioredoxin (H.W. Hellinga, J.P. Caradonna, F.M. Richards. J. *Mol. Biol.* **222**, 787-803, 1991). His work in protein engineering has been aided by a program which he has developed called DEZYMER. Over the past several years, he has extended his research and his successes in the area of protein engineering to include the introduction of a FeS_4 cubiodal cluster into thioredoxin and the introduction an extraneous iron superoxide dismutase active site into thioredoxin. The title of his talk will be "Structure- Based Design of Protein Function". Dr. Hellinga is an ...ciate professor at Duke University.

STRUCTURAL BIOLOGY

In the last decade, structural biology has transformed the face of biomedical research. Due to the improvement of equipment and experimental methodology, the advent of cloning, and the availability of rapid, inexpensive computers, the solving of macromolecular structures by NMR and X-ray crystallography has become commonplace. These techniques are allowing us to view the molecules that make up a living organism at an extraordinary resolution, sometimes better than 1 Angstrom (atomic level resolution). However, this is not to say that it is easy. To solve a novel macromolecular structure (be it protein, DNA, RNA, etc.) by NMR involves specific isotopic labelling of that molecule, collection of data, and assignment of resonances within that data, followed by initial structure generation and refinement of that structure. This process can take several months to several years to complete. Likewise, to solve a novel macromolecular structure by X-ray crystallography involves crystallization of the macromolecule in question, introduction of specific heavy-atoms into the protein crystal, collection of data in-house or at a synchrotron, location of the heavy-atom sites, and initial generation of electron density maps; this is followed by initial structure generation and refinement. As in the case of NMR, structure solution can take up to several years to complete, with the average being approximately 1-2 years. While this time is not prohibitive in a typical research effort where attention is directed towards a single protein, current cutting-edge research is demanding a more rapid approach. In conjunction with the massive genomic sequencing efforts underway, structural biologists are initiating structural genomic efforts where representative proteins from each genome are solved and used-in conjunction with molecular modeling-to predict the structure (or at least the fold) of homologous proteins within the genome in question and in other genomes as well.

Protein and DNA

A healthy mind and body require the coordinated action of billions of tiny molecular workers called proteins. Our genes contain the DNA scripts for manufacturing these proteins. Some proteins build our cells and other proteins work to allow us to think, smell, eat and breathe. Proteins are indispensable molecules in our bodies, and each has a unique three-dimensional shape that is well suited for its particular job; and if the shape of even one protein happens to go awry, there can be major consequences for human health. Misshapen proteins, especially those that make up part of the cell surface or cell membranes, are the culprits behind many diseases (cystic fibrosis, Alzheimer's disease, etc.). Membrane proteins which account for about 30% of the proteins in a cell also turn out to be one of the most important classes of proteins for health. They are major drug targets for the development of disease treatments.

There is a need of development of rapid, efficient, and dependable methods to produce protein samples that scientists can use to determine the three-dimensional structure, or shape, of a protein. The new effort are on the way to catalyse what is currently a hit-or-miss process into a organized, coordinated, systematic and streamlined routine, helping researchers clarify the role of protein shape in health and disease.

Many membrane proteins are complex biological machines consisting of several interlocking proteins working together. To understand how these machines work- and to learn how to fix them when they don't- researchers need to view the protein complexes in several different orientations, mimicking the way thes[illegible] assemblies twist and bend inside living cells. Extremely large amounts [illegible] quality protein are needed to perform these difficult experiments. It is necessary to determine structures for complexes that are formed between membrane proteins and their soluble partner proteins (for example, complexes of G-proteins with GPCRs) or membrane proteins and proteins of the intracellular and extracellular matrix.

The structural genomics helps to investigate the structure of the proteins coded within the genome of hypothermic archebacteria (*Pyrobaculum aerophilum*). Dr. Terwilliger has written a wonderful program called SOLVE which rapidly completes analysis of the diffraction data and generation of interpretable electron density maps.

BIOINFORMATICS/COMPUTATIONAL GENOMICS

Bioinformatics is the combination of biology and information technology. It is the branch of science that deals with the computer-based analysis of large biological data sets. Bioinformatics incorporates the development of databases to store and search data, and of statistical tools and algorithms to analyse and determine relationships between biological data sets, such as macromolecular sequences, structures, expression profiles and biochemical pathways. It is mainly used for map genomes and identifies genes, to determine protein structure and simulate protein interactions, to discover new therapeutic targets and design medicines aimed at the targets and to assess the effects of virtual mutations on gene function. Computers are required in bioinformatics for their processing speed (allowing repetitive tasks to be carried out quickly and systematically) and for their problem-solving power. It requires expert human input, and the integrity and quality of the source data. They are also critical because bioinformatics technology uses computational tools provided by the information technology revolution, such as statistical software, graphics simulation and database management, to organize and analyse information about biological systems, which for biotechnology, is information about cells and biological molecules. Using another product of the information revolution, the Internet, scientists broadcast this information around the world.

The last five years has seen an unforeseen amount of activity and success in the area of genomic sequencing, where researchers determine the actual DNA sequence of a genome of an entire organism with the help of bioinformatics.

At last count over 269 genomes are currently being sequenced or have been completed, spread between viruses (more than 186), procaryotes (65), and eucaryotes (18). It is anticipated that the sequencing of the human genome has been completed in year 2002. Of course, this data is worth little without the tools to interpret and mine it. This is a daunting task and program suites have been written to identify putative genes (also known as Open Reading Frames or ORFs), promoter regions, and other significant landmarks within this sequence data.

Translation and Open Reading Frame Search: Regions of DNA that encode proteins are first transcribed into messenger RNA and then translated into protein. By examining the DNA sequence alone we can determine the sequence of amino acids that will appear in the final protein. In translation codons of three nucleotides determine which amino acid will be added next in the growing protein chain. It is important then to decide which nucleotide to start translation, and when to stop, this is called an open reading frame.

Open Reading Frame

Open Reading frames where successive nucleotide triplets can be read as codons specifying amino acids and where the sequence of these triplets is not interrupted by stop codons. A reading frame, in biology, consists of three-nucleotide codon sets in either DNA or RNA that are contiguous and non-overlapping. An open reading frame (ORF) is a similar sequence that can be translated into a protein or a polypeptide. In any open reading frame, the start-code sequence or initiation codon that begins the protein is methionine ATG, and then stop-code sequence or termination codon ends it. The stop sequence is coded by what is termed a nonsense codon, or a codon that does not have an RNA match. There are only three nonsense codons: amber (UAG) ochre (UAA) and opal (UGA). As you can see, each one contains a "U" nucleotide, not normal to DNA.

Any time a geneticist finds a long open reading frame, it is an indication that there is probably a gene near it. Sometimes short open reading frames occur outside the genes, simply by chance, and these are generally examined and discarded. Open reading frames are crucial to genetic sequencing.

Once a gene has been sequenced it is important to determine the correct open reading frame (ORF). Every region of DNA has six possible reading frames, three in each direction. The reading frame that is used determines which amino acids will be encoded by a gene. Typically only one reading frame is used in translating a gene (in eukaryotes), and this is often the longest open reading frame. Once the open reading frame is known the DNA sequence can be translated into its corresponding amino acid sequence. An open reading frame starts with an atg (Met) in most species and ends with a stop codon (taa, tag or tga).

For example, the following sequence of DNA can be read in six reading frames. Three is the forward and three in the reverse direction. The three reading frames in the forward direction are shown with the translated amino acids below each DNA sequence. Frame **1** starts with the "a", Frame **2** with the "t" and Frame **3** with the "g". Stop codons are indicated by an "*" in the protein sequence. The longest ORF is in Frame **1**.

5'3'

Atgcccaagctgaatagcgtagaggggttttcatcatttgaggacgatgtataa

1 atg ccc aag ctg aat agc gta gag ggg ttt tca tca ttt gag gac gat gta taa

M P K L N S V E G F S S F E D D V *

2 tgc cca agc tga ata gcg tag agg ggt ttt cat cat ttg agg acg atg tat

C P S * I A * R G F H H L R T M Y

3 gcc caa gct gaa tag cgt aga ggg gtt ttc atc att tga gga cga tgt ata

A Q A E * R R G V F I I * G R C I

The interpretation of the data from these Human Genome projects is of significant interest to pharma, biotechnology concerns, and academic biomedical research, as each ORF might represent a significant biological target. The genomic databases allows the description and evaluation of the secondary/tertiary fold and active site of all identified ORFs within a genome, with an eye towards discovering suitable biological targets for structure-based drug design.

After development of neural networks it is possible to predict the three-dimensional protein fold of a putative gene, discriminating between introns (non-coding regions) and exons (coding regions) within putative genes (ORFs), and identification of putative RNA genes in genomic data. This approach allows a significant and comprehensive mining of any genomic data.

CHAPTER 40

New Therapeutic Approaches

Despite important advances in understanding its pathophysiology, therapy for various diseases with the help of different tools and techniques has been developed. There are several approaches of new therapies, however, a few of them have been described here.

Brain Tumours

The blood-brain barrier is a substantial obstacle for delivering anticancer agents to brain tumours, and new strategies for bypassing it are greatly needed for brain-tumor therapy. Intranasal delivery provides a practical, non-invasive method for delivering therapeutic agents to the brain and could provide an alternative to intravenous injection and convection-enhanced delivery. The rates were treated bearing intracerebral human tumor xenografts intranasally with GRN163 (Intranasal delivery of telomerase inhibitor GRN163). It is an oligonucleotide N3'→ P5'thio-phosphoramidate telomerase inhibitor. 3'-Fuorescein isothiocyanate (FITC)-labelled GRN163 was administered intranasal every 2 min as 6 μl drops into alternating sides of the nasal cavity over 22 min. FITC-labelled GRN163 was present in tumor cells at all time points studied, and accumulation of GRN163 peaked at 4 h after delivery. Moreover, GRN163 delivered intranasal, daily for 12 days, significantly prolonged the median survival from 35 days in the control group to 75.5 days in the GRN163-treated group. Thus, intranasal delivery of GRN163 readily bypassed the blood-brain barrier, exhibited favourable tumor uptake, and inhibited tumor growth, leading to a prolonged lifespan for treated rats compared to controls. This delivery approach appears to kill tumor cells selectively, and no toxic effects were noted in normal brain tissue. These data support further development of intranasal delivery of tumor-specific therapeutic agents for brain tumor patients.

Parkinson's Disease

The incidence of Parkinson's disease (PD) and other neurodegenerative disorders increases with old age although the reason(s) why is not clear. There is a direct link between aging and neurodegeneration as well as a potential new treatment against PD. The inhibitory action of SIRT2- a member of the sirtuin family is linked to aging and prevents the toxicity of the protein aggregates that are believed to be behind the neuronal death characteristic of PD. Interestingly, and contrarily to "classic" approaches that try to eliminate these aggregates, SIRT2 inhibition appears to work by "fusing" many small protein aggregates into larger (apparently less neuro-toxic) ones. The research by showing that aging and neurodegeneration can be linked opens the door to a total new type of therapeutic approaches not only towards PD, but also other neurodegenerative diseases linked to aberrant protein deposits in the brain and known to become worst with age.

In many neurodegenerative disorders, including PD, toxic aggregates of aberrant proteins found in the brain tissue of patients are believed to be behind the neural death characteristic of these diseases. In PD, the disorder seems to result from deposition of alpha-synuclein - a protein of unknown functions – which leads to neuron death in the substantia nigra (literally the black substance), a part of the brain responsible for motor control. Neurons within the substantia nigra produce the neurotransmitter dopamine (and so are called dopaminergic neurons) that acts as messenger between this "control centre" and other neurons in the body for a correct regulation of the body movements. If the substantia nigra degenerates, like it happens in PD, dopamine is no longer correctly produced and the patient starts showing the disease characteristic symptoms, which include tremors, rigidity, slowness of movements as well as problems in balance and coordination. Nevertheless, and contrary to many other neurodegenerative diseases, PD has a number of effective treatments that by increasing dopamine levels in patients' brain can efficiently alleviate the symptoms. But there is no cure and, as disease progresses, treatments become less and less efficient and full deterioration is inevitable.

Cardiovascular Disease

The increase in cardiovascular disease and its clinical complications occurs despite extensive use of effective drugs and, thus, development of new and complementary therapeutic approaches is urged.

Alzheimer's Disease (AD)

Research into Alzheimer's disease (AD) pathology has identified several underlying disease processes that are potential targets for drug discovery and development. One strategy targets glutamatergic neurotransmission mediated by the N-methyl-D-aspartate (NMDA) receptor. Therapeutic intervention with high-affinity NMDA receptor antagonists, such as phencyclidine (PCP) and MK-801, is not practical due to adverse side effects; however, a low-moderate affinity, uncompetitive and strongly voltage-dependent NMDA receptor antagonist, memantine (NamendaTM), is well tolerated and recently has been approved by the U.S. Food and Drug Administration for the treatment of moderate to severe AD. Clinical results support NMDA receptor antagonism as a viable therapeutic strategy for AD and suggest that this novel pharmacologic approach, either alone or in combination with other drugs, is likely to significantly impact the current AD treatment paradigm.

It is known that oxidized lipids produced by the body initiate a cascade of responses in the vascular system that cause an inflammatory response and can lead to the development of atherosclerosis. However, the reason that these oxidized lipids cause an inflammatory response has previously been unknown. The oxidized lipids can activate **tight junctions** - which are like gates that connect the cells in the body - and it is this opening and closing of the tight junctions between cells that can cause an inflammatory response. The interactions between oxidized lipids and tight junctions of cells have developed the prospect for a new therapeutic approach to cardiovascular disease.

Fibrillar proteins, such as collagens type I and III, and elastin are components of the extracellular matrix. They form an intricate widespread network that provides a basis for maintaining the physical structure of the heart and vessels and also play an important role in determining cardiovascular function. Physiologically, collagen and elastin fibres are enzymatically cross-linked to form matrix. In addition to these enzymatically formed cross-links, collagen fibres may be linked non-enzymatically, most notably by formation of AGEs (**advanced glycation end-products**).

AGEs are formed by a reaction between reducing sugars and body proteins;

they are formed increasingly in diabetes mellitus and hypertension and they accumulate with aging. There are several mechanisms whereby AGEs may affect cardiovascular structure and function. These include increased myocardial and vascular stiffness and (upon reaction with their receptors) inflammatory reactions, release of growth factors and cytokines, and increased oxidative stress. Therefore breaking AGEs appears as a promising tool in the therapy of cardiovascular injury related to diabetes, hypertension and aging. Breakers of AGE cross-links have been developed and one of them, alagebrium, has been extensively studied. This brief review discusses the formation of AGEs, their role in mediating cardiovascular injury, as well as the results of experimental and clinical studies involving alagebrium.

Lipid Treatment

High levels of LDL cholesterol (LDL-c) are causally related to the development of atherosclerosis. A large body of epidemiological evidence supports a direct relationship between the level of serum LDL-c and the risk of cardiovascular disease, and clinical trials have shown that lowering LDL-c with statins leads to a 30% decrease in the risk of cardiovascular events. However, many patients taking statins continue to suffer cardiovascular disease, highlighting the need both to improve the number of patients reaching LDL-C therapeutic goals and to develop new therapeutic targets. The apoB/apoA-I ratio has been shown to be a major contributor to the risk of myocardial infarction and hence increasing HDL-cholesterol (HDL-C) offers a promising approach to further decrease cardiovascular risk. Clinical trial data with fibrates and nicotinic acid, together with preliminary data on potent new treatments, support HDL-c as a therapeutic target. The cardio protective effect of HDL-c may not solely be related to its role in reverse cholesterol transport.

Statin therapy: Reduction of low-density lipoprotein cholesterol (LDL-C) with statin therapy is currently identified in treatment guidelines as the primary focus for patients with or at risk of coronary heart disease (CHD). Yet despite effective statin therapy there is still an unacceptably high residual coronary risk. A substantial proportion of patients with CHD have mixed dyslipidemia, including low levels of high-density lipoprotein cholesterol (HDL-C), an independent and predictive risk factor for CHD. Although effective in reducing LDL-C, statin therapy has only modest effects in raising HDL-C.

Fibrate therapy: Fibrate therapy is an alternative lipid-modifying strategy, and is effective in reducing CHD mortality and morbidity, with the magnitude of clinical benefit similar to statin therapy. Multi-drug therapy with complementary mechanisms of action has been proposed as a means of improving lipid-modifying efficacy. **Nicotinic acid** is the most potent agent for increasing HDL-C and also substantially reduces LDL-C and triglycerides. Addition of nicotinic acid to statin therapy would be a logical management approach, given the potential for complementary therapeutic benefit. The clinical benefits of this combination are supported by the results of the HDL Atherosclerosis Treatment Study, which showed reduction of 60-90% in the incidence of major coronary events when both agents were administered. In addition, combination treatment led to angiographic regression of stenosis, compared with placebo, rather than slowed progression as previously reported with statin mono-therapy. Given that the prevalence of low HDL-C, particularly amongst individuals with CHD, is higher than previously anticipated, combining nicotinic acid and a statin represents an innovative approach to further reducing CHD risk.

Cholesteryl ester transfer protein: There is substantial interest in the concept of inhibition of the cholesteryl ester transfer protein (CETP) as a novel strategy for raising HDL cholesterol levels, as well as reducing levels of atherogenic lipoproteins.

ATP binding cassette transporter (ABCA1): Cholesterol is a major cardiovascular disease risk factor, therapeutic interventions to lower plasma cholesterol levels have had limited success in reducing coronary events, underscoring the need for other treatment strategies. A promising therapeutic target is an ATP binding cassette transporter called ABCA1, a cell membrane protein that is the gatekeeper for secretion of excess cholesterol from macrophages into the high-density lipoprotein (HDL) metabolic pathway. Mutations in ABCA1 cause Tangier disease, a severe HDL-deficiency syndrome characterized by accumulation of cholesterol in tissue macrophages and prevalent atherosclerosis. Thus, ABCA1-activating drugs have the potential to mobilize cholesterol from macrophages of atherosclerotic lesions, making them powerful agents for preventing and reversing cardiovascular disease.

C-reactive protein (CRP): Circulating concentrations of C-reactive protein (CRP), the classical acute phase protein, are strongly predictive of atherothrombotic events and of outcome following myocardial infarction. CRP binds modified LDL, activates complement, and is deposited in atherosclerotic lesions, suggesting that it may contribute to atherogenesis and plaque destabilisation. CRP is also deposited, together with activated complement, in all acute myocardial infarction lesions, and injection of human CRP increases myocardial infarction size after coronary artery ligation in rats, via a complement-dependent mechanism. Inhibition of CRP-binding and its effects is therefore an attractive therapeutic target in acute myocardial infarction and possibly also to prevent atheroma and atherothrombosis. We report here a novel high throughput screening assay for inhibitors of CRP-binding to enzyme-modified LDL, a potentially critical pathophysiological ligand in cardiovascular disease.

Liver X receptors (LXRs)

Liver X receptors (LXRs) are cholesterol sensors within cells and play a critical role in the modulation of lipid and lipoprotein metabolism. Reverse cholesterol transport, bile acid synthesis and intestinal cholesterol absorption are all positively affected by LXR activation. In mice, LXR agonists have demonstrated beneficial effects on atherosclerosis, but also resulted in unfavorable effects on triglyceride metabolism, mainly due to the activation of LXRα in the liver. Activation of LXRß reduces atherosclerosis and increases HDL cholesterol, without raising plasma triglyceride levels, making it an attractive therapeutic target.

Psoriasis

Chronic plaque psoriasis is an immune-mediated, inflammatory skin disease with a heavy burden on quality of life of patients. The disease has a chronic relapsing course and may be lifelong. Comorbid disorders include psoriatic arthritis, obesity, dyslipidemia, hypertension and an increased rate of cardiovascular disease. Conventional systemic treatments include methotrexate, cyclosporine and acitretin, which are associated with end organ toxicity that precludes long term therapy. Biological drugs are designed to selectively interfere with the immune mechanisms that induce psoriasis. Efalizumab is effective for skin psoriasis but not psoriatic arthritis. Anti-TNF- agents (etanercept, infliximab and adalimumab) are active on both psoriasis and psoriatic arthritis. Infliximab is the most effective and rapid agent, but its safety profile may be less favourable. Moreover, efficacy can reduce over time. Etanercept is moderately active but has a better safety profile, and can be discontinued and re-used without loss of efficacy. The long term safety of all these agents has not been established.

Thrombosis

Low-molecular-weight (LMW) heparins are increasingly used in the prevention and treatment of thromboembolic disease because, compared to unfractionated heparins, they show a higher benefit/risk ratio and better clinical convenience. LMW heparins is used in the prophylaxis of venous thromboembolism after surgery and in nonsurgical situations, as well as in the treatment of established venous thrombosis.

Asthma

Nicotinic agonists (such as nicotine) are a family of molecules that stimulate nicotinic receptors present on inflammatory and smooth muscle cells. Stimulation of these receptors provokes the down-regulation of these cells. The treatment of asthma, a disease characterised by airway inflammation and broncho-constriction, has hardly changed over the past 20 years. There is definitely still a need for new therapeutic approaches for this highly prevalent disease. Nicotinic agonist, dimethylphenylpiperazinium (DMPP), can prevent the development of asthma in mice. The interest of DMPP and other agonists of similar nature is that they are hydrosoluble and therefore, in contrast to nicotine, do not readily cross the blood brain barrier and should not cause addiction.

Chronic Fatigue Syndrome (CFS)

Chronic fatigue syndrome (CFS) is a disorder of unknown etiology, consisting of prolonged, debilitating fatigue, and a multitude of symptoms including neurocognitive dysfunction, flu-like symptoms, myalgia, muscle weakness, arthralgia, low- grade fever, sore throat, headache, sleep disturbances, and swelling and tenderness of lymph nodes. No effective treatment for CFS is known. NADH (ENADA®) may be an effective therapy in the management of Chronic fatigue syndrome, however, it requires further clinical trials be performed to establish its efficacy in this clinically perplexing disorder.

Drug Abuse

After a period of forced exposure to alcohol, rats develop a preference for alcohol when given the choice to drink either an alcohol solution or water. The development of alcohol preference was dependent on the alcohol concentration but independent of the presence of an alcohol withdrawal phase. The highest alcohol consumption was found with a 3% solution. The **serotonin 5-HT$_2$** antagonist ritanserin reduced alcohol intake and alcohol preference in a dose-related manner. Changes in test procedures and injection schedules did not affect the effectiveness of ritanserin. The activity of ritanserin started at doses ≥0.04 mg/kg and in the active dose range ritanserin reduced alcohol intake by 46 to 67% and alcohol preference by 33 to 42%. At any time during treatment, the decrease in alcohol drinking was compensated by an increase in water consumption.

CHAPTER 41

Kidney and Liver Function Tests

KIDNEY FUNCTION TESTS

The kidneys regulate the amount of water and salts that we have in our bodies. They do this by filtering the blood through millions of structures called nephrons. The kidneys also pass out certain waste products from the body. Urine is made up of the excess water, salts and waste products passed out by the kidneys down to the bladder.

Elevated creatinine and blood urea nitrogen (BUN) levels are likely evidence of decreased kidney function. Creatinine and urea nitrogen are waste products that your kidneys normally remove from your blood. If your kidneys are not working properly, these substances may build up in your blood. Kidney function tests include blood tests that measure creatinine and urea nitrogen.

If you have elevated creatinine and BUN, your doctor will likely recommend additional tests to further evaluate your kidney function. Tests may include:

Kidney function tests are a collective term for a variety of individual tests and procedures that can be done to evaluate how well the kidneys are functioning.

The kidneys, the body's natural filtration system, perform many vital functions, including removing metabolic waste products from the bloodstream, regulating the body's water balance, and maintaining the pH (acidity/alkalinity) of the body's fluids. Approximately one and a half quarts of blood per minute are circulated through the kidneys, where waste chemicals are filtered out and eliminated from the body (along with excess water) in the form of urine. Kidney function tests help to determine if the kidneys are performing their tasks adequately.

Important Terms

- **Blood urea nitrogen (BUN):** The nitrogen portion of urea in the bloodstream. Urea is a waste product of protein metabolism in the body.
- **Creatinine:** The metabolized by-product of creatine, an organic acid that assists the body in producing muscle contractions. Creatinine is found in the bloodstream and in muscle tissue. It is removed from the blood by the kidneys and excreted in the urine.
- **Osmolality:** A measurement of urine concentration that depends on the number of particles dissolved in it. Values are expressed as milliosmols per kilogram (mOsm/kg) of water.
- **Urea:** A by-product of protein metabolism that is formed in the liver. Because urea contains ammonia, which is toxic to the body, it must be quickly filtered from the blood by the kidneys and excreted in the urine.

A complete history should be taken prior to kidney function tests to assess the patient's food and drug intake. A wide variety of prescription and over-the-counter medications can affect blood and urine kidney function test results, as can some food and beverages.

Several conditions can affect the ability of the kidneys to carry-out their vital functions. Some lead to a rapid (acute) decline in kidney function; others lead to a gradual (chronic) decline in function. Both result in a build-up of toxic waste substances in the blood. A number of clinical laboratory tests that measure the levels of substances normally regulated by the kidneys can help determine the cause and extent of kidney dysfunction. These tests are done on urine samples, as well as on blood samples.

Urine Tests

There are a variety of urine tests that assess kidney function. A simple, inexpensive screening test, called a routine urinalysis, is often the first test administered if kidney problems are suspected. A small, randomly collected urine sample is examined physically for things like colour, odour, appearance, and concentration (specific gravity); chemically for substances such as protein, glucose, and pH (acidity/ alkalinity); and microscopically for the presence of cellular elements (red blood cells, white blood cells, and epithelial cells), bacteria, crystals, and casts (structures formed by the deposit of protein, cells, and other substances in the kidneys' tubules). If results indicate a possibility of disease or impaired kidney function, one or more of the following additional tests is usually performed to more specifically diagnose the cause and the level of decline in kidney function.

1. **Creatinine clearance test**. This test evaluates how efficiently the kidneys clear a substance called creatinine from the blood. Creatinine, a waste product of muscle energy metabolism, is produced at a constant rate that is proportional to the muscle mass of the individual. Because the body does not recycle it, all of the creatinine filtered by the kidneys in a given amount of time is excreted in the urine, making creatinine clearance a very specific measurement of kidney function. The test is performed on a timed urine specimen-a cumulative sample collected over a two to twenty-four hour period. Determination of the blood creatinine level is also required to calculate the urine clearance.
2. **Urea clearance test**. Urea is a waste product that is created by protein metabolism and excreted in the urine. The urea clearance test requires a blood sample to measure the amount of urea in the bloodstream and two urine specimens, collected one hour apart, to determine the amount of urea that is filtered, or cleared, by the kidneys into the urine.
3. **Urine osmolality test**. Urine osmolality is a measurement of the number of dissolved particles in urine. It is a more precise measurement than specific gravity for evaluating the ability of the kidneys to concentrate or dilute the urine. Kidneys that are functioning normally will excrete more water into the urine as fluid intake is increased, diluting the urine. If fluid intake is decreased, the kidneys excrete less water and the urine becomes more concentrated. The test may be done on a urine sample collected first thing in the morning, on multiple timed samples, or on a cumulative sample collected over a twenty-four hour period. The patient will typically be prescribed a high-protein diet for several days before the test and asked to drink no fluids the night before the test.
4. **Urine protein test**. Healthy kidneys filter all proteins from the bloodstream and then reabsorb them, allowing no protein, or only slight amounts of protein, into the urine. The persistent presence of significant amounts of protein in the urine, then, is an important indicator of kidney disease. A positive screening test for protein (included in a routine urinalysis) on a random urine sample is

usually followed-up with a test on a 24-hour urine sample that more precisely measures the quantity of protein.

Blood Tests

There are also several blood tests that can aid in evaluating kidney function. These include:

Blood Urea Nitrogen Test (BUN)

Urea is a by-product of protein metabolism. This waste product is formed in the liver, then filtered from the blood and excreted in the urine by the kidneys. Urea nitrogen also is produced from the breakdown of food protein. A normal BUN level is between 7 and 20 mg/dL. As kidney function decreases, the BUN level increases.

The BUN test measures the amount of nitrogen contained in the urea. High BUN levels can indicate kidney dysfunction, but because blood urea nitrogen is also affected by protein intake and liver function, the test is usually done in conjunction with a blood creatinine, a more specific indicator of kidney function.

Creatinine Test

Serum creatinine: Creatinine (kree-AT-uh-nin) is a waste product that comes from meat protein in the diet and from the normal wear and tear on muscles of the body. Creatinine levels in the blood can vary, and each laboratory has its own normal range. In many labs the normal range is 0.6 to 1.2 mg/dL. Higher levels may be a sign that the kidneys are not working properly. As kidney disease progresses, the level of creatinine in the blood increases.

This test measures blood levels of creatinine, a by-product of muscle energy metabolism that, like urea, is filtered from the blood by the kidneys and excreted into the urine. Production of creatinine depends on an individual's muscle mass, which usually fluctuates very little. With normal kidney function, then, the amount of creatinine in the blood remains relatively constant and normal. For this reason, and because creatinine is affected very little by liver function, an elevated blood creatinine is a more sensitive indication of impaired kidney function than the BUN.

Creatinine Clearance

A creatinine clearance test compares the creatinine in a 24-hour sample of urine to the creatinine level in the blood, to show how many milliliters of blood the kidneys are filtering out each minute (mL/min). The creatinine clearance can also be estimated accurately from the serum creatinine alone using well-established prediction equations.

Other Blood Tests

Measurement of the blood levels of other elements regulated in part by the kidneys can also be useful in evaluating kidney function. These include sodium, potassium, chloride, bicarbonate, calcium, magnesium, phosphorus, protein, uric acid, and glucose.

Preparation

Patients will be given specific instructions for collection of urine samples, depending on the test to be performed. Some timed urine tests require an extended collection period of up to 24 hours, during which time the patient collects all urine voided and transfers it to a specimen container. Refrigeration and/or preservatives are typically required to maintain the integrity of such urine specimens. Certain dietary and/or medication restrictions may be imposed for some of the blood and urine tests. The patient may also be instructed to avoid exercise for a period of time before a test.

Dissolved Salts

Dissolved salts that are routinely measured are sodium, potassium, chloride and bicarbonate. They are sometimes referred to as 'electrolytes'. Abnormal blood levels of any of these may be due to a kidney problem. (Some other conditions may also alter the salt balance in the blood.)

Other Tests of Kidney Function

The routine kidney blood test is a general marker of kidney function. If the blood test is abnormal it cannot say what is causing the kidney problem. Therefore, if you have an abnormal result you may need further tests to find the cause of a kidney problem. Urine tests to look for increased protein, red blood cells and casts in your urine as indicators of impaired kidney function. Glomerular filtration rate should be tested to determine your kidney function as a per cent of normal. The other important testes are other blood tests, scans, x-rays, kidney biopsy, etc.

Aftercare

If medication was discontinued prior to a urine kidney function test, it may be resumed once the test is completed.

Risks

Risks for these tests are minimal, but may include slight bleeding from a blood-drawing site, hematoma (accumulation of blood under a puncture site), or fainting or feeling light-headed after venipuncture. In addition, suspension of medication or dietary changes imposed in preparation for some blood or urine tests may trigger side-effects in some individuals.

Normal Results

Normal values for many tests are determined by the patient's age and sex. Reference values can also vary by laboratory, but are generally within the ranges that follow.

Urine Tests

1. **Creatinine clearance:** For a 24-hour urine collection, normal results are 90-139 ml/min for adult males less than 40 years old, and 80-125ml/min for adult females less than 40 years old. For people over 40, values decrease by 6.5 ml/min for each decade of life.
2. **Urea clearance:** With maximum clearance, normal is 64-99 ml/min.
3. **Urine osmolality:** With restricted fluid intake (concentration testing), osmolality should be greater than 800 mOsm/kg of water. With increased fluid intake (dilution testing), osmolality should be less than 100 mOSm/kg in at least one of the specimens collected.
4. **Urine protein:** A 24-hour urine collection should contain no more than 150 mg of protein.

Blood Tests

1. Blood urea nitrogen (BUN). 8-20 mg/dl.
2. Creatinine. 0.8-1.2 mg/dl for males and 0.6-0.9 mg/dl for females.

Abnormal Results

Low clearance values for creatinine and urea indicate diminished ability of the kidneys to filter these waste products from the blood and excrete them in the urine. As clearance levels decrease, blood levels of creatinine and urea nitrogen increase.

Since it can be affected by other factors, an elevated BUN, by itself, is suggestive, but not diagnostic, for kidney dysfunction. An abnormally elevated blood creatinine, a more specific and sensitive indicator of kidney disease than the BUN, is diagnostic of impaired kidney function.

Inability of the kidneys to concentrate the urine in response to restricted fluid intake, or to dilute the urine in response to increased fluid intake during osmolality testing may indicate decreased kidney function. Because the kidneys normally excrete almost no protein in the urine, its persistent presence, in amounts that exceed the normal 24-hour urine value, usually indicates some type of kidney disease as well.

LIVER FUNCTION TESTS

The liver is the chemical factory of the body. It performs many important following functions:

1. Storing some nutrients.
2. Converting fats to energy when the body needs it.
3. Producing bile.
4. Making blood proteins.
5. Helping the blood to clot.
6. Making other chemicals the body needs.
7. Breaking down harmful substances.
8. A blood test can show problems with these processes, and indicate what could be causing the problems.

Liver function tests are necessary as part of a general health check when you are first diagnosed with cancer, or many other illnesses. They also help to diagnose the cause of symptoms, such as jaundice. This is yellowing of the skin and whites of the eyes. Liver function tests may help the doctor find the cause. Such testes are necessary before treatment starts. Many drugs affect the liver so your doctor will want to check on how well your liver is working before during and after your treatment and to check for secondary cancer. It wouldn't be the only test you need, but helps the doctors find out whether the cancer is affecting your liver. Other tests you may need include an ultrasound scan, CT scan.

Liver function tests (LFT) are a helpful screening tool, which are an effective modality to detect hepatic dysfunction. Since the liver performs a variety of functions so no single test is sufficient to provide complete estimate of function of liver. Often clinicians are faced with reports that do not tally with the clinical condition of the patient and they face difficulty in interpreting the LFT. An attempt is being made to study and understand the LFT and simplify their interpretation with algorithms.

Liver function tests represent a broad range of normal functions performed by the liver. The diagnosis of liver disease depends upon a complete history, complete physical examination, and evaluation of liver function tests and further invasive and non-invasive tests. Many patients become confused regarding the meaning of a liver function test. This section is designed to describe the basic liver function tests and the meaning for patients.

The hepatobiliary tree represents hepatic cells and biliary tract cells. Inflammation of the hepatic cells results in elevation in the alanine aminotransferase (ALT), aspartate aminotransferase (AST) and possibly the bilirubin. Inflammation of the biliary tract cells results predominantly in an elevation of the alkaline phosphatase. In liver disease there are crossovers between purely biliary disease and hepatocellular disease. To interpret these, the physician will look at the entire picture of the hepatocellular disease and biliary tract disease to determine which is the primary abnormality.

Alanine Aminotransferase (Alt)

ALT is the enzyme produced within the cells of the liver. The level of ALT abnormality is increased in conditions where cells of the liver have been inflamed or undergone cell death. As the cells are damaged, the ALT leaks into the bloodstream leading to a rise in the serum levels. Any form of hepatic cell damage can result in an elevation in the ALT. The ALT level may or may not correlate with the degree of cell death or inflammation. ALT is the most sensitive marker for liver cell damage.

Aspartate Ainotransferase (AST) and Alanine Aminotransferase (ALT)

This enzyme also reflects damage to the hepatic cell. It is less specific for liver disease. It may be elevated and other conditions such as a myocardial infarct (heart attack). Although AST is not a specific for liver as the ALT, ratios between ALT and AST are useful to physicians in assessing the etiology of liver enzyme abnormalities.

AST levels indicate the degree of inflammation. The aminotransferases are enzymes that are present in the liver cells (hepatocytes). They leak into the blood stream when the liver cells are damaged. These values are usually high in hepatitis – possibly twenty to fifty times higher than normal. The ALT value is more specific to the liver than the AST value. The AST value may also give an indication of muscle damage elsewhere in the body. Ratios of these enzymes can be helpful in NASH and alcohol-related liver disease.

Alkaline Phosphatase

Alkaline phosphatase (ALP), an enzyme found mainly in the bile ducts of the liver. Increases in ALP and another liver enzyme called **Gamma GT** (GGT) can indicate obstructive or cholestatic liver disease, where bile is not properly transported from the liver because of obstruction (blockage) of the bile duct. GGT is tested with ALP to make sure that ALP increases are coming from your liver. The GGT is also a potential indicator of your alcohol usage. It is not specific to the biliary tract. It is also found in bone and the placenta. Renal or intestinal damage can also cause the alkaline phosphatase to rise. If the alkaline phosphatase is elevated, biliary tract damage and inflammation should be considered. However, considering the above other etiologies must also be entertained. One way to assess the etiology of the alkaline phosphatase is to perform a serologic evaluation called isoenzymes. Another more common method to asses the etiology of the elevated alkaline phosphatase is to determine whether the GGT is elevated or whether other function tests are abnormal (such as bilirubin).

Alkaline phosphatase may be elevated in primary biliary cirrhosis, alcoholic hepatitis, PSC, gallstones in choledocholithiasis.

Gamma Glutamic Transpeptidase (GGT)

This enzyme is also produced by the bile ducts. However, it is not very specific to the liver or bile ducts. It is used often times to confirm that the alkaline phosphatase is of the hepatic etiology. Certain GGT levels, as an isolated finding, reflect rare forms of liver disease. Medications commonly cause GGT to be elevated. Liver toxins such as alcohol can cause increases in the GGT.

Bilirubin

Bilirubin, formed from haemoglobin and the main pigment in bile (a yellow/green substance made by your liver). An increase of bilirubin causes jaundice, a yellowing of the eyes and skin in liver disease. Hemoglobin is derived from red cells that have outlived their natural life and subsequently have been removed by the spleen. During splenic degradation of red blood cells, hemoglobin (the part of the red blood cell that carries oxygen to the tissues) is separated out from iron and cell membrane components. Hemoglobin is transferred to the liver where it undergoes further metabolism in a process called conjugation. Conjugation allows hemoglobin to become more water-soluble. The water solubility of bilirubin allows the bilirubin to be excreted into bile. Bile then is used to digest food.

As the liver becomes irritated, the total bilirubin may rise. It is then important to understand the difference between total bilirubin, which has undergone conjugation (that is hepatic cell metabolism), and at portion of bilirubin which has not been metabolized. These two components are called total bilirubin and direct bilirubin. The direct bilirubin fraction is that portion of bilirubin that has undergone metabolism by the liver. When this fraction is elevated, the cause of elevated bilirubin (hyperbilirubinemia) is usually outside the liver. These types of causes are typically gallstones. This type of abnormality is usually treated with surgery (such as a gallbladder removal or choleycystectomy).

If the direct bilirubin is low, while the total bilirubin is high, this reflects liver cell damage or bile duct damage within the liver itself.

Albumin

Albumin, a very important protein that helps keep fluid pressures in the body stable and carries many substances in the body. Albumin may decrease in chronic liver disease, particularly if the disease is getting worse, but may be decreased for other reasons such as a lack (deficiency) of protein, for example malnutrition.

Albumin is the major protein present within the blood. Albumin is synthesised by the liver. As such, it represents a major synthetic protein and is a marker for the ability of the liver to synthesise proteins. It is only one of many proteins that are synthesised by the liver. However, since it is easy to measure, it represents a reliable and inexpensive laboratory test for physicians to assess the degree of liver damage present in the in any particular patient. When the liver has been chronically damaged, the albumin may be low. This would indicate that the synthetic function of the liver has been markedly diminished. Such findings suggest a diagnosis of cirrhosis. Malnutrition can also cause low albumin (hypoalbuminemia) with no associated liver disease.

Prothrombin Time (PT)

Another measure of hepatic synthetic function is the prothrombin time. Prothrombin time is affected by proteins synthesised by the liver. Particularly, these proteins are associated with the incorporation of vitamin K metabolites into a protein. This allows normal coagulation (clotting of blood). Thus, in patients who have prolonged prothrombin times, liver disease may be present. Since a prolonged PT is not a specific test for liver disease, confirmation of other abnormal liver tests is essential. This may include reviewing other liver function tests or radiology studies of the liver. Diseases such as malnutrition, in which decreased vitamin K ingestion is present, may result in a prolonged PT time. An indirect test of hepatic synthetic function includes administration of vitamin K (10mg) subcutaneously over three days. Several days later, the prothrombin time may be measured. If the prothrombin time becomes normal, then hepatic synthetic function is intact. This test does not indicate that there is no liver disease, but is suggestive that malnutrition may coexist with (or without) liver disease.

Platelet Count

Platelets are cells that form the primary mechanism in blood clots. They're also the smallest of blood cells. They derived from the bone marrow from the larger cells known as megakaryocytes. Individuals with liver disease develop a large spleen. As this process occurs platelets are trapped with in the sinusoids (small pathways within the spleen) of the spleen. While the trapping of platelets is a normal function for the spleen, in liver disease it becomes exaggerated because of the enlarged spleen (splenomegaly). Subsequently, the platelet count may become diminished.

Serum Protein Electrophoresis

This is an evaluation of the types of proteins that are present with in a patient's serum. By using an electrophoretic gel, major proteins can be separated out. This results in four major types of proteins. These are 1) albumin, 2) alpha globulins, 3) beta globulins and 4) gammaglobulins. This test is useful for evaluation of patients who have abnormal liver function tests since it allows a direct quantification of multiple different serum proteins. If the gamma globulin fraction is elevated, autoimmune hepatitis may be present. In addition a deficiency in the alpha globulin fraction can result in the diagnosis, or a clinical clue, to A. alpha-1 antitrypsin deficiency. This is a simple blood test that is commonly performed by hepatologists.

CHAPTER 42

Alkaloids

Alkaloid is the basic nitrogen compounds (mostly heterocyclic) occurring mostly in the plant kingdom (but not excluding those of animal origin). They have also been isolated from such diverse organisms as frogs, ants (pheromones), butterflies (defense), marine bacteria, sponges, fungi, spiders (venom neurotoxin), beetles (defense), and mammals, although is not yet clear whether de novo alkaloid biosynthesis occurs in each organism.

Amino acids, peptides, proteins, nucleotides, nucleic acids, amino sugars, and antibiotics are not normally regarded as alkaloids. By extension, certain neutral compounds biogenetically related to basic alkaloids are included. One of the largest groups of chemical arsenals produced by plants is the alkaloid. Many of these metabolic by-products are derived from amino acids and include an enormous number of bitter, nitrogenous compounds. More than 10,000 different alkaloids have been discovered in species from over 300 plant families. The use of Cinchona bark to treat malaria was first reported in Europe in 1633. The first alkaloid was isolated originally from *Cinchona succirubra*, quinine is one of 31 alkaloids with related structures, and the principal antimalarial compound, in the plant.

Alkaloids contain nitrogen - usually derived from an amino acid. They are bitter in taste, generally white solids (exception - nicotine is a brown liquid). They give a precipitate with heavy metal iodides. Most alkaloids are precipitated from neutral or slightly acidic solution by Mayer's reagent (potassiomercuric iodide solution). Cream coloured precipitate. Dragendorff's reagent (solution of potassium bismuth iodide) gives orange coloured precipitate with alkaloids. Caffeine, a purine derivative, does not precipitate like most alkaloids. Alkaloids are basic - they form water soluble salts. Most alkaloids are well-defined crystalline substances which unite with acids to form salts. In plants, they may exist in the free state, as salts or as N-oxides. They occur in a limited number of plants. Nucleic acid exists in all plants, whereas, morphine exists in only one plant species.

PHYSICAL AND CHEMICAL PROPERTIES

Physical

Alkaloids are colourless, crystalline, non-volatile solids that are insoluble in water but soluble in organic solvents such as ethanol, ether, chloroform, etc. Their Molecular weight ranges from 100-900. Those alkaloids which are without oxygen atoms in the structure are usually liquid at ordinary temperature (e.g. nicotine, sparteine, and coniine) while those with oxygen atoms are crystalline (e.g. berberine to yellow).

Chemical Properties

Alkaloids often contain one or more rings of carbon atoms, usually with a nitrogen atom in the ring. The position of the nitrogen atom in the carbon ring varies with different alkaloids and with different plant families. In some alkaloids, such as mescaline, the nitrogen atom is not within a carbon ring. In fact, it is the precise position of the nitrogen atom that effects the properties of these alkaloids.

Most are optically active (usually laevorotatory), but not those from the purine group. They are basicity depends on the availability of lone pairs of electrons on the N atom and the type of hetero cycle and substituents. They have the ability to form salts with mineral acids (e.g. with HCl, H_2SO_4, HNO_3) or organic acids (e.g. tartrates, sulphamates, maleates).

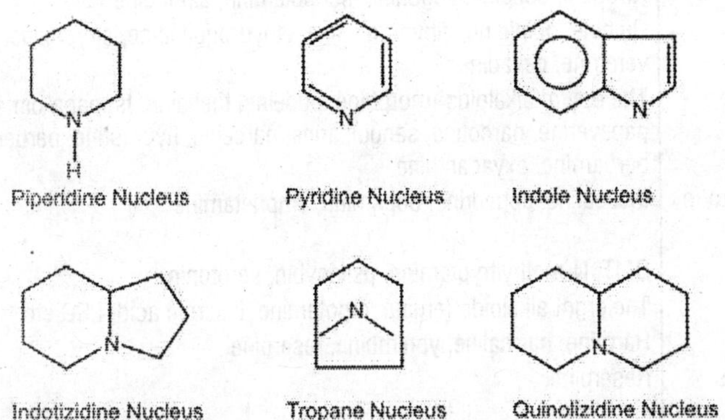

FIG. 42.1: Nuclei of several alkaloids.

Although they undoubtedly existed long before humans, some alkaloids have remarkable structural similarities with neurotransmitters in the central nervous system of humans, including dopamine, serotonin and acetylcholine. The amazing effect of these alkaloids on humans has led to the development of powerful pain-killer medications, spiritual drugs, and serious addictions by people who are ignorant of the properties of these powerful chemicals water or mild acid and then recovered as crystalline material by treatment with base.

CLASSIFICATION OF ALKALOIDS

There are three main types of alkaloids: (1) True alkaloids, (2) Proto alkaloids, and (3) Pseudo alkaloids. True and proto alkaloids are derived from amino acids while pseudo alkaloids are not derived from amino acids. True and pseudo alkaloids have heterocyclic ring with nitrogen while proto alkaloid does not have heterocyclic ring with nitrogen. Pseudo-alkaloids can be derived from Terpenoids or Purines.

Alkaloids are sub-classified on the basis of the chemical type of their nitrogen containing ring. Alkaloids are formed as metabolic by-products. However, their characteristic bitter taste and accompanying toxicity generally help to repel insects and herbivores. Complicated ecological interrelationships have developed between several insects and alkaloids. For example, the consumption of specific pyrrolizidine alkaloids is necessary for some danoid butterflies to synthesise their pheromones.

Alkaloids can be classified in terms of their biological activity, chemical structure (nucleus containing nitrogen), and biosynthetic pathway (the way they are produced in the plant). Alkaloids are grouped under the names of known compounds, even some non-nitrogenous ones (since those molecules' structures appear in the finished product; the opium alkaloids are sometimes called phenanthrenes, for example), or by the plants they are isolated from. When more is learned about a certain alkaloid, the grouping is changed to reflect the new knowledge, usually taking the name of a biologically-important amine that stands out in the synthesis process.

The major groups of alkaloids are as follows:

TABLE 42.1: Various types of alkaloid groups and their types

Gruop	Type
Pyridine group	Piperine, coniine, trigonelline, arecaidine, guvacine, pilocarpine, cytisine, nicotine, sparteine, pelletierine
Pyrrolidine group	Hygrine, cuscohygrine, nicotine
Tropane group	Atropine, cocaine, ecgonine, scopolamine, catuabine
Quinoline group	Quinine, quinidine, dihydroquinine, dihydroquinidine, strychnine, brucine, veratrine, cevadine
Isoquinoline group	The opium alkaloids (morphine, codeine, thebaine, Isopapa-dimethoxy-aniline, papaverine, narcotine, sanguinarine, narceine, hydrastine, berberine), emetine, berbamine, oxyacanthine
Phenethylamine group	Mescaline, ephedrine, dopamine, amphetamine
Indole group	
Tryptamines	DMT, N-methyltryptamine, psilocybin, serotonin
Ergolines	The ergot alkaloids (ergine, ergotamine, lysergic acid, LSD etc.)
Beta-carbolines:	Harmine, harmaline, yohimbine, reserpine
Rauwolfia alkaloids	Reserpine
Purine group	
Xanthines	Caffeine, theobromine, theophylline
Terpenoid group	
Aconite alkaloids	Aconitine
Steroids	Solanine, samandaris (quaternary ammonium compounds): muscarine, choline, neurine
Vinca alkaloids	Vinblastine, vincristine
Miscellaneous	Capsaicin, cynarin, phytolaccine, phytolaccotoxin

BIOSYNTHESIS

Alkaloids belong to the broad category of secondary metabolites. This class of molecule has historically been defined as naturally occurring substances that are not vital to the organism that produces them. Alkaloid any of a class of organic compounds composed of carbon, hydrogen, nitrogen, and usually oxygen that are often derived from plants. They are natural products of non-peptidic origin. Marine organisms have proved to be an extremely fertile source of very unusual and highly complex alkaloids.

With the introduction of molecular biology into the plant alkaloid field, induction of alkaloid biosynthesis in response to exposure to wounding or to elicitors can be analysed at the level of gene activation, and gene expression patterns in the plant can be determined and interpreted as a first indication of possible function.

Their basic character is reflected in the name derived from alkaline which means basic. Plants produce unique natural products as a result of gene mutation and subsequent adaptation of metabolic pathways to create new secondary metabolites. However, their biosynthesis and accumulation remains remarkably under the control of the biotic and abiotic environments. Alkaloid biosynthesis requires the adaptation of cellular activities to perform specialized metabolism without compromising general homeostasis, is accomplished by restricting product biosynthesis and accumulation to particular cells and to defined times of plant development. The cell and developmental biology of alkaloid biosynthesis, which is remarkably complex, evolved in part by recruiting pre-existing enzymes to perform new functions.

The biosynthetic precursors of most alkaloids are amino acids. At a very basic level it is possible to distinguish between aromatic and aliphatic amino acid precursors.

Aliphatic Amino Acids

(L) Ornithine five carbons

(L) Lysine six carbons

Aromatic Amino Acids

Cocaine

Coniine

Noradrenalin

Mescaline

Even in complex alkaloids the aromatic amino acid portion can easily be identified.

Strychnine

LSD

(L) -Tryptophane

N-norlaudanosing

Morphine

L-Tyrosine

The amino acid building blocks of alkaloids are modified by following:

a) decarboxylation,
b) aldol condensation,
c) Reductive amination or methylation.

Successive alterations using these relatively few biosynthetic transformations.

Decarboxylation

Decarboxylation reactions are important in biological systems because intermediates which are chemically disposed for decarboxylation, such as ß-keto acids, occur frequently in primary and secondary metabolism.

Mild heat

$-CO_2$

α-Keto acids are chemically not predisposed towards decarboxylation. This is reflected in much higher temperatures required to effect the above transformation. Nature uses enzymes for this reaction which carry PLP as co-factor. The scheme below shows the decarboxylation of a α-amino acid.

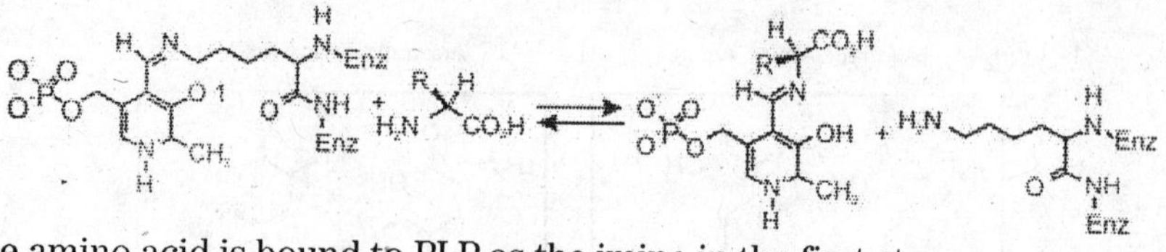

The amino acid is bound to PLP as the imine in the first step.

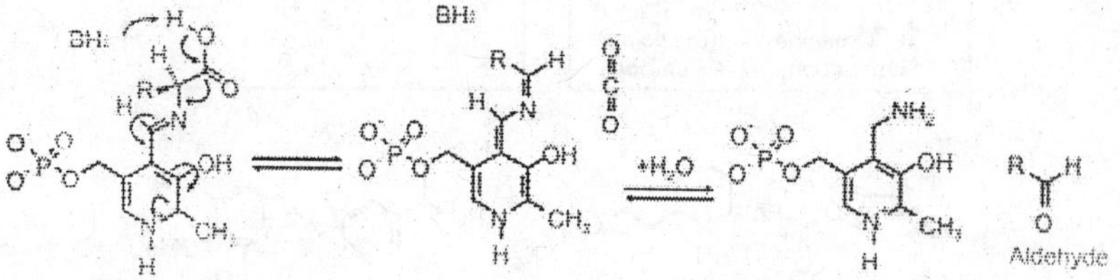

In the actual decarboxylation step, the electronic effects are the same: the pyridine nitrogen acts as an electron-withdrawing group, this time facilitating deprotonation of the carboxylic acid group. Loss of carbon dioxide and hydrolysis of the imine gives the reaction products.

Aldol Condensation

An Aldol condensation is an organic reaction in which an enolate ion reacts with a carbonyl compound to form a ß-hydroxyaldehyde or ß-hydroxyketone, followed by dehydration to give a conjugated enone.

It provides a good way to form carbon–carbon bonds. It involves the nucleophilic addition of a ketone enolate to an aldehyde to form a ß-hydroxy ketone, or aldol (aldehyde + alcohol), a structural unit found in many naturally occurring molecules and pharmaceuticals. It includes three steps:

Nature's equivalents of enolates are enamines and coenzyme A. These are co-factors of aldolase enzymes.

Enamines: The side chain of the amino acid lysine carries an amino group. Reaction with carbonyl compounds leads to imines which tautomerise to give enamines. Enamines are enolate equivalents and react with carbonyl compounds through nucleophilic attack via their ß-carbon. They are used in a very similar way in organic chemistry as shown below for the reaction of a secondary amine (pyrrolidin) with a ketone.

Aldol reactions require several levels of control are (1) Enol versus carbonyl component, (2) Regioselectivity and (3) Stereoselectivity.

Reductive Amination

It is also known as reductive alkylation, is a chemical reaction which involves the conversion of a carbonyl group to an amine, via an intermediate imine. The carbonyl group is most commonly a ketone or an aldehyde.

STEP 1: First, an acid-base reaction. Hydroxide functions as a base and removes the acidic a-hydrogen giving the reactive enolate.

STEP 2: The nucleophilic enolate attacks the aldehyde at the electrophilic carbonyl C in a nucleophilic addition type process giving an intermediate alkoxide.

STEP 3: An acid-base reaction. The alkoxide deprotonates a water molecule creating hydroxide and the β hydroxyaldehydes or aldol product.

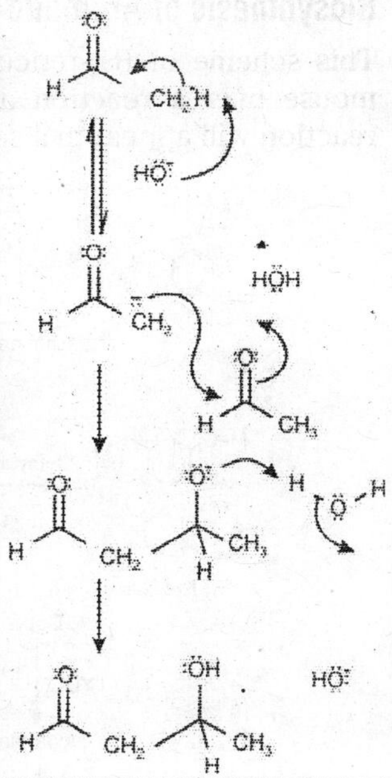

Methylation

Methylation is a term used in the chemical sciences to denote the attachment or substitution of a methyl group on various substrates. Methylation more specifically refers to the replacement of a hydrogen atom with the methyl group. In biological systems, methylation is catalysed by enzymes; such methylation can be involved in modification of heavy metals, regulation of gene expression, regulation of protein function, and RNA metabolism. Methylation of heavy metals can also occur outside of biological systems. Chemical methylation of tissue samples is also one method for reducing certain histological staining artifacts.

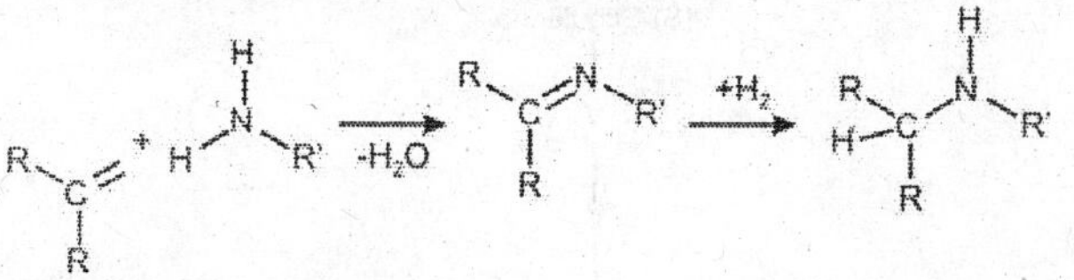

Biosynthesis of Alicyclic Alkaloids

This scheme of cocaine biosynthesis is a clickable map. If you move your mouse over a reaction arrow you can click on a hyperlink and details of the reaction will appear in a separate window.

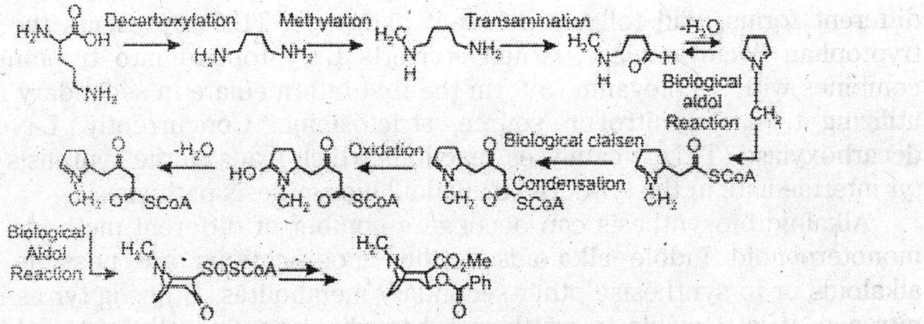

Biosynthesis of Aromatic Alkaloids

This scheme of (S)-reticuline biosynthesis is a clickable map. If you move your mouse over a reaction arrow you can click on a hyperlink and details of the reaction will appear in a separate window.

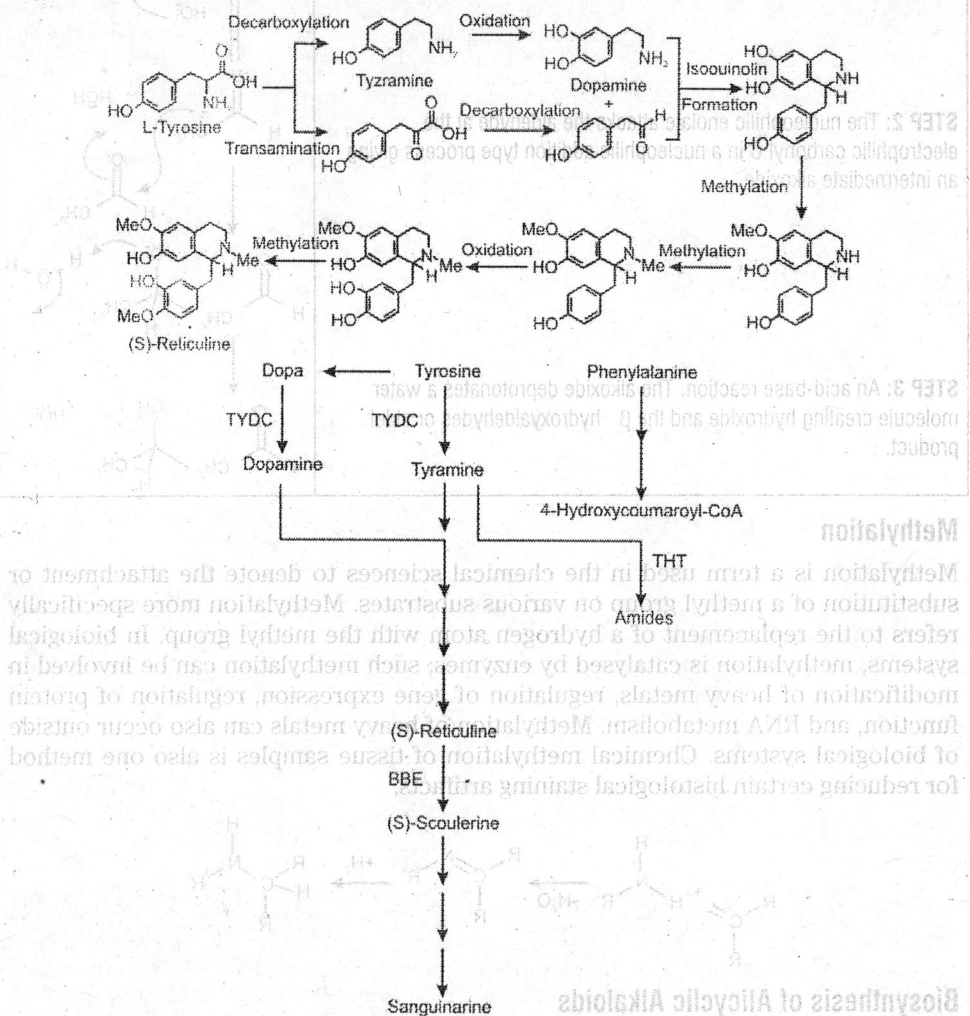

Biosynthesis of Opiate Alkaloids in *P. somniferum*

The alkaloid compounds synthesised by opium poppy can take a number of different forms, and follows different pathways. TDC represents the enzyme L-tryptophan decarboxylase, which converts L-tryptophan into tryptamine, which combines with secologanin to form the first intermediate in secondary metabolism utilizing trp as a nitrogen source, strictosidine. Concurrently, L-tyrosine/dopa decarboxylase (TYDC) catalyses reactions which leads to the synthesis of the first tyr intermediate in the tyr-derived alkaloid biosynthesis pathway.

Alkaloid biosynthesis can occur in a number of different methods. Within the monoterpenoid indole alkaloids further modifications are possible to modify alkaloids or to synthesise other secondary metabolites. Utilizing tyr as a source of nitrogen, it is possible to synthesise benzylisoquinoline alkaloids, which contain benzene rings. It can undergo further modifications to form morphine and codeine, which are also benzylisoquinoline alkaloids. Tyramine can also combine with coumarin derivatives to give rise to cell wall-bound amides. (S)-nococlaurine is recognized as the common precursor for alkaloids, with over 2500 known benzylisoquinoline compounds being derived from it.

The versatiliy of alkaloid biosynthesis is one of the reasons that a diverse number of secondary metabolites exist. Alkaloid biosynthesis incorporates a number of different secondary metabolites to form many of the compounds that are found in the opium poppy and in other plants. The opium poppy utilizes unique biosynthetic pathways that enable it to produce compounds such as sanguinarine, morphine, and codeine.

In the second figure, BBE represents the berberine bridge enzyme, which is responsible for converting the last common intermediate (S)-reticuline into (S)-scoulerine in the morphine and sanguinarine biosynthetic pathways. The synthesis of (S)-scoulerine represents the first committed step in sanguinarine biosynthesis. Since the conversion of (S)-reticuline to (S)-scoulerine is not a reversible reaction, BBE must have a regulatory function to moderate the resources that the opium poppy will use for sanguinarine biosynthesis. THT represents tyramine hydroxycinnamoyl transferase, and is utilized to synthesise amides. The activity of THT combines a phenolic compound, 4-hydroxycoumaroyl-CoA with tyramine, a proto-benzylisoquinoline alkaloid to form amides.

The pathways involved in alkaloid biosynthesis utilizing tyr as a nitrogen source are responsible for a wide range of compounds that are used by humans today. The pathways all seem to have a primary to secondary intermediates involved, and may be rate-limiting steps in the formation of the intermediate compound between primary and secondary metabolite. All of the opiates and their derivatives are alkaloids derived from the amino acid tryrosine. The biosynthesis of dopamine from L-dopa, through the action of dopa decarboxylase and 4-hydroxyphenylacetylaldehyde are also important steps, as they are needed to form the first committed alkaloid intermediate (S)-norcoclaurine. The activity of TYDC, DODC, and BBE must be highly regulated so the opium poppy does not expend too much energy on the biosynthesis of benzylisoquinoline compounds when they are not needed. A study conducted on opium poppy cell cultures elicited with *Botrytis* ssp. mycelia responded with sanguinarine accumulation between five to ten hours after the addition of the elicitor. The fungal pathogen also caused the mRNA levels of BBE and TYDC to increase to a much higher level than was present in the controls. This suggests that there is a molecule in the fungus that induces the synthesis of sanguinarine. The BBE and TYDC induction pathways have not yet been elucidated, however, mycorrhizal fungi most likely do not cause an upregulation of BBE and TYDC gene transcription, so there must be a compound that is only present in pathogenic fungi that causes the induction.

Biosynthesis of Tropane

The starting compound of this synthesis is ornithine, methylornithine is the first intermediate.

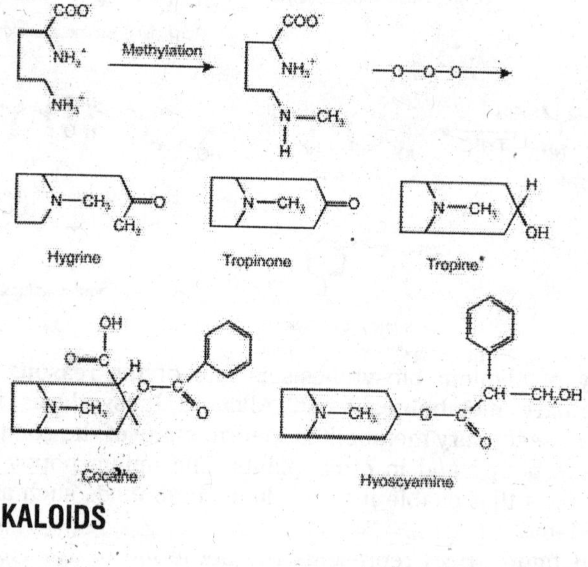

Methylation

Hygrine Tropinone Tropine

Cocaine Hyoscyamine

IMPORTANT ALKALOIDS

Ergot Alkaloids

One of the most amazing stories about naturally-occurring alkaloids in fungi concerns ergot (*Claviceps purpurea*), a fungus that infects grains of rye and related grasses. One of the psychoactive components of ergot fungus is the alkaloid ergine (d-lysergic acid amide), better known as natural LSD. The more potent synthetic LSD, (d-lysergic acid diethylamide), also known as LSD 25, is one of the most powerful psychoactive drugs known. LSD 25 was originally synthesised from natural pshchoactive alkaloids in ergot. According to Lewis and Lewis (1977), it is 4,000 times more powerful than mescaline. Natural LSD (ergine) is also found in the seeds of two species of Mexican morning glory vines which are still ingested by native Indians in an important medicinal and religious ritual.

In both cases, death usually follows and outbreaks of ergotism caused 11,000 deaths in Russia as late as 1926. Today the problem is recognized and controlled. Some of the ergot alkaloids have been used to treat migraine headaches and sexual disorders in clinical applications. The most famous of these alkaloids is lysergic acid diethylamide, LSD, a powerful hallucinogen that is a synthetic derivative of the natural products. Similar alkaloids, particularly ergine, are also found in Mexican morning glories, such as *Ipomeoa tricolor*.

LSD

Morphine Alkaloids

The morphine alkaloids, derived from the opium poppy, *Papaver somniferum*, are powerful pain relievers and narcotics. The narcotic activity of *P. somniferum* was noted on Sumerian tablets in 3500 B.C., making it one of the oldest drugs known. Opium is the dried latex of the seed heads of *P. somniferum*, and has been used as an analgesic (eliminates or relieves pain) and narcotic (induces sleep or drowsiness) in preparations such as laudanum and paregoric. Morphine is the principal alkaloid.

Morphine Heroin

It was widely used for pain relief. In an attempt to make morphine less addictive, Bayer chemists acetylated the hydroxyl groups to produce diacetylmorphine and is marketed as a non-addictive pain reliever under the trade name Heroin. Other derivatives of morphine have been developed and found use as opiate antagonists or as animal tranquilizers.

Tropane Alkaloids

A common property of tropane alkaloids is a methylated nitrogen atom $N\text{-}CH_3$ at one end of the molecule (see illustration at left). This chemical structure is also found in the neurotransmitter acetylcholine, which transmits impulses between nerves in the brain and neuromuscular junctions.

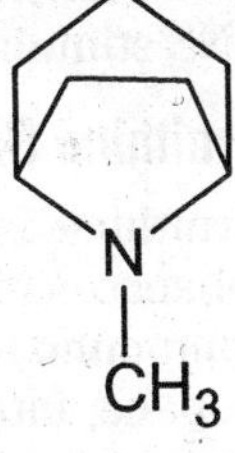

Tropane is a nitrogenous bicyclic organic compound. It is mainly known for a group of alkaloids derived from it (called tropane alkaloids), which include, among others, atropine and cocaine. Both alkaloids contain tropinone from which tropane is a derivate. Tropane alkaloids occur in plants of the families Solanaceae (mandrake, henbane, deadly nightshade, Datura, Potato, Tomato species) and Erythroxylaceae (coca).

The tropane alkaloids have the 8-azabicyclo[3.2.1]octane nucleus, are commonly found in plants of three families, the Solanaceae, Erythroxylaceae, and Convolvulaceae families. The alkaloids isolated from plants of these families, while having several legitimate medicinal uses, are probably best known for their toxic properties. This can be a major problem since the plants produce very attractive berries which are tempting to small children. As few as three berries of henbane (*Hyoscyamus niger*) or deadly nightshade (*Atropa belladonna*) can cause death in infants.

Many of the plants in the Solanaceae family contain tropane alkaloids, which are responsible for the toxic effects of the plants.

The mandrake (*Mandragora officinarum*) was reputed to possess aphrodisiac properties and was prized for these properties. However, the roots also contain large quantities of the tropane alkaloid hyoscine (scopolamine), making the plant highly toxic. *Datura stramonium*, a plant common to Virginia, and other Datura species are widely distributed throughout the world. Extracts of *Datura metel*, sometimes called Hindu Datura, were used as knockout drops to lure virgiins into prostitution and then by the prostitutes to sedate their clients.

The Solanaceae alkaloids derived from these plants, while very toxic, are often important medicinal agents. Atropine, the racemic form of (-)-hyoscyamine, was first isolated from *Atropa belladonna* in 1833. (-)-Hyoscyamine was also isolated in 1833 from *Hyoscyamus niger*. This tropane alkaloid is a muscarinic receptor antagonist. Muscarinic receptors are involved in constriction of the pupil, vasodilation, moderating the heartbeat, and stimulating secretions. The most familiar use of atropine, as the sulphate salt, is as a mydriatic to dilate the pupil of the eye during an opthamological examination. Atropine has also found use as a CNS stimulant and is useful in treating poisoning cases, particularly anticholinesterase poisoning induced by organophosphorous insecticides and nerve gas and poisoning induced by the toxic principles of the mushroom Amanita muscaria.

Scopolamine, (-)-hyoscine, which was first isolated in 1881 from *Hyoscyamus muticus* and *Scopolia atropides* is a CNS depressant at high doses, and has been used in combination with morphine to induce "twilight sleep" before and during labour, and in smaller doses, to prevent motion sickness.

In recent years, a tropane alkaloid isolated from several Erythroxylaceae species has become more notorious. This, of course, is cocaine, which was first isolated from coca, the dried leaves of *Erythroxylum coca* (*Bolivian coca*) or *Erythroxylum truxillense* (*Peruvian coca*), in 1862 by Wohler. The cocaine has anesthetic properties. The hydrochloride of cocaine is used as a topical anesthetic in opthamology as 1% solution (maximum). Cocaine is toxic in large enough doses when taken internally, particularly when taken by injection, but is also a potent CNS stimulant which has led to its abuse.

Ornithine Derivatives

Ornithine is a precursor of the cyclic pyrrolidines that occur in the alkaloids of tobacco (nicotine, nornicotine) and other Solanaceae. Nicotine is a starting compound of numerous further tobacco alkaloids. During the biosynthesis of tropane, intermediates are produced that are at the same time starting compounds for cocaine and hyoscamine.

Vincristine

Vincristine, one of the most potent antileukemic drugs in use today, was isolated in a search for diabetes treatments from *Vinca rosea* (now *Catharanthus roseus*) in the 1950's along with vinblastine, a homologue in which the N-methyl group is oxidized to an aldehyde moiety. This is such a complex structure that it is still isolated from the plant (*Madagascan periwinkle*) today rather than prepared by synthesis. The small change in structure, however, causes a significant change in pharmacological efficacy. Vincristine (leurocristine, VCR) is most effective in treating childhood leukemias and non-Hodgkin's lymphomas, where vinblastine (vincaleukoblastine, VLB) is used to treat Hodgkin's

HO
N
OAc
N
NH
CO_2CH_3
CO_2CH_3
H_3CO
N
OH
CH_3
Vincristine

disease.

Putrescine

Ornithine

Nicotine

Nicotinic acid

Ephedrine Alkaloids

Ephedra is the common name for three principal species: *Ephedra sinica*, *Ephedra equisentina*, and *Ephedra intermedia*. The active compounds in the plant's stem (about 1.32% by weight) are the phenylalanine-derived alkaloids ephedrine, pseudoephedrine, phenylpropanolamine (norephedrine), and cathine (norpseudoephedrine).

Ephedrine

Alkaloids of the Lythraceae

Alkaloid content and composition vary by species and growth conditions; total alkaloid content can vary from 0.5% to 2.3%. Ephedrine, the most potent alkaloid, can account for up to 90% of the total alkaloid content and pseudoephedrine can account for up to 27%. The pharmacologic activity of an *Ephedra* sample depends on its alkaloid composition. North American *Ephedra* species, such as *E. nevadensis* (known as Mormon tea), contain little or no ephedrine or other alkaloids.

Serotonin **Psilocybin**

Indole Alkaloid

Indole alkaloids contain the indole carbon-nitrogen ring which is also found in the fungal alkaloids ergine and psilocybin, the neurotransmitter serotonin, and the mind-altering drug LSD. [See illustration at left showing 5-sided ring with N at the bottom.] These alkaloids may interfere or compete with the action of serotonin in the brain.

It is a complex group, founded on tryptamine, and including many of pharmacological interest. They include the 'animal alkaloids', adrenaline, noradrenaline, serotonin (5-hydroxytryptamine or 5-HT) and their ilk, the tranquillizing alkaloids of passion flower, the ophthalmic alkaloids related to physostigmine from the calabar bean, the uterine stimulants ergotamine and ergometrine from the fungus ergot of rye, and the man-made derivative of these, lysergic acid diethylamide (LSD). Also included are the alkaloids of Indian snakeroot (Rauwolfia serpentaria), including reserpine, having powerful hypotensive (and in isolation from the whole plant, depressive) effects. There are several notable central nervous stimulants: strychnine, yohimbine and psilocybin.

All these alkaloids have their effect on the neuromuscular system by interacting with adrenergic receptors. We might also note two new orthodox discoveries, the antileukaemic medications vincristine and vinblastine isolated from the Madagascar periwinkle (*Catharanthus rosea*).

Indole alkaloids is a group of alkaloids forming one of the major groups of hallucinogens and psychedelic drugs, including bufotenin, DMT, harmine, LSD, and psilocybin. The indole alkaloids are a class of marine natural products that show unique promise in the development of new drug leads. This report reviews the literature on indole alkaloids of marine origin and also highlights our own research. Specific biological activities of indole alkaloids presented here include: cytotoxicity, antiviral, antiparasitic, anti-inflammatory, serotonin antagonism, Ca-releasing, calmodulin antagonism, and other pharmacological activities.

Pyrrolidine Alkaloids

Originating from the amino-acid ornithine, this group includes the tropane alkaloids, atropine, hyoscine and hyoscyamine from the nightshade family (henbane, datura, Jimson weed, belladonna and bittersweet, for example). As a group these block parasympathetic nerve activity. The group also includes the 'truth medication' scopolamine and cocaine.

Pyridine and Piperidine Alkaloids

Derived from the B vitamin, nicotinic acid, this group includes the alkaloids of lobelia and tobacco (especially nicotine). They act to stimulate and then to block all autonomic nervous activity.

Pyrrolizidine and quinolizidine alkaloids: Both have notable toxic reputations. The first is derived from ornithine and linked with liver damage: it is found in the ragworts (potential trouble for grazing stock), borage, comfrey and coltsfoot (where evidence for toxicity is less clear).

The second group is derived from the amino-acid lysine, and provides the alkaloids that make the consumption of some plants of the pea family, notably laburnum, lupins and brooms, so dangerous.

Quinoline Alkaloids

Derived from phenylalanine, this group includes quinine and other alkaloids from cinchona bark, and also the alkaloids found in Angostura Bitters.

Isoquinoline Alkaloids

Related to the above, this is one of the most significant groups. It includes several subclasses:

1. **Simple isoquinolines:** Alkaloids of mescaline cactus (*Lopophora williamsii*), notably mescaline.
2. **Benzyl isoquinolines:** Precursors of many below, including the opium poppy's papaverine.
3. **Phthalide isoquinolines:** Including narcotine.
4. **Protopines:** Exclusive to the poppy family, including protopine.
5. **Protoberberines:** Berberine, hydrastine, canadine, etc., from *Berberis* spp., and golden seal.

6. **Morphine alkaloids:** Including morphine, codeine and thebaine from the opium poppy.
7. **Ipecac alkaloids:** Including the emetic alkaloid emetine from ipecacuanha.

Purine Alkaloids

Notably including caffeine, theobromine, theophylline and aminophylline from cocoa, tea and coffee, these are derived from the purine nucleotides adenine and guanine (hence their collective name: xanthines). They act to prolong the effective life of many hormones, notably adrenaline.

Terpenoid Alkaloids

Having a terpene skeleton, this group includes the poisons of aconite and the delphiniums, the tranquillizing alkaloid of valerian, and a range of interesting new discoveries in the plant world.

APPENDIX 1

Uric Acid: A Uremic Toxin?

Uric acid might often be regarded as a simple marker of renal disease. Although it is well known that hyperuricemia causes gout which is associated with renal insufficiency and cardiovascular disease, one might think that it could attribute to the intrarenal urate crystal, but not to uric acid per se.

Ammonia is linked to carbon dioxide and a phosphate at the expense of energy in the form of ATP. This new nitrogen compound, carbamoyl phosphate, joins with the non-protein amino acid ornithine to form citrulline. The citrulline combines with another amino acid, aspartate to form the larger molecule arginosuccinate. This large compound breaks down into the Krebs cycle intermediate fumarate and the amino acid arginine. Splitting arginine releases urea and regenerates ornithine. Here are the key points to remember. Urea is formed by a cyclic pathway called the urea cycle and the urea cycle is linked to the Krebs cycle by shared intermediates.

Bases to Uric Acid: Both adenine and guanine nucleotides converge at the common intermediate **xanthine**. Hypoxanthine, representing the original adenine, is oxidized to xanthine by the enzyme xanthine oxidase. Guanine is deaminated, with the amino group released as ammonia, to xanthine. If this process is occurring in tissues other than liver, most of the ammonia will be transported to the liver as glutamine for ultimate excretion as urea.

$H_2N-C(=O)-NH_2$

Urea

Xanthine, like hypoxanthine, is oxidized by oxygen and xanthine oxidase with the production of hydrogen peroxide. In man, the urate is excreted and the hydrogen peroxide is degraded by catalase. Xanthine oxidase is present in significant concentration only in liver and intestine. The pathway to the nucleosides, possibly to the free bases, is present in many tissues.

Uremia: Uremia is a term used to loosely describe the illness accompanying kidney failure (also called renal failure). In kidney failure, urea and other waste products, which are normally excreted into the urine, are retained in the blood. Early symptoms include anorexia and lethargy, and late symptoms can include decreased mental acuity and coma. It is usually diagnosed in kidney dialysis patients when the glomerular filtration rate, a measure of kidney function, is below 50% of normal.

Azotemia is another word that refers to high levels of urea, but is used primarily when the abnormality can be measured chemically but is not yet so severe as to produce symptoms. Uremia can also result in fibrinous pericarditis. There are many dysfunctions caused by uremia affecting many systems of the body, such as blood (lower levels of erythropoietin), sex (lower levels of testosterone/estrogen) and bones (osteoporosis and metastatic calcifications).

Patients with chronic renal failure who undergo hemodialysis experience accelerated atherosclerosis and premature death. Since the end- metabolite, oxalic acid accumulates in plasma in proportion to the severity of renal failure.

It is caused by renal failure, the level of urea in the blood can also be increased by increased production of urea in the liver, due to high protein diet, increased protein breakdown (surgery, infection, trauma, cancer), gastrointestinal bleeding, drugs (e.g. tetracyclines and corticosteroids), decreased elimination of urea. It occurs due to decreased blood flow through kidney (e.g. hypotension, cardiac failure). The urinary outflow obstruction may be due to dehydration, kidney failure and chronic infection of the kidney, such as chronic polynephritis.

Glossary

A form. A duplex DNA structure with right-handed twisting in which the planes of the base pairs are tilted about 70° with respect to the helix axis.

Acetal. The product formed by the successive condensation of two alcohols with a single aldehyde. It contains two ether-linked oxygen attached to a central carbon atom.

Acetyl CoA. Acetyl-coenzyme A is a high-energy ester of acetic acid that is important both in the tricarboxylic acid cycle and in fatty acid biosynthesis.

Actin. A protein found in combination with myosin in muscle and also found as filaments constituting an important part of the cytoskeleton in many eukaryotic cells.

Actinomycin D. An antibiotic, that binds to DNA and inhibits RNA chain elongation.

Activated complex. The highest free energy state of a complex in going from reactants to products.

Active site. The region of an enzyme molecule that contains the substrate binding site and the catalytic site for converting the substrate(s) into product(s).

Active transport. The energy-dependent transport of a substance across a membrane.

Adenine Dinucleotides. Major electron carrying compounds in the oxidation of fuel molecules.

Adenine. A purine base found in DNA or RNA.

Adenosine Diphosphate (ADP). The nucleotide formed by adding a pyrophosphate group to the 5'-OH group of adenosine.

Adenosine triphosphate (ATP). The nucleotide formed by adding yet another phosphate group to the pyrophosphate group on ADP.

Adenosine. A purine nucleoside found in DNA, RNA, and many cofactors.

Adenylate cyclase. The enzyme that catalyzes the formation of cyclic 3',5' adenosine monophosphate (cAMP) from ATP.

Adipocyte. A specialized cell that functions as a storage depot for lipid.

Aerobe. An organism that utilizes oxygen for growth.

Affinity chromatography. A column chromatographic technique that employs attached functional groups that have a specific affinity for sites on particular proteins.

Alcohol. A molecule with a hydroxyl group attached to a carbon atom. all alcohols contain the hydroxyl group, **-oh**, usually formed as a byproduct by the fermentation of sugars or starch.

Aldehyde. A molecule containing a doubly bonded oxygen and a hydrogen attached to the same carbon atom.

Alginate. Component of the cell walls of many rhodophytes and kelps. Alginates have an affinity for water, and so help to slow dessication when the algae are exposed to the air; they are commercially important in the production of paper, toothpaste, beer, and frozen foods.

Alipathic hc. do not contain a benzene group, or benzene ring;

Aliphatic aa. amino acids in which the shorter side chains are

Aliphatic hydroxylic amino acid. Amino acids with highly hydrophilic.

Alleles. Alternative forms of a gene.

Allosteric enzyme. An enzyme whose active site can be altered by the binding of a small molecule at a nonoverlapping site.

Alosteric effect. These are interactions between spatially distinct sites; i.e.. a conformational change induced by the binding of a molecule to one site in a protein that alters other sites of the same protein (a molecular switch for receiving, integration, and transmission of signals); as in the case of hemoglobin, phosphofructokinase, and ribonucleotide reductase.

Alpha helix. A tightly coiled polypeptide main inner-chain forming the inner part of the rod, and the side chains extended outward in a helical array.

Amino acid. Unit molecule from which proteins are constructed by polymerization. A peptide; the basic building block of proteins or polypeptides equipped w/ a carboxyl group COOH.

Anabolic pathway. ATP-consuming;

Anomers. The C-1 carbon of a sugar ring (either pyranose or furanose form) to which the -OH group is attached. The sugar isomers that differ in configuration about the carbonyl carbon atom. This carbon atom is called the anomeric carbon atom of the sugar.

Antibiotic. A natural product that inhibits bacterial growth (is bacteriostatic) and sometimes results in bacterial death.

Antibody. A specific protein that interacts with a foreign substance (antigen) in a specific way.

Anticodon. A sequence of three bases on the transfer RNA that pair with the bases in the corresponding codon on the messenger RNA.

Antigen. A foreign substance that triggers antibody formation and is bound by the corresponding antibody.

Apoactivator. A regulatory protein that stimulates transcription from one or more genes in the presence of a coactivator molecule.

Asymmetric carbon. A carbon that is covalently bonded to four different groups.

ATP. "Adenosine triphosphate". A relatively stable, high energy molecule used to fuel chemical reactions within cells.

Attenuator. A provisional transcription stop signal.

Autoradiography. The technique of exposing film in the presence of disintegrating radioactive particles. Used to obtain information on the distribution of radioactivity in a gel or a thin cell section.

Autoregulation. The process in which a gene regulates its own expression.

Autotroph. An organism that can form its organic constituents from CO2.

Auxotroph. A mutant that cannot grow on the minimal medium on which a wild-type member of the same species can grow.

B cell. One of the major types of cells in the immune system. B cells can differentiate to form memory cells or antibody-forming cells.

B form. The most common form of duplex DNA, containing a right-handed helix and about 10 (10.5 exactly) base pairs per turn of the helix axis.

Base analog. A compound, usually a purine or a pyrimidine, that differs somewhat from a normal nucleic acid base.

Base stacking. The close packing of the planes of base pairs, commonly found in DNA and RNA structures.

Base. The adenine, guanine, cytosine or thymine group attached to a nucleotide or nucleoside. Also may be used to refer to a nucleic acid unit within a polynucleotide chain, as when a gene is said to be 2000 bases long.

Bidirectional replication. Replication in both directions away from the origin, as opposed to replication in one direction only (unidirectional replication).

Bilayer. A double layer of lipid molecules with the hydrophilic ends oriented outward, in contact with water, and the hydrophobic parts oriented inward.

Biochemical pathway. A series of enzyme-catalyzed reactions that results in the conversion of a precursor molecule into a product molecule.

Biochemistry. The study of those molecules used and manufactured by living things.

Bioluminescence. The production of light by a chemical reaction within an organism. The process occurs in many bacteria and protists, as well as certain animals and fungi.

Blastoderm. The stage in embryogenesis when a unicellular layer at the surface surrounds the yolk mass.

Bond energy. The energy required to break a bond.

Bonds between amino acid. A link between two amino acids in which a byproduct is released (always endothermic). Bonds requires an input of free energy, whereas their hydrolysis is thermodynamically downhill.

Buffer. A conjugate acid-base pair that is capable of resisting changes in ph when acid or base is added to the system. This tendency will be maximal when the conjugate forms are present in equal amounts.

Calcium carbonate. a "salt" used by many marine invertebrates, such as corals and echinoderms, and by protists, such as coccolithophorids, to construct their exoskeletons.

cAMP. 3',5' cyclic adenosine monophosphate. The camp molecule plays a key role in metabolic regulation.

CAP. Catabolite activator protein, an alosteric protein which binds first to a camp (cyclic adenosine monophosphate) before it can dock onto the DNA, enabling the RNA polymerase to join it, triggering the mRNA synthesis.

Carbohydrates. A class of biochemical compounds which includes sugars, starch, chitin, and steroids.

Carboxylic acid. A molecule containing a carbon atom attached to a hydroxyl group and to an oxygen atom by a double bond.

Carcinogen. A chemical that can cause cancer.

Carotenoids. Lipid-soluble pigments that are made from isoprene units.

Catabolic pathway. ATP-generating;

Catabolism. That part of metabolism that is concerned with degradation reactions.

Catabolite repression. The general repression of transcription of genes associated with catabolism that is seen in the presence of glucose.

Catalyst. A compound that lowers the activation energy of a reaction without itself being consumed.

Catalytic site. The site of an enzyme involved in the catalytic process.

Catenane. An interlocked pair of circular structures, such as covalently closed DNA molecules.

Catenation. The linking of molecules without any direct covalent bonding between them, as when two circular DNA molecules interlock. like the links in a chain.

cDNA. Complementary DNA, made in vitro from the mRNA by the enzyme reverse transcriptase using deoxyribonucleotide triphosphates. Unlike mRNA, cDNA can be easily propagated and sequenced.

Cell cycle. All of those stages that a cell passes through from one cell generation to the next.

Cell line. An established clone originally derived from a whole organism through a long process of cultivation.

Cell lineage. The pedigree of cells resulting from binary fission.

Cell wall. A tough outer coating found in many plant, fungal, and bacterial cells that accounts for their ability to withstand mechanical stress or abrupt changes in osmotic pressure. Cell walls always contain a carbohydrate component and frequently also a peptide and a lipid component.

Cellulose. carbohydrate polymer of the simple sugar glucose. It is found in the cell walls of plants and green algae, as well as dinoflagellates. Cellulose is the most abundant compound on earth that is manufactured by living things.

Chelate. A molecule that contains more than one binding site and frequently binds to another molecule through more than one binding site at the same time.

Chemiosmotic coupling. The coupling of ATP synthesis to an electrochemical potential gradient across a membrane.

Chimeric DNA. Recombinant DNA whose components originate from two or more different sources.

Chitin. A carbohydrate polymer found in the cell walls of fungi and in the exoskeletons of arthropods, which provides strength for support and protection.

Chlorophyll. The green-colored pigment that absorbs light during photosynthesis, often found in plants, algae, and some bacteria; it includes a porphyrin ring, and often has a long hydrophobic tail.

Chloroplast. A chlorophyll-containing photosynthetic organelle, found in eukaryotic cells, that can harness light energy.

Chromatin. The nucleoprotein fibers of eukaryotic chromosomes.

Chromatography. A procedure for separating chemically similar molecules. Segregation is usually carried out on paper or in glass or metal columns with the help of different solvents. The paper or glass columns contain porous solids with functional groups that have limited affinities for the molecules being separated.

Chromosome puff. A swollen region of a giant chromosome; the swelling reflects a high degree of transcription activity.

Chromosome. A thread-like structure, visible in the cell nucleus during metaphase, that carries the hereditary information.

Cis dominance. Property of a sequence or a gene that exerts a dominant effect on a gene to which it is linked.

Cistron. A genetic unit that encodes a single

polypeptide chain.

Citric acid cycle. See tricarboxylic acid (TCA) cycle.

Clone. One of a group of genetically identical cells or organisms derived from a common ancestor.

Cloning vector. A self-replicating entity to which foreign DNA can be covalently attached for purposes of amplification in host cells.

Coactivator. A molecule that functions together with a protein apoactivator. For example, camp is a coactivator of the cap protein.

Codon. In a messenger RNA molecule, a sequence of three bases that represents a particular amino acid.

Coenzyme. A nonprotein organic molecule that combines with an apoenzyme to form the functioning holoenzyme; it Aids in enzyme-catalyzesd reactions, often by acting as an electron carrier (donor or acceptor).

Cofactor. A small molecule required for enzyme activity. It could be organic in nature, like a coenzyme, or inorganic in nature, like a metallic cation.

Collagen. long proteins whose structure is wound into a triple helix. The resulting fibers have a high tensile strength. Collagen is a primary component of mammalian hair.

Complementary base sequence. For a given sequence of nucleic acids, the nucleic acids that are related to them by the rules of base pairing.

Configuration. The spatial arrangement in which atoms are covalently linked in a molecule.

Constitutive enzymes. Enzymes synthesized in fixed amounts, regardless of growth conditions.

Cooperative binding. A situation in which the binding of one ligand to a macromolecule favors the binding of another. For example, DNA cooperatively binds histone molecules, and hemoglobin cooperatively binds oxygen molecules.

Coordinate induction. The simultaneous expression of two or more genes.

Cosmid. A DNA molecule with cos ends from lambda-bacteriophage that can be packaged in vitro into a virus for infection purposes.

Cytidine. A pyrimidine nucleoside found in DNA and RNA.

Cytochrome. Heme-containing Proteins with an iron-containing poryphyrin ring prosthetic groups (heme) attached to them. They undergo oxidation and reduction through loss or gain of a single electron by the iron atom at the center of the cytochrome.

Cytokinin. A plant hormone produced in root tissue.

Cytoplasm. The contents enclosed by the plasma (or cytoplasmic) membrane, excluding the nucleus.

Cytosine. A pyrimidine base found in DNA and RNA.

Cytosine. pairs with guanine.

Cytoskeleton. The filamentous skeleton, formed in the eukaryotic cytoplasm, that is largely responsible for controlling cell shape.

Cytosol. The liquid portion of the cytoplasm, including the macromolecules but not including the larger structures like subcellular organelles or cytoskeleton.

D loop. An extended loop of single-stranded DNA displaced from a duplex structure by an oligonucleotide.

Dalton. A unit of mass equivalent to the mass of a hydrogen atom (1.66 x 10-24 g)

Dark reactions. Reactions that can occur in the dark, in a process that is usually associated with light, such as the dark reactions of photosynthesis.

De novo pathway. A biochemical pathway that starts from elementary substrates and ends in the synthesis of a biochemical.

Deamination. The enzymatic removal of an amine group, as in the deamination of an amino acid to an alpha keto acid.

Dehydrogenase. An enzyme that catalyzes the removal of a pair of electrons (and usually one or two protons) from a substrate molecule.

Denaturation. The disruption of the native folded structure of a nucleic acid or protein molecule; may be due to heat, chemical treatment, or change in ph.

Density-gradient centrifugation. The separation, by centrifugation, of molecules according to their density, in a gradient varying in solute concentration.

Dialysis. Removal of small molecules from a macromolecule preparation by allowing them to pass across a semipermeable membrane.

Diauxic growth. Biphasic growth on a mixture of two carbon sources in which one carbon source is used up before the other one. For example, in the presence of glucose and lactose, *e. Coli* will utilize the glucose before the lactose.

Difference spectra. Plots comparing the absorption spectra of a molecule or an assembly of molecules in different states, for example, those of mitochondria under oxidizing or reducing conditions.

Differential centrifugation. Separation of molecules and/or organelles by sedimentation rate.

Differentiation. A change in the form and pattern of a cell and the genes it expresses as a result of growth and replication, usually during development of a multicellular organism. Also occurs in microorganisms.

Dimer. Structure resulting from the association of two subunits.

Dinosteranes/dinosteroids. chemicals found in dinoflagellates, which have been useful in documenting their existence early in the fossil record.

Dipeptide bond. two amino acids joined together.

Diploid cell. A cell that contains two chromosomes (2n) of each type.

Dipole. A separation of charge within a single molecule.

Directed mutagenesis. In a DNA sequence, an intentional alteration that can be genetically inherited.

Dissociation constant. An equilibrium constant for the dissociation of a molecule into two parts (e.g., dissociation of acetic acid into acetate anion and a proton); kd.

DNA cloning. The propagation of individual segments of DNA as clones.

DNA double helix. two right-handed interlocking helixes, constituting a b-DNA type helix, joined by hydrogen bonds between the pairs purine-pyrimidine bases (a pairs w/ t and g w/ c); together 2nm in diameter. The helical structure makes a 360° twist after each 10 residues of each chain.

DNA library. A mixture of clones, each containing a cloning vector and a segment of DNA from a source of interest.

DNA major groove 1.2nm wide and slightly deeper then the **minor groove**, 0.6nm wide; these result due to the non-glycosidic bonding of opposing pairs of the centrally located purine and pyrimidine pairs and the phosphate-sugar backbone on the outside.

DNA mutation. (l. Mutare, to change) A permanent change in chemical structure, organization, or amount of DNA; produces a gene or a chromosome set differing from the wild type, resulting in a faulty protein (loss or gain of function; gains and selection are the tools of evolution); e.g.. UV-radiation (*xerodermy pigmentosum*), etc.

DNA polymerase. An enzyme that catalyzes the formation of 3'-5' phosphodiester bonds from deoxyribonucleotide triphosphates.

DNA sequence. The linear assembly of purine-pyrimidine nucleotides (a pairs w/ t and g w/ c) along a DNA strand.

DNA. Deoxyribonucleic acid. A polydeoxyribonucleotide in which the sugar is deoxyribose; the main repository of genetic information in all cells and most viruses. A double chain of linked nucleotides; composed of a base, purine or pyrimidine, and a phosphate group, PO_3, having deoxyribose as their sugars) ; the fundamental substance of which genes are composed (see genetics for image); in **eukaryota**. DNA wrapped around histones, forming nucleosomes on solenoids; i.e. Chromosomes. In **prokaryota**. DNA is circular and supercoiled, do not have chromatin (histones etc.).

Domain. A segment of a folded protein structure showing conformational integrity. A domain could include the entire protein or just a fraction of the protein. Some proteins, such as antibodies, contain many structural domains.

Dominant. Describing an allele whose phenotype is expressed regardless of whether the organism is homozygous or heterozygous for that allele.

Double helix. A structure in which two helically-twisted polynucleotide strands are held together by hydrogen bonding and base stacking.

Ecdysone. A hormone that stimulates the molting process in insects.

Edman degradation. A systematic method of sequencing proteins, proceeding by stepwise removal of single amino acids from the amino terminus of a polypeptide chain.

Eicosanoid. Any fatty acid with 20 carbons.

Electron transport chain. The energy bucket brigade - the voltage gradient across the mitochondrial wall, drives electrons along with hydrogen ions to the oxygen to generate water.

Electrophoresis. The movement of particles in an electrical field. A commonly-used technique for analysis of mixtures of molecules in solution according to their electrophoretic mobilities.

Elongation factors. Protein factors uniquely required during the elongation phase of protein synthesis.

Eluate. The fluid that has passed through (eluted from) a chromatographic column.

Enantiomorphs. Isomers that are mirror images of one another.

Endergonic reaction. A reaction with a positive standard free energy change.

Endocrine glands. Specialized tissues whose function is to synthesize and secrete hormones.

Endonuclease. An enzyme that breaks a phosphodiester linkage at some point within a polynucleotide chain.

Endopeptidase. An enzyme that breaks a polypeptide chain at an inteRNAl peptide linkage.

Endoplasmic reticulum. A system of double membranes in the cytoplasm that is involved in the synthesis of transported proteins. The rough endoplasmic reticulum has ribosomes associated with it. The smooth endoplasmic reticulum does not.

End-product (feedback) inhibition. The inhibition of the first enzyme in a pathway by the end product of that pathway.

Energy charge. The fractional degree to which the amp-adp-atp system is filled with high-energy phosphates (phosphoryl groups).

Energy harvest. The aerobic pathway breaks down glucose to co2 and h2o, consumes o2 as a final electron acceptor, and produces a total of 36 atp per molecule of glucose.

Enhancer. A DNA sequence that can stimulate transcription at an appreciable distance from the site where it is located. It acts in either orientation and either upstream or downstream from the promoter.

Entropy. The randomness of a system.

Enzyme. Usually a protein functioning as a catalyst in living organisms, which promotes *specific* reactions or groups of reactions; i.e.. lowers activation energy without being consumed or affected by the process with out altering reaction equilibria.

Epimers. Two stereoisomers with more than one chiral center that differ in configuration at one of their chiral centers.

Equilibrium. The point at which the concentrations of two compounds are such that the interconversion of one compound into the other compound does not result in any change in free energy.

Ether. A molecule containing two carbons linked by an oxygen atom.

Eukaryote. A cell or organism that has a membrane-bound nucleus.

Excision repair. DNA repair in which a damaged region is replaced.

Excited state. An energy-rich state of an atom or a molecule, produced by the absorption of radiant energy.

Exergonic reaction. A chemical reaction that takes place with a negative change in standard free energy.

Exon. Any non-intron section of the coding sequence of a gene (in eukaryota only); exons spliced together constitute the mRNA and are translated into proteins (compare intron).

Exonuclease. An enzyme that breaks a phosphodiester linkage at one or the other end of a polynucleotide chain so as to release single or small nucleotide residues.

F factor. A large bacterial plasmid, known as the sex-factor plasmid because it permits mating between F^+ and F^- bacteria.

Facultative aerobe. An organism that can use molecular oxygen in its metabolism but that also can live anaerobically.

Fatty acid. A long-chain hydrocarbon containing a carboxyl group at one end.

Fermentation. The energy-generating breakdown of glucose or related molecules by a process that does not require molecular oxygen.

Fingerprinting. The characteristic two-dimensional paper chromatogram obtained from the partial hydrolysis of a protein or a nucleic acid.

Fischer representation. A two-dimensional representation of the molecule in which the c-atoms are aligned in a linear fashion, usually in a open-chain manner.

Flagellin. Protein which is the primary component of prokaryotic flagella.

Flavoprotein. Proteins containing a derivative of roboflavin; the flavon portion, which is the prosthetic group that is alteRNAtely reduced as it accepts h-atoms and oxidizes when electrons are passed on. It is Commonly found in the electron transport chain during the synthesis of ATP. Riboflavin, also called Vitamin b2, is a required growth factor for some organisms.

Fluorescence. The emission of light by an excited molecule in the process of making the transition from the excited state to the ground state.

Free energy. That part of the energy of a system that is available to do useful work.

Fucoxanthin. Yellowish-brown pigment found in some members of the chromista, including kelps and diatoms.

Functional group. It is a group of atoms that is largely responsible for the chemical behavior of the parent molecule.

Furanose. A sugar that contains a five-membered ring as a result of intramolecular hemiacetal formation.

Gene amplification. The duplication of a particular gene within a chromosome two or more times.

Gene splicing. The cutting and rejoining of DNA sequences.

Gene. A segment of the genome that codes for a functional product.

Genetic map. The arrangement of genes or other identifiable sequences on a chromosome.

Genome. The total genetic content of a cell or a virus.

Genotype. The genetic characteristics of an organism (distinguished from its observable characteristics, or phenotype).

Globular protein. A folded protein that adopts an approximately globular shape. May also be called soluble proteins.

Gluconeogenesis. The production of sugars from nonsugar precursors such as lactate or amino acids. Applies more specifically to the production of free glucose by vertebrate livers.

Glucose. Simple sugar, and the primary product of photosynthesis. It is polymerized to make cellulose and chitin.

Glycogen. A polymer of glucose residues in 1,4 linkage, with 1,6 linkages at branch points.

Glycogenic. Describing amino acids whose metabolism may lead to gluconeogenesis.

Glycolipid. A lipid containing a carbohydrate group.

Glycolysis. A series of reactions in the cytoplasms of a cell, that converts glucose to pyruvate w/ the concomitant production of a small amount of ATP. It is the catabolic conversion of glucose to pyruvate with the production of ATP.

Glycoprotein. A membrane-bound protein linked to an oligosaccharide or a polysaccharide and attached branching carbohydrates. These may function in cell-cell recognition, such as in human blood groups and immune system response, as well as in resisting compression of cells.

Glycosidic bond. The bond between a sugar and an alcohol. Also the bond that links two sugars in disaccharides, oligosaccharides, and polysaccharides.

Glyoxylate cycle. A pathway that uses some of the enzymes of the TCA cycle and some enzymes whereby acetate can be converted into succinate and carbohydrates.

Glyoxysome. An organelle containing some enzymes of the glyoxylate cycle.

Golgi apparatus. A complex series of double-membrane structures that interact with the endoplasmic reticulum and that serve as a transfer point for proteins destined for other organelles, the plasma membrane, or extracellular transport.

Gram molecular weight. For a given compound, the weight in grams that is numerically equal to its molecular weight.

Ground state. The lowest electronic energy state of an atom or a molecule.

Growth factor. A substance that must be present in the growth medium to permit eucaryotic cell proliferation.

Growth fork. The region on a DNA duplex molecule where synthesis is taking place. It resembles a fork in shape, since it consists of a region of duplex DNA connected to a region of unwound single strands.

GTP. Guanosinetriphosphate, high energy molecule similar to ATP that participates in several energy-requiring processes, i.e. Peptide bond formation; as with ATP, GTP can acquire the lower energy levels as GDP and GMP.

Guanine. A purine base found in DNA or RNA.

Guanosine. A purine nucleoside found in DNA and RNA.

Gyrase. An energy-transducing enzyme; it converts free energy of atp into torsional energy for supercoiling.

Hairpin loop. A single-stranded complementary region of DNA or RNA that folds back on itself and base-pairs into a double helix.

Half-life. The time required for the disappearance of one half of a substance.

Haploid cell. A cell containing only one chromosome of each type.

Helicase. An ATP-driven enzyme actively involved in the separation of the complimentary nucleotides of DNA. It separates the double helix into separate parental strands at room temperature, which would otherwise take a min. Temperature of 90° C.

Helix. A spiral structure with a repeating pattern.

Heme group. an iron protophyrine portion of many respiration pigments.

Heme. An iron-porphyrin complex found in hemoglobin and cytochromes.

Hemoglobin (Hg). The oxygen carrying pigment of the erythrocytes, formed by the developing erythrocyte in bone marrow. It is a complex protein composed of four heme groups and four globin polypeptide chains. It is a protein complex found in the blood of most chordates and the roots of certain legumes. It binds oxygen molecules, and in chordates serves as the means by which the oxygen is supplied to the cells of the body.

Hexose. A sugar with a six-carbon backbone.

High-energy compound. A compound that undergoes hydrolysis with a high negative standard free energy change.

Histone. A type of basic protein that forms the unit around which DNA is coiled in the nucleosomes of eukaryotic chromosomes, allowing extreme long DNA molecules to be packed into a cell nucleus.

Holoenzyme. An intact enzyme containing all of its subunits and any necessary cofactors with full enzymatic activity.

Homologous chromosomes. Chromosomes that carry the same pattern of genes, but not necessarily the same alleles.

Hormone receptor. A protein that is located on the cell membrane or inside the responsive cell and that interacts specifically with the hormone.

Hormone. A chemical substance made in one cell and secreted so as to influence the metabolic activity of a select group of cells located at other sites in the organism.

Host cell. A cell used for growth and reproduction of a virus.

Hybrid (or chimeric) plasmid. A plasmid that contains DNA from two different organisms.

Hydrocarbons. these molecules mainly consist of two elements, hydrogen and carbon.

Hydrogen bond. A weak, noncovalent, attractive force between one electronegative atom and a hydrogen atom that is covalently linked to a second electronegative atom.

Hydrolysis. The cleavage of a molecule by the addition of water. Hydrophilic. Preferring to be in contact with water.

Hydrophilic. "Water loving". Hydrophilic compounds dissolve easily in water, and are usually polar.

Hydrophobic effect. The noncovalent association of nonpolar groups with each other in aqueous solution.

Hydrophobic. "Water fearing". Hydrophobic compounds do not dissolve easily in water, and are usually non-polar. Oils and other long hydrocarbons are hydrophobic.

Hydroxyapatite. A calcium phosphate gel used, in the case of nucleic acids, to selectively absorb duplex DNA-RNA from a mixture of single-stranded and duplex nucleic acids.

Imine. A molecule containing a nitrogen atom attached to a carbon atom by a double bond. The nitrogen is also covalently linked to a hydrogen.

Immunoglobulin. A protein made in a b plasma cell and usually secreted; it interacts specifically with a foreign agent. Synonymous with antibody. It is composed of two heavy and two light chains linked by disulfide bonds. Immunoglobulins can be divided into five classes (Igg, Igm, Iga, Igd, and Ige) based on their heavy-chain component.

Inducers. Molecules that cause an increase in a protein activity when added to cells.

Inducible proteins. Those which are synthesized in different amounts depending on cellular signals.

Inhibition of enzyme activity. It is brought about by small molecules and ions; serves as a major control mechanism.

Initiation factors. Those protein factors that are specifically required during the initiation phase of protein synthesis.

Integrin. Adhesive protein of the extracellular matrix in animals.

Interferon. One of a family of proteins that are liberated by special host cells in the mammal in response to viral infection. The interferons attach to an infected cell, where they stimulate antiviral protein synthesis.

Intron (gk. Intervening sequence). A segment of largely unknown function (non-coding) within a gene of eukaryota which is initially transcribed but is not found in the functional mRNA - cut out (compare exon).

Inverted repeat. A chromosome segment that is identical to another segment on the same chromosome except that it is oriented in the opposite direction.

Irreversible inhibitor. Tightly bound to the target enzyme (either covalently or non-covalently). Therefore,

Isomer. one of two or more compounds that contain the same number of the same atoms in different arrangements.

Isomerase. An enzyme that catalyzes an intramolecular rearrangement.

Isomerization. Rearrangement of atomic groups within the same molecule without any loss or gain of atoms.

Isozymes. Multiple forms of an enzyme that differ from one another in one or more of the properties.

Ketogenic. Describing amino acids that are metabolized to acetoacetate and acetate.

Ketone bodies. Refers to acetoacetate, acetone, and -hydroxybutyrate made from acetyl-coa in the liver and used for energy in? nonhepatic tissue.

Ketone. A functional group of an organic compound in which a carbon atom is double-bonded to an oxygen. Neither of the other substituents attached to the carbon is a hydrogen. Otherwise the group would be called an aldehyde.

Ketosis. A condition in which the concentration of ketone bodies in the blood or urine is unusually high.

Kilobase. One thousand bases in a DNA molecule.

Kinase. An enzyme catalyzing phosphorylation of an acceptor molecule, usually with atp serving as the phosphate (phosphoryl) donor.

Kinetochore. A structure that attaches laterally to the centromere of a chromosome; it is the site of chromosome tubule attachment.

Krebs cycle. A series of eight major reactions following glycolysis, in which acetate residues within mitochondria are degraded to CO_2 and NADH.

Lagging strand. the strand that is synthesized apparently in the 3' to 5' direction, by ligating short fragments synthesized individually in the 5' to 3' direction.

Laminarin. a beta-glucan polysaccharide produced by many chromists through photosynthesis.

Lampbrush chromosome. Giant diplotene chromosome found in the oocyte nucleus. The loops that are observed are the sites of extensive gene expression.

Law of mass action. The finding that the rate of a chemical reaction is a function of the product of the concentrations of the reacting species.

Leader region. The region of an mRNA between the 5' end and the initiation codon for translation of the first polypeptide chain.

Leader sequence. An n-terminal signal sequence that directs secretion and processing of proteins.

Leading strand. The strand that is made in the 5' to 3' direction by continuous polymerization at the 3' growing tip.

Lectins. Agglutinating proteins usually extracted from plants.

Ligand. A (usually small) molecule that binds to another, such as oxygen when it binds to myoglobin.

Ligase. an ATP-driven enzyme that can rejoin a broken phosphodiester bond in a nucleic acid, primarily used in the lagging strand to bond Okazaki fragments together (3' with 5' terminus), as well as bonding strands after the repair of mismatched bases. It is an enzyme that catalyzes the joining of two molecules together. In DNA it joins 5'-oh to 3' phosphates.

Linkage. The tendency of markers to be inherited together. Linkage of two markers is an indication that they are close to one another in the genome.

Linkers. Short oligonucleotides that can be ligated (connected) to larger DNA fragments, then cleaved (cut) to yield overlapping cohesive (sticky) ends, suitable for ligation to other DNAs that contain comparable cohesive ends.

Linking number. The net number of times one polynucleotide chain crosses over another polynucleotide chain. By convention, right-handed crossovers are given a plus designation.

Lipid bilayer. Model for the structure of the cell membrane based on the interaction between the hydrophobic regions of phospholipids.

Lipid. A biological molecule that is soluble in organic solvents. Lipids include steroids, fatty acids, prostaglandins, terpenes, and waxes.

Lipopolysaccharide. Usually refers to a unique glycolipid found in gram negative bacteria.

Luciferase. enzyme which activates luciferin to produce bioluminescence.

Luciferin. Compound whose activated form emits light.

Lyase. An enzyme that catalyzes the removal of a group to form a double bond, or the reverse reaction.

Lysosome. An organelle that contains hydrolytic enzymes designed to break down proteins that are targeted to that organelle.

Lytic infection. A virus infection that leads to the iysis of the host cell, yielding progeny virus particles.

M phase. That period of the cell cycle when mitosis takes place.

Marker. A "landmark" that can be localized to a specific region of the genome.

Meiosis. Process in which diploid cells undergo division to form haploid sex cells.

Membrane protein. A protein that is associated with a membrane, rather than found free in the cell. A membrane protein may be integral (embedded or buried) in the membrane, or peripheral (attached more loosely, by interactions with either lipid or intergral membrane proteins).

Membrane transport. The facilitated transport of a molecule across a membrane.

Membrane. A sheet-like composite of protein and lipid that is the boundary of cells and organelles.

Merodiploid. An organism that is diploid for some but not all of its genes.

Mesosome. An invagination of the bacterial cell membrane.

Messenger RNA (mRNA). The template RNA carrying the message for protein synthesis.

Metabolic turnover. A measure of the rate at which already existing molecules of the given species are replaced by newly-synthesized molecules of the same type. Usually isotopic labeling is required to measure turnover.

Metabolism. The sum total of the enzyme-catalyzed reactions that occur in a living organism.

Metamorphosis. A change of form, especially the conversion of a larval form to an adult form.

Metaphase. That stage in mitosis or meiosis when all of the chromosomes are lined up on the equator (i.e., an imaginary line that bisects the cell).

Micelle. An aggregate of lipids in which the polar head groups face outward and the hydrophobic tails face inward; no solvent is trapped in the center.

Michaelis constant (km). The substrate concentration at which an enzyme-catalyzed reaction proceeds at one-half of the maximum velocity.

Michaelis-menten equation (also known as the Henri-Michaelis-Menten equation). An equation relating the reaction velocity to the substrate concentration of an enzyme.

Microtubules. Thin tubules, made from globular proteins, that serve multiple purposes in eukaryotic cells.

Mismatch repair. The replacement of a base in a heteroduplex structure by one that forms a watson-crick base pair.

Missense mutation. A change in which a codon for one amino acid is replaced by a codon for another amino acid.

Mitochondrion. An organelle, found in eukaryotic cells, in which oxidative phosphorylation takes place. It contains its own genome and unique ribosomes to carry out protein syn thesis of only a fraction of the proteins located in this organelle.

Mitosis. The process whereby replicated chromosomes segregate equally toward opposite poles prior to cell division.

Mobile genetic element. A segment of the genome that can move as a unit from one location on the genome to another, without any requirement for sequence homology.

Monolayer. A single layer of oriented lipid molecules.

Monomer. One unit of a protein or other structure.

Mutagen. An agent that can bring about a heritable change (mutation) in an organism.

Mutagenesis. A process that leads to a change in the genetic material that is inherited in later generations.

Mutant. An organism that carries an altered gene or change in its genome.

Mutarotation. The change in optical rotation of a sugar that is observed immediately after it is dissolved in aqueous solution, as the result of the slow approach of equilibrium of a pyranose or a furanose in its alpha and beta forms.

Mutation. The genetically inheritable alteration of a gene or group of genes.

Myofibril. A unit of thick and thin filaments in a muscle fiber.

Myoglobin. (gk. *Myo* = muscle) an iron containing protophyrin-globin complex found in muscle; serves as A reservoir for oxygen and gives some muscles their red or pink color; the oxygen binding site allows the Attachment of a distal and proximal histidine molecule. Mg has a far higher O_2-affinity at low partial Pressures than Hg.

Myosin. The main protein of the thick filaments in a muscle myofibril. It is composed of two coiled subunits (mr about 220,000) that can aggregate to form a thick filament, which is globular at each end.

Nascent RNA. The initial transcripts of RNA, before any modification or processing.

Negative control. Repression of biological activity by the presence of a specific molecule.

Negatively supercoiling. Allows a more compact packing than a relaxed twisted DNA circle, by twisting the

Nernst equation. An equation that relates the redox potential to the standard redox potential and the concentrations of the oxidized and reduced form of the couple.

Neurotoxin. Poison which interferes with nerve function, usually by affecting the flow of ions through the cell membrane.

Nitrogen cycle. The passage of nitrogen through various valence states, as the result of reactions carried out by a wide variety of different organisms.

Nitrogen fixation. Conversion of atmospheric nitrogen into a form that can be converted by biochemical reactions to an organic form. This reaction is carried out by a very limited number of microorganisms.

Nitrogenous base. An aromatic nitrogen-containing molecule with basic properties. Such bases include purines and pyrimidines.

Noncompetitive inhibitor. An inhibitor of enzyme activity whose effect is not reversed by increasing the concentration of substrate molecule.

Nuclease. An enzyme that cleaves phosphodiester bonds of nucleic acids.

Nucleic acid. A class of biochemical compounds which includes DNA and RNA. They are among the largest molecules known. These are polymers of the ribonucleotides or deoxyribonucleotides.

Nucleohistone. A complex of DNA and histone.

Nucleolus, A spherical structure visible in the nucleus during interphase. The nucleolus is associated with a site on the chromosome that is involved in ribosomal RNA synthesis.

Nucleophilic group. An electron-rich group that tends to attack an electron-deficient nucleus.

Nucleoside. a purine or pyrimidine base bounded to a sugar.

Nucleoside. An organic molecule containing a purine or pyrimidine base and a five-carbon sugar (ribose or deoxyribose).

Nucleosome. The basic unit of eukaryotic chromosome structure; a ball of eight histone molecules wrapped around by two coils of DNA; it is the main protagonist in packing the DNA strand; can easily be disturbed by UV-exposure (easily absorbs wavelengths of about 260 nm).

Nucleotide. An organic molecule containing a purine or pyrimidine base, a five-carbon sugar (ribose or deoxyribose), and one or more phosphate groups. A phosphoester of a nucleoside.

Nucleotide. Unit from which nucleic acids are constructed by polymerization. It contains a sugar, a phosphate group, and an organic base. ATP is a nucleotide.

Nucleus. In eukaryotic cells, the centrally-located organelle that encloses most of the chromosomes. Minor amounts of chromosomal substance are found in some other organelles, most notably the mitochondria and the chloroplasts.

Okazaki fragment. A short segment of single-stranded DNA that is an intermediate in DNA synthesis. In bacteria, Okazaki fragments are 1000-2000 bases in length; in eukaryotes, 100-200

bases in length. Each of the short discontinued segments in the 3'-5'direction of the lagging strand made by DNA polymerase-iii - about 1500 bases in eu-, 150 bases in prokaryota.

Oligonucleotide. A polynucleotide containing a small number of nucleotides. The linkages are the same as in a polynucleotide; the only distinguishing feature is the small size.

Oligosaccharide. A molecule containing a small number of sugar residues joined in a linear or a branched structure by glycosidic bonds.

Oncogene. A gene of cellular or viral origin that is responsible for rapid, unruly growth of animal cells. A cancer-causing gene.

Operon. A group of contiguous genes that are coordinately regulated by two cis-acting elements, a promoter and an operator. Found only in prokaryotic cells.

Optical activity. The property of a molecule that leads to rotation of the plane of polarization of plane-polarized light when the latter is transmitted through the substance. Chirality is a necessary and sufficient property for optical activity.

Organelle. A subcellular membrane-bounded body with a well-defined function.

Oxidation. The loss of electrons from a compound.

Oxidative phosphorylation. The formation of atp as the result of the transfer of electrons to oxygen.

Oxido-reductase. An enzyme that catalyzes oxidation-reduction reactions.

PCR. Polymerase chain reaction. A method for amplifying DNA sequences.

Pentose phosphate pathway. The pathway involving the oxidation of glucose-6-phosphate to pentose phosphates and further reactions of pentose phosphates.

Pentose. A sugar with five carbon atoms.

Peptide mapping. Same as fingerprinting.

Peptide unit. a unit of a rigid planar array of n, h, c, and a o-atom;

Peptide. An organic molecule in which a covalent amide bond is formed between the amino group of one amino acid and the carboxyl group of another amino acid, with the elimination of a water molecule. The resulting connection is called a peptide bond.

Peptidoglycan. carbohydrate polymer cross-linked by proteins. It is found in the cell wall of gram positive bacteria, where it stains with the dye crystal-violet.

Peptidoglycan. The main component of the bacterial cell wall, consisting of a two-dimensional network of heteropolysaccharides running in one direction, cross-linked with polypeptides running in the perpendicular direction.

Periplasm. The region between the inner (cytoplasmic) membrane and the cell wall or outer membrane of a bacterium.

Permease. A protein that catalyzes the transport of a specific small molecule across a membrane.

Peroxisomes. Subcellular organelles that contain flavin-requiring oxidases and that regenerate oxidized flavin by reaction with oxygen.

Phenotype. The observable trait(s) that result from the genotype in cooperation with the environment.

Phenylketonuria. A human disease caused by a genetic deficiency in the enzyme that converts phenylalanine to tyrosine. The immediate cause of the disease is an excess of phenylalanine, which can be alleviated by a diet low in phenylalanine.

Pheromone. A hormone-like substance that acts as an attractant.

Phosphate. an ion consisting of a phosphorus atom and four oxygen atoms. Among other things, it is used in the constuction of nucleic acids.

Phosphodiester. A molecule containing two alcohols esterified to a single molecule of phosphate. For example, the backbone of nucleic acids is connected by 5'-3' phosphodiester linkages between the adjacent individual nucleotide residues.

Phosphogluconate pathway. Another name for the pentose phosphate pathway. This name derives from the fact that 6-phosphogluconate is an intermediate in the formation of pentoses from glucose.

Phospholipid. A lipid containing charged hydrophilic phosphate groups; a component of cell membranes.

Phosphorylation. The formation of a phosphate derivative of a biomolecule.

Photoreactivation. DNA repair in which the damaged region is repaired with the help of light and an enzyme. The lesion (break) is repaired without excision (cutting out) from the DNA.

Photosynthesis. The biosynthesis that directly harnesses the chemical energy resulting from the absorption of light.

Frequently used to refer to the formation of carbohydrates from co2 that occurs in the chloroplasts of plants or the plastids of photosynthetic microorganisms.

Phycocyanin. blue, water-soluble pigment found in the cyanobacteria and the red algae.

Phycoerythrin. Red, water-soluble pigment found in the cyanobacteria and red algae.

Pigment. Any colorful compound, used by living things to absorb or block sunlight, and in sexual displays.

Plasma membrane. The membrane that surrounds the cytoplasm.

Plasmid. A circular DNA duplex that replicates autonomously in bacteria. Plasmids that integrate into the host genome are called episomes. Plasmids differ from viruses in that they never form infectious nucleoprotein particles.

Polar group. A hydrophilic (water-loving) group.

Polar mutation. A mutation in one gene that reduces the expression of a gene or genes distal to the promoter in the same operon.

Polarimeter. An instrument for determining the rotation of polarization of light as the light passes through a solution containing an optically-active substance.

Polyamine. A hydrocarbon containing more than two amino groups.

Polycistronic messenger RNA. In prokaryotes, an RNA that contains two or more cistrons; note that only in prokaryotic mRNAs can more than one cistron be utilized by the translation system to generate individual proteins.

Polymer. A large molecule constructed from many smaller identical units. These include proteins, nucleic acids, and starches.

Polymerase chain reaction (PCR). a method used to amplify a specific DNA sequence in vitro by repeated cycles of synthesis using specific primers and DNA polymerase.

Polymerase. An enzyme that catalyzes the synthesis of a polymer from monomers. The enzymes that synthesizes new DNA strands (from 5' to 3') involved in the polymerization (formation) of large molecules out of monomeric units (building blocks), using a DNA template.

Polynucleotide phosphorylase. An enzyme that polymerizes ribonucleotide diphosphates. No template is required.

Polynucleotide. A chain structure containing nucleotides linked together by phosphodiester (5'-3') bonds. The polynucleotide chain has a directional sense with a 5' and a 3' end.

Polypetide bond. Several amino acids linked together by peptide bonds determining the primary structure.

Polyribosome (polysome). A complex of an mRNA and two or more ribosomes actively engaged in protein synthesis.

Polysaccharide. A linear or branched chain structure containing many sugar molecules linked by glycosidic bonds.

Porphyrin. A complex planar structure containing four substituted pyrroles covalently joined in a ring and frequently containing a central metal atom. For example, heme is a porphyrin with a central iron atom.

Positive control. A system that is turned on by the presence of a regulatory protein.

Posttranslational modification. The covalent bond changes that occur in a polypeptide chain after it leaves the ribosome and before it becomes a mature protein.

Primary structure. In a polymer, the sequence of monomers and the covalent bonds. In proteins, it refers to the amino acid sequence.

Primase. This specializes RNA polymerase joins the prepriming complex in a multisubunit assembly called primosome.

Primer. A structure that serves as a growing point for polymerization. Short primers of DNA are often used in sequencing and mutagenesis procedures.

Primosome. A multiprotein complex that catalyzes synthesis of RNA primer at various points along the DNA template.

Prochiral molecule. A nonchiral molecule that lacks handedness and is optically inactive, but would become chiral by a change in one of the substituents at the chiral center. A prochiral molecule may react with an enzyme so that two groups that have a mirror-image relationship to each other are treated differently.

Prokaryote. A unicellular organism that contains a single chromosome, no nucleus, no membrane-bound organelles, and has characteristic ribosomes and biochemistry.

Promoter (initation). The site on DNA where RNA polymerase binds and begins transcription, i.e.. a regulator region just shortly off the 5' end of a gene. RNA polymerase itself is unable to start transcription at promoter site, rather it needs the subunit to make this holoenzyme complete and to make the promoter site recognizable. Once found, RNA polymerase unwinds the template DNA by nearly 2 turns. It is that region of the gene that signals RNA polymerase binding and the initiation of transcription.

Prophage. The silent phage genome. Some prophages integrate into the host genome; others replicate autonomously. The prophage state is maintained by a phage-encoded repressor.

Proprotein. A protein that is made in an active form, so that it requires processing to become functional.

Prostaglandin. An oxygenated eicosanoid that has a hormonal function. Prostaglandins are unusual hormones in that they usually have effects only in that region of the organism where they are synthesized.

Prosthetic group. The tightly bound, non-protein portion of an enzyme but essential for its function; they differ from coenzymes in that they are more firmly attached (usually permanently) to the enzyme protein; e.g.. the heme group present in cytochromes.

Protamines. Highly basic, arginine-rich proteins found complexed to DNA in the sperm of many invertebrates and fish.

Protein structure. folding of the protein structure is achieved with the help of chaperones. Protein function Arises from conformation, which is the 3d arrangement of atoms in a structure.

Protein subunit. One of the components or monomers of a multicomponent protein.

Protein. (gk. Proteios, primary). A complex organic compound composed of many (about 100) aminoacids joined together by peptide bonds. It is a class of biochemical compounds constructed from amino acids. Proteins may be structural, such as those that make up hair and cartilage, or they may be reactive, such as the enzymes.

Proteinaceous. describes any structure which is composed of protein.

Proteoglycan. A protein-linked heteropolysaccharide in which the heteropolysaccharide is usually the major component.

Protist. A relatively undifferentiated organism that can survive as a single cell.

Proton acceptor. A functional group capable of accepting a proton from a proton donor molecule.

Proton motive force. The thermodynamic driving force for proton translocation.

Proto-oncogene. A cellular gene that can undergo modification to a cancer-causing gene (oncogene).

Pseudocycle. A sequence of reactions that can be arranged in a cycle but that usually do not function simultaneously in both directions. Also called a futile cycle, since the net result of simultaneous functioning in both directions would be the expenditure of energy without accomplishing any useful work.

Pulse-chase. An experiment in which a short labeling period is followed by the addition of an excess of the same, unlabeled compound to dilute out the labeled material. Useful for observing time-dependent behavior of compounds.

Purine. A heterocyclic ring structure with varying functional groups. The purines adenine and guanine are found in both DNA and RNA.

Puromycin. An antibiotic that inhibits polypeptide synthesis by competing with aminoacyl-tRNA for the ribosomal binding site a.

Pyranose. A simple sugar containing the six-membered pyran ring.

Pyrimidine. A heterocyclic six-membered ring structure. Cytosine and uracil are the main pyrimidines found in RNA, and cytosine and thymine are the main pyrimidines found in DNA.

Pyrimidine. A type of single CN-ring base.

Pyrophosphate. A molecule formed by two phosphates in anhydride linkage.

Quaternary structure. In a protein, the way in which the different folded subunits interact to form the multisubunit protein.

R group. Shorthand for the side chain of an amino acid.

R loop. A triple-stranded structure in which RNA displaces a DNA strand by DNA-RNA hybrid formation in a region of the DNA.

Rapid-start complex. The complex that RNA polymerase forms at the promoter site just before initiation.

Recombination. The transfer to offspring of genes not found together in either of the parents.

Redox couple. An electron donor and its corresponding oxidized form.

Redox potential (e). The relative tendency of a pair of molecules to release or accept an electron.

Regulatory enzyme. An enzyme in which the active site is subject to regulation by factors other than the enzyme substrate. The enzyme frequently contains a nonoverlapping site for binding the regulatory factor that affects the activity of the active site.

Regulatory gene. A gene whose principal product is a protein designed to regulate the synthesis of other genes.

Renaturation. The process of returning a denatured structure to its original native structure, as when two single strands of DNA are reunited to form a regular duplex, or an unfolded polypeptide chain is returned to its normal folded three-dimensional structure.

Repair synthesis. DNA synthesis following excision (cutting out) of damaged DNA.

Repetitive DNA. A DNA sequence that is present in many copies per genome.

Replica plating. A technique in which an impression of a culture is taken from a master plate and transferred to a fresh plate. The impression can be of bacterial clones or phage plaques.

Replication fork. The point at which the two strands of DNA are separated to allow replication of each strand moving from the 3' to the 5' end of the parental sense (coding, upper or + strand), see polymerase.

Replicon. A genetic element that behaves as an autonomous replicating unit. It can be a plasmid, phage, or bacterial chromosome.

Repressor. A regulatory protein that inhibits transcription from one or more genes. It can combine with an inducer (resulting in specific enzyme induction) or with an operator element (resulting in repression).

Resonance hybrid. A molecular structure that is a hybrid of two structures that differ in the locations of some of the electrons. For example, the benzene ring can be drawn in two ways, with double bonds in different positions. The actual structure of benzene is in-between these two equivalent structures.

Restriction-modification system. A pair of enzymes found in most bacteria (but not eukaryotic cells). The restriction enzyme recognizes a certain sequence in duplex DNA and makes one cut in each unmodified DNA strand at or near the recognition sequence. The modification enzyme methylates (or modifies) the same sequence, thus protecting it from the action of the restriction enzyme.

Reverse transcriptase. An enzyme that synthesizes DNA from an RNA template, using deoxyribonucleotide triphosphates.

Reversible inhibitor. characterized by a rapid dissociation of the enzyme-inhibitor complex. These feedback loops are necessary to control the product concentration; i.e. halt production when sufficient product material is available. This effect can either be:

Rho-factor. protein factor of prokaryota required to recognize certain transcription termination signals. It is a protein involved in the termination of transcription of some messenger RNAs.

Ribose. The five-carbon sugar found in RNA.

Ribosomal RNA (rRNA). The RNA parts of the ribosome.

Ribosomes. Small cellular particles made up of ribosomal RNA and protein. They are the site, together with mRNA, of protein synthesis.

RNA (ribonucleic acid). A polynucleotide in which the sugar is ribose.

RNA polymerase. An enzyme that catalyzes the formation of RNA from ribonucleotide triphosphates, using DNA as a template.

RNA polymerase. Enzyme that catalyzes the synthesis of an RNA strand from a DNA template (does not require a primer, i.e. A core enzyme and the sigma factor - together form the holoenzyme). RNA polymerase is under control of camp and cap; transcription is hindered (starvation) if only few glucose molecules are present whereas many glucose molecules degrade camp, facilitating transcription (*stop 'n go* mechanism). Synthesis of RNA (transcription) starts with initiation, elongation, and termination; transcription is based on the non-coding or complementary DNA strand, which makes the synthesized RNA identical to the coding strand (except for the uracil-base).

RNA splicing. The excision of a segment of RNA, followed by a rejoining of the remaining fragments.

RNA. "Ribonucleic acid". The nucleic acid which carries the DNA message into parts of the cell where it is interpreted and used. The 18s ribosomal RNA sequence has been used in many groups of organisms to reconstruct phylogeny.

Rolling circle replication. A mechanism for the replication of circular DNA. A nick in one strand allows the 3' end to be extended, displacing the strand with the 5' end, which is also replicated, to generate a double-stranded tail that can become larger than the unit size of the circular DNA.

rRNA (ribosomal RNA, 80% of total RNA). A class of small (21 different proteins + an extra RNA molecule) and large (34 different proteins + 2 extra RNA molecules) subunit-RNA molecules, coded in the nuclear organizer, that have an integral role in ribosome structure and function. Eukaryotic and prokaryotic mRNA (include their subunits) just differ slightly in their molecular weight.

Rubisco. Protein which fixes carbon in photosynthetic organisms. It binds molecules of carbon dioxide to a five-carbon molcule. Rubisco is the most common protein on earth.

Salting in. The increase in solubility that is displayed by typical globular proteins upon the addition of small amounts of certain salts, such as ammonium sulfate.

Salting out. The decrease in protein solubility that occurs when salts such as ammonium sulfate are present at high concentrations.

Salvage pathway. A family of reactions that permits, for instance, nucleosides as well as purine and pyrimidine bases resulting from the partial breakdown of nucleic acids to be re-utilized in nucleic acid synthesis.

Satellite DNA. A DNA fraction whose base composition differs from that of the main component of DNA, as revealed by the fact that it bands at a different density in a csci gradient. Usually repetitive DNA or organelle DNA.

Saxitoxin. Neurotoxin found in a variety of dinoflagellates. If ingested, it may cause respiratory failure and cardiac arrest.

Scaffold. The central framework of a chromosome to which the DNA solenoid is attached as loops; composed largely of topoisomerase.

Schiff base. A nucleophile attacks a carbonyl group to form a tetrahedral intermediate which dehydrates afterwards generating the Schiff base; serves as an e—sink and is considered a potent e-acceptor.

Second messenger. A diffusible small molècule, such as camp, that is formed at the inner surface of the plasma membrane in response to a hormonal signal.

Secondary structure. In a protein or a nucleic acid, any repetitive folded pattern that results from the interaction of the corresponding polymeric chains.

Self splicing introns. A ribosome oh-group attacks a 5' splice site. The newly formed 3'-oh terminus of the upstream exon then attacks the 3' splice site to form a phosphodiester bond with the downstream exon.

Semiconservative replication. Duplication of DNA in which the daughter duplex carries one old strand and one new strand.

Semiconservative replication. The established model of DNA replication in which each double-stranded molecule is composed of on parental strand and one newly polymerized strand; i.e.. parental strand determines the sequence of the complimentary strand (after meselson and stahl). DNA synthesis is mediated by several enzymes.

Sigma factor. A subunit of RNA polymerase that recognizes specific sites on DNA for initiation of RNA synthesis.

Signal sequence. A (usually n-terminal) sequence of a protein that directs its processing or localization within the cell.

Silica. amorphous silicon dioxide (glass). It is a structural component in many organisms, such as diatoms and horsetails.

Single-copy DNA. A region of the genome whose sequence is present only once per haploid complement.

Site-directed mutagenesis. An intentional alteration in a DNA sequence that can be genetically inherited.

Somatic cell. Any cell of an organism that cannot contribute its genes to a later generation.

Sos system. A set of DNA repair enzymes and regulatory proteins that regulate their synthesis so that maximum synthesis occurs when the DNA is damaged.

Southern blotting. A method for detecting a specific DNA restriction fragment, developed by edward southern. DNA from a gel electrophoresis pattern is blotted onto nitrocellulose paper; then the DNA is denatured and fixed on the paper. Subsequently the pattern of specific sequences in the southern blot can be determined by hybridization to a suitable probe and autoradiography. A northern blot is similar, except that RNA is blotted instead onto the nitrocellulose paper.

Splicing. The reaction that removes introns and joins together exons in eukaryotic RNA.

Spongin. Proteinacous compound of which the spicules in demospongiae are composed.

Sporulation. Formation from vegetative cells of metabolically inactive cells that can resist extreme environmental conditions.

Stacking energy. The energy of interaction that favors the face-to-face packing of purine and pyrimidine base pairs.

Starch. A complex polymer of glucose, used by plants and green algae to store surplus sugar for later use.

Steady state. In enzyme-kinetic analysis, the time interval when the rate of reaction is approximately constant with time. The term is also used to describe the state of a living cell where the concentrations of many molecules are approximately constant because of a balancing between their rate of synthesis and breakdown.

Stem cell. A cell from which other cells stem or arise by differentiation.

Stereoisomers. Isomers that are nonsuperimposable mirror images of each other.

Steroids. Compounds that are derivatives of a tetracyclic structure composed of a cyclopentane ring fused to a substituted phenanthrene nucleus.

Structural domain. An element of protein tertiary structure that forms an independent folding unit.

Structural gene. A gene encoding the amino acid sequence of a polypeptide chain.

Structural protein. A protein that serves a structural function.

Substitution reaction. a reaction in which an atom (or a group of atoms) replaces an atom in the original molecule; or in complexes; reaction typical of alkanes in which the displacement atom is a hydrogen atom.

Substrate. The molecule undergoing reaction with an enzyme.

ubunit. Individual polypeptide chains in a protein.

ugar. Any of several small carbohydrates, such as glucose, which are "sweet" to the taste.

Sulfuric side chain. hydrophobic sidechain (cys, met);

Supercoil. A closed double stranded DNA molecule that is twisted on itself in prokaryotes. Kinking of specific base sequences allows bending of discrete sites; packaging is even further increased by the slightly twisted base pairs, they are not co-planar.

Supercoiled DNA. Supertwisted, covalently-closed duplex DNA.

Suppressor gene. A gene that can reverse the phenotype of a mutation in another gene.

Suppressor mutation. A mutation that restores a function lost by an initial mutation and that is located at a site different from the initial mutation.

Svedberg unit (S). The unit used to express the sedimentation constant (S = 10^{-13} sec). The sedimentation constant s is proportional to the rate of sedimentation of a molecule in a given centrifugal field and is related to the size and shape of the molecule.

Synapse. The chemical connection for communication between two nerve cells or between a nerve cell and a target cell such as a muscle cell.

Synapsis. The pairing of homologous chromosomes, seen during the first meiotic prophase.

Tandem duplication. A duplication in which the repeated regions are immediately adjacent to one another.

Template. A molecular "mold" that shapes the structure or sequence of another molecule; e.g. The nucleotide sequence of DNA acts as a template to control the nucleotide sequence of RNA during transcription.

Termination factors. Proteins that are exclusively involved in the termination reactions of protein synthesis on the ribosome.

Termination. A stop signal in the form of a hairpin loop followed by several uracil residues.

Terpenes. A diverse group of lipids made from isoprene precursors.

Tertiary structure. In a protein or nucleic acid, the final folded form of the polymer chain.

Tetramer. Structure resulting from the association of four subunits.

Tetrapeptide bond. Four amino acids joined together to form a linear chain.

Thioester. An ester of a carboxylic acid with a thiol or mercaptan.

Thymidine. One of the four nucleosides found in DNA.

Thymine. A pyrimidine base found in DNA.

Thymine. Pairs with adenine; in DNA only.

Topoisomerase. An enzyme that changes the extent of supercoiling of a DNA duplex. It is an enzyme unwinding the tightly coiled DNA arrangement, for DNA-replication (see below).

Transcription bubble (elongation). during elongation of the RNA transcript, duplex DNA is unwound at the forward end of RNA polymerase and rewound at its rear end.

Transcription. RNA synthesis that occurs on a DNA template.

Transduction. Genetic exchange in bacteria that is mediated via phage.

Transfection. An artificial process of infecting cells with naked viral DNA.

Transfer RNA (tRNA). Any of a family of low-molecular weight RNAs that transfer amino acids from the cytoplasm to the template for protein synthesis on the ribosome.

Transferase. An enzyme that catalyzes the transfer of a molecular group from one molecule to another.

Transformation. Genetic exchange in bacteria that is mediated via purified DNA. In somatic cell genetics the term is also used to indicate the conversion of a normal cell to one that grows like a cancer cell.

Transgenic. Describing an organism that contains transfected DNA in the germ line.

Transition state. The activated state in which a molecule is best suited to undergoing a chemical reaction.

Translation. The process of reading a messenger RNA sequence for the specified amino acid sequence it contains.

Transport protein. A protein whose primary function is to transport a substance from one part of the cell to another, from one cell to another, or from one tissue to another.

Tricarboxylic acid (TCA) cycle. The cyclical process whereby acetate is completely oxidized to CO_2 and water, and electrons are transferred to NAD^+ and flavine. The TCA cycle is localized to the mitochondria in eukaryotic cells and to the plasma membrane in prokaryotic cells. Also called the Krebs or citric acid cycle.

Tripeptide bond. three amino acids joined together to form a linear chain.

tRNA. (Transfer RNA, 15% of total RNA). A small cloverleaf (schematic; skeletal model is l-shaped) adapter molecules that bear specific amino acids (at the 3'-end = CCA) to the ribosome during translation. The attachment of the amino acid to the specific tRNA is catalyzed by specific *aminoacetyl-tRNA-synthetases* (activation enzymes), this process requires the presence of the aminoacetyl and an ATP.

Trypsin. A proteolytic enzyme that cleaves (cuts) peptide chains next to the basic amino acids arginine and iysine.

Tryptic peptide mapping. The technique of generating a chromatographic profile characteristic of the fragments resulting from trypsin enzyme cleavage of the protein.

Tumorigenesis. The mechanism of tumor formation.

Turnover number. The maximum number of molecules of substrate that can be converted to product per active site per unit time.

Ultracentrifuge. A high-speed centrifuge that can attain speeds up to 60,000 rpm and centrifugal fields of 500,000 times gravity. Useful for characterizing and/or separating macromolecules.

Unwinding proteins. Proteins that help to unwind double-stranded DNA during DNA replication.

Uracil. In place of thymine (found in RNA only) that pairs with adenine.

Urea cycle. A metabolic pathway in the liver that leads to the synthesis of urea from amino groups and co2. The function of the pathway is to convert the ammonia resulting from catabolism to a nontoxic form, which is then secreted.

UV irradiation. Electromagnetic radiation with a wavelength shorter than that of visible light (200-390 nm). Causes damage to DNA (mainly by forming pyrimidine dimers).

Van der Waals forces. Refers to the combined effect of two types of interactions, one attractive and one repulsive. The attractive forces are due to favorable interactions among the induced instantaneous dipole moments that arise from fluctuations in the electron charge densities of neighboring nonbonded atoms. Repulsive forces arise when noncovalently bonded atoms come too close together.

Viroids. Pathogenic agents, mostly of plants, that consist of short (usually circular) RNA molecules.

Virus. A complex of nucleic-acid and protein, that can infect and replicate inside a specific host cell to make more virus particles.

Vitamin. A trace organic substance required in the diet of some species. Many vitamins are precursors of coenzymes.

Watson-Crick base pairs. The type of hydrogen-bonded base pairs found in DNA, or comparable base pairs found in RNA. The base pairs are A-T, G-C, and A-U.

Western blot. Similar in principle to a southern blot, but where the species adsorbed to the nitrocellulose filter is a protein, and the detection makes use of specific antibodies.

Wild-type gene. The form of a gene (allele) normally found in nature.

Wobble. A proposed explanation for base pairing that is not of the Watson-Crick type and that often occurs between the 3' base in the codon and the 5' base in the anticodon.

X-ray crystallography. A technique for determining the structure of molecules from the x-ray diffraction patterns that are produced by crystalline arrays of the molecules.

Ylid. A compound in which adjacent, covalently-bonded atoms, both having an electronic octet, have opposite charges.

Z form. A duplex DNA structure in which there is the usual type of hydrogen bonding between the base pairs but in which the helix formed by the two polynucleotide chains is left-handed rather than right-handed.

Zwitterion. A dipolar ion with spatially-separated positive and negative charges.

Zygote. A cell that results from the union of haploid male and female sex cells. Zygotes are diploid.

Zymogen. An inactive precursor of an enzyme. For example, trypsin exists in the inactive form trypsinogen before it is converted to its active form, trypsin.